Boling Guo, Liming Ling, Yansheng Ma, and Hui Yang
Infinite-Dimensional Dynamical Systems

Also of Interest

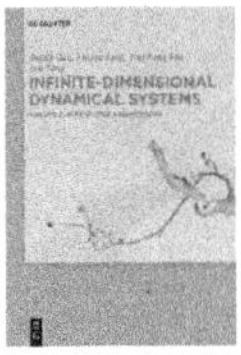

Infinite-Dimensional Dynamical Systems. Volume 2: Attractor and Methods
Boling Guo, Liming Ling, Yansheng Ma, Hui Yang, 2018
ISBN 978-3-11-058699-2, e-ISBN (PDF) 978-3-11-058726-5, e-ISBN (EPUB) 978-3-11-058708-1

Solitons
Boling Guo, Xiao-Feng Pang, Yu-Feng Wang, Nan Liu, 2018
ISBN 978-3-11-054924-9, e-ISBN (PDF) 978-3-11-054963-8, e-ISBN (EPUB) 978-3-11-054941-6

Rogue Waves. Mathematical Theory and Applications in Physics
Boling Guo, Lixin Tian, Zhenya Yan, Liming Ling, Yu-Feng Wang, 2017
ISBN 978-3-11-046942-4, e-ISBN (PDF) 978-3-11-047057-4, e-ISBN (EPUB) 978-3-11-046969-1

Vanishing Viscosity Method. Solutions to Nonlinear Systems
Boling Guo, Dongfen Bian, Fangfang Li, Xiaoyu Xi, 2016
ISBN 978-3-11-049528-7, e-ISBN (PDF) 978-3-11-049427-3, e-ISBN (EPUB) 978-3-11-049257-6

Stochastic PDEs and Dynamics
Boling Guo, Hongjun Gao, Xueke Pu, 2016
ISBN 978-3-11-049510-2, e-ISBN (PDF) 978-3-11-049388-7, e-ISBN (EPUB) 978-3-11-049243-9

Boling Guo, Liming Ling, Yansheng Ma, and
Hui Yang

Infinite-Dimensional Dynamical Systems

Volume 1: Attractors and Inertial Manifolds

DE GRUYTER

Mathematics Subject Classification 2010
76D03, 35Q35, 35E15, 35A01, 35A02

Authors
Prof. Boling Guo
Laboratory of Computational Physics
Institute of Applied Physics and
Computational Mathematics
6 Huayuan Road
Haidian District
100088 Beijing
People's Republic of China
gbl@iapcm.ac.cn

Prof. Liming Ling
South China University of Technology
School of Mathematics
Wushan RD., Tianhe District 381
510640 Guangzhou
People's Republic of China
greven@163.com

Dr Yansheng Ma
Northeast Normal University
School of Mathematics and Statistics
5268 Renmin Street
Jilin Province
130024 Changchun
People's Republic of China
mays538@nenu.edu.cn

Prof. Hui Yang
Yunnan Normal University
School of Mathematics
768 Junxian Road
Yunnan Province
650500 Kuming
People's Republic of China
yh_m1026@aliyun.com

ISBN 978-3-11-054925-6
e-ISBN (PDF) 978-3-11-054965-2
e-ISBN (EPUB) 978-3-11-054942-3

Library of Congress Control Number: 2018934551

Bibliographic information published by the Deutsche Nationalbibliothek
The Deutsche Nationalbibliothek lists this publication in the Deutsche Nationalbibliografie; detailed bibliographic data are available on the Internet at http://dnb.dnb.de.

Typesetting: VTeX UAB, Lithuania
Printing and binding: CPI books GmbH, Leck
Cover image: Chong Guo

www.degruyter.com

Preface

This book introduces the mathematical theory, research methods and results for infinite-dimensional dynamical systems. In 1993, the author gave a brief introduction to the conceptual framework, methods and research progress on infinite-dimensional dynamical systems in his monograph "Nonlinear evolution equations". However he didn't have an extensive and in-depth discussion on the topic due to limited space. As numerous new and important research achievements began to accumulate during the past decades, the authors have made a decision to write a monograph about infinite-dimensional dynamical systems.

The aim of this book is to introduce the rudimentary knowledge, some interesting problems and important and new results including the authors', cooperators' and other scholars' recent research on infinite-dimensional dynamical systems through some concise but heuristic methods.

The main emphasis in the first volume is on the mathematical analysis of attractors and inertial manifolds. In Chapter 1 "Attractor and its dimension estimation" we mainly introduce the global attractor and estimation of Hausdorff and fractal dimensions for some dissipative nonlinear evolution equations in modern physics. Chapter 2 "Inertial manifold" deals with inertial manifolds for a range of generalized differential equations and the spectral gap conditions. Moreover, we study the existence, smoothness and normal hyperbolic properties of inertial manifolds. Chapter 3 "Approximate inertial manifold" constructs the inertial manifolds and investigates the convergence of approximate inertial manifolds, which provides a constructive method to establish existence of inertial manifolds.

The second volume devotes to the modern analytical tools and methods in infinite dimensional dynamic system. In Chapter 1 "Discrete attractor and approximate calculation" we introduce results on the existence of discrete attractor and approximate calculation which are closely related to infinite-dimensional dynamical systems. Employing numerical calculations, we provide images of global attractors and approximate inertial manifolds. Chapter 2 "Some properties of a global attractor" introduces some properties of a global attractor, including oscillatory properties and asymptotic behavior. The asymptotic behavior of inertial manifolds can be determined only by the properties of a few points and it is closely related to the unstable manifold of the hyperbolic fixed point. We estimate an upper bound of Hausdorff length of level sets through the geometric measure method and provide a new method to give a lower bound on the dimensional estimate for the attractor. In Chapter 3 "Structures of small dissipative dynamical systems" we mainly introduce the structure of stable and unstable manifolds with small perturbations and the chaotic behaviors by employing the geometric singular perturbation theory, the center manifold theory in infinite dimensional setting and Melnikov method. The structure of stable and unstable manifolds is related to the first Chern number on the fiber bundle. Chapter 4 "Existence and sta-

https://doi.org/10.1515/9783110549652-201

bility of solitary waves" uses the concentration-compactness principle to study the existence of solitary wave solutions and discusses the nonlinear stability, instability and asymptotic stability for the solitary waves by the energy functional method and spectral analysis.

As the content of infinite-dimensional dynamical systems is quite rich and extensive, it is closely related to many subjects, such as fluid mechanics, functional analysis, topology, geometric measure theory, numerical mathematics, and so on. There are numerous new methods and results due to the quick developments of infinite-dimensional dynamical systems. Owing to the limited time and knowledge of the authors, there must be some inadvertent errors and omissions in the book. Any suggestions and comments are welcomed.

Beijing, China
July 2017

Boling Guo

Contents

1 Attractor and its dimension estimation

1.1 Global attractor and estimation of Hausdorff and fractal dimensions

In this section, a very important concept of a global attractor is introduced in the infinite-dimensional dynamical system. Moreover, the existence theorem of a global attractor and the estimates of the Hausdorff and fractal dimensions are given.

Definition 1.1.1. Assume that E is a Banach space and $S(t)$ is a continuous semigroup of operators, i. e., $S(t) : E \to E$, $S(t+\tau) = S(t) \cdot S(\tau)$, for any $t, \tau \geq 0$, $S(0) = I$ (identity operator). If a compact set $\mathscr{A} \subset E$ is
(i) invariant, i. e., $S(t)\mathscr{A} = \mathscr{A}$ for any $t \geq 0$;
(ii) attractive, i. e., for any bounded set $B \subset E$,

$$\mathrm{dist}(S(t)B, \mathscr{A}) = \sup_{x \in B} \inf_{y \in \mathscr{A}} \|S(t)x - y\|_E \to 0, \quad t \to \infty;$$

and, in particular, if all the paths $S(t)u_0$ departed from the initial data u_0 converge to the set $\mathscr{A}$, i. e.,

$$\mathrm{dist}(S(t)u_0, \mathscr{A}) \to 0, \quad t \to \infty, \tag{1.1.1}$$

then the compact set $\mathscr{A}$ is called the global attractor.

The structure of the global attractor is rather complicated. The semigroup $S(t)$ generated from an initial problem of nonlinear evolution equation

$$\frac{\mathrm{d}u(t)}{\mathrm{d}t} = F(u(t)), \tag{1.1.2}$$

$$u(0) = u_0, \tag{1.1.3}$$

may involve, besides the simple equilibrium point (maybe a multiple solution), also a time-periodic orbit, quasi-periodic orbit, a fractal and strange attractor, and so on. The semigroup is likely not to be a smooth manifold, possessing non-integer dimension.

To give the existence theorem of a global attractor, we need to introduce the concept of an attractive set.

Definition 1.1.2. If a bounded set $B_0 \subset E$ is such that for any bounded set $B \subset E$ there exists a $t_0(B) > 0$ such that

$$S(t)B \subset B_0, \quad \forall t \geq t_0(B), \tag{1.1.4}$$

then the set B_0 is called an attractive set in the set E.

https://doi.org/10.1515/9783110549652-001

Theorem 1.1.1. *Suppose E is a Banach space and $S(t), t \geq 0$ is a semigroup of operators, $S(t) : E \to E, S(t+\tau) = S(t)S(\tau), t, \tau \geq 0, S(0) = I$, where I is the identity operator. Assume the following conditions:*

(i) *The semigroup $S(t)$ is uniformly bounded in the set E, i. e., for any $R > 0$, there exists a constant $C(R)$ such that when $\|u\|_E \leq R$, it follows that*

$$\|S(t)u\|_E \leq C(R), \quad \forall t \in [0, \infty). \tag{1.1.5}$$

(ii) *There exists a bounded attractive set B_0 in the set E.*

(iii) *When $t > 0$, the operator $S(t)$ is completely continuous.*

Then the semigroup $S(t)$ possesses a compact global attractor $\mathscr{A}$.

Remark 1.1.1. If we replace the bounded attractive set B_0 of the condition (ii) with the compact attractive set B_0, then the completely continuous property of semigroup $S(t)$ can be replaced with that of being a continuous operator. Theorem 1.1.1 will remain valid.

Remark 1.1.2. Furthermore, we can prove that the global attractor $\mathscr{A}$ is the ω limiting set of the attractive set B_0, i. e.,

$$\mathscr{A} = \omega(B_0) = \bigcap_{s \geq 0} \overline{\bigcup_{t \geq s} S(t)B_0} \tag{1.1.6}$$

where the closure is taken in the set E.

Another frequently-used existence theorem of an attractor is the following:

Theorem 1.1.2. *Suppose E is a Banach space and a semigroup $S(t)$ is continuous. Assume that there exists an open set $\mathscr{U} \subset E$ and a bounded set $\mathscr{B}$ in $\mathscr{U}$ such that the set $\mathscr{B}$ is absorptive in $\mathscr{U}$. Also the following conditions are satisfied:*

(1) *The operator $S(t)$ is uniformly compact for any t large enough, i. e., for every bounded set $\mathscr{B}$, there exists $t = t_0(\mathscr{B})$ such that the set*

$$\bigcup_{t \geq t_0} S(t)\mathscr{B} \tag{1.1.7}$$

is relatively compact in the set E. Or

(2) *$S(t) = S_1(t) + S_2(t)$, where the operator $S_1(\cdot)$ is uniformly compact for t large enough (i. e., satisfies condition (1.1.7)), the operator $S_2(t)$ is a continuous map, $S_2(t) : E \to E$, and for every bounded set $B \subset E$,*

$$r_B(t) = \sup_{\phi \in B} \|S_2(t)\phi\|_E \to 0, \tag{1.1.8}$$

then the ω limiting set of $\mathscr{B}$ is a compact attractor, which attracts the bounded set in $\mathscr{U}$. It is the largest attractor in $\mathscr{U}$, and when $\mathscr{U}$ is convex and connected, $\mathscr{A}$ is connected.

Therefore, to prove the existence of a global attractor, we merely need to verify whether the conditions in Theorem 1.1.1 or 1.1.2 are valid. The most important is that:

(i) The existence and continuity of semigroup $S(t)$;

(ii) There exists a bounded or compact absorbing set;

(iii) The semigroup $S(t)$ $(t \geq 0)$ is a completely continuous operator or satisfies condition (1.1.7) (or condition (1.1.8)).

To portray the geometric property of a global attractor in the simplest way, we can estimate the Hausdorff and fractal dimension.

Definition 1.1.3. The Hausdorff measure of a set X is

$$\mu_H(X, d) = \lim_{\epsilon \to 0} \mu_H(X, d, \epsilon) = \sup_{\epsilon > 0} \mu_H(X, d, \epsilon) \tag{1.1.9}$$

where

$$\mu_H(X, d, \epsilon) = \inf \sum_i r_i^d \tag{1.1.10}$$

where inf is taken over all coverings of X by balls with radius $r_i \leq \epsilon$. If there exists a number $d = d_H(X) \in [0, +\infty]$ such that

$$\mu_H(X, d) = 0, \quad d > d_H(X), \tag{1.1.11}$$

$$\mu_H(X, d) = \infty, \quad d < d_H(X), \tag{1.1.12}$$

then the number $d_H(X)$ is called the Hausdorff dimension of the set X.

Definition 1.1.4. The fractal dimension of a set X is

$$d_F(X) = \lim_{\epsilon > 0} \sup \frac{\lg n_X(\epsilon)}{\lg \frac{1}{\epsilon}} \tag{1.1.13}$$

where $n_X(\epsilon)$ is the smallest number of balls with radius $\leq \epsilon$ in a covering of X. It is readily seen that

$$d_F(X) = \inf\{d > 0, \ \mu_F(X, d) = 0\} \tag{1.1.14}$$

where

$$\mu_F(X, d) = \lim_{\epsilon \to 0} \sup \epsilon^d n_X(\epsilon). \tag{1.1.15}$$

Since $\mu_F(X, d) \geq \mu_H(X, d)$, then

$$d_H(X) \leq d_F(X). \tag{1.1.16}$$

In the following, we consider the initial problem

$$\frac{du(t)}{dt} = F(u(t)), \quad t > 0, \tag{1.1.17}$$
$$u(0) = u_0, \tag{1.1.18}$$

where $F(u)$ is a determined function, $F(u) : E \to E$, and E is a Hilbert space. Suppose that for any $u_0 \in E$, there exists a global solution $u(t) \in E$, denoted by $u(t) = S(t)u_0$, where the map $S(t) : E \to E$ is the semigroup of the initial problem (1.1.17)–(1.1.18).

Suppose $F : E \to E$ is Fréchet differentiable and the linear initial problem

$$\frac{dU(t)}{dt} = F'(S(t)u_0)U(t), \tag{1.1.19}$$
$$U(0) = \xi, \tag{1.1.20}$$

is solvable for every u_0 and $\xi \in E$. Finally, suppose $S(t)$ is differentiable, with the derivative $L(t, u_0)$, i. e.,

$$L(t, u_0)\xi = U(t), \quad \forall \xi \in E, \tag{1.1.21}$$

and $U(t)$ is the solution of (1.1.19) and (1.1.20). Since (1.1.19) is the first order variational equation of (1.1.17), the above assumptions are natural and can be readily verified.

For a fixed $u_0 \in E$, solutions $\xi_1, \xi_2, \dots, \xi_J$ are J elements of E, $U_1(t), U_2(t), \dots, U_J(t)$ denote J solutions of linearized equation (1.1.19) with the initial data

$$U_1(0) = \xi_1, \quad U_2(0) = \xi_2, \quad \dots, \quad U_J(0) = \xi_J.$$

Through direct calculation we arrive at

$$\frac{d}{dt}\|U_1(t) \wedge \cdots \wedge U_J(t)\|_{\wedge E}^2 - 2\operatorname{tr}(F'(u(t)) \cdot Q_J)\|U_1(t) \wedge \cdots \wedge U_J(t)\|_{\wedge E}^2 = 0, \tag{1.1.22}$$

where $F'(u(t)) = F'S(t)u_0$ is a linear map $U \to F'(u(t))U$, $u(t) = S(t)u_0$ is the solution of (1.1.17) and (1.1.18), $\wedge$ denotes the wedge product, tr denotes the trace of operator, and Q_J represents the orthogonal projection from E to the subspace spanned by $U_1(t), U_2(t), \dots, U_J(t)$. The J-dimensional volume $\bigwedge_{j=1}^{J} \xi_j$ is

$$\omega_J(t) = \sup_{u_0 \in A} \sup_{\xi_i \in E} \|U_1(t) \wedge \cdots \wedge U_J(t)\|_{\wedge^J E}^2, \tag{1.1.23}$$

where A is the invariant set of the semigroup $\{S(t)\}_{t \geq 0}$. It is readily verified that $\omega_j(t)$ is subexponential about t, i. e.,

$$\omega_j(t + t') \leq \omega_j(t)\omega_j(t'), \quad \forall t, t' \geq 0. \tag{1.1.24}$$

Therefore,

$$\lim_{t \to \infty} \omega_j(t)^{\frac{1}{t}} = \Pi_j, \quad \forall j, \ 1 \leq j \leq J \tag{1.1.25}$$

exists. By (1.1.22) we have

$$\Pi_J \le \exp q_J \tag{1.1.26}$$

where

$$q_J = \limsup_{t\to\infty} q_J(t), \tag{1.1.27}$$

$$q_J(t) = \sup_{u_0\in A,\, \xi_j\in E} \sup_{\|\xi_j\|_E\le 1} \left\{ \frac{1}{t} \int_0^t \operatorname{tr}(F'(S(\tau)u_0)Q_J(\tau))\mathrm{d}\tau \right\}. \tag{1.1.28}$$

Definition 1.1.5. If a group of sequence $\Lambda_1, \Lambda_2, \dots, \Lambda_m$ is defined as

$$\Lambda_1 = \Pi_1, \quad \Lambda_1\Lambda_2 = \Pi_2, \quad \dots, \quad \Lambda_1 \cdots \Lambda_m = \Pi_m,$$

or

$$\Lambda_1 = \Pi_1, \quad \Lambda_m = \frac{\Pi_m}{\Pi_{m-1}}, \quad m \ge 2, \tag{1.1.29}$$

$$\Lambda_m = \lim_{t\to\infty} \left(\frac{\omega_m(t)}{\omega_{m-1}(t)} \right)^{\frac{1}{t}}, \quad m \ge 2,$$

then Λ_m is the global (or uniform) Lyapunov number in the set A and

$$\mu_m = \lg \Lambda_m, \quad m > 1$$

is the corresponding Lyapunov index. Due to equation (1.1.26), we have

$$\mu_1 + \mu_2 + \cdots + \mu_J \le q_J. \tag{1.1.30}$$

Theorem 1.1.3. *Under the assumptions of initial problems* (1.1.17)–(1.1.18) *and* (1.1.19)–(1.1.20), *if we have*

$$q_J(t) \le -\delta < 0, \quad \forall t \ge t_0, \tag{1.1.31}$$

for some J and $t_0 > 0$, then the volume element $\|U_1(t) \wedge \cdots \wedge U_J(t)\|_{\wedge^J E}$ exponentially decays as $t \to \infty$. For $u_0 \in A, \xi_1, \xi_2, \dots, \xi_J \in E$, it follows that

$$\|U_1(t) \wedge \cdots \wedge U_J(t)\|_{\wedge^m E} \le \|U_1(t_0) \wedge \cdots \wedge U_m(t_0)\|_{\wedge^J E} \exp(-\delta(t - t_0))$$

uniformly. If A is the bounded functional invariant set of the semigroup $S(t)$, then for some j the inequality

$$q_j < 0 \tag{1.1.32}$$

is valid. From

$$\Pi_j = \Lambda_1 \Lambda_2 \cdots \Lambda_j < 1, \quad \mu_1 + \mu_2 + \cdots + \mu_j < 0,$$

it follows that

$$\Lambda_j < 1, \tag{1.1.33}$$

i. e.,

$$\mu_j < 0. \tag{1.1.34}$$

Theorem 1.1.4. *Assume that there exists a global attractor of the initial problem of the nonlinear evolution equation* (1.1.17)–(1.1.18)*, which is bounded in* $H^1(\Omega)$*. Suppose the linear initial problem* (1.1.19)–(1.1.20) *is solvable and the semigroup* $S(t)$ *determined by the initial problem* (1.1.17)–(1.1.18) *is differentiable. If*

$$q_j < 0 \tag{1.1.35}$$

is determined by (1.1.27) *for some* j*, then the Hausdorff dimension of the global attractor* A *is finite and* $\leq j$*. Its fractal dimension is less than or equal to*

$$j\left(1 + \max_{1 \leq l \leq j-1} \frac{(q_l)_+}{|q_j|}\right). \tag{1.1.36}$$

We can weaken the Fréchet differentiability condition of $S(t)$, i. e., we can merely demand that the operator $S(t)$ is uniformly differentiable in the compact invariant subset, i. e., for every $u \in X$, there exists a linear operator $L(t,u) \in \mathscr{L}(\mathscr{E})$, and

$$\sup_{u,v \in X,\, 0<|u-v|\leq \epsilon} \frac{|S(t)u - S(t)v - L(t,u)(u-v)|}{|u-v|} \to 0, \quad \epsilon \to 0, \tag{1.1.37}$$

$$\sup_{u \in X} |L(t,u)|_{\mathscr{L}(\mathscr{E})} < +\infty, \tag{1.1.38}$$

$$\sup_{u \in X} \omega_d(L(t,u)) < 1, \quad \text{for some real number } d > 0, \tag{1.1.39}$$

where $\omega_d(L) = \omega_n^{1-s}(L)\omega_{n+1}^s(L)$, $d = n + s$. Then we have

Theorem 1.1.5. *Under the assumptions* (1.1.37)–(1.1.39)*, the Hausdorff dimension of* X *is finite and less than or equal to* d*.*

1.2 Kuramoto–Sivashinsky equation

The Kuramoto–Sivashinsky equation (KS equation) [30, 181]

$$\varphi_t + \frac{1}{2}(\varphi_x)^2 + \nu\varphi + \alpha\varphi_{xx} + \gamma\varphi_{xxxx} = 0 \tag{1.2.1}$$

was proposed by Kuramoto [157] in the study of a reaction diffusion in 1978 and independently by Sivashinsky [193] who researched combustion flame propagation in 1977, where α, γ and ν are positive constants. This model also appears when studying a membrane vibration [3] and bifurcation solutions of Navier–Stokes equation [192]. Nicolaenko et al. [183] made an in-depth and systematic research on the global attractor and bifurcation solution of the one-dimensional KS equation. B. Nicolaenko [179] proposed a generalized KS-type equation (involving a high-dimensional KS equation). Guo et al. [101] studied the asymptotic behavior as $t \to \infty$, the structure of the traveling solution, the similarity solution by Lie group and infinitesimal transformation, and the approximate solution by the spectral method. Guo [88] proved the existence of a global attractor of the KS equation and its finite dimension. In 1993, Guo and Su [109] gave the first proof of the existence of a global attractor of the high-dimensional KS equation and estimated the Hausdorff and fractal dimensions.

Setting $u = \varphi_x$, equation (1.2.1) can be converted into

$$u_t + uu_x + \nu u + \alpha u_{xx} + \gamma u_{xxxx} = 0. \tag{1.2.2}$$

We consider the following generalized KS-type equation:

$$u_t + \alpha\triangle u + \gamma\triangle^2 u + \nabla\cdot f(u) + \triangle\varphi(u) = g(u) + h(x), \tag{1.2.3}$$

and periodic initial condition

$$u(x,t) = u(x + 2de_i, t), \quad x \in \Omega,\ t \geq 0,\ i = 1,2,\dots,n, \tag{1.2.4}$$

$$u(x,0) = u_0(x), \quad x \in \Omega, \tag{1.2.5}$$

where $\Omega \subset \mathbf{R}^n$ is the n dimensional cube with side length $2d$, i. e.,

$$\overline{\Omega} = \{x = (x_1,\dots,x_n) \mid |x_i| \geq d,\ i = 1,2,\dots,n\}$$

where $x + 2de_i = (x_1,\dots,x_{i-1},x_i + 2d, x_{i+1},\dots,x_n)$, $i = 1,2,\dots,n$; $\alpha \geq 0$; $\gamma > 0$; $\nabla\cdot f(u) = \sum_{k=1}^n \frac{\partial f_k(u)}{\partial x_k}$; $\triangle u = \sum_{k=1}^n \frac{\partial^2 u}{\partial x_k^2}$. If $\alpha = 0$ and $f(u) = 0$, equation (1.2.3) is the inhomogeneous Cahn–Hilliard equation.

To prove the existence of a global attractor, the following priori estimates are given:

Lemma 1.2.1. *Suppose*

(1) $\varphi'(u) \leq \varphi_0$;

(2) $\gamma > \frac{\alpha+\varphi_0}{2}$;

(3) $g(0) = 0, g'(u) \leq g_0, g_0 < -\frac{\alpha+\varphi_0-1}{2}$;

(4) $h(x) \in L^2(\Omega)$, $u_0(x) \in L^2(\Omega)$.

Then the smooth solution of problem (1.2.3)–(1.2.5) *satisfies the following estimate:*

$$\|u(\cdot,t)\|^2 \le e^{(2g_0+\alpha-1-\varphi_0)t}\|u_0(x)\|^2 + \frac{1}{|2g_0+\alpha+\varphi_0+1|}(1-e^{(2g_0-\alpha-1-\varphi_0)t})\|h(x)\|^2, \quad 0<t<\infty. \tag{1.2.6}$$

Moreover, we have

$$\limsup_{t\to\infty}\|u(\cdot,t)\|^2 \le \frac{\|h(x)\|^2}{|2g_0+\alpha+1+\varphi_0|} = E_0, \tag{1.2.7}$$

$$\limsup_{t\to\infty}\frac{1}{t}\int_0^t \|\Delta u(\cdot,t)\|^2 \mathrm{d}t \le \frac{1}{[2\gamma-(\alpha+\varphi_0)]}[\|u_0(x)\|^2+\|h(x)\|^2]. \tag{1.2.8}$$

Proof. Taking the inner product of (1.2.3) and u, it follows that

$$(u, u_t + \alpha\Delta u + \gamma\Delta^2 u + \nabla\cdot f(u) + \Delta\varphi(u) - g(u) - h(x)) = 0, \tag{1.2.9}$$

where

$$\begin{aligned}(\nabla\cdot f(u), u) &= \int_\Omega \sum_{k=1}^n \frac{\partial f_k(u)}{\partial x_k} u \mathrm{d}x\\ &= -\sum_{k=1}^n \int_\Omega f_k(u) u_{x_k} \mathrm{d}x\\ &= -\sum_{k=1}^n \int_\Omega \Phi_k(u)_{x_k}\mathrm{d}x = 0, \quad \Phi_k(u) = \int_0^u f_k(u)\mathrm{d}u,\end{aligned}$$

$$(\Delta\varphi(u), u) = -(\nabla\varphi(u), \nabla u) = -(\varphi'(u), |\nabla u|^2) \ge -\varphi_0\|\nabla u\|^2.$$

By Gagliardo–Nirenberg inequality, we obtain

$$\|\nabla u\|^2 \le \|\Delta u\|\|u\| \le \frac{1}{2}(\|\Delta u\|^2 + \|u\|^2).$$

It follows that

$$\begin{aligned}&|(u, \alpha\nabla u)| = \alpha\|\nabla u\|^2 \le \frac{\alpha}{2}(\|\nabla u\|^2 + \|u\|^2),\\ &(g(u), u) \le g_0\|u\|^2,\\ &(h(x), u) \le \frac{1}{2}(\|u\|^2 + \|h(x)\|^2).\end{aligned}$$

Thus, through (1.2.9), we arrive at

$$\begin{aligned}&\frac{1}{2}\frac{\mathrm{d}}{\mathrm{d}t}\|u(\cdot,t)\|^2 + \left(\gamma - \frac{\alpha+\varphi_0}{2}\right)\|\Delta u(\cdot,t)\|^2\\ &\le \left(g_0 + \frac{\alpha+\varphi+1}{2}\right)\|u(\cdot,t)\|^2 + \frac{1}{2}\|h(x)\|^2.\end{aligned} \tag{1.2.10}$$

Finally, equations (1.2.6), (1.2.7) and (1.2.8) can be deduced from (1.2.10) using the Gronwall inequality. □

Lemma 1.2.2. *Under the conditions of Lemma* 1.2.1, *assume that*

(1) $\max_{k=1,\dots,n} |f_k(u)| \le A|u|^p$, $1 \le p \le 1 + \frac{6}{n}$;
(2) $|\varphi'(u)| \le B|u|^q$, $0 \le q < \frac{4}{n}$;
(3) $h(x) \in L^2(\Omega)$, $u_0(x) \in H^1(\Omega)$, $\Omega \subset \mathbf{R}^n$, $1 \le n \le 6$.

Then the solution of problem (1.2.3)–(1.2.5) *satisfies*

$$\begin{aligned}\|\nabla u(\cdot,t)\|^2 &\le e^{2g_0 t}\|\nabla u_0(x)\|^2 \\ &\quad + \frac{1}{|g_0|}(1 - e^{2g_0 t})(C_6\|h(x)\|^2 + C_7), \quad 0 \le t < \infty, \end{aligned} \tag{1.2.11}$$

where functions $C_6(\cdot)$ *and* $C_7(\cdot)$ *depend on* $\|u(\cdot,t)\|$. *Moreover, we have*

$$\begin{aligned}&\overline{\lim_{t\to\infty}} \|\nabla u(\cdot,t)\|^2 \\ &\quad \le \frac{1}{|g_0|}\Big[\max_{t\ge0} C_6(\|u(\cdot,t)\|)\|h(x)\|^2 + \max_{t\ge0} C_7(\|u(\cdot,t)\|)\Big] = E_1,\end{aligned} \tag{1.2.12}$$

$$\overline{\lim_{t\to\infty}} \frac{1}{t}\int_0^t \|\nabla\Delta u(\cdot,t)\|^2 dt \le \frac{6}{\gamma}\Big[\|u_0(x)\|_{H^1}^2 + \max_{t\ge0} C_6\|h\|^2 + \max_{t\ge0} C_7\Big]. \tag{1.2.13}$$

Proof. Taking the inner product of (1.2.3) and Δu, we arrive at

$$(\Delta u, u_t + \alpha\Delta u + \gamma\Delta^2 u + \nabla\cdot f(u) + \Delta\varphi(u) - g(u) - h(x)) = 0, \tag{1.2.14}$$

where

$$\begin{aligned}(\Delta u, -g(u)) &= (\nabla g(u), \nabla u) = (g'(u)\nabla u, \nabla u) \\ &\le g_0\|\nabla u\|^2 |(\Delta u, \Delta\varphi(u))| \\ &= |(\nabla\varphi(u), \nabla\Delta u)| = |(\varphi'(u)\nabla u, \nabla\Delta u)| \\ &\le \frac{\gamma}{6}\|\nabla\Delta u\|^2 + \frac{3}{2\gamma}\|\varphi'(u)\nabla u\|^2.\end{aligned}$$

By virtue of $|\varphi'(u)| \le B|u|^q$, it follows that

$$\frac{3}{2\gamma}\|\varphi'(u)\nabla u\|^2 \le \frac{3}{2\gamma}\|\varphi'(u)\|_\infty^2\|\nabla u\|^2 \le \frac{3}{2\gamma}B^2\|u\|_\infty^{2q}\|\nabla u\|^2.$$

Due to the Sobolev interpolation inequality,

$$\begin{aligned}&\|u\|_\infty \le C_1\|\nabla\Delta u\|^{\frac{11}{6}}\|u\|^{\frac{6-n}{6}} + C_1'\|u\|, \\ &\|\nabla u\| \le C_2\|\nabla\Delta u\|^{\frac{1}{3}}\|u\|^{\frac{2}{3}} + C_2'\|u\|, \quad 1 \le n \le 6,\end{aligned}$$

hence it follows that

$$\frac{3}{2\gamma}\|\varphi'(u)\nabla u\|^2 \le \frac{3}{2\gamma}B^2C_1^{2q}C_2^2\|\nabla\triangle u\|^{\frac{nq+2}{3}}\|u\|^{((16-n)q+1)/3}$$
$$\le \frac{\gamma}{6}\|\nabla\triangle u\|^2 + C_3(C_1, C_2, q, \|u\|), \quad 0 < q < \frac{4}{n},$$
$$|(\nabla\cdot f(u), \triangle u)| = \left|\sum_{k=1}^{n}\int_\Omega f_k(u)\frac{\partial}{\partial x_k}\triangle u dx\right|$$
$$\le \sum_{k=1}^{n}\|f_k(u)\|\|\nabla\triangle u\|$$
$$\le \frac{\gamma}{6}\|\nabla\triangle u\|^2 + \frac{3}{2\gamma}\sum_{k=1}^{n}\|f_k(u)\|^2,$$

where

$$\frac{3}{2\gamma}\|f_k(u)\|^2 \le \frac{3}{2\gamma}A^2\|u\|_p^{2p} \le \frac{3}{2\gamma}A^2\|u\|_\infty^{2p-2}\|u\|^2$$
$$\le \frac{3}{2\gamma}A^2C^{2p-2}\|\nabla\triangle u\|^{\frac{n(p-1)}{3}}\|u\|^{\frac{n+(6-n)p}{3}},$$
$$1 \le p < 1 + \frac{6}{n} \le \frac{3}{2\gamma}\|\nabla\triangle u\|^2 + C_4(\gamma, A, \|u\|),$$
$$\alpha\|\nabla\|^2 \le \alpha C_2\|\nabla\triangle u\|^{\frac{1}{3}}\|u\|^{\frac{2}{3}}$$
$$\le \frac{\gamma}{6}\|\nabla\triangle u\|^2 + C_5(\alpha, C_2, \|u\|),$$
$$(\triangle u, -g(u)) = (g'(u)\nabla u, \nabla u) \le g_0\|\nabla u\|^2$$
$$(\triangle u, -h(x)) \le \frac{\gamma}{12}\|\nabla\triangle u\|^2 + C_6\|h(x)\|^2.$$

Finally, due to equation (1.2.14), we obtain

$$\frac{1}{2}\frac{\mathrm{d}}{\mathrm{d}t}\|\nabla u\|^2 + \frac{\gamma}{12}\|\nabla\triangle u\|^2 \le g_0\|\nabla u\|^2 + C_6\|h\|^2 + C_7(\|u\|),$$

which implies

$$\|\nabla u\|^2 \le e^{2g_0t}\|\nabla u_0(x)\|^2 + \frac{1}{|g_0|}(1 - e^{-g_0t})(C_6\|h\|^2 + C_8),$$

as well as (1.2.12) and (1.2.13). □

Lemma 1.2.3. *Under the conditions of Lemma* 1.2.2, *suppose that*

(1) $\max_{k=1,\dots,n}\|f_k'(u)\| \le A|u|^{p-1}$, $|\varphi''(u)| \le B|u|^{q-1}$;

(2) $|g''(u)| \le C|u|^l$, $0 \le l < \frac{40}{3n} - \frac{4}{3}$;

(3) $u_0(x) \in H^2(\Omega)$.

Then we have

$$\|\triangle u(\cdot,t)\|^2 \le e^{2g_0 t}\|\triangle u_0(x)\|^2 + \frac{1}{|g_0|}(1-e^{2g_0 t})\Big(\frac{3}{\gamma}\|h\|^2 + C_{12}\Big), \tag{1.2.15}$$

where the function C_{12} depends on $\|u(\cdot,t)\|_{H^1}$. Moreover, we have

$$\overline{\lim_{t\to\infty}}\,\|\triangle u(\cdot,t)\|^2 \le \frac{1}{|g_0|}\Big[\frac{3}{\gamma}\|h(x)\|^2 + \max_{t\ge 0} C_{12}(\|u(\cdot,t)\|_{H^1})\Big] = E_2, \tag{1.2.16}$$

$$\overline{\lim_{t\to\infty}}\,\frac{1}{t}\int_0^t \|\nabla\triangle u(\cdot,t)\|^2 dt \le \frac{2}{\gamma}\Big[\|u_0(x)\|^2 + \frac{3}{\gamma}\|h\|^2 + \max_{t\ge 0} C_{12}\Big]. \tag{1.2.17}$$

Proof. Taking the inner product of (1.2.3) and $\triangle^2 u$, we arrive at

$$(\triangle^2 u, u_t + \alpha\triangle u + \gamma\triangle^2 u + \nabla\cdot f(u) + \triangle\varphi(u) - g(u) - h(x)) = 0, \tag{1.2.18}$$

where

$$\begin{aligned}
|(\triangle^2 u, \nabla\cdot f(u))| &\le \sum_{k=1}^n \|f_k'(u)\|_\infty \|\nabla u\|\|\triangle^2 u\| \\
&\le nA\|u\|_\infty^{p-1}\|\nabla u\|\|\triangle^2 u\| \le nAC_1\|\nabla\triangle u\|^{\frac{(p-1)n}{6}}\|u\|^{\frac{(p-1)(6-n)}{6}}\|\nabla u\|\|\triangle^2 u\| \\
&\le nAC_1C_\star^{\frac{n(p-1)}{6}}\|\nabla u\|^{\frac{(p-1)n}{18}}\|\nabla^2 u\|^{\frac{(p-1)n}{9}}\|u\|^{\frac{(p-1)(6-n)}{6}}\|\nabla u\|\|\triangle^2 u\| \\
&\le \frac{\gamma}{6}\|\triangle^2 u\|^2 + C_3(\|u\|_{H^1}).
\end{aligned}$$

In the above equation, we use the following Sobolev interpolation inequalities:

$$\begin{aligned}
\|\nabla\triangle u\| &\le C_\star\|\nabla u\|^{\frac13}\|\triangle^2 u\|^{\frac23} + C\|u\|, \\
|(\triangle\varphi(u), \triangle^2 u)| &\le \|\varphi'(u)\triangle u + \varphi''(u)(\nabla u)^2\|\|\triangle^2 u\| \\
&\le [\|\varphi'(u)\|_\infty\|\triangle u\| + \|\varphi''(u)\|_\infty\|\nabla u\|_\infty\|\nabla u\|], \\
\|\triangle^2 u\| &\le (B\|u\|_\infty^q C_5\|\triangle^2 u\|^{\frac13}\|u\|^{\frac23} + B\|u\|_\infty^{q-1}C_6\|\triangle^2 u\|^{\frac{11}{6}}\|\nabla u\|^{1-\frac{n}{6}}\|\triangle u\|)\|\triangle^2 u\| \\
&\le [BC_4^q C_5\|\triangle^2 u\|^{\frac{qn}{8}}\|u\|^{q(1-\frac{n}{8})}\|\triangle^2 u\|^{\frac13}\|u\|^{\frac23} \\
&\quad + BC_4^{q-1}C_6\|\triangle^2 u\|^{\frac{(q-1)n}{8}}\|u\|^{(q-1)(1-\frac{n}{8})}\|\triangle^2 u\|^{\frac{n}{6}}\|\nabla u\|^{2-\frac{n}{6}}], \\
\|\triangle^2 u\| &\le C_7(\|\triangle^2 u\|^{\frac{3nq+8}{24}} + \|\triangle^2 u\|^{\frac{(3q+1)n}{24}})\|\triangle^2 u\| \\
&\le \frac{\gamma}{6}\|\triangle^2 u\|^2 + C_8(\|u\|_{H^1}).
\end{aligned}$$

In the above we used the following Sobolev interpolation inequalities:

$$\|u\|_\infty \le C_4\|\triangle^2 u\|^{\frac{n}{8}}\|u\|^{1-\frac{n}{8}} + C_1'\|u\|,$$

$$
\begin{aligned}
\|\Delta u\| &\le C_5\|\Delta^2 u\|^{\frac{1}{3}}\|\Delta u\|^{\frac{2}{3}} + C_5'\|u\|,\\
\|\nabla u\|_\infty &\le C_6\|\Delta^2 u\|^{\frac{n}{6}}\|\nabla u\|^{1-\frac{n}{6}} + C_6'\|u\|,\\
(\Delta^2 u, g(u)) &= (\Delta u, \Delta g(u)) = (\Delta u, g'(u)\Delta u + g''(u)(\nabla u)^2)\\
&\le g_0\|\Delta u\|^2 + \|g''(u)\|_\infty\|\nabla u\|\|\nabla u\|_\infty\|\Delta u\|\\
&\le g_0\|\Delta u\|^2 + C\|u\|_\infty^l\|\nabla u\|\|\nabla u\|_\infty\|\Delta u\|\\
&\le g_0\|\Delta u\|^2 + CC_4^l\|\Delta^2 u\|^{\frac{ln}{8}}\|u\|^{l(1-\frac{n}{8})}C_6\|\Delta^2 u\|^{\frac{n}{6}}\|\nabla u\|^{\frac{8}{3}-\frac{n}{6}}C_5\|\Delta^2 u\|^{\frac{1}{3}}\\
&\le g_0\|\Delta u\|^2 + C_9\|\Delta^2 u\|^{\frac{(3l+4)n+8}{24}}\\
&\le g_0\|\Delta u\|^2 + \frac{\gamma}{6}\|\Delta^2 u\|^2 + C_{10}(\|u\|_{H^1}),\\
|(\Delta^2 u, h)| &\le \frac{\gamma}{12}\|\Delta^2 u\|^2 + \frac{3}{\gamma}\|h\|^2,\\
\alpha\|\nabla\Delta u\|^2 &\le \alpha C_*\|\nabla u\|^{\frac{1}{3}}\|\Delta^2 u\|^{\frac{2}{3}} \le \frac{\gamma}{6}\|\Delta^2 u\|^2 + C_{11}(\|u\|_{H^1}).
\end{aligned}
$$

Thus, by using (1.2.18), it follows that

$$
\frac{1}{2}\frac{\mathrm{d}}{\mathrm{d}t}\|\Delta u\|^2 + \frac{\gamma}{4}\|\Delta^2 u\|^2 \le g_0\|\Delta u\|^2 + \frac{3}{\gamma}\|h(x)\|^2 + C_{12}(\|u(\cdot,t)\|_{H^1}),
$$

which yields

$$
\|\Delta u(\cdot,t)\|^2 \le e^{2g_0 t}\|\Delta u_0(x)\|^2 + \frac{1}{|g_0|}(1 - e^{2g_0 t})\Big(\frac{3}{\gamma}\|h\|^2 + C_{13}\Big),
$$

making equations (1.2.15), (1.2.16) and (1.2.17) valid. □

Lemma 1.2.4. *Under the conditions of Lemma* 1.2.3, *suppose that*

(1) $f(u) \in C^2, \varphi(u) \in C^3, g(u) \in C^1$,

$$
|\varphi''(u)| + |\varphi'''(u)| \le K, \quad K > 0, \quad 4 \le n < 5,
$$
$$
|g'(u)| \le C|u|^{l-1}, \quad 0 \le l < \frac{40}{3n} - \frac{4}{3};
$$

(2) $u_0(x) \in H^2(\Omega), h(x) \in H^1(\Omega)$.

Then the smooth solution of the problem (1.2.3)–(1.2.5) *satisfies the following estimate:*

$$
\|\nabla\Delta u(\cdot,t)\| \le \frac{E_3}{t}, \quad t > 0, \tag{1.2.19}
$$

where the constant E_3 *depends on* $\|u_0(x)\|_{H^2}$ *and* $\|h(x)\|_{H^1}$.

Proof. Taking the inner product of (1.2.3) and $t^2\Delta^3 u$, we have

$$
(t^2\Delta^3 u, u_t + \alpha\Delta u + \gamma\Delta^2 u + \nabla\cdot f(u) + \Delta\varphi(u) - g(u) - h(u)) = 0. \tag{1.2.20}
$$

By direct calculation we have:

$$
\begin{aligned}
(t^2\triangle^3 u, u_t) &= -\frac{1}{2}\frac{\mathrm{d}}{\mathrm{d}t}\|t\nabla\triangle u\|^2 + \|t^{\frac{1}{2}}\nabla\triangle u\|^2, \\
(t^2\triangle^3 u, \alpha\triangle u) &= \alpha t^2\|\triangle^2 u\|^2, \\
\gamma(t^2\triangle^3 u, \triangle^2 u) &= -\gamma\|t\nabla\triangle^2 u\|^2, \\
|(\nabla\cdot f(u), \triangle^3 u)| &= |(\nabla(\nabla\cdot f(u)), \nabla\triangle^2 u)| \\
&\le C_1\Big[\max_{k=1,\dots,n}\|f_k''(u)\|_\infty\|\nabla u\|_\infty\|\nabla u\| + \|f_k'(u)\|_\infty\|\triangle u\|\Big]\|\nabla\triangle^2 u\|.
\end{aligned}
$$

When $n < 4$, using the following interpolation inequalities:

$$
\begin{aligned}
\|u\|_\infty &\le C_2\|\triangle u\|^{\frac{n}{4}}\|u\|^{1-\frac{n}{4}} + C_2' \le \text{const}, \\
\|\nabla u\|_\infty &\le C_3\|\triangle u\|^{1-\frac{n}{8}}\|\nabla\triangle^2 u\|^{\frac{n}{8}} + C_3',
\end{aligned}
$$

we obtain

$$
|(\nabla\cdot f(u), t^2\triangle^3 u)| \le \frac{\gamma}{6}\|t\cdot\nabla\triangle^2 u\|^2 + C_4.
$$

When $4 \le n < 6$, due to the Kato interpolation inequality [148], we obtain

$$
\begin{aligned}
\|u\|_\infty &\le C_5\|u\|_{H^2}^{\frac{10-n}{6}}\|\nabla\triangle u\|^{\frac{n-4}{6}}, \\
\|\nabla u\|_\infty &\le C_6\|u\|_{H^2}^{\frac{8-n}{6}}\|\nabla\triangle u\|^{\frac{n-2}{6}}.
\end{aligned}
$$

Moreover, we have the following estimate:

$$
\begin{aligned}
|(\nabla\cdot f(u), t^2\triangle^3 u)| &\le C_7[\|\nabla\triangle^2 u\|^{\frac{4n-12}{3n}} + \|\nabla\triangle^2 u\|^{\frac{n-4}{4}}]\|\nabla\triangle^2 u\| \\
&\le \frac{\gamma}{6}\|\nabla\triangle^2 u\|^2 + C_8.
\end{aligned}
$$

Using a similar procedure, when $n < 4$, we have

$$
\begin{aligned}
|(\triangle\varphi(u), \triangle^3 u)| &= |(\nabla\triangle\varphi(u), \nabla\triangle^2 u)| \\
&= |(\nabla(\varphi'(u))\triangle u + \varphi''(u)(\nabla u), \nabla\triangle^2 u)| \\
&\le C[\|\varphi''(u)\|_\infty\|\nabla u\|_\infty\|\triangle u\| \\
&\quad + \|\varphi'(u)\|_\infty\|\nabla\triangle u\| + \|\varphi''(u)\|_\infty\|\nabla u\|_\infty^2\|\nabla u\|]\|\nabla\triangle^2 u\| \\
&\le \frac{\gamma}{6}\|\nabla\triangle^2 u\|^2 + C_9.
\end{aligned}
$$

When $4 \le n < 5$, we have

$$
\begin{aligned}
|(\triangle\varphi(u), \triangle^3 u)| &\le C[\|\nabla\triangle^2 u\|^{\frac{n-2}{6}} + \|\nabla\triangle^2 u\|^{\frac{7n-16}{6n}} + \|\nabla\triangle^2 u\|^{\frac{n-2}{3}}]\|\nabla\triangle^2 u\| \\
&\le \frac{\gamma}{6}\|\nabla\triangle^2 u\|^2 + C_{10},
\end{aligned}
$$

$$\begin{aligned}
\alpha t^2\|\triangle^2 u\|^2 &\le \frac{\gamma}{6}\|t\nabla\triangle^2 u\|^2 + C_{11},\\
|(g(u),\triangle^3 u)| &= \|(g'(u)\nabla u,\nabla\triangle^2 u)\| \le \|g'(u)\|_\infty\|\nabla u\|\|\nabla\triangle^2 u\|\\
&\le \frac{\gamma}{6}\|\nabla\triangle^2 u\|^2 + C_{12},\\
|(g(u),\triangle^3 u)| &= C_{13}\|\nabla\triangle^2 u\|^{\frac{(40-n)(n-4)}{18n}}\|\nabla\triangle^2 u\|\\
&\le \frac{\gamma}{6}\|\nabla\triangle^2 u\|^2 + C_{14},\\
|(h(x),\triangle^3 u)| &\le |(\nabla h,\nabla\triangle^2 u)|\\
&\le \frac{\gamma}{6}\|\nabla\triangle^2 u\|^2 + \frac{3}{2\gamma}\|\nabla h\|^2.
\end{aligned}$$

Therefore, from (1.2.20) we have

$$\frac{1}{2}\frac{\mathrm{d}}{\mathrm{d}t}\|t\nabla\triangle u\|^2 + \frac{\gamma}{6}\|t\nabla\triangle^2 u\|^2 \le C_{15}[\|\nabla\triangle u\|^2 + \|\nabla h\|^2 + 1].$$

Using the above equation, we obtain

$$\|\nabla\triangle u(\cdot,t)\| \le \frac{E_3}{t}, \quad t > 0,$$

where the parameter E_3 merely depends on $\|u_0(x)\|_{H^2}$, $\|h(x)\|_{H^1}$ and t, $0 \le t \le T$. □

Now we prove the existence and uniqueness of the global smooth solution of periodic initial problem (1.2.3)–(1.2.5) by Galerkin method. Suppose $\omega_j(x)$, $j = 1, 2, \ldots$, are the eigenfunctions of the equation $\triangle u + \lambda u = 0$ with periodic boundary condition at $\lambda = \lambda_j$; $\{\omega_j\}$ is the standard orthogonal basis. Suppose an approximate solution of problem (1.2.3)–(1.2.5) can be represented as

$$u_N(x,t) = \sum_{j=1}^{N} \alpha_{jN}(t)\omega_j(x) \tag{1.2.21}$$

where $\alpha_{jN}(t)$, $j = 1, 2, \ldots, N$; $N = 1, 2, \ldots$, are undetermined function coefficients for $t \in \mathbf{R}^+$. According to the Galerkin method, the coefficient $\alpha_{jN}(t)$ should satisfy the following first order nonlinear ordinary differential equation [137]

$$(u_{N,t} + \alpha\triangle u_N + \gamma\triangle^2 u_N + \triangle\cdot f(u_N) + \triangle\varphi(u_N) - g(u_N) - h(x), \omega_j(x)) = 0, \tag{1.2.22}$$

$j = 1, 2, \ldots, N$, with the initial value condition

$$(u_N(x,0),\omega_j(x)) = (u_0(x),\omega_j(x)), \quad j = 1, 2, \ldots, N. \tag{1.2.23}$$

Obviously, we obtain

$$\begin{aligned}
(u_{N,t}(x,t),\omega_j(x)) &= \alpha'_{j,N}(t),\\
(u_{N,t}(x,0),\omega_j(x)) &= \alpha_{j,N}(0),
\end{aligned}$$

and $u_{0,j}(x) = (u_0(x), \omega_j(x))$, $j = 1, 2, \dots, N$, are the coefficients of the approximate expansion $\sum_{j=1}^{N} u_{0,j}\omega_j$ of the function $u_0(x)$.

Similarly as in the proofs of Lemmas 1.2.1–1.2.4, we can establish uniform estimates of the approximate solutions by Galerkin method. These uniform estimates ensure the existence of the global solution $\alpha_{j,N}(t)$, $j = 1, 2, \dots, N$; $0 \le t \le T$, for the problem (1.2.22)–(1.2.23). And we can prove that the approximate solutions $u_N(x,t)$ of problem (1.2.22)–(1.2.23) converge to the global solution of problem (1.2.3)–(1.2.5).

Lemma 1.2.5. *Suppose the following conditions are valid:*

(1) $\varphi'(u) \le \varphi_0, \gamma > \frac{\alpha+\varphi_0}{2}$;

(2) $g(0) = 0, g'(u) \le g_0$;

(3) $|f^{(k)}(u)| \le A|u|^{p-k}, k = 0, 1, 1 \le p \le \frac{6}{n} + 1,$
$|\varphi^{(k)}(u)| \le B|u|^{q-k-1}, k = 1, 2, 0 \le q < 1 + \frac{4}{n}, n < 4,$
$|\varphi'(u)| \le B|u|^q, 0 \le q < \frac{4}{n}, n > 4,$
$|\varphi''(u)| + |\varphi'''(u)| \le K, K > 0,$
$|g^{(k)}(u)| \le C|u|^{l+2-k}, k = 1, 2, 0 < l < \frac{40}{3n} - \frac{4}{3};$

(4) $f(u) \in C^2, \varphi(u) \in C^3, g(u) \in C^2$;

(5) $h(x) \in H^1(\Omega), u_0(x) \in H^3(\Omega)$.

Then the solutions of problem (1.2.22)–(1.2.23) *satisfy the following estimate:*

$$\sup_{0 \le t \le T} \|u_N(\cdot,t)\|^2_{H^3(\Omega)} \le K_3, \tag{1.2.24}$$

where the constant K_3 depends on $\|u_0(x)\|_{H^3(\Omega)}$ and $\|h(x)\|_{H^1(\Omega)}$, and does not depend on N.

Lemma 1.2.6. *Suppose the conditions of Lemma* 1.2.5 *are satisfied and*

(1) $f(u) \in C^{2m-1}, \varphi(u) \in C^{2m}, g(u) \in C^{2m}, m > 2$;

(2) $u_0(x) \in H^{2m}(\Omega), h(x) \in H^{2(m-1)}(\Omega)$.

Then the solutions of problem (1.2.22)–(1.2.23) *satisfy the following estimate:*

$$\sup_{0 \le t \le T} \|\triangle^m u_N(\cdot,t)\| \le K_m \tag{1.2.25}$$

where the constant K_m depends on $\|u_0(x)\|_{H^{2m}}$ and $\|h(x)\|_{H^{2m-2}}$, and does not depend on N.

Proof. From the equation $\triangle\omega_j + \lambda_j\omega_j = 0$, it follows that

$$\triangle^{2m}\omega_j - \lambda_j^{2m}\omega_j = 0, \quad m > 2.$$

Multiplying (1.2.22) by $-\lambda_j^{2m}\alpha_{jN}(t)$, and summing over all $j = 1, 2, \dots, N$,

$$(u_{N,t} + \alpha\triangle u_N + \gamma\triangle^2 u_N + \triangle\cdot f(u_N) + \triangle\varphi(u_N) - g(u_N) - h(x), \triangle^{2m}u_N) = 0, \tag{1.2.26}$$

where

$$(u_{N,t}, \triangle^{2m} u_N) = \frac{1}{2}\frac{\mathrm{d}}{\mathrm{d}t}\|\triangle^m u_N\|^2,$$
$$\alpha(\triangle u_N, \triangle^{2m} u_N) = (-1)^{2m-1}\alpha\|\nabla\triangle^m u_N\|^2,$$
$$\gamma(\triangle^{2m} u_N, \triangle^2 u_N) = \gamma\|\triangle^{m-1} u_N\|^2.$$

By the Lemma 1.2.5 and Sobolev imbedding theorem, one deduces

$$\sup_{0\le t\le T}\|u_N(\cdot,t)\|_\infty \le K_4, \quad 1\le n<5,$$

where the constant K_4 does not depend on N. On account of the following equalities:

$$\|D^s f(u)\| \le C_s(\|u\|_\infty)\|D^s u(s)\|, \quad f(u)\in C^s,$$

one deduces that

$$\begin{aligned}
|(\triangle\cdot f(u_N), \triangle^{2m} u_N)| &= |(\triangle^{m+1} u_N, \triangle^{m-1}\nabla\cdot f(u_N))| \\
&\le \frac{\gamma}{6}\|\triangle^{m+1} u_N\|^2 + C_{2m-1}(\|u_N\|_\infty\|\nabla^{2m-1} u_N\|^2), \\
|(\triangle^{2m} u_{2N}, \triangle\varphi(u_N))| &= |(\triangle^{m+1} u_N, \triangle^{m-1}(\varphi'(u_N)\triangle u_N + \varphi''(u_N)(\nabla u_N)^2))| \\
&\le C^0_{2m-2}[\|\triangle^{m-1}\varphi'(u_N)\|\|\triangle u_N\| + \|\triangle^m u_N\|\|\varphi'(u_N)\| \\
&\quad +\|\triangle^{m-1}\varphi''(u_N)\|\|\nabla u_N\|^2 + \|\triangle^{m-1}(\nabla u_N)^2\|\|\varphi''(u_N)\|]\|\triangle^{m-1} u_N\| \\
&\le C^0_{2m-2}[C^1_{2m-2}\|\triangle^{m-1} u_N\|\|\triangle u_N\| + \|\triangle^m u_N\|\|\varphi'(u_N)\| \\
&\quad +C^2_{2m-2}\|\triangle^{m-1} u_N\|\|\nabla u_N\|_\infty\|\nabla u_N\| + 2(\|\nabla^{2m-1} u_N\|\|\nabla u_N\|_\infty \\
&\quad +\|\triangle u_N\|_\infty\|\nabla^{2m-2} u_N\|)\|\varphi''(u_N)\|]\|\triangle^{m-1} u_N\| \\
&\le C_1[\|\triangle^{m-1} u_N\|^{\frac{2m-5}{2m-1}+\frac{n}{2(2m+1)}} + \|\triangle^{m-1} u_N\|^{\frac{2m-5}{2m-1}+\frac{n}{4m}}]\|\triangle^{m-1} u_N\| \\
&\le \frac{\gamma}{6}\|\triangle^{m+1} u_N\|^2 + C_2.
\end{aligned}$$

The above inequality utilizes the following Sobolev interpolation inequalities:

$$\|\nabla u_N\|_\infty \le C\|\triangle^{m-1} u_N\|^{\frac{n}{2(2m+1)}}\|\nabla u_N\|^{1-\frac{n}{2(2m-1)}} + C',$$
$$\|\triangle^{m-1} u_N\| \le C\|\triangle^{m+1} u_N\|^{\frac{2m-5}{2m-1}}\|\nabla\triangle u_N\|^{1-\frac{2m-5}{2m-1}} + C',$$
$$\|\triangle u_N\| \le C\|\triangle^{m-1} u_N\|^{\frac{n}{4m}}\|\triangle u_N\|^{1-\frac{n}{4m}} + C'.$$

In a similar way, we obtain

$$\begin{aligned}
|(\triangle^{2m} u_N, g(u_N))| &= |(\triangle^m u_N, \triangle^m g(u_N))| \\
&\le |(\triangle^m u_N, g'(u_N)\nabla^m u_N)| \\
&\quad +|(\triangle^m u_N, \nabla^{2m-1}(g'(u_N)\nabla u_N) - g'(u_N)\triangle^m u_N)| \\
&\le g_0\|\triangle^m u_N\|^2 + \frac{\gamma}{6}\|\triangle^{m-1} u_N\|^2 + C_3,
\end{aligned}$$

$$
\begin{aligned}
|(\triangle^{2m}u_N, h)| &= |(\triangle^{m-1}u_N, \triangle^{m-1}h)| \\
&\le \frac{\gamma}{12}\|\triangle^{m-1}u_N\|^2 + \frac{3}{\gamma}\|\triangle^{m-1}h\|^2, \\
\alpha\|\nabla^{2m+1}u_N\|^2 &\le \frac{\gamma}{6}\|\triangle^{m-1}u_N\|^2 + C_4,
\end{aligned}
$$

then, on account of (1.2.26), we deduce

$$\frac{1}{2}\frac{\mathrm{d}}{\mathrm{d}t}\|\triangle^m u_N\|^2 + \frac{\gamma}{4}\|\triangle^{m+1}u_N\|^2 \le g_0\|\triangle^m u_N\|^2 + \frac{3}{\gamma}\|\triangle^{m-1}h\|^2 + C_5.$$

Moreover, we get

$$\sup_{0\le t\le T}\|\triangle^m u_N\|^2 \le K_m,$$

where the constant K_m does not depend on N. □

Theorem 1.2.1. *Under the conditions of Lemma* 1.2.6, *there exists a unique global smooth solution* $u(x,t)$ *of the periodic initial value problem* (1.2.3)–(1.2.5) *such that*

$$
\begin{aligned}
u(x,t) &\in L^\infty(0,T;H^{2m}(\Omega)), \\
u_t(x,t) &\in L^\infty(0,T;H^{2m-4}(\Omega)), \quad m > 1.
\end{aligned}
$$

Proof. On account of Lemmas 1.2.5 and 1.2.6, we have

$$\sup_{0\le t\le T}\|u_N(\cdot,t)\|_{H^{2m}(\Omega)} \le K_m,$$

where the constant K_m does not depend on N. Thus we can choose a subsequence $\{u_{N_i}(x,t)\}$ from the approximate sequence $\{u_N(x,t)\}$ such that there exists a function $u(x,t) \in L^\infty(0,T;H^{2m}(\Omega))$ which satisfies the following conditions:

- $u_{N_i}(x,t) \to u(x,t)$ is weak $*$ convergent in $L^\infty(0,T;H^{2m}(\Omega))$,
- $\triangle u_{N_i}(x,t) \to \triangle u(x,t)$ is strongly and almost surely convergent in $L^\infty(0,T;L_2(\Omega))$ as $N_i \to \infty$.

Using the equation

$$\left(\frac{\partial u_N}{\partial t}, \frac{\partial u_N}{\partial t} + \alpha\triangle u_N + \gamma\triangle^2 u_N + \nabla f(u_N) + \triangle\varphi(u_N) - g(u_N) - h(x)\right) = 0,$$

we obtain

$$\left\|\frac{\partial u_N}{\partial t}\right\| \le K'_m,$$

where the constant K'_m does not depend on N. Therefore, we have

- $\frac{\partial u_{N_i}}{\partial t} \to \frac{\partial u}{\partial t}$ is weak $*$ convergent in $L^\infty(0,T;L_2(\Omega))$ as $N_i \to \infty$.

It follows that the function $u(x,t)$ almost always satisfies equation (1.2.3) and the periodic initial value conditions (1.2.4) and (1.2.5). So the existence of a smooth solution of problem (1.2.3)–(1.2.5) is proved. The uniqueness of the smooth solution can be readily obtained by the energy method. □

To prove the existence of a global attractor for periodic initial value problem (1.2.3)–(1.2.5), we use the following result by Babin–Vishik [7]:

Theorem 1.2.2. *Suppose E is a Banach space, $\{S_t\}_{t\geq 0}$ is a semigroup of operators $S_t : E \to E$:*

$$S_t \cdot S_\tau = S_{t+\tau}, \quad S_0 = I,$$

where I is the identity operator, and S_t satisfies the following conditions:

(1) *Operator S_t is bounded uniformly, i. e., if $\forall R > 0$, $\|u\|_E \leq R$, then there exists a constant $C(R)$ such that*

$$\|S_t u\|_E \leq C(R), \quad t \in [0,\infty);$$

(2) *There exists a bounded absorbing set $B_0 \subset E$, i. e., for any bounded set $B \subset E$, there exists a constant T such that*

$$S_t B \subset B_0, \quad t \geq T;$$

(3) *S_t is a completely continuous operator for all $t > 0$.*

Then the semigroup of operators S_t possesses a compact global attractor.

Theorem 1.2.3. *Suppose the problem* (1.2.3)–(1.2.5) *possesses a global smooth solution and satisfies:*

(1) $f(u) \in C^2, \varphi(u) \in C^3, g(u) \in C^2$,

$$|f^{(k)}(u)| \leq A|u|^{p-k}, \quad k = 0,1,\ 1 \leq p < 1 + \frac{6}{n},$$

$$|\varphi^{(k)}(u)| \leq B|u|^{q-k+1}, \quad k = 1,2,\ 0 \leq q < 1 + \frac{4}{n},\ n < 4,$$

$$|\varphi'(u)| \leq B|u|^q, \quad 0 \leq q < \frac{4}{n},\ n \geq 4;$$

$$|\varphi''(u)| + |\varphi'''(u)| \leq K, \quad K > 0;$$

(2) $\varphi'(u) \leq \varphi_0, \gamma > \frac{\alpha+\varphi_0}{2}$;

(3) $g(0) = 0, g'(u) \leq g_0, g_0 < -\frac{\alpha+\varphi_0+1}{2}$,

$$|g^{(k)}| \leq C|u|^{l+2-k}, \quad k = 1,2,\ 0 \leq l < \frac{40}{3n} - \frac{4}{3};$$

(4) $u_0(x) \in H^2(\Omega), h(x) \in H^1(\Omega), \Omega \subset \mathbf{R}^n, 1 \leq n \leq 5$.

Then there exists a global attractor of the periodic initial value problem (1.2.3)–(1.2.5), *i. e., there exists a set* $\mathscr{A} \subset H^2(\Omega)$ *such that*

(1) $S_t\mathscr{A} = \mathscr{A}$, $t \in \mathbf{R}^+$;

(2) $\lim_{t\to\infty} \operatorname{dist}(S_tB, \mathscr{A}) = 0$, *where* B *is an arbitrary bounded set* $\subset H^2(\Omega)$,

$$\operatorname{dist}(X, Y) = \sup_{x\in X} \inf_{y\in Y} \|x - y\|_E,$$

where S_t *is the semigroup operator generated by problem* (1.2.3)–(1.2.5).

Proof. We will verify conditions (1)–(3) of Theorem 1.2.2.

Under the assumptions of Theorem 1.2.3, we know that there exists a semigroup operator generated by problem (1.2.3)–(1.2.5). Denote the Banach space $E = H^2(\Omega)$, $S_t : H^2(\Omega) \to H^2(\Omega)$. Using Lemmas 1.2.1–1.2.3, and letting $B \subset H^2(\Omega)$ be contained in the ball $\{\|u\|_{H^2} \le R\}$, we have

$$\|S_tu_0\|_{H^2}^2 = \|u(\cdot, t)\|_{H^2}^2 \le \|u_0(x)\|_{H^2}^2 + C_1\|h(x)\|^2 + C_2 \le R^2 + C', \quad t \ge 0, \quad u_0 \in B,$$

where C_1, C_2 and C' are constants. This means that $\{S_t\}$ is uniformly bounded in $H^2(\Omega)$. Secondly, using the mentioned lemmas, we deduce that

$$\|S_tu_0\|_{H^2} = \|u(\cdot, t)\|_{H^2}^2 \le 2(E_0 + E_1 + E_2), \quad t \ge t_0 = t_0(R, \|h(x)\|). \tag{1.2.27}$$

Thus $\bar{A} = \{u(\cdot, t) \in H^2(\Omega), \|u\|_{H^2} \le 2(E_0 + E_1 + E_2)\}$ is a bounded absorbing set of the semigroup operator S_t. By Lemma 1.2.4 we have

$$\|\nabla\triangle u(\cdot, t)\| \le \frac{E(R, t)}{t}, \quad t > 0,$$

where $\|u_0\|_{H^2} \le R$. By compact imbedding, $H^3(\Omega) \hookrightarrow H^2(\Omega)$, we obtain that the semigroup operator S_t is completely continuous when $t > 0$. So the theorem is proved. □

Remark 1.2.1. As stated in Theorem 1.1.1 of [197], the global attractor $\mathscr{A}$ obtained in Theorem 1.2.3 is the ω limit of absorbing set $\bar{A}$, i. e.,

$$\mathscr{A} = \omega(\bar{A}) = \bigcap_{\tau\ge 0} \overline{\bigcup_{t\ge\tau} S_t\bar{A}}.$$

To establish estimates of the Hausdorff and fractal dimensions of the global attractor for the problem (1.2.3)–(1.2.5), we consider the following linear variational problem of (1.2.3)–(1.2.5):

$$v_t + L(u(t))v = 0, \tag{1.2.28}$$

$$v(0) = v_0, \tag{1.2.29}$$

where $L(u(t))v = \alpha\triangle v + \gamma\triangle^2 v + \sum_{k=1}^n (f_k(u)v)_{x_k} + \triangle(\varphi'(u)v) - g'(u)v$.

Since the solutions of problem (1.2.3)–(1.2.5) are smooth enough, one can readily prove that the linear problem (1.2.28)–(1.2.29) possesses a global smooth solution when the initial data $v_0(x)$ is smooth enough, i. e., there exists a solution operator G_t such that $v(t) = G_t v_0$. Also it is readily proved that the semigroup operator $S_t u_0$ is differentiable in $L_2(\Omega)$, i. e., the Fréchet differential $S'_t u_0$ exists, and $G_t v_0 = S'_t u_0$. Actually, setting

$$\omega(t) = S_t(u_0 + v_0) - S_t(u_0) - G_t(u_0)v_0 = u_1(t) - u(t) - v(t),$$

we have

$$\begin{aligned} \partial_t \omega(t) &= L_1(u_1(t)) - L_1(u(t)) + L(u(t))v(t) \\ &= L_1(u(t) + v(t) + \omega(t)) - L_1(u(t)) + L(u(t))v(t), \end{aligned} \tag{1.2.30}$$

$$\omega(0) = 0, \tag{1.2.31}$$

where $u_t = L_1(u)$ is the operator form of equation (1.2.3). Thus equation (1.2.30) can be rewritten as

$$\partial_t \omega(t) + L(u(t))\omega = \Lambda_0(u, v, \omega), \tag{1.2.32}$$

where

$$\Lambda_0(u, v, \omega) = L_1(u(t) + v(t) + \omega(t)) - L_1(u(t)) + L(u(t))(v + \omega). \tag{1.2.33}$$

Using the theory of linear partial differential equations, we can obtain the following L_2 estimate:

$$\|\omega(t)\| \le C\|v_0\|^2, \tag{1.2.34}$$

which infers that the semigroup operator S_t is differentiable in $L_2(\Omega)$.

Let $v_1(t), v_2(t), \dots, v_J(t)$ be the solutions of the linear equation corresponding to the initial values $v_1(0) = \xi_1, \dots, v_J(0) = \xi_J$, where $\xi = (\xi_1, \xi_2, \dots, \xi_J) \in L_2$. Using the results of [197], we arrive at

$$\frac{\mathrm{d}}{\mathrm{d}t}\|v_1(t) \wedge \cdots \wedge v_J(t)\|^2 + 2\,\mathrm{tr}(L(u(t)) \cdot Q_J)\|v_1(t) \wedge \cdots \wedge v_J(t)\|^2 = 0, \tag{1.2.35}$$

where $L(u(t)) = L(S_t u_0)$ is the linear map $v \to L(u(t))v$, "$\wedge$" represents the wedge, tr represents the trace of an operator, $Q_J(t)$ is the orthogonal projection from the space $L_2(\Omega)$ to the subspace $\mathrm{span}\{v_1(t), \dots, v_J(t)\}$. Thus, by equation (1.2.31), we obtain that the volume of a J-dimensional cube is bounded by

$$\begin{aligned} \omega_J(t) &= \sup_{u_0 \in \mathscr{A}} \sup_{\xi_j \in L_2, |\xi_j| \le 1} \|v_1(t) \wedge \cdots \wedge v_J(t)\|^2_{\Lambda^J, L_2} \\ &\le \sup_{u_0 \in \mathscr{A}} \exp\left(-\int_0^t \inf(\mathrm{tr}\, L(S_\tau u_0) \cdot Q_J(\tau)) d\tau \right). \end{aligned} \tag{1.2.36}$$

Notice that by the results of [197], $\omega_j(t)$ is log-subadditive, i. e.,

$$\omega_j(t+t') \le \omega_j(t)\omega_j(t'), \quad t, t' \ge 0. \tag{1.2.37}$$

Therefore, we have

$$\lim_{t\to\infty} \omega_j(t)^{\frac{1}{t}} = \Pi_j, \quad 1 \le j \le J, \tag{1.2.38}$$

$$\Pi_j \le \exp(-q_J), \tag{1.2.39}$$

where

$$q_J = \lim_{t\to\infty} \sup\left(\inf_{u_0\in\mathscr{A}} \frac{1}{t} \int_0^t \inf(\operatorname{tr} L(S(\tau)u_0) \cdot Q_J(\tau))d\tau \right). \tag{1.2.40}$$

Definition 1.2.1. The Hausdorff measure [55, 56] of a set X is defined as

$$n_H(X,d) = \lim_{\epsilon\to 0} n_H(X,d,\epsilon) = \sup_{\epsilon>0} n_H(X,d,\epsilon)$$

where

$$n_H(X,d,\epsilon) = \inf \sum_i r_i^d,$$

and inf is taken over all coverings of X by balls with radius $r_i \le \epsilon$. If there exists a number $d = d_H(X) \in [0,+\infty]$ such that

$$\mu_H(X,d) = 0, \quad d > d_H(X),$$
$$\mu_H(X,d) = \infty, \quad d < d_H(X),$$

then the number $d_H(X)$ is called the Hausdorff dimension of the set X.

Definition 1.2.2. The fractal dimension of a set X is defined as the number

$$d_F(X) = \lim \sup_{\epsilon>0} \frac{\lg n_X(\epsilon)}{\lg \frac{1}{\epsilon}}$$

where $n_X(\epsilon)$ denotes the minimum number of balls with radius $\le \epsilon$ in a covering of the set X. By the results in [197],

$$d_F(X) = \inf\{d > 0,\ n_F(X,d) = 0\}$$

where $n_F(X,d) = \lim\sup_{\epsilon\to 0} \epsilon^d n_X(\epsilon)$.

Theorem 1.2.4 ([33]). *Suppose $\mathscr{A}$ is the attractor of a nonlinear evolution equation (e. g., Navier–Stokes equation, equation* (1.2.3), *etc.). If it is bounded in $H^1(\Omega)$, then for a certain J, the Hausdorff dimension of $\mathscr{A} \le J$. Its fractal dimension is no more than*

$$J\left(1 + \max_{1\le i\le J} \frac{-q_i}{q_J}\right). \tag{1.2.41}$$

Lemma 1.2.7 (Generalized Sobolev–Lieb–Thirring inequality, [197]). *Suppose $\Omega \subset \mathbf{R}^n$ is a bounded domain, $\{\varphi_1, \varphi_2, \dots\}$, $\varphi_j \in H^1(\Omega)$ is an orthogonal basis in $L_2(\Omega)$. Then we have the estimate*

$$\int_\Omega \left(\sum_{j=1}^N |\varphi_j(x)|^2 \right)^{1+\frac{2}{n}} \mathrm{d}x \le k_0 \sum_{j=1}^N \int_\Omega |\mathrm{grad}\, \varphi_j|^2 \mathrm{d}x. \tag{1.2.42}$$

Theorem 1.2.5. *Under the conditions of Theorem 1.2.3, the Hausdorff and fractal dimensions of the global attractor for problem (1.2.3)–(1.2.5) is finite and*

$$d_H(\mathscr{A}) \le J_0, \quad d_F(\mathscr{A}) \le 2J_0,$$

where J_0 is the minimum integer satisfying the following equations:

$$\begin{aligned} J_0 \ge{}& \frac{1}{2\gamma c'} [(\alpha + \|\varphi'(u)\|_\infty) c'^{\frac{1}{2}} + (\alpha + \|\varphi'(u)\|_\infty)^2 c' \\ &+ 4\gamma c' [k_0 c'^{\frac{1}{4}} \|f''(u)\|_\infty (\|\nabla u\| + \|\nabla u\|_{L_{\frac{5}{2}}}) \\ &+ \|\varphi'''(u)\|_\infty \|\nabla u\|_\infty^2 + \|\varphi''(u)\|_\infty \|\Delta u\|_\infty]^{\frac{1}{2}}], \quad n = 2, \end{aligned} \tag{1.2.43}$$

$$J_0 \ge \left(\frac{4\nu_1}{c'\gamma} \right)^{\frac{3}{7}}, \quad n = 3, \tag{1.2.44}$$

$$\begin{aligned} \nu_1 ={}& \frac{2}{7} \left(\frac{5c'\gamma}{28} \right)^{-\frac{5}{2}} [c'^{\frac{1}{2}} (\alpha + \|\varphi'(u)\|_\infty)]^{\frac{2}{7}} \\ &+ \frac{9}{14} \left(\frac{5c'\gamma}{56} \right)^{-\frac{5}{9}} k_0 \|f''(u)\|_\infty c'^{\frac{1}{4}} (\|\nabla u\| + \|\nabla u\|_{L_{\frac{5}{2}}})^{\frac{14}{9}} \\ &+ \frac{4}{7} \left(\frac{3c'\gamma}{28} \right)^{-\frac{3}{4}} (\|\varphi'''(u)\|_\infty \|\nabla u\|_\infty^2 + \|\varphi''(u)\|_\infty \|\Delta u\|_\infty)^{\frac{7}{4}}, \end{aligned} \tag{1.2.45}$$

where c' and k_0 are absolute constants, and

$$\|f''(u)\|_\infty = \max_{k=1,\dots,n} \|f_k''(u)\|_\infty.$$

Proof. Based on Theorem 1.2.4, we must estimate the lower bound of $\mathrm{tr}(L(u(t)) \cdot Q_J)$, which we do as follows:

$$\begin{aligned} &\mathrm{tr}(L(u(t)) \cdot Q_J) \\ &= \sum_{j=1}^J \left[\left(\alpha \Delta \varphi_j + \gamma \Delta^2 \varphi_j + \sum_{k=1}^n (f_k'(u)\varphi_j)_{x_k} - \Delta(\varphi'(u)\varphi_j) - g'(u)\varphi_j, \varphi_j \right) \right] \\ &= \sum_{j=1}^J \left[\left(\gamma \|\Delta \varphi_j\|^2 - \alpha \|\nabla \varphi_j\|^2 + \left(\sum_{k=1}^n f_k'(u)\varphi_j \right)_{x_k} - \Delta(\varphi'(u)\varphi_j) - g'(u)\varphi_j, \varphi_j \right) \right] \end{aligned} \tag{1.2.46}$$

where

$$\left\|\sum_{j=1}^{J}\left(\left(\sum_{k=1}^{n} f_k'(u)\varphi_j\right)_{x_k}, \varphi_j\right)\right\| = \left\|\sum_{j=1}^{J}\left(\sum_{k=1}^{n} f_k'(u)\varphi_j\right), \varphi_{jx_k}\right\|$$

$$= \frac{1}{2}\left|\left(\sum_{k=1}^{n} f_k''(u)u_{x_k}, \sum_{j=1}^{J}\varphi_j^2\right)\right| = \frac{1}{2}\left|\left(\sum_{k=1}^{n} f_k''(u)u_{x_k}, \rho(x)\right)\right|,$$

$$\rho(x) = \sum_{j=1}^{J}\varphi_j^2.$$

By Lemma 1.2.7, we have

$$\left|\sum_{k=1}^{n}(f_k''(u)u_{x_k}, \rho(x))\right| \le \left\|\sum_{k=1}^{n} f_k''(u)u_{x_k}\right\|\|\rho(x)\|$$

$$\le \|f_k''(u)\|_\infty\|\nabla u\|k_0\left(\sum_{j=1}^{J}\int|\nabla\varphi_j|^2\mathrm{d}x\right)^{\frac{1}{2}}, \quad n = 2,$$

$$\left|\left(\sum_{k=1}^{n} f_k''(u)u_{x_k}, \rho(x)\right)\right| \le \left(\int_\Omega\left|\sum_{k=1}^{n} f_k''(u)u_{x_k}\right|^{\frac{5}{2}}\mathrm{d}x\right)^{\frac{2}{5}}\left(\int_\Omega|\rho(x)|^{\frac{5}{3}}\mathrm{d}x\right)^{\frac{3}{5}}$$

$$\le \|f''(u)\|_\infty\|\nabla u\|_{L^{\frac{5}{2}}}k_0\left(\sum_{j=1}^{J}\int|\nabla\varphi_j|^2\mathrm{d}x\right)^{\frac{1}{2}}, \quad n = 3,$$

$$\begin{aligned}
|-(\triangle(\varphi'(u)\varphi_j), \varphi_j)| &= |(\nabla\varphi'(u)\varphi_j, \nabla\varphi_j)| \\
&= |(\varphi''(u)\nabla u\varphi_j, \nabla\varphi_j) + (\varphi'(u)\nabla\varphi_j, \nabla\varphi_j)| \\
&= \left|-\frac{1}{2}(\nabla\varphi''(u)\nabla u, \varphi_j^2) + (\varphi'(u)\nabla\varphi_j, \nabla\varphi_j)\right| \\
&= \left|-\frac{1}{2}(\varphi'''(u)(\nabla u)^2, \varphi_j^2) - \frac{1}{2}(\varphi''(u)\triangle u, \varphi_j^2) + (\varphi'(u)\nabla\varphi_j, \nabla\varphi_j)\right| \\
&\le \|\varphi'''(u)\|_\infty\|\nabla u\|_\infty^2\|\varphi_j\|^2 \\
&\quad + \|\varphi''(u)\|_\infty\|\triangle u\|_\infty\|\varphi_j\|^2 + \|\varphi'(u)\|_\infty\|\varphi_j\|^2
\end{aligned}$$

Now select $\varphi_j(x)$s which are the eigenfunctions corresponding to eigenvalues λ_j, $j = 1, 2, \dots$, for equation $-\triangle u = \lambda u$ with the periodic boundary conditions and such that

$$\|\nabla\varphi_j\|^2 = \lambda_j, \quad \|\triangle\varphi_j\|^2 = \lambda_j^2, \quad \|\varphi_j\|^2 = 1.$$

As in [197], we estimate the eigenvalues λ_j as follows:

$$\lambda_j \ge \left[\frac{(j-1)^{\frac{1}{n}}}{2} - 1\right]^2 \sim cj^{\frac{2}{n}}.$$

Hence, from (1.2.46)

$$\begin{aligned}
\operatorname{tr}(L(u(t))\cdot Q_J) &\ge \lambda\sum_{j=1}^{J}\lambda_j^2-\alpha\sum_{j=1}^{J}\lambda_j-\|f''(u)\|_\infty(\|\nabla u\|_{L^{\frac52}}+\|\nabla u\|)k_0\Big(\sum_{j=1}^{J}\lambda_j\Big)^{\frac12}\\
&\quad -J(\|\varphi''(u)\|_\infty\|\nabla u\|_\infty^2+\|\varphi''(u)\|_\infty\|\nabla u\|_\infty)-\|\varphi'(u)\|_\infty\sum_{j=1}^{J}\lambda_j\\
&\ge \gamma\sum_{j=1}^{J}\lambda_j^2-(\alpha+\|\varphi'(u)\|_\infty)J^{\frac12}\Big(\sum_{j=1}^{J}\lambda_j^2\Big)^{\frac12}\\
&\quad -k_0\|f''(u)\|_\infty(\|\nabla u\|+\|\nabla u\|_{L^{\frac52}})J^{\frac14}\Big(\sum_{j=1}^{J}\lambda_j^2\Big)^{\frac14}\\
&\quad -J(\|\varphi'''(u)\|_\infty\|\nabla u\|^2+\|\varphi''(u)\|_\infty\|\triangle u\|_\infty).
\end{aligned}$$

From the inequality $\lambda_j \ge c' j^{\frac{2}{n}}$, when $n = 2$, and

$$\begin{aligned}
J \ge \frac{1}{2\gamma c'}\Big[(\alpha+\|\varphi'(u)\|_\infty)c'^{\frac12}&+\big[(\alpha+\|\varphi''(u)\|_\infty)^2c'\\
&+4\gamma c'(k_0c'^{\frac14}\|f''(u)\|_\infty(\|\nabla u\|+\|\nabla u\|_{L^{\frac52}})\\
&+\|\varphi'''(u)\|_\infty\|\nabla u\|_\infty^2+\|\varphi''(u)\|_\infty\|\triangle u\|_\infty)\big]^{\frac12}\Big]=J_0,
\end{aligned}$$

we have

$$\operatorname{tr}(L(u(t))\cdot Q_J) > 0.$$

By the generalized Young inequality

$$ab \le \epsilon\frac{a^p}{p}+\epsilon^{1-p'}\frac{b^{p'}}{p'},\quad \frac1p+\frac1{p'}=1, \tag{1.2.47}$$

we get

$$\begin{aligned}
&c'^{\frac12}(\alpha+\|\varphi''(u)\|_\infty)J^{\frac53}\le\frac{c'\gamma}{4}J^{\frac72}+\frac27\Big(\frac{5c'\gamma}{28}\Big)^{-\frac52}[c'^{\frac12}(\alpha+\|\varphi'(u)\|_\infty)]^{\frac72},\\
&k_0\|f''(u)\|_\infty c'^{\frac14}(\|\nabla u\|+\|\nabla u\|_{L^{\frac52}})J^{\frac56}\\
&\quad\le\frac{c'\gamma}{4}J^{\frac73}+\frac{9}{14}\Big(\frac{5c'\gamma}{56}\Big)^{-\frac59}[k_0\|f''(u)\|_\infty c'^{\frac14}(\|\nabla u\|+\|\nabla u\|_{L^{\frac52}})]^{\frac{14}{9}},\\
&(\|\varphi'''(u)\|_\infty\|\nabla u\|_\infty^2+\|\varphi''(u)\|_\infty\|\triangle u\|_\infty)J\\
&\quad\le\frac{c'\gamma}{4}J^{\frac73}+\frac47\Big(\frac{3c'\gamma}{28}\Big)^{-\frac34}(\|\varphi''(u)\|_\infty\|\nabla u\|^2+\|\varphi''(u)\|_\infty\|\triangle u\|_\infty)^{\frac74}.
\end{aligned}$$

So when $n = 3$ and

$$J > \left(\frac{4\nu_1}{c'\gamma}\right)^{\frac{3}{7}},$$

we have

$$\mathrm{tr}(L(u(t)) \cdot Q_J) > 0.$$

From the definition of q_l it is easy to verify that

$$-\frac{q_l}{q_J} \le 1, \quad l \le J_0 - 1.$$

So by Theorem 1.2.4 we get

$$d_H(\mathscr{A}) \le J_0, \quad d_F(\mathscr{A}) \le 2J_0. \qquad \square$$

1.3 A type of nonlinear viscoelastic wave equation

Consider the following nonlinear wave equation with viscoelastic item:

$$u_{tt} = \alpha u_{xx} + \sigma(u_x)_x - f(u) + g(x), \quad x \in (0,1), \quad t \in (0,\infty), \tag{1.3.1}$$

with the initial conditions

$$u(0) = u_0, \quad u_t(0) = u_1, \tag{1.3.2}$$

and the boundary conditions

$$u(0,t) = u(1,t) = 0. \tag{1.3.3}$$

The problem comes from the homogeneous beam longitudinal motion, $u(x,t)$ describes the displacement of the beam cross-section at time t. If we use $T(x,t)$ to express the cross-section stress at time t, then equation (1.3.1) can be written as

$$\rho_0 u_{tt} = T_x - f(u) + g(x), \quad x \in (0,1), \quad t \in (0,\infty), \tag{1.3.4}$$

where ρ_0 means the density of the beam. Denote $\rho_0 = 1$. In equation (1.3.4), we make structural assumptions on the stress $T(x,t)$, namely

$$T(x,t) = \alpha u_{xt}(x,t) + \sigma(u_x). \tag{1.3.5}$$

Then we obtain equation (1.3.1), where $\sigma(s)$ is a smooth function, having the property

$$\sigma(0) = 0, \quad \sigma'(s) \ge \gamma_0 > 0, \quad \forall s \in R, \tag{1.3.6}$$

where α and γ_0 are positive constants.

Problem (1.3.1)–(1.3.3) has been studied by many mathematicians and physicists. When $f = g = 0$, in 1968–1969 Greenberg et al. studied the existence and stability of the global classical solution in [81, 80]. In 1981 Fitzgibben proved the existence of a global solution $(u, u_t) \in W^{1,\infty} \times W^{1,2}$ in [57]. In 1984 Chang and Guo constructed its difference scheme in [22] and proved the convergence of the difference solution and the existence of a global solution.

When $\sigma'(s) = 1, \sigma(u_x)_x = u_{xx}, g(x) = 0, f'(u) \le C$, Massat [175] proved the existence of a global attractor in $X^\alpha \times X^\beta$ and its finite dimension in 1983, where $X^\alpha = D(A^\alpha)$, $0 \le \beta \le \alpha < 1, A = -\partial_{xx}$. When $\sigma(s)$ is nonlinear, Berhaliev [15, 16] proved the existence of a global attractor in (E_0, E_1) in 1985, where $E_0 = E_1$. In 1994, Guo et al. [110] proved the existence of a compact global attractor in the space $\mathscr{H} = D(A) \times L^2(\Omega)$ for the problem (1.3.1)–(1.3.3) and obtained dimensional estimates.

Due to the semigroup method, we can readily establish the local existence of a global solution for the problem (1.3.1)–(1.3.3).

Theorem 1.3.1. *Suppose $\sigma(s)$ and $f(u)$ are smooth functions, $\sigma(0) = 0, g(x) \in L^2(\Omega), \Omega = [0, 1]$. Then for $(u_0, u_1) \in \mathscr{H} = D(A) \times L^2(\Omega)$ there exist $t_0 = t_0(u_0, u_1) > 0$ and $u(t, u_0, u_1)$ such that $(u(t), u_t(t))$ from $[0, t_0]$ to $\mathscr{H}$ is continuous, $(u(t), u_t(t)) \in D(A) \times D(A)$, (u_t, u_{tt}) exists for almost all $t \in (0, t_0)$ and $(u(t), u_t(t))$ satisfies* (1.3.1)–(1.3.3).

We provide a priori estimate in the following. For the nonlinear functions, we make the following assumptions:

(i)

$$\lim_{|s|\to\infty} \inf \frac{F(s)}{|s|^2} \ge 0; \tag{1.3.7}$$

(ii) There exists $\omega > 0$ such that

$$\lim_{|s|\to\infty} \frac{sf(s) - F(s)}{|s|^2} \ge 0, \tag{1.3.8}$$

where $F(s) = \int_0^1 f(s)ds$. For (i), suppose $0 < \omega \le \frac{1}{2}$, and let

$$\|v\| = \left(\int_\Omega |v|^2 dx\right)^{\frac{1}{2}}, \quad (u, v) = \int_\Omega u \cdot v dx,$$

$$\|(u, v)\| = (\|u\|_{H^2}^2 + \|v\|^2)^{\frac{1}{2}}.$$

Lemma 1.3.1. *Suppose $\sigma(s)$ satisfies condition* (1.3.6), *$f(s)$ satisfies conditions* (1.3.7) *and* (1.3.8). *Then for a solution of problem* (1.3.1)–(1.3.3) *there exist constants $C_1 = C_1(\|u_0, u_1\|)$ and $C_2 = C_2(\|g\|)$ such that*

$$\|u_x\|^2 + \|u\|^2 + \|u_t\|^2 \le C_1(\|(u_0, u_1)\|)e^{-\omega t} + C_2(\|g\|) \tag{1.3.9}$$

is valid.

Proof. Let $v = u_t + \rho u$, where ρ is a sufficiently small undetermined positive number. Then (1.3.1) becomes

$$v_t - \alpha v_{xx} - \rho v - \sigma(u_x)_x + \alpha\rho u_{xx} + \rho^2 u + f(u) = g(x). \tag{1.3.10}$$

Taking the inner product of (1.3.10) and v, we have

$$\begin{aligned}
&\frac{1}{2}\frac{d}{dt}\left\{\frac{1}{2}\|v\|^2 + \int_\Omega E(u_x)dx - \frac{\alpha\rho}{2}\|u_x\|^2\right.\\
&\left.+\frac{\rho^2}{2}\|u\|^2 + \int_\Omega F(u)dx - (g,u)\right\}\\
&+\left\{\alpha\|v_x\|^2 - \rho\|v\|^2 + \rho\int_\Omega \sigma(u_x)u_x dx - \alpha\rho^2\|u_x\|^2\right.\\
&\left.+\rho^3\|u\|^2 + \rho\int_\Omega f(u)udx - \rho(g,u)\right\} = 0,
\end{aligned} \tag{1.3.11}$$

where $E(s) = \int_0^s \sigma(s)ds$, $F(s) = \int_0^s f(s)ds$. Let

$$H_1(u,v) = \frac{1}{2}\|v\|^2 + \int_\Omega E(u_x)dx - \frac{\alpha\rho}{2}\|u_x\|^2 + \frac{\rho^2}{2}\|u_x\|^2 + \int_\Omega F(u)dx - (g,u), \tag{1.3.12}$$

$$\begin{aligned}
K_1(u,v) = {}&\alpha\|v_x\|^2 - \rho\|v\|^2 + \rho\int_\Omega \sigma(u_x)u_x dx\\
&- \alpha\rho^2\|u_x\|^2 + \rho^3\|u\|^2 + \rho\int f(u)udx - \rho(g,u).
\end{aligned} \tag{1.3.13}$$

Then (1.3.11) can be written as

$$\frac{d}{dt}H_1(u,v) + K_1(u,v) = 0. \tag{1.3.14}$$

Now we estimate $H_1(u,v)$ and $K_1(u,v)$. Select ρ sufficiently small such that

$$\rho < \min\left\{\frac{2}{3}\pi^2\alpha, \frac{\gamma_0}{8\alpha}\right\}. \tag{1.3.15}$$

Since $\sigma(s)$ satisfies (1.3.6), we have

$$\sigma(s)s \ge E(s) \ge \frac{\gamma_0}{2}s^2. \tag{1.3.16}$$

On the other hand, since $f(u)$ satisfies (1.3.7) and (1.3.8), there exist constants C_{11} and C_{12} such that

$$\int_\Omega F(u)dx \ge -\frac{\rho^2}{2}\|u\|^2 - C_{11}, \tag{1.3.17}$$

$$\int_\Omega [f(u)u - \omega F(u)]dx \geq -\frac{\pi\gamma_0}{16}\|u\|^2 - C_{12}. \tag{1.3.18}$$

Hence, since $0 < \omega \leq \frac{1}{2}$, using (1.3.16)–(1.3.18), we have

$$\begin{aligned}
K_1(u,v) - \omega\rho H_1(u,v) = {} & \alpha\|v_x\|^2 - \left(\rho + \frac{\rho\omega}{2}\right)\|v\|^2 + \rho\int_\Omega [\sigma(u_x)u_x - \omega E(u_x)]dx \\
& - \left(\alpha\rho^2 - \frac{\alpha\omega\rho^2}{2}\right)\|u_x\|^2 + \left(\rho^3 - \frac{\rho^3\omega}{2}\right)\|u\|^2 \\
& + \rho\int [f(u)u - \omega F(u)]dx + \rho\omega(g,u) \\
\geq {} & \alpha\|v_x\|^2 - \left(\rho + \frac{\omega\rho}{2}\right)\|v\|^2 + \frac{\rho}{2}\int \sigma(u_x)u_x dx \\
& - \left(\alpha\rho^2 - \frac{\alpha\omega\rho^2}{2}\right)\|u_x\|^2 \\
& + \left(\rho^3 - \frac{\rho^3\omega}{2}\right)\|u\|^2 - \frac{\pi^2\rho\gamma_0}{16}\|u\|^2 - C_{12}\rho - \rho\|g\|\|u\| \\
\geq {} & \left(\alpha\pi^2 - \frac{3\rho}{2}\right)\|v\|^2 + \frac{\gamma_0\rho}{4}\|u_x\|^2 \\
& - \alpha\rho^2\|u_x\|^2 - \frac{\pi^2\rho\gamma_0}{8}\|u^2\| - C_{12}\rho - C(\|g\|) \geq -C(\|g\|),
\end{aligned} \tag{1.3.19}$$

$$\begin{aligned}
H_1(u,v) & \geq \frac{1}{2}\|v\|^2 + \frac{\gamma_0}{2}\|u_x\|^2 - \frac{\alpha\rho}{2}\|u_x\|^2 + \frac{\rho^2}{4}\|u\|^2 - C(\|g\|) \\
& \geq \frac{1}{2}\|v\|^2 + \frac{\gamma_0}{4}\|u_x\|^2 + \frac{\rho^2}{4}\|u\|^2 - C(\|g\|).
\end{aligned} \tag{1.3.20}$$

From (1.3.14) and (1.3.19) we have

$$\frac{d}{dt}H_1(u,v) + \rho\omega H_1(u,v) \leq C(\|g\|). \tag{1.3.21}$$

Gronwall inequality yields

$$H_1(u,v) \leq H_1(u,v)(0)e^{-\omega\rho t} + C(\|g\|)(1 - e^{-\omega\rho t}). \tag{1.3.22}$$

Since $u_0 \in H^2 \cap H_0^1(\Omega)$, $\sigma(s)$ is continuous, by the definition of $H_1(u,v)$ and the embedding theorem, we arrive at $|H_1(u,v)| \leq C(\|(u_0,u_1)\|)$ and from (1.3.20) obtain the conclusion of the lemma. □

Now we estimate the uniform boundedness of $\|u_{xx}\|$. Taking the inner product of (1.3.1) and u_{xx}, we get

$$\frac{\mathrm{d}}{\mathrm{d}t}\left\{\frac{\alpha}{2}\|u_{xx}\|^2 - (u_t, u_{xx})\right\} + (\sigma'(u_x)u_{xx}, u_{xx}) = \|u_{xt}\|^2 + (f(u) - g(x), u_{xx}). \tag{1.3.23}$$

Multiplying (1.3.23) by α and adding equation (1.3.14), we notice that

$$\|u_{xt}+\rho u_x\|^2 \geq \frac{1}{2}\|u_{xt}\|^2 - \rho^2\|u_x\|^2.$$

The latter implies

$$\frac{d}{dt}H_2(u,u_t) + K_2(u,u_t) \leq 0, \tag{1.3.24}$$

where

$$H_2(u,u_t) = \frac{\alpha^2}{2}\|u_{xx}\|^2 - \alpha(u_t,u_{xx}) + H_1(u,v) \tag{1.3.25}$$

$$\begin{aligned} K_2(u,u_t) &= \alpha\int_\Omega \sigma'(u_x)|u_{xx}|^2 dx - \alpha(f(u)-g,u_{xx}) \\ &\quad - \alpha\rho\|u_x\|^2 - \rho\|v\|^2 + \rho\int_\Omega \sigma(u_x)u_x dx \\ &\quad - \alpha\rho^2\|u_x\|^2 + \rho^3\|u\|^2 + \rho\int_\Omega f(u)u dx - \rho(q,u). \end{aligned} \tag{1.3.26}$$

Choosing ρ as in (1.3.15), we now get

$$\frac{\alpha}{2}\int_\Omega \sigma'(u_x)|u_{xx}|^2 dx - \frac{\alpha^2\rho}{2}\|u_{xx}\|^2 \geq \left(\frac{\alpha\gamma_0}{2} - \frac{\alpha^2\rho}{2}\right)\|u_{xx}\|^2 \geq 0,$$

$$\alpha\rho|(u_t,u_{xx})| \leq \frac{\alpha\gamma_0}{8}\|u_{xx}\|^2 + C\|u_t\|^2,$$

$$\alpha|(f(u)-g,u_{xx})| \leq \frac{\alpha\gamma_0}{8}\|u_{xx}\|^2 + C(\|f(u)\|^2 + \|g\|^2).$$

Thus there exist C_{21} and C_{22} such that

$$\begin{aligned} K_2(u,u_t) - \rho H_2(u,u_t) &\geq -C_{21}(\|u_t\|^2 + \|u\|^2 + \|u_x\|^2 + \|F(u)\|_\infty + \|f(u)\|_\infty) - C_{22}(\|g\|) \\ &\geq -C_3(\|(u_0,u_1\|) - C_4(\|g\|). \end{aligned} \tag{1.3.27}$$

By Lemma 1.3.1, (1.3.27) and (1.3.24), we get

$$\frac{d}{dt}H_2(u,u_t) + \rho H_2(u,u_t) \leq C_3\|(u_0,u_1)\| + C_4(\|g\|). \tag{1.3.28}$$

Hence

$$H_2(u,u_t) \leq H_2(u,u_t)(0)e^{-\rho t} + \rho^{-1}(C_3(\|(u_0,u_1)\|) + C_4). \tag{1.3.29}$$

Since $H_2(u,u_t)(0) \leq C(\|(u_0,u_1)\|)$, through formula (1.3.29) we know that $H_2(u,v)$ is uniformly bounded. On the other hand,

$$H_2(u,u_t) \geq \frac{\alpha^2}{4}\|u_{xx}\|^2 - C(\|u_x\|^2 + \|u\|^2 + \|u_t\|^2 + \|F(u)\|_\infty + \|g\|). \tag{1.3.30}$$

By the embedding theorem and Lemma 1.3.1, we know that $\|u_{xx}\|$ is uniformly bounded. Thus we have

Theorem 1.3.2. *Suppose $\sigma(s)$ and $f(u)$ satisfy* (1.3.6) *and* (1.3.8), *respectively. Then for any $(u_0, u_1) \in \mathscr{H} = D(A) \times L^2(\Omega)$, problem* (1.3.1)–(1.3.3) *has a global solution $(u, u_t) \in C(0, \infty; \mathscr{H})$.*

From Theorems 1.3.1 and 1.3.2 we know that problem (1.3.1)–(1.3.3) generates a nonlinear semigroup $S(t)$ on $\mathscr{H}$, $S(t)(u_0, u_1) = (u(t), u_t(t))$, where $u(t)$ is the unique solution of problem (1.3.1)–(1.3.3), and semigroup $S(t)$ is continuous in t and (u_0, u_1). We prove that $S(t)$ has bounded absorbing sets in $\mathscr{H}$.

Proposition 1.3.1. *Under the assumption of Theorem 1.3.2, there exists a constant M_0 such that for any solution $(u(t), u_t(t))$ of problem* (1.3.1)–(1.3.3) *with the initial value $(u_0, u_1) \in \mathscr{H}$ satisfying $\|(u_0, u_1)\| \le R$ there exists a $T = T(R) > 0$ such that*

$$\|S(t)(u_0, u_1)\| = \|(u(t), u_t(t))\| \le M_0, \quad t \ge T(R).$$

In other words, $B = \{(u_1, u_2) \in \mathscr{H} \mid \|(u_1, u_2)\| \le M_0\}$ is a bounded absorbing set of $S(t)$ in $\mathscr{H}$.

Proof. From Lemma 1.3.1 we know that there exist constants $\rho_{0,\infty} = \rho_{0,\infty}(\|g\|)$ and $T_0 = T_0(R)$ such that

$$\|u_x\|^2 + \|u\|^2 + \|u_t\|^2 \le \rho_{0,\infty}^2, \quad t \ge T_0(R). \tag{1.3.31}$$

By Sobolev embedding theorem, $\|u\|_\infty \le \|u_x\| \le \rho_{0,\infty}$, $t \ge T_0(R)$. Via equation (1.3.29) we have

$$K_2(u, u_t) - \rho H_2(u, u_t) \ge -C\rho_{0,\infty}^2 - C\|g\|^2 \equiv \rho_{1,\infty}^2, \quad t \ge T_0(R). \tag{1.3.32}$$

From equations (1.3.24) and (1.3.31),

$$\frac{\mathrm{d}}{\mathrm{d}t} H_2(u, u_t) + \rho H_2(u, u_t) \le \rho_{1,\infty}^2, \quad t \ge T_0(R), \tag{1.3.33}$$

thus

$$H_2(u, u_t) \le H_2(u, u_t)(0) e^{-\rho t} + \rho^{-1} \rho_{1,\infty}^2, \quad t \ge T_0(R). \tag{1.3.34}$$

From equation (1.3.30) we know that there exists a constant $\rho_{2,\infty}$ such that

$$\|u_{xx}\|^2 \le \rho_{2,\infty}^2, \quad t \ge T_1(R). \tag{1.3.35}$$

Using equations (1.3.31) and (1.3.35), we finish the proof. □

In order to prove the existence of a compact global attractor of $S(t)$, we must decompose $S(t)$.

Proposition 1.3.2. *The semigroup $S(t)$ defined by the problem (1.3.1)–(1.3.3) can be decomposed as*

$$S(t) = C(t) + V(t), \tag{1.3.36}$$

where $V(t)$, when $t > t_0$, is consistent relatively compact; $C(t)$ is a continuous mapping from $\mathscr{H}$ to $\mathscr{H}$, and for any bounded set $B_1 \subset \mathscr{H}$ it satisfies

$$r_c(t) = \sup_{\phi \in B_1} \|C(t)\phi\| \to 0, \quad t \to \infty. \tag{1.3.37}$$

Proof. Suppose $(u_0, u_1) \in B_1 = \{(u_0, u_1) \mid \|(u_0, u_1)\| \le R\}$.

Let $\overline{v}(x,t) \in L^\infty(0,\infty; L^2(\Omega)) \cap L^2(0,T; H_0^1(\Omega))$, $\forall T > 0$ be the unique solution of the following problem:

$$(P_1) \qquad \begin{cases} \overline{v}_t - \alpha \overline{v}_{xx} = 0, \\ \overline{v}(x,0) = u_1, \\ \overline{v}(0,t) = \overline{v}(1,t) = 0. \end{cases}$$

Also let $v(x,t)$ be the unique solution of the following problem:

$$(P_2) \qquad \begin{cases} v_t - \alpha v_{xx} = (\sigma(u_x))_x - f(u) + g, \\ v(x,0) = 0, \\ v(0,t) = v(1,t) = 0. \end{cases}$$

By the uniqueness of the solution of (P_2), we deduce that $u_1 = \overline{v} + v$. On the other hand, from (P_1) we have

$$\|\overline{v}(t)\|^2 \le \|u_1\|^2 e^{-\alpha\pi^2 t}. \tag{1.3.38}$$

Since $(u, u_t) \in L^\infty(0,\infty; L^2(\Omega))$, by Proposition 1.3.1 we know that $\overline{G}(x,t) = \sigma(u_x)_x - f(u) + g(x) \in L^\infty(0,\infty; L^2(\Omega))$ and

$$\|\overline{G}(x,t)\| \le \|\sigma'(u_x)\|_\infty \|u_{xx}\| + \|f(u)\| + \|g\| \le C(M_0), \quad t \ge T_0(R).$$

From the problem (P_2) we know that $v(x,t) \in L^\infty(0,\infty; H_0^1(\Omega))$ and

$$\|v\|^2 + \|v_x\|^2 \le C\|G\|^2 (1 - e^{-\frac{\alpha\pi^2}{2} t}) \le C(R), \quad t > T. \tag{1.3.39}$$

Set $\overline{\omega}(x,t) \in L^\infty(0,\infty; H^2 \cap H_0^1)$ to be the solution of the following problem:

$$(P_3) \qquad \begin{cases} \overline{\omega}_{tt} - \alpha \overline{\omega}_{xxt} + \alpha^{-1}\sigma'(u_x)(\overline{\omega}_t - \alpha \overline{\omega}_{xx}) \\ \quad = \alpha^{-1}\sigma'(u_x)\overline{v}(x,t) \triangleq \overline{F}(x,t), \\ \overline{\omega}(0) = u_0, \quad \overline{\omega}_t(0) = u_1, \\ \overline{\omega}(0,t) = \overline{\omega}(1,t) = 0, \end{cases}$$

where $\overline{v}(x,t)$ is the solution of (P_1). It is easy to prove that the solution of (P_3) is unique. In fact, let $\overline{\theta}(x,t)$ be the unique solution of the following problem:

$$(P_4)\qquad \begin{cases} \dfrac{\partial\overline{\theta}}{\partial t} + \alpha^{-1}\sigma'(u_x)\overline{\theta} = \overline{F}(x,t), \\ \overline{\theta}(x,0) = u_1 - \alpha u_{0xx} \in L^2. \end{cases}$$

Then $\overline{\theta}$ can be expressed as

$$\begin{aligned} \overline{\theta}(x,t) &= e^{-\alpha^{-1}\int_0^t \sigma'(u_x(x,\tau)d\tau)}(u_1 - \alpha u_{0xx}) \\ &\quad + e^{-\alpha^{-1}\int_0^t \sigma'(u_x(x,\tau)d\tau)} \cdot \int_0^t \overline{F}(x,\tau)e^{\alpha^{-1}\int_0^t \sigma'(u_x(x,s)ds)}d\tau \end{aligned} \tag{1.3.40}$$

From formula (1.3.40), we know that $\overline{\theta} \in L^\infty(0,\infty;L^2(\Omega))$, $\overline{\theta}_t \in L^\infty(0,\infty;L^2(\Omega))$, and there exist constants $\beta > 0$ and $C > 0$ such that

$$\begin{aligned} &\|\overline{\theta}\|^2 \le C(\|u_1\|^2 + \|u_{0xx}\|^2)e^{-\beta t}, \\ &\|\overline{\theta}_t\|^2 \le C(\|u_1\|^2 + \|u_{0xx}\|^2)e^{-\beta t}. \end{aligned} \tag{1.3.41}$$

Now, we solve the following problem

$$(P_5)\qquad \begin{cases} \overline{\omega}_t - \alpha\overline{\omega}_{xx} = \overline{\theta}(x,t); \\ \overline{\omega}(x,0) = u_0; \\ \overline{\omega}(0,t) = \overline{\omega}(1,t) = 0. \end{cases}$$

then $\overline{\omega}(x,t) \in L^\infty(0,\infty;H^2\cap H_0^1)$, $\overline{\omega_t}(x,t) \in L^\infty(0,\infty;L^2(\Omega))$, $\overline{\omega}(x,t)$ is the solution of problem (P_3). Obviously, the problem (P_3) has a unique solution. By the virtue of (1.3.41) and (P_5), there exist constants β_2 and $C = C(\|(u_0,u_1)\|)$, s. t.

$$\|(\overline{\omega},\overline{\omega}_t)\| \le C(\|(u_0,u_1)\|)e^{-\beta_2 t} \le C(R)e^{-\beta_2 t}. \tag{1.3.42}$$

Finally, we consider the following problem:

$$(P_6)\qquad \begin{cases} (\omega_t - \alpha\omega_{xx})_t + \alpha^{-1}\sigma'(u_x)(\omega_t - \alpha\omega_{xx}) = F; \\ \omega(x,0) = 0, \quad \omega_t(x,0) = 0; \\ \omega(0,t) = \omega(1,t) = 0. \end{cases}$$

where

$$\begin{aligned} F(x,t) &= \alpha^{-1}\sigma'(u_x)u_t - \alpha^{-1}\sigma'(u_x)\overline{v} - f(u) + g \\ &= \alpha^{-1}\sigma'(u_x)v(x,t) - f(u) + g \end{aligned}$$

Since $(u,u_t) \in \mathscr{H}$, $v(x,t) \in L^\infty(0,\infty;H_0^1)$, and it satisfies (1.3.39), we can get $F(x,t) \in L^\infty(0,\infty;H^1)$, and

$$\|F(x,t)\|^2 + \|F(x,t)\|^2 \le C(R_0). \tag{1.3.43}$$

Defining

$$\theta(x,t) = e^{-\alpha^{-1}\int_0^t \sigma'(u_x(x,\tau))d\tau} \cdot \int_0^t F(x,t)e^{-\alpha^{-1}\int_0^t \sigma'(u_x(x,s))ds} d\tau \tag{1.3.44}$$

we know $\theta(x,t) \in L^\infty(0,\infty;H^1)$, $\theta_t(x,t) \in L^\infty(0,\infty;H^1)$, and

$$\|\theta\|_{H^1}^2 + \|\theta_t\|_{H^1}^2 \le C(R).$$

Now we solve the following problem

$$\begin{cases} \omega_t - \alpha\omega_{xx} = \theta(x,t); \\ \omega(x,0) = 0, \\ \omega(0,t) = \omega(1,t) = 0. \end{cases}$$

Then $\omega(x,t) \in H^3 \cap H_0^1$ is the unique solution of (P_5), and we have

$$\|\omega(x,t)\|_{H^3} \le C(R). \tag{1.3.45}$$

By the uniqueness of solution of problem (P_6), we get $u(x,t) = \overline{\omega}(x,t) + \omega(x,t)$. Now we define

$$C(t)(u_0,u_1) = (\overline{\omega},\overline{v}),$$
$$U(t)(u_0,u_1) = (\omega,v).$$

Then using (1.3.38), (1.3.39), (1.3.42) and (1.3.45), we obtain $S(t) = C(t) + U(t)$, where $C(t)$ and $U(t)$ satisfy the required properties of Proposition 1.3.2. □

From Propositions 1.3.1, 1.3.2 and Theorem 1.1.1 in [197], we get

Theorem 1.3.3. *The semigroup operator $S(t)$ defined by problem (1.3.1)–(1.3.3) has a global attractor in $\mathscr{H}$, which attracts all the bounded sets of $\mathscr{H}$.*

In the following, we estimate the dimension of the attractor $\mathscr{A}$. Set $\xi_0 = (u_0,u_1) \in \mathscr{H}$, and let $u(t)$ be the solution of the corresponding problem (1.3.1)–(1.3.3), that is, $S(t)\xi_0 = (u(t),u_t(t))$. It is easy to prove the consistent differentiability of $S(t)$ in a similar way.

Proposition 1.3.3. *Set $\xi_0^1 = (u_0^1,u_1^1)$, $\xi_0^2 = (u_0^2,u_1^2) \in \mathscr{H}$, and $\|\xi_0^1\| \le R$, $\|\xi_0^2\| \le R$. Then for all T, R, $0 < T, R < +\infty$, there exists a constant $C = C(R,T)$ such that*

$$\|S(t)\xi_0^1 - S(t)\xi_0^2\|^2 \le C(R,T)\|\xi_0^1 - \xi_0^2\|, \quad 0 \le t \le T. \tag{1.3.46}$$

Now we consider the linear variational problem for problem (1.3.1)–(1.3.3):

$$U_{tt} - U_{xxt} - (\sigma'(u_x)U_x)_x + f'(u)U = 0, \tag{1.3.47}$$

$$U(x,0) = \omega_0, \quad U_t(x,0) = \omega_1, \tag{1.3.48}$$

$$U(0,t) = U(1,t) = 0, \tag{1.3.49}$$

where $(u, u_t) = S(t)\xi_0$ and $\eta_0 = (\omega_0, \omega_1) \in \mathscr{H}$. Since $S(t)\xi_0 \in C(R^+; \mathscr{H})$, the linear problem (1.3.47)–(1.3.49) has a unique solution $(U(t), U_t(t)) \in C(R^+; \mathscr{H})$. We can prove that

$$(DS(t)\xi_0)\eta_0 = (U(t), U_t(t)), \tag{1.3.50}$$

where $S(t)$ is consistently differentiable at ξ_0.

Proposition 1.3.4. *For any R and T, $0 < R, T < \infty$, there is a constant $C(R,T)$ such that for any $\xi_0 = (u_0, u_1)$, $h_0 = (h_{01}, h_{02})$ satisfying $\|\xi_0\| \le R$, $\|\xi_0 + h_0\| \le R$, $t \le T$, we have*

$$\|S(t)(\xi_0 + h_0) - S(t)\xi_0 - (DS(t)\xi_0)h_0\| \le C(R,T)\|h_0\|^2. \tag{1.3.51}$$

Take $\eta_0^1, \eta_0^2, \ldots, \eta_0^m \in \mathscr{H}$ and study the evolution of Gram determinant

$$\|\eta^1(t) \wedge \eta^2(t) \wedge \cdots \wedge \eta^m(t)\|_{\wedge^m(\mathscr{H})} = \det_{1\le i,j\le m}((\eta^i, \eta^j)), \tag{1.3.52}$$

where $\eta^j(t) = (DS(t)\xi_0)\eta_0^j$, and $((\cdot,\cdot))$ means the inner product in $\mathscr{H}$. We will prove, for sufficiently large m and when $t \to \infty$, that the determinant (1.3.52) exponentially decays to zero.

Theorem 1.3.4. *Set $f'(u) \ge 0$. If $\mathscr{A}$ is the global attractor of problem* (1.3.1)–(1.3.3) *then there are constants $u > 0$, $C_1 > 0$ and $C_2 > 0$ such that for any $\xi_0 \in \mathscr{A}$, $m \ge 1$ and $t \ge 0$, we have*

$$\begin{aligned}&\|(DS(t)\xi_0)\eta^1 \wedge (DS(t)\xi_0)\eta^2 \wedge \cdots \wedge (DS(t)\xi_0)\eta^m\|_{\wedge^m(\mathscr{H})}\\ &\quad\le \|\eta_0^1 \wedge \eta_0^2 \wedge \cdots \wedge \eta_0^m\|_{\wedge^m(\mathscr{H})} C_1^m \exp(C_2\sqrt{m} - \mu m)t, \quad \forall \eta_0^i \in \mathscr{H}.\end{aligned} \tag{1.3.53}$$

Proof. Firstly, we consider an equivalent norm on $\mathscr{H}$. Let $\xi_0 \in \mathscr{A}$, $(u, u_t) = S(t)\xi_0$ and $\eta(t) = (U(t), U_t(t)) = (DS(t)\xi_0)\eta_0$, where $\eta_0 = (\omega_0, \omega_1) \in \mathscr{H}$. Then $U(t)$ satisfies (1.3.46)–(1.3.49), setting $\omega(t) = e^{\mu t}U(t)$, where μ is a positive number to be determined, and $\omega(t)$ satisfies

$$\omega_{tt} - \alpha\omega_{xxt} - 2\mu\omega_t - (\sigma'(u_x)\omega_x)_x + \alpha\mu\omega_{xx} + \mu^2\omega + f'(u)\omega = 0. \tag{1.3.54}$$

Multiplying (1.3.54) by ω_t and integrating over Ω, we have

$$\begin{aligned}&\frac{d}{dt}\left\{\frac{1}{2}\|\omega_t\|^2 + \frac{1}{2}\int_\Omega \sigma'(u_x)|\omega_x|^2 dx - \frac{\alpha\mu}{2}\|\omega_x\|^2 + \frac{\mu^2}{2}\|\omega\|^2 + \frac{1}{2}\int_\Omega f'(u)|\omega|^2 dx\right\}\\ &= -\alpha\|\omega_{xt}\|^2 + 2\mu\|\omega_t\|^2 + \frac{1}{2}\int_\Omega \sigma''(u_x)u_{xx}|\omega_x|^2 dx - \frac{1}{2}\int_\Omega f''(u)u_t|\omega|^2 dx.\end{aligned} \tag{1.3.55}$$

Multiplying (1.3.54) by ω_{xx} and integrating over Ω, we obtain

$$\begin{aligned}\frac{\mathrm{d}}{\mathrm{d}t}&\left\{\frac{\alpha}{2}\|\omega_{xx}\|^2-(\omega_x,\omega_{xx})-\frac{\mu}{2}\|\omega_x\|^2\right\}\\&=\|\omega_{xt}\|^2-\int_\Omega\sigma'(u_x)|\omega_{xx}|^2\mathrm{d}x-\int_\Omega\sigma''(u_x)u_{xx}\omega_x\omega_{xx}\mathrm{d}x\\&\quad+\alpha\mu\|\omega_{xx}\|^2-\mu^2\|\omega_x\|^2-\int_\Omega f'(u)|\omega_x|^2\mathrm{d}x-\int_\Omega f''(u)u_x\omega\omega_x\mathrm{d}x.\end{aligned}\tag{1.3.56}$$

Multiplying (1.3.55) by a constant k and then adding (1.3.56) yields

$$\frac{\mathrm{d}}{\mathrm{d}t}J(\xi(t))=\frac{\mathrm{d}}{\mathrm{d}t}J(\omega,\omega_t)=K(\xi(t))=K(\omega,\omega_t),\tag{1.3.57}$$

where $\xi(t)=(\omega,\omega_t)\in\mathscr{H}$,

$$\begin{aligned}J(\xi)=&\frac{\alpha}{2}\|\xi_{1xx}\|^2-(\xi_2,\xi_{1xx})-\frac{\mu}{2}\|\xi_{1x}\|^2\\&+\frac{k}{2}\|\xi_2\|^2+\frac{k}{2}\left[\int_\Omega\sigma'(u_x)\xi_{1x}-\alpha\mu\|\xi_{1x}\|^2\right]\\&+\frac{\mu^2k}{2}\|\xi_1\|^2+\frac{k}{2}\int_\Omega f'(u)|\xi_1|^2\mathrm{d}x,\end{aligned}\tag{1.3.58}$$

$$\begin{aligned}K(\xi)=&(-\alpha k+1)\|\xi_{2x}\|^2+2\mu k\|\xi_2\|^2-\int_\Omega\sigma'(u_x)|\xi_{1xx}|^2\mathrm{d}x\\&+\alpha\mu\|\xi_{1xx}\|^2-\mu^2\|\xi_{1x}\|^2+\frac{2}{k}\int_\Omega\sigma''(u_x)u_{xx}|\xi_{1x}|^2\mathrm{d}x\\&+\frac{k}{2}\int_\Omega f''(u)u_t|\xi_1|^2\mathrm{d}x-\int_\Omega f'(u)|\xi_{1x}|^2\mathrm{d}x\\&-\int_\Omega f''(u)u_x\xi_1\xi_{1x}\mathrm{d}x-\int_\Omega\sigma''(u_x)u_{xx}\xi_{1x}\xi_{1xx}\mathrm{d}x\end{aligned}\tag{1.3.59}$$

for $\xi=(\xi_1,\xi_2)\in\mathscr{H}$ and $\xi_{2x}\in L^2(\Omega)$. Now we take $k=\frac{1}{\alpha}$ and μ such that

$$0<u<\min\left(\frac{\gamma_0}{2\alpha},\frac{\alpha\pi^2}{4},1\right).\tag{1.3.60}$$

Since $f'(u)\geq 0$, we have

$$\begin{aligned}J(\xi)\geq&\frac{\alpha}{2}\|\xi_{1xx}\|^2-\frac{\alpha}{4}\|\xi_{1xx}\|^2-\frac{1}{2}\|\xi_2\|^2+\frac{k}{2}\|\xi_2\|^2\\&+\frac{k\gamma_0}{4}\|\xi_{1x}\|^2-\frac{\mu}{2}\|\xi_{1x}\|^2-\frac{k\alpha\mu}{4}\|\xi_{1x}\|^2\end{aligned}\tag{1.3.61}$$

$$+\frac{\mu^2 k}{2}\|\xi_1\|^2+\frac{k}{2}\int_\Omega f'(u)|\xi_1|^2\mathrm{d}x$$
$$\geq\frac{\alpha}{4}\|\xi_{1xx}\|^2+\frac{1}{\alpha}\|\xi_2\|^2+\frac{\gamma_0}{4\alpha}\|\xi_{1x}\|^2+\frac{2\mu^2}{\alpha}\|\xi_1\|^2. \tag{1.3.62}$$

Since $S(t)\xi_0=(u(t),u_t(t))\in\mathscr{A}$, the norms $\|u_x\|_\infty$, $\|u\|_\infty$, $\|u_{xx}\|$ and $\|u\|$ are uniformly bounded. By the definitions (1.3.58) and (1.3.61), there exist positive constants k_0 and k_1 such that

$$k_0\|\xi_0\|\leq J(\xi)\leq k_1\|\xi\|^2. \tag{1.3.63}$$

So $J(\xi)^{\frac{1}{2}}$ is an equivalent norm of $\mathscr{H}$. As for $K(\xi)$, we have

$$\left|\frac{k}{2}\int_\Omega\sigma''(u_x)u_{xt}|\xi_{1x}|^2\mathrm{d}x\right|$$
$$=\left|-\frac{k}{2}\int_\Omega\sigma'''(u_x)u_{xx}u_t|\xi_{1x}|^2\mathrm{d}x-k\int_\Omega\sigma''(u_x)u_t\xi_{1x}\xi_{1xx}\mathrm{d}x\right|$$
$$\leq C(\|\xi_{1x}\|_\infty^2)+\|\xi_{1x}\|\|\xi_{1xx}\|$$
$$\leq C(\|\xi_1\|_{H^2}^2+\|\xi_2\|^2)^{\frac{3}{4}}\|\xi_{1x}\|^{\frac{1}{2}}.$$

The three terms on the right-hand side of $K(\xi)$ can be estimated similarly. Since k and μ are selected in (1.3.63), the first three terms of $K(\xi)\leq 0$, i. e.,

$$(-\alpha k+1)\|\xi_{2x}\|^2+2\mu k\|\xi_2\|^2-\int_\Omega\sigma'(u_x)|\xi_{1xx}|^2\mathrm{d}x+\alpha\mu|\xi_{1xx}|^2\mathrm{d}x\leq 0.$$

Thus we have

$$K(\xi)\leq C(\|\xi_1\|_{H^2}^2+\|\xi_2\|^2)^{\frac{3}{4}}\|\xi_{1x}\|^{\frac{1}{2}}. \tag{1.3.64}$$

To estimate the variation of the m-dimensional volume, we need the following lemma:

Lemma 1.3.2. *Suppose $\phi(\cdot,\cdot)$ and $\phi_1(\cdot,\cdot)$ are two inner products in a Hilbert space, which are continuous and equivalent, i. e.,*

$$\alpha\phi(\xi,\xi)\leq\phi_1(\xi,\xi)\leq\beta\phi(\xi,\xi),\quad\forall\xi\in H. \tag{1.3.65}$$

Then we have

$$\alpha^m\det_{1\leq i,j\leq m}\phi(\xi^i,\xi^j)\leq\det_{1\leq i,j\leq m}\phi_1(\xi^i,\xi^j)\leq\beta^m\det_{1\leq i,j\leq m}\phi(\xi^i,\xi^j),\quad\forall\xi^i\in H,\ i=1,2,\dots,m. \tag{1.3.66}$$

Set $\omega^i(t) = e^{\mu t}U^i(t)$, $i = 1, 2, \ldots, m$. Then $(U^i(t), U^i_t(t)) = (DS(t)\xi_0)\eta^i_0$. Let $\overline{\eta^i}(t) = (\omega^i(t), \omega^i_t - \mu\omega^i)$, $\xi^i(t) = (\omega^i(t), \omega^i_t(t))$, and then

$$\begin{aligned}
&\|\eta^l(t) \wedge \eta^2(t) \wedge \cdots \wedge \eta^m(t)\|^2_{\wedge^m(\mathscr{H})} \\
&= e^{-2\mu mt}\|\overline{\eta^l}(t) \wedge \overline{\eta^2}(t) \wedge \cdots \wedge \overline{\eta^m}(t)\|_{\wedge^m(\mathscr{H})} \\
&= e^{-2\mu mt} \det_{1\le i,j\le m}((\overline{\eta^i}(t), \overline{\eta^j}(t))) \\
&= e^{-2\mu mt} \det_{1\le i,j\le m} \phi_1(\xi^i(t), \xi^j(t)),
\end{aligned} \tag{1.3.67}$$

where $\phi_1(\xi, \eta) = ((\xi_1, \eta_1))_{2s} + (\xi_2 - \mu\xi_1, \eta_2 - \mu\eta_1)$, $\xi = (\xi_1, \xi_2)$, $\eta = (\eta_1, \eta_2) \in \mathscr{H}$. Since $\mu < 1$, we know that $\phi_1(\cdot, \cdot)$ satisfies the condition of Lemma 1.3.2. Hence there exists a constant C such that

$$\det_{1\le i,j\le m} \phi_1(\xi^i(t), \xi^j(t)) \le C \det_{1\le i,j\le m}((\xi^i(t), \xi^j(t))). \tag{1.3.68}$$

In order to estimate $\det_{1\le i,j\le m}((\xi^i(t), \xi^j(t)))$, we introduce the following bilinear form on $\mathscr{H}$, for any $\xi = (\xi_1, \xi_2), \eta = (\eta_1, \eta_2) \in \mathscr{H}$:

$$\begin{aligned}
\phi(\xi, \eta) &= \frac{\alpha}{2}(\xi_{1xx}, \eta_{1xx}) \\
&= \frac{1}{2}[(\xi_2, \eta_{1xx}) + (\eta_2, \xi_{1xx})] - \frac{\mu}{2}(\xi_{1x}, \eta_{1x}) \\
&\quad + \frac{k}{2}(\eta_2, \xi_2) + \frac{k}{2}\int_\Omega \sigma'(u_x)\xi_{1x}\eta_{1x}\mathrm{d}x - \frac{\alpha\mu k}{2}(\xi_{1x}, \eta_{1x}) \\
&\quad + \frac{\mu^2 k}{2}(\xi_1, \eta_1) + \frac{k}{2}\int_\Omega f'(u)\xi_1\eta_1\mathrm{d}x,
\end{aligned} \tag{1.3.69}$$

and then $\phi(\eta, \eta) = J(\eta)$. From equation (1.3.63), we know that $\phi(\eta, \eta)^{\frac{1}{2}}$ is an equivalent norm. Hence, by Lemma 1.3.2, we have

$$k_0^m \det_{1\le i,j\le m}((\xi^i(t), \xi^j(t))) \le \det_{1\le i,j\le m} \phi(\xi^i(t), \xi^j(t)) \le k_1^m \det_{1\le i,j\le m}((\xi^i(t), \xi^j(t))). \tag{1.3.70}$$

Moreover, we need to estimate $H_m(t) = \det_{1\le i,j\le m} \phi(\xi^i(t), \xi^j(t))$. Similarly as in the proof in [74], we have

$$\frac{dH_m(t)}{\mathrm{d}t} = H_m(t)\sum_{l=1}^m \max_{F\subset R^m, \dim F=l,} \min_{x\in F, x\ne 0} \frac{K(\sum_{j=1}^m x_j\xi^j(t))}{J(\sum_{j=1}^m x_j\xi^j(t))}. \tag{1.3.71}$$

From equations (1.3.63) and (1.3.64), we have

$$\frac{dH_m(t)}{\mathrm{d}t} \le \frac{C}{k_0}H_m(t)\sum_{l=1}^m \max_{F\subset R^m, \dim F=l,} \min_{x\in F, x\ne 0} \frac{\|\sum_{j=1}^m x_j\xi^j(t)\|_H^{\frac{3}{2}}}{\|\sum_{j=1}^m x_j\xi^j(t)\|_H^2}\left\|\sum_{j=1}^m x_j\xi^j(t)\right\|^{\frac{1}{2}}$$

$$\leq CH_m(t)\sum_{l=1}^{m}\max_{F\subset R^m,\dim F=l,}\min_{x\in F,x\neq 0}\frac{\|\sum_{j=1}^{m}x_j\xi^j(t)\|^{\frac{1}{2}}}{\|\sum_{j=1}^{m}x_j\xi^j(t)\|_H^{\frac{1}{2}}}$$

$$\leq CH_m(t)\sum_{l=1}^{m}\frac{1}{\lambda_l^{\frac{1}{4}}}\leq C\sqrt{m}H_m(t). \tag{1.3.72}$$

Since an eigenvalue of $\mathscr{A}$ is $\lambda_l = \pi^2 l^2$, we have

$$H_m(t)\leq H_m(0)e^{C\sqrt{m}t}. \tag{1.3.73}$$

Now using (1.3.67), (1.3.70) and (1.3.72), we complete the proof of the theorem. □

As a corollary of Theorem 1.3.4, we have

Theorem 1.3.5. *If $f'(u)\geq 0$ then the attractor determined by Theorem* 1.3.4 *has finite fractal and Hausdorff dimensions.*

Proof. With the aid of Theorem 1.3.4 we know that $\forall\xi_0\subset\mathscr{A}$,

$$\omega_m(DS(t)\xi_0)\leq C_1^m\exp(C_2\sqrt{m}-\mu m)t,$$

where ω_m is a Lyapunov index. So $\overline{\omega}_m(\mathscr{A}) < 1$. If m is large enough, say $m > (\frac{C_2}{\mu})^2$, then $\overline{\omega}_m(\mathscr{A})$ is a consistent Lyapunov index of $\mathscr{A}$. Using Lemma V.3.1 in [197], we can deduce that $\mathscr{A}$ has finite fractal and Hausdorff dimensions. □

1.4 Coupled KdV equations

Now we consider the following coupled Korteweg–de Vries (KdV) equations:

$$u_t = u_{xxx}+6uu_x+2vv_x, \tag{1.4.1}$$

$$v_t = 2(uv)_x, \tag{1.4.2}$$

which were proposed by Kupershmidt [156] in 1985 and could describe the interaction of two long waves. Ito proposed the operator method to give the infinite symmetries and motion constants in [144]. In 1991, Guo and Tan proved the existence and uniqueness of a global smooth solution for the above equation [113]. In 1996, Guo and Yang considered the corresponding dissipation equation and proved the existence and finiteness of dimension of the global attractor [128, 127]. And the other similar resutls on the well-posedness can refer to the reference [129, 130, 131, 132].

Now we consider the following dissipation coupled KdV equations with periodic initial value problem:

$$u_t+f(u)_x-\alpha u_{xx}+\beta u_{xxx}+2vv_x = G_1(u,v)+h_1(x), \tag{1.4.3}$$

$$v_t-\gamma v_{xx}+2(uv)_x = G_2(u,v)+h_2(x),\quad x\in\mathbf{R},\quad t\geq 0, \tag{1.4.4}$$

$$u(x+D,t)=u(x-D,t),\quad v(x+D,t)=v(x-D,t),\quad \forall x\in\mathbf{R},\quad t\geq 0, \tag{1.4.5}$$
$$u(x,0)=u_0(x),\quad v(x,0)=v_0(x),\quad x\in\mathbf{R}, \tag{1.4.6}$$

where $D>0$, $\alpha>0$, $\beta\neq 0$ and $\gamma>0$ are constants. We first make a uniform priori estimates for problem (1.4.3)–(1.4.6) which are independent of t, and then prove the global existence of the attractor. Finally, we estimate an upper bound of its dimension.

Lemma 1.4.1. *Suppose*
(1) $G_i(0,0)=0$, $i=1,2$, *and for any* $(\xi,\eta)\in\mathbf{R}^2$,

$$(\xi,\eta)\begin{pmatrix}-G_{1u} & -G_{1v}\\ -G_{2u} & -G_{2v}\end{pmatrix}\begin{pmatrix}\xi\\ \eta\end{pmatrix}\geq b_0(\xi^2+\eta^2);$$

where $b_0>0$ *is a constant;*
(2) $u_0(x),v_0(x)\in L^2(\Omega)$, $h_i(x)\in L^2(\Omega)$, $i=1,2$, $\Omega=(-D,D)$;

then the following estimate for problem (1.4.3)–(1.4.6) *holds:*

$$\|u\|^2+\|v\|^2\leq e^{-b_0t}(\|u_0\|^2+\|v_0\|^2)+\frac{1}{b_0^2}(1-e^{-b_0t})(\|h_1\|^2+\|h_2\|^2). \tag{1.4.7}$$

Furthermore, we have

$$\overline{\lim_{t\to\infty}}\left(\|u(\cdot,t)\|^2+\|v(\cdot,t)\|^2\right)\leq\frac{1}{b_0^2}(\|h_1\|^2+\|h_2\|^2)\equiv E_0, \tag{1.4.8}$$

$$\overline{\lim_{t\to\infty}}\frac{1}{t}\int_0^t[\alpha\|u_x(\cdot,\tau)\|^2+\gamma\|v_x(\cdot,\tau)\|^2]d\tau\leq\frac{1}{2b_0}(\|h_1\|^2+\|h_2\|^2). \tag{1.4.9}$$

Proof. Taking the inner product of equation (1.4.3) and u, as well as of (1.4.3) and v, we get

$$(u,u_t+f(u)_x-\alpha u_{xx}+\beta u_{xxx}+2vv_x)=(u,G_1(u,v)+h_1), \tag{1.4.10}$$
$$(v,v_t-\gamma u_{xx}+2(uv)_x)=(v,G_2(u,v)+h_2), \tag{1.4.11}$$

where $(u,\omega)=\int_{-D}^{D}u(x,t)\omega(x,t)dx$ and

$$(u,f(u)_x)=0,\quad (u,-\alpha u_{xx})=\alpha\|u_x\|^2,\quad (u,u_{xxx})=0,$$
$$(u,2vv_x)+(u,2(uv)_x)=0,$$
$$(u,G_1(u,v))+(v,G_2(u,v))\leq -b_0(\|u\|^2+\|v\|^2).$$

From equations (1.4.10) and (1.4.11) we get

$$\frac{1}{2}\frac{d}{dt}(\|u\|^2+\|v\|^2)+\alpha\|u_x\|^2+\gamma\|v_x\|^2+\frac{b_0}{2}(\|u\|^2+\|v\|^2)\leq\frac{1}{2b_0}((\|h_1\|^2+\|h_2\|^2)) \tag{1.4.12}$$

Inequality (1.4.12) now yields (1.4.8) and (1.4.9). □

Lemma 1.4.2. *Under the conditions of Lemma* 1.4.1, *assume that*
(1) $f \in C^2, G_i \in C^1, i = 1, 2$ *and*

$$|f(u)| \le A|u|^{5-\delta}, \quad \delta > 0,\ A > 0,$$
$$|G_i(u, v)| \le B_i(|u|^5 + |v|^5), \quad B_i > 0,\ i = 1, 2;$$

(2) *if* $u_{0x}, v_{0x} \in L^2(\Omega)$, *then*

$$\begin{aligned}\|u_x\|^2 + \|v_x\|^2 &\le 2e^{-2b_0 t}\left(\|u_{0x}\|^2 + \|v_{0x}\|^2 - \frac{2}{\beta}\int F(u_0(x))\mathrm{d}x\right)\\ &\quad + 2e^{-2b_0 t}\int_0^t C_1 e^{2b_0 s}\mathrm{d}s\\ &\quad + \frac{1}{b_0}(1 - e^{-2b_0 t})\left(\frac{4}{\gamma}\|h_2\|^2 + \frac{3}{\alpha}\|h_1\|^2\right) + C_2 \end{aligned} \tag{1.4.13}$$

where the functions C_1 *and* C_2 *depend on* $\|u\|$ *and* $\|v\|$. *Moreover, we obtain that*

$$\begin{aligned}&\overline{\lim_{t\to\infty}}\left(\|u_x(\cdot, t)\|^2 + \|v_x(\cdot, t)\|^2\right)\\ &\quad\le \frac{1}{b_0}\max_{t\ge 0} C_1 + \frac{1}{b_0}\left(\frac{4}{\gamma}\|h_2\|^2 + \frac{3}{\alpha}\|h_1\|^2\right) + \max_{t\ge 0} C_2 \triangleq E_1, \end{aligned} \tag{1.4.14}$$

$$\begin{aligned}&\overline{\lim_{t\to\infty}}\frac{1}{t}\int_0^t [\alpha\|u_{xx}(\cdot, \tau)\|^2 + \gamma\|v_{xx}(\cdot, \tau)\|^2]\mathrm{d}\tau\\ &\quad\le \max_{t\ge 0}(b_0 C_2 + C_1) + \frac{4}{\gamma}\|h_2\|^2 + \frac{3}{\alpha}\|h_1\|^2. \end{aligned} \tag{1.4.15}$$

Proof. Taking the inner product of (1.4.3) and u_{xx}, we obtain

$$(u_{xx}, u_t + f(u)_x - \alpha u_{xx} + \beta u_{xx} + 2vv_x) = (u_{xx}, G_1(u, v) + h_1(x)), \tag{1.4.16}$$

where

$$\begin{aligned}(u_{xx}, f(u)_x) &= -(u_{xxx}, f(u))\\ &= \frac{1}{\beta}(u_t + f(u)_x - \alpha u_{xx} + 2vv_x - G_1(u, v) - h_1, f(u))\\ &= \frac{1}{\beta}\frac{\mathrm{d}}{\mathrm{d}t}\int F(u(x, t))\mathrm{d}x - \frac{\alpha}{\beta}(u_{xx}, f(u))\\ &\quad + \frac{2}{\beta}(vv_x, f(u)) - \frac{1}{\beta}(G_1 + h_1, f)\end{aligned}$$

and $F(u) = \int_0^u f(s)\mathrm{d}s$. By Sobolev interpolation inequality, we have

$$\begin{aligned}|(u_{xx}, f(\omega))| &\le \|u_{xx}\|\|f(u)\| \le A\|u_{xx}\|\|u\|_{2(5-\delta)}^{5-\delta}\\ &\le \frac{|\beta|}{12}\|u_{xx}\|^2 + C(\|u\|),\end{aligned}$$

$$\begin{aligned}
\left|\frac{2}{\beta}(vv_x, f(u))\right| &\le \frac{2}{|\beta|}\|v\|_4\|v_x\|_4\|f(u)\|_2 \\
&\le \frac{\alpha}{12}\|u_{xx}\|^2 + \frac{\gamma}{8}\|v_{xx}\|^2 + C(\|u\|, \|v\|), \\
\left|\frac{1}{\beta}(G_1, f)\right| &\le \frac{AB_1}{|\beta|}[\|u\|_{10-\delta}^{10-\delta} + \|u\|_{2(5-\delta)}^{5-\delta}\|v\|_{10}^5] \\
&\le \frac{\alpha}{12}\|u_{xx}\|^2 + \frac{\gamma}{8}\|v_{xx}\|^2 + C(\|u\|, \|v\|), \\
\left|\frac{1}{\beta}(h_1, f(u))\right| &\le \frac{1}{|\beta|}\|h_1\|\|f\| \le \frac{\alpha}{12}\|u_{xx}\|^2 + C, \\
|(u_{xx}, h_1)| &\le \|u_{xx}\|\|h_1\| \le \frac{\alpha}{12}\|u_{xx}\|^2 + \frac{3}{\alpha}\|h_1\|^2.
\end{aligned}$$

Due to equation (1.4.16), we have

$$\begin{aligned}
&\frac{1}{2}\frac{\mathrm{d}}{\mathrm{d}t}\left[\|u_x\|^2 - \frac{2}{\beta}\int F(u)\mathrm{d}x\right] + \frac{7\alpha}{12}\|u_{xx}\|^2 - 2(u_{xx}, vv_x) \\
&\quad \le -(u_{xx}, G_1(u,v)) + \frac{\gamma}{4}\|v_{xx}\|^2 - \frac{3}{\alpha}\|h_1\|^2 + C, \qquad (1.4.17)
\end{aligned}$$

where C depends on $\|u\|$, $\|v\|$ and $\|h_1\|$. Taking the inner product of (1.4.4) and v_{xx} gives

$$(v_{xx}, v_t - \gamma v_{xx} + 2(uv)_x) = (v_{xx}, G_2(u,v) + h_2),$$

hence we obtain

$$\frac{1}{2}\frac{\mathrm{d}}{\mathrm{d}t}\|v_x\|^2 + \gamma\|v_{xx}\|^2 - 2(v_{xx}, (uv)_x) = -(v_{xx}, G_2) - (v_{xx}, h_2). \qquad (1.4.18)$$

From equations (1.4.17) and (1.4.18) it follows that

$$\begin{aligned}
&\frac{1}{2}\frac{\mathrm{d}}{\mathrm{d}t}\left[\|u_x\|^2 + \|v_x\|^2 - \frac{2}{\beta}\int F(u)\mathrm{d}t\right] + \frac{7\alpha}{12}\|u_{xx}\|^2 + \frac{3\gamma}{4}\|v_{xx}\|^2 \\
&\quad \le -(u_{xx}, G_1) - (v_{xx}, G_2) + 2[(u_{xx}, vv_x) + (v_{xx}, (uv)_x)] + (v_{xx}, h_2) + \frac{3}{\alpha}\|h_1\|^2 + C,
\end{aligned}$$

where

$$\begin{aligned}
-(u_{xx}, G_1) - (v_{xx}, G_2) &= \int [u_x(G_{1u}u_x + G_{1v}v_x) + v_x(G_{2u}u_x + G_{2v}v_x)]\mathrm{d}x \\
&\le -b_0[\|u_x\|^2 + \|v_x\|^2] \\
|(v_{xx}, h_2)| &\le \frac{\gamma}{8}\|v_{xx}\|^2 + \frac{4}{\gamma}\|h_2\|^2, \\
|2(u_{xx}, vv_x) + 2(v_{xx}, (uv)_x)| &= 3\left|\int u_x v_x^2 \mathrm{d}x\right| \le 3\|u_x\|\|v_x\|_4^2 \\
&\le \frac{\alpha}{24}\|u_{xx}\|^2 + \frac{\gamma}{8}\|v_{xx}\|^2 + C,
\end{aligned}$$

$$\left|\frac{2b_0}{\beta}\int F(u)\mathrm{d}x\right| \le C\|u\|_{6-\sigma}^{6-\sigma} \le \frac{\alpha}{24}\|u_{xx}\|^2 + C.$$

Let $\phi(t) = \|u_x\|^2 + \|v_x\|^2 - \frac{2}{\beta}\int F(u)\mathrm{d}x$. Then we have

$$\frac{\mathrm{d}\phi(t)}{\mathrm{d}t} + \alpha\|u_{xx}\|^2 + \gamma\|v_{xx}\|^2 + 2b_0\phi(t) \le \frac{3}{\alpha}\|h_1\|^2 + \frac{4}{\gamma}\|h_2\|^2 + C.$$

Integrating the above inequality, we obtain

$$\phi(t) \le e^{-2b_0 t}\phi(0) + e^{-2b_0 t}\int_0^t Ce^{2b_0 s}\mathrm{d}s + \frac{1}{2b_0}(1-e^{-2b_0 t})\left(\frac{4}{\gamma}\|h_2\|^2 + \frac{3}{\alpha}\|h_1\|^2\right).$$

Notice that

$$\left|\frac{2}{\beta}\int F(u)\mathrm{d}x\right| \le A\|u\|_{6-\sigma}^{6-\sigma} \le \frac{1}{2}\|u_{xxx}\|^2 + C,$$

hence we get

$$\begin{aligned}\|u_x\|^2 + \|v_x\|^2 &\le 2e^{-2b_0 t}\phi(0) + 2e^{-2b_0 t}\int_0^t C_1 e^{2b_0 s}\mathrm{d}s\\ &\quad + \frac{1}{b_0}(1-e^{-2b_0 t})\left(\frac{4}{\gamma}\|h_2\|^2 + \frac{3}{\alpha}\|h_1\|^2\right) + C_2, \end{aligned} \tag{1.4.19}$$

where the functions C_1 and C_2 depend on $\|u\|$ and $\|v\|$. From equation (1.4.19), we get (1.4.14) and (1.4.15). □

Lemma 1.4.3. *Under the conditions of Lemma* 1.4.2, *assume that*
(1) $f(u) \in C^2, G_i(u,v) \in C^2, i = 1,2;$
(2) $u_0(x), v_0(x) \in H^2(\Omega), h_i(x) \in H^1(\Omega), i = 1,2.$

Then for the smooth solution of problem (1.4.3)–(1.4.6) *we have*

$$\begin{aligned}\|u_{xx}\|^2 + \|v_{xx}\|^2 &\le e^{-2b_0 t}[\|u_{0xx}\|^2 + \|v_{0xx}\|^2]\\ &\quad + \frac{1-e^{-2b_0 t}}{2b_0}\left[\frac{5}{\alpha}\|h_{1x}\|^2 + \frac{2}{\gamma}\|h_{2x}\|^2\right]\\ &\quad + e^{-2b_0 t}\int_0^t C_3 e^{2b_0 s}\mathrm{d}s, \end{aligned} \tag{1.4.20}$$

where the function C_3 depends on $\|u\|_{H^1}$ and $\|v\|_{H^1}$. Furthermore, we have

$$\overline{\lim_{t\to\infty}}\,[\|u_{xx}\|^2 + \|v_{xx}\|^2] \le \frac{1}{2b_0}\max_{t\ge 0} C_3 + \frac{1}{2b_0}\left[\frac{5}{\alpha}\|h_{1x}\|^2 + \frac{2}{\gamma}\|h_{2x}\|^2\right] \tag{1.4.21}$$

$$\overline{\lim_{t\to\infty}}\,\frac{1}{t}\int_0^t [\alpha\|u_{xx}\|^2 + \|v_{xxx}\|^2]\mathrm{d}x \le \max_{t\ge 0} C_3 + \left[\frac{5}{\alpha}\|h_{1x}\|^2 + \frac{2}{\gamma}\|h_{2x}\|^2\right]. \tag{1.4.22}$$

Proof. First, by Lemmas 1.4.1, 1.4.2 and Sobolev embedding theorem, we have

$$\|u\|_\infty^2 + \|v\|_\infty^2 \le C^*(\|u\|_{H^1}^2 + \|v\|_{H^1}^2),$$

where C^* is the Sobolev embedding constant. Moreover,

$$\overline{\lim_{t\to\infty}}\,[\|u(\cdot,t)\|_\infty^2 + \|v(\cdot,t)\|_\infty^2] \le C^*(E_0 + E_1).$$

Taking the inner product of (1.4.3) and u_{xxxx}, we obtain

$$(u_{xxxx}, u_t + f(u)_x - \alpha u_{xx} + \beta u_{xxx} + 2vv_x) = (u_{xxxx}, G_1(u,v) + h_1), \tag{1.4.23}$$

where

$$\begin{aligned}
|(u_{xxxx}, f(u)_x)| &= |(u_{xxx}, f''(u)u_x^2 + f'(u)u_{xx})| \\
&\le \|f'(u)\|_\infty \|u_{xxx}\|\|u_{xx}\| + \|f''(u)\|_\infty \|u_{xxx}\|\|u_x\|_4^2 \\
&\le \frac{\alpha}{10}\|u_{xxx}\|^2 + C, \\
|(u_{xxxx}, 2vv_x)| &\le |(u_{xxx}, 2v_x^2 + 2vv_{xx})| \\
&\le \frac{\alpha}{10}\|u_{xxx}\|^2 + \frac{\gamma}{8}\|v_{xxx}\|^2 + C, \\
|(u_{xxxx}, h_1)| &\le \|u_{xxx}\|\|h_{1x}\| \le \frac{\alpha}{10}\|u_{xxx}\|^2 + \frac{5}{\alpha}\|h_{1x}\|^2.
\end{aligned}$$

Taking the inner product of (1.4.3) and v_{xxxx}, we have

$$(v_{xxxx}, v_t - \gamma v_{xx} + (2uv)_x) = (v_{xxxx}, G_2(u,v) + h_2). \tag{1.4.24}$$

Notice that

$$|(v_{xxxx}, (2uv)_x)| \le \frac{\alpha}{10}\|u_{xxx}\|^2 + \frac{\gamma}{8}\|v_{xxx}\|^2 + C$$

and

$$\begin{aligned}
&(u_{xxxx}, G_1(u,v)) + (v_{xxxx}, G_2(u,v)) \\
&\quad = (u_{xx}, G_{1u}u_{xx} + G_{1v}v_{xx} + G_{1uu}u_x^2 + 2G_{1uv}u_xv_x + G_{1vv}v_x^2) \\
&\qquad + (v_{xx}, G_{2u}u_{xx} + G_{2v}v_{xx} + G_{2uu}u_x^2 + 2G_{2uv}u_xv_x + G_{2vv}v_x^2) \\
&\quad \le -b_0(\|u_{xx}\|^2 + \|v_{xx}\|^2) + \frac{\alpha}{10}\|u_{xxx}\|^2 + \frac{\gamma}{8}\|v_{xxx}\|^2 + C,
\end{aligned}$$

where the function C depends on $\|u\|_H^1$ and $\|v\|_H^1$. Moreover, from (1.4.23) and (1.4.24), we have

$$\begin{aligned}
&\frac{d}{dt}[\|u_{xx}\|^2 + \|v_{xx}\|^2] + \alpha\|u_{xxx}\|^2 + \gamma\|v_{xxx}\|^2 + 2b_0[\|u_{xx}\|^2 + \|v_{xx}\|^2] \\
&\quad \le \frac{5}{\alpha}\|h_{1x}\|^2 + \frac{2}{\gamma}\|h_{2x}\|^2 + C.
\end{aligned}$$

Thus we get

$$\|u_{xx}\|^2 + \|v_{xx}\|^2 \le e^{-2b_0 t}[\|u_{0xx}\|^2 + \|v_{0xx}\|^2] + e^{-2b_0 t}\int_0^t C_4 e^{2b_0 s} ds$$
$$+ \frac{1}{2b_0}(1 - e^{-2b_0 t})\Big(\frac{5}{\alpha}\|h_{1x}\|^2 + \frac{2}{\gamma}\|h_{2x}\|^2\Big), \tag{1.4.25}$$
$$\overline{\lim_{t\to\infty}}\,[\|u_{xx}\|^2 + \|v_{xx}\|^2] \le \frac{1}{2b_0}\Big[\max_{t\ge 0} C_4 + \frac{5}{\alpha}\|h_{1x}\|^2 + \frac{2}{\gamma}\|h_{2x}\|^2\Big] \equiv E_2. \quad \square$$

Lemma 1.4.4. *Under the conditions of Lemma* 1.4.3, *assume that*
(1) $f(u) \in C^3$, $G_i(u, v) \in C^2$, $i = 1, 2$;
(2) $u_0(x) \in H^2(\Omega)$, $h_i(x) \in H^2(\Omega)$, $v_0(x) \in H^2(\Omega)$, $i = 1, 2$.

Then for the smooth solution of problem (1.4.3)–(1.4.6), *we have the following estimate:*

$$\|u_{xxx}\|^2 + \|v_{xxx}\|^2 \le \frac{E_3}{t}, \quad t > 0, \tag{1.4.26}$$

where constant E_3 *depends on* $\|u_0\|_{H^2}$, $\|v_0\|_{H^2}$, $\|h_i\|_{H^2}$ *and* t.

Proof. Taking the inner product of (1.4.3) and $t^2\frac{\partial^6 u}{\partial x^6}$, and respectively the inner product of (1.4.3) and $t^2\frac{\partial^6 v}{\partial x^6}$, we have

$$\Big(t^2\frac{\partial^6 u}{\partial x^6}, u_t - \alpha u_{xx} + \beta u_{xxx} + 2vv_x + f(u)_x\Big) = \Big(t^2\frac{\partial^6 u}{\partial x^6}, G_1 + h_1(x)\Big),$$
$$\Big(\frac{\partial^6 v}{\partial x^6}, v_t - \gamma v_{xx} + 2(uv)_x\Big) = \Big(t^2\frac{\partial^6 v}{\partial x^6}, G_2 + h_2(x)\Big).$$

Since

$$\Big(t^2\frac{\partial^6 u}{\partial x^6}, u_t\Big) = -\frac{1}{2}\frac{d}{dt}\|tu_{xxx}\|^2 + \|\sqrt{t}u_{xxx}\|^3,$$
$$\Big(t^2\frac{\partial^6 u}{\partial x^6}, -\alpha u_{xx}\Big) = -\alpha\|tu_{xxxx}\|^2,$$
$$\Big|\Big(t^2\frac{\partial^6 u}{\partial x^6}, f(u)_x\Big)\Big| = \Big|\Big(t^2\frac{\partial^4 u}{\partial x^4}, f(u)_{xxx}\Big)\Big|$$
$$\le \frac{\alpha}{12}\|tu_{xxxx}\|^2 + C(\|tu_{xxx}\|^2 + 1),$$
$$\Big|\Big(t^2\frac{\partial^6 u}{\partial x^6}, 2vv_x\Big)\Big| \le \frac{\alpha}{12}\|tu_{xxxx}\|^2 + C(\|tv_{xxx}\|^2 + 1),$$
$$\Big|\Big(t^2\frac{\partial^6 u}{\partial x^6}, G_1(u, v)\Big)\Big| \le \frac{\alpha}{12}\|tu_{xxxx}\|^2 + C(\|tu_{xxx}\|^2 + \|tv_{xxx}\|^2 + 1),$$
$$\|\sqrt{t}u_{xxx}\|^2 = t\|u_{xxx}\|^2 \le \frac{\alpha}{12}\|tu_{xxxx}\|^2 + C,$$
$$\Big|\Big(t^2\frac{\partial^6 u}{\partial x^6}, h_1\Big)\Big| \le \frac{\alpha}{12}\|tu_{xxxx}\|^2 + C,$$

where the constant C depends on $\|u_0\|_{H^2}$, $\|v_0\|_{H^2}$, and t, we obtain that

$$\frac{d}{dt}[\|tu_{xxx}\|^2 + \|tv_{xxx}\|^2] + \alpha\|tu_{xxxx}\|^2 - \gamma\|tv_{xxxx}\|^2 \leq C(\|tu_{xxx}\|^2 + \|tv_{xxx}\|^2 + 1).$$

Hence

$$\|u_{xxx}\|^2 + \|v_{xxx}\|^2 \leq E_3/t, \quad t > 0. \qquad \square$$

By Galerkin method we prove the existence of a global smooth solution for problem (1.4.3)–(1.4.6). Set $\omega_j(x)$, $j = 1, 2, \ldots$ to be the eigenfunctions for the equation $u_{xx} + \alpha u = 0$ with periodic boundary conditions with corresponding eigenvalues α_j, $j = 1, 2, \ldots$. Then $\{\omega_j\}$ forms the standard orthonormal basis. The approximate solutions $u_N(x, t)$ and $v_N(x, t)$ of problem (1.4.3)–(1.4.6) possess the following form:

$$u_N(x, t) = \sum_{j=1}^{N} \alpha_{jN}(t)\omega_j(t), \quad v_N(x, t) = \sum_{j=1}^{N} \beta_{jN}(t)\omega_j(t), \tag{1.4.27}$$

where α_{jN}, β_{jN}, $j = 1, 2, \ldots, N$; $N = 1, 2, \ldots$, are the coefficient functions of $t \in \mathbf{R}^+$. According to the Galerkin method, the coefficients α_{jN}, β_{jN} must satisfy the following first order nonlinear ordinary differential equations:

$$\begin{aligned} &(u_{Nt} + f(u_N)_x - \alpha u_{Nxx} + \beta u_{Nxxx} - G_1(u_N, v_N) - h_1, \omega_j) = 0, \\ &(v_{Nt} - \gamma v_{Nxx} + 2(u_N v_N)_x - G_2(u_N, v_N) - h_2, \omega_j) = 0, \\ &j = 1, 2, \ldots, N, \end{aligned} \tag{1.4.28}$$

with the initial conditions

$$\begin{aligned} \alpha_{jN}(0) &= (u_N(x, 0), \omega_j(x)) = (u_0(x), \omega_j(x)), \\ \beta_{jN}(0) &= (v_N(x, 0), \omega_j(x)) = (v_0(x), \omega_j(x)). \end{aligned} \tag{1.4.29}$$

From the solution existence theory for the ordinary differential equations, we get the existence of a local smooth solution for problems (1.4.28)–(1.4.29). Similarly as in the proof of Lemmas 1.4.1–1.4.4, we can establish consistent integral estimates of approximate solutions $\{u_N(x, t)\}$ and $\{v_N(x, t)\}$, which depend on N. The uniform priori estimates not only ensure the existence of a global solution α_{jN}, β_{jN} for the problem (1.4.28)–(1.4.29), but can also prove that the approximate solutions $\{u_N(x, t)\}$ and $\{v_N(x, t)\}$ converge to the global solution of problem (1.4.3)–(1.4.6). Then we have

Theorem 1.4.1. *Suppose the following conditions are satisfied:*
(1) $f(u) \in C^m$, $G_i \in C^{m-1}$, $i = 1, 2$, *and*

$$|f(u)| \leq A|u|^{5-\sigma}, \quad |G_i(u, v)| \leq B(|u|^5 + |v|^5), \quad B > 0,\ i = 1, 2;$$

(2) $G_i(0,0) = 0$, $i = 1,2$, *and* $\forall(\xi,\eta) \in \mathbf{R}^2$,

$$(\xi,\eta)\begin{pmatrix} -G_{1u} & -G_{1v} \\ -G_{2u} & -G_{2v} \end{pmatrix}\begin{pmatrix} \xi \\ \eta \end{pmatrix} \geq -b_0(\xi^2+\eta^2),$$

where $b_0 > 0$ *is a constant;*

(3) *if* $u_0, v_0 \in H^m(\Omega)$, $h_i(x) \in H^m(\Omega)$, $i = 1,2$, *then there exists a unique global solution for problem* (1.4.3)–(1.4.6), $u(x,t), v(x,t) \in L^\infty(0,T;H^m(\Omega))$.

In order to prove the existence of a global attractor for the periodic initial value problem (1.4.3)–(1.4.6), we use the following theorem:

Theorem 1.4.2 ([8]). *Suppose E is a Banach space and* $\{S_t,\ t \leq 0\}$ *is a set of semigroup operators,* $S_t : E \longrightarrow E$, $S_t \cdot S_\tau = S_{t+\tau}$, $S_0 = I$, *where I is the identity operator. Assume that* S_t *satisfies:*

(1) S_t *is bounded, i. e., for any* $R > 0$, *if* $\|u\|_E \leq R$, *then there exists a constant* $C(R)$, *such that*

$$\|S_t u\|_E \leq C(R), \quad t \in [0,\infty);$$

(2) *There exists a bounded absorbing set* $B_0 \subset E$, *i. e., for any bounded set* $B \subset E$, *there exists a constant T such that*

$$S_t B \subset B_0, \quad t \geq T;$$

(3) S_t *is a completely continuous operator for* $t > 0$.

Then the semigroup operator S_t *possesses a compact global attractor.*

Theorem 1.4.3. *Suppose there exists a unique global smooth solution for problem* (1.4.3)–(1.4.6) *and the conditions of Lemma* 1.4.4 *are satisfied. Then the periodic initial value problem* (1.4.3)–(1.4.6) *has a global attractor* $\mathscr{A}$, *which possess the following properties:*

(1) $S_t\mathscr{A} = \mathscr{A}$, $\forall t \in \mathbf{R}^+$.

(2) $\lim_{t\to\infty} \operatorname{dist}(S_t B, \mathscr{A}) = 0$, $\forall$ *bounded set* $B \subset H^2(\Omega)$, *where*

$$\operatorname{dist}(X,Y) = \sup_{x\in X}\inf_{y\in Y}\|x-y\|_E$$

and $S_t u_0$ *is the semigroup generated by problem* (1.4.3)–(1.4.6).

Proof. By Theorem 1.4.2, we merely need to verify its conditions for the problem (1.4.3)–(1.4.6). Under the assumptions of Theorem 1.4.3, we know that problem (1.4.3)–(1.4.6) generates a semigroup S_t. Denote the Banach space $E = H^2 \times H^2$,

$(u, v) \in E$, $\|(u, v)\|_E^2 = \|u\|_{H^2}^2 + \|v\|_{H^2}^2$, $S_t : E \longrightarrow E$. By Lemmas 1.4.1–1.4.3, if $B \subset E$ is the ball $\{\|(u, v)\|_E \le R\}$, then we have

$$\begin{aligned}\|S_t(u_0, v_0)\|_E^2 &= \|(u, v)\|_E^2 = \|u\|_{H^2}^2 + \|v\|_{H^2}^2 \\ &\le C(\|u_0\|_{H^2}^2 + \|v_0\|_{H^2}^2, \|h_i\|_{H^1}^2) \\ &\le C(R^2, \|h_1\|_{H^1}^2 + \|h_2\|_{H^1}^2), \quad t \ge 0, (u_0, v_0) \in B.\end{aligned}$$

This means that $\{S_t\}$ is consistently bounded in E. Furthermore, by using the above lemmas, we have

$$\|S_t(u_0, v_0)\|_E^2 = \|u\|_{H^2}^2 + \|v\|_{H^2}^2 \le 2(E_0 + E_1 + E_2), \quad t \ge t_0(R, \|h_1\|_{H^1}^2 + \|h_2\|_{H^1}^2). \tag{1.4.30}$$

Thus,

$$\overline{A} = \{(u(\cdot, t), v(\cdot, t)) \in E, \|(u, v)\|_E^2 \le 2(E_0 + E_1 + E_2)\}$$

is a bounded absorbing set of the semigroup operator S_t. From Lemma 1.4.4, we know

$$\|u_{xxx}\|^2 + \|v_{xxx}\|^2 \le \frac{E_3}{t}, \quad \forall t > 0, \|(u_0, v_0)\|_E \le R.$$

By use of the compact embedding $H^3(\Omega) \hookrightarrow H^2\Omega$, we know that the semigroup operator $S_t : E \to E$ is continuous for $t > 0$. This proves the theorem. □

Remark 1.4.1. As [197] points out, we get the attractor $\mathscr{A}$ in Theorem 1.4.3 as the limit set of an absorbing set:

$$\mathscr{A} = \omega(\overline{A}) = \bigcap_{r \ge 0} \bigcup_{t \ge r} S_t A. \tag{1.4.31}$$

To establish upper bounds of Hausdorff and fractal dimensions of the global attractor for the initial value problem (1.4.3)–(1.4.6), we need to consider the variational problem for (1.4.3)–(1.4.6):

$$v_t + L(u, v)v = 0, \tag{1.4.32}$$

$$v(0) = v_0, \tag{1.4.33}$$

where

$$v = \begin{pmatrix} \eta \\ \xi \end{pmatrix}, \quad v_0 = \begin{pmatrix} \eta_0 \\ \xi_0 \end{pmatrix},$$

$$L(u, v)v = \begin{pmatrix} -\alpha\eta_{xx} + \beta\eta_{xxx} + (f'(u)\eta)_x + 2(v\xi)_x - G_{1u}\eta - G_{1v}\xi \\ -\gamma\xi_{xx} + 2(u\xi + v\eta)_x - G_{2u}\eta - G_{2v}\xi \end{pmatrix}.$$

Since the solution of problem (1.4.3)–(1.4.6) is sufficiently smooth, we can prove that the linear problem (1.4.32)–(1.4.33) possesses a global smooth solution, as long as the initial value v_0 can be reasonably smooth, i. e., there exists G_t such that $v(t) = G_t v_0$. The semigroup operator $S_t U_0$ is differentiable in $L^2(\Omega)$, which is equivalent to the existence of the Fréchet derivative $S_t U_0$, and $G_t v_0 = S_t U_0$, $U_0 = (u_0, v_0)^T$. Indeed, denote

$$\omega(t) = S_t(U_0 - v_0) - S_t(U_0) - G_t(U_0)v_0 = U_1(t) - U(t) - v(t),$$

where

$$U(t) = \begin{pmatrix} u \\ v \end{pmatrix}.$$

It follows that

$$\partial_t \omega(t) = L_1(U_1) - L_1(U) + L(U(t))v(t) = L_1(U + v + \omega) - L_1(U) + L(U(t))v(t), \quad (1.4.34)$$

$$\omega(0) = 0, \quad (1.4.35)$$

where $U_1 = L_1(U)$ is the subform of problem (1.4.3)–(1.4.6). Thus (1.4.34) can be written as

$$\partial_t \omega - L(U)\omega = \Lambda_0(U, v, \omega), \quad (1.4.36)$$

where

$$\Lambda_0(U, v, \omega) = L_1(U, v, \omega) - L_1(U) + L(U)(v + \omega)(t). \quad (1.4.37)$$

Applying the linear PDE theory, we have the following L_2 estimate:

$$\|\omega(t)\| \le C\|v_0\|^2. \quad (1.4.38)$$

This yields that the semigroup operator S_t is differentiable in $L^2(\Omega)$.

Denote by $v_1(t), v_2(t), \ldots, v_J(t)$ the solutions of the linear equation with initial values $v_1(0) = \xi_1, v_2(0) = \xi_2, \ldots, v_J(0) = \xi_J$, respectively. Here $\xi_j \in L^2(\Omega) \times L^2(\Omega)$, $j = 1, 2, \ldots, J$. By a simple calculation, we have

$$\frac{\mathrm{d}}{\mathrm{d}t}\|v_1(t) \wedge v_2(t) \wedge \cdots \wedge v_J(t)\|^2 + 2\operatorname{tr}(L(U(t))Q_J)\|v_1(t) \wedge v_2(t) \wedge \cdots \wedge v_J(t)\|^2 = 0, \quad (1.4.39)$$

where $L(U(t)) = L(S_t U_0)$ is the linear mapping $v \longrightarrow L(U(t))v$, $\wedge$ denotes the cross-product, tr denotes the trace of an operator, $Q_j(t)$ means the representation space $L^2(\Omega) \times L^2\Omega)$ to $v_1(t), v_2(t), \ldots, v_J(t)$ into the subspace orthogonal projection. From formula (1.4.39), we can obtain the variation of volume of the J-dimensional cube $\bigwedge_{j=1}^{J} \xi_j$ as

$$\begin{aligned}\omega_J(t) &= \sup_{u_0 \in \mathscr{A}} \sup_{\xi \in L^2, |\xi_j| \le 1} \|v_1(t) \wedge v_2(t) \wedge \cdots \wedge v_J(t)\|^2_{\mathscr{L}^J L^2} \\ &\le \sup_{u_0 \in \mathscr{A}} \exp\left(-2\int_0^t \operatorname{tr}(L(S_r U_0)Q_J(r))\mathrm{d}r\right), \quad \text{where } \mathscr{A} \text{ is an attractor.}\end{aligned}$$

Noticing the result in [197], we know that $\omega_J(t)$ is log-subadditive, that is,

$$\omega_J(t+t_1) \le \omega_J(t)\omega_J(t_1), \quad t, t_1 \ge 0. \tag{1.4.40}$$

Therefore, we obtain

$$\lim_{t\to\infty} \omega_J(t)^{\frac{1}{t}} = \pi_J \le \exp(-2q_J) \tag{1.4.41}$$

where

$$q_J = \lim_{t\to\infty} \sup_{u_0\in\mathscr{A}} \left(\sup_{\xi\in L^2, |\xi_j|\le 1} \frac{1}{t} \int_0^t \operatorname{tr}(L(S_\tau U_0)Q_J(\tau))\mathrm{d}\tau \right). \tag{1.4.42}$$

By use of the following theorem and lemma

Theorem 1.4.4 ([33]). *Let $\mathscr{A}$ be an attractor for a nonlinear evolution equation. Assume it is bounded in $H^1(\Omega)$. If $q_J > 0$, for some J, then the Hausdorff dimension of the attractor $\le J$ and its fractal dimension $\le J(1+\max_{1\le j\le J}(-\frac{q_j}{q_J}))$.*

Theorem 1.4.5 (Generalized Sobolev–Lieb–Thirring inequality [197]). *Let $\Omega \subset \mathbf{R}^n$ be a bounded set, and $\{\phi_1, \phi_2, \dots, \phi_N\}$ an orthogonal basis in $L^2(\Omega)$, $\phi_i \in H^m$, $i = 1, 2, \dots, N$, and for almost all $x \in \Omega$, $\rho(x) = \sum_{j=1}^N |\phi_j(x)|^2$. Then for almost $x \in \Omega$, $\rho(x) = \sum_{j=1}^N |\phi_j|^2$, the following estimate holds:*

$$\int \rho(x)^{1+2m/n}\mathrm{d}x \le \frac{k_0}{|\Omega|^{2m/n}} \int \rho(x)\mathrm{d}x + k_0 \sum_{j=1}^N \int |D^m\phi_j|^2 \mathrm{d}x, \tag{1.4.43}$$

where the constant k_0 depends on m, n and Ω, but is independent of N and ϕ_j.

Theorem 1.4.6. *Under the conditions of Theorems 1.4.1 and 1.4.3, the global attractor has finite Hausdorff and fractal dimensions for the periodic initial value problem (1.4.3)–(1.4.6), which can be bounded as*

$$d_H(\mathscr{A}) \le J_0, \quad d_F(\mathscr{A}) \le J_0\left(1 + \frac{2b\sqrt{(b/(3a))}}{3(aJ_0^3 - bJ_0)}\right),$$

where J_0 is the smallest integer such that

$$J_0 > \left[k_0 + \frac{4k_0D^2}{\min\{\alpha,\gamma\}}\left(\frac{1}{\sqrt{2D}}\left(\frac{1}{2}\|f''(u)\|_\infty + 1\right)(\|u_x\| + \|v_x\|) - b_0\right)\right]^{\frac{1}{2}} \ge J_0 - 1$$

and

$$a = \frac{\min\{\alpha,\gamma\}}{4k_0D^2},$$

$$b = \frac{\min\{\alpha,\gamma\}}{4D^2} + \frac{1}{\sqrt{2D}}\left(\frac{1}{2}\|f''(u)\|_\infty + 1\right)(\|u_x\| + \|v_x\|) - b_0.$$

Proof. By Theorem 1.4.4, we only need to estimate the lower bound of $\mathrm{tr}(L(U)Q_J)$. Set $\{\Phi_1, \Phi_2, \dots, \Phi_J\}$, $\Phi_j = (\phi_j, \psi_j)^T$ as the orthonormal basis of the subspace $Q_J(L^2 \times L^2)$. We have

$$\begin{aligned}
\mathrm{tr}(L(U(t))Q_J) &= \sum_{j=1}^{J} [(-\alpha\phi_{jxx} - \beta\phi_{jxxx} + (f'(u)\phi_j)_x \\
&\quad + (2v\phi_j)_x - G_{1u}(u,v)\phi_j - G_{1v}\psi_j, \phi_j) \\
&\quad + (-\gamma\psi_{jxx} + 2(u\psi_j + v\phi_j)_x - G_{2u}\phi_j - G_{2v}\psi_j, \psi_j)] \\
&= \sum_{j=1}^{J}\Big[\alpha\|\phi_{jx}\|^2 + \gamma\|\psi_{jx}\|^2 + \frac{1}{2}\int f''(u)u_x\phi_j^2\mathrm{d}x + \int u_x\psi_j^2\mathrm{d}x \\
&\quad + 2\int v_x\phi_j\psi_j\mathrm{d}x - \int(G_{1u}\phi_j^2 + G_{1v}\phi_j\psi_j + G_{2u}\phi_j\psi_j + G_{2v}\psi_j^2)\mathrm{d}x\Big] \\
&\geq \sum_{j=1}^{J}\Big[\min\{\alpha,\gamma\}(\|\phi_{jx}\|^2 + \|\psi_{jx}\|^2) \\
&\quad - \int\Big(\frac{1}{2}|f''(u)u_x| + |u_x| + |v_x|\Big) \\
&\quad \times(\phi_j^2 + \psi_j^2)\mathrm{d}x + b_0\int(\phi_j^2 + \psi_j^2)\mathrm{d}x\Big] \\
&\geq \min\{\alpha,\gamma\}\Big[\frac{1}{k}\int\rho(x)^3\mathrm{d}x - \frac{1}{(2D)^2}J\Big] + b_0J \\
&\quad - \Big\|\Big(\frac{1}{2}|f''(u)u_x| + |u_x| + |v_x|\Big)\Big\|_{3,2}\Big(\int\rho(x)^3\mathrm{d}x\Big)^{\frac{1}{3}} \\
&\geq \min\{\alpha,\gamma\}\frac{1}{k_0}\int\rho(x)^3\mathrm{d}x + \Big(b_0 - \frac{\min\{\alpha,\gamma\}}{4D^2}\Big)J \\
&\quad - \Big[\Big(\frac{1}{2}\|f''(u)\|_\infty + 1\Big)(2D)^{\frac{1}{6}}(\|u_x\| + \|v_x\|)\Big]\Big(\int\rho(x)^3\mathrm{d}x\Big)^{\frac{1}{3}}. \qquad (1.4.44)
\end{aligned}$$

Since

$$J = \int_\Omega \rho(x)\mathrm{d}x \leq \Big(\int\rho(x)^3\mathrm{d}x\Big)^{\frac{1}{3}}(2D)^{\frac{2}{3}},$$

it follows that

$$\int\rho(x)^3\mathrm{d}x \geq \frac{1}{(2D)^2}J^3. \qquad (1.4.45)$$

Thus from equations (1.4.44) and (1.4.45), we get

$$\begin{aligned}
\mathrm{tr}(U(t)\cdot Q_J) &\geq \frac{\min\{\alpha,\gamma\}}{4k_0D^2}J^3 + \Big(b_0 - \frac{\min\{\alpha,\gamma\}}{4D^2}\Big)J \\
&\quad - \Big(\frac{1}{2}\|f''(u)\|_\infty + 1\Big)(\|u_x\| + \|v_x\|)\frac{1}{\sqrt{2D}}J > 0.
\end{aligned}$$

If

$$J > \left(\frac{b}{a}\right)^{\frac{1}{2}},$$

$$a = \frac{\min\{\alpha, \gamma\}}{4k_0 D^2},$$

$$b = \frac{\min\{\alpha, \gamma\}}{4D^2} + \frac{1}{\sqrt{2D}}\left(\frac{1}{2}\|f''(u)\|_\infty + 1\right)(\|u_x\| + \|v_x\|) - b_0$$

and

$$J_0 - 1 \le \left(\frac{b}{a}\right)^{\frac{1}{2}} \le J_0,$$

$$-\frac{q_l}{q_{J_0}} \le \frac{bl - al^3}{aJ_0^3 - bJ_0} \le \frac{2b\sqrt{b/3a}}{3(aJ_0^3 - bJ_0)},$$

then by Theorem 1.4.6, we have

$$d_H(\mathscr{A}) \le J_0,$$

$$d_F(\mathscr{A}) \le J_0\left(1 + \frac{2b\sqrt{b/3a}}{3(aJ_0^3 - bJ_0)}\right)$$

and the claim is proved. □

1.5 Davey–Stewartson equation

We consider the following Davey–Stewartson (DS) equation [29, 62, 206]:

$$\frac{\partial A}{\partial t} - \alpha\frac{\partial^2 A}{\partial x^2} - b\frac{\partial^2 A}{\partial y^2} = \chi A - \beta|A|^2 A + \gamma QA, \quad t > 0,\ (x,y) \in \Omega, \tag{1.5.1}$$

$$\frac{\partial^2 Q}{\partial x^2} + \frac{\partial^2 Q}{\partial y^2} = \frac{\partial^2}{\partial y^2}(|A|^2), \tag{1.5.2}$$

with boundary condition

$$A(t,x,y) = 0, \quad Q(t,x,y) = 0, \quad t \ge 0,\ (x,y) \in \partial\Omega \tag{1.5.3}$$

and initial value condition

$$A(0,x,y) = A_0(x,y), \quad (x,y) \in \Omega, \tag{1.5.4}$$

where $a = a_1 + ia_2$, $b = b_1 + ib_2$, $\beta = \beta_1 + i\beta_2$, $\gamma = \gamma_1 + i\gamma_2$, $\chi = \chi_1 + i\chi_2$ are complex numbers, $\Omega \subset \mathbf{R}^2$ is a smooth bounded area. This system of equations was proposed by Davey et al. [39] while studying planar Poiseuille flow in a nonlinear three-dimensional disturbance evolution. Here $A(t,x,y)$ stands for complex amplitude, and $Q(t,x,y)$ describes real speed disturbance.

If we set $Q = |A|^2 - \frac{\partial \varphi}{\partial x}$, together with equation (1.5.2), we will have

$$\Delta Q = \Delta |A|^2 - \frac{\partial}{\partial x}\Delta\varphi = \frac{\partial^2}{\partial y^2}|A|^2,$$

where $\Delta = \frac{\partial^2}{\partial x^2} + \frac{\partial^2}{\partial y^2}$, so we will obtain

$$\Delta\varphi = \frac{\partial}{\partial x}|A|^2. \tag{1.5.5}$$

Hence the amplitude equation becomes

$$\frac{\partial A}{\partial t} - a\frac{\partial^2 A}{\partial x^2} - b\frac{\partial^2 A}{\partial y^2} = \chi A - \overline{\beta}|A|^2 A - \gamma A\frac{\partial\varphi}{\partial x} \tag{1.5.6}$$

where $\overline{\beta} = \beta + \gamma$. The system of equations (1.5.5)–(1.5.6) was first proposed in [38, 48] by Davey and Stewartson. Ablowitz and Haberman [2] studied two-dimensional completely integrable nonlinear Schrödinger equation systems and obtained some special cases of the above equations. Many mathematical physicists did a series of studies about the equations, for instance, investigated the existence of a generalized local or global solution, the stability of a plane wave solution, studied the properties of the solitary wave solutions, solution of the singularity development, etc.; for details we could refer to Davey, Hocking and Stewartson [39], Holmes [142], Ghidaglia and Saut [75], Anker and Freeman [4], Ablowitz and Fokas [1], Tsutsumi [201], Hayashi and Saut [138], Linares and Ponce [165], and so on. In 1997, Guo and Yang [210] proved the existence of global smooth solutions for a class of DS equations, Guo and Li [105] proved the existence of the attractor for a class of DS equations, and showed finite fractal dimension.

From equations (1.5.2) and (1.5.3) we can solve for Q as a function of A:

$$Q = -(-\Delta)^{-1}\frac{\partial^2}{\partial y^2}|A|^2 \triangleq E(|A|^2), \tag{1.5.7}$$

where $(-\Delta)^{-1}$ is the inverse of Laplace operator which has Dirichlet boundary conditions. We can reduce equations (1.5.1) and (1.5.2) into the following nonlinear Schrödinger (Ginzburg–Landau) equation:

$$\frac{\partial A}{\partial t} - a\frac{\partial^2 A}{\partial x^2} - b\frac{\partial^2 A}{\partial y^2} = \chi A - \beta|A|^2 A - \gamma A E(|A|^2), \quad t > 0,\ (x,y) \in \Omega, \tag{1.5.8}$$

$$A(t,x,y) = 0, \quad t \geq 0,\ (x,y) \in \partial\Omega, \tag{1.5.9}$$

$$A(0,x,y) = A_0(x,y), \quad (x,y) \in \Omega. \tag{1.5.10}$$

From Sobolev inequality, there exists a $C(p) > 0$, $1 < p < \infty$, such that

$$\left\|\frac{\partial^2 u}{\partial y^2}\right\|_p \leq C(p)\|\Delta u\|_p, \quad u \in C_0(\Omega),$$

where $\|\cdot\|_p$ denotes the norm of $L^p(\Omega)$. Hence

$$\|E(u)\|_p \le C(p)\|u\|_p, \quad u \in C_0^\infty(\Omega) \tag{1.5.11}$$

and $E = (-\Delta)^{-1}\frac{\partial^2}{\partial y^2}$ can be extended to a bounded linear operator $(1 < p < \infty)$ in $L^p(\Omega)$ with norm $C(p)$. Actually, $\frac{\partial^2}{\partial y^2}$ and Δ commute in $W^{2,p}(\Omega)$, $(-\Delta)^{-1}$ is a linear bounded operator in $W^{2,p}(\Omega)$, thus $\bar{E} = \frac{\partial^2}{\partial y^2}(-\Delta)^{-1}$ is the extension of E in $L^p(\Omega)$. And $\|\bar{E}\|_{\mathscr{A}(L^p(\Omega))} = C(p)$.

We will prove the following theorem:

Theorem 1.5.1. *Suppose*

$$[H] \quad a_1 > 0,\ b_1 > 0,\ \beta_1 > 0,\ \beta_1 + C(2)\gamma_1 > 0,\ \chi_1 > 0.$$

Then the semigroup operator generated by equations (1.5.8) and (1.5.9) has a global compact attractor in $H_0^1(\Omega)$, and its Hausdorff and fractal dimensions are finite.

Lemma 1.5.1. *Suppose condition $[H]$ is satisfied, $A_0 \in L^2(\Omega)$. Then $A \in L^\infty(R^-, L^2(\Omega))$ satisfies*

$$\|A(t)\|^2 \le \|A_0\|^2 e^{-t} + C, \quad \forall t \ge 0.$$

Thus there exists a $t_1(R) > 0$ such that

$$\|A(t)\| \le C, \quad \forall t \ge t_1(R),$$

where $\|A_0\| \le R$, $C = C(\beta_1, \gamma_1, \chi_1, \Omega)$. Moreover, for all $r > 0$, we have

$$\int_t^{t+r} (\|\nabla A(s)\|^2 + \|A(s)\|_4^4)\mathrm{d}s \le C(r, \beta_1, \gamma_1, \chi_1, \Omega), \quad \forall t \ge t_1(R).$$

Proof. Multiplying equation (1.5.8) by $\overline{A}$, integrating over Ω, and taking its real component, we obtain

$$\begin{aligned} &\frac{1}{2}\frac{\mathrm{d}}{\mathrm{d}t}\|A\|^2 + a_1\|A_x\|^2 + b_1\|A_y\|^2 \\ &\quad = \chi_1\|A\|^2 - \beta_1\|A\|_4^4 - \gamma_1\int |A|^2 E(|A|^2)\mathrm{d}x\mathrm{d}y. \end{aligned} \tag{1.5.12}$$

Noticing that

$$\begin{aligned} &\int |A|^2 E(|A|^2)\mathrm{d}x\mathrm{d}y = \int |[(-\Delta)^{\frac{1}{2}}|A|^2]_y|^2 \mathrm{d}x\mathrm{d}y \ge 0, \\ &\int |A|^2 E(|A|^2)\mathrm{d}x\mathrm{d}y \le \||A|^2\|\|E(|A|^2)\| \\ &\qquad \le C(2)\||A|^2\|^2 = C(2)\|A\|_4^4, \end{aligned} \tag{1.5.13}$$

we have

$$-\beta_1\|A\|_4^4-\gamma_1\int|A|^2E(|A|^2)\mathrm{d}x\mathrm{d}y\le-K\|A\|_4^4,\tag{1.5.14}$$

where $K=\beta_1>0$ when $\gamma_1>0$ and $K=\beta_1+C(2)\gamma_1>0$ when $\gamma_1<0$. By Hölder inequality,

$$\chi_1\|A\|^2\le\frac{1}{2}K\|A\|_4^4+C,\tag{1.5.15}$$

where C only depends on β_1,γ_1,χ_1 and Ω. Combining with equations (1.5.12)–(1.5.12), we obtain

$$\frac{1}{2}\frac{\mathrm{d}}{\mathrm{d}t}\|A\|^2+\alpha_0\|\nabla A\|^2+\frac{1}{2}K\|A\|_4^4\le C.\tag{1.5.16}$$

Similar to equation (1.5.15), $\|A\|^2\le\frac{1}{2}K\|A\|_4^4+C$, we have

$$\frac{\mathrm{d}}{\mathrm{d}t}\|A\|^2+\|A\|^2\le C.$$

By Gronwall inequality we deduce that

$$\|A\|^2\le\|A_0\|^2e^{-t}+C(1-e^{-t}).\tag{1.5.17}$$

Hence there exists a $t_1(R)>0$ such that $\|A\|^2\le1+C,t\ge t_1(R),\|A_0\|\le R$. From equations (1.5.16) and (1.5.17), we obtain

$$\begin{aligned}2\alpha_0\int_t^{t+r}\|\nabla A(s)\|^2\mathrm{d}s+K\int_t^{t+r}\|A(s)\|_4^4\mathrm{d}s&\\ \le Cr-\|A(s)\|^2\Big|_{s=t}^{s=t+r}&\\ \le Cr+\|A_0\|^2e^{-t}+C(1-e^{-t})\le Cr+C,&\end{aligned}$$

$\forall t>t_1(R)$, $r>0$, and the lemma is proved. □

Lemma 1.5.2. *Suppose condition [H] is satisfied, $A_0\in H_0^1(\Omega)$. Then there exist $A\in L^\infty(R^+,L_0^1(\Omega))$ and $t_2(R)>0$, $C>0$ such that*

$$\|\nabla A(t)\|^2\le C,\quad\forall t\ge t_2(R),$$

where $\|A_0\|_{H^1}\le R$.

Proof. Multiplying equation (1.5.8) by $-\Delta\bar{A}$, integrating over Ω, and taking its real component yields

$$\mathrm{Re}\int aA_{xx}\Delta\bar{A}\mathrm{d}x\mathrm{d}y=a_1\|A_{xx}\|^2+a_1\|A_{xy}\|^2,$$

$$\mathrm{Re}\int aA_{yy}\Delta\bar{A}\mathrm{d}x\mathrm{d}y = b_1\|A_{yy}\|^2 + b_1\|A_{xy}\|^2,$$

and we have

$$\frac{1}{2}\frac{\mathrm{d}}{\mathrm{d}t}\|\nabla A\|^2 + a_1\|A_{xx}\|^2 + b_1\|A_{yy}\|^2 + (a_1+b_1)\|A_{xy}\|^2$$
$$= \chi_1\|\nabla A\|^2 - \mathrm{Re}\,\beta\int |A|^2A\Delta\bar{A}\mathrm{d}x\mathrm{d}y - \mathrm{Re}\,\gamma\int AE(|A|^2)\Delta\bar{A}\mathrm{d}x\mathrm{d}y. \tag{1.5.18}$$

Noticing that

$$a_1\|A_{xx}\|^2 + (a_1+b_1)\|A_{xy}\|^2 + b_1\|A_{yy}\|^2 \geq \alpha_0\|\Delta A\|^2,$$

and using Gagliardo–Nirenberg inequality

$$\|u\|_6 \leq C\|u\|^{1/3}\|\nabla u\|^{2/3}, \quad \forall u \in H_0^1(\Omega),$$

we get the following estimates:

$$\begin{aligned}
\left|\mathrm{Re}\,\beta\int |A|^2A\Delta\bar{A}\mathrm{d}x\mathrm{d}y\right| &\leq |\beta|\|A\|_6^3\|\Delta A\| \\
&\leq C\|A\|\|\nabla A\|^2\|\Delta A\| \\
&\leq \frac{1}{6}\alpha_0\|\Delta A\|^2 + C\|A\|^2\|\nabla A\|^4, \\
\left|\mathrm{Re}\,\gamma\int AE(|A|^2)\Delta\bar{A}\mathrm{d}x\mathrm{d}y\right| &\leq |\gamma|\|A\|_6\|E(|A|^2)\|_3\|\Delta A\| \\
&\leq |\gamma|\|A\|_5^3\|\Delta A\| \leq C\|A\|\|\nabla A\|^2\|\Delta A\| \\
&\leq \frac{1}{6}\alpha_0\|\Delta A\|^2 + C\|A\|^2\|\nabla A\|^4, \\
\chi_1\|\nabla A\|^2 &\leq \frac{1}{6}\alpha_0\|\Delta A\|^2 + C\|A\|^2.
\end{aligned}$$

Consequently we obtain

$$\frac{1}{2}\frac{\mathrm{d}}{\mathrm{d}t}\|\nabla A\|^2 + \frac{1}{2}\alpha_0\|\Delta A\|^2 \leq C\|A\|^2 + C\|A\|^2\|\nabla A\|^4. \tag{1.5.19}$$

In order to prove the claim of the lemma, i. e.,

$$\left\|\nabla A(t)\right\|^2 \leq C, \quad \forall t \geq t_2(R),$$

we need the following lemma

Lemma 1.5.3 (Uniform Gronwall inequality). *Let $g(t)$, $h(t)$ and $y(t) \geq 0$ be such that*

$$y'(t) \leq g(t)y(t) + h(t), \quad \forall t \geq s.$$

If $\int_t^{t+r} g(s) \leq k_1$, $\int_t^{t+r} h(s)\mathrm{d}s \leq k_2$, $\int_t^{t+r} y(s)\mathrm{d}s \leq k_3$, $r > 0$, then

$$y(t+r) \leq \left(\frac{k_3}{r} + k_2\right)e^{k_1}, \quad \forall t \geq s.$$

Now let $y(t) = \|\nabla A(t)\|^2$, $g(t) = C\|A(t)\|^2\|\nabla A(t)\|^2$, $h(t) = C\|A(t)\|^2$. Then by Lemma 1.5.2, for any $r > 0$, $t \geq t_1(R)$,

$$\int_t^{t+r} y(s)\mathrm{d}s \leq Cr + C \triangleq k_3,$$

$$\int_t^{t+r} g(s)\mathrm{d}s \leq C \int_t^{t+r} \|\nabla A\|^2\mathrm{d}s \leq Cr + C \triangleq k_1,$$

$$\int_t^{t+r} h(s)\mathrm{d}s \leq Cr \triangleq k_2,$$

where $\|A_0\|_{H_0^1(\Omega)} \leq R$. By the uniform Gronwall inequality, we have

$$y(t+r) = \|\nabla A(t+r)\| \leq \left(\frac{C}{r} + C + Cr\right)e^{Cr+C}, \quad t \geq t_1(R), \quad r > 0.$$

Thus Lemma 1.5.3 is proved. Furthermore, from equation (1.5.19) we have

$$\int_t^{t+r} \|\Delta A\|^2\mathrm{d}s \leq C(r), \quad \forall t \geq 0.$$

□

Lemma 1.5.4. *Suppose the conditions of Lemma* 1.5.2 *are satisfied. Then we get*

$$\|\Delta A\|^2 \leq C + \frac{C}{t}, \quad \forall t > 0.$$

Proof. Acting Δ on both sides of equation (1.5.8), multiplying by $-t\Delta\bar{A}$, integrating in Ω over x, and taking its real part, we obtain

$$\begin{aligned}&\frac{1}{2}\frac{\mathrm{d}}{\mathrm{d}t}(t\|\Delta A\|^2) - \frac{1}{2}\|\Delta A\|^2 + t(a_1\|\Delta A_x\|^2 + b_1\|\Delta A_y\|^2)\\ &= \chi_1 t\|\Delta A\|^2 + \operatorname{Re}\beta t\int \Delta(|A|^2A)\Delta\bar{A}\mathrm{d}x\mathrm{d}y + \operatorname{Re}\gamma t\int \Delta(AE(|A|^2))\Delta\bar{A}\mathrm{d}x\mathrm{d}y.\end{aligned} \tag{1.5.20}$$

Noticing that

$$\begin{aligned}a_1\|\Delta A_x\|^2 + b_1\|\Delta A_y\|^2 &\geq \alpha_0\|\nabla\Delta A\|^2,\\ \|\nabla A\|_4 &\leq C\|A\|_4^{\frac{3}{5}}\|\nabla\Delta A\|^{\frac{2}{3}},\\ \|\Delta A\|_4 &\leq C\|A\|_4^{\frac{1}{5}}\|\nabla\Delta A\|^{\frac{4}{5}},\end{aligned}$$

we get

$$\begin{aligned}&\left|\operatorname{Re}\beta\int \Delta(|A|^2A)\Delta\bar{A}\mathrm{d}x\mathrm{d}y\right|\\ &\leq 2|\beta|\int(|A|^2|\Delta A|^2 + |\nabla A|^2|A||\Delta A|)\mathrm{d}x\mathrm{d}y\end{aligned}$$

$$\leq C(\|A\|_4^2\|\Delta A\|_4^2 + \|\nabla A\|_4^2\|A\|_4\|\Delta A\|_4)$$
$$\leq C\|A\|_4^{\frac{12}{5}}\|\nabla\Delta A\|_4^{\frac{8}{5}}$$
$$\leq \frac{1}{4}\alpha_0\|\nabla\Delta A\|^2 + C\|A\|_{H^1}^{12},$$

$$\left|\mathrm{Re}\,\gamma\int \Delta(AE(|A|^2))\Delta\bar{A}\mathrm{d}x\mathrm{d}y\right|$$
$$\leq |\gamma|\|\Delta A\|_4^2\|E(|A|^2)\| + 2|\gamma|\|\nabla A\|_4\|\nabla E(|A|^2)\|\|\Delta A\|_4$$
$$+ |\gamma|\|A\|_4\|\Delta A\|_4\|\Delta E(|A|^2)\|$$
$$\leq C\|A\|_4^2\|\Delta A\|_4^2 + C\|\nabla A\|_4\|\nabla|A|^2\|\|\Delta A\|_4 + C\|A\|_4\|\Delta A\|_4\left\|\frac{\partial^2|A|^2}{\partial y^2}\right\|$$
$$\leq C\|A\|_4\|\Delta A\|_4^2 + C\|\nabla A\|_4^2\|A\|_4\|\Delta A\|_4 + C\|A\|_4\|\Delta A\|_4\left(2\|A\|_4\left\|\frac{\partial^2 A}{\partial y^2}\right\|_4 + 2\left\|\frac{\partial A}{\partial y}\right\|_4^2\right)$$
$$\leq C\|A\|_4^{\frac{12}{5}}\|\nabla\Delta A\|^{\frac{8}{5}}$$
$$\leq \frac{1}{4}\alpha_0\|\nabla\Delta A\|^2 + C\|A\|_{H^1}^{12}.$$

From equation (1.5.20) and the above estimate, we obtain

$$\frac{1}{2}\frac{\mathrm{d}}{\mathrm{d}t}(t\|\Delta A\|^2) + \frac{1}{2}\alpha_0\lambda_1 t\|\Delta A\|^2 \leq Ct\|A\|_{H^1}^{12} + \frac{1}{2}\|\Delta A\|^2 \leq Ct + \frac{1}{2}\|\Delta A\|^2,$$

where $\lambda_1\|u\|^2 \leq \|\nabla u\|^2$, $u \in H_0^1(\Omega)$ and λ_1 is the first eigenvalue with Dirichlet boundary conditions of $-\Delta$. Set $\alpha_0 = \min\{a, b\}$. If $\alpha_0 > 0$, then

$$a_1\|u_x\|^2 + b_1\|u_y\|^2 \geq \alpha_0\lambda_1\|u\|^2.$$

By $\int_0^r \|\Delta A(t)\|^2\mathrm{d}t \leq C(r)$, $\forall r > 0$, and Gronwall inequality, we get

$$t\|\Delta A\|^2 \leq G(t) \triangleq \int_0^t e^{-\alpha_0\lambda_1(t-s)}(Cs + \|\Delta A(s)\|^2)\mathrm{d}s. \tag{1.5.21}$$

Thus we have

$$\|\Delta A\|^2 \leq \frac{C(r)}{t}, \quad \forall 0 < t \leq r. \tag{1.5.22}$$

From equation (1.5.21) we have

$$\|\Delta A\|^2 \leq \frac{1}{t}G(t), \quad \forall t > 0. \tag{1.5.23}$$

Differentiating $G(t)$ with respect to t, using formula (1.5.23), and inserting into the above equation, we have

$$\frac{\mathrm{d}}{\mathrm{d}t}G(t) = Ct + \|\Delta A\|^2 - \alpha_0\lambda_1 G(t) \leq Ct - \left(\alpha_0\lambda_1 - \frac{1}{t}\right)G(t).$$

Taking $r_* > \frac{2}{\alpha_0\lambda_1}$, we have

$$\frac{\mathrm{d}}{\mathrm{d}t}G(t) + \frac{1}{2}\alpha_0\lambda_1 G(t) \le Ct, \quad \forall t \ge r_*.$$

From Gronwall inequality we obtain

$$\begin{aligned} G(t) &\le G(r_*)e^{-\frac{1}{2}\alpha_0\lambda_1(t-r_*)} + \int_{r_*}^{t} e^{-\frac{1}{2}\alpha_0\lambda_1(t-r)}Cs\mathrm{d}s \\ &\le G(r_*) + C\int_0^t se^{-\frac{1}{2}\alpha_0\lambda_1(t-s)}\mathrm{d}s \\ &\le G(r_*) + \frac{2C}{\alpha_0\lambda_1}t + \frac{4C}{\alpha_0^2\lambda_1^2}, \quad \forall t \ge r_*. \end{aligned}$$

Hence

$$\|\Delta A\|^2 \le \frac{C(r_*)}{t} + C, \quad \forall t \ge r_*. \tag{1.5.24}$$

From equations (1.5.22) and (1.5.24) we get the claim. □

Now we rewrite (1.5.8) and (1.5.9) using a functional form

$$A_t = F(A) \triangleq LA + f(A), \quad A|_{t=0} = A_0, \tag{1.5.25}$$

where $L = P(D) = a\frac{\partial^2}{\partial x^2} + b\frac{\partial^2}{\partial y^2}$ is a differential operator on $X = L^2(\Omega)$, $D(L) = W^{2,p} \cap W_0^{1,p}(\Omega)$, $1 < p < \infty$, and the linear mapping $f(A) = \chi A - \beta|A|^2 A - \gamma E(|A|^2)$. Obviously $0 \in \rho(L)$, where $\rho(L)$ is the resolvent equation of ρ. The symbol for $L = P(D)$ is $P(\xi) = a\xi_1^2 + b\xi_2^2$, its real component $\mathrm{Re}\,P(\xi) \ge \alpha_0|\xi|^2$. So $P(\xi)$ is a strong elliptic polynomial, and $L = P(D)$ can generate a bounded analytic semigroup on X. So we can define the fractal power of $-L$ as $(-L)^\alpha$, its domain is $D((-L)^\alpha) = X^\alpha$. In particular, when $p = 2\alpha = \frac{1}{2}$, we see $D((-L)^{\frac{1}{2}}) = H_0^1(\Omega)$, the nonlinear mapping $f(A)$ is locally continuous from $X^{\frac{1}{2}}$ to $X^{\frac{1}{2}}$, that is,

$$\begin{aligned} &\|f(A_1) - f(A_2)\|_X \le C(R)\|A_1 - A_2\|_{X^{1/2}}, \\ &\forall A_1, A_2 \in X^{\frac{1}{2}} = H_0^1(\Omega), \quad \|A_k\| \le R, \quad k = 1, 2. \end{aligned}$$

Hence, for $A_0 \in X^{1/2}$, and formula has the unique local solution

$$A \in C([0, t_0), X^{1/2}) \cap C^1((0, t_0), X).$$

By the priori estimates of Lemmas 1.5.1–1.5.4, we know that the local solution can be extended to the global solution. So we have

Theorem 1.5.2. *Suppose condition [H] is satisfied and $A_0 \in H_0^1(\Omega)$. Then the system* (1.5.8)–(1.5.9) *has a unique global solution*

$$A \in C([0,+\infty);\ H_0^1(\Omega)) \cap C^1((0,+\infty), L^2(\Omega)).$$

The solution operator $S(t) : A_0 \longmapsto A(t)$ forms a continuous semigroup in $H_0^1(\Omega)$, and possesses an absorbing bounded set $\mathscr{B} \subset H_0^1(\Omega)$.

By Theorem 1.5.2 and Lemma 1.5.4, for any bounded set $B \subset H_0^1(\Omega)$, $\overline{\bigcup_{t\geq 1} S(t)B}$ is bounded in $H_0^1(\Omega)$. That is, $S(t)$ is compact for sufficiently large t. By the results in [197] we know

Theorem 1.5.3. *Suppose condition [H] is met, $S(t)$ is the semigroup generated by problem* (1.5.8)–(1.5.9) *and $\mathscr{B}$ is the absorbing set of $S(t)$ in $H_0^1(\Omega)$. Then the ω limit set of $\mathscr{B}$ is*

$$\mathscr{A} = \bigcap_{s\geq 0} \overline{\bigcup_{t\geq s} S(t)\mathscr{B}},$$

where closure is taken under the topology of $H_0^1(\Omega)$, and it satisfies:
(1) *$S(t)\mathscr{A} = \mathscr{A}$, $\forall t \geq 0$ (invariance property).*
(2) $\lim_{t\to\infty} \operatorname{dist}(S(t)B, \mathscr{A}) \triangleq \lim_{t\to\infty} \sup_{A_0\in B} \operatorname{dist}(S(t)A_0, \mathscr{A}) = 0$, $\forall B \subset H_0^1(\Omega)$ *absorbing.*
(3) *$\mathscr{A}$ is compact in $H_0^1(\Omega)$.*

We will prove that $\mathscr{A}$ has finite Hausdorff and fractal dimensions in the following.

From Theorem 1.5.2 and Lemma 1.5.4 we know that for $A_0 \in H_0^1(\Omega)$, equation (1.5.25) has a unique solution

$$A \in C([0,+\infty); H_0^1(\Omega)) \cap L^2_{\text{loc}}([0,+\infty); H^2(\Omega)) \cap L^\infty((t_*,+\infty); H^2(\Omega)), \quad \forall t_* > 0.$$

Let $U(t,\cdot,\cdot)$ is the solution of the following variational problem:

$$\begin{aligned}
\frac{\mathrm{d}U}{\mathrm{d}t} &= F'(A(t))U \\
&= aU_{xx} + bU_{yy} + \chi U - 2\beta|A|^2U - \beta A^2\overline{U} - \gamma E(|A|^2)U - \gamma E(\overline{A}U + A\overline{U})A, \\
&\qquad t > 0,\ (x,y) \in \Omega, \\
U(t,x,y) &= 0, \quad t \geq 0,\ (x,y) \in \partial\Omega, \\
U(0,\cdot,\cdot) &= U_0 \in H_0^1(\Omega), \quad (x,y) \in \Omega.
\end{aligned}$$

It is easy to verify that the above problem has a unique solution

$$U \in C_b(R^+; H_0^1(\Omega)) \cap L^2_{\text{loc}}(R^+; H^2(\Omega)),$$

this means that $S(t)$ is Fréchet differentiable in $H_0^1(\Omega)$. Also $U(t,\cdot,\cdot) = (DS(t)A_0)U_0$ is the Fréchet derivative of $S(t)$ at $A_0 \in H_0^1(\Omega)$. Then we have

$$\|S(t)(A_0+U_0) - S(t)A_0 - (DS(t)A_0)U_0\|_{H_0^1(\Omega)} \le C(R,T)\|U_0\|^2_{H_0^1(\Omega)}.$$

For $0 \le t \le T$, we have $A_0, U_0 \in H_0^1(\Omega)$, $\|A_0\|_{H_0^1(\Omega)} \le R$. Suppose that $U_{k_0} \in H_0^1(\Omega)$ is linearly independent, $U_k = (DS(t)A_0)U_{k0}$, $1 \le k \le m$, then

$$\begin{aligned}&\|U_1(t)\wedge U_2(t)\wedge\cdots\wedge U_m(t)\|^2_{\wedge^m H_0^1(\Omega)}\\ &\quad= \|U_{10}(t)\wedge U_{20}(t)\wedge\cdots\wedge U_{m0}(t)\|^2_{\wedge^m H_0^1(\Omega)} \exp\int_0^t \operatorname{Re}\operatorname{tr}(F'(S(\tau)A_0)\cdot Q_m(\tau))d\tau,\end{aligned} \tag{1.5.26}$$

where $(\cdot,\cdot)_{H_0^1(\Omega)}$ is the inner product in $H_0^1(\Omega)$ and $(\cdot,\cdot)$ is the inner product in $L^2(\Omega)$. We omit the variable τ and get

$$\begin{aligned}\operatorname{Re}(F'(A)\phi_j,\phi_j)_{H_0^1(\Omega)} &= -a_1\|\nabla\phi_{jx}\|^2 - b_1\|\nabla\phi_{jy}\|^2 + \chi\|\nabla\phi_j\|\\ &\quad- 2\operatorname{Re}\beta(\nabla(|A|^2\phi_j),\nabla\phi_j) - \operatorname{Re}\beta(\nabla(A^2\overline{\phi_j}),\nabla\phi_j)\\ &\quad- \operatorname{Re}\gamma(\nabla(E(|A|^2)\phi_j),\nabla\phi_j) - \operatorname{Re}\gamma(\nabla(E(\overline{A}\phi_j + A\overline{\phi_j})A),\nabla\phi_j).\end{aligned}$$

Using Gagliardo–Nirenberg inequality, we get

$$\begin{aligned}|2\operatorname{Re}\beta(\nabla(|A|^2\phi_j),\nabla\phi_j)| &\le C(\|A\|_\infty\|\nabla A\|_3\|\phi_j\|_6 + \|A\|^2_\infty\|\nabla\phi_j\|)\|\nabla\phi_j\|\\ &\le C(\|\nabla A\|_3^2\|\nabla\phi_j\|^2) \le C\|A\|^{2/3}\|\Delta A\|^{4/3}\|\nabla\phi_j\|^2,\end{aligned}$$

Similarly, we have

$$\begin{aligned}|\operatorname{Re}(\nabla(A^2\overline{\phi_j}),\nabla\phi_j)| &\le C\|A\|^{2/3}\|\Delta A\|^{4/3}\|\nabla\phi_j\|^2,\\ |\operatorname{Re}\gamma(\nabla(E(|A|^2)\phi_j),\nabla\phi_j)| &\le |\gamma|\|E(|A|^2)\|_\infty\|\nabla\phi_j\|^2 + C|\gamma|\|\nabla|A|^2\|_3\|\phi_j\|_6\|\nabla\phi_j\|\\ &\le C|\gamma|\|\nabla E(|A|^2)\|_3\|\nabla\phi_j\|^2 + C|\gamma|\|A\|_\infty\|\nabla|A|\|_3\|\nabla\phi_j\|^2\\ &\le C\|A\|^{2/3}\|\Delta A\|^{4/3}\|\nabla\phi_j\|^2,\\ |\operatorname{Re}\gamma(\nabla E(\overline{A}\phi_j + A\overline{\phi_j}),\nabla\phi_j)| &\le |\gamma|\|E(\overline{A}\phi_j + A\overline{\phi_j})\|_6\|\nabla A\|_3\|\nabla\phi_j\|\\ &\quad+ |\gamma|\|\nabla E(\overline{A}\phi_j + A\overline{\phi_j}\|\|A\|_\infty\|\nabla\phi_j\|\\ &\le C|\gamma|\|A\|_\infty\|\phi_j\|_6\|\nabla A\|_3\|\nabla\phi_j\|\\ &\le C\|A\|^{2/3}\|\Delta A\|^{4/3}\|\nabla\phi_j\|^2.\end{aligned}$$

Thus we get

$$\begin{aligned}\operatorname{Re}(F'(A)\phi_j,\phi_j)_{H_0^1(\Omega)} &\le -\alpha_0\|\Delta\phi_j\|^2 + \chi_1\|\nabla\phi_j\|^2 + C\|A\|^{2/3}\|\Delta A\|^{4/3}\|\nabla\phi_j\|^2\\ &\le -\alpha_0\|\Delta\phi_j\|^2 + \chi_1\|\nabla\phi_j\|^2 + C_0(\|A\|^2 + \|\Delta A\|^2)\|\nabla\phi_j\|^2,\end{aligned}$$

where constant C_0 only depends on β, γ and Ω_0. Since $\|\nabla\phi_j\| = 1$, we have

$$\operatorname{Re}\operatorname{tr}(F'(A)\cdot Q_m) \le -\alpha_0 \sum_{j=1}^{m} \|\Delta\phi_j\|^2 + m(\chi_1 + C_0\|A\|^2) + C_0 m\|\Delta A\|^2. \tag{1.5.27}$$

Set $\rho_m(x,y) = \sum_{i=1}^{m} |\nabla\phi_j(x,y)|^2$, $\sigma_m(x,y) = \sum_{j=1}^{m} |\Delta\phi_j(x,y)|^2$. Then $\int_\Omega \rho_m(x,y)\mathrm{d}x\mathrm{d}y = m$. By the generalized Sobolev–Lieb–Thirring inequality

$$\int_\Omega \rho_m^2(x,y)\mathrm{d}x\mathrm{d}y \le k_0 \int_\Omega \sigma_m(x,y)\mathrm{d}x\mathrm{d}y,$$

where k_0 only depends on Ω, so by Hölder inequality, we have

$$m = \int_\Omega \rho_m(x,y)\mathrm{d}x\mathrm{d}y \le |\Omega|^{1/2}\Big(\int_\Omega \rho_m^2(x,y)\mathrm{d}x\mathrm{d}y\Big)^{1/2} \le k_0|\Omega|^{1/2}\Big(\int_\Omega \sigma_m(x,y)\mathrm{d}x\mathrm{d}y\Big)^{1/2}.$$

Hence

$$\sum_{j=1}^{m} \|\Delta\phi_j\|^2 = \int_\Omega \sigma_m(x,y)\mathrm{d}x\mathrm{d}y \ge \frac{m^2}{k_0^2|\Omega|}. \tag{1.5.28}$$

For $A_0 \in \mathscr{A}$(global attractor), $A(t) = S(t)A_0$, and we have

$$A(t) \le C_1, \quad \int_0^t \|\Delta A(s)\|^2 \mathrm{d}s \le C_2 t + C_3, \quad \forall t \ge 0.$$

Hence, by equations (1.5.27) and (1.5.28), we obtain

$$\begin{aligned}
&\int_0^t \operatorname{Re}\operatorname{tr}(F'(S(\tau)A_0)\cdot Q_m(\tau))\mathrm{d}\tau \\
&\quad \le -\frac{\alpha_0 m^2}{k_0^2|\Omega|} t + m(\chi_1 + C_0C_1)t + C_0C_2 mt + C_0C_3 m.
\end{aligned}$$

Thus we get

$$\begin{aligned}
q_m &= \limsup_{t\to\infty} \sup_{A_0\in\mathscr{A}} \sup_{\|U_{j0}\|\le 1} \frac{1}{t}\int_0^t \operatorname{Re}\operatorname{tr}(F'(S(\tau)A_0)\cdot Q_m(\tau))\mathrm{d}\tau \\
&\le -\frac{\alpha_0 m^2}{k_0^2|\Omega|} + m(\chi_1 + C_0C_1 + C_0C_2).
\end{aligned}$$

Set m_0 to be the smallest integer satisfying

$$m_0 > \frac{\chi_1 + C_0C_1 - C_0C_2}{\alpha_0}|\Omega|k_0. \tag{1.5.29}$$

Then we have $q_m < 0$, $m \ge m_0$. Then using the results of [197] we obtain

Theorem 1.5.4. *Suppose $\mathscr{A}$ is the attractor of problem* (1.5.8)–(1.5.9) *in $H_0^1(\Omega)$ and m_0 is determined by* (1.5.29). *Then we have the following estimates:*

(1) *the Hausdorff dimension of $\mathscr{A} \le m_0$;*

(2) *the fractal dimension of $\mathscr{A} \le 2m_0$.*

1.6 Derivative Ginzburg–Landau equation

Derivative Ginzburg–Landau equation appears in many physical problems, for example, Rayleigh–Benard convection, Taylor–Couette flow in fluid mechanics, the dissipation in the plasma drift flow, chemical reaction of turbulent flow, see [19, 46, 102, 103, 133, 134, 135].

The generalized (with a derivative term) derivative Ginzburg–Landau equation in a one dimensional space has the following form:

$$\frac{\partial u}{\partial t} = \alpha_0 u + \alpha_1 u_{xx} + \alpha_2 |u|^2 u + \alpha_3 |u|^2 u_x + \alpha_4 u^2 u_x^* + \alpha_5 |u|^{2\sigma} u, \tag{1.6.1}$$

where $\sigma > 0$, $\alpha_k = a_k + ib_k$, $\alpha_0 = a_0 > 0$, a_k, b_k are real constants and u_x^* is the complex conjugate of u_x. In 1992, Duan, Holmes and Titi [51] proved the global existence and uniqueness of a solution for equation (1.6.1) in a bounded area for $\sigma = 2$. In 1994, Duan, Holmes and Titi [50] proved the well-posedness of Cauchy problem (1.6.1). Guo and Gao proved the existence of the global attractor for the periodic initial value problem (1.6.1) and showed the finiteness of its fractal and Hausdorff dimensions [97].

If $\alpha_2 = \alpha_3 = \alpha_4 = 0$, then equation (1.6.1) reduces to

$$\frac{\partial u}{\partial t} = \rho u + (1 + i\nu)\Delta u - (1 + i\mu)|u|^{2\sigma} u. \tag{1.6.2}$$

In 1990, Bartuccelli et al. [10] studied the "hard" and "soft" states. In 1994, Doering et al. [49] proved the existence of weak and strong solutions in a two-dimensional space.

Now we consider the following generalized equation in a two-dimensional bounded area:

$$\frac{\partial u}{\partial t} = \rho u + (1 + i\nu)\Delta u - (1 + i\mu)|u|^{2\sigma} u + \alpha\lambda_1 \cdot \nabla(|u|^2 u) + \beta(\lambda_2 \cdot \nabla u)|u|^2 \tag{1.6.3}$$

with initial conditions

$$u(x, 0) = u_0(x), \quad x \in \Omega \tag{1.6.4}$$

and boundary conditions

$$\Omega = (0, L_1) \times (0, L_1), \quad u \text{ is periodic in } \Omega, \tag{1.6.5}$$

where $u(x,t)$ is an unknown complex variable function, $\sigma > 0$, $\rho > 0$, v, μ, α and β all are real constants and λ_1, λ_2 are real vectors.

In 1996, Guo and Wang [114] proved that if $u_0 \in H^2_{\text{per}}(\Omega)$ and $\exists\sigma > 0$ such that

$$\epsilon \leq \sigma \leq \frac{1}{\sqrt{1+(\frac{u-v\delta^2}{1+\delta^2})^2}-1}, \tag{1.6.6}$$

then there exists a global unique solution $u(x,t)$ of problem (1.6.1)–(1.6.5),

$$u(x,t) \in L^\infty(0,T;H^2(\Omega)) \cap L^2(0,T;H^3(\Omega)), \quad \forall T > 0,$$

and the equation has a global attractor having finite Hausdorff and fractal dimensions.

Now we establish a consistent a priori estimate for problem (1.6.1)–(1.6.5).

Lemma 1.6.1. *Suppose* $u_0 \in H^2_{\text{per}}(\Omega)$. *Then for a solution of* (1.6.1)–(1.6.5), *we have*

$$\|u(t)\|^2 \leq C_1, \quad \forall t \geq t_1.$$

Proof. Taking the inner product of (1.6.3) and u in $L^2(\Omega)$, and then the real component, we obtain

$$\begin{aligned}\frac{1}{2}\frac{\mathrm{d}}{\mathrm{d}t}\|u\|^2 &= \rho\int_\Omega |u|^2\mathrm{d}x - \|\nabla u\|^2 - \int_\Omega |u|^{2\sigma+2}\mathrm{d}x \\ &\quad + \alpha\,\mathrm{Re}\int_\Omega (\lambda_1\cdot\nabla(|u|^2u))u^*\mathrm{d}x + \beta\,\mathrm{Re}\int_\Omega(\lambda_2\cdot\nabla u)|u|^2u^*\mathrm{d}x. \end{aligned}\tag{1.6.7}$$

Denoting $\lambda_2 = (a,b)$, we have

$$\begin{aligned}&\mathrm{Re}\int_\Omega(\lambda_2\cdot\nabla u)|u|^2u^*\mathrm{d}x \\ &= \mathrm{Re}\int_\Omega\left(a\frac{\partial u}{\partial x_1}u^* + b\frac{\partial u}{\partial x_2}u^*\right)|u|^2\mathrm{d}x \\ &= \frac{1}{2}\int_\Omega a|u|^2\frac{\partial}{\partial x_1}|u|^2\mathrm{d}x + \frac{1}{2}\int_\Omega b|u|^2\frac{\partial}{\partial x_2}|u|^2\mathrm{d}x \\ &= \frac{1}{2}\int_0^{L_2}\mathrm{d}x_2\int_0^{L_1}a|u|^2\frac{\partial}{\partial x_1}|u|^2\mathrm{d}x_1 + \frac{1}{2}\int_0^{L_1}\mathrm{d}x_1\int_0^{L_2}b|u|^2\frac{\partial}{\partial x_2}|u|^2\mathrm{d}x_2 = 0, \end{aligned}\tag{1.6.8}$$

$$\begin{aligned}&\mathrm{Re}\int_\Omega(\lambda_1\cdot\nabla(|u|^2u))u^*\mathrm{d}x \\ &= \mathrm{Re}\int_\Omega(\lambda_1\cdot\nabla|u|^2)|u|^2\mathrm{d}x + \mathrm{Re}\int_\Omega(\lambda_1\cdot\nabla u)|u|^2u^*\mathrm{d}x\end{aligned}$$

$$= \int_{\Omega} (\lambda_1 \cdot \nabla |u|^2)|u|^2 \mathrm{d}x + \frac{1}{2} \int_{\Omega} (\lambda_1 \cdot \nabla |u|^2)|u|^2 \mathrm{d}x = 0, \tag{1.6.9}$$

thus from (1.6.7)–(1.6.9) we obtain

$$\begin{aligned} \frac{1}{2}\frac{\mathrm{d}}{\mathrm{d}t}\|u\|^2 + \|\nabla u|^2 + \int_{\Omega} |u|^{2\sigma+2}\mathrm{d}x &= \rho \int_{\Omega} |u|^2 \mathrm{d}x \\ &\le \frac{1}{2}\int_{\Omega} |u|^{2\sigma+2}\mathrm{d}x + C \quad \text{(by Young inequality)}, \end{aligned} \tag{1.6.10}$$

where the constant C depends on parameters σ and ρ. In the following, C will possibly denote a different constant and will depend on parameters (σ, ρ, ν, μ). From equation (1.6.10), we have

$$\frac{1}{2}\frac{\mathrm{d}}{\mathrm{d}t}\|u\|^2 + \|\nabla u\|^2 + \frac{1}{2}\int_{\Omega} |u|^{2\sigma+2}\mathrm{d}x \le C.$$

It follows that

$$\frac{\mathrm{d}}{\mathrm{d}t}\|u\|^2 + \int_{\Omega} |u|^{2\sigma+2}\mathrm{d}x \le C. \tag{1.6.11}$$

Using Young inequality we obtain

$$|u|^2 = |u|^2 \cdot 1 \le |u|^{2\sigma+2} + C. \tag{1.6.12}$$

Integrating equation (1.6.12), we get

$$\|u\|^2 \le \int_{\Omega} |u|^{2\sigma+2}\mathrm{d}x + C. \tag{1.6.13}$$

From equations (1.6.11) and (1.6.13), we have

$$\frac{\mathrm{d}}{\mathrm{d}t}\|u\|^2 + \|u\|^2 \le C.$$

Moreover, by Gronwall inequality we have

$$\begin{aligned} \|u(t)\|^2 &\le \|u_0\|^2 e^{-t} + C \le R^2 e^{-t} + C, \quad \forall t \ge 0, \\ &\le 2C, \quad \forall t \ge t_*, \end{aligned}$$

where $t_* = \ln \frac{R^2}{C}$. The lemma has been proved. □

Lemma 1.6.2. *Suppose the conditions of Lemma* 1.6.1 *are satisfied, and there exists* $\delta > 0$ *such that equation* (1.6.6) *is valid. Then* $\forall \varepsilon > 0$ *we have*

$$\frac{1}{2(1+\sigma)}\frac{\mathrm{d}}{\mathrm{d}t}\int_{\Omega} |u|^{2\sigma+2}\mathrm{d}x \le -\frac{1}{2}\int_{\Omega} |u|^{4\sigma+2}\mathrm{d}x + \varepsilon\|\Delta u\|^2$$

$$+ \varepsilon\|\nabla u\|^4 - \frac{1}{4}\int_\Omega |u|^{2\sigma-2}((1+2\sigma)|\nabla|u|^2|^2$$
$$- 2\nu\sigma\nabla|u|^2 \cdot i(u\nabla u^* - u^*\nabla u) + |u\nabla u^* - u^*\nabla u|^2)\mathrm{d}x$$
$$+ C_3 - C_2(\varepsilon)(|\alpha\lambda_1| + |\beta\lambda_2|)^{\frac{8(1+\sigma)(1+2\sigma)}{6\sigma^2-11\sigma-7}}, \quad \forall t \geq t_2,$$

where the constant C_3 depends on parameters (σ, ρ, ν, μ), $C_2(\varepsilon)$ depends on ε, t_2 depends on R, and $\|u_0\| \leq R_0$.

Proof. Taking the inner product of (1.6.3) and $|u|^{2\sigma}u$ in L^2, we have

$$\int_\Omega \frac{\partial u}{\partial t}|u|^{2\sigma}u^*\mathrm{d}x = \rho\int_\Omega |u|^{2\sigma+2}\mathrm{d}x + (1+i\nu)\int_\Omega \Delta u \cdot |u|^{2\sigma}u^*\mathrm{d}x - (1+i\mu)\int_\Omega |u|^{4\sigma-2}\mathrm{d}x$$
$$+ \alpha\int_\Omega (\lambda_1 \cdot \nabla(|u|^2u))|u|^{2\sigma}u^*\mathrm{d}x + \beta\int_\Omega (\lambda_2 \cdot \nabla u)|u|^{2\sigma+2}u^*\mathrm{d}x. \quad (1.6.14)$$

Since

$$(1+i\nu)\int_\Omega \Delta u \cdot |u|^{2\sigma}u^*\mathrm{d}x = -(1+i\nu)\int_\Omega |\nabla u|^2|u|^{2\sigma}\mathrm{d}x$$
$$- (1+i\nu)\int_\Omega \sigma|u|^{2\sigma-2}u^*\nabla u \cdot \nabla|u|^2\mathrm{d}x, \quad (1.6.15)$$

taking the real component of equation (1.6.14), we get

$$\frac{1}{2(1+\sigma)}\frac{\mathrm{d}}{\mathrm{d}t}\int_\Omega |u|^{2\sigma+2}\mathrm{d}x = \rho\int_\Omega |u|^{2\sigma-2}\mathrm{d}x - \int_\Omega |\nabla u|^2|u|^{2\sigma}\mathrm{d}x - \frac{\sigma}{2}\int_\Omega |u|^{2\sigma-2}||\nabla u|^2|^2\mathrm{d}x$$
$$+ \frac{1}{2}\nu\sigma\int_\Omega |u|^{2\sigma-2}\nabla|u|^2 \cdot i(u\nabla u^* - u^*\nabla u)\mathrm{d}x$$
$$- \int_\Omega |u|^{4\sigma+2}\mathrm{d}x + \operatorname{Re}\alpha\int_\Omega (\lambda_1 \cdot (|u|^2u))|u|^{2\sigma}u^*\mathrm{d}x$$
$$+ \operatorname{Re}\beta\int_\Omega (\lambda_2 \cdot \nabla u)|u|^{2\sigma-2}u^*\mathrm{d}x. \quad (1.6.16)$$

Since

$$|u|^2|\nabla u|^2 = \frac{1}{4}|\nabla|u|^2|^2 + \frac{1}{4}|u\nabla u^* - u^*\nabla u|^2, \quad (1.6.17)$$

we obtain

$$-\int_\Omega |\nabla u|^2|u|^{2\sigma}\mathrm{d}x - \frac{\sigma}{2}\int_\Omega |u|^{2\sigma-2}||\nabla u|^2|^2\mathrm{d}x + \frac{1}{2}\nu\sigma\int_\Omega |u|^{2\sigma-2}\nabla|u|^2 \cdot i(u\nabla u^* - u^*\nabla u)\mathrm{d}x$$

$$= -\frac{1}{4}\int_\Omega |u|^{2\sigma-2}((1+2\sigma)|\nabla|u|^2|^2 - 2\nu\sigma\nabla|u|^2 \cdot i(u\nabla u^* - u^*\nabla u) + |u\nabla u^* - u^*\nabla u|^2)\mathrm{d}x \tag{1.6.18}$$

$$\rho\int_\Omega |u|^{2\sigma+2}\mathrm{d}x = \rho\int_\Omega |u|^{2\sigma+1}\cdot|u|\mathrm{d}x \le \frac{1}{6}\int_\Omega |u|^{4\sigma+2}\mathrm{d}x + \frac{3}{2}\rho^2\int_\Omega |u|^2\mathrm{d}x$$
$$\le \frac{1}{6}\int_\Omega |u|^{4\sigma+2}\mathrm{d}x + C, \quad \forall t \ge t_1, \tag{1.6.19}$$

$$\left|\mathrm{Re}\,\beta\int_\Omega(\lambda_2\cdot\nabla u)|u|^{2\sigma+2}u^*\mathrm{d}x\right| \le |\beta\lambda_2|\int_\Omega |\nabla u||u|^2|u|^{2\sigma+1}\mathrm{d}x$$
$$\le 3|\beta\lambda_2|^2\int_\Omega |\nabla u|^2|u|^4\mathrm{d}x + \frac{1}{12}\int_\Omega |u|^{4\sigma+2}\mathrm{d}x$$
$$< 3|\beta\lambda_2|^2\|\nabla u\|_4^2\|u\|_8^4 + \frac{1}{12}\int_\Omega |u|^{4\sigma+2}. \tag{1.6.20}$$

By the Sobolev interpolation inequality,

$$\|u\|_4 \le K\|u\|_{H^1}^{\frac{1}{2}}\|u\|^{\frac{1}{2}}, \quad \forall u \in H^1(\Omega), \tag{1.6.21}$$
$$\|u\|_8 \le K\|u\|_{H^2}^{\theta}\|u\|_q^{1-\theta}, \quad \forall u \in H^2(\Omega), \tag{1.6.22}$$

where

$$\theta = \frac{8-q}{4q+8} \quad \text{when } 1 < q < 8; \quad \theta = 0 \quad \text{when } q \ge 8,$$

and then we have

$$\|\nabla u\|_4^2 \le C\|\nabla u\|_{H^1}\|\nabla u\| \le C\|u\|_{H^2}\|u\|_{H^1}. \tag{1.6.23}$$

From equations (1.6.20)–(1.6.23), when $q > 3$ and $0 < \gamma \le 1$, we have

$$\left|\mathrm{Re}\,\beta\int_\Omega(\lambda_2\cdot\nabla u)|u|^{2\sigma+2}u^*\mathrm{d}x\right|$$
$$\le C|\beta\lambda_2|^2\|u\|_{H^2}^{4\theta+1}\|u\|_{H^1}\|u\|_q^{4(1-\theta)} + \frac{1}{12}\int_\Omega |u|^{4\sigma+2}\mathrm{d}x$$
$$\le \gamma\|u\|_{H^2}^2 + C(\gamma)|\beta\lambda_2|^{\frac{4}{1-4\theta}}\|u\|_{H^1}^{\frac{2}{1-4\theta}}\|u\|_q^{\frac{8(1-\theta)}{1-4\theta}} + \frac{1}{12}\int_\Omega |u|^{4\delta+2}\mathrm{d}x.$$

When $q > \frac{14}{3}$, we have

$$\left|\mathrm{Re}\,\beta\int_\Omega(\lambda_2\cdot\nabla u)|u|^{2\sigma+2}u^*\mathrm{d}x\right|$$
$$\le \gamma\|u\|_{H^2}^2 + \gamma\|u\|_{H^1}^4 + C(\gamma)|\beta\lambda_2|^{\frac{8}{1-8\theta}}\|u\|_q^{\frac{16(1-\theta)}{1-8\theta}} + \frac{1}{12}\int_\Omega |u|^{4\sigma+2}\mathrm{d}x.$$

When $q > \frac{34}{3}$, we have

$$\left|\operatorname{Re}\beta\int_\Omega(\lambda_2\cdot\nabla u)|u|^{2\sigma+2}u^*\mathrm{d}x\right|$$
$$\le \gamma\|u\|_{H^2}^2+\gamma\|u\|_{H^1}^4+\frac{1}{12}\|u\|_q^q+C(\gamma)|\beta\lambda_2|^{\frac{8q}{q-8q\theta-16-16\theta}}+\frac{1}{12}\int_\Omega|u|^{4\sigma-2}\mathrm{d}x.$$

Since $\sigma\ge 3$, we have $q=4\sigma+2>\frac{34}{3}$ and so

$$\left|\operatorname{Re}\beta\int_\Omega(\lambda_2\cdot\nabla u)|u|^{2\sigma+2}u^*\mathrm{d}x\right|$$
$$\le \gamma\|u\|_{H^2}^2+\gamma\|u\|_{H^1}^4+\frac{1}{12}\|u\|_{4\sigma+2}^{4\sigma+2}+C(\gamma)|\beta\lambda_2|^{\frac{8(1+\sigma)(1+2\sigma)}{6\sigma^2-11\sigma-7}}+\frac{1}{12}\int_\Omega|u|^{4\sigma-2}\mathrm{d}x,$$

$$\left|\operatorname{Re}\beta\int_\Omega(\lambda_2\cdot\nabla u)|u|^{2\sigma+2}u^*\mathrm{d}x\right|$$
$$\le \gamma\|u\|_{H^2}^2+\gamma\|u\|_{H^1}^4+\frac{1}{6}\int\Omega|u|^{4\sigma+2}\mathrm{d}x+C(\gamma)|\beta\lambda_2|^{\frac{8(1+\sigma)(1+2\sigma)}{6\sigma^2-11\sigma-7}}$$
$$\le \gamma C\|\Delta u\|^2+8\gamma\|\nabla u\|^4+\frac{1}{6}\int_\Omega|u|^{4\sigma+2}\mathrm{d}x+C+C(\gamma)|\beta\lambda_2|^{\frac{8(1+\sigma)(1+2\sigma)}{6\sigma^2-11\sigma-7}}. \tag{1.6.24}$$

Since

$$\nabla(|u|^2u)=|u|^2\nabla u+u\nabla|u|^2=2|u|^2\nabla u+u^2\nabla u, \tag{1.6.25}$$

we similarly have

$$\left|\operatorname{Re}\alpha\int_\Omega(\lambda_1\cdot\nabla(|u|^2u))|u|^{2\sigma}u^*\mathrm{d}x\right|$$
$$\le 3|\alpha\lambda_1|\int_\Omega|\nabla u||u|^2\cdot|u|^{2\sigma+1}\mathrm{d}x\le\gamma C\|\Delta u\|^2$$
$$+8\gamma\|\nabla u\|^4+\frac{1}{6}\int_\Omega|u|^{4\sigma+2}\mathrm{d}x+C(\gamma)|\alpha\lambda_1|^{\frac{8(1+\sigma)(1+2\sigma)}{6\sigma^2-11\sigma-7}}. \tag{1.6.26}$$

From equations (1.6.16), (1.6.18), (1.6.19), (1.6.24) and (1.6.26), we get

$$\frac{1}{2(1+\sigma)}\frac{\mathrm{d}}{\mathrm{d}t}\int_\Omega|u|^{2\sigma+2}\mathrm{d}x\le-\frac{1}{2}\int_\Omega|u|^{4\sigma+2}\mathrm{d}x+\gamma C\|\Delta u\|^2$$
$$+16\gamma\|\nabla u\|^4+C+C(\gamma)\big(|\alpha\lambda_1|+|\beta\lambda_2|\big)^{\frac{8(1+\sigma)(1+2\sigma)}{6\sigma^2-11\sigma-7}}$$
$$-\frac{1}{4}\int_\Omega|u|^{2\sigma-2}\big((1+2\sigma)|\nabla|u|^2|^2$$

$$-2\nu\sigma\nabla|u|^2\cdot i(\nabla u^*-u^*\nabla u)+|u\nabla u^*-u^*\nabla u|^2)\mathrm{d}x,$$

when γ is small enough, and the lemma is proved. □

Lemma 1.6.3. *Suppose the condition of Lemma 1.6.2 is satisfied. Then we have*

$$\|\nabla u\|^2\le C_3+C_3(|\alpha\lambda_1|+|\beta\lambda_2|)^{\frac{8(1+\sigma)(1+2\sigma)}{6\sigma^2-11\sigma-7}},\quad \forall t\ge t_3, \tag{1.6.27}$$

where constant C_3 depends on data parameters; t_3 depends on data parameters and R; $\|u_0\|_{H^1}\le R$.

Proof. Taking the L^2 inner product of equation (1.6.3) and Δu, we get

$$\frac{1}{2}\frac{\mathrm{d}}{\mathrm{d}t}\|\nabla u\|^2+\|\Delta u\|^2=\rho\|\nabla u\|^2+\operatorname{Re}(1+i\mu)\int_\Omega|u|^{2\sigma}u\Delta u^*\mathrm{d}x-\alpha\operatorname{Re}\int_\Omega(\lambda_1\cdot\nabla(|u|^2u))\Delta u^*\mathrm{d}x$$
$$-\beta\operatorname{Re}\int_\Omega(\lambda_2\cdot\nabla u)|u|^2\Delta u^*\mathrm{d}x, \tag{1.6.28}$$

$\forall\varepsilon>0$, and, by Young inequality,

$$\rho\|\nabla u\|^2\le\varepsilon\|\nabla u\|^4+C(\varepsilon), \tag{1.6.29}$$

$$\begin{aligned}
&\operatorname{Re}(1+i\mu)\int_\Omega|u|^{2\sigma}u\Delta u^*\mathrm{d}x\\
&=-\operatorname{Re}(1+i\mu)\int_\Omega|u|^{2\sigma}|\nabla u|^2\mathrm{d}x-\operatorname{Re}(1+i\mu)\int_\Omega\sigma|u|^{2\sigma-2}u\Delta u^*\cdot\nabla|u|^2\mathrm{d}x\\
&=-\int_\Omega|u|^{2\sigma}|\nabla u|^2\mathrm{d}x-\frac{\sigma}{2}\int_\Omega|u|^{2\sigma-2}|\nabla|u|^2|^2\mathrm{d}x\\
&\quad+\frac{1}{2}\mu\sigma\int_\Omega|u|^{2\sigma-2}\nabla|u|^2\cdot i(u\cdot\nabla u-u\nabla u^*)\mathrm{d}x\\
&=-\frac{1}{4}\int_\Omega|u|^{2\sigma-2}((1+2\sigma)|\nabla|u|^2|^2\\
&\quad-2\mu\sigma\nabla|u|^2\cdot i(u^*\nabla u-u\nabla u^*)+|u^*\nabla u-u\nabla u^*|^2)\mathrm{d}x\quad\text{(by (1.6.25))}
\end{aligned}\tag{1.6.30}$$

$$\begin{aligned}
&\left|-\beta\operatorname{Re}\int_\Omega(\lambda_2\cdot\nabla u)|u|^2\Delta u^*\mathrm{d}x\right|\\
&\le|\beta\lambda_2|\int_\Omega|\nabla u||u|^2|\Delta u|\mathrm{d}x\\
&\le|\beta\lambda_2|\|\Delta u\|\|\nabla u\|_4\|u\|_8^2\quad\text{(by Hölder inequality)}\\
&\le C|\beta\lambda_2|\|\Delta u\|\|\nabla u\|_{H^1}^{\frac{1}{2}}\|\nabla u\|^{\frac{1}{2}}\|u\|_{H^2}^{2\theta}\|u\|_q^{2(1-\theta)}\\
&\le C|\beta\lambda_2|\|u\|_{H^2}^{2\theta-\frac{3}{2}}\|\nabla u\|^{\frac{1}{2}}\|u\|_q^{2(1-\theta)}\\
&\le\gamma\|u\|_{H^2}^2+C(\gamma)|\beta\lambda_2|^{\frac{4}{1-4\theta}}\|\nabla u\|^{\frac{2}{1-4\theta}}\|u\|_q^{\frac{8(1-\theta)}{1-4\theta}}\quad\text{(when } q>3,\ 0<\gamma\le 1),
\end{aligned}$$

$$\left|-\beta\operatorname{Re}\int_\Omega(\lambda_2\cdot\nabla u)|u|^2\Delta u^*\,dx\right|$$
$$\le\gamma\|u\|_{H^2}^2+\gamma\|\nabla u\|^4+C(\gamma)|\beta\lambda_2|^{\frac{8}{1-8\theta}}\|u\|_q^{\frac{16(1-\theta)}{1-8\theta}}\quad\left(\text{when } q>\frac{14}{3}\right),$$
$$\left|-\beta\operatorname{Re}\int_\Omega(\lambda_2\cdot\nabla u)|u|^2\Delta u^*\,dx\right|$$
$$\le\gamma\|u\|_{H^2}^2+\gamma\|\nabla u\|^4+\gamma\|u\|_q^q+C(\gamma)|\beta\lambda_2|^{\frac{8q}{q-8q\theta-16+16\theta}}\quad\left(\text{when } q>\frac{34}{3}\right),$$
$$\left|-\beta\operatorname{Re}\int_\Omega(\lambda_2\cdot\nabla u)|u|^2\Delta u^*\,dx\right|$$
$$\le\gamma\|\Delta u\|^2+\gamma\|\nabla u\|^4+\gamma\|u\|_q^q+C(\gamma)|\beta\lambda_2|^{\frac{8q}{q-8q\theta-16+16\theta}}\le\gamma\|\Delta u\|^2$$
$$+\gamma\|\nabla u\|^4+\gamma\|u\|_{4\sigma+2}^{4\sigma+2}+C(\gamma)|\beta\lambda_2|^{\frac{8(1+\sigma)(1+2\sigma)}{6\sigma^2-11\sigma-7}}\quad\left(\text{when } \sigma\ge 3,\ q=4\sigma+2>\frac{34}{3}\right),$$
$$\left|-\beta\operatorname{Re}\int_\Omega(\lambda_2\cdot\nabla u)|u|^2\Delta u^*\,dx\right|$$
$$\le\frac{1}{2}\varepsilon\|\Delta u\|^2+\frac{1}{2}\varepsilon\|\nabla u\|^4+\frac{1}{2}\varepsilon\|u\|_{4\sigma+2}^{4\sigma+2}+C(\varepsilon)|\beta\lambda_2|^{\frac{8(1+\sigma)(1+2\sigma)}{6\sigma^2-11\sigma-7}}.\tag{1.6.31}$$

Similarly, by equation (1.6.25) we have

$$\left|-\alpha\operatorname{Re}\int_\Omega(\lambda_1\cdot\nabla(u|u|^2))\Delta u^*\,dx\right|$$
$$\le\frac{1}{2}\varepsilon\|\Delta u\|^2+\frac{1}{2}\varepsilon\|\nabla u\|^4+\frac{1}{2}\varepsilon\|u\|_{4\sigma+2}^{4\sigma+2}+C(\varepsilon)|\alpha\lambda_1|^{\frac{8(1+\sigma)(1+2\sigma)}{6\sigma^2-11\sigma-7}}.\tag{1.6.32}$$

From equations (1.6.28)–(1.6.32), we have

$$\frac{1}{2}\frac{d}{dt}\|\nabla u\|^2+\|\Delta u\|^2\le\epsilon\|\Delta u\|^2+2\varepsilon\|\nabla u\|^4+\varepsilon\|u\|_{4\sigma+2}^{4\sigma+2}$$
$$+C(\varepsilon)+C(\varepsilon)(|\alpha\lambda_1|+|\beta\lambda_2|)^{\frac{8(1+\sigma)(1+2\sigma)}{6\sigma^2-11\sigma-7}}-\frac{1}{4}\int_\Omega|u|^{2\sigma-2}((1+2\sigma)|\nabla|u|^2|^2$$
$$-2\mu\sigma\nabla|u|^2\cdot i(u^*\cdot\nabla u-u\nabla u^*)+|u^*\cdot\nabla u-u\nabla u^*|^2)dx.\tag{1.6.33}$$

Using equation (1.6.33) and Lemma 1.6.2, we arrive at

$$\frac{1}{2}\frac{d}{dt}\left(\|\nabla u\|^2+\frac{\delta^2}{1+\sigma}\int_\Omega|u|^{2\sigma+2}dx\right)+\|\Delta u\|^2+\frac{\delta^2}{2}\int_\Omega|u|^{4\sigma+2}dx$$
$$\le\varepsilon(1+\delta^2)\|\Delta u\|^2+\varepsilon(2+\delta^2)\|\nabla u\|^4$$
$$+\varepsilon\int_\Omega|u|^{4\sigma-2}dx+C(\varepsilon)+C(\varepsilon)(|\alpha\lambda_1|+|\beta\lambda_2|)^{\frac{8(1+\sigma)(1+2\sigma)}{6\sigma^2-11\sigma-7}}$$

$$
\begin{aligned}
&- \frac{1}{4} \int_\Omega |u|^{2\sigma-2}((1+2\sigma)(1+\delta^2)\big||\nabla u|^2\big|^2 \\
&+ 2\sigma(\nu\delta^2-\mu)\nabla|u|^2 \cdot i(u^*\nabla u - u\nabla u^*) \\
&+ (1+\delta^2)|u^*\nabla u - u\nabla u^*|^2)\mathrm{d}x. \qquad (1.6.34)
\end{aligned}
$$

Since

$$
\|\nabla u\|^2 = (-\Delta u, u) \le \|\Delta u\|\|u\| \le K\|\Delta u\|, \qquad (1.6.35)
$$

where K depends only on data parameters, but not on ε, we have

$$
\begin{aligned}
&\varepsilon(1+\sigma^2)\|\Delta u\|^2 + \varepsilon(2+\delta^2)\|\nabla u\|^4 \\
&\le \varepsilon K_0\|\Delta u\|^2 \quad (K_0 \text{ is independent of } \varepsilon) \\
&\le \frac{1}{2}\|\Delta u\|^2. \qquad (1.6.36)
\end{aligned}
$$

Inspecting condition (1.6.6), we deduce that the matrix

$$
\begin{pmatrix} (1+2\sigma)(1+\sigma^2) & \sigma(\nu\delta^2-\mu) \\ \sigma(\nu\delta^2-\mu) & 1+\sigma^2 \end{pmatrix}
$$

is negative definite. So the last term on the right of (1.6.34) is positive. Choose $\varepsilon \le \frac{\delta^2}{4}$ small enough so that equation (1.6.34) is valid. By equations (1.6.34) and (1.6.36), we get

$$
\begin{aligned}
&\frac{1}{2}\frac{\mathrm{d}}{\mathrm{d}t}\left(\|\nabla u\|^2 + \frac{\delta^2}{1+\sigma}\int_\Omega |u|^{2\sigma+2}\mathrm{d}x\right) + \frac{1}{2}\|\Delta u\|^2 + \frac{1}{4}\delta^2 \int_\Omega |u|^{4\sigma+2}\mathrm{d}x \\
&\le C + C(|\alpha\lambda_1| + |\beta\lambda_2|)^{\frac{8(1+\sigma)(1+2\sigma)}{6\sigma^2-11\sigma-7}}. \qquad (1.6.37)
\end{aligned}
$$

From equation (1.6.35), we get

$$
\|\nabla u\|^2 \le K\|\Delta u\| \le \|\Delta u\|^2 + \frac{1}{4}K^2, \qquad (1.6.38)
$$

$$
\begin{aligned}
\frac{\delta^2}{2(1+\sigma)}\int_\Omega |u|^{2\sigma+2}\mathrm{d}x &= \frac{\delta^2}{2(1+\sigma)}\int_\Omega |u|^{2\sigma+1}|u|\mathrm{d}x \\
&\le \frac{\delta^2}{4}\int_\Omega |u|^{4\sigma+2}\mathrm{d}x + C\int_\Omega |u|^2\mathrm{d}x \le \frac{\delta^2}{4}\int_\Omega |u|^{4\sigma+2}\mathrm{d}x + C. \qquad (1.6.39)
\end{aligned}
$$

From equations (1.6.37)–(1.6.39), we get

$$
\begin{aligned}
&\frac{\mathrm{d}}{\mathrm{d}t}\left(\|\nabla u\|^2 + \frac{\delta^2}{1+\sigma}\int_\Omega |u|^{2\sigma+2}\mathrm{d}x\right) + \|\nabla u\|^2 + \frac{\delta^2}{1+\sigma}\int_\Omega |u|^{2\sigma+2}\mathrm{d}x \\
&\le C + C(|\alpha\lambda_1| + |\beta\lambda_2|)^{\frac{8(1+\sigma)(1+2\sigma)}{6\sigma^2-11\sigma-7}}, \quad \forall t \ge t_*,
\end{aligned}
$$

where $t_* = \max\{t_1, t_2\}$ and t_1, t_2 are described in Lemmas 1.6.1 and 1.6.2.

Using Gronwall inequality, we deduce

$$
\begin{aligned}
&\|\nabla u(t)\|^2 + \frac{\delta^2}{1+\sigma}\int_\Omega |u|^{2\sigma+2}\mathrm{d}x \\
&\le \Big(\|\nabla u(t_*)\|^2 + \frac{\delta^2}{1+\sigma}\int_\Omega |u(t_*)|^{2\sigma+2}\mathrm{d}x\Big)e^{-(t-t_*)} + C \\
&+ C(|\alpha\lambda_1| + |\beta\lambda_2|)^{\frac{8(1+\sigma)(1+2\sigma)}{6\sigma^2-11\sigma-7}}, \quad \forall t \ge t_*.
\end{aligned} \tag{1.6.40}
$$

From the existence of a global solution for equation (1.6.3)–(1.6.5), we can readily obtain

$$
\|\nabla u(t_*)\|^2 + \frac{\delta^2}{1+\sigma}\int_\Omega |u(t_*)|^{2\sigma+2}\mathrm{d}x \le C(R), \tag{1.6.41}
$$

where $C(R)$ depends on data parameters and R, and $\|u_0\|_{H^2} \le R$.

Then, by equations (1.6.40)–(1.6.41), we get

$$
\begin{aligned}
&\|\nabla u(t)\|^2 + \frac{\delta^2}{1+\sigma}\int_\Omega |u(t)|^{2\sigma-2}\mathrm{d}x \\
&\le C(R)e^{-(t-t_*)} + C + C(|\alpha\lambda_1| + |\beta\lambda_2|)^{\frac{8(1+\sigma)(1+2\sigma)}{6\sigma^2-11\sigma-7}}, \quad \forall t \ge t_*' \\
&\le 2C + C(|\alpha\lambda_1| + |\beta\lambda_2|)^{\frac{8(1+\sigma)(1+2\sigma)}{6\sigma^2-11\sigma-7}}, \quad \forall t \ge t_*',
\end{aligned}
$$

where

$$
t_*' = \max\{t_*, t_* + \ln C(R) - \ln(C + C(|\alpha\lambda_1| + |\beta\lambda_2|)^{\frac{8(1+\sigma)(1+2\sigma)}{6\sigma^2-11\sigma-7}})\}
$$

which yields the claim of this lemma. □

Lemma 1.6.4. *Suppose the condition of Lemma* 1.6.2 *is satisfied. Then we have*

$$
\|\Delta u(t)\|^2 \le C_4 + C_4(|\alpha\lambda_1| + |\beta\lambda_2|)^{\frac{16(1+\sigma)(1+2\sigma)(7+8\sigma)}{3(6\sigma^2-11\sigma-7)}} + C_4(|\alpha\lambda_1| + |\beta\lambda_2|)^{8+\frac{240(1+\sigma)(1+2\sigma)}{6\sigma^2-11\sigma-7}}, \quad \forall t \ge t_4
$$

where C_4 depends on data parameters, t_4 depends on data parameters and R, and $\|u_0\|_{H^2} \le R$.

Proof. Taking the inner product equation (1.6.3) and $\Delta^2 u$, and then its the real component, we arrive at

$$
\begin{aligned}
\frac{1}{2}\frac{\mathrm{d}}{\mathrm{d}t}\|\Delta u\|^2 &= \rho\|\Delta u\|^2 - \|\nabla\Delta u\|^2 \\
&\quad - \operatorname{Re}(1+i\mu)\int_\Omega |u|^{2\sigma}u\Delta^2 u^*\,\mathrm{d}x
\end{aligned}
$$

$$+\alpha\,\mathrm{Re}\int_\Omega(\lambda_1\cdot\nabla(|u|^2u))\Delta^2u^*\,dx$$
$$+\beta\,\mathrm{Re}\int_\Omega(\lambda_2\cdot\nabla u)|u|^2\Delta^2u^*\,dx. \tag{1.6.42}$$

Applying Lemmas 1.6.1 and 1.6.3, we get $\forall t\ge t_*$,

$$\begin{aligned}&\frac{d}{dt}\|\Delta u\|^2+\|\nabla\Delta u\|^2+\|\Delta u\|^2\\&\le C+C(|\alpha\lambda_1|+|\beta\lambda_2|)^{\frac{16(1+\sigma)(1+2\sigma)(7+8\sigma)}{3(6\sigma^2-11\sigma-7)}}+C(|\alpha\lambda_1|+|\beta\lambda_2|)^{8+\frac{240(1+\sigma)(1+2\sigma)}{6\sigma^2-11\sigma-7}}.\end{aligned}\tag{1.6.43}$$

Using Gronwall lemma, we obtain

$$\begin{aligned}\|\Delta u(t)\|^2&\le\|\Delta u(t_*)\|^2e^{-(t-t_*)}+C\\&+C(|\alpha\lambda_1|+|\beta\lambda_2|)^{\frac{16(1+\sigma)(1+2\sigma)(7+8\sigma)}{3(6\sigma^2-11\sigma-7)}}+C(|\alpha\lambda_1|+|\beta\lambda_2|)^{8+\frac{240(1+\sigma)(1+2\sigma)}{6\sigma^2-11\sigma-7}},\quad\forall t\ge t_*.\end{aligned}\tag{1.6.44}$$

With the aid of the priori estimate of the global solution, we have

$$\|\Delta u(t_*)\|^2\le C(R),$$

where $C(R)$ depends on data parameters and R, and $\|u_0\|_{H^2}\le R$. When t is large enough, we get

$$\|\Delta u(t)\|^2\le 2C+2C(|\alpha\lambda_1|+|\beta\lambda_2|)^{\frac{16(1+\sigma)(1+2\sigma)(7+8\sigma)}{3(6\sigma^2-11\sigma-7)}}+2C(|\alpha\lambda_1|+|\beta\lambda_2|)^{8+\frac{240(1+\sigma)(1+2\sigma)}{6\sigma^2-11\sigma-7}},$$

which proves the claim. □

Notice that

$$\begin{aligned}\|u\|_{H^2}^2&\le C(\|\Delta u(t)\|+\|u\|)^2\\&\le C+C(|\alpha\lambda_1|+|\beta\lambda_2|)^{\frac{16(1+\sigma)(1+2\sigma)(7+8\sigma)}{3(6\sigma^2-11\sigma-7)}}+C(|\alpha\lambda_1|+|\beta\lambda_2|)^{8+\frac{240(1+\sigma)(1+2\sigma)}{6\sigma^2-11\sigma-7}},\quad\forall t\ge t_*,\end{aligned}\tag{1.6.45}$$

where C merely depends on data parameters and (σ,ρ,ν,μ). Also

$$\begin{aligned}\|u\|_\infty^2&\le C(\|u\|\|u\|_{H^2})\\&\le C+C(|\alpha\lambda_1|+|\beta\lambda_2|)^{\frac{16(1+\sigma)(1+2\sigma)(7+8\sigma)}{3(6\sigma^2-11\sigma-7)}}+C(|\alpha\lambda_1|+|\beta\lambda_2|)^{8+\frac{240(1+\sigma)(1+2\sigma)}{6\sigma^2-11\sigma-7}},\quad\forall t\ge t_*.\end{aligned}\tag{1.6.46}$$

Lemma 1.6.5. *Suppose the condition of Lemma* 1.6.2 *is satisfied. Then we have*

$$\|\nabla\Delta u(t)\|^2\le K,\quad\forall t\ge t_*,$$

where K depends on data parameters $(\sigma,\rho,\nu,\mu,\alpha,\beta,\lambda_1,\lambda_2,\Omega)$; t_5 depends on data parameters $(\sigma,\rho,\nu,\mu,\alpha,\beta,\lambda_1,\lambda_2,\Omega)$ and R, and $\|u_0\|_{H^2}\le R$.

Proof. Taking the inner product of equation (1.6.3) and $\Delta^3 u$, and then its the real component, we deduce

$$\begin{aligned}\frac{1}{2}\frac{\mathrm{d}}{\mathrm{d}t}\|\nabla\Delta u\|^2 = \rho\|\nabla\Delta u\|^2 - \|\Delta^2 u\|^2 \\ + \operatorname{Re}(1+i\mu)\int_\Omega |u|^{2\sigma} u\Delta^3 u^*\,\mathrm{d}x \\ - \operatorname{Re}\alpha\int_\Omega (\lambda_1\cdot\nabla(|u|^2 u))\Delta^3 u^*\,\mathrm{d}x \\ - \operatorname{Re}\beta\int_\Omega (\lambda_2\cdot\nabla u)|u|^2\Delta^3 u^*\,\mathrm{d}x.\end{aligned} \tag{1.6.47}$$

Applying Lemmas 1.6.1, 1.6.3 and 1.6.4, after further complicated operations, we get

$$|\operatorname{Re}(1+i\mu)\int_\Omega |u|^{2\sigma} u\Delta^3 u^*\,\mathrm{d}x| \le K\|\Delta^2 u\|$$

$$\le \frac{1}{6}\|\Delta^2 u\|^2 + K \tag{1.6.48}$$

$$\times\left|\alpha\int_\Omega (\lambda_1\cdot\nabla(|u|^2 u))\Delta^3 u^*\,\mathrm{d}x\right|$$

$$\le \frac{1}{6}\|\Delta^2 u\|^2 + K\|\nabla\Delta u\|^2 + K \tag{1.6.49}$$

$$\times\left|\beta\int_\Omega (\lambda_2\cdot\nabla u)|u|^2\Delta^3 u^*\,\mathrm{d}x\right|$$

$$\le \frac{1}{6}\|\Delta^2 u\|^2 + K\|\nabla\Delta u\|^2 + K \tag{1.6.50}$$

From (1.6.47)–(1.6.50), we have

$$\begin{aligned}\frac{1}{2}\frac{\mathrm{d}}{\mathrm{d}t}\|\nabla\Delta u\|^2 &\le \rho\|\nabla\Delta u\|^2 - \|\Delta^2 u\|^2 + \frac{1}{2}\|\Delta^2 u\|^2 + K\|\nabla\Delta u\|^2 + K \\ &\le (K+\rho)\|\nabla\Delta u\|^2 + K,\end{aligned}$$

thus arriving at

$$\frac{\mathrm{d}}{\mathrm{d}t}\|\nabla\Delta u\|^2 \le K\|\nabla\Delta u\|^2 + K. \tag{1.6.51}$$

Due to equation (1.6.43), we have

$$\frac{\mathrm{d}}{\mathrm{d}t}\|\Delta u\|^2 + \|\nabla\Delta u\|^2 \le K, \quad \forall t \ge t_*. \tag{1.6.52}$$

Integrating equation (1.6.52) from t to $t+1$, we have

$$\|\Delta u(t+1)\|^2 - \|\Delta u(t)\|^2 + \int_t^{t+1} \|\nabla\Delta u\|^2\mathrm{d}t \le K, \quad \forall t \le t_*.$$

Applying Lemma 1.6.4, we obtain

$$\int_t^{t+1} \|\nabla\Delta u\|^2 dt \le K, \quad \forall t \ge t_*. \tag{1.6.53}$$

Using equations (1.6.51) and (1.6.53) and consistent Gronwall lemma, we get

$$\|\nabla\Delta u(t)\|^2 \le K, \quad t \ge t_* + 1,$$

which shows that the lemma has been proved. □

Now we establish the global existence of the attractor and its Hausdorff and fractal dimension estimates for the problem (1.6.3)–(1.6.5). From (1.6.45), we infer that the ball

$$B = \{u \in H^2(\Omega) : \|u\|_{H^2} \le K_0\}$$

is the absorbing set of $S(t)$ in $H^2(\Omega)$. Lemma 1.6.5 shows that the semigroup $S(t)$ in $H^2(\Omega)$ is compact for sufficiently large t. So we can get the global existence of the attractor by using results of [197]. We get the following:

Theorem 1.6.1. *Suppose* (1.6.6) *is valid. Then the* ω *limit set*

$$\mathscr{A} = \omega(B) = \bigcap_{s\ge 0}\overline{\bigcup_{t\ge s} S(t)B}$$

is the compact attractor of $S(t)$ *in* $H^2(\Omega)$*. Here the closure is taken in* $H^2(\Omega)$*.*

Next, we prove that the dimension of the global attractor $\mathscr{A}$ is finite. For this, rewrite equation (1.6.3) in the abstract form

$$\frac{du}{dt} = F(u), \tag{1.6.54}$$

where $F(u)$ means the right-hand side of equation (1.6.3).

We consider a variational equation for problem (1.6.3)–(1.6.5), namely

$$v_t = F'(u(t))v \tag{1.6.55}$$

with the initial value

$$v(0) = v_0 \in H, \tag{1.6.56}$$

where

$$\begin{aligned} F'(u(t))v = {} & \rho v + (1 + i\nu)\Delta v - (1 + i\mu)(1 + \sigma)|u|^{2\sigma} v \\ & - (1 + i\mu)\sigma|u|^{2\sigma-2}u^2 v^* + 2\alpha(\lambda_1 \cdot \nabla(|u|^2 v)) \end{aligned}$$

$$+\alpha(\lambda_1\cdot\nabla(u^2v^*))+\beta(\lambda_2\cdot\nabla v)|u|^2$$
$$+\beta(\lambda_2\cdot\nabla u)(vu^*+uv^*),$$

$u(t)=S(t)u_0$ is the solution of problem (1.6.3)–(1.6.5), and $u_0\in\mathscr{A}$.

We know that, for $u_0\in\mathscr{A}$, we have $S(t)u_0\in H^2(\Omega)$. Using standard methods, $\forall v_0\in H$ we can prove that the linear initial value problem (1.6.55)–(1.6.56) has a unique solution $v(t)$ such that

$$v(t)\in L^2(0,T;H^1(\Omega))\cap L^\infty(0,T;H),\quad \forall T>0. \tag{1.6.57}$$

For $\forall v_0\in H$, let $G(t)v_0$ denote the solution of (1.6.55)–(1.6.56), and through complex calculations and energy estimation, we can prove that for $\forall\omega_0, u_0\in\mathscr{A}$,

$$\frac{\|S(t)\omega_0-S(t)u_0-G(t)(\omega_0-u_0)\|^2}{\|\omega_0-u_0\|^2}\le K\|\omega_0-u_0\|,\quad \forall 0\le t\le T,$$

where K depends on the data parameters $(\sigma,\rho,\nu,\mu,\alpha,\beta,\lambda_1,\lambda_2,\Omega)$, T and R; $\|u_0\|_H^2\le R$. This inequality shows that the semigroup $S(t)$ is differentiable on $\mathscr{A}$, and the differential operator $L(t,u_0): v_0\in H\to G(t)v_0\in H$. We consider that $v_0=v_{01},\dots,v_{0m}$ are m elements in H. The corresponding solutions of (1.6.55)–(1.6.56) are $v(t)=v_1(t),\dots,v_m(t)$. Then by the results of [197] we have

$$|v_1(t)\wedge\cdots\wedge v_m(t)|_{\wedge^m H}=|v_{01}\wedge\cdots\wedge v_{0m}|_{\wedge^m H}\exp\int_0^t \operatorname{Re}\operatorname{tr}F'(u(\tau))\cdot Q_m(\tau)\mathrm{d}\tau, \tag{1.6.58}$$

where

$$Q_m(\tau)=Q_m(\tau,u_0,v_{01},\dots,v_{0m})$$

is the orthogonal projection of $\{v_1(\tau),\dots,v_m(\tau)\}$. Suppose for an given time $\tau,\varphi_j(\tau)$, $j\in N$ is the orthogonal basis of H, and $Q_m(\tau)H=\operatorname{span}\{v_1(\tau),\dots,v_m(\tau)\}$, $v_j(\tau)\in H^1(\Omega)$. Then

$$\begin{aligned}&\operatorname{Re}\operatorname{tr}F'(u(\tau))\cdot Q_m(\tau)\\&\quad=\sum_{j=1}^m \operatorname{Re}(F'(u(\tau))\cdot Q_m(\tau)\varphi_j(\tau),\varphi_j(\tau))\\&\quad=\sum_{j=1}^m \operatorname{Re}(F'(u(\tau)\varphi_j(\tau)),\varphi_j(\tau)).\end{aligned} \tag{1.6.59}$$

Omitting parameter τ, we get

$$\operatorname{Re}(F'(u)\varphi_j,\varphi_j)=\rho\|\varphi_j\|^2-\|\nabla\varphi_j\|^2-(1+\sigma)\int_\Omega |u|^{2\sigma}|\varphi_j|^2\mathrm{d}x$$

$$
\begin{aligned}
&- \operatorname{Re}(1+i\mu)\sigma \int_\Omega |u|^{2\sigma-2}u^2(\varphi_j^*)^2 \mathrm{d}x + \operatorname{Re} 2\alpha \int_\Omega (\lambda_1 \cdot \nabla(|u|^2\varphi_j))\varphi_j^* \mathrm{d}x \\
&+ \operatorname{Re}\alpha \int_\Omega (\lambda_1 \cdot \nabla(|u|^2\varphi_j^*))\varphi_j^* \mathrm{d}x + \operatorname{Re}\beta \int_\Omega (\lambda_2 \cdot \nabla\varphi_j)|u|^2\varphi_j^* \mathrm{d}x \\
&+ \operatorname{Re}\beta \int_\Omega (\lambda_2 \cdot \nabla u)(u^*|\varphi_j|^2 + u(\varphi_j^*)^2)\mathrm{d}x.
\end{aligned}
\tag{1.6.60}
$$

Now we estimate the right-hand side of (1.6.60):

$$
\begin{aligned}
&- \operatorname{Re}(1+i\mu)\sigma \int_\Omega |u|^{2\sigma-2}u^2(\varphi_j^*)^2 \mathrm{d}x \\
&\le \sigma|1+i\mu|\|u\|_\infty^{2\sigma}\|\varphi_j\|^2 \\
&\le C\|\varphi_j\|^2 + C\|\varphi_j\|^2(|\alpha\lambda_1| + |\beta\lambda_2|)^{\frac{8\sigma(1+\sigma)(1+2\sigma)(7+8\sigma)}{3(6\sigma^2-11\sigma-7)}} \\
&\quad + C\|\varphi_j\|^2(|\alpha\lambda_1| + |\beta\lambda_2|)^{4\sigma+\frac{120\sigma(1+\sigma)(1+2\sigma)}{6\sigma^2-11\sigma-7}},
\end{aligned}
\tag{1.6.61}
$$

$$
\begin{aligned}
&\operatorname{Re} 2\alpha \int_\Omega (\lambda_1 \cdot \nabla(|u|^2\varphi_j))\varphi_j^* \mathrm{d}x \\
&= -2\alpha \operatorname{Re} \int_\Omega (\lambda_1 \cdot \nabla\varphi_j^*)|u|^2\varphi_j \mathrm{d}x \\
&\le 2|\alpha\lambda_1|\|u\|_\infty^2\|\nabla\varphi_j\|\|\varphi_j\| \le \frac{1}{8}\|\nabla\varphi_j\|^2 + C|\alpha\lambda_1|\|u\|_\infty^4\|\varphi_j\|^2 \\
&\le \frac{1}{8}\|\nabla\varphi_j\|^2 + C\|\varphi_j\|_\infty^2(|\alpha\lambda_1| + |\beta\lambda_2|)^2 + C\|\varphi_j\|^2(|\alpha\lambda_1| - |\beta\lambda_2|)^{2+\frac{16(1+\sigma)(1+2\sigma)(7+8\sigma)}{3(6\sigma^2-11\sigma-7)}} \\
&\quad + C\|\varphi_j\|^2(|\alpha\lambda_1| - |\beta\lambda_2|)^{10+\frac{240(1+\sigma)(1+2\sigma)}{6\sigma^2-11\sigma-7}},
\end{aligned}
\tag{1.6.62}
$$

$$
\begin{aligned}
&\operatorname{Re}\alpha \int_\Omega (\lambda_1 \cdot \nabla(u^2\varphi_j^*))\varphi_j^* \mathrm{d}x = -\alpha \operatorname{Re} \int_\Omega (\lambda_1 \cdot \nabla\varphi_j^*)u^2\varphi_j^* \mathrm{d}x \\
&\le |\alpha\lambda_1|\|u\|_\infty^2\|\nabla\varphi_j\|\|\varphi_j\| \\
&\le \frac{1}{8}\|\nabla\varphi_j\|^2 + C\|\varphi_j\|^2(|\alpha\lambda_1| + |\beta\lambda_2|)^2 + C\|\varphi_j\|^2(|\alpha\lambda_1| + |\beta\lambda_2|)^{2+\frac{16(1+\sigma)(1+2\sigma)(7+8\sigma)}{3(6\sigma^2-11\sigma-7)}} \\
&\quad + C\|\varphi_j\|^2(|\alpha\lambda_1| + |\beta\lambda_2|)^{10+\frac{240(1+\sigma)(1+2\sigma)}{6\sigma^2-11\sigma-7}},
\end{aligned}
\tag{1.6.63}
$$

$$
\begin{aligned}
&\operatorname{Re}\beta \int_\Omega (\lambda_2 \cdot \nabla\varphi_j)|u|^2\varphi_j^* \mathrm{d}x \\
&\le |\beta\lambda_2|\|u\|_\infty^2\|\nabla\varphi_j\|\|\varphi_j\| \\
&\le \frac{1}{8}\|\nabla\varphi_j\|^2 + C\|\varphi_j\|^2(|\alpha\lambda_1| + |\beta\lambda_2|)^2 + C\|\varphi_j\|^2(|\alpha\lambda_1| + |\beta\lambda_2|)^{2+\frac{16(1+\sigma)(1+2\sigma)(7+8\sigma)}{3(6\sigma^2-11\sigma-7)}} \\
&\quad + C\|\varphi_j\|^2(|\alpha\lambda_1| + |\beta\lambda_2|)^{10+\frac{240(1+\sigma)(1+2\sigma)}{6\sigma^2-11\sigma-7}},
\end{aligned}
\tag{1.6.64}
$$

$$\begin{aligned}
&\operatorname{Re}\beta\int_{\Omega}(\lambda_2\cdot\nabla u)(u^*|\varphi_j|^2+u(\varphi_j^*)^2)\mathrm{d}x\\
&\le 2|\beta\lambda_2|\|u\|_\infty\|\nabla u\|\|\varphi_j\|_4^2\\
&\le C|\beta\lambda_2|\|u\|_\infty\|\nabla u\|\|\varphi_j\|\|\varphi_j\|_{H^1}\\
&\le \frac{1}{8}\|\varphi_j\|_{H^1}^2+C|\beta\lambda_2|^2\|u\|_\infty^2\|\nabla u\|^2\|\varphi_j\|^2\\
&\le \frac{1}{8}\|\varphi_j\|_{H^1}^2+C|\beta\lambda_2|^2\|u\|_\infty^2(-\Delta u,u)\|\varphi_j\|^2\\
&\le \frac{1}{8}\|\varphi_j\|_{H^1}^2+C|\beta\lambda_2|^2\|u\|\|u\|_{H^2}\|\Delta u\|\|u\|\|\varphi_j\|^2\\
&\le \frac{1}{8}\|\varphi_j\|_{H^1}^2+C|\beta\lambda_2|^2\|u\|_{H^2}^2\|\varphi_j\|^2\\
&\le \frac{1}{8}\|\nabla\varphi_j\|^2+\frac{1}{8}\|\varphi_j\|^2+C\|\varphi_j\|^2(|\alpha\lambda_1|+|\beta\lambda_2|)^2\\
&\quad+C\|\varphi_j\|^2(|\alpha\lambda_1|+|\beta\lambda_2|)^{2+\frac{16(1+\sigma)(1+2\sigma)(7+8\sigma)}{3(6\sigma^2-11\sigma-7)}}\\
&\quad+C\|\varphi_j\|^2(|\alpha\lambda_1|+|\beta\lambda_2|)^{10+\frac{240(1+\sigma)(1+2\sigma)}{6\sigma^2-11\sigma-7}}.
\end{aligned}\tag{1.6.65}$$

By (1.6.61)–(1.6.65), we have

$$\operatorname{Re}\operatorname{tr}F'(u(\tau))\cdot Q_m(\tau)\le -\frac{1}{2}\sum_{j=1}^m\|\nabla\varphi_j\|^2+E\sum_{j=1}^m\|\varphi_j\|^2,\tag{1.6.66}$$

where

$$\begin{aligned}
E&=C+C(|\alpha\lambda_1|+|\beta\lambda_2|)^2\\
&\quad+C(|\alpha\lambda_1|+|\beta\lambda_2|)^{\frac{8\sigma(1+\sigma)(1+2\sigma)(7+8\sigma)}{3(6\sigma^2-11\sigma-7)}}\\
&\quad+C(|\alpha\lambda_1|+|\beta\lambda_2|)^{4\sigma+\frac{120\sigma(1+\sigma)(1+2\sigma)}{6\sigma^2-11\sigma-7}}\\
&\quad+C(|\alpha\lambda_1|+|\beta\lambda_2|)^{2+\frac{16(1+\sigma)(1+2\sigma)(7+8\sigma)}{3(6\sigma^2-11\sigma-7)}}\\
&\quad+C(|\alpha\lambda_1|+|\beta\lambda_2|)^{10+\frac{240\sigma(1+\sigma)(1+2\sigma)}{6\sigma^2-11\sigma-7}}.
\end{aligned}\tag{1.6.67}$$

Set

$$\eta=\eta(x,\tau)=\sum_{j=1}^m|\varphi_j|^2.\tag{1.6.68}$$

Since $\{\varphi_j;\ j=1,\dots,m\}$ is an orthonormal set in H, we have

$$\sum_{j=1}^m\|\varphi_j\|^2=\int_{\Omega}\eta\mathrm{d}x=m.\tag{1.6.69}$$

By Sobolev–Lieb–Thirring inequality, we have

$$\int_{\Omega}\eta^2\mathrm{d}x\le C_0\int_{\Omega}\eta\mathrm{d}x+C_0\sum_{j=1}^m\|\nabla\varphi_j\|^2,\tag{1.6.70}$$

where C_0 depends only on the shape of Ω.

By Hölder inequality we have

$$\left(\int_\Omega \eta \mathrm{d}x\right)^2 \leq |\Omega| \int_\Omega \eta^2 \mathrm{d}x. \tag{1.6.71}$$

Due to equations (1.6.66)–(1.6.71), we have

$$\begin{aligned} \operatorname{Re}\operatorname{tr} F'(u(\tau)) \cdot Q_m(\tau) &\leq -\frac{m^2}{2C_0|\Omega|} + \frac{m}{2} + Em \\ &\leq -\frac{m^2}{4C_0|\Omega|} + C_0|\Omega|\left(E + \frac{1}{2}\right)^2. \end{aligned} \tag{1.6.72}$$

For $i = 1, \ldots, m$ and $v_{0i} \in H$, we define

$$q_m(t) = \sup_{u_0 \in \mathscr{A}} \sup_{\|v_{0i}\| \leq 1} \left(\frac{1}{t}\int_0^t \operatorname{Re}\operatorname{tr} F'(S(\tau)u_0) \cdot Q_m(\tau)\mathrm{d}\tau\right),$$

$$q_m(t) = \limsup_{t \to \infty} q_m(t).$$

From (1.6.72), we obtain

$$q_m(t) \leq -\frac{m^2}{4C_0|\Omega|} + C_0|\Omega|\left(E + \frac{1}{2}\right)^2.$$

Hence, if m satisfies

$$m - 1 < \sqrt{8}C_0|\Omega|\left(E + \frac{1}{2}\right) \leq m, \tag{1.6.73}$$

then $q_m < 0$. From these considerations we have

Lemma 1.6.6. *Suppose $\mathscr{A}$ is the global attractor of problem* (1.6.3)–(1.6.5). *Then the dimension of $\mathscr{A} \leq m$, and its fractal dimension $\leq 2m$, where m is determined by* (1.6.73).

1.7 Ginzburg–Landau model in superconductivity

In this section, we consider the evolutional Ginzburg–Landau (GL) model in superconductivity

$$\eta\Psi_t + i\eta k\Phi\Psi + \left(\frac{i}{k}\nabla + A\right)^2 \Psi - \Psi + |\Psi|^2\Psi = 0, \quad (x,t) \in \Omega \times \mathbf{R}^+, \tag{1.7.1}$$

$$A_t + \operatorname{grad}\Phi + \operatorname{curl}^2 A + \frac{i}{2k}(\Psi^* \operatorname{grad}\Psi - \Psi \operatorname{grad}\Psi^*) + |\Psi|^2 A = 0, \tag{1.7.2}$$

with the boundary conditions

$$\nabla\Psi \cdot n = 0, \quad \left(\frac{i}{k}\nabla\Psi + A\Psi\right) \times n = 0, \quad (x,t) \in \partial\Omega \times \mathbf{R}^+ \tag{1.7.3}$$

and the initial conditions

$$\Psi(x,0) = \Psi_0(x), \quad A(x,0) = A_0(x), \quad x \in \Omega, \tag{1.7.4}$$

where $\Omega \subset \mathbf{R}^N$, $N = 2, 3$, is a domain with a smooth boundary $\partial\Omega$; $\mathbf{R}^+ = [0, \infty)$; η, k are positive constants related to some physical quantities; $i = -\sqrt{-1}$, n is the unit outward normal vector of $\partial\Omega$. In equations (1.7.1)–(1.7.2), Ψ, A, Φ are unknown functions, Ψ is a complex function and serves as an order parameter in GL theory, Ψ^* denotes the complex conjugate of Ψ, $|\Psi|^2$ means the density of superconducting charge carriers. Function Φ is scalar and real, representing the electrical potential. The vector function A is real and represents the magnetic potential, i. e., $H = \operatorname{curl} A$ is the magnetic field. In 1950, this model was first proposed by Ginzburg and Landau based on the second order phase transition theory of steady state conditions in fluids. In 1968, Gorkov and Eliashberg obtained equations (1.7.2)–(1.7.3) from Bardeen–Cooper–Schrieffer (BCS) theory.

The discovery of high-temperature superconductivity has brought potential commercial use. The Ginzburg–Landau model of superconductivity has drawn interest of many people. Since 1984, Berger, Chapman and Yang, among others, did some studies on the existence, uniqueness and regularity of a global solution [14, 23, 208]. But regarding the asymptotical behavior when $t \to \infty$, little research is done. In 1995, Guo and Wu [125] studied the long time asymptotic behavior and the existence of an attractor. The other results for the similar model can refer to the reference [58, 67, 72, 73, 92, 149, 150, 151, 153, 154, 155, 162, 164, 166, 167, 168, 176, 187, 200, 202, 204, 207].

From equations (1.7.2)–(1.7.3) we know that the number of unknown functions is larger than the number of equations. In order to guarantee the uniqueness of a solution, usually we must attach three types of gauge transformation:

(1) The Coulomb gauge. For this gauge, A is taken to be divergence-free, that is, $\operatorname{div} A = 0$, and it can be converted into

$$\eta\Psi_t + i\eta k\Phi\Psi + \left(\frac{i}{k}\nabla + A\right)^2 \Psi - \Psi + |\Psi|^2\Psi = 0, \quad (x,t) \in \Omega \times \mathbf{R}^+, \tag{1.7.5}$$

$$A_t + \operatorname{grad}\Phi + \operatorname{curl}^2 A + \frac{i}{2k}(\Psi^* \operatorname{grad}\Psi - \Psi \operatorname{grad}\Psi^*) + |\Psi|^2 A = 0, \tag{1.7.6}$$

$$-\Delta\Phi = \operatorname{div}\left[\frac{i}{2k}(\Psi^* \operatorname{grad}\Psi^*) + |\Psi|^2 A\right].$$

(2) The instantaneous gauge, $\Phi = \operatorname{div} A$. In this case, this model becomes

$$\eta\Psi_t + i\eta k \operatorname{div} A\Psi + \left(\frac{i}{k}\nabla + A\right)^2 \Psi - \Psi + |\Psi|^2\Psi = 0, \tag{1.7.7}$$

$$A_t + \Delta A + \frac{i}{2k}(\Psi^* \operatorname{grad}\Psi - \Psi \operatorname{grad}\Psi^*) + |\Psi|^2 A = 0, \quad (x,t) \in \Omega \times \mathbf{R}^+. \tag{1.7.8}$$

(3) Zero potential gauge, $\Phi(x) = 0$. Then the model can be rewritten as

$$\eta\Psi_t + \left(\frac{i}{k}\nabla + A\right)^2\Psi - \Psi + |\Psi|^2\Psi = 0, \quad (x,t) \in \Omega \times \mathbf{R}^+, \tag{1.7.9}$$

$$A_t + \operatorname{curl}^2 A + \frac{i}{2k}(\Psi^* \operatorname{grad}\Psi) + |\Psi|^2 A = 0, \quad (x,t) \in \Omega \times \mathbf{R}^+. \tag{1.7.10}$$

The existence of a global solution for all kinds of gauge has been obtained. But regarding the asymptotical behavior, we cannot obtain satisfactory results under any gauges, since it is difficult to get the uniformly bounded estimate under various kinds of gauge. Here we get the H^2-estimate of a solution under instantaneous gauge. Assume $H^m(\Omega)$ is a standard Sobolev space and $\mathscr{H}^m$ denotes the corresponding complex function space

$$H_n^1(\Omega) = \{A \in H^1(\Omega),\ A \cdot n = 0,\ x \in \partial\Omega\} \tag{1.7.11}$$

with the norm $(\|\operatorname{div} A\|^2 + \|\operatorname{curl} A\|^2)^{\frac{1}{2}}$, $A \in H_n^1(\Omega)$, and the norm $(\|\operatorname{div} A\|^2 + \|\operatorname{curl} A\|^2)^{\frac{1}{2}}$ is equivalent to the norm $\|\nabla A\|$, i. e.,

$$\|u\| \le K_1\|\nabla u\|, \quad u \in H_n^1(\Omega). \tag{1.7.12}$$

In the following we establish an a priori estimate. Set $D_A = \frac{i}{k}\nabla + A$. Then equations (1.7.1)–(1.7.2) can be written as for the gauge $\Phi = \operatorname{div} A$, namely

$$\eta\Psi_t + i\eta k \operatorname{div} A\Psi + D_A^2\Psi - \Psi + |\Psi|^2\Psi = 0, \quad \Omega \times \mathbf{R}^+, \tag{1.7.13}$$

$$A_t - \Delta A = \frac{1}{2}[\Psi^* D_A\Psi + \Psi(D_A\Psi)^*], \quad \Omega \times \mathbf{R}^+. \tag{1.7.14}$$

It is easy to see that, for $\Psi, \Phi \in \mathscr{H}^1$, $A \in H^1$, D_A satisfies

$$(D_A^2\Psi, \Phi^*) = (D_A\Psi, D_A\Phi^*), \tag{1.7.15}$$

$$(D_A\Psi)_t = D_A\Psi_t + A_t\Psi. \tag{1.7.16}$$

In the following, we always assume $|\Psi_0(x)| \le 1$, $x \in \Omega_0$. Then we know $|\Psi(x,t)| \le 1$, $(x,t) \in \Omega\times\mathbf{R}^+$. Multiplying equation (1.7.13) by Ψ^*, integrating by parts, and taking the real part, we obtain

$$\eta\frac{d}{dt}\|\Psi\|^2 + 2\|D_A\Psi\|^2 - \|\Psi\|^2 + \|\Psi\|_4^4 = 0. \tag{1.7.17}$$

Multiplying both sides of equation (1.7.13) by their conjugates, and integrating by parts, we arrive at

$$\begin{aligned}\eta^2\|\Psi_t\|^2 + \|D_A^2\Psi\|^2 - \eta\frac{d}{dt}\|D_A\Psi\|^2 &= \eta(A_t\Psi_t, (D_A\Psi)^*) + \eta(A_t\Psi^*, D_A\Psi)\\ &\le 2\|D_A\Psi\|\|A_t\| \le \|D_A\Psi\|^2 + \|A_t\|^2.\end{aligned} \tag{1.7.18}$$

Multiplying both sides of equation (1.7.14) with themselves, and integrating by parts, we obtain

$$\|A_t\|^2 + \|\Delta A\|^2 + \frac{\mathrm{d}}{\mathrm{d}t}[\|\operatorname{div} A\|^2 + \|\operatorname{curl} A\|^2] = \frac{1}{4}\int_\Omega [\Psi^* D_A\Psi + \Psi(D_A\Psi)^*]^2 \mathrm{d}x \le \|D_A\Psi\|^2. \tag{1.7.19}$$

The last inequality follows from $|\Psi(0)| \le 1$, $x \in \Omega$. Then we have

$$\frac{\mathrm{d}}{\mathrm{d}t}[\|\operatorname{div} A\|^2 + \|\operatorname{curl} A\|^2 + \eta\|\Psi\|^2] + \|A_t\|^2 + \|\Delta A\|^2 + \|D_A\Psi\|^2 \le C|\Omega|. \tag{1.7.20}$$

Moreover, we multiply (1.7.13) by $2\Psi_t^*$, equation (1.7.14) by A_t, and add them together to get

$$\begin{aligned}&\frac{\mathrm{d}}{\mathrm{d}t}\left[\|\operatorname{div} A\|^2 + \|\operatorname{curl} A\|^2\frac{1}{2} + \|\Psi\|_4^4 - \|\Psi\|^2 + \|D_A\Psi\|^2\right]\\ &\le -2\eta\|\Psi_t\|^2 - 2\|A_t\|^2 + i\eta k\int_\Omega \operatorname{div} A(\Psi^*\Psi_t - \Psi\Psi_t^*)\mathrm{d}t\\ &\le \frac{4k^2}{\eta}\|\operatorname{div} A\|^2 - \eta\|\Psi_t\|^2 - 2\|A_t\|^2.\end{aligned} \tag{1.7.21}$$

Noting that

$$\begin{aligned}&\|\nabla A\| \le C\|A\|^{\frac{1}{2}}\|\Delta A\|^{\frac{1}{2}} \le C\|\nabla A\|^{\frac{1}{2}}\|\Delta A\|^{\frac{1}{2}}\\ &\|\operatorname{div} A\|^2 + \|\operatorname{curl} A\|^2 \le C\|\nabla A\|,\end{aligned}$$

we get

$$(\|\operatorname{div} A\|^2 + \|\operatorname{curl} A\|^2) \le K_2\|\Delta A\|.$$

This, together with equations (1.7.17), (1.7.19) and (1.7.21), implies that

$$\begin{aligned}&\frac{\mathrm{d}}{\mathrm{d}t}\left[C\eta\|\Psi\|^2 + C_1\|\operatorname{div} A\|^2 + \|\operatorname{curl} A\|^2 + \frac{1}{2}\|\Psi\|_4^4 - \|\Psi\|^2 + \|D_A\Psi\|^2\right]\\ &+ C_2\left[C\eta\|\Psi\|^2 + C_1\|\operatorname{div} A\|^2 + \|\operatorname{curl} A\|^2 + \frac{1}{2}\|\Psi\|_4^4 - \|\Psi\|^2 + \|D_A\Psi\|^2\right]\\ &+ C_3[\eta\|\Psi_t\|^2 + \|A_t\|^2] \le C|\Omega|.\end{aligned} \tag{1.7.22}$$

By Gronwall inequality, we have

$$\begin{aligned}&C\eta\|\Psi\|^2 + C_1\|\operatorname{div} A\|^2 + \|\operatorname{curl} A\|^2 + \frac{1}{2}\|\Psi\|_4^4 - \|\Psi\|^2 + \|D_A\Psi\|^2\\ &\le e^{-C_2 t}\left[C\eta\|\Psi_0\|^2 + C_1\|\nabla A_0\|^2 + \frac{1}{2}\|\Psi_0\|_4^4 - \|\Psi_0\|^2 + \|D_A\Psi_0\|^2\right] + C|\Omega|(1 - e^{-C_2 t}).\end{aligned} \tag{1.7.23}$$

Choosing t_0 large enough, for $t \geq t_0$ we have

$$C\eta\|\Psi\|^2 + C_1\|\mathrm{div}\, A\|^2 + \|\mathrm{curl}\, A\|^2 + \frac{1}{2}\|\Psi\|_4^4 - \|\Psi\|^2 + \|D_A\Psi\|^2 \leq C|\Omega|, \tag{1.7.24}$$

where the constant C is independent of Ψ_0 and A_0. Hence, for $t \geq t_0$, we get $(\|\mathrm{div}\, A(t)\|^2 + \|\mathrm{curl}\, A(t)\|^2)^{\frac{1}{2}} \leq K_3$. Noticing that

$$\|D_A\Psi\| \geq \frac{1}{k}\|\nabla\Psi\| - \|A\|,$$

we obtain

$$\frac{1}{k}\|\nabla\Psi\| \leq \|D_A\Psi\| + \|A\|, \quad \forall t \geq t_0.$$

Thus

$$\|\nabla\Psi\| \leq K_4.$$

Furthermore, from equation (1.7.22) we have

$$\int_t^{t+1} [\|\Psi_s(s)\|^2 + \|A_s(s)\|^2]\mathrm{d}s \leq K_5, \quad t \geq t_0. \tag{1.7.25}$$

Then we see that

$$B = \{(\Psi, A) \in \mathcal{H}^1 \times H^1 \mid \|\Psi\| \leq 1,\ \|\nabla\Psi\| \leq K_4,\ (\|\mathrm{div}\, A(t)\|^2 + \|\mathrm{curl}\, A\|^2)^{\frac{1}{2}} \leq K_3\}$$

is an absorbing set of equation (1.7.13)–(1.7.14) in $\mathcal{H}^1 \times H^1$. In order to prove that the ω-limit set of B is the attractor of (1.7.13)–(1.7.14) and (1.7.3)–(1.7.4) in $\mathcal{H}^1 \times H^1$, $|\Psi| \leq 1$, we must prove the compactness of this absorbing set. Noting that

$$(D_A^2\Psi, \Psi^*) = (D_A\Psi, (D_A\Psi^*)),$$
$$(D_A^2\Psi)_t = D_A^2\Psi_t + D_A(A_t\Psi) + A_tD_A\Psi,$$

we differentiate (1.7.13) with respect to t, multiply it by Ψ_t^*, and take the real part to get

$$\frac{\eta}{2}\frac{\mathrm{d}}{\mathrm{d}t}\|\Psi_t\|^2 - \eta k\,\mathrm{Im}\int_\Omega (\mathrm{div}\, A_t)\Psi\Psi_t^*\,\mathrm{d}x + \|D_A\Psi_t\|^2$$
$$+ \int_\Omega A_t\Psi(D_A\Psi_t)^*\,\mathrm{d}x + \int_\Omega (D_A\Psi)(A_t\Psi_t)^*\,\mathrm{d}x - \|\Psi_t\|^2 + \int_\Omega |\Psi|^2|\Psi_t|^2\mathrm{d}x \leq 0, \tag{1.7.26}$$

which yields

$$\frac{\eta}{2}\frac{\mathrm{d}}{\mathrm{d}t}\|\Psi_t\|^2 + \|D_A\Psi_t\|^2 + \int \Omega|\Psi|^2|\Psi_t|^2\mathrm{d}x$$
$$\leq \|\Psi_t\|^2 + \eta k\|\nabla A_t\|\|\Psi_t\| - \int_\Omega A_t\Psi(D_A\Psi_t)^*\,\mathrm{d}x - \int_\Omega (D_A\Psi)(A_t\Psi_t)^*\,\mathrm{d}x. \tag{1.7.27}$$

In the same way, we differentiate (1.7.14) with respect to t, multiply it by A_t, and integrate in by parts to obtain

$$\frac{1}{2}\frac{\mathrm{d}}{\mathrm{d}t}\|A_t\|^2 + \|\nabla A_t\|^2 = -\operatorname{Re}\int_\Omega (D_A\Psi)^*(A_t\Psi_t)\mathrm{d}x - \operatorname{Re}\int_\Omega (A_t\Psi)^*(D_A\Psi + A_t\Psi)\mathrm{d}x$$
$$\leq \|D_A\Psi\|\|A_t\Psi_t\| + \|A_t\Psi\|\|D_A\Psi_t\| - \|A_t\Psi\|^2, \tag{1.7.28}$$

which implies

$$\frac{1}{2}\frac{\mathrm{d}}{\mathrm{d}t}\|A_t\|^2 + \|\nabla A_t\|^2 + \|A_t\Psi\|^2 \leq \|D_A\Psi\|\|A_t\Psi_t\| + \|A_t\Psi\|\|D_A\Psi_t\|, \tag{1.7.29}$$
$$\frac{\mathrm{d}}{\mathrm{d}t}[\eta\|\Psi_t\|^2 + \|A_t\|^2] + \|D_A\Psi_t\|^2 + \|\nabla A_t\|^2 + \|A_t\Psi\|^2 \leq C(\|A_t\|^2 + \|\Psi_t\|^2). \tag{1.7.30}$$

Using the uniform Gronwall inequality and equation (1.7.20), we get

$$\|A_t\| \leq K_6, \quad \|\Psi_t\| \leq K_6, \quad t > 1, \tag{1.7.31}$$

which in turn ensures the bounds on the H^2-norm

$$\|A\|_{H^2} \leq K_7, \quad \|\Psi\|_{H^2} \leq K_7, \quad t > 1. \tag{1.7.32}$$

Set the semigroup $S(t)$ to be

$$S(t) : \mathscr{H}^1 \times H^1 \to \mathscr{H}^1 \times H^1 \quad \text{where } S(t)(\Psi_0, A_0) = (\Psi(t), A(t)),$$

$(\Psi(t), A(t))$ is the solution of problem (1.7.13)–(1.7.14) and (1.7.3)–(1.7.4) with the initial value (Ψ_0, A_0). By the Sobolev embedding theorem, we know that $\bigcup_{t>1} S(t)B$ is compact in $\mathscr{H}^1 \times H^1$, and the ω-limit sets of B, namely

$$\mathscr{A} = \omega(B) = \bigcap_{s\geq 0}\bigcup_{t\geq s}\overline{S(t)B},$$

is the attractor of $|\Psi| \leq 1$ in $\mathscr{H}^1 \times H^1$. Then we have

Theorem 1.7.1. *Assume that* $|\Psi_0| \leq 1$, $x \in \Omega$. *Under the instantaneous gauge* $\Phi = \operatorname{div} A$, *the problem* (1.7.1)–(1.7.4) *possesses a unique attractor in* $\mathscr{H} \times H^1$.

Now we estimate the dimension of the attractor $\mathscr{A}$. Set $u(t) = (\Psi(t), A(t))$, $u_0 = (\Psi(0)_0, A(0)_0)$. The variational equations for (1.7.13)–(1.7.14) are

$$V_t + L(u)V = 0, \quad V(x,0) = V_0(x), \quad (x,t) \in \Omega \times \mathbf{R}^+, \quad V \in \mathscr{H}^1 \times H^1 \tag{1.7.33}$$

with the initial condition

$$V(x,0) = V_0(x), \quad x \in \Omega, \tag{1.7.34}$$

where $V = (\Psi, E)$, $L(u)V = (L_1, L_2)$,

$$L_1 = ik(\operatorname{div} A)\Psi - ik(\operatorname{div} E)\Psi$$
$$+ \frac{1}{\eta}\left[D_A\Psi + \frac{i}{k}\nabla(E\Psi) + \frac{i}{k}E\cdot\Psi + 2A\cdot E\Psi - \Psi + |\Psi|^2\Psi + 2\Psi^2\Psi^* + 2|\Psi|^2\Psi\right], \quad (1.7.35)$$

$$L_2 = \frac{1}{2}[\Psi^* D_A\Psi + \Psi^* D_A\Psi + E\Psi + D_A\Psi^* + \Psi D_A^*\Psi + E\Psi^*]. \quad (1.7.36)$$

We can see that, when $V_0(x) \in \mathscr{H}^1 \times H^1$, there exists a unique global solution of problem (1.7.13) such that

$$V(x,t) \in L^\infty([0,\infty), \mathscr{H}^1 \times H^1) \cap L^\infty((0,\infty); \mathscr{H}^2 \times H^2).$$

Let $V_i(t)$ be the solution of equation (1.7.33) with the initial value $V_i(0) = \xi_i$, $i = 1,2,\dots,N$, where $\xi_1, \xi_2, \dots, \xi_N \in L^2$ are linearly independent. Also $Q_N(t)$ means the orthogonal projection from L^2 to the linear span of $V_1(t), V_2(t), \dots, V_N(t)$. Set

$$q_N = \lim_{t\to\infty} \sup_{u_0\in\mathscr{A}} \left(\sup_{\xi_i\in L^2, |\xi_j|=1} \frac{1}{t}\int_0^t \operatorname{tr}(L(u(s))Q_N(s))\mathrm{d}s \right),$$

where tr means the trace of an operator. To estimate the dimension of the global attractor $\mathscr{A}$, we need the following lemma:

Lemma 1.7.1 ([33]). *Suppose $\mathscr{A}$ is the global attractor of* (1.7.13), (1.7.3), *and* (1.7.4). *If $q_N > 0$ for some N, then the Hausdorff dimension of $\mathscr{A}$ satisfies*

$$d_H(\mathscr{A}) \le N,$$

while the fractal dimension of $\mathscr{A}$ satisfies

$$d_F(\mathscr{A}) \le N\left(1 + \max_{1\le j\le N} -\frac{q_j}{q_N}\right).$$

Lemma 1.7.2 ([197]). *Let $\Phi_j \in H^m$, $1 \le j \le N$ be orthogonal in L^2, for almost $x \in \Omega$, and set*

$$\rho(y) = \sum_{j=1}^N |\Psi_j(x)|^2.$$

Then for any p satisfying

$$1 < p \le 1 + \frac{1}{2m},$$

there exists a K such that

$$\left(\int_\Omega \rho(x)^{\frac{p}{p-1}}\mathrm{d}x\right)^{2m(p-1)} \le K\sum_{j=1}^N \int_\Omega |D^m\Phi_j|^2\mathrm{d}x,$$

where the constant K depends on m, p, but not on Φ_j and N.

We estimate $\operatorname{tr}(L(u)Q_N(t))$. Set $\phi_1, \phi_2, \dots, \phi_N \in \mathscr{H}^1 \times H^1$ to be the orthonormal basis of the linear span of

$$\{V_1(t), V_2(t), \dots, V_N(t)\}.$$

Then we have

$$\begin{aligned}
\operatorname{tr}(L(u)Q_N(t)) &= \sum_{j=1}^{N} (L(u(t))\phi_j, \phi_j) \\
&\geq \sum_{j=1}^{N} \left\{ \left(\frac{1}{\eta k^2} \cdot \|\nabla \phi_j\|^2 + \|\operatorname{div} E_j\|^2 + \|\operatorname{curl} E_j\|^2 \right) \right. \\
&\quad - k\|\nabla E_j\|\|\phi_j\| - \left(2k + \frac{1}{k\eta}\right)\|A\|_\infty \|\phi_j\|\|\nabla \phi_j\| \\
&\quad - \frac{1}{k\eta}\|E_j\|\|\nabla \phi_j\| - \frac{1}{k\eta}\|A\|_\infty \|\nabla \phi_j\|\|\phi_j\| \\
&\quad - \frac{1}{k\eta}\|E_j\|\|\nabla \phi_j\|_4 \|\phi_j\|_4 - \frac{2}{\eta}\|A\|_\infty \|E_j\|\|\phi_j\| \\
&\quad - \frac{1}{\eta}\|A\|_\infty^2 \|\phi_j\|^2 + \frac{5}{\eta}\|\phi_j\|^2 - \frac{1}{k}\|\nabla \Psi\|_4 \|\phi_j\|_4 \|E_j\|_4 \\
&\quad \left. - \frac{1}{k}\|\nabla \Psi_j\|\|E_j\| - 2\|\phi_j\|_4 \|A\|_4 \|E_j\| - 2\|\phi_j\|\|A\|_\infty \|E_j\| \right\} \\
&> \frac{1}{2}\sum_{j=1}^{N} \left(\frac{1}{\eta k^2}\|\nabla \phi_j\|^2 + \|\operatorname{div} E_j\|^2 + \|\operatorname{curl} E_j\|^2 \right) \\
&\quad - C \sum_{j=1}^{N} \left(\frac{1}{\eta k^2}\|\phi_j\|^2 + \|E_j\|^2 \right).
\end{aligned} \tag{1.7.37}$$

Set $\rho(x) = \sum_{j=1}^{N} (\frac{1}{\eta k^2}\|\Phi_j\|^2 + \|E_j\|^2)$. Applying Lemma 1.7.2, for $N = 3$, we have

$$\int_\Omega \rho(x)^{\frac{2}{5}} \mathrm{d}x \leq k_1 \left\{ \sum_{j=1}^{N} \left(\frac{1}{\eta k^2}\|\nabla \Phi_j\|^2 + \|\operatorname{div} E_j\|^2 + \|\operatorname{curl} E_j\|^2 \right) \right\} \tag{1.7.38}$$

and

$$\begin{aligned}
\operatorname{tr}(L(u)Q_N(t)) &\geq \frac{1}{2}\sum_{j=1}^{N} \left(\frac{1}{\eta k^2}\|\nabla \Psi_j\|^2 + \|\operatorname{div} E_j\|^2 + \|\operatorname{curl} E_j\|^2 \right) - C\int_\Omega \rho(x)\mathrm{d}x \\
&\geq \frac{1}{4}\sum_{j=1}^{N} \left(\frac{1}{\eta k^2}\|\nabla \Psi_j\|^2 + \|\operatorname{div} E_j\|^2 + \|\operatorname{curl} E_j\|^2 \right) - C(\eta, k, A_1, A_2, A_3) \\
&= \sum_{j=1}^{N} \left(\frac{1}{\eta k^2}(\alpha_1 + \alpha_2 + \cdots + \alpha_N) + \mu_1 + \mu_2 + \cdots + \mu_N \right) - C(\eta, k, A_1, A_2, A_3),
\end{aligned} \tag{1.7.39}$$

where α_i, μ_i, $i = 1, 2, \ldots, N$, are the eigenvalues in $\mathscr{H}$ and H_n^1, respectively. Choose N_0 large enough so that

$$\operatorname{tr}(L(u)Q_N(t)) \geq \sum_{j=1}^{N}\left(\frac{1}{\eta k^2}(\alpha_1 + \alpha_2 + \cdots + \alpha_N) + (\mu_1 + \mu_2 + \cdots + \mu_N)\right) - C(\eta, k, A_1, A_2, A_3) \geq 0. \tag{1.7.40}$$

Then by a theorem in [197], for the problem (1.7.1)–(1.7.4), its Hausdorff and fractal dimensions are bounded for the instantaneous gauge as follows:

$$d_H(\omega(B)) \leq N_c, \quad d_F(\omega(B)) \leq 2N_0. \tag{1.7.41}$$

Finally, we have the following theorem:

Theorem 1.7.2. *Under the assumptions of Theorem* 1.7.1, *the attractor for the problem* (1.7.1)–(1.7.4) *has bounded Hausdorff and fractal dimensions under the instantaneous gauge. In this case, the long time behavior of the solution of problem* (1.7.1)–(1.7.4) *is determined by finite many parameters.*

1.8 Landau–Lifshitz–Maxwell equation

In 1935, Landau and Lifshitz [160] put forward the following ferromagnetic chain coupling equations of electromagnetic field [110]:

$$z_t = \lambda_1 z \times (\Delta z + H) - \lambda_2 z \times (z \times (\Delta z + H)), \tag{1.8.1}$$

$$\nabla \times H = \frac{\partial E}{\partial t} + \sigma E, \tag{1.8.2}$$

$$\nabla \times E = -\frac{\partial H}{\partial t} - \beta \frac{\partial z}{\partial t}, \tag{1.8.3}$$

$$\nabla \cdot H + \beta \nabla \cdot z = 0, \quad \nabla \cdot E = 0, \tag{1.8.4}$$

where $\lambda_1, \lambda_2, \sigma, \beta$ are constants and $\lambda_2 \geq 0$, $\sigma \geq 0$. The unknown vector-valued function $z(x,t) = (z_1(x,t), z_2(x,t), z_3(x,t))$ models magnetization, $H(x,t) = (H_1(x,t), H_2(x,t), H_3(x,t))$ means the magnetic field, $E(x,t) = (E_1(x,t), E_2(x,t), E_3(x,t))$ is the electromagnetic field, $H^e = \Delta z + H$ represents the effective magnetic field, $\Delta = \sum_{i=1}^{n} \frac{\partial^2}{\partial x_i^2}$, $\nabla = (\frac{\partial}{\partial x_1}, \frac{\partial}{\partial x_2}, \ldots, \frac{\partial}{\partial x_n})$, and $\times$ means vector cross-product.

If $H = 0$, $E = 0$, then we can get Landau–Lifshitz–Maxwell equation system with Gilbert term:

$$z_t = \lambda_1 z \times \Delta z - \lambda_2 z \times z \times (z \times \Delta z) \tag{1.8.5}$$

where $\lambda_2 > 0$ is Gilbert damping constant. Guo and Hong in 1993, Guo and Wang in 1995, Chen and Guo in 1996, Guo and Ding in 1997 systematically researched the properties of the solutions of equation (1.8.5); see respective references [100, 116, 24, 95, 96].

Especially, in [100] the authors found a close relationship between the harmonic mapping on a Riemann manifold and the solutions of (1.8.5). When $\lambda_2 = 0$, equation (1.8.5) can be reduced to

$$z_t = \lambda_1 z \times \Delta z. \tag{1.8.6}$$

In a one-dimensional space, it is an integrable system, and has the soliton solution. Nakamura, Lakshmanan, and Zakharov, among others, studied the interaction of soliton, infinite conservation law, and inverse scattering method in detail, respectively in [178, 159, 211]. Since 1982, Zhou and Guo [216] did a systematic and deep research on the kinds of determining solution problems (including the space of one-dimensional and multidimensional initial value problem, linear and nonlinear boundary value problems) for the equation (1.8.6). In particular, in 1991, Zhou, Guo and Tan [217] proved the existence and uniqueness of a smooth solution (in one spacial dimension).

Here we consider the global existence of an attractor and its dimension estimation problem with periodic initial conditions of the system (1.8.1)–(1.8.4) as in [110],

$$\begin{aligned} &z(x+2De_i,t) = z(x,t), \quad H(x+2De_i,t) = H(x,t), \\ &E(x+2De_i,t) = E(x,t), \quad x \in \Omega,\ t \geq 0,\ i = 1,2,\dots,n, \end{aligned} \tag{1.8.7}$$

$$\begin{aligned} &z(x,0) = z_0(x), \quad H(x,0) = H_0(x), \\ &E(x,0) = E_0(x), \quad x \in \Omega, \end{aligned} \tag{1.8.8}$$

where $x + 2De_i = (x_1,\dots,x_{i-1},x_i + 2D,x_{i+1},\dots,x_n)$, $i = 1,2,\dots,n$, $D > 0$, $\Omega \subset \mathbf{R}^n$ is an n-dimensional cube with side length $2D$.

In the following we establish a priori estimate

Lemma 1.8.1. *Let* $|z_0(x) = 1|$. *Then for the smooth solution of the periodic initial value problem* (1.8.1)–(1.8.4) *and* (1.8.7)–(1.8.8), *we have*

$$|z(x,t)| = 1, \quad x \in \Omega,\ t \geq 0. \tag{1.8.9}$$

Proof. Taking the dot product of equation (1.8.1) with z, we arrive at

$$\frac{\partial}{\partial t}|z(x,t)|^2 = 0,$$

which yields the claim. □

Lemma 1.8.2. *Let* $\lambda_2 > 0$, $\beta > 0$, $\sigma \geq 0$, $\nabla z_0(x) \in L_2(\Omega)$, $E_0(x) \in L_2(\Omega)$, $H_0(x) \in L_2(\Omega)$. *Then we have the following estimates:*

$$\sup_{0\leq t<\infty} [\|\nabla z(\cdot,t)\|_2^2 + \|E(\cdot,t)\|_2^2 + \|H(\cdot,t)\|_2^2] \leq K_1,$$

$$\int_0^\infty \|z \times (\Delta z + H)\|_2^2 dt \leq K_2, \tag{1.8.10}$$

where constants K_1 *and* K_2 *depend on* $\|\nabla z_0(x)\|_2$, $\|E_0(x)\|_2$ *and* $\|H_0(x)\|_2$.

Proof. Taking the dot product of equation (1.8.2) with E, and that of equation (1.8.2) with $-H$, and then adding them together, we get

$$(\nabla \times H) \cdot E - (\nabla \times E) \cdot H = \frac{\partial E}{\partial t} \cdot E + \sigma |E|^2 + \frac{\partial H}{\partial t} \cdot H + \beta \frac{\partial z}{\partial t} \cdot H. \tag{1.8.11}$$

Using the vector formula

$$(\nabla \times H) \cdot E - (\nabla \times E) \cdot H = \nabla \cdot (H \times E) \tag{1.8.12}$$

and integrating equation (1.8.11) over Ω, we obtain

$$\frac{1}{2} \frac{\mathrm{d}}{\mathrm{d}t} (\|E\|_2^2 + \|H\|_2^2) + \sigma \|E\|_2^2 + \beta \int_\Omega \frac{\partial z}{\partial t} \cdot H \mathrm{d}x = 0. \tag{1.8.13}$$

Taking the dot product of $(\Delta z + H)$ with (1.8.1), we have

$$(\Delta z + H) \cdot \frac{\partial z}{\partial t} = -\lambda_2 (\Delta z + H) \cdot [z \times (z \times (\Delta z + H))], \tag{1.8.14}$$

where

$$-(\Delta z + H) \cdot [z \times (z \times (\Delta z + H))] = |z \times (\Delta z + H)|^2. \tag{1.8.15}$$

Integrating (1.8.14) over Ω, we get

$$\begin{aligned} \int_\Omega (\Delta z + H) \cdot \frac{\partial z}{\partial t} \mathrm{d}x &= \lambda_2 \int_\Omega |z \times (\Delta z + H)|^2 \mathrm{d}x, \\ \int_\Omega \frac{\partial z}{\partial t} \cdot H \mathrm{d}x &= -\int_\Omega \Delta z \cdot \frac{\partial z}{\partial t} \mathrm{d}x + \lambda_2 \int_\Omega |z \times (\Delta z + H)|^2 \mathrm{d}x \\ &= \frac{1}{2} \frac{\mathrm{d}}{\mathrm{d}t} \|\nabla z\|_2^2 + \lambda_2 \int_\Omega |z \times (\Delta z + H)|^2 \mathrm{d}x. \end{aligned} \tag{1.8.16}$$

Using (1.8.13) and (1.8.16), we get

$$\frac{1}{2} \frac{\mathrm{d}}{\mathrm{d}t} (\|E\|_2^2 + \|H\|_2^2) + \sigma \|E\|_2^2 + \frac{\beta}{2} \frac{\mathrm{d}}{\mathrm{d}t} \|\nabla z\|_2^2 + \beta \lambda_2 \|z \times (\Delta z + H)\|_2^2 = 0. \tag{1.8.17}$$

Integrating the above equation with respect to $t \in [0, T]$, we get

$$\begin{aligned} \varepsilon(t) &\triangleq \frac{1}{2} (\|E(\cdot, t)\|_2^2 + \|H(\cdot, t)\|_2^2) + \sigma \int_0^t \|E(\cdot, t)\|_2^2 + \frac{\beta}{2} \|\nabla z(\cdot, t)\|_2^2 + \beta \lambda_2 \int_0^t \|z \times (\Delta z + H)\|_2^2 \\ &= \varepsilon(0) = \frac{1}{2} (\|E_0(x)\|_2^2 + \|H_0(x)\|_2^2) + \sigma \|E_0(x)\|_2^2 + \frac{\beta}{2} \|\nabla z_0(x)\|_2^2. \end{aligned}$$

Thus the estimate (1.8.10) is established. □

In order to get the uniform priori estimate in t for $(z, H, E) \in (H^2(\Omega), H^1(\Omega), H^1(\Omega))$, we first transform equations (1.8.2)–(1.8.3) into the equivalent second order wave equations.

In fact, acting by $\nabla\times$ on (1.8.2) and (1.8.3), we get

$$\begin{aligned}\nabla \times (\nabla \times H) &= \frac{\partial}{\partial t}\nabla \times E + \sigma\nabla \times E,\\ \nabla \times (\nabla \times E) &= -\frac{\partial}{\partial t}\nabla \times H - \beta\frac{\partial}{\partial t}\nabla \times z.\end{aligned}$$

From the vector formulas

$$\begin{aligned}\nabla \times (\nabla \times H) &= \nabla(\nabla \cdot H) - \Delta H = -\beta\nabla(\nabla \cdot z) - \Delta H,\\ \nabla \times (\nabla \times E) &= \nabla(\nabla \cdot E) - \Delta E = -\Delta E,\end{aligned}$$

we have

$$\begin{aligned}-\beta\nabla(\nabla \cdot z) - \Delta H &= \frac{\partial}{\partial t}\nabla \times E + \sigma\nabla \times E = -H_{tt} - \beta z_{tt} - \sigma H_t - \sigma\beta z_t,\\ -\Delta E &= -\frac{\partial}{\partial t}\nabla \times H - \beta\frac{\partial}{\partial t}\nabla \times z = -E_{tt} - \sigma E_t - \beta(\nabla \times z)_t,\end{aligned}$$

i. e.,

$$H_{tt} - \Delta H + \beta z_{tt} + \sigma H_t + \sigma\beta z_t - \beta\nabla(\nabla \cdot z) = 0, \tag{1.8.18}$$

$$E_{tt} - \Delta E + \sigma E_t + \beta(\nabla \times z)_t = 0. \tag{1.8.19}$$

Getting the uniform priori estimate in t of the solution $(z, E, H) \in (H^2 \times H^2 \times H^1)$ directly from (1.8.1), (1.8.18) and (1.8.19) is very difficult. In the following, we apply the small parameter method of the energy functional.

Define the Lyapunov energy functional as

$$e(t) = \frac{1}{2}(\|H_t\|_2^2 + \|\nabla H\|_2^2 + \|E_t\|_2^2 + \|\nabla E\|_2^2 + \|\Delta z\|_2^2) + \eta_1(H, H_t) + \eta_2(E, E_t), \tag{1.8.20}$$

where η_1, η_2 are undetermined constants. We can prove that $e(t)$ satisfies the following differential inequalities:

$$\frac{\mathrm{d}e(t)}{\mathrm{d}t} + ae(t) \le K, \tag{1.8.21}$$

where the constants $a > 0$, K are independent of t. This gives a priori estimate of $e(t)$, and then we get the estimate of $(H, E, z) \in (H^1, H^1, H^1)$.

In fact, from (1.8.18)–(1.8.19) we have

$$\begin{aligned}\frac{\mathrm{d}e(t)}{\mathrm{d}t} &= (H_t, H_{tt}) + (\nabla H, \nabla H_t) + (E_t, E_{tt}) + (\nabla E, \nabla E_t) + (\Delta z, \Delta z_t) + \eta_1(H_t, H_t)\\ &\quad + \eta_1(H, H_{tt}) + \eta_2(E_t, E_t) + \eta_2(E, E_{tt}),\end{aligned} \tag{1.8.22}$$

where

$$
\begin{aligned}
(H_t, H_{tt}) &= (H_t, \Delta H) - \beta(H_t, z_{tt}) - \sigma(H_t, H_t) - \sigma\beta(H_t, z_t) + \beta(H_t, \nabla(\nabla \cdot z)),\\
(\nabla H, \nabla H_t) &= -(\Delta H, H_t),\\
(E_t, E_{tt}) &= (E_t, \Delta E) - \sigma(E_t, E_t) - \beta(E_t, (\nabla \times z)_t),\\
(\nabla E, \nabla E_t) &= (-\Delta E, E_t),\\
(\Delta z, \Delta z_t) &= \lambda_1(\Delta(z \times \Delta z), \Delta z) + \lambda_1(\Delta(z \times H), \Delta z)\\
&\quad - \lambda_2(\Delta(|\nabla z|^2 z), \Delta z) + \lambda_2(\Delta^2 z, \Delta z)\\
&\quad + \lambda_2(\Delta H, \Delta z) - \lambda_2(\Delta(z \cdot H)z, \Delta z),\\
\eta_1(H, H_{tt}) &= -\eta_1\|\nabla H\|_2^2 - \beta\eta_1(H, z_{tt}) - \sigma\eta_1(H, H_t)\\
&\quad - \sigma\beta\eta_1(H, z_t) + \eta_1\beta(H, \nabla(\nabla \cdot z)),\\
\eta_2(E, E_{tt}) &= -\eta_2\|\nabla E\|_2^2 - \eta_2\sigma(E, E_t) - \eta_2\beta(E, (\nabla \times z)_t),
\end{aligned}
$$

thus we obtain

$$
\begin{aligned}
\frac{de(t)}{dt} &= -(\sigma - \eta_1)\|H_t\|_2^2 - (\sigma - \eta_2)\|E_t\|_2^2\\
&\quad - \eta_1\|\nabla H\|_2^2 - \eta_2\|\nabla E\|_2^2 - \lambda_2\|\nabla \Delta z\|_2^2\\
&\quad - \beta\eta_1(H, z_{tt}) - \eta_1\sigma(H, H_t) - \sigma\beta\eta_1(H, z_t)\\
&\quad - \eta_2\sigma(E, E_t) - \eta_2\beta(E, (\nabla \times z)_t)\\
&\quad + \beta(H, \nabla(\nabla \cdot z)) + \eta_1\beta(H, \nabla(\nabla \cdot z))\\
&\quad - \beta(H_t, z_{tt}) - \sigma\beta(H_t, z_t) - \beta(E_t, (\nabla \times z)_t)\\
&\quad + \lambda_1(\Delta(z \times \Delta z), \Delta z) + \lambda_1(\Delta(z \times H), \Delta z)\\
&\quad + \lambda_2(\Delta(|\nabla z|^2 z), \Delta z) + \lambda_2(\Delta z, \Delta H) - \lambda_2(\Delta(z \cdot H)z, \Delta z),
\end{aligned}
\tag{1.8.23}
$$

where

$$
\begin{aligned}
(H, z_{tt}) &= (H, z_t)_t - (H_t, z_t),\\
(H_t, z_{tt}) &= (H_t, z_t)_t - (z_t, \Delta H - \beta z_{tt} - \sigma H_t - \sigma\beta z_t + \beta\nabla(\nabla \cdot z))\\
&= (H_t, z_t)_t + \frac{\beta}{2}(z_t, z_t)_t - (z_t, \Delta H) + \sigma(z_t, H_t) + \sigma\beta\|z_t\|_2^2 - \beta(z_t, \nabla(\nabla \cdot z)),\\
\beta(E_t, \nabla \times z)_t) &= \beta(E_t, \nabla \times z)_t + \frac{\beta^2}{2}(\nabla \times z, \nabla \times z)_t + \sigma\beta(E_t, \nabla \times z) + \beta(\nabla E, \nabla(\nabla \times z)),\\
\eta_2\beta(E, (\nabla \times z)_t) &= \eta_2\beta(E, \nabla \times z)_t - \eta_2\beta(E_t, \nabla \times z).
\end{aligned}
$$

Let

$$
e_1(t) = \frac{1}{2}G(t) + R(t),
$$

where

$$
\begin{aligned}
G(t) &\triangleq \|E_t\|_2^2 + \|H_t\|_2^2 + \|\nabla E\|_2^2 + \|\nabla H\|_2^2 + \|\Delta z\|_2^2\\
&\triangleq 2e(t) - 2\eta_1(H, H_t) - 2\eta_2(E, E_t),
\end{aligned}
$$

$$\begin{aligned}R(t) \triangleq\ & \beta(z_t, H_t) + \frac{\beta^2}{2}\|z_t\|^2 + \beta(E_t, \nabla \times z)\\ &+ \frac{\beta^2}{2}\|\nabla \times z\|_2^2 - \eta_2\beta(E, \nabla \times z) + \frac{1}{2}\sigma\eta_1\|H\|_2^2\\ &+ \frac{1}{2}\sigma\eta_2\|E\|_2^2 + \eta_2\beta(E, \nabla \times z) + \eta_2(E, E_t) + \eta_1(H, vh_t),\end{aligned} \tag{1.8.24}$$

which yields

$$\begin{aligned}&\frac{de_1(t)}{dt} + (\sigma - \eta_1)\|H_t\|_2^2 + (\sigma - \eta_2)\|E_t\|_2^2 + \eta_1\|\nabla H\|_2^2 + \eta_2\|\nabla E\|_2^2 + \lambda_2\|\nabla\Delta z\|_2^2 + \sigma\beta^2\|z_t\|^2\\ &\quad = -(2\sigma\beta - \eta_1\beta)(z_t, H_t) + \beta(z_t, \Delta H) - (\sigma - \eta_2)\beta(E_t, \nabla \times z) - \beta(\nabla E, \nabla(\nabla \times z))\\ &\qquad + \lambda_1(\Delta(z \times \Delta z), \Delta z) + \lambda_1(\Delta(z \times H), \Delta z)\\ &\qquad + \lambda_2(\Delta(|\nabla z|^2 z), \Delta z) + \lambda_2(\Delta z, \Delta H) - \lambda_2(\Delta(z \cdot H)z, \Delta z)\\ &\qquad - \sigma\beta\eta_1(H, z_t) + \beta(H_t, \nabla(\nabla \cdot z)) + \eta_1\beta(H, \nabla(\nabla \cdot z)) + \beta^2(z_t, \nabla(\nabla \cdot z)).\end{aligned} \tag{1.8.25}$$

To estimate the right-hand side terms of (1.8.25), we need to use Sobolev interpolation inequality repeatedly.

(1) Through (1.8.25), we have

$$\begin{aligned}|z_t|^2 &= \lambda_1^2|z \times (\Delta z + H)|^2 + \lambda_2^2|z \times (z \times (\Delta z + H))|^2\\ &\le (\lambda_1^2 + \lambda_2^2)|\Delta z + H|^2\\ &\le 2(\lambda_1^2 + \lambda_2^2)(|\Delta z|^2 + |H|^2).\end{aligned}$$

With the aid of Sobolev interpolation inequality and Hölder inequality, we obtain

$$\begin{aligned}|-(2\sigma\beta + \eta_1\beta)(z_t, H_t)| &\le \varepsilon_1\|H_t\|_2^2 + C(\varepsilon_1, \sigma, \beta, \eta_1)\|z_t\|_2^2\\ &\le \varepsilon_1\|H_t\|_2^2 + 2C(\lambda_1^2 + \lambda_2^2)\|\Delta z\|_2^2 + d_1\\ &\le \varepsilon_1\|H\|_2^2 + \frac{\lambda_2}{l}\|\nabla\Delta z\|_2^2 + d_2,\end{aligned}$$

where ε_1, l are undetermined positive coefficients.

(2) Acting with ∇ on (1.8.1), we get

$$\begin{aligned}\nabla z_t &= \lambda_1\nabla z \times (\Delta z + H) + \lambda_1 z \times \nabla\Delta z\\ &\quad - \lambda_2\nabla z \times (z \times (\Delta z + H)) - \lambda_2 z \times (\nabla z \times (\Delta z + H))\\ &\quad - \lambda_2 z \times (z \times \nabla\Delta z) - \lambda_2 z \times (z \times \nabla H),\\ \beta(z_t, \Delta H) &= -\beta(\nabla z_t, \nabla H)\\ &= -\beta\lambda_1(\nabla z \times (\Delta z + H), \nabla H) - \beta\lambda_1(z \times \nabla\Delta z, \nabla H)\\ &\quad + \lambda_2\beta(\nabla z \times (z \times \Delta z), \nabla H) + \lambda_2\beta(\nabla z \times (z \times H), \nabla H)\\ &\quad + \lambda_2\beta(z \times (\nabla z \times \Delta z), \nabla H) + \lambda_2\beta(z \times (\nabla z \times H), \nabla H)\\ &\quad + \lambda_2\beta(z \times (z \times \nabla\Delta z), \nabla H) - \lambda_2\beta\|z \times \nabla H\|_2^2\end{aligned}$$

$$
\begin{aligned}
&\le -\lambda_2\beta\|z\times\nabla H\|_2^2 + (|\lambda_1| + 2\lambda_2)\beta\|\nabla z\|_\infty\|\Delta z\|_2\|\nabla H\|_2 \\
&\quad + (|\lambda_1| + 2\lambda_2)\beta\|\nabla z\|_\infty\|H\|_2\|\nabla H\|_2\|\nabla H\|_2 \\
&\quad + (|\lambda_1| + \lambda_2)\beta\|\nabla\Delta z\|_2\|\nabla H\|_2 \\
&\le C_1(|\lambda_1| + 2\lambda_2)\beta\|\nabla z\|_2^{\frac{3}{2}-\frac{n}{4}}\|\nabla\Delta z\|_2^{\frac{1}{2}+\frac{n}{4}}\|\nabla H\|_2 \\
&\quad + C_2(|\lambda_1| + 2\lambda_2)\beta\|\nabla\Delta z\|_2^{\frac{n}{4}}\|\nabla H\|_2 \\
&\quad + \frac{(|\lambda_1| + |\lambda_2|)}{4}\|\nabla\Delta z\|_2^2 + (|\lambda_1| + \lambda_2)\beta^2\|\nabla H\|_2^2 \\
&\le \frac{C_0^2}{\lambda_2}(|\lambda_1| + 2\lambda_2)^2\|\nabla z\|_2^{3-\frac{n}{2}}\|\nabla\Delta z\|^{1+\frac{n}{2}} + C_3\|\nabla\Delta z\|^{\frac{n}{2}} \\
&\quad + \frac{(|\lambda_1| + \lambda_2)}{4}\|\nabla\Delta z\|_2^2 + 2(|\lambda_1| + \lambda_2)\beta^2\|\nabla H\|_2^2 \\
&\le \begin{cases} (\frac{\lambda^2}{l} + \frac{(|\lambda_1|+\lambda_2)}{4})\|\nabla\Delta z\|_2^2 \\ \quad + 2(|\lambda_1| + \lambda_2)\beta^2\|\nabla H\|_2^2 + C, \quad \text{when } n = 1, \\ (\frac{\lambda_2}{l} + \frac{(|\lambda_1|+\lambda_2)}{4}) + \frac{C_0^2}{\lambda_2}(|\lambda_1| + 2\lambda_2)^2\|\nabla z\|_2^2\|\nabla\Delta z\|_2^2 \\ \quad + 2(|\lambda_1| + \lambda_2)\beta^2\|\nabla H\|_2^2 + C, \quad \text{when } n = 2, \end{cases}
\end{aligned}
$$

(3)

$$
\big|-(\sigma + \eta_2)\beta(E_t, \nabla\times z)\big| \le \varepsilon_2\|E_t\|_2^2 + C_1\|\nabla z\|_2^2 \le \varepsilon_2\|E_t\|_2^2 + C_2,
$$

(4)

$$
\big|-\beta(\nabla E, \nabla(\nabla\times z))\big| \le \frac{\lambda_2}{l}\|\nabla\Delta z\|_2^2 + C_1\|E\|_2^2 \le \frac{\lambda_2}{l}\|\nabla\Delta z\|_2^2 + C_2,
$$

(5)

$$
\begin{aligned}
\big|\lambda_1(\Delta(z\times\Delta z), \Delta z)\big| &\le |\lambda_1|\big|(\nabla z\times\Delta z, \nabla\Delta z)\big| \\
&\le |\lambda_1|\|\Delta z\|_\infty\|\Delta z\|_2\|\nabla\Delta z\|_2^2 \\
&\le C_0|\lambda_1|\|\nabla z\|_2^{\frac{3}{2}-\frac{n}{4}}\|\nabla\Delta z\|_2^{\frac{3}{2}+\frac{n}{4}} \\
&\le \begin{cases} \frac{\lambda_2}{l}\|\nabla\Delta z\|_2^2 + C, & n = 1, \\ C_0|\lambda_1|\|\nabla z\|_2\|\nabla\Delta z\|_2^2, & n = 2, \end{cases}
\end{aligned}
$$

(6)

$$
\begin{aligned}
\big|\lambda_1(\Delta(z\times H), \Delta z)\big| &= \big|\lambda_1(\nabla(z\times H), \nabla\Delta z)\big| \\
&\le \|\nabla z\|_\infty|\lambda_1|\|H\|_2\|\nabla\Delta z\|_2 + |\lambda_1|\|\nabla H\|_2\|\nabla\Delta z\|^2 \\
&\le \frac{\lambda_2}{4}\|\nabla\Delta z\|_2^2 + \frac{2\lambda_1^2}{\lambda_2}\|\nabla H\|_2^2 + C,
\end{aligned}
$$

(7)

$$
\big|\lambda_2(\Delta(|\nabla z|^2 z), \Delta z)\big| = \big|-\lambda_2(\nabla(|\nabla z|^2 z), \nabla\Delta z)\big|
$$

$$
\begin{aligned}
&\le \lambda_2\|\nabla z\|_6^3\|\nabla\Delta z\|_2 + C\lambda_2\|\nabla z\|_\infty\|\nabla\Delta z\|_2 \\
&\le C\lambda_2\left(\|\nabla z\|^{3-\frac{n}{4}}\|\nabla\Delta z\|_2^{1+\frac{n}{4}} + \|\nabla z\|_2^{\frac{3}{2}-\frac{n}{4}}\|\nabla\Delta z\|_2^{\frac{3}{2}+\frac{n}{4}}\right) \\
&\le \begin{cases} \frac{\lambda_2}{l}\|\nabla\Delta z\|_2^2 + C, & n = 1, \\ C\lambda_2(\|\Delta z\|_2^2 + \|\nabla z\|_2)\|\nabla\Delta z\|_2^2, & n = 2, \end{cases}
\end{aligned}
$$

(8)

$$
\left|\lambda_2(\Delta z, \Delta H)\right| \le \frac{\lambda_2}{4}\|\nabla\Delta z\|_2^2 + \lambda_2\|\nabla H\|_2^2,
$$

(9)

$$
\begin{aligned}
\left|\lambda_2(\Delta(z\cdot H)z, \Delta z)\right| &= \lambda_2\left|(\nabla(z\cdot H)z, \nabla^3 z)\right| \\
&\le \lambda_2(2\|H\|_2\|\nabla z\|_\infty\|\nabla\Delta z\|_2 + \|\nabla H\|_2\|\nabla\Delta z\|_2) \\
&\le \frac{\lambda_2}{8}\|\nabla\Delta z\|_2^2 + 3\lambda_2\|\nabla H\|_2^2 + C,
\end{aligned}
$$

(10)

$$
\left|-\sigma\beta\eta_1(H, z_t)\right| \le C_1\|H\|_2\|z_t\|_2 \le C_2\|\Delta z\|_2 + d_1 \le \frac{\lambda_2}{l}\|\nabla\Delta z\|_2^2 + d_2,
$$

(11)

$$
\left|\beta(H_t, \nabla(\nabla\cdot z))\right| \le \beta\|H_t\|_2\|\Delta z\|_2 \le \varepsilon_3\|H_t\|_2 + \frac{\lambda_2}{l}\|\nabla\Delta z\|_2^2 + C,
$$

(12)

$$
\left|\eta_1\beta(H, \nabla(\nabla\cdot z))\right| \le \eta_1\beta\|H\|_2\|\Delta z\|_2 \le \frac{\lambda_2}{l}\|\nabla\Delta z\|_2^2 + C.
$$

Combining (1)–(12), and using equation (1.8.25), we arrive at the following bounds:

- when $n = 1$,

$$
\begin{aligned}
&\frac{\mathrm{d}e_1(t)}{\mathrm{d}t} + (\sigma - \eta_1 - \varepsilon_1 - \varepsilon_3)\|H_t\|_2^2 + (\sigma - \eta_2 - \varepsilon_2)\|E_t\|_2^2 \\
&\quad + \eta_2\|\nabla E\|_2^2 + \left(\eta_1 - \frac{(4+3\beta^2)(\lambda_2^2 + 2\lambda_1^2)}{\lambda_2}\right)\|\nabla H\|_2^2 \\
&\quad + \left(\frac{1}{8} - \frac{7}{l}\right)\lambda_2\|\nabla\Delta z\|_2^2 \le C_1;
\end{aligned} \tag{1.8.26}
$$

- when $n = 2$,

$$
\begin{aligned}
&\frac{\mathrm{d}e_1(t)}{\mathrm{d}t} + (\sigma - \eta_1 - \varepsilon_1 - \varepsilon_3)\|H_t\|_2^2 + (\sigma - \eta_2 - \varepsilon_2)\|E_t\|_2^2 \\
&\quad + \eta_2\|E_t\|_2^2 + \left(\eta_1 - \frac{(4+3\beta^2)(\lambda_2^2 + 2\lambda_1^2)}{\lambda_2}\right)\|\nabla H\|_2^2
\end{aligned}
$$

$$+\left[\left(\frac{1}{8}-\frac{6}{l}\right)\lambda_2 - C(|\lambda_1|+\lambda_2)\|\nabla z\|_2\right.$$
$$\left.-\frac{C}{\lambda_2}\left(\lambda_2^2+(|\lambda_1|+2\lambda_2)^2\|\nabla z\|_2^2\right)\right]\|\nabla\Delta z\|_2^2 \le C_2, \tag{1.8.27}$$

where constants C_1, C_2 are independent of t, and all the parameters ε_1, ε_2, ε_3, η_1, η_2 and l can be selected as follows:

$$\sigma > \frac{(4+3\beta^2)(\lambda_2^2+2\lambda_1^2)}{\lambda_2} \tag{1.8.28}$$

and

(i) $\frac{(4+3\beta^2)(\lambda_2^2+2\lambda_1^2)}{\lambda_2} < \eta_1 < \sigma$,

(ii) $\eta_2 = \varepsilon_2 = \frac{\sigma}{4}$,

(iii) $n = 1, l > 56$; $n = 2, l = 60$.

Suppose $\|\nabla z_0(x)\|$, $\|E_0(x)\|$ and $\|H_0(x)\|$ are sufficiently small. Using Lemma 1.8.2 and

$$C(|\lambda_1|+\lambda_2)\|\nabla z\|_2 + \frac{C}{\lambda_2}[\lambda_2^2 + (|\lambda_1|+2\lambda_2^2)^2]\|\nabla z\|_2^2$$
$$\le C(|\lambda_1|+\lambda_2)\frac{1}{\sqrt{\beta}}[\|E_0(x)\| + \|H_0(x)\| + \sqrt{2}\sigma\|E_0(x)\| + \|\nabla z_0(x)\|]$$
$$+\frac{C}{\lambda_2}\left\{\lambda_2^2 + (|\lambda_1|+2\lambda_2)^2\frac{1}{\beta}\left[\left(1+\frac{\sigma}{\beta}\right)\|E_0(x)\|^2\right.\right.$$
$$\left.\left.+\|H_0(x)\|^2 + \beta\|\nabla z_0(x)\|^2\right]\right\} < \frac{\lambda_2}{40}, \tag{1.8.29}$$

there exists a constant $a > 0$ such that

$$\frac{\mathrm{d}e_1(t)}{\mathrm{d}t} + a(\|H_t\|_2^2 + \|E_t\|_2^2 + \|\nabla H\|_2^2 + \|\nabla E\|_2^2 + \|\nabla\Delta z\|_2^2) \le C, \tag{1.8.30}$$

where the constant C is independent of t, for $1 \le n \le 2$.

Since $\Delta z(x,t)$ is a periodic function of x, $\int_0^{2D}\Delta z\mathrm{d}x = 0$. Using Poincaré inequality, we have

$$\|\Delta z\|_2^2 \le \delta\|\nabla\Delta\|_2^2,$$

where we choose $\delta_0 = \min\{a, \frac{a}{\delta}\}$. From equation (1.8.30), we have

$$\frac{\mathrm{d}e_1(t)}{\mathrm{d}t} + \delta_0(\|H_t\|_2^2 + \|E_t\|_2^2 + \|\nabla H\|_2^2 + \|\nabla E\|_2^2 + \|\Delta z\|_2^2) \le C. \tag{1.8.31}$$

Inequality (1.8.31) can be written as

$$\frac{\mathrm{d}e_1(t)}{\mathrm{d}t} + 2\delta_0 e_1(t) \le C + 2\delta_0 R(t) \le C + 2\delta_0 \sup_t |R(t)|. \tag{1.8.32}$$

Hence

$$e_1(t)e^{2\delta_0 t} \le e_1(0) + \left(\frac{C}{2\delta_0} + \sup_t |R(t)|\right)(e^{2\delta_0 t} - 1),$$
$$e_1(t) \le e_1(0) + \frac{C}{2\delta_0} + \sup_t |R(t)|,$$

that is,

$$\frac{1}{2}G(t) + R(t) \le C_0 + \sup_t |R(t)|, \quad C_0 \triangleq e_1(0) + \frac{C}{2\delta_0},$$
$$G(t) \le 2C_0 + 2\Big(\sup_t |R(t)| - R(t)\Big) \le 2C_0 + 4\sup_t |R(t)|. \tag{1.8.33}$$

By equation (1.8.24), we have

$$\begin{aligned}|R(t)| &\le \beta\|z_t\|\|H_t\| + \frac{\beta^2}{2}\|z_t\|_2^2 + \beta\|z_t\|_2\|\nabla z\|_2 \\ &\quad + \eta_1\|H\|_2\|\nabla E\|_2 + \eta_2\|E\|_2\|E_t\|_2 + \frac{\beta^2}{2}\|\nabla z\|_2^2 \\ &\quad + \frac{1}{2}\sigma\eta_1\|H\|_2^2 + \frac{1}{2}\sigma\eta_2\|E\|_2^2 + \eta_2\beta\|E\|_2\|\nabla z\|_2 \\ &\le \frac{\beta+\beta^2}{2}\|z_t\|_2^2 + \frac{\beta}{2}(\|H_t\|_2^2 + \|E_t\|_2^2 + \|\nabla E\|_2^2) + C_1 \\ &\le (\beta+\beta^2)(\lambda_1^2+\lambda_2^2)\|\Delta z\|_2^2 + \frac{\beta}{2}(\|H_t\|_2^2 + \|E_t\|_2^2 + \|\nabla E\|_2^2) + C_2.\end{aligned} \tag{1.8.34}$$

Set $\beta < \frac{1}{2}$, $(\beta+\beta^2)(\lambda_1^2+\lambda_2^2) < \frac{1}{4}$, then we can take

$$a_0 = 4\max\left\{\frac{\beta}{2}, (\lambda_1^2+\lambda_2^2)(\beta+\beta^2)\right\} < 1.$$

Due to equation (1.8.34), we have

$$|R(t)| \le \frac{1}{4}a_0(\|H_t\|_2^2 + \|E_t\|_2^2 + \|\nabla E\|_2^2 + \|\Delta z\|_2^2) \le \frac{1}{4}a_0 G(t). \tag{1.8.35}$$

Inserting equation (1.8.35) into equation (1.8.33), it follows that

$$G(t) \le 2C_0 + a_0 \sup_t G(t),$$
$$\sup_t G(t) \le \frac{2C_0}{1-a_0} \triangleq d_0. \tag{1.8.36}$$

Thus we have the following result:

Lemma 1.8.3. *Suppose* $(z(x,t), H(x,t), E(x,t))$ *is a smooth solution of periodic initial value problem* (1.8.1)–(1.8.4) *and* (1.8.7)–(1.8.8). *If the initial data satisfies* $(z_0(x), H_0(x), E_0(x)) \in (H^2(\Omega), H^1(\Omega), H^1(\Omega))$, $\Omega \subset \mathbf{R}^n$, $1 \le n \le 2$, $|z_0(x)| = 1$, *and the following conditions:*

(1) $\lambda_2 > 0, \sigma > \frac{(4+3\beta^2)\lambda_2^2+2\lambda_1^2}{\lambda_2}$;
(2) $0 < \beta < \frac{1}{2}, (\beta + \beta^2)(\lambda_1^2 + \lambda_2^2) \le \frac{1}{4}$;
(3) *when* $n = 2$,

$$\|\nabla z_0\|_{L^2(\Omega)}^2 + \|H_0\|_{L^2(\Omega)}^2 + \|E_0\|_{L^2(\Omega)}^2 \le \lambda,$$

where $\lambda = \lambda(\lambda_1, \lambda_2, \beta)$ *is an appropriately small constant,*

then we have the estimate

$$\sup_{t\in[0,\infty)} \left(\|z(\cdot,t)\|_{H^2(\Omega)}^2 + \|H(\cdot,t)\|_{H^1(\Omega)}^2 + \|E(\cdot,t)\|_{H^1(\Omega)}^2\right) \le K,$$

where the constant K *depends on* $\|z_0(x)\|_{H^2(\Omega)}$, $\|H_0(x)\|_{H^1(\Omega)}$ *and* $\|E_0(x)\|_{H^1(\Omega)}$.

Using the theorems from [194], we obtain the following theorem:

Theorem 1.8.1. *Assume that the constants* $\lambda_2 > 0, \beta \ge 0, \sigma \ge 0$, *and the initial function*

$$(z_0(x), H_0(x), E_0(x)) \in (H^k(\Omega), H^{k-1}(\Omega), H^{k+1}(\Omega))$$

for $k \ge 1 + [\frac{n}{2}]$, *where* $\Omega \subset \mathbf{R}^n$, $1 \le n \le 2$, *is a bounded domain, and* $|z_0(x)| = 1$, $\nabla \cdot E_0 = 0$, $\nabla \cdot (H_0 + \beta z_0) = 0$, *when* $n = 2$, *are such that*

$$\|\nabla z_0\| + \|E_0\| + \|H_0\| \le \delta,$$

where δ *is an appropriately small constant. Then the periodic initial value problem* (1.8.7)–(1.8.8) *for the Landau–Lifshitz–Maxwell equation system* (1.8.1)–(1.8.4) *has a unique global smooth solution. Moreover,*

$$|z(x,t)| = 1, \quad x \in \Omega,\ t \in \mathbf{R}^+,$$

$$z(x,t) \in \bigcap_{s=0}^{[\frac{k}{2}]} W_\infty^s(0,T; H^{k-2s}(\Omega)),$$

$$H(x,t) \in \bigcap_{s=0}^{k-1} W_\infty^s(0,T; H^{k-1-s}(\Omega)),$$

$$E(x,t) \in \bigcap_{s=0}^{k-1} W_\infty^s(0,T; H^{k-1-s}(\Omega)),$$

Together with the above established a priori estimate, we get

Theorem 1.8.2. *Under the conditions of Theorem* 1.8.1 *and Lemma* 1.8.3, *the problem* (1.8.1)–(1.8.4) *and* (1.8.7)–(1.8.8) *has a unique attractor* $\mathscr{A}$, *that is, the set* $\mathscr{A}$ *has the following properties:*
(1) $\mathscr{A}$ *is weak compact in* $H^2(\Omega) \times H^1(\Omega) \times H^1(\Omega)$;

(2) $S(t)\mathscr{A} = \mathscr{A}$;
(3) $\lim_{t\to\infty} \operatorname{dist}(S(t)B, \mathscr{A}) = 0$, $\forall B \subset H^2 \times H^1 \times H^1$ *bounded, where*

$$\operatorname{dist}(X, Y) = \sup_{x\in X} \inf_{y\in Y} \|x - y\|,$$

where $S(t)(z_0, H_0, E_0)$ *is the semigroup operator generated by problem* (1.8.1)–(1.8.4) *and* (1.8.7)–(1.8.8).

Proof. Using Theorem 1.8.1, the problem (1.8.1)–(1.8.4) and (1.8.7)–(1.8.8) generates a continuous semigroup operator $S(t)(z_0, H_0, E_0)$. Taking

$$E = \{(z, H, E) \in H^2(\Omega) \times H^1(\Omega) \times H^1(\Omega) : |z(x, t)| = 1,\ \nabla \cdot E = 0,\ \nabla \cdot (H + \beta z) = 0\}$$

and its subset

$$\begin{aligned} B = \{&|z(x, t)| = 1,\ \nabla \cdot E = 0, \nabla(H + \beta z) = 0,\\ &z(\cdot, t) \in H^2(\Omega),\ H(\cdot, t) \in H^1(\Omega),\ E(\cdot, t) \in H^1(\Omega)\\ &\|z(\cdot, t)\|^2_{H^2(\Omega)} + \|H(\cdot, t)\|^2_{H^1(\Omega)} + \|E(\cdot, t)\|^2_{H^1(\Omega)} \le \varepsilon_0 + \delta_0\}\end{aligned}$$

as a bounded absorbing set in E, we see that B is weakly compact in E. Then we know that the set $\mathscr{A} = \omega(B)$ is the weakly compact attractor of the periodic initial value problem (1.8.1)–(1.8.4) and (1.8.7)–(1.8.8). □

In the following, we estimate the upper bound of Hausdorff and fractal dimensions of the attractor $\mathscr{A}$. Now consider the linear variational problem corresponding to (1.8.1)–(1.8.4) and (1.8.7)–(1.8.8):

$$\begin{aligned} z_t &= \lambda_2 \Delta z + 2\lambda_2(\nabla z, \nabla z)z + \lambda_2|\nabla z|^2 z\\ &\quad + \lambda_1 z \times \Delta z + \lambda_1 z \times h - \lambda_1(\Delta z + H) \times z\\ &\quad + \lambda_2 h - \lambda_2(z \cdot H)z - \lambda_2(z \cdot h)z - \lambda_2(z \cdot H)z, && (1.8.37)\\ e_t &= \nabla \times h - \sigma e, && (1.8.38)\\ h_t &= -\nabla \times e - \beta z_t, && (1.8.39)\\ z(0) &= z_0, \quad h(0) = h_0, \quad e(0) = e_0, && (1.8.40)\\ z(x, t),&\quad e(x, t), \quad h(x, t) \quad \text{are periodic in } x \text{ with period } 2D. && (1.8.41)\end{aligned}$$

Then equations (1.8.37)–(1.8.41) can be written in operator form as

$$v_t = -L(u)v, \quad v_0 = v(0), \qquad (1.8.42)$$

where $u = (z, e, h)$ is the solution of problem (1.8.1)–(1.8.4) and (1.8.7)–(1.8.8) with $v = (z, E, H)$ and $v_0 = (z_0, e_0, h_0)$.

Since the problem (1.8.1)–(1.8.4) and (1.8.7)–(1.8.8) possesses a smooth solution, the coefficients of the linear variational equations (1.8.37)–(1.8.41) are smooth. When the initial value v_0 is sufficiently smooth, equations (1.8.37)–(1.8.41) possess a unique smooth solution, i. e., there exists an operator $G(t)$ such that $v_t = G(t)v_0$. Furthermore, we prove that the semigroup operator $S(t)$ is differentiable in $L^2(\Omega)$, and its Fréchet derivative is $S'(t)u_0 = G(t)v_0$.

Lemma 1.8.4. *The smooth solution of the periodic initial value problem* (1.8.1)–(1.8.4) *and* (1.8.7)–(1.8.8) *continuously depends on the initial conditions.*

Proof. Set $(z_i(x,t), H_i(x,t), E_i(x,t))$, $i = 1,2$, to be the smooth solution of problem (1.8.1)–(1.8.4) with initial values $z_i(x,0) = z_{0i}(x)$, $H_i(x,0) = H_{0i}(x)$, $E_i(x,0) = E_{0i}(x)$, $i = 1,2$. Set $z(x,t) = z_2(x,t)-z_1(x,t)$, $H(x,t) = H_2(x,t)-H_1(x,t)$ and $E(x,t) = E_2(x,t)-E_1(x,t)$. Then $(z(x,t), H(x,t), E(x,t))$ satisfies

$$\begin{aligned} z_t &= \lambda_1 z \times \Delta z + \lambda_1 z_1 \times \Delta z + \lambda_1 z \times H_2 + \lambda_1 z_1 \times H \\ &\quad + \lambda_2 \Delta z + \lambda_2 |\Delta z_2|^2 z + \lambda_2 (\nabla z, \nabla(z_1 + z_2)) z_1 \\ &\quad + \lambda_2 H - \lambda_2 (z_2 \cdot H_2) z - \lambda_2 (z_2 \cdot H + H_1 \cdot z) z_1, \end{aligned} \tag{1.8.43}$$

$$E_t + \sigma E = \nabla \times H, \tag{1.8.44}$$

$$H_t + \beta z_t = -\nabla \times E, \tag{1.8.45}$$

$$\begin{gathered} \nabla \cdot (H + \beta z) = 0, \quad \nabla \cdot E = 0, \\ z(x,t), \quad H(x,t), \quad E(x,t) \quad \text{are periodic in } x \text{ with period } 2D, \\ z(x,0) = z_0(x), \quad H(x,0) = H_0(x), \\ E(x,0) = E_0(x), \quad |z_0(x)| = 1, \\ \nabla \cdot (H_0 + \beta z_0) = 0, \quad \nabla E_0 = 0. \end{gathered}$$

Then we can establish the following inequality:

$$\begin{aligned} &\sup_{0 \le t \le T} [\|\nabla z(\cdot,t)\|_{H^1}^2 + \|H(\cdot,t)\|_{L^2}^2 + \|E(\cdot,t)\|_{L^2}^2] \\ &\quad \le C(\|z_0(x)\|_{H^1}^2 + \|H_0(x)\|_{L^2}^2 + \|E_0(x)\|_{L^2}^2), \end{aligned}$$

where C is an absolute constant. Obviously, if inequality (1.8.44) is established, then Lemma 1.8.4 is proved.

Indeed, taking the inner product of (1.8.43) and z, we get

$$\frac{1}{2}\frac{d}{dt}\int_\Omega |z|^2 dx + \lambda_2 \|\nabla z\|^2 \le C_1 [\|\nabla z\|^2 + \|H\|^2 + \|E\|^2].$$

Multiplying (1.8.43) by Δz, and integrating it with respect to x over Ω, we arrive at

$$\begin{aligned} \frac{1}{2}\frac{d}{dt}\int_\Omega |\nabla z|^2 dx + \lambda_2 \|\Delta z\|^2 &= -\lambda_1 \int_\Omega z \times \Delta z \cdot \Delta z dx \\ &\quad - \lambda_1 \int_\Omega z \times H_2 \cdot \Delta z dx - \lambda_1 \int_\Omega z_2 \times H \cdot \Delta z dx \end{aligned}$$

$$
\begin{aligned}
&-\lambda_2\int_\Omega |\nabla z_2|^2 z\cdot\Delta z\mathrm{d}x-\lambda_2\int_\Omega\big(\nabla z\cdot\nabla(z_1+z_2)z_1\big)\cdot\Delta z\mathrm{d}x\\
&+\lambda_2\int_\Omega H\cdot\Delta z\mathrm{d}x+\lambda_2\int_\Omega(z_2\cdot H_2)z\cdot\Delta z\mathrm{d}x\\
&+\lambda_2\int_\Omega(z_2\cdot H+H_1\cdot z)z_1\cdot\Delta z\mathrm{d}x,
\end{aligned}\tag{1.8.46}
$$

where

$$
\begin{aligned}
\left|-\lambda_1\int_\Omega z\times\Delta z_2\cdot\Delta z\mathrm{d}x\right| &= \left|\lambda_1\int_\Omega z\times\nabla\Delta z_2\cdot\nabla z\mathrm{d}x\right|\\
&\le|\lambda_1|\|\nabla\Delta z_2\|_\infty\|z\|\|\nabla z\|\le C|\lambda_1|(\|z\|^2+\|\nabla z\|^2),\\
\left|-\lambda_1\int_\Omega z\times H_2\cdot\Delta z\mathrm{d}x\right| &= \left|\lambda_1\int_\Omega z\times\nabla H_2\cdot\Delta z\mathrm{d}x\right|\\
&\le|\lambda_1|\|\nabla H_2\|_\infty\|z\|\|\nabla z\|\le C|\lambda_1|(\|z\|^2+\|\nabla z\|^2),\\
\left|-\lambda_2\int_\Omega |\nabla z_2|^2 z\cdot\Delta z\mathrm{d}x\right| &\le \lambda_2\|\nabla z_2\|_\infty^2\|z\|\|\Delta z\|\\
&\le\frac{\lambda_2}{K}\|\Delta z\|^2+C(K)\lambda_2\|z\|^2,
\end{aligned}
$$

where K is an unknown constant. Furthermore,

$$
\begin{aligned}
\left|-\lambda_2\int_\Omega \nabla z\cdot\nabla(z_1+z_2)z\cdot\Delta z\mathrm{d}x\right| &\le \lambda_2\|\nabla(z_1+z_2)\|_\infty\|\nabla z\|\|\Delta z\|\\
&\le\frac{\lambda_2}{K}\|\Delta z\|^2+C(K)\lambda_2\|\nabla z\|^2,\\
\left|\lambda_2\int_\Omega H\cdot\Delta z\mathrm{d}x\right| &\le\frac{\lambda_2}{K}\|\Delta z\|^2+C(K)\lambda_2\|H\|^2,\\
\left|\lambda_2\int_\Omega(z_2\cdot H_2)z\cdot\Delta z\mathrm{d}x\right| &\le\lambda_2\|H\|\|\Delta z\|\\
&\le\lambda_2\|H_2\|_\infty\|z\|\|\Delta z\|\le\frac{\lambda_2}{K}\|\Delta z\|^2+C(K)\lambda_2\|z\|^2,\\
\left|\lambda_2\int_\Omega(z_2\cdot H+H_1\cdot z)z_1\cdot\Delta z\mathrm{d}x\right| &\le\frac{\lambda_2}{K}\|\Delta z\|^2+C(K)\lambda_2(\|H\|^2+\|z\|^2),
\end{aligned}
$$

and from equation (1.8.46) we have

$$
\frac{1}{2}\frac{\mathrm{d}}{\mathrm{d}t}\|\nabla z\|^2+\lambda_2\|\Delta z\|^2+\lambda_1\int_\Omega z_1\times H\cdot\Delta z\mathrm{d}x\le\frac{5}{K}\lambda_2\|\Delta z\|^2+C(K)\big(\|\nabla z\|^2+\|H\|^2+\|z\|^2\big).\tag{1.8.47}
$$

Multiplying equations (1.8.10) and (1.8.11) by E and H, respectively, and integrating in Ω (for $\beta > 0$), we get

$$\frac{1}{2\beta}\frac{\mathrm{d}}{\mathrm{d}t}\int_\Omega (|E|^2 + |H|^2)\mathrm{d}x + \frac{\sigma}{\beta}\|z\|^2 = -\int_\Omega z_t \cdot H\mathrm{d}x. \tag{1.8.48}$$

From equations (1.8.47) and (1.8.48), we get

$$\begin{aligned}\frac{1}{2}\frac{\mathrm{d}}{\mathrm{d}t}&\int_\Omega \Big(|E|^2 + |H|^2 + \frac{1}{\beta}(|E|^2 + |H|^2)\Big)\mathrm{d}x + \lambda_2\|\Delta z\|^2 \\ &\le -\int_\Omega (z_t \cdot H + \lambda_1 z_1 \times H \cdot \Delta z)\mathrm{d}x \\ &\quad + \frac{5}{K_2}\lambda_2\|\Delta z\|^2 + C(K)(\|\nabla z\|^2\|z\|^2 + \|H\|^2).\end{aligned} \tag{1.8.49}$$

Taking the inner product of equation (1.8.43) and H, we have

$$\begin{aligned}&\int_\Omega (z_t \cdot H + \lambda_1 z_1 \times H \cdot \Delta z)\mathrm{d}x \\ &\quad \le |\lambda_1|\Big|\int_\Omega (z \times \Delta z_2 + z \times H_2)\cdot H\mathrm{d}x\Big| + \lambda_2\Big|\int_\Omega \Delta z \cdot H\mathrm{d}x\Big| \\ &\qquad + \lambda_2\Big|\int_\Omega |\nabla z|^2 z \cdot H\mathrm{d}x\Big| + \lambda_2\Big|\int_\Omega \nabla z \cdot \nabla(z + z_2)z_1 \cdot H\mathrm{d}x\Big| \\ &\qquad + \lambda_2\|H\|^2 + \lambda_2\Big|\int_\Omega (z_2 \cdot H_2)z \cdot H\mathrm{d}x\Big| \\ &\qquad + \lambda_2\Big|\int_\Omega (z_2 \cdot H)(z_1 \cdot H)\mathrm{d}x\Big| + \lambda_2\Big|\int_\Omega (H_1 \cdot z)(z_1 \cdot H)\mathrm{d}x\Big| \\ &\quad \le \frac{\lambda_2}{K}\|\Delta z\|^2 + C(K)(\|\nabla z\|^2 + \|z\|^2 + \|H\|^2).\end{aligned} \tag{1.8.50}$$

From equation (1.8.47)–(1.8.50), taking $K \ge 6$, we get

$$\frac{\mathrm{d}}{\mathrm{d}t}\int_\Omega \Big[|\nabla z|^2 + |z|^2 + \frac{1}{\beta}(|E|^2 + |H|^2)\Big]\mathrm{d}x \le C(K, \lambda_1, \lambda_2)[\|\nabla z\|^2 + \|z\|^2 + \|H\|^2],$$

which yields the claim. □

In order to prove that the semigroup $S(t)$ is Fréchet differentiable, now we consider the linear variational problem (1.8.1)–(1.8.4) and (1.8.7)–(1.8.8).

Set $DS(t)(z_{01}, H_{01}, E_{01}) = (\omega(t), I(t), F(t))$ to be Fréchet differentiable at (z_0, H_0, E_0) of semigroup operator. We have

$$\omega_t(t) = \lambda_1\omega \times (\Delta z_1 + H_1) + \lambda_1 z_1 \times (\Delta\omega + I)$$

$$\begin{aligned} &- \lambda_2 \omega \times (z_1 \times (\Delta z_1 + H_1)) - \lambda_2 z_1 \times (\omega \times (\Delta z_1 + H_1)) \\ &- \lambda_2 z_1 \times (z_1 \times (\Delta \omega + I)), \end{aligned} \tag{1.8.51}$$

$$\nabla \times I = F + \sigma F, \tag{1.8.52}$$

$$\nabla \times f = -i_t - \beta \omega_t, \tag{1.8.53}$$

$$\nabla \cdot (i + \rho \omega) = 0, \quad \nabla f = 0, \tag{1.8.54}$$

$$(\omega, I(t), F(t))_{t=0} = (z_0, H_0, E_0), \tag{1.8.55}$$

where $(z_1, H_1, E_1) = S(t)(z_{01}, H_{01}, E_{01})$ is the solution having initial value (z_{01}, H_{01}, E_{01}) of problem (1.8.1)–(1.8.4) and (1.8.7)–(1.8.8). Let

$$\begin{aligned} (\tilde{z}, \widetilde{H}, \widetilde{E}) &= (z, H, E) - (\omega, I, F) \\ &= S(t)(z_{01}, H_{01}, E_{01}) - DS(t)(z_{01}, H_{01}, E_{01})(z_0, H_0, E_0). \end{aligned} \tag{1.8.56}$$

Then we have

$$\begin{aligned} z_t = {} & \lambda_1 [z + (\Delta z_2 + H_2) - \omega \times (\Delta z_1 + H_1)] \\ & + \lambda_1 [z_1 \times (\Delta z + H) - z_1 \times (\Delta \omega + I)] \\ & - \lambda_2 [z \times (z_2 \times (\Delta z_2 + H_2)) - z \times (z_1 \times (\Delta z_1 + H_1)] \\ & - \lambda_2 [z_1 \times (z \times (\Delta z_2 + H_2)) - z_1 \times (\omega \times (\Delta z_1 + H_1)] \\ & - \lambda_2 [z_1 \times (z_1 \times (\Delta z + H)) - z_1 \times (z_1 \times (\Delta \omega + I)], \end{aligned} \tag{1.8.57}$$

$$\nabla \times \widetilde{H} = \widetilde{E_t} + \sigma \widetilde{E}, \tag{1.8.58}$$

$$\nabla \times \widetilde{E} = -\widetilde{H_t} - \beta \widetilde{Z_t}, \tag{1.8.59}$$

$$\nabla \cdot (\widetilde{H} + \beta \widetilde{Z}) = 0, \quad \nabla \cdot \widetilde{E} = 0, \tag{1.8.60}$$

$$(\widetilde{Z}, \widetilde{H}, \widetilde{E})|_{t=0} = 0. \tag{1.8.61}$$

Thus equation (1.8.57) can be rewritten as

$$\begin{aligned} \widetilde{Z_t} = {} & \lambda_1 [\widetilde{Z} + (\Delta z_1 + H_1) + z \times (\Delta z + H)] \\ & + \lambda_1 [z_1 \times (\Delta \widetilde{Z} + \widetilde{H})] - \lambda_2 [z \times (z \times (\Delta z_2 + H_2)] \\ & + z \times (z_1 \times (\Delta z + H)) + \widetilde{Z} \times [z_1 \times (\Delta z_1 + H_1)] \\ & - \lambda_2 [z_1 \times (\tilde{z} \times (\Delta z_1 + H_1))] + z_1 \times (z \times (\Delta z + H)) \\ & - \lambda_2 [z_1 \times (z_1 \times (\Delta \widetilde{Z} + \widetilde{H}))]. \end{aligned} \tag{1.8.62}$$

With the aid of equation (1.8.62), we have

$$\frac{1}{2} \frac{\mathrm{d}}{\mathrm{d}t} \|\widetilde{Z}\|^2 \le C_1 \|\widetilde{Z}\|^2 + C_2 (\|z\| + \|H\| + \|E\|)^4,$$

$$\|\widetilde{Z}(t)\|^2 \le \|\widetilde{Z}(0)\| e^{C_1 t} + \int_0^t e^{C_1 (t-s)} C_2 (\|z(s)\| + \|H(s)\| + \|E(s)\|)^4 \mathrm{d}s$$

$$= \int_0^t e^{C_1(t-s)} C_2(\|z(s)\| + \|H(s)\| + \|E(s)\|)^4 \mathrm{d}s.$$

Using Lemma 1.8.4, we obtain

$$\|\widetilde{Z}(t)\| \le C(T,K)(\|z_0\| + \|H_0\| + \|E_0\|)^2, \quad 0 \le t \le T.$$

We can get similar estimates of $\|H(t)\|$ and $\|E(t)\|$. Finally, we have

Lemma 1.8.5. *If the solutions for problem* (1.8.1)–(1.8.4) *and* (1.8.7)–(1.8.8) *are sufficiently smooth, then* $S(t) : (z_0, H_0, E_0) \to \{z(t), H(t), E(t)\}$ *is uniformly differentiable. If it is differentiable at* $(z_0, H_0, E_0) \in \mathscr{A}$, *then the mapping*

$$DS(t)(z_0, H_0, E_0) = (\omega(t), I(t), F(t))$$

is the solution of problem (1.8.51)–(1.8.55).

Let $v_1(t), v_2(t), \dots, v_J(t)$ be the solutions of problem (1.8.37)–(1.8.41) with the initial values $v_1(0) = \xi_1, v_2(0) = \xi_2, \dots, v_J(0) = \xi_J$, respectively, where $\xi_i \in L^2(\Omega)$, $i = 1, 2, \dots, J$. By explicit calculation, we arrive at

$$\frac{\mathrm{d}}{\mathrm{d}t}\|v_1(t) \wedge \cdots \wedge v_J(t)\|_2^2 + 2\,\mathrm{tr}(L(u(t)) \cdot Q_J)\|v_1(t) \wedge \cdots \wedge v_J(t)\|_2^2 = 0, \tag{1.8.63}$$

where $L(u(t)) = L(S(t)u_0)$ is the linear mapping $v \to L(u(t))v$; $\wedge$ means the cross-product, tr is the trace of an operator. Also Q_J is the orthogonal projection from $L^2(\mathbf{R})$ to the subspace spanned by $\{v_1(t), v_2(t), \dots, v_J(t)\}$. From equation (1.8.63) we get the variation of the volume for the J-dimensional cube $\bigwedge_{j=0}^J \xi_j$:

$$\begin{aligned} \omega_J(t) &= \sup_{u_0 \in A} \sup_{\xi \in L^2, |\xi_j| \le 1} \|v_1(t) \wedge \cdots \wedge v_J(t)\|_{L^2(\Omega)}^2 \\ &\le \sup_{u_0 \in A} \exp\Big(-\int_0^t \inf(\mathrm{tr}(L(S(\tau)u_0) \cdot Q_J(\tau)))\mathrm{d}\tau\Big). \end{aligned} \tag{1.8.64}$$

Now we rewrite equations (1.8.37)–(1.8.39) as

$$z_t + f(z, \nabla z, \Delta z, h; Z, \nabla Z, \Delta Z, H) = 0, \tag{1.8.65}$$

$$e_t + \sigma e - \nabla \times h = 0, \tag{1.8.66}$$

$$h_t + \beta z_t + \nabla \times e = 0, \tag{1.8.67}$$

where

$$\begin{aligned} &f(z, \nabla z, \Delta z, h; Z, \nabla Z, \Delta Z, H) \\ &\quad = -\lambda_2 \Delta z - 2\lambda_2(\nabla z \cdot \nabla Z)Z - \lambda_2|\nabla Z|^2 z - \lambda_1 Z \times \Delta z \end{aligned}$$

$$\begin{aligned} & - \lambda_1 Z \times h + \lambda_1(\Delta Z + H) \times z - \lambda_2 h \\ & + \lambda_2(Z \cdot H)z + \lambda_2(Z \cdot h)zZ + \lambda_2(z \cdot H)Z. \end{aligned} \tag{1.8.68}$$

Now selecting a periodical orthogonal basis $\{\varphi_j, e_j(t), h_j(t)\}$ which satisfies:

(1) $\Delta\varphi_j = -\lambda_j^2\varphi_j$;

(2) $\|\varphi_j\|_2 = \|e_j\| = \|h_j\|_2 = 1$;

we obtain that $\|\nabla\varphi_j\|_2 = |\lambda_j|$, $\|\Delta\varphi_j\|_2 = \lambda_j^2$. By definition, it follows that

$$\begin{aligned} \operatorname{tr}\{L(u(t)) \cdot Q_J(t)\} = \sum_{j=1}^{J} & \big[(f(\varphi_j, \nabla\varphi_j, \Delta\varphi_j, h_j; Z, \nabla Z, \Delta Z, H), \varphi_j) \\ & + \sigma(e_j, e_j) - (\nabla \times h_j, e_j) + (\nabla \times e_j, h_j) \\ & - \beta(f(\varphi_j, \nabla\varphi_j, \Delta\varphi_j, h_j; Z, \nabla Z, \Delta Z, H), h_j)\big]. \end{aligned} \tag{1.8.69}$$

Since

$$(\nabla \times e_j, h_j) - (\nabla \times h_j, e_j) = \int_\Omega \nabla \cdot (e_j \times h_j) \mathrm{d}x = 0,$$

we only need to estimate in (1.8.69) the quantity

$$(f(\varphi_j, \nabla\varphi_j, h_j; H), \varphi_j)$$

and

$$-\beta(f(\varphi_j, \nabla\varphi_j, \Delta\varphi_j, h_j; Z, \nabla Z, \Delta Z, H), \varphi_j).$$

By equation (1.8.68),

$$\begin{aligned} & (f(\varphi_j, \nabla\varphi_j, \Delta\varphi_j, h_j; Z, \nabla Z, \Delta Z, H), \varphi_j) \\ & \quad = -\lambda_2(\Delta\varphi_j, \varphi_j) - 2\lambda_2((\nabla Z, \nabla\varphi_j)Z, \varphi_j) - \lambda_2(|\nabla Z|^2\varphi_j, \varphi_j) - \lambda_1(Z \times \Delta\varphi_j, \varphi_j) \\ & \quad\quad - \lambda_1(Z \times h_j, \varphi_j) - \lambda_2(h_j, \varphi_j) + \lambda_2((\varphi_j \cdot H)Z, \varphi_j) \\ & \quad\quad + \lambda_2((Z \cdot h_j)Z, \varphi_j) + \lambda_2((Z \cdot H)\varphi_j, \varphi_j), \end{aligned}$$

where

$$\begin{aligned} -\lambda_2(\Delta\varphi_j, \varphi_j) &= \lambda_2\lambda_j^2, \\ \big|-2\lambda_2((\nabla Z \cdot \nabla\varphi_j)Z, \varphi_j)\big| &\le 2\lambda_2\|\nabla\varphi_j\|_2\|\varphi_j\|_2\|\nabla Z\|_\infty = 2\lambda_2|\lambda_j|\|\nabla Z\|_\infty, \\ \big|-\lambda_2(|\nabla Z|^2\varphi_j, \varphi_j)\big| &\le \lambda_2|\nabla Z|_\infty^2\|\varphi_j\|_2^2 = \lambda_2|\nabla Z|_\infty^2, \\ \big|-\lambda_1(Z \times \Delta\varphi_j, \varphi_j)\big| &= \big|\lambda_1(\nabla\varphi_j, \nabla Z \times \varphi_j)\big| \\ &\le |\lambda_1|\|\nabla\varphi_j\|_2\|\varphi_j\|_2|\nabla Z|_\infty = |\lambda_1\lambda_j|\|\nabla Z\|_\infty, \\ \big|-\lambda_1(Z \times h_j, \varphi_j)\big| &\le |\lambda_1|\|h_j\|_2\|\varphi_j\|_2 = |\lambda_1|, \end{aligned}$$

$$
\begin{aligned}
\left|-\lambda_2(h_j,\varphi_j)\right| &\leq \lambda_2,\\
\left|\lambda_2((\varphi_j\cdot H)Z,\varphi_j\right| &\leq \lambda_2\|H\|_\infty\|\varphi_j\|_2^2 = \lambda_2\|H\|_\infty,\\
\left|\lambda_2((Z\cdot h_j)Z,\varphi_j\right| &\leq \lambda_2\|h_j\|_2\|\varphi_j\|_2 = \lambda_2,\\
\left|\lambda_2((Z\cdot H)\varphi_j,\varphi_j\right| &\leq \lambda_2\|H\|_\infty,
\end{aligned}
$$

thus

$$
\begin{aligned}
&(f(\varphi_j,\nabla\varphi_j,\triangle\varphi_j,h_j;Z,\nabla Z,\Delta Z,H),\varphi_j)\\
&\quad\geq \lambda_2\lambda_j^2-(2\lambda_2+|\lambda_1|)|\lambda_j|\|\nabla Z\|_\infty-\lambda_2\|\nabla Z\|_\infty^2-(2\lambda_2+|\lambda_1|)-2\lambda_2\|H\|_\infty.
\end{aligned}
\tag{1.8.70}
$$

On account of equation (1.8.70)

$$
\begin{aligned}
&-\beta(f(\varphi_j,\nabla\varphi_j,\triangle\varphi_j,h_j;Z,\nabla Z,\Delta Z,H),\varphi_j)+\lambda_2\beta(\Delta\varphi_j,h_j)+2\lambda_2\beta((\nabla Z\cdot\nabla\varphi_j)Z,h_j)\\
&\quad+\lambda_2\beta(|\nabla Z|^2\varphi_j,h_j)+\lambda_1\beta(Z\times\Delta\varphi_j,h_j)-\lambda_1\beta((\Delta Z+H)\times\varphi_j,h_j)+\lambda_2\beta(h_j,h_j)\\
&\quad-\lambda_2\beta((\varphi_j\cdot H)Z,h_j)-\lambda_2\beta((Z\cdot h_j)Z,h_j-\lambda_2\beta((Z\cdot H)\varphi_j,h_j),
\end{aligned}
$$

where

$$
\begin{aligned}
\left|\lambda_2\beta(\Delta\varphi_j,h_j)\right| &\leq \lambda_2\beta\|\Delta\varphi_j\|_2\|h_j\|_2=\lambda_2\beta\lambda_j^2,\\
\left|2\lambda_2\beta((\nabla Z\cdot\nabla\varphi_j)Z,h_j)\right| &\leq 2\lambda_2\beta\|\nabla\varphi_j\|_2\|h_j\|_2\|\nabla Z\|_\infty=2\lambda_2\beta|\lambda_j|\|\nabla Z\|_\infty,\\
\left|\lambda_2\beta(|\nabla Z|^2\varphi_j,h_j)\right| &\leq \lambda_2\beta\|\nabla Z\|_\infty^2\|h_j\|_2\|\varphi_j\|_2=\lambda_2\beta\|\nabla Z\|_\infty^2,\\
\left|\lambda_1\beta(Z\times\Delta\varphi_j,h_j)\right| &\leq |\lambda_1|\|\beta\|\|\Delta\varphi_j\|_2\|h_j\|_2=|\lambda_1|\beta\lambda_j^2,\\
\left|-\lambda_1\beta((\Delta Z+H)\times\varphi_j,h_j)\right| &\leq |\lambda_1|\beta|\Delta Z+H|_\infty\|\varphi_j\|_2\|h_j\|_2\\
&=|\lambda_1|\beta\|\Delta Z+H\|_\infty,\\
\lambda_2\beta(h_j,h_j) &= \lambda_2\beta,\\
\left|-\lambda_2\beta((\varphi_j\cdot H)Z,h_j)\right| &\leq \lambda_2\|H\|_\infty,\\
\left|-\lambda_2\beta((Z\cdot h_j)Z,h_j)\right| &\leq \lambda_2\beta,\\
\left|-\lambda_2\beta((Z\cdot H)\varphi_j,h_j)\right| &\leq \lambda_2\|H\|_\infty,
\end{aligned}
$$

we then have

$$
\begin{aligned}
&-\beta(f(\varphi_j,\nabla\varphi_j,\Delta\varphi_j,h_j;Z,\nabla Z,\Delta Z,H),\varphi_j)\\
&\quad\geq -|\lambda|\beta\lambda_j^2-\lambda_2\beta\lambda_j^2-2\lambda_2\beta|\lambda_j|\|\nabla Z\|_\infty\\
&\qquad-\left[\lambda_2\beta\|\nabla Z\|_\infty^2+|\lambda_1|\beta\|\Delta Z+H\|_\infty+2\lambda_2\beta\|H\|_\infty\right].
\end{aligned}
\tag{1.8.71}
$$

Substituting equations (1.8.70) and (1.8.71) into (1.8.69), we arrive at

$$\begin{aligned}\operatorname{tr}\{L(u(t))\cdot Q_J(t)\} \geq{}& (\lambda_2 - (\lambda_2 + |\lambda_1|\beta))\sum_{j=1}^{J}\lambda_j^2\\ &- (2\lambda_2 + 2\lambda_2\beta + |\lambda_1|)\sum_{j=1}^{J}|\lambda_j|\|\nabla Z\|_\infty\\ &+ (\sigma - \lambda_2(1+\beta))\|\nabla Z\|_\infty^2 - 2\lambda_2(1+\beta)\|H\|_\infty\\ &- |\lambda_1|\beta\|\Delta Z + H\|_\infty J.\end{aligned} \tag{1.8.72}$$

Choose parameter β sufficiently small so that

$$\lambda_2 > (\lambda_2 + |\lambda_1|)\beta,$$

that is,

$$0 < \beta < \frac{\lambda_2}{\lambda_2 + |\lambda_1|}. \tag{1.8.73}$$

Setting

$$\begin{aligned}&\delta = \lambda_2 - (\lambda_2 + |\lambda_1|)\beta, \quad \chi = \left(\sum_{j=1}^{J}\lambda_j^2\right)^{\frac{1}{2}},\\ &a = (2\lambda_2 + 2\lambda_2\beta + |\lambda_1|)\|\Delta Z\|_\infty,\\ &b = \sigma - \lambda_2(1+\beta)\|\nabla Z\|_\infty^2 - 2\lambda_2(1+\beta)\|H\|_\infty - |\lambda_1|\beta\|\Delta Z + H\|_\infty,\end{aligned}$$

and noting that

$$\sum_1^J|\lambda_j| \leq \left(\sum_1^J\lambda_j^2\right)^{\frac{1}{2}} J^{\frac{1}{2}},$$

we can rewrite equation (1.8.72) as

$$\operatorname{tr}\{L(u(t))\cdot Q_J(t)\} \geq \sigma\chi^2 - aJ^{\frac{1}{2}}\chi + bJ = \delta\left(\chi - \frac{aJ^{\frac{1}{2}}}{2\delta}\right)^2 + \frac{4\delta b - a^2}{4\delta}J. \tag{1.8.74}$$

From equation (1.8.74), we have

(1) $4\delta b - a^2 \leq 0$,

$$\operatorname{tr}\{L(u(t))\cdot Q_J(t)\} \geq \delta\left(\chi - \frac{a + \sqrt{a^2 - 4\delta b}}{2\delta}J^{\frac{1}{2}}\right)\left(\chi - \frac{a - \sqrt{a^2 - 4\delta b}}{2\delta}J^{\frac{1}{2}}\right).$$

Choosing J such that

$$\chi > \frac{a + \sqrt{a^2 - 4\delta b}}{2\delta}J^{\frac{1}{2}} \tag{1.8.75}$$

for λ_j, we have the estimate

$$\lambda_j^2 \geq \left[\frac{(j-1)^{\frac{1}{n}}}{2} - 1\right]^2 = \frac{1}{4}(j-1)^{\frac{2}{n}} - (j-1)^{\frac{1}{n}} + 1,$$

that is,

$$\lambda_j^2 \geq \begin{cases} \frac{1}{4}(j-1)^2 - j + 2 = \frac{1}{4}j^2 + \frac{3}{2}j + \frac{9}{4}, & \text{when } n = 1, \\ \frac{1}{4}(j-1) - (j-1)^{\frac{1}{2}} + 1 = \frac{1}{4}j + \frac{3}{4} - (j-1)^{\frac{1}{2}}, & \text{when } n = 2. \end{cases}$$

Now consider several cases:

(i) When $n = 1$,

$$\begin{aligned} \sum_{j=1}^{J} \lambda_j^2 &\geq \frac{1}{4}\sum_{1}^{J} j^2 - \frac{3}{2}\sum_{1}^{J} j + \frac{9}{4}J \\ &= \frac{1}{24}(J+1)(2J+1)J - \frac{3}{4}(J+1)J + \frac{9}{4}J \\ &= \frac{1}{12}J^3 - \frac{5}{8}J^2 + \frac{37}{24}J. \end{aligned}$$

Therefore this ensures that equation (1.8.75) is valid. Choose J_0 such that

$$\begin{aligned} \frac{1}{12}J_0^3 - \frac{5}{8}J_0^2 + \left(\frac{37}{24} - \frac{(a+\sqrt{a^2-4\delta b})^2}{4\delta^2}\right)J_0 &> 0, \\ 2J_0^2 - 15J_0 + 37 - \frac{6}{\delta^2}(a+\sqrt{a^2-4\delta b})^2 &> 0, \\ \left(J_0 - \frac{15}{4}\right)^2 + \frac{71}{16} - \frac{3}{\delta^2}(a+\sqrt{a^2-4\delta b})^2 &> 0. \end{aligned}$$

When

$$a + \sqrt{a^2 - 4\delta b} < 2\delta,$$

we can choose $J_0 = 1$. When

$$a + \sqrt{a^2 - 4\delta b} \geq 2\delta,$$

we can choose

$$J_0 > \sqrt{\frac{3(a+\sqrt{a^2-4\delta b})}{\delta^2} - \frac{71}{16}} + \frac{15}{4}.$$

(ii) When $n = 2$,

$$\sum_{j=1}^{J} \lambda_j^2 \geq \frac{1}{4}\sum_{j=1}^{J} j + \frac{3}{4}J - \sum_{j=1}^{J}(j-1)^{\frac{1}{2}} = \frac{1}{8}(J-1)J + \frac{3}{4}J - \sum_{j=1}^{J-1} j^{\frac{1}{2}}$$

$$\geq \frac{J^2+7J}{8} - \left(\sum_{j=1}^{J-1} j\right)^{\frac{1}{2}} \sqrt{J-1}$$
$$\geq \frac{J^2+7J}{8} - \frac{1}{\sqrt{2}} J^{\frac{3}{2}}.$$

To guarantee that (1.8.75) is true, we choose J_0 such that

$$\frac{J_0+7J}{8} - \frac{1}{\sqrt{2}} J_0^{\frac{1}{2}} > \frac{(a+\sqrt{a^2-4\delta b})^2}{4\delta^2},$$

that is,

$$J_0 > \left\{2\sqrt{2} + \left[1 + 2\left(\frac{a+\sqrt{a^2-4\delta b}}{\delta}\right)^2\right]^{\frac{1}{2}}\right\}^2.$$

Lastly, we get the following theorem:

Theorem 1.8.3. *Assume that $\Omega \subset \mathbf{R}^n$, $1 \leq n \leq 2$, is a bounded domain, and the following conditions are satisfied:*

(1) $\lambda_2 > 0$, $\sigma > \frac{(4+3\beta^2)\lambda_2^2+2\lambda_1^2}{\lambda_2}$;

(2) $0 < \beta < \min\{\frac{1}{2}, \frac{\lambda_2}{\lambda_2+|\lambda_1|}\}$;

(3) $(\beta+\beta^2)(\lambda_1^2+\lambda_2^2) < \frac{1}{4}$,

(4) *when $n = 2$,*

$$\|\nabla Z_0\|_2 + \|H_0\|_2 + \|E_0\|_2 \leq v$$

where $v = v(\lambda_1, \lambda_2, \beta)$ is a sufficiently small constant. Then the periodic initial value problem (1.8.1)–(1.8.4) and (1.8.7)–(1.8.8) possesses a unique attractor $\mathscr{A} = \omega(B)$, where

$$B = \{(Z,H,E) \in (H^2(\Omega), H^1(\Omega), H^1(\Omega)) \mid \|Z\|_{H^2}^2 + \|E\|_{H^1}^2 + \|H\|_{H^2}^2 \leq \varepsilon_0 + d_0\}$$

is an absorbing set. The Hausdorff and fractal dimensions of the attractor $\mathscr{A}$ are bounded and satisfy:

(1) *if $a^2 - 4\delta b < 0$,*

$$d_H(\mathscr{A}) \leq 1, \quad d_F(\mathscr{A}) \leq 2;$$

(2) *if $a^2 - 4\delta b \geq 0$,*

(i) *for $n = 1$, if*

$$a + \sqrt{a^2-4\delta b} < 2\delta,$$

then

$$d_H(\mathscr{A}) \leq 1, \quad d_F(\mathscr{A}) \leq 2;$$

If

$$a + \sqrt{a^2 - 4\delta b} \geq 2\delta,$$

then

$$d_H(\mathscr{A}) \leq J_1, \quad d_F(\mathscr{A}) \leq 2J_1,$$

where J_1 *is the smallest integer, satisfying*

$$J_1 > \sqrt{\frac{3(a + \sqrt{a^2 - 4\delta b})}{\delta^2} - \frac{71}{16} + \frac{15}{4}};$$

(ii) *for* $n = 2$,

$$d_H(\mathscr{A}) \leq J_2, \quad d_F(\mathscr{A}) \leq 2J_2,$$

where J_1 *is the smallest integer, satisfying*

$$J_2 > \left\{2\sqrt{2} + \left[1 + 2\left(\frac{a + \sqrt{a^2 - 4\delta b}}{\delta}\right)^2\right]^{\frac{1}{2}}\right\}^2,$$

thus

$$\delta = \lambda_2 - (\lambda_2 + |\lambda_1|)\beta,$$
$$a = (2\lambda_2 + 2\lambda_2\beta + |\lambda_1|)\|\nabla Z\|_\infty,$$
$$b = \sigma - \lambda_2(1 + \beta)\|\nabla Z\|_\infty^2 + 2\lambda_2(1 + \beta)\|H\|_\infty - |\lambda_1\beta|\|\Delta Z + H\|_\infty.$$

1.9 Nonlinear Schrödinger–Boussinesq equations

Consider the initial boundary value problem for the following dissipation of nonlinear Schrödinger–Boussinesq equations [99]

$$i\varepsilon_t + \Delta\varepsilon - n\varepsilon - \beta|\varepsilon|^2\varepsilon + i\gamma\varepsilon = g(x), \tag{1.9.1}$$
$$n_t = \Delta\varphi, \tag{1.9.2}$$
$$\varphi_t = n + f(n) + \mu n_t - \lambda\Delta n + |\varepsilon|^2 - \alpha\varphi, \tag{1.9.3}$$
$$\varepsilon(0) = \varepsilon_0, \quad n(0) = n_0, \quad \varphi(0) = \varphi_0, \tag{1.9.4}$$
$$\varepsilon\,|_{\partial\Omega} = n\,|_{\partial\Omega} = \varphi\,|_{\partial\Omega} = 0, \tag{1.9.5}$$

where $x \in \Omega \subset \mathbf{R}^N$, $t \in \mathbf{R}^+$, $\alpha, \beta, \gamma, \mu, \lambda > 0$ are constants, $i = \sqrt{-1}$, $\varepsilon(x,t) = (\varepsilon_1(x,t), \varepsilon_2(x,t), \dots, \varepsilon_J(x,t))$ is an unknown complex function vector, $n(x,t)$, $\varphi(x,t)$ are unknown real functions. This system of equations appears in the nonlinear interaction of laser and plasma, where ε represents electric field, n denotes density disturbance, φ denotes potential function; see [169, 184, 212]. Constants $\gamma > 0$, $\mu > 0$, $\alpha > 0$ describe the dissipation effect. When $\gamma = \mu = \alpha = 0$, this system is integrable and has the soliton solution. The global existence and uniqueness of a smooth solution for equation (1.9.1)–(1.9.5) was first obtained by Guo [83, 108] in 1983.

In what follows, we study the existence of a global attractor and the estimation of dimension for the problem (1.9.1)–(1.9.5).

For simplicity, we consider one-dimensional space, $\Omega = [0, L]$. Suppose $f \in C^{\infty}(\mathbf{R})$ and it satisfies

(i)

$$\lim_{|s|\to\infty} \inf \frac{F(s)}{|s|^2} \geq 0; \tag{1.9.6}$$

(ii) there exists $\omega \geq 0$ such that

$$\lim_{|s|\to\infty} \frac{sf(s) - \omega F(s)}{|s|^2} \geq 0, \tag{1.9.7}$$

where $F(s) = \int_0^s f(\delta)\mathrm{d}\delta$. Without loss of generality, we set $\omega \leq 1$.

Denote the norm $\|u\| = \|u\|_{L^2} = (\int |u|^2 \mathrm{d}x)^{\frac{1}{2}}$, $H^m(\Omega) = W^{m,2}(\Omega)$

$$A = -\partial_{xx}.$$

Lemma 1.9.1. *Assume that $\varepsilon_0(x) \in L^2(\Omega)$, $g(x) \in L^2(\Omega)$. Then for a solution of (1.9.1)–(1.9.5), we have*

$$\|\varepsilon(t)\|^2 \leq \|\varepsilon_0\|^2 e^{-\gamma t} + \frac{\|g\|^2}{\gamma^2}(1 - e^{-\gamma t}). \tag{1.9.8}$$

Proof. Multiplying the equation (1.9.1) by $\bar{\varepsilon}$ and integrating it over Ω, we get

$$(i\varepsilon_t, \varepsilon) + (\varepsilon_{xx}, \varepsilon) - (n\varepsilon, \varepsilon) - \beta(|\varepsilon|^2\varepsilon, \varepsilon) + i\gamma(\varepsilon, \varepsilon) = (g, \varepsilon), \tag{1.9.9}$$

where $(\cdot, \cdot)$ means the inner product in $L^2(\Omega)$. Taking the imaginary part of equation (1.9.9) yields

$$\frac{1}{2}\frac{\mathrm{d}}{\mathrm{d}t}\|\varepsilon\|^2 + \gamma\|\varepsilon\|^2 = \mathrm{Im}(g, \varepsilon),$$

then, using Cauchy and Gronwall inequalities, we get (1.9.8). □

Taking the real part of equation (1.9.9), we obtain

$$-\mathrm{Im}(\varepsilon_t, \varepsilon) - \|\varepsilon_x\|^2 - \int n|\varepsilon|^2 \mathrm{d}x - \beta \int |\varepsilon|^4 \mathrm{d}x = \mathrm{Re}(g, \varepsilon). \tag{1.9.10}$$

Multiplying equation (1.9.1) by $\overline{\varepsilon_t}$ and integrating it over Ω, we get

$$i(\varepsilon_t, \varepsilon_t) + (\varepsilon_{xx}, \varepsilon_t) - (n\varepsilon, \varepsilon_t) - \beta(|\varepsilon|^2\varepsilon, \varepsilon_t) + i\gamma(\varepsilon, \varepsilon_t) = (g, \varepsilon_t). \tag{1.9.11}$$

On account of

$$\frac{d}{dt}\|\varepsilon_x\|^2 = -2\,\mathrm{Re}(\varepsilon_{xx}, \varepsilon_t),$$
$$\frac{d}{dt}\int n|\varepsilon|^2 dx = 2\,\mathrm{Re}(n\varepsilon, \varepsilon_t) + \int n_t|\varepsilon|^2 dx,$$
$$\frac{1}{4}\frac{d}{dt}|\varepsilon|^4 dx = \mathrm{Re}(|\varepsilon|^2\varepsilon, \varepsilon_t),$$

and after taking the real part of equation (1.9.11), we arrive at

$$\frac{d}{dt}\left[\frac{1}{2}\|\varepsilon_x\|^2 + \frac{1}{2}\int n|\varepsilon|^4 dx + \frac{\beta}{4}\int |\varepsilon|^2 dx\right] - \frac{1}{2}\int n_t|\varepsilon|^2 dx + \mathrm{Im}\,\gamma(\varepsilon, \varepsilon_t) = -\,\mathrm{Re}(g, \varepsilon_t). \tag{1.9.12}$$

Combining equations (1.9.10) and (1.9.12), we have

$$\frac{d}{dt}\left[\frac{1}{2}\|\varepsilon_x\|^2 + \frac{1}{2}\int n|\varepsilon|^2 dx + \frac{\beta}{4}\int |\varepsilon|^2 dx + \mathrm{Re}(g, \varepsilon)\right] + \gamma\left[\|\varepsilon_x\|^2 + \int n|\varepsilon|^2 dx + \beta\int |\varepsilon|^4 dx + \mathrm{Re}(g, \varepsilon)\right] = \frac{1}{2}\int n_t|\varepsilon|^2 dx. \tag{1.9.13}$$

Now we estimate $\frac{1}{2}\int n_t|\varepsilon|^2 dx$. Setting $m = n_t + \rho n$, where ρ is small enough, taking the inner product of equation (1.9.3) with m, we arrive at

$$(\varphi_t, m) = (n + f(n) + \mu n_t - \lambda n_{xx} + |\varepsilon|^2 - \alpha\varphi, m).$$

Since $A\varphi_t = -n_{tt}$, we obtain

$$(m_t + (\alpha - \rho)m + \mu Am + [\lambda A^2 + A - \mu\rho A - \rho(\alpha - \rho)]n + Af(n) + A|\varepsilon|^2, A^{-1}m) = 0.$$

Then we have

$$\begin{aligned}&\frac{d}{dt}\left\{\frac{1}{2}\|A^{-\frac{1}{2}}m\|^2 + \frac{1}{2}\lambda\|A^{\frac{1}{2}}n\|^2 + \frac{1}{2}\|n\|^2 + \int F(n)dx\right\}\\ &+ (\alpha - \rho)\|A^{-\frac{1}{2}}m\|^2 + \lambda\rho\|A^{\frac{1}{2}}m\|^2 + \rho\|n\|^2\\ &+ \mu\|m\|^2 + \rho\int f(n)n dx - \mu\rho(n, m)\\ &- \rho(\alpha - \rho)(n, A^{-1}m) + \rho\int n|\varepsilon|^2 dx + \int n_t|\varepsilon|^2 dx = 0.\end{aligned} \tag{1.9.14}$$

On account of

$$\mu\rho|(n, m)| \le \mu\rho\|n\|\|m\| \le \frac{1}{8}\rho\|n\|^2 + 2\mu^2\|m\|^2,$$

$$\rho(\alpha-\rho)|(n, A^{-1}m)| \leq \frac{1}{8}\rho\|n\|^2 + 2\rho(\alpha-\rho)^2\|A^{-1}m\|^2$$
$$\leq \frac{1}{8}\rho\|n\|^2 + 2\rho(\alpha-\rho)^2 c_0\|m\|^2,$$

where c_0 depends on the first eigenvalue of A, if we choose ρ small enough so that

$$\rho \leq \gamma, \quad \rho \leq \frac{1}{2}\alpha, \quad 2\mu^2\rho + 2\rho(\alpha-\rho)^2 c_0 < \frac{1}{2}\mu, \tag{1.9.15}$$

then equation (1.9.14) becomes

$$\begin{aligned}&\frac{d}{dt}\left\{\frac{1}{2}\|A^{-\frac{1}{2}}m\|^2 + \frac{1}{2}\lambda\|A^{\frac{1}{2}}n\|^2 + \frac{1}{2}\|n\|^2 + \int F(n)dx\right\}\\ &\quad + \frac{1}{2}\alpha\|A^{-\frac{1}{2}}m\|^2 + \lambda\rho\|A^{\frac{1}{2}}n\|^2 + \frac{3}{4}\rho\|n\|^2 + \frac{1}{2}\mu\|m\|^2\\ &\quad + \rho\int f(n)n dx + \rho\int n|\varepsilon|^2 dx + \int n_t|\varepsilon|^2 dx \leq 0.\end{aligned} \tag{1.9.16}$$

Denote

$$\begin{aligned}H(\varepsilon, n, \varphi) &= \|A^{\frac{1}{2}}\varepsilon\|^2 + \int n|\varepsilon|^2 dx + \frac{1}{2}\beta\int |\varepsilon|^4 dx\\ &\quad + 2\operatorname{Re}(g, \varepsilon) + \frac{1}{2}\|A^{-\frac{1}{2}}m\|^2 + \frac{1}{2}\lambda\|A^{\frac{1}{2}}n\|^2 + \frac{1}{2}\|n\|^2 + \int F(n)dx,\end{aligned} \tag{1.9.17}$$

$$\begin{aligned}K(\varepsilon, n, \varphi) &= 2\gamma\left[\|A^{\frac{1}{2}}\varepsilon\|^2 + \int n|\varepsilon|^2 dx + \beta\int |\varepsilon|^4 dx + \operatorname{Re}(g, \varepsilon)\right]\\ &\quad + \frac{1}{2}\alpha\|A^{\frac{1}{2}}\varepsilon\|^2 + \lambda\rho\|A^{\frac{1}{2}}n\|^2\\ &\quad + \frac{3}{4}\rho\|n\|^2 + \rho\int f(n)n dx + \rho\int n|\varepsilon|^2 dx,\end{aligned} \tag{1.9.18}$$

then from equations (1.9.14) and (1.9.15), we get

$$\frac{d}{dt}H(\varepsilon, n, \varphi) + K(\varepsilon, n, \varphi) + \frac{1}{2}\mu\|m\|^2 \leq 0. \tag{1.9.19}$$

Now we estimate the terms $H(\varepsilon, n, \varphi)$ and $K(\varepsilon, n, \varphi)$.

Lemma 1.9.2. *For any small $\theta > 0$, there exist constants c_1 and c_2, which only depend on θ, such that*

$$\begin{aligned}H(\varepsilon, n, \varphi) &\geq (1-\theta)\|A^{\frac{1}{2}}\varepsilon\|^2 + \frac{1}{2}\beta\int |\varepsilon|^4 dx\\ &\quad + \frac{1}{2}\|A^{-\frac{1}{2}}m\|^2 + \frac{1}{2}\lambda\|A^{\frac{1}{2}}n\|^2\\ &\quad + \frac{1}{4}\|n\|^2 - \|\varepsilon\|^2 - c_1\|\varepsilon\|^6 - \|g\|^2 - c_2.\end{aligned} \tag{1.9.20}$$

Proof. By Young and Sobolev inequalities, $\forall\theta_i > 0, i = 1,2,3$, there exist $c(\theta_i), i = 1,2,3$, such that

$$\left|\int n|\varepsilon|^2 dx\right| \le \theta_1\|n\|^2 + c(\theta_1)\|\varepsilon\|_{L^4}^4 \le \theta_1\|n\|^2 + \theta_2\|\varepsilon_x\|^2 + c(\theta_1)c(\theta_2)\|\varepsilon\|^6,$$

$$\left|\int F(n)dx\right| \le \theta_3\|n\|^2 + c(\theta_3).$$ □

By the definition of H, and choosing $\theta_1 = \theta_3 = \frac{1}{8}, \theta_2 = 0 \in (0,1)$, inequality (1.9.20) follows.

Lemma 1.9.3. *If ρ is small enough and such that* (1.9.16) *is valid, then there exist constants c_3, c_4 such that*

$$\rho\omega H - K \le \rho(\|g\|^2 + \|\varepsilon\|^2) + c_3\|\varepsilon\|^6 + c_4. \tag{1.9.21}$$

Proof. Firstly, we have

$$\begin{aligned}
&\rho\omega H(\varepsilon,n,\varphi) - K(\varepsilon,n,\varphi)\\
&= -(2\gamma - \omega\rho)\|A^{\frac{1}{2}}\varepsilon\|^2 - \left(2\gamma - \frac{1}{2}\rho\omega\right)\beta\int|\varepsilon|^4 dx\\
&\quad - \left(\frac{1}{2}\alpha - \frac{1}{2}\rho\omega\right)\|A^{-\frac{1}{2}}m\|^2 - \left(1 - \frac{1}{2}\omega\right)\lambda\rho\|A^{-\frac{1}{2}}n\|^2\\
&\quad - \left(\frac{3}{4} - \frac{1}{2}\omega\right)\rho\|n\|^2 - \rho(\omega - 1)\int n|\varepsilon|^2 dx\\
&\quad - 2(\gamma - \omega\rho)(g,\omega) - \rho\int[f(n)n - \omega F(n)]dx.
\end{aligned} \tag{1.9.22}$$

With the aid of equation (1.9.7), $f(n)n - \omega F(n) \ge -\frac{1}{8}n^2 - c$, and then we have

$$-\int[f(n)n - \omega F(n)]dx \le \frac{1}{8}\|n\|^2 + c. \tag{1.9.23}$$

From equations (1.9.22) and (1.9.23), together with $\omega \le 1$ and ρ, which satisfies equation (1.9.15), we get

$$\begin{aligned}
\rho\omega H(\varepsilon,n,\varphi) - K(\varepsilon,n,\varphi) &\le -(2\gamma - 2\rho)\|A^{\frac{1}{2}}\varepsilon\|^2\\
&\quad - \left(2\gamma - \frac{1}{2}\rho\right)\beta\int|\varepsilon|^4 dx - \left(\frac{1}{2}\alpha - \frac{1}{2}\rho\right)\|A^{-\frac{1}{2}}m\|^2\\
&\quad + \frac{1}{2}\rho\|g\|^2 - \frac{1}{2}\rho\|\varepsilon\|^2 + c_3\|\varepsilon\|^6 + c_4\\
&\le \rho\|g\|^2 + \rho\|\varepsilon\|^2 + c_3\|\varepsilon\|^6 + c_4.
\end{aligned}$$ □

It follows that

$$\frac{d}{dt}H(\varepsilon,n,\varphi) + \omega\rho H(\varepsilon,n,\varphi) + \frac{1}{2}\mu\|m\|^2 \le \rho\|g\|^2 + \rho\|\varepsilon\|^2 + c_3\|\varepsilon\|^6 + c_4.$$

By Gronwall inequality, we have

$$H(\varepsilon,n,\varphi) \le H(\varepsilon_0,n_0,\varphi_0)e^{-\rho\omega t} + e^{-\rho\omega t}\int_0^t (\rho\|g\|^2 + \rho\|\varepsilon\|^2 + c_3\|\varepsilon\|^6 + c_4)e^{\rho\omega\tau}\,d\tau, \tag{1.9.24}$$

then Lemma 1.9.1 yields

Proposition 1.9.1. *Let $\varepsilon(t), n(\cdot,t), \varphi(\cdot,t) \in H_0^1(\Omega)$ be the solution of problem* (1.9.1)–(1.9.5). *Suppose equations* (1.9.6)–(1.9.15) *are valid, then we have*

$$H(\varepsilon,n,\varphi) \le H(\varepsilon_0,n_0,\varphi_0)e^{-\rho\omega t} + a_\infty(1-e^{-\rho\omega t}) + e^{-\rho\omega t}c(\|g\|^2,\|\varepsilon_0\|^2), \quad \forall t \ge 0, \tag{1.9.25}$$

where a_∞ depends on $\|g_0\|^2$, $c(\|g\|^2,\|\varepsilon_0\|^2)$, $\|g\|^2$ and $\|\varepsilon_0\|^2$.

By use of Galerkin method and the above a priori estimate, it is easy to prove

Theorem 1.9.1. *Suppose $\varepsilon_0(x), n_0(x), \varphi_0(x) \in H_0^1(\Omega)$, $g(x) \in L^2(\Omega)$, and f satisfies equations* (1.9.6)–(1.9.7). *Then there exists a unique global solution $(\varepsilon(\cdot,t), n(\cdot,t), \varphi(\cdot,t)) \in L^\infty(0,\infty;H_0^1(\Omega)) \cap C(0,\infty;H_0^1(\Omega))$, $n_t \in L^2(0,\infty;L^2(\Omega))$ of problem* (1.9.1)–(1.9.5). *When $t > 0$, the map $(\varepsilon_0,n_0,\varphi_0) \to (\varepsilon(t),n(t),\varphi(t))$ is continuous in $H_0^1 \times H_0^1 \times H_0^1$.*

Let $S(t)$ be the operator semigroup generated by problem (1.9.1)–(1.9.5), that is,

$$S(t)u_0 = S(t)(\varepsilon_0,n_0,\varphi_0) = (\varepsilon(\cdot,t), n(\cdot,t), \varphi(\cdot,t)).$$

By Theorem 1.9.1, we know that $S(t)u_0$ is continuous in $H_0^1 \times H_0^1 \times H_0^1$. Through Lemma 1.9.2, we have

Proposition 1.9.2. *Suppose* (1.9.6) *and* (1.9.7) *are valid, and there exists a constant ρ_1 such that for any $R > 0$, there exists $t_1(R) > 0$ such that for any $t \ge t_1(R)$, whenever $\varepsilon_0, n_0, \varphi_0 \in H_0^1$ and*

$$\|\varepsilon_0\|_{H_0^1}^2 + \|n_0\|_{H_0^1}^2 + \|\varphi_0\|_{H_0^1}^2 \le R^2 \tag{1.9.26}$$

for the initial value $(\varepsilon_0,n_0,\varphi_0)$, the solution $(\varepsilon(\cdot,t), n(\cdot,t), \varphi(\cdot,t))$ of problem (1.9.1)–(1.9.5) *satisfies*

$$\|\varepsilon(\cdot,t)\|_{H_0^1}^2 + \|n(\cdot,t)\|_{H_0^1}^2 + \|\varphi(\cdot,t)\|_{H_0^1}^2 \le \rho_1^2, \tag{1.9.27}$$

that is, there exists a bounded absorbing set for the problem (1.9.1)–(1.9.5) *in H^2.*

Now we consider the bounded absorbing set in H^2.

Proposition 1.9.3. *Assume that* (1.9.6)–(1.9.7) *are valid and there exists a constant $\rho_2 \ge 0$ such that for any $R > 0$, there exists $t_2 = t_2(R) > 0$ such that for any $\varepsilon_0, n_0, \varphi_0 \in D(A)$,*

$$\|\varepsilon_0(x)\|_{H^2(\Omega)}^2 + \|n_0(x)\|_{H^2(\Omega)}^2 + \|\varphi_0(x)\|_{H^2(\Omega)}^2 \le R^2. \tag{1.9.28}$$

Then solutions of problem (1.9.1)–(1.9.5) *satisfy*

$$\|\varepsilon(\cdot,t)\|_{H^2}^2 + \|n(\cdot,t)\|_{H^2}^2 + \|\varphi(\cdot,t)\|_{H^2}^2 \le \rho_2^2, \quad t \ge t_2. \tag{1.9.29}$$

Proof. We have a uniform estimate of solution using H^1 norm. Now for t large enough, we estimate $\|\varepsilon_{xx}\|^2$, $\|n_{xx}\|^2$ and $\|\varphi_{xx}\|^2$. Taking the inner product of(1.9.1) with $A\varepsilon_t + \gamma A\varepsilon$, integrating it and taking real part, we arrive at

$$\operatorname{Re}(\varepsilon_{xx}, A\varepsilon_t + \gamma A\varepsilon) - \operatorname{Re}(n\varepsilon, A\varepsilon_t + \gamma A\varepsilon) - \operatorname{Re}(\beta|\varepsilon|^2 + g, A\varepsilon_t + \gamma A\epsilon) = 0. \tag{1.9.30}$$

Since

$$\begin{aligned}
\operatorname{Re}(\beta|\varepsilon|^2\varepsilon, \varepsilon_{xxt}) &= \frac{\mathrm{d}}{\mathrm{d}t}\operatorname{Re}\int\Big[\beta(|\varepsilon|_x^2)\varepsilon_x + \frac{1}{2}|\varepsilon|^2|\varepsilon_x|^2 + (\operatorname{Re}(\varepsilon\cdot\overline{\varepsilon_x}))^2\Big]\mathrm{d}x \\
&\quad + \operatorname{Re}\int\beta\Big[(|\varepsilon|_x^2)\varepsilon_t - \frac{1}{2}|\varepsilon_x|^2|\varepsilon|_t^2 \\
&\quad + 2\operatorname{Re}(\varepsilon_x,\overline{\varepsilon_t})\operatorname{Re}(\varepsilon,\varepsilon_x) + 2\operatorname{Re}(\varepsilon,\overline{\varepsilon_t})|\varepsilon_x|^2\Big]\mathrm{d}x \\
&= \frac{\mathrm{d}h_1}{\mathrm{d}t} + h_2,
\end{aligned} \tag{1.9.31}$$

where

$$h_1 = \operatorname{Re}\beta\int\Big[(|\varepsilon|_x^2)\overline{\varepsilon_x} + \frac{1}{2}|\varepsilon|^2|\varepsilon_x|^2 + (\operatorname{Re}(\varepsilon\cdot\overline{\varepsilon_{tx}}))^2\Big]\mathrm{d}x \tag{1.9.32}$$

$$h_2 = \beta\int\Big[\operatorname{Re}(|\varepsilon|^2)_x\varepsilon_t - \frac{1}{2}|\varepsilon_x|^2|\varepsilon|_t^2 + 2\operatorname{Re}(\varepsilon_x\overline{\varepsilon_t}) + 2\operatorname{Re}(\varepsilon\overline{\varepsilon_x})|\varepsilon_x|^2\Big]\mathrm{d}x \tag{1.9.33}$$

and

$$(g, A\varepsilon_t) = \frac{\mathrm{d}}{\mathrm{d}t}(g, A\varepsilon),$$

using equation (1.9.30), we obtain

$$\begin{aligned}
&\frac{\mathrm{d}}{\mathrm{d}t}\Big[\frac{1}{2}|A\overline{\varepsilon}|^2 - h_1 + \operatorname{Re}(g, A\varepsilon)\Big] + \gamma|A\varepsilon|^2 - h_2 \\
&\quad + \beta\gamma\operatorname{Re}\int(|\varepsilon|^2\varepsilon)_x\overline{\varepsilon_x}\mathrm{d}x + \operatorname{Re}\gamma(g, A\varepsilon) \\
&\quad + \gamma\operatorname{Re}\int(n\overline{\varepsilon})_x\overline{\varepsilon_x}\mathrm{d}x + \operatorname{Re}(n\varepsilon, A\varepsilon_t) = 0.
\end{aligned} \tag{1.9.34}$$

On the other hand, setting $m = n_t + \rho n$, and taking the inner product of (1.9.3) with m, we have

$$\begin{aligned}
&\frac{\mathrm{d}}{\mathrm{d}t}\Big[\frac{1}{2}\|m\|^2 + \frac{1}{2}\lambda\|An\|^2 + \frac{1}{2}\|A^{\frac{1}{2}}n\|^2 + \frac{1}{2}\int f'(n)|A^{\frac{1}{2}}n|^2\mathrm{d}x\Big] \\
&\quad + (\alpha-\rho)\|m\|^2 + \mu\|A^{\frac{1}{2}}m\|^2 - \rho(\alpha-\rho)(n,m) - \mu\rho(An,m) + \rho\|A^{\frac{1}{2}}n\|^2 + \rho\lambda\|An\|^2
\end{aligned}$$

$$+\int\left[\rho|\varepsilon|_x^2 n_x - \frac{1}{2}f''(n)n_t|n_x|^2 + \rho f'(n)|n_x|^2\right]\mathrm{d}x = -(A|\varepsilon|^2, n_t). \tag{1.9.35}$$

Choosing ρ small enough, such that (1.9.15) is valid, we then get

$$\begin{aligned}&\frac{\mathrm{d}}{\mathrm{d}t}\left[\frac{1}{2}|m|^2 + \frac{1}{2}\lambda\|An\|^2 + \frac{1}{2}\|A^{\frac{1}{2}}n\|^2 + \frac{1}{2}\int f'(n)|A^{\frac{1}{2}}n|^2\mathrm{d}x\right]\\&+(\alpha-\rho)\|m\|^2 + \frac{3}{4}\rho\|A^{\frac{1}{2}}n\|^2 + \rho\lambda\|An\|^2 + \frac{1}{2}\mu\|A^{\frac{1}{2}}m\|^2\\&+\int\left[\rho|\varepsilon|_x^2 n_x - \frac{1}{2}f''(n)n_t|n_x|^2 + \rho f'(n)|n_x|^2\right]\mathrm{d}x \le -(A|\varepsilon|^2, n_t).\end{aligned} \tag{1.9.36}$$

Since

$$\begin{aligned}&(-A|\varepsilon|^2, n_t) - 2\operatorname{Re}(n\varepsilon, A\varepsilon_t)\\&= \frac{\mathrm{d}}{\mathrm{d}t}\operatorname{Re}\int[-n|\varepsilon_x|^2 - n_x\varepsilon\overline{\varepsilon_x}]\mathrm{d}x + 2\operatorname{Re}\int n_x\varepsilon_t\overline{\varepsilon_x}\mathrm{d}x + n_t|\varepsilon_x|^2\mathrm{d}x,\end{aligned} \tag{1.9.37}$$

taking (1.9.34) multiplied by 2, together with (1.9.36), we get

$$\begin{aligned}&\frac{\mathrm{d}}{\mathrm{d}t}\left[\|A\varepsilon\|^2 + \frac{1}{2}\|m\|^2 + \frac{1}{2}\lambda\|An\|^2 - 2\operatorname{Re}(g, A\varepsilon) - 2h_1 + h_3\right]\\&+2\gamma\|\Lambda\varepsilon\|^2 + (\alpha-\rho)\|m\|^2 + \rho\lambda\|An\|^2 + \frac{1}{2}\mu\|A^{\frac{1}{2}}m\|^2\\&-2h_2 + h_1 + 2\operatorname{Re}(g, A\varepsilon) \le 0,\end{aligned} \tag{1.9.38}$$

where h_1 and h_2 are defined by (1.9.32) and (1.9.33), respectively,

$$h_3 = \frac{1}{2}\|A^{\frac{1}{2}}n\|^2 + \frac{1}{2}\int f'(n)|A^{\frac{1}{2}}n|^2\mathrm{d}x + \operatorname{Re}\int[n|\varepsilon_x|^2 + 2n_x\varepsilon\overline{\varepsilon_x}]\mathrm{d}x, \tag{1.9.39}$$

$$\begin{aligned}h_4 &= \frac{3}{4}\rho\|A^{\frac{1}{2}}n\|^2 + 2\beta\gamma\operatorname{Re}\int(|\varepsilon|^2\varepsilon)_x\overline{\varepsilon_x}\mathrm{d}x\\&\quad+2\gamma\operatorname{Re}\int(n\varepsilon)_x\overline{\varepsilon_x}\mathrm{d}x - 2\operatorname{Re}\int n_x\varepsilon_t\overline{\varepsilon_x}\mathrm{d}x + \int n_t|\varepsilon_x|^2\mathrm{d}x.\end{aligned} \tag{1.9.40}$$

Set

$$\widetilde{H}(\varepsilon, n, \varphi) = \|A\varepsilon\|^2 + \frac{1}{2}\|m\|^2 + \frac{1}{2}\lambda\|An\|^2 - 2\operatorname{Re}(g, A\varepsilon) - 2h_1 + h_3, \tag{1.9.41}$$

$$\widetilde{K}(\varepsilon, n, \varphi) = 2\gamma\|A\varepsilon\|^2 + (\alpha-\rho)\|m\|^2 - \rho\lambda\|An\|^2 + 2\operatorname{Re}(g, A\varepsilon) - 2h_2 + h_4. \tag{1.9.42}$$

Then equation (1.9.38) can be converted into

$$\frac{\mathrm{d}}{\mathrm{d}t}\widetilde{H}(\varepsilon, n, \varphi) + \widetilde{K}(\varepsilon, n, \varphi) + \frac{1}{2}\mu\|A^{\frac{1}{2}}m\|^2 \le 0. \tag{1.9.43}$$

Now we estimate $\widetilde{H}(\varepsilon, n, \varphi)$ and $\widetilde{K}(\varepsilon, n, \varphi)$. By Proposition 1.9.2, there exists a constant ρ_1' such that

$$\|\varepsilon(\cdot, t)\|_{H_0^1}^2 + \|n(\cdot, t)\|_{H_0^1}^2 + \|\varphi(\cdot, t)\|_{H_0^1}^2 \le \rho_1'^2.$$

On account of the embedding theorem, there exists a constant ρ_1'', which depends on $\|g\|$, such that

$$\|\varepsilon(\cdot,t)\|_{L^\infty(\Omega)},\ \|n(\cdot,t)\|_{L^\infty(\Omega)},\ \|\varphi(\cdot,t)\|_{L^\infty(\Omega)} \le \rho_1''.$$

By the definition of h_1 and h_3, we have

$$|h_1| \le C\rho_1'' \int |\varepsilon_x|^2 dx \le \rho_1^{(3)}, \quad t \ge T_1(R), \tag{1.9.44}$$

$$|h_3| \le C\rho_1' + C(\rho_1'') \le \rho_1^{(4)}, \quad t \ge T_2(R), \tag{1.9.45}$$

where $\rho_1^{(3)}$ and $\rho_1^{(4)}$ only depend on $\|g\|$. As with h_2 and h_4, we have

$$|h_2| \le C\rho_1'' \int |\varepsilon_x|^2|\varepsilon_t| dx, \quad t \ge T(R), \tag{1.9.46}$$

$$|h_4| \le C\rho_1' + C_2 \int |\varepsilon_x||n_x||\varepsilon_t| dx + C_3 \int |n_t||\varepsilon_x|^2 dx. \tag{1.9.47}$$

Denote $\varepsilon_t = i\Delta\varepsilon + b_1$ for $b_1 = -in\varepsilon - i\beta|\varepsilon|^2 - \gamma\varepsilon - ig \in L^\infty(0,\infty;L^2)$. Substituting ε_t into (1.9.46), we have

$$\begin{aligned}|h_2| &\le C\rho_1'' \int |\varepsilon_x|^2|\Delta\varepsilon| dx + C\rho_1'' \int |b_1||\Delta\varepsilon_x|^2 dx \\ &\le C\|\Delta\varepsilon\|\|\varepsilon_x\|_{L^4}^2 + C\|b_1\|\|\varepsilon_x\|_{L^4}^2 \le C\|\Delta\varepsilon\|^{\frac{3}{2}}\|\varepsilon_x\|^{\frac{3}{2}} + C\|b_1\|\|\Delta\varepsilon\|^{\frac{1}{2}}\|\varepsilon_x\|^{\frac{3}{2}} \\ &\le \theta\|\Delta\varepsilon\|^2 + C(\theta), \quad \forall\theta \ge 0,\end{aligned} \tag{1.9.48}$$

where we use the embedding theorem, and C depends only on $\|g\|$. Similarly, since $m = n_t + \rho n$, we have

$$\begin{aligned}|h_4| &\le C + C\int |\varepsilon_x||n_x|^2|\Delta\varepsilon| dx + C\int |\varepsilon_x||n_x||b_1| dx \\ &\quad + C\int |m||\varepsilon_x|^2 dx + C\int |n||\varepsilon_x|^2 dx \\ &\le C + C\|A\varepsilon\|\|\varepsilon_x\|_{L^4}\|n_x\|_{L^4} + C\|b_1\|\|\varepsilon_x\|_{L^4}\|n_x\|_{L^4} + C\|m\|\|\varepsilon_x\|_{L^4}^2 \\ &\le C + C\|A\varepsilon\|\|A\varepsilon\|^{\frac{1}{4}}\|An\|^{\frac{1}{4}}\|\varepsilon_x\|^{\frac{3}{4}}\|n_x\|^{\frac{3}{4}} \\ &\quad + C|b_1|\|A\varepsilon\|^{\frac{1}{4}}\|An\|^{\frac{1}{4}}\|\varepsilon_x\|^{\frac{3}{4}}\|n_x\|^{\frac{3}{4}} + C\|m\|\|A\varepsilon\|^{\frac{1}{2}}\|\varepsilon_x\|^{\frac{3}{2}} \\ &\le C + \theta_1\|A\varepsilon\|^2 + \theta_2\|m\|^2 + \theta_3\|An\|^2, \quad \forall\theta_1,\theta_2,\theta_3 \ge 0,\end{aligned} \tag{1.9.49}$$

where C depends only on θ_i, $i = 1,2,3$, and $\|g\|$. If we select θ, θ_i, $i = 1,2,3$, small enough, then by the definition of $\widetilde{H}$ and $\widetilde{K}$, there exist constants $C_0 \ge 0$ and $C_1 \ge 0$ such that $C_0\widetilde{H} - \widetilde{K} \le C_1$. From equation (1.9.43), we have

$$\frac{d\widetilde{H}}{dt} + C_0\widetilde{K} + \frac{1}{2}\mu\|Am\|^2 \le C_1, \quad \forall t \ge T_1(R). \tag{1.9.50}$$

By Gronwall inequality, we have

$$\widetilde{H}(\varepsilon,n,\varphi)\le e^{-C_0(t-T_1)}\widetilde{H}(\varepsilon,n,\varphi)(T_1)+C_1(1-e^{-C_0(t-T_1)}). \tag{1.9.51}$$

Since also

$$\widetilde{H}\ge(1-\theta)\|A\varepsilon\|^2+\frac{1}{2}\|m\|^2+\frac{\lambda}{2}\|An\|^2-C,$$

where $0<\theta<1$ and C depends only on $\|g\|$ and ρ_1', this yields the claim. □

Now consider the long time behavior of the semigroup $S(t)u_0$. As a corollary of Proposition 1.9.2, we have

Corollary 1.9.1. *The set*

$$B_1=\{(\varepsilon,n,\varphi)\in V^3=H_0^1\times H_0^1\times H_0^1,\ \|\varepsilon(\cdot,t)\|_{H_0^1}^2+\|n(\cdot,t)\|_{H_0^1}^2+\|\varphi(\cdot,t)\|_{H_0^1}^2\le\rho_1^2\}$$

is the bounded absorbing set of $S(t)$ in V^3, and the set

$$B_2=\{(\varepsilon,n,\varphi)\in D(A),\ \|\varepsilon(\cdot,t)\|_{H^2}^2+\|n(\cdot,t)\|_{H^2}^2+\|\varphi(\cdot,t)\|_{H^2}^2\le\rho_2^2\}$$

is the bounded absorbing set of $S(t)$ in $D(A)$.

Now we consider the ω-limit set of B_1 in V^3 and B_2 in $D(A)$. Set

$$\omega^\omega(B_2)=\bigcap_{s\ge0}\overline{\bigcup_{t\ge s}S(t)B_2}$$

where the closures are taken in weak topology of $D(A)^3$. Let

$$\omega^\omega(B_1)=\bigcap_{s\ge0}\overline{\bigcup_{t\ge s}S(t)B_1}$$

where the closures are in the weak topology of V^3. Based on the related theorems in [197], we have

Theorem 1.9.2. *The set $\mathscr{A}=\omega^\omega(B_2)$ satisfies:*

(1) *$\mathscr{A}$ is weakly compact and is bounded in $D(A)$. Also*

$$S(t)\mathscr{A}=\mathscr{A},\quad \forall t\in\mathbf{R}.$$

(2) *For any bounded set in $D(A)$, it follows that*

$$\lim_{t\to\infty}d_H^\omega(S(t)B,\mathscr{A})=\lim_{t\to\infty}\sup_{x\in B}\inf_{y\in\mathscr{A}}d^\omega(S(t)x,y)=0.$$

(3) *Under the weak topology of $D(A)$, $\mathscr{A}$ is connected.*

Corollary 1.9.2. *For any bounded set B in $D(A)$, the set $S(t)B$ converges to $\mathscr{A}$ in V^3.*

We now investigate the dimension of the invariant set in $D(A)$.

Assume that $\varepsilon_0 = (\xi_0, n_0, \varphi_0) \in V \times V \times V$ is given, and let $S(t)\xi_0 = (\varepsilon(\cdot, t), n(\cdot, t), \varphi(\cdot, t))$ be the semigroup generated by the solution of problem (1.9.1)–(1.9.5). We consider the following variational equations:

$$iU_t + U_{xx} - nU - V\varepsilon - \beta|\varepsilon|^2 U - 2\beta \operatorname{Re}(\bar{\varepsilon}U)\varepsilon + i\gamma U = 0, \tag{1.9.52}$$

$$V_t = W_{xx}, \tag{1.9.53}$$

$$W_t = V + f'(n)V + \mu V - \lambda\Delta V - \alpha W + 2\operatorname{Re}(\bar{\varepsilon}U) \tag{1.9.54}$$

with boundary values

$$U(0) = u_0, \quad V(0) = v_0, \quad W(0) = \omega_0, \tag{1.9.55}$$

$$U(0,t) = U(L,t) = V(0,t) = V(L,t) = W(0,t) = W(L,t) = 0. \tag{1.9.56}$$

It is easy to prove that, when $(u_0, v_0, \omega_0) \in V \times V \times V$, the linear problem (1.9.52)–(1.9.56) possesses a unique solution $\widetilde{U}(t) = (U(t), V(t), W(t)) \in C(\mathbf{R}, V^3)$. By using the energy estimation method, we can prove

Proposition 1.9.4. *The semigroup operator $S(t)\xi_0 = (\varepsilon(t), n(t), \varphi(t))$ is differentiable in $V \times V \times V$, and its differential is the linear mapping $\eta_0 = (u_0, v_0, \omega_0) \in V^3 \to \widetilde{U}(t) = (U(t), V(t), W(t))$, that is, $DS(t)\xi_0 = \widetilde{U}(t)$.*

Set $u = e^{\gamma t}U$, $v = e^{\gamma t}V$, $\omega = e^{\gamma t}W$. Then (1.9.52)–(1.9.54) can be rewritten as

$$iu_t + u_{xx} - v\varepsilon - nv - \beta|\varepsilon|^2 u - 2\beta(\varepsilon u)\varepsilon = 0, \tag{1.9.57}$$

$$v_t = \omega_{xx} + \gamma v, \tag{1.9.58}$$

$$\omega_t = v + f'(n)v - \mu v_t - \mu\gamma v - \lambda\Delta v - (\alpha - \lambda)\omega + 2\operatorname{Re}(\bar{\varepsilon}u). \tag{1.9.59}$$

Multiplying equation (1.9.57) by $\bar{u}$, integrating it over Ω, and taking the real part, we obtain

$$\begin{aligned} &\frac{\mathrm{d}}{\mathrm{d}t}\left[\frac{1}{2}\|u_x\|^2 + \frac{1}{2}\int n|u|^2\mathrm{d}x + \beta\int |\varepsilon|^2|u|^2\mathrm{d}x + \beta\int [\operatorname{Re}(\varepsilon u)]^2\mathrm{d}x + \int v\varepsilon\bar{u}\mathrm{d}x\right] \\ &\quad - \int \frac{1}{2}n_t|u|^2\mathrm{d}x - \beta\int (|\varepsilon|_t^2)|u|^2\mathrm{d}x \\ &\quad - \beta\int \operatorname{Re}(\bar{\varepsilon}u)\operatorname{Re}(\bar{\varepsilon}_t u)\mathrm{d}x - \operatorname{Re}\int [v\varepsilon_t u + \varepsilon u v_t]\mathrm{d}x = 0. \end{aligned} \tag{1.9.60}$$

Multiplying equation (1.9.58) by v_t, equation (1.9.59) by ω_t, integrating them over Ω, and adding together, produces

$$\begin{aligned} &\frac{\mathrm{d}}{\mathrm{d}t}\left[\frac{1}{2}\lambda\|v_x\|^2 + \frac{1}{2}\|\omega_x\|^2 + \int\left(\frac{1}{2}(1 + f'(n))|v|^2 - \gamma(v, \omega)\right)\mathrm{d}x\right] \\ &\quad - \frac{1}{2}\int f''(n)n_t|v|^2\mathrm{d}x + \mu\|v_t\|^2 - (\alpha - 2\gamma)(\omega, v_t) + 2\operatorname{Re}\int v_t\varepsilon u\mathrm{d}x = 0. \end{aligned} \tag{1.9.61}$$

Through (1.9.60) and (1.9.61), we obtain

$$\frac{\mathrm{d}}{\mathrm{d}t}\Big\{\|u_x\|^2 + \frac{1}{2}\lambda\|v_x\|^2 + \frac{1}{2}\|\omega_x\|^2 - \frac{1}{2}\int [n|u|^2 + 2\beta|\varepsilon|^2|u|^2 + 2\beta(\mathrm{Re}(\bar{\varepsilon}u))^2 + 2\,\mathrm{Re}\; v\varepsilon\bar{u}]$$
$$+ \int\Big[\frac{1}{2}(1+f'(n))|v|^2 - \frac{1}{2}\mu\gamma|v|^2 - \gamma(v,\omega)\Big]\mathrm{d}x\Big\}$$
$$+ \int\Big[-n_t|u|^2 - 2\beta(|\varepsilon|^2)_t|u|^2 - 2\beta\,\mathrm{Re}(\bar{\varepsilon}_t u) - 2v\varepsilon_t u - \frac{1}{2}f''(n)n_t|v|^2\Big]\mathrm{d}x$$
$$+ \mu\|v_t\|^2 - (\alpha - 2\gamma)(\omega, v_t) = 0. \tag{1.9.62}$$

Let

$$J(\eta) = J(u,v,\omega) = \frac{1}{2}\|u_x\|^2 + \frac{1}{2}\lambda\|v_x\|^2 + \frac{1}{2}\|\omega_x\|^2 + J_1, \tag{1.9.63}$$

where

$$J_1 = \int\Big\{\frac{1}{2}n|u|^2 + \beta|\varepsilon|^2|u|^2 + \beta(\mathrm{Re}(\bar{\varepsilon}u))^2 + \mathrm{Re}(v\varepsilon\bar{u}) +$$
$$+ \frac{1}{2}(1+f'(n))|v|^2 - \frac{1}{2}\mu\gamma|v|^2 - \gamma(v,\omega)\Big\}\mathrm{d}x. \tag{1.9.64}$$

Setting

$$I(\eta) = I(u,v,\omega) = -\mu\|v_t\|^2 + (\alpha - 2\lambda)(\omega, v_t) + I_1, \tag{1.9.65}$$

where

$$I_1(\eta) = \int\Big\{n_t|u|^2 + 2\beta(|\varepsilon|^2)_t|u|^2 + 2\beta\,\mathrm{Re}(\bar{\varepsilon}u)\,\mathrm{Re}(\bar{\varepsilon_t}u) + 2\,\mathrm{Re}(v\varepsilon_t\bar{u}) + \frac{1}{2}f''(n)n_t|v|^2\Big\}\mathrm{d}x, \tag{1.9.66}$$

equation (1.9.62) becomes

$$\frac{\mathrm{d}}{\mathrm{d}t}J(\eta) = I(\eta). \tag{1.9.67}$$

On the other hand,

$$\frac{\mathrm{d}}{\mathrm{d}t}\|u\|^2 = 2\,\mathrm{Im}\int \varepsilon\bar{u}v\mathrm{d}x + 2\beta\int \mathrm{Re}(\varepsilon u)\,\mathrm{Im}(\varepsilon\bar{u})\mathrm{d}x \tag{1.9.68}$$

$$\frac{\mathrm{d}}{\mathrm{d}t}\Big(\frac{1}{2}\lambda\|v\|^2 + \frac{1}{2}\|\omega\|^2\Big) = -\mu\|\omega_x|^2 + \lambda y\|v\|^2$$
$$- (\alpha - \gamma)\|\omega\|^2 - 2\,\mathrm{Re}(\varepsilon\bar{u}, \omega) + \int (I + f'(n))v\omega\mathrm{d}x. \tag{1.9.69}$$

Letting

$$J_\sigma(\eta) = J_\sigma(u,v,\omega) = J(\eta) + \sigma\|u\|^2 + \sigma(\lambda\|v\|^2 + \|\omega\|^2), \tag{1.9.70}$$

$$I_\sigma(\eta) = I_\sigma(u, v, \omega) = I(\eta) - \sigma I_2 - 2\sigma\mu\|\sigma\omega\|^2, \tag{1.9.71}$$

where

$$\begin{aligned} I_2 &= 2((1 + f'(n))v, \omega) + 2\lambda\gamma\|v\|^2 - 2(\alpha - \lambda)\|\omega\|^2 \\ &\quad - 4\operatorname{Re}(\varepsilon\bar{u}, \omega) + 2\operatorname{Im}\int \varepsilon\bar{u}v\mathrm{d}x + 2\beta\int \operatorname{Re}(\varepsilon\bar{u})\operatorname{Im}(\bar{\varepsilon}u)\mathrm{d}x, \end{aligned} \tag{1.9.72}$$

we get the energy equality

$$\frac{\mathrm{d}J_\sigma(\eta)}{\mathrm{d}t} = I_\sigma(\eta). \tag{1.9.73}$$

Assume that X is a bounded invariant set of $S(t)$ in $D(A)^3$, that is,

$$S(t)X = X, \quad \forall t \geq 0.$$

Also X is bounded in $D(A)^3$, $\xi_0 \in X$, $S(t)\xi_0 = (\varepsilon(t), n(t), \varphi(t)) \in X$ and

$$|X|_\infty = \sup_{\xi_0 \in X}\big[|\varepsilon(t)|_\infty + |n(t)|_\infty + |\varphi(t)|_\infty\big] < \infty,$$

where $|\cdot|_\infty = \sup_t \|\cdot\|_{L^\infty(\Omega)}$. By the definition of J_σ in equation (1.9.71), we know that there exist constants $\sigma \geq 0, C_0, C_1 \geq 0$ such that

$$\begin{aligned} C_0\big(\|u\|_V^2 + \|v\|_V^2 + \|\omega\|_V^2\big) &\leq J_\sigma(u, v, \omega) \\ &\leq C_1\big(\|u\|_V^2 + \|v\|_V^2 + \|\omega\|_V^2\big), \quad \forall (u, v, \omega) \in V \times V \times V. \end{aligned} \tag{1.9.74}$$

Thus $J_\sigma(u, v, \omega)$ is an equivalent norm in $V \times V \times V$. Now we introduce **R**-linear inner product in V^3. Suppose that $\eta = (\eta_1, \eta_2, \eta_3), \xi = (\xi_1, \xi_2, \xi_3) \in V \times V \times V$. Define

$$\begin{aligned} \Psi(\eta, \xi) &= \operatorname{Re}\int\Big[\eta_{1x}\overline{\xi_{1x}} + \frac{1}{2}\lambda\eta_{2x}\overline{\xi_{2x}} + \frac{1}{2}\eta_{3x}\overline{\xi_{3x}} + \frac{1}{2}n\eta_1\overline{\xi_1} \\ &\quad + \beta\|\varepsilon\|^2\eta_1\overline{\xi_1} + \beta\operatorname{Re}(\bar{\varepsilon}\eta_1)\operatorname{Re}(\varepsilon\xi_1) + \frac{1}{2}(1 + f'(n))\eta_2\overline{\xi_2} - \frac{1}{2}\mu\gamma\eta_2\overline{\xi_2}\Big]\mathrm{d}x \\ &\quad + \frac{1}{2}\int \operatorname{Re}(\eta_2\varepsilon\xi_2 + \eta_1\varepsilon\xi_1)\mathrm{d}x - \frac{1}{2}\gamma\int(\eta_2\xi_3 + \eta_3\xi_2)\mathrm{d}x \\ &\quad + \sigma\operatorname{Re}\int \eta_1\overline{\xi_1}\mathrm{d}x + \sigma\int \operatorname{Re}[\lambda\eta_2\overline{\xi_2} + \eta_3\overline{\xi_3}]\mathrm{d}x. \end{aligned} \tag{1.9.75}$$

It is easy to verify that $\Psi(\eta, \xi)$ is bilinear in $V^3 \times V^3$, and $\Psi(\eta, \eta) = J_\sigma(\eta_1, \eta_2, \eta_3)$. By equation (1.9.74), we know that it is equivalent. Hence $\Psi(\eta, \xi)^{\frac{1}{2}}$ is an equivalent norm in $V \times V \times V$.

Suppose that $\xi_0^j = (u_{0j}, v_{0j}, \omega_{0j})$, $j = 1, 2, \ldots, m$, are m elements in $V \times V \times V$. Assume $\xi^j(t)$, $j = 1, 2, \ldots, m$, are solutions of problem (1.9.52)–(1.9.56), that is, $\xi^j(t) = (u^j(t), v^j(t), \omega^j(t)) = (DS(t)\xi_0)\xi_0^j$. Let $\eta^i(t) = e^{\gamma_t}\xi^i(t)$, $i = 1, 2, \ldots, m$. Then

$\eta^i(t) = (u_i(t), v_i(t), \omega_i(t))$ is the solution of (1.9.57)–(1.9.59) with the initial value $\eta^i(0) = (u_i(0), v_i(0), \omega_i(0)) = \xi_0^i$. Now we consider the variation of the volume:

$$|\xi^1(t) \wedge \xi^2(t) \wedge \cdots \wedge \xi^m(t)|_{\wedge^m(V^3)} = \det_{1\le i,j\le m} (\xi^i, \xi^j)_{V^3} \tag{1.9.76}$$

where $(\cdot,\cdot)_{V^3}$ is the inner product of $\psi(\cdot,\cdot)$ determined by equation (1.9.75). Thus we have the following theorem:

Theorem 1.9.3. *Assume that X is a bounded invariant set in $D(A)^3$, then there exist constants $C_1, C_2 \ge 0$ such that for any $\xi_0 = (\varepsilon_0, n_0, \varphi_0) \in X$, $\xi^i(t) = (DS(t)\xi_1)\xi_0^i$ satisfies*

$$\begin{aligned} &|\xi^1(t) \wedge \xi^2(t) \wedge \cdots \wedge \xi^m(t)|_{\wedge^m(V^3)} \\ &\quad \le |\xi_0^1(t) \wedge \xi_0^2(t) \wedge \cdots \wedge \xi_0^m(t)|_{\wedge^m(V^3)} \cdot C_1^m e^{(C_2-m\gamma)t} \end{aligned} \tag{1.9.77}$$

where constants C_1, C_2 depend on coefficients $\alpha, \beta, \gamma, \mu$ and λ. Also $m \ge 1, t \ge 0$.

Proof. First, notice that

$$\begin{aligned} &|\xi^1(t) \wedge \xi^2(t) \wedge \cdots \wedge \xi^m(t)|^2_{\wedge^m(V^3)} \\ &\quad = e^{-2\gamma mt} |\eta^1(t) \wedge \eta^2(t) \wedge \cdots \wedge \eta^m(t)|_{\wedge^m(V^3)} \\ &\quad = e^{-2\gamma mt} \det_{1\le i,j\le m} \Psi(\eta^i(t), \eta^j(t)). \end{aligned} \tag{1.9.78}$$

So we need to estimate $\det_{1\le i,j\le m} \Psi(\eta^i(t), \eta^j(t))$. Let

$$\begin{aligned} H_m(t) &= \det_{1\le i,j\le m} \Psi(\eta^i(t), \eta^j(t)), \\ \frac{\mathrm{d}}{\mathrm{d}t} H_m(t) &= \frac{\mathrm{d}}{\mathrm{d}t} \det_{1\le i,j\le m} \Psi(\eta^i(t), \eta^j(t)) - \sum_{l=1}^m \det_{1\le i,j\le m} \Psi(\eta^i(t), \eta^j(t))_l, \end{aligned} \tag{1.9.79}$$

where

$$\Psi(\eta^i(t), \eta^j(t))_l = (1 - \delta_{jl})\Psi(\eta^i, \eta^j) + \delta_{jl} \frac{\mathrm{d}}{\mathrm{d}t} \Psi(\eta^i(t), \eta^j(t)). \tag{1.9.80}$$

The mapping $\Psi(\cdot,\cdot)$ is an $\mathbf{R}$-linear inner product in V^3. Hence

$$\Psi(\eta^i(t), \eta^j(t)) = \frac{1}{4}[\Psi(\eta^i + \eta^j, \eta^i + \eta^j) - \Psi(\eta^i - \eta^j, \eta^i - \eta^j)],$$

and then, by equation (1.9.75), we have

$$\begin{aligned} \frac{\mathrm{d}}{\mathrm{d}t} \Psi(\eta^i(t), \eta^j(t)) &= \frac{1}{4} \frac{\mathrm{d}}{\mathrm{d}t} J_\sigma(\eta^i(t) + \eta^j(t)) \\ &\quad - \frac{1}{4} \frac{\mathrm{d}}{\mathrm{d}t} J_\sigma(\eta^i(t) - \eta^j(t)) = \frac{1}{4}[I_\sigma(\eta^i + \eta^j) - I_\sigma(\eta^i - \eta^j)]. \end{aligned}$$

By Lemma 1.3.2 in [74], from (1.9.79)–(1.9.80), we have

$$\frac{dH_m(t)}{dt} = H_m(t)\sum_{l=1}^{m} \max_{F\subset \mathbf{R}^m, \dim F=l} \min_{x\neq 0, x\in F} \frac{I_\sigma(\sum_{j=1}^m x_j\eta^j(t))}{J_\sigma(\sum_{l=1}^m x_j\eta^j(t))}. \tag{1.9.81}$$

Using the definition in (1.9.71), we obtain

$$\begin{aligned} I_\sigma\Big(\sum x_j\eta^j(t)\Big) &= I_\sigma\Big(\sum x_j u_j, \sum x_j v_j, \sum x_j\omega_j\Big) \\ &= -\mu\Big\|\sum x_j v_{jt}\Big\|^2 + (\alpha - 2\gamma)\Big(\sum_{j=1}^m x_j\omega_j, \sum x_j v_{jt}\Big) \\ &\quad - I_1\Big(\sum x_j\eta^j(t)\Big) - \mu\sigma\Big\|\sum x_j\omega_{jt}(t)\Big\|^2 + \sigma I_2\Big(\sum x_j\eta^j(t)\Big) \\ &\leq C\Big\|\sum x_j\omega_j(t)\Big\|^2 + \Big|I_1\Big(\sum x_j\eta^j(t)\Big)\Big| + \sigma\Big|I_2\Big(\sum x_j\eta^j(t)\Big)\Big|. \end{aligned}$$

On account of equations (1.9.74) and (1.9.72), there exists $C_3 \geq 0$ such that

$$I_\sigma\Big(\sum x_j\eta^j(t)\Big) \leq C_3\Big\|\sum x_j\eta^j(t)\Big\|_{V^3}^{\frac{1}{2}}\Big\|\sum x_j\eta^j(t)\Big\|^{\frac{3}{2}}. \tag{1.9.82}$$

On the other hand, by (1.9.74), we have

$$I_\sigma\Big(\sum x_j\eta^j(t)\Big) \geq C_0\Big\|\sum x_j\eta^j(t)\Big\|_{V^3}^2. \tag{1.9.83}$$

By (1.9.81), (1.9.82) and (1.9.83), we have

$$\begin{aligned} \frac{d}{dt}H_m(t) &\leq \frac{C_3}{C_0}H_m(t)\sum_{l=1}^m \max_{F\subset\mathbf{R}^m, \dim F=t} \min_{x\neq 0, x\in F} \frac{\|\sum x_j\eta^j(t)\|^{\frac{3}{2}}}{\|\sum x_j\eta^j(t)\|_{V^3}^{\frac{3}{2}}} \\ &\leq \frac{C_3}{C_0}H_m(t) \max_{F\subset\mathbf{R}^m, \dim F=t} \min_{x\neq 0, x\in F}\left(\frac{\|\sum x_j u_j\|}{\|\sum x_j u_j\|_V} + \frac{\|\sum x_j v_j\|}{\|\sum x_j v_j\|_V} + \frac{\|\sum x_j\omega_j\|}{\sum x_j\omega_j\|_V}\right)^{\frac{3}{2}} \\ &\leq \frac{C_3}{C_0}H_m(t)\sum_{l=1}^m\left(\frac{3}{\sqrt{\lambda_l}}\right)^{\frac{3}{2}} \\ &\leq \frac{\sqrt{27}}{C_0}C_3\sum_{l=1}^m\frac{1}{\lambda_l^{\frac{3}{4}}}H_m(t), \end{aligned} \tag{1.9.84}$$

where we use the maximum and minimum principles, λ_l is the lth eigenvalue of $A = -\partial_{xx}$. We know that $\lambda_l \sim Cl^2$ as $l \to \infty$. Hence, there exists constant a C_2 such that

$$\frac{dH_m(t)}{dt} \leq C_2 H_m(t). \tag{1.9.85}$$

Hence, using $H_m(t) \leq e^{2C_2 t}H_m(0)$, $t \geq 0$, together with equation (1.9.78), we get equation (1.9.77). □

Using Theorem 1.9.3, we have

Theorem 1.9.4. *The global attractor $\mathscr{A}$ determined by Theorem 1.9.2 possesses bounded Hausdorff and fractal dimensions.*

Proof. If we introduce the inner product $\Psi(\cdot,\cdot)$ in V^3, then by Theorem 1.9.3, we get

$$\omega(DS(t)\xi_0) \leq C^m \exp(C_2 - \gamma m)t, \quad \forall \xi_0 \in \mathscr{A},$$

where ω_m is the Lyapunov index. When m is large enough, $\overline{\omega_m}(\mathscr{A}) < 1$, here $\overline{\omega}(\mathscr{A})$ is the uniform Lyapunov index of $\mathscr{A}$. Using Theorem V.31 of [197], we know that $\mathscr{A}$ possesses bounded Hausdorff and fractal dimensions. Since the strongness of equivalent, the attractor has the same Hausdorff and fractal dimension. Theorem 1.9.4 is proved. □

1.10 A new method to prove existence of a strong topology attractor

In 1995, Guo and Wu [124] studied the Benjamin–Ono (BO) global attractor in an unbounded region with the linear dissipation equation, proving the existence of a strongly compact attractor on $H^1(\mathbf{R})$. An important method which can make a weakly compact attractor become a strong one was proposed, i. e., for a class of nonlinear evolution equations with a generalized energy conservation integral, they proved that the weak convergence of a solution sequence is actually equivalent to the convergence in norm; moreover, the existence of a strong compact attractor of BO equation was proved. This is consistent with the idea that Ball and Ghidaglia [27] used to prove the existence of a strongly compact attractor of KdV equation, and it was found almost at the same time. By now this method has received much attention and is widely used, for instance, for the two-dimensional Davey–Stewartson equations some very good results have been obtained.

Now we consider the following Cauchy problem for Benjamin–Ono equation (BO) with weak dissipation equation [98]:

$$u_t + Hu_{xx} + (u^2)_x + C_0 u = f, \quad x \in \mathbf{R}, t > 0, \tag{1.10.1}$$

$$u(x,0) = u_0(x), \quad x \in \mathbf{R}, \tag{1.10.2}$$

where $C_0 > 0$ is a constant, $C_0 u$ means the zeroth order dissipation effect, f is the external force, which is independent of t, and H is Hilbert transform, that is,

$$H(\phi) = \frac{1}{\pi} P \int_{-\infty}^{\infty} \frac{\phi(y)}{y-x} dy,$$

where P denotes Cauchy integral principal value. If $C_0 = 0$, equation (1.10.1) is the well-known Benjamin–Ono equation, which can be used to describe the propagation of inner wave of long wavelength in the stream physically, it has a soliton solution and lots of important features, see [147]. For the BO problem and its generalization, the existence and uniqueness of a global smooth solution has been established, see, e. g., [215, 143].

In order to prove the existence of a weakly compact attractor for the BO problem, we must now establish a uniform a priori estimate. First, for the Hilbert transform and for any $f, g \in L^2(\mathbf{R})$, we have:

$$H^2 f = -f, \quad H(fg) = H(HfHg) + fHg + gHf, \tag{1.10.3}$$

$$(f, Hg) = -(Hf, g), \quad (Hf, f) = 0, \tag{1.10.4}$$

$$(Hf, Hg) = (f, g), \quad \|Hf\| = \|f\|, \tag{1.10.5}$$

$$Hf_x = (Hf)_x, \quad \forall f \in H^1(\mathbf{R}). \tag{1.10.6}$$

Multiplying (1.10.1) with u, and integrating by parts over $\mathbf{R}$, we get

$$\frac{1}{2}\frac{\mathrm{d}}{\mathrm{d}t}\|u\|^2 + C_0\|u\|^2 = (f, u). \tag{1.10.7}$$

Moreover, we obtain

$$\|u(t)\|^2 \le e^{-C_0 t}\|u_0\|^2 - \frac{1}{C_0^2}(1 - e^{-C_0 t})\|f\|^2. \tag{1.10.8}$$

Hence, for $t \ge t_0 = \frac{2}{C_0}\ln\frac{C_0\|u_0\|}{\|f\|}$, we have

$$\|u(t)\|^2 \le \frac{2}{C_0^2} = A_1. \tag{1.10.9}$$

In order to get an estimate of the H^1 norm, we multiply (1.10.1) by u_{xx}, and integrate by parts, and then get

$$\frac{1}{2}\frac{\mathrm{d}}{\mathrm{d}t}\|u_x\|^2 + C_0\|u_x\|^2 = (f_x, u_x) + \int (u^2)_x u_{xx}\mathrm{d}x. \tag{1.10.10}$$

Multiplying (1.10.1) by u^3 and integrating over $\mathbf{R}$ with respect to x, we have

$$\frac{1}{4}\frac{\mathrm{d}}{\mathrm{d}t}\|u\|_{L^4}^4 + C_0\|u\|_{L^4}^4 + \int u^3 Hu_{xx}\mathrm{d}x + \int (u^2)_x u^3\mathrm{d}x = (f, u^3). \tag{1.10.11}$$

By (1.10.4), the above equation can be reduced to

$$\frac{1}{4}\frac{\mathrm{d}}{\mathrm{d}t}\|u\|_{L^4}^4 + C_0\|u\|_{L^4}^4 = 3\int u^2 u_x Hu_x\mathrm{d}x + (f, u^3). \tag{1.10.12}$$

Furthermore,

$$\begin{aligned}\frac{\mathrm{d}}{\mathrm{d}t}\int u^2 Hu_x\mathrm{d}x &= 2\int uu_t Hu_x\mathrm{d}x + u^2 Hu_{xx}\mathrm{d}x \\ &= -2\int uHu_x[Hu_{xx} + C_0 u + (u^2)_x - f]\mathrm{d}x - \int Hu^2 u_x\mathrm{d}x\end{aligned}$$

$$\begin{aligned}&= 2\int fuHu_x\mathrm{d}x - 2C_0\int u^2Hu_x\mathrm{d}x - 2\int u(u^2)_xHu_x\mathrm{d}x - 2\int uHu_xHu_{xx}\mathrm{d}x\\&\quad + \int Hu^2[Hu_{xx} + (u^2)_{xx} + C_0u_x - f_x]\mathrm{d}x\\&= 2\int fuHu_x\mathrm{d}x - 3C_0\int u^2Hu_x\mathrm{d}x - 4\int u^2Hu_x\mathrm{d}x + \int (u)_x^2u_{xx}\mathrm{d}x\\&\quad + \int u_x(Hu_x)^2\mathrm{d}x + \int u^2Hf_x\mathrm{d}x.\end{aligned} \tag{1.10.13}$$

Noting that

$$H(u_xHu_x) = H(Hu_xH^2u_x) - u_xH^2u_x + (Hu_x)^2, \tag{1.10.14}$$

we get

$$(Hu_x)^2 = u_x^2 + 2H(u_xHu_x). \tag{1.10.15}$$

Hence

$$\begin{aligned}\int u_x(Hu_x)^2\mathrm{d}x &= 2\int u_xH(u_xHu_x)\mathrm{d}x + \int u_x^3\mathrm{d}x\\&= -2\int u_x(Hu_x)^2\mathrm{d}x - 2\int uu_xu_{xx}\mathrm{d}x,\end{aligned} \tag{1.10.16}$$

through which we arrive at

$$\int u_x(Hu_x)^2\mathrm{d}x = -\frac{1}{3}\int (u^2)_xu_{xx}\mathrm{d}x, \tag{1.10.17}$$

$$\begin{aligned}\frac{\mathrm{d}}{\mathrm{d}t}\int u^2Hu_x\mathrm{d}x &= 2\int fuHu_x\mathrm{d}x - 3C_0\int u^2Hu_x\mathrm{d}x\\&\quad - 4\int u^2u_xHu_x\mathrm{d}x + \int u^2Hf_x\mathrm{d}x + \frac{2}{3}\int (u^2)_xu_{xx}\mathrm{d}x.\end{aligned} \tag{1.10.18}$$

Combining with equations (1.10.9), (1.10.10) and (1.10.18), we have

$$\begin{aligned}&\frac{\mathrm{d}}{\mathrm{d}t}\left[\|u_x\|^2 - 3\int u^2Hu_x\mathrm{d}x - \|u\|_{L^4}^4 + 2C_0\|u_x\|^2 - 4C_0\|u\|_{L^4}^4\right]\\&= 2(f_x, u_x) - 6\int fuHu_x\mathrm{d}x + 9C_0\int u^2Hu_x\mathrm{d}x - 3\int u^2Hf_x\mathrm{d}x - 4(f, u^3).\end{aligned} \tag{1.10.19}$$

Using Gagliardo–Nirenberg inequality, we have

$$\|u\|_{L^4}^4 \le C\|u\|^{\frac{3}{4}}\|u_x\|^{\frac{1}{4}}, \quad \|u\|_{L^6} \le C\|u\|^{\frac{2}{3}}\|u_x\|^{\frac{1}{3}}, \quad \forall u \in H^1(\mathbf{R}). \tag{1.10.20}$$

Hence

$$\int u^2Hu_x\mathrm{d}x \le \|u\|_{L^4}^2\|u_x\| \le \|u\|^{\frac{3}{2}}\|u_x\|^{\frac{3}{2}}, \tag{1.10.21}$$

$$\int fuHu_x\mathrm{d}x \le \|f\|_{L^4}\|u\|_{L^4}\|u_x\| \le C\|f\|_{H^1}\|u\|^{\frac{1}{4}}\|u_x\|^{\frac{5}{4}}, \tag{1.10.22}$$

$$\int u^2 H f_x \mathrm{d}x \le C\|f_x\|\|u\|^{\frac{3}{2}}\|u_x\|^{\frac{1}{2}}. \tag{1.10.23}$$

Therefore

$$\begin{aligned}
&\frac{\mathrm{d}}{\mathrm{d}t}\left[\|u_x\|^2 - 3\int u^2 H u_x \mathrm{d}x - \|u\|_{L^4}^4\right] + 2C_0\left[\|u_x\|^2 - 3\int u^2 H u_x \mathrm{d}x - \|u\|_{L^4}^4\right]\\
&= 2(f_x, u_x) + 2C_0\|u\|_{L^4}^4 + 3C_0\int u^2 H u_x \mathrm{d}x - 3\int u^2 H f_x \mathrm{d}x - 6\int f u H u_x \mathrm{d}x - 4(f, u^3)\\
&\le C[\|f_x\|^2 + \|u\|^6 + \|f_x\|^{\frac{4}{3}}\|u\|^2 + \|f\|^{\frac{8}{3}}\|u\|^2 + \|f\|^2\|u\|^2],
\end{aligned} \tag{1.10.24}$$

which yields

$$\begin{aligned}
&\|u_x\|^2 - 3\int u^2 H u_x \mathrm{d}x - \|u\|_{L^4}^4\\
&\le \left(\|u_{0x}\|^2 - 3\int u_0^2 H u_{0x} \mathrm{d}x - \|u_0\|_{L^4}^4\right)e^{-C_0 t}\\
&\quad + C\int_0^t [\|f_x\|^2 + \|u\|^6 + \|f_x\|^{\frac{4}{3}}\|u\|^2 + \|f\|^{\frac{8}{3}}\|u\|^2 + \|f\|^2\|u\|^4]e^{-C_0(t-s)}\mathrm{d}s.
\end{aligned} \tag{1.10.25}$$

For $t \le t_0$, we have

$$\begin{aligned}
&\|u_x\|^2 - 3\int u^2 H u_x \mathrm{d}x - \|u\|_{L^4}^4\\
&\le \left(\|u_{0x}\|^2 - 3\int u_0^2 H u_{0x} \mathrm{d}x - \|u_0\|_{L^4}^4\right)\\
&\quad + C\int_0^{t_0} [\|f_x\|^2 + \|u\|^6 + \|f_x\|^{\frac{4}{3}}\|u\|^2 + \|f\|^{\frac{8}{3}}\|u\|^2 + \|f\|^2\|u\|^4]e^{-C_0(t-s)}\mathrm{d}s\\
&\le C(t_0, \|u_0\|_{H^1}, \|f\|_{H^1}).
\end{aligned} \tag{1.10.26}$$

For $t \ge t_0$, we have

$$\begin{aligned}
&\|u_x\|^2 - 3\int u^2 H u_x \mathrm{d}x - \|u\|_{L^4}^4\\
&\le \left(\|u_x(t_0)\|^2 - 3\int u^2 H u_x(t_0) \mathrm{d}x - \|u(t_0)\|_{L^4}^4\right)e^{-C_0(t-t_0)}\\
&\quad + C\left[\|f_x\|^2 + \frac{2}{C_0^2}\|f\|^2 + \frac{2}{C_0^2}\|f\|^2\|f_x\|^{\frac{1}{3}} + \frac{2}{C_0^2}\|f\|^{\frac{14}{3}} + \frac{4}{C_0^2}\|f\|^4\right]\\
&\le C\|f\|_{H^1}^6 + \left(\|u_x(t_0)\|^2 - 3\int u^2 H u_x(t_0) \mathrm{d}x - \|u(t_0)\|_{L^4}^4\right)e^{-C_0(t-t_0)}.
\end{aligned} \tag{1.10.27}$$

Therefore there exists $t_1 > t_0$ such that

$$\|u_x(t)\|^2 \le C\|f\|_{H^1}^6. \tag{1.10.28}$$

Thus

$$\|u_x(t)\|^2 \le C\|f\|_{H^1}^6 + 3\int u^2 Hu_x dx + \|u\|_{L^4}^4$$
$$\le C\|f\|_{H^1}^6 + C[\|u\|^{\frac{3}{2}}\|u_x\|^{\frac{3}{2}} + \|u\|^3\|u_x\|] \le C\|f\|_{H^1}^6 + \frac{1}{2}\|u_x\|^2 + C\|u\|^6.$$

When $t > t_1$, we obtain that

$$\|u_x\|^2 \le C\|f\|_{H^1}^6 = A_2. \tag{1.10.29}$$

Setting

$$B = \{\phi(x) \mid \phi(x) \in H^1,\ \|\phi\| \le A_1,\ \|\phi_x\| \le A_2\},$$

it is easy to see that B is the absorbing set of the BO problem. Define the ω limit set by

$$\omega^{\omega}(B) = \bigcap_{s>0}\overline{\bigcup_{t\ge s} S(t)B^{\omega}} \tag{1.10.30}$$

where the closure is taken with respect to the weak topology of $H^1(\mathbf{R})$. Since B is a bounded set in $H^1(\mathbf{R})$, it is weakly compact in $H^1(\mathbf{R})$. Since $H^1(\mathbf{R})$ is a separable topological space, we know that B is metrizable. Define d^{ω} to be its distance. Under the effect of semigroup $S(t)$, a bounded set A attracts a bounded set B relative to the measure d^{ω} if and only if

$$\lim_{t\to\infty} d_{H^1}^{\omega}(S(t)B, A) = \lim_{t\to\infty} \sup_{x\in B} \inf_{y\in A} d^{\omega}(S(t)x, y). \tag{1.10.31}$$

Following the idea in [197], we have

Proposition 1.10.1. *The set $\omega^{\omega}(B)$ is the nonempty weakly compact attractor of the BO problem.*

Now we prove that this global weakly compact attractor $\omega^{\omega}(B)$ is also a global strongly compact attractor in H^1.

Since $\omega^{\omega}(B)$ is invariant under the action of the semigroup $S(t)$, we prove that for any bounded set $A \in H^1(\mathbf{R})$, we have

$$\lim_{t\to\infty} d_{H^1}(S(t)A, \omega^{\omega}(B)) = 0. \tag{1.10.32}$$

To do so, we only need to prove that if $\forall t_n \to \infty$, $\phi_n \in A$ such that

$$S(t_n)\phi_n \to \overline{u}(x) \quad \text{weakly in } H^1(\mathbf{R}) \tag{1.10.33}$$

where $\overline{u}(x) \in \omega^{\omega}(B)$, then

$$S(t_n)\phi_n \to \overline{u}(x) \quad \text{strongly in } H^1(\mathbf{R}). \tag{1.10.34}$$

In fact, $\forall T > 0$, $\exists n_j \to \infty$, $v(x) \in \omega^{\omega}(B)$ such that for $t_{n_j} > T$ we have

$$v_{n_j}(x) = S(t_{n_j} - T)\phi_{n_j} \to v(x) \quad \text{weakly in } H^1(\mathbf{R}).$$

Set $W_{n_j} = S(t_{n_j} + t - T)\phi_{n_j}$. Since $S(t) : H^1(\mathbf{R}) \to H^1(\mathbf{R})$ is bounded, we have

$$\begin{aligned} &S(t) : H^1(\mathbf{R}) \to L^2(\mathbf{R}) \quad \text{is continuous,} \\ &S(t) : H^1(\mathbf{R}) \to H^1(\mathbf{R}) \quad \text{is weakly continuous.} \end{aligned}$$

Hence, we have

$$W_{n_j}(t) \to S(t)v \quad \text{weakly in } H^1(\mathbf{R}), \quad S(T)v = \overline{u}.$$

Since

$$\frac{1}{2}\frac{\mathrm{d}}{\mathrm{d}t}\|W_{nj}(t)\|^2 + C_0\|W_{n_j}(t)\|^2 = (f, W_{nj}(t)),$$

we have

$$\|W_{n_j}(t)\|^2 = \|v_{n_j}(t)\|^2 e^{-2C_0 t} + \frac{1}{2C_0}\int_0^t (f, W_{n_j}(s))e^{-2C_0(t-s)}\mathrm{d}s,$$

$$\|W_{n_j}(t)\|^2 = e^{-2C_0 t}\|v_{n_j}(t)\|^2 + \frac{1}{2C_0}\int_0^t (f, W_{n_j}(s))e^{-2C_0(T-s)}\mathrm{d}s.$$

Since $\phi_{n_j} \in B$, where B is the absorbing set in $H^1(\mathbf{R})$, we can find $C > 0$ such that

$$(f, W_{n_j}(s)e^{-2C_0(t-s)} + C) > 0.$$

Therefore

$$\begin{aligned} \limsup_{n_j\to\infty}\|W_{n_j}(T)\|^2 + \frac{CT}{2C_0} &\le e^{-2C_0 T}\limsup_{n_j\to\infty}\|v_{n_j}\|^2 \\ &\quad + \limsup_{n_j\to\infty}\frac{1}{2C_0}\int_0^T [(f, W_{n_j}(s))e^{-2C_0(T-s)} + C]\mathrm{d}s. \end{aligned} \tag{1.10.35}$$

Noting that

$$W_{n_j}(t) \to S(t)v \quad \text{weakly in } H^1(\mathbf{R}), \quad \forall t > 0,$$

we have

$$(f, W_{n_j}(s)) \to (f, S(t)v), \quad \forall t > 0.$$

Together with Fatou lemma, this implies

$$\limsup_{n_j\to\infty}\int_0^T[(f,W_{n_j}(s))e^{-2C_0(T-s)}+C]ds \le \int_0^T[(f,S(s)v)e^{-2C_0(T-s)}+C]ds \tag{1.10.36}$$

and

$$\limsup_{n_j\to\infty}\|W_{n_j}(T)\|^2+\frac{CT}{2C_0} \le e^{2C_0T}\limsup_{n_j\to\infty}\|v_{n_j}\|^2+\frac{1}{2C_0}\int_0^T(f,S(s)v)e^{-2C_0(T-s)}ds+\frac{CT}{2C_0}. \tag{1.10.37}$$

On the other hand, since $\overline{u}=S(T)v$, we have

$$\|\overline{u}\|^2=e^{-2C_0T}\|v\|^2+\frac{1}{2C_0}\int_0^T(f,S(s)v)e^{-2C_0(T-s)}ds. \tag{1.10.38}$$

Hence

$$\limsup_{n_j\to\infty}\|W_{n_j}(T)\|^2\le\|\overline{u}\|^2+e^{-2C_0T}\left[\limsup_{n_j\to\infty}\|v_{n_j}\|^2-\|v\|^2\right]. \tag{1.10.39}$$

Noting that B is the bounded absorbing set in $H^1(\mathbf{R})$, for $t_j>T$, we have $\{u_{n_j}\subset A\}$, $\|u_{n_j}\|^2\le C$, $\|v_{n_j}\|^2\le C$, where C is independent of T and n_j. So

$$\limsup_{n_j\to\infty}\|W_{n_j}(T)\|^2\le\|\overline{u}\|^2+2Ce^{-2C_0T}. \tag{1.10.40}$$

On account of $W_{n_j}(T)=S(t_{n_j})\phi_{n_j}$, we have

$$\limsup_{n_j\to\infty}\|S_{n_j}\phi_{n_j}\|^2\le\|\overline{u}\|^2-2Ce^{-2C_0T}.$$

Letting $T\to\infty$, we get

$$\limsup_{n_j\to\infty}\|S(t_{n_j})\phi_{n_j}\|^2\le\|\overline{u}\|^2. \tag{1.10.41}$$

On the other hand, since $S_{n_j}\phi_{n_j}\to\overline{u}$ weakly in $H^1(\mathbf{R})$,

$$\limsup_{n_j\to\infty}\|S(t_{n_j})\phi_{n_j}\|^2\ge\|\overline{u}\|^2. \tag{1.10.42}$$

From equations (1.10.41) and (1.10.42), we have

$$\limsup_{n_j\to\infty} \|S(t_{n_j})\phi_{n_j}\|^2 = \|\overline{\mu}\|^2. \tag{1.10.43}$$

Together with the fact that $S(t_{n_j})\phi_{n_j} \to \overline{u}$ (weakly), we get

$$S(t_{n_j})\phi_{n_j} \to \overline{u} \quad \text{strongly in } L^2(\mathbf{R}).$$

Since for any subsequences t_{n_j} and ϕ_{n_j} the above limit is unique, we obtain

$$S(t_{n_j})\phi_n \to \overline{u} \quad \text{strongly in } L^2(\mathbf{R}).$$

Noting that, if $u_n \to u$ weakly in $H^1(\mathbf{R})$, and $u_n \to u$ strongly in $L^2(\mathbf{R})$, then $u_n \to u$ strongly in $L^p(\mathbf{R})$ for any $p > 2$.

From equation (1.10.24), we get

$$\begin{aligned}
&\|u_x(t)\|^2 - 3\int n^2 Hu_x \mathrm{d}x - \|u(t)\|_{L^4}^4 \\
&= e^{-2C_0 t}\left[\|u_{0x}\|^2 - 3\int u_0^2 Hu_{0x}\mathrm{d}x - \|u_0\|_{L^4}^4\right] \\
&+ \frac{1}{2C_0}\left[\int_0^t e^{-2C_0(t-s)}[2(f_x, u_x(s)) + 2C_0\|u(s)\|_{L^4}^4]\right. \\
&\left. + 3C_0\int u^2 Hu_x \mathrm{d}x - 3\int u^2 Hf_x \mathrm{d}x - 4(f, u^3)\right]\mathrm{d}s.
\end{aligned} \tag{1.10.44}$$

Similar to the proof of the L^2-convergence, we can prove

$$S(t_n)\phi_n \to \overline{u} \quad \text{strongly in } H^1(\mathbf{R}).$$

Hence, we have

Theorem 1.10.1. *Assume that $u_0 \in H^1, f \in H^1$, and $C_0 > 0$ is a constant, then the BO problem has a unique strongly compact global attractor in $H^1(\mathbf{R})$.*

We can also prove that the semigroup $S(t)$ is continuous in H^1. From the above proof, we know that on the global attractor, the weak H^1-convergence induces the strong H^1-convergence. Thus we get

Theorem 1.10.2. *The semigroup operator $S(t)$ is H^1-continuous on the global attractor.*

From the dimension estimation of the global attractor, we get

Theorem 1.10.3. *Suppose the assumption of Theorem 1.10.1 is valid. Then the attractor has bounded fractal dimension.*

1.11 Nonlinear KdV–Schrödinger equation

We consider the following periodic initial problem of nonlinear KdV–Schrödinger equation with weak damping

$$i\varepsilon_t + \varepsilon_{xx} - bn\varepsilon + i\gamma\varepsilon = g_1(x), \tag{1.11.1}$$

$$n_t + \frac{1}{2}f(n)_x + \frac{\beta}{2}n_{xxx} + \nu n + \frac{1}{2}|\varepsilon|_x^2 = g_2(x), \tag{1.11.2}$$

$$\varepsilon(x,t) = \varepsilon(x+L,t), \quad n(x,t) = n(x+L,t), \quad \forall x \in \mathbf{R}, t \geq 0, \tag{1.11.3}$$

$$\varepsilon(x,0) = \varepsilon_0(x), \quad n(x,0) = n_0(x), \tag{1.11.4}$$

where $\varepsilon(x,t) = (\varepsilon^1(x,t), \varepsilon^2(x,t), \dots, \varepsilon^N(x,t))$ is an unknown complex vector function; $n(x,t)$ is a real function; $f(n)$ is a nonlinear function; b, γ, β, ν are constants; $L > 0$. The problem (1.11.1)–(1.11.4) appears in the studies of laser and plasma physics, including electric field ε, where n denotes a density disturbance. When there is no damping, $\gamma = \nu = 0$, $g_1 = g_2 = 0$, Appert, Vaclavik, Makhankov, Gibbons, among others, [5, 170, 78] did lots of explicit studies on the soliton solution of equations (1.11.1)–(1.11.4). In 1983, Guo [84] first proved the existence and uniqueness of a global solution for the periodic initial value problem (1.11.1)–(1.11.4). In 1997, Guo and Chen [93] proved the existence and boundedness of dimension of the global attractor for the problem (1.11.1)–(1.11.4). The other related problems on the well-posedness can refer to the reference [111, 112].

In the following, suppose that $\nu > \gamma > 0$, $b\beta < 0$ and $f(s) \in C^{(4)}(\mathbf{R})$ is such that there exist A_0 and $B > 0$ so that

$$|f(n)| \leq A_0 n^2 + B, \quad \forall n \in \mathbf{R}. \tag{1.11.5}$$

Suppose that m is a non-negative integer, L is a positive real number, H_L^m means the Sobolev space H^m with period L. Also denote

$$IH_L^m = H_L^m \times H_L^m \times \cdots \times H_L^m,$$

$$\|v\|_m^2 = \sum_{k=0}^{m} L^{2k}|v|_k^2, \quad |v|_k^2 = \int_0^L |\frac{\mathrm{d}^k v}{\mathrm{d}x^k}|^2 \mathrm{d}x,$$

$$\|\varepsilon\|_m = \left[\sum_{j=1}^{N} \|\varepsilon_j\|_m^2\right]^{\frac{1}{2}}, \quad \varepsilon = (\varepsilon^1, \dots, \varepsilon^N) \in IH_L^m.$$

Letting

$$X_k = IH_L^k \times H_L^k, \quad k = 1, 2, \dots, m,$$

$$\|\eta\|_k = (\|\eta_1\|_k^2 + \|\eta_2\|_k^2)^{\frac{1}{2}},$$

$$\eta = (\eta_1, \eta_2) \in X_k, \quad k = 1, 2, \dots, m,$$

we obtain the Sobolev inequality

$$|v|_{L^\infty} = \sup_{0\le x\le L} |v(x)| \le |v|_0^{\frac{1}{2}}\left(2|v|_1 + \frac{1}{L}|v|_0\right)^{\frac{1}{2}}, \quad \forall v \in H_L^1. \tag{1.11.6}$$

Following the method in [84], we have

Theorem 1.11.1. *Suppose $f(n)$ satisfies condition* (1.11.5). *Let $b\beta < 0$, $v > \gamma > 0$, $g = (g_1, g_2) \in X_m$, $(\varepsilon_0(x), n_0(x)) \in X_m$, $m \ge 2$. Then for any T, $0 < T < \infty$, there exists a global unique solution $(\varepsilon(x,t), n(x,t)) \in C(0,T;X_m)$ for the problem* (1.11.1)–(1.11.4), *and it is continuous with respect to the initial value.*

The semigroup $S(t)\zeta_0 = S(t)(\varepsilon_0, n_0) = (\varepsilon(t), n(t))$ is generated by the solution of problem (1.11.1)–(1.11.4), where $S(t)$ is defined on $\mathbf{R}^+$, and is continuous in t. But from the bound on the solution, we know that for any $R > 0$, $0 < T < \infty$, there exists a constant $C_0(R,T)$ such that

$$\sup_{t\in(0,T]} \|S(t)(\varepsilon_0, n_0)\|_2 \le C_0(R,T), \quad \text{provided } \|(\varepsilon_0, n_0)\|_2 \le R. \tag{1.11.7}$$

Lemma 1.11.1. *For $t \in \mathbf{R}$, the map $S(t)$ is continuous with respect to X_1 on the bounded set X_2.*

Proof. Assume that $(\varepsilon_{10}, n_{10}), (\varepsilon_{20}, n_{20}) \in X_2$, $\|\varepsilon_{k0}\|_2 + \|n_{k0}\|_2 \le R$, $k = 1,2$, and $(\varepsilon_1(t), n_1(t))$, $(\varepsilon_2(t), n_2(t))$ stand for the corresponding solutions $(\varepsilon_k(t), n_k(t)) = S(t)(\varepsilon_{k0}, n_{k0})$, $k = 1,2$, respectively. Let $\varepsilon(t) = \varepsilon_1(t) - \varepsilon_2(t)$, $n(t) = n_1(t) - n_2(t)$. Then $(\varepsilon(t), n(t))$ satisfies

$$i\varepsilon_t + \varepsilon_{xx} - b(n_1\varepsilon_1 - n_2\varepsilon_2) + i\gamma\varepsilon = 0, \tag{1.11.8}$$

$$n_t + \frac{1}{2}(f(n_1) - f(n_2))_x + \frac{\beta}{2}n_{xxx} + vn + \frac{1}{2}(|\varepsilon_1|_x^2 - |\varepsilon_2|_x^2) = 0. \tag{1.11.9}$$

Multiplying equation (1.11.8) by $\bar{\varepsilon}$ and integrating over $(0,L)$, we get

$$(i\varepsilon_t, \varepsilon)t + (\varepsilon_{xx}, \varepsilon) - b\int_0^L (n_1|\varepsilon|^2 + n_1\varepsilon_2\bar{\varepsilon})dx + i\gamma|\varepsilon|_0^2 = 0. \tag{1.11.10}$$

Taking the imaginary part of the above equation, we obtain

$$\frac{1}{2}\frac{d}{dt}|\varepsilon|_0^2 + \gamma|\varepsilon|_0^2 - b\,\mathrm{Im}\int_0^L \varepsilon_2 n\bar{\varepsilon}dx = 0. \tag{1.11.11}$$

On the other hand, taking the real part of equation (1.11.10), we get

$$\mathrm{Im}(\varepsilon_t, \varepsilon) + |\varepsilon_x|_0^2 + b\int_0^t n_1|\varepsilon|^2 dx + b\,\mathrm{Re}\int_0^L n\varepsilon\bar{\varepsilon}dx = 0. \tag{1.11.12}$$

Multiplying equation (1.11.8) by $\overline{\varepsilon}$, integrating over $(0, L)$, and taking its real part, we get

$$\frac{\mathrm{d}}{\mathrm{d}t}|\varepsilon_x|_0^2 + \operatorname{Re} b(n_1\varepsilon + \varepsilon_2 n, \varepsilon_t) + \gamma \operatorname{Im}(\varepsilon, \varepsilon_t) = 0. \tag{1.11.13}$$

Combining with equations (1.11.12) and (1.11.13), we have

$$\frac{1}{2}\frac{\mathrm{d}}{\mathrm{d}t}|\varepsilon_x|_0^2 + \gamma|\varepsilon_x|_0^2 + \gamma b \int_0^L n_1|\varepsilon|^2 \mathrm{d}x + \gamma b \operatorname{Re}\int_0^L n\varepsilon_2\overline{\varepsilon}\mathrm{d}x + \operatorname{Re} b(n_1\varepsilon + \varepsilon_2 n, \varepsilon) = 0. \tag{1.11.14}$$

Multiplying equation (1.11.9) by n and integrating over $[0, L]$, we get

$$\frac{1}{2}\frac{\mathrm{d}}{\mathrm{d}t}|n|_0^2 + \nu|n|_0^2 - \frac{1}{2}\int_0^1 f'(\tau n + (1-\tau)n_2)\mathrm{d}\tau \cdot nn_x - \frac{1}{2}(\varepsilon_1\overline{\varepsilon} + \overline{\varepsilon_2}\varepsilon, n_x) = 0. \tag{1.11.15}$$

Multiplying equation (1.11.9) by n_{xx} and integrating over $[0, L]$, we arrive at

$$\frac{\mathrm{d}}{\mathrm{d}t}\left\{\frac{1}{2}|n_x|_0^2 - \frac{1}{\beta}(\varepsilon_1\overline{\varepsilon} + \overline{\varepsilon_2}\varepsilon, n)\right\} + \phi(\varepsilon, n) = 0, \tag{1.11.16}$$

where

$$\begin{aligned}
\phi(\varepsilon, n) = &-\frac{3}{4}\int_0^L\left(\int_0^1 f'(\tau n_1 + (1-\tau)n_2)\mathrm{d}\tau\right)_x n_x^2\mathrm{d}x \\
&-\frac{1}{2}\int_0^L\left(\int_0^1 f'(\tau n_1 + (1-\tau)n_2)\mathrm{d}\tau\right)_{xx} nn_x\mathrm{d}x \\
&+\frac{1}{\beta}(\varepsilon_1\overline{\varepsilon} + \overline{\varepsilon_2}\varepsilon_t, n) + \frac{1}{\beta}(\varepsilon_{1t}\varepsilon + \overline{\varepsilon_{2t}}\varepsilon, n) \\
&-\frac{1}{\beta}\left(\varepsilon_1\overline{\varepsilon} + \overline{\varepsilon_2}\varepsilon, \frac{1}{2}\left(\int_0^1 f'(\tau n_1 + (1-\tau)n_2)\mathrm{d}\tau \cdot n\right)_x + \nu n\right).
\end{aligned} \tag{1.11.17}$$

Together with equation (1.11.16) and (1.11.11), (1.11.14), (1.11.15), we have

$$\begin{aligned}
&\frac{\mathrm{d}}{\mathrm{d}t}\left\{\frac{1}{2}|n_x|_0^2 + \frac{1}{2}|\varepsilon_x|_0^2 + \sigma|n|_0^2 + \sigma|\varepsilon|_0^2 - \frac{1}{\beta}(\varepsilon_1\overline{\varepsilon} + \overline{\varepsilon_2}\varepsilon, n)\right\} \\
&\quad + \nu|\varepsilon_x|_0^2 + \gamma b\int_0^L n_1|\varepsilon|^2\mathrm{d}x + \gamma b\operatorname{Re}\int_0^L n\varepsilon_2\overline{\varepsilon}\mathrm{d}x \\
&\quad + \operatorname{Re}\, b(n_1\varepsilon + \varepsilon_2 n, \varepsilon_t) + \sigma\gamma|\varepsilon|_0^2 - b\sigma\operatorname{Im}\int_0^L \varepsilon_2 n\overline{\varepsilon}\mathrm{d}x \\
&\quad + \sigma\nu|n|_0^2 - \frac{\sigma}{2}\left(\int_0^1 f'(\tau n_1 + (1-\tau)n_2)\mathrm{d}\tau \cdot n, n_x\right) \\
&\quad - \frac{\sigma}{2}(\varepsilon_1\overline{\varepsilon} + \overline{\varepsilon_2}\varepsilon, n_x) + \phi(\varepsilon, n) = 0,
\end{aligned} \tag{1.11.18}$$

where $\sigma \geq \frac{2C}{\beta}$, $|\varepsilon_k(t)|_{L^\infty} \leq C(R,T)$, $|n_k(t)|_{L^\infty} \leq C(R,T)$, $k = 1,2$. Using equation (1.11.1) and integrating by parts, from equation (1.11.18) we have

$$\frac{\mathrm{d}}{\mathrm{d}t}\left\{\frac{1}{2}|n_x|_0^2 + \frac{1}{2}|\varepsilon_x|_0^2 + \sigma|n|_0^2 + \sigma|\varepsilon|_0^2 - \frac{1}{\beta}(\varepsilon_1\bar{\varepsilon} + \overline{\varepsilon_2}\varepsilon, n)\right\}$$
$$\leq C_1\left\{\frac{1}{2}|n_x|_0^2 + \frac{1}{2}|\varepsilon_x|_0^2 + \sigma|n|_0^2 + \sigma|\varepsilon|_0^2 - \frac{1}{\beta}(\varepsilon_1\bar{\varepsilon} + \overline{\varepsilon_2}\varepsilon, n)\right\}.$$

Hence, we get

$$\frac{1}{2}|n_x|_0^2 + \frac{1}{2}|\varepsilon_x|_0^2 + \sigma|n|_0^2 + \sigma|\varepsilon|_0^2 - \frac{1}{\beta}(\varepsilon_1\bar{\varepsilon} + \overline{\varepsilon_2}\varepsilon, n) \leq C_2(\|(n(0), \varepsilon(0))\|_1^2)e^{C_1 t},$$

proving the lemma. □

Proposition 1.11.1. *For any $t \in \mathbf{R}^+$, the semigroup operator $S(t)$ determined by problem* (1.11.1)–(1.11.4) *is weakly continuous from X_2 to X_2.*

Proof. Let ζ^n be a weakly convergent sequence in X_2, and let ξ be its limit. For a fixed $t \in \mathbf{R}^+$, from equation (1.11.7) we can infer that $\{S(t)\zeta^n\}$ is bounded in X_2. Since it is weakly compact, we can extract a subsequence $\{S(t)u^n\}$, which weakly converges to $v \in X_2$. On the other hand, by the compactness of X_2 embedded into X_1, $\{u^n\}$ strongly converges to u in X_1. In virtue of Lemma 1.11.1, $\{S(t)u^n\}$ strongly converges to $S(t)u$, $v = S(t)u$. Hence, the sequence $\{S(t)u^n\}$ weakly converges to $S(t)u$. The proposition has been proved. □

In the following we prove the existence of an absorbing set in X_1 and X_2.

Proposition 1.11.2. *Assume that $v > \gamma > 0$, $b\beta < 0$ and let $f(n)$ satisfy equation* (1.11.5), *where $g = (g_1, g_2) \in X_1$. Then there exists a constant $\rho_1 = \rho_1(L, v, \gamma, \|g\|_1)$ such that for any $R > 0$, there is $T_1 = T_1(R) > 0$ such that for every (ε_0, n_0) in X_1, $\|(\varepsilon_0, n_0)\|_1^2 \leq R^2$, we have*

$$\|S(t)(\varepsilon_0, n_0)\|_1^2 \leq \rho_1^2, \quad \forall t \geq T_1(R). \tag{1.11.19}$$

To prove Proposition 1.11.2, we need the following two lemmas.

Lemma 1.11.2. *Assume that $\varepsilon_0(x) \in L^2(\Omega)$, $\Omega = [0, L]$. Then for any solution $(\varepsilon(t), n(t))$ of problem* (1.11.1)–(1.11.4), *we have*

$$\|\varepsilon(t)\|_0^2 \leq |\varepsilon|_0^2 e^{-\gamma t} + \frac{|g_1|_0^2}{\gamma^2}(1 - e^{-\gamma t}). \tag{1.11.20}$$

Proof. Taking the inner product of equation (1.11.4) with $\bar{\varepsilon}$, and then its imaginary part, we obtain the result. □

Lemma 1.11.3. *Assume that $\nu > \gamma$, $(\varepsilon_0, n_0) \in X_1$. Then for any solution $(\varepsilon(t), n(t))$ of problem* (1.11.1)–(1.11.4), *there exist constants $C_1 = C_1(|n_0|_0^2, |\varepsilon_0|_1^2)$, $C_2 = C_2(\|g_1\|_1^2, |g_2|_0)$ such that*

$$\|n\|_0^2 \le |b|^{-1}|\varepsilon|_0|\varepsilon_x|_0 + C_1 e^{-\gamma t} + C_2(1 - e^{-2\gamma t}). \tag{1.11.21}$$

Proof. From equations (1.11.1) and (1.11.2), we deduce that

$$\frac{1}{2}\frac{\mathrm{d}}{\mathrm{d}t}\int_0^L [|b|n^2 + \operatorname{Sign} b \operatorname{Im}(\varepsilon\bar{\varepsilon}_x)]\mathrm{d}x + \nu|b||n|_0^2$$
$$+\gamma\int_0^L \operatorname{Sign} b \operatorname{Im}(\varepsilon\bar{\varepsilon}_x)\mathrm{d}x - |b|(g_2, n) + \operatorname{Sign} b(g_1 x, \varepsilon) = 0. \tag{1.11.22}$$

From equation (1.11.22), when $\nu > \gamma$ we know that

$$\frac{\mathrm{d}}{\mathrm{d}t}\int_0^L [|b|n^2 + \operatorname{Sign} b \operatorname{Im}(\varepsilon\bar{\varepsilon}_x)]\mathrm{d}x$$
$$+ 2\gamma\int_0^L [|b|n^2 + \operatorname{Sign} b \operatorname{Im}(\varepsilon\bar{\varepsilon}_x)]\mathrm{d}x \le \frac{|b||g_2|_0^2}{2(\nu-\gamma)} + 2|g_{1x}|_0|\varepsilon|_0.$$

Using Gronwall inequality and Lemma 1.11.2, we deduce that

$$\|n_0\|_0^2 \le |b|^{-1}|\varepsilon|_0|\varepsilon_x|_0 + C_1 e^{-\gamma t} + C_2(1 - e^{-\gamma t}),$$

where

$$C_1 = |n_0|_0^2 + |b|^{-1}|\varepsilon|_0|\varepsilon_{0x}|_0 + \gamma^{-1}|\varepsilon|_0^2,$$
$$C_2 = |b|^{-1}\left(\frac{|g_{1x}|_0^2}{2\gamma} + \frac{|g|_0^2}{2\gamma^3} + \frac{|b||g_2|_0^2}{4\gamma(\nu-\gamma)}\right). \qquad \square$$

Using Lemmas 1.11.2 and 1.11.3, we now prove Proposition 1.11.2. Taking the inner product of (1.11.2) with n, we have

$$\frac{1}{2}\frac{\mathrm{d}}{\mathrm{d}t}|n|_0^2 + \frac{1}{2}(|\varepsilon|_x^2, n) + \nu|n|_0^2 + (g_2, n) = 0. \tag{1.11.23}$$

On the other hand, by direct calculation we deduce from equation (1.11.1) that

$$\frac{\mathrm{d}}{\mathrm{d}t}\left\{|\varepsilon_x|_0^2 + b\int_0^L n|\varepsilon|^2\mathrm{d}x + 2\operatorname{Re}(g_1, \varepsilon) - \frac{b\beta}{2}|n_x|_0^2 + b\int_0^L F(n)\mathrm{d}x\right\}$$
$$+ 2\gamma\left[|\varepsilon_x|_0^2 + b\int_0^L n|\varepsilon|^2\mathrm{d}x + 2\operatorname{Re}(g_1, \varepsilon)\right] - \nu b\beta|n_x|_0^2$$
$$+ \nu b\int_0^L f(n)n\mathrm{d}x + b\nu\int_0^L n|\varepsilon|^2\mathrm{d}x - b(g_2, f(n) - n_x - |\varepsilon|^2) = 0, \tag{1.11.24}$$

where $F(n) = \int_0^n f(s)\mathrm{d}s$. Combining equations (1.11.22) and (1.11.2), we get

$$\frac{\mathrm{d}}{\mathrm{d}t}H_1(\varepsilon, n) + K_1(\varepsilon, n) = 0, \tag{1.11.25}$$

where

$$H_1(\varepsilon, n) = |\varepsilon_x|_0^2 + b\int_0^L n|\varepsilon|^2\mathrm{d}x + 2\,\mathrm{Re}(g_1, \varepsilon) - \frac{b\beta}{2}|n_x|_0^2 + b\int_0^L F(n)\mathrm{d}x + \lambda|n|_0^2, \tag{1.11.26}$$

$$\begin{aligned} K_1(\varepsilon, n) = {} & 2\gamma\left[|\varepsilon_x|_0^2 + b\int_0^L n|\varepsilon|^2\mathrm{d}x + \mathrm{Re}(g_1, \varepsilon)\right] \\ & - \nu b\beta|n_x|_0^2 + \nu b\int_0^L f(n)n\mathrm{d}x + b\beta\int_0^L n|\varepsilon|^2\mathrm{d}x \\ & - b(g_2, f(n) - n_x - |\varepsilon|^2) + 2\nu\lambda|n|_0^2 + \lambda\int_0^L n|\varepsilon|_x^2\mathrm{d}x - \lambda(g_2, n), \end{aligned} \tag{1.11.27}$$

and λ is an unknown positive constant.

According to Lemmas 1.11.2 and 1.11.3, there exists a constant $\rho_0 = \rho_0(\|g_1\|_2^2, \|g_2\|_0^2)$ such that for any solution $(\varepsilon(t), n(t))$ of the problem (1.11.1)–(1.11.4) with the initial value $(\varepsilon_0, n_0) \in X_1$, $\|(\varepsilon_0, n_0)\|_1^2 \le R^2$, there exists $T_0 = T_0(R) > 0$ such that

$$|\varepsilon|_0^2 \le \frac{2}{\gamma^2}|g_1|_0^2, \quad t \ge T_0(R), \tag{1.11.28}$$

$$|n|_0^2 \le |b|^{-1}|\varepsilon_0||\varepsilon_x|_0 + \rho_0, \quad t \ge T_0(R). \tag{1.11.29}$$

Hence, when $t \le T_0(R)$, we have

$$\begin{aligned} \int_0^L |n|^3\mathrm{d}x & \le |n|_\infty|n|_0^2 \le \left(2|n|_1 + \frac{1}{L}|n|_0\right)|n|_0^{\frac{5}{2}} \\ & \le \left(2|n|_1 + \frac{1}{L}|n|_0\right)^{\frac{1}{2}}\left(\frac{\sqrt{2}|g_1|_0}{\gamma|b|}|\varepsilon|_1 + \rho_0\right)^{\frac{5}{4}} \\ & \le \theta_1\|\varepsilon_x\|_0^2 + \theta_2|n_x|_0^2 + \frac{\theta_2}{2L}|n|_0^2 + C(\theta_1, \theta_2, |g|_1^2), \end{aligned} \tag{1.11.30}$$

where

$$C(\theta_1, \theta_2, \|g_1\|_1^2) = \frac{\gamma^2|b|^2\rho_0^2}{2|g_1|_0^2}\theta_1 + \frac{5^5|g_1|_0^{10}}{2^8\gamma^{10}|b|^{10}\theta_1^5\theta_2^2}, \quad \theta_1, \theta_2 > 0.$$

Hence, by assumption (1.11.5) on $f(n)$, we have

$$|b|\left|\int_0^L F(n)\mathrm{d}x\right| \le \frac{|b|A_0}{3}\int_0^L |n|^3\mathrm{d}x + |b|B\int_0^L |n|\mathrm{d}x$$
$$\le \frac{1}{4}|\varepsilon_x|_0^2 + \frac{|b\beta|}{8}|n_x|_0^2 + \left(\frac{|b\beta|}{16L} + \frac{|b|}{2}\right)|n|_0^2 + C(L,\nu,\gamma,\|g_1\|_1). \tag{1.11.31}$$

On the other hand,

$$\left|b\int_0^L n|\varepsilon|^2\mathrm{d}x\right| \le \frac{|b|}{2}|n|_0^2 + \frac{1}{4}|\varepsilon|_0^2 + C(|g_1|_0^2), \quad t \ge T_0(R). \tag{1.11.32}$$

Hence, when $t \ge T_0(R)$, there exists a constant $C_3(\|g\|_1^2)$ such that

$$H_1(\varepsilon,n) \ge \frac{1}{2}|\varepsilon_x|_0^2 + \frac{|b\beta|}{4}|n_x|_0^2 + \left(\lambda - \frac{|b\beta|}{16L} - |b|\right)|n|_0^2 - C_3(\|g\|_1^2), \tag{1.11.33}$$

$$H_1(\varepsilon,n) \le \frac{3}{2}|\varepsilon_x|_0^2 + \frac{3|b\beta|}{4}|n_x|_0^2 + \left(\lambda + \frac{|b\beta|}{16L} + |b|\right)|n|_0^2 + C_3(\|g\|_1^2). \tag{1.11.34}$$

As to $K_1(\varepsilon,n)$, we have the following lemma:

Lemma 1.11.4. *Suppose the conditions of Proposition* 1.11.1 *are satisfied. Then there exists a constant* $C_4 = C_1(\|g_1\|)$ *such that*

$$K_1(\varepsilon,n) - \gamma H_1(\varepsilon,n) \ge \frac{1}{2}\left(\gamma\lambda - (2\nu+\gamma)|b| - \frac{|b\gamma|\gamma}{2} - 2|b||g|_{L^\infty}A_0\right)|n|_0^2 - C_4. \tag{1.11.35}$$

Proof. First,

$$\begin{aligned}K_1(\varepsilon,n) - \gamma H_1(\varepsilon,n) &= \gamma|\varepsilon_x|_0^2 + (2\nu-\gamma)|b\beta||n_x|_0^2\\ &\quad + (2\nu-\gamma)\lambda|n|_0^2 + (\gamma+\nu)b\int_0^L n|\varepsilon|^2\mathrm{d}x\\ &\quad + \nu b\int_0^L [f(n)n - F(n)]\mathrm{d}x - b(g_2, f(n) - n_x - |\varepsilon|^2 + \lambda n)\\ &\quad + \lambda\int_0^L n|\varepsilon|_x^2\mathrm{d}x.\end{aligned} \tag{1.11.36}$$

Similar to equations (1.11.30) and (1.11.32), when $t \ge T_0(R)$, we have

$$(\gamma+\nu)b\int_0^L n|\varepsilon|^2\mathrm{d}x \le \frac{(\nu+\gamma)|b|}{2}|n_0^2 + \frac{\gamma}{4}|\varepsilon_x|_0^2 + C(|g_1|_0^2) \tag{1.11.37}$$

$$\left|\nu b\int_0^L [f(n)n - F(n)]\mathrm{d}x\right| \le |\nu b|\int_0^L \frac{4A_0}{3}|n|^3 + B_0|n|\mathrm{d}x$$
$$\le \frac{\gamma}{4}|\varepsilon_x|_0^2 + \frac{|b\beta|\gamma}{4}|n_x|_0^2 + \left(\frac{|b\beta|\gamma}{8L} + \frac{|\nu b|}{2}\right)|n|_0^2$$
$$+ C(\|g\|_1^2), \tag{1.11.38}$$
$$|b(g_2, f(n) - n_x - |\varepsilon|^2 + \lambda n)| \le \frac{|b\beta|\gamma}{4}|n_x|_0^2 + \frac{\lambda\gamma}{4}|n|_0^2$$
$$+ |b||g_2|_\infty A_0|n|_0^2 + C(\|g\|_0^2). \tag{1.11.39}$$

For $\lambda\int_0^L n|\varepsilon|_x^2\mathrm{d}x$ we have

$$\left|\lambda\int_0^L n|\varepsilon|_x^2\mathrm{d}x\right| = \left|\lambda\int_0^L n_x|\varepsilon|^2\mathrm{d}x\right| \le \lambda|n_x|_0\|\varepsilon\|_{L^4}^2$$
$$\le \frac{|b\beta|\gamma}{4}|\varepsilon_x|_0^2 + \frac{\lambda\gamma}{4}|n|_0^2 - C(\|g_1\|_0^2). \tag{1.11.40}$$

From equations (1.11.36)–(1.11.40) we know that there exists a constant $C_4 = C_4(\|g\|_1)$ such that equation (1.11.35) is valid. □

Now if we choose λ large enough, namely

$$\lambda > \max\left\{\frac{|b\beta|}{16L} + |b|, \left(2\nu + \gamma + \frac{|b\beta|\gamma}{4} - 2|g_2|_\infty|b|\gamma^{-1}\right)\right\}, \tag{1.11.41}$$

then from equations (1.11.25) and (1.11.35) we get

$$\frac{\mathrm{d}}{\mathrm{d}t}H_1(\varepsilon, n) + \gamma H_1(\varepsilon, n) \le C_4, \quad t \ge T_0(R). \tag{1.11.42}$$

Hence, by Gronwall inequality, we have

$$H_1(\varepsilon, n) \le H_1(\varepsilon, n)(T_0)e^{-\gamma(t-T_0)} + \frac{C_4}{\gamma}(1 - e^{-\gamma(t-T_0)}), \quad t \ge T_0. \tag{1.11.43}$$

From equations (1.11.33), (1.11.34) and (1.11.43) we can deduce the claim of Theorem 1.11.2.

Proposition 1.11.3. *Assume that* $\nu > \gamma > 0$, $b\beta < 0$, f *satisfies equation* (1.11.5) *and* $g = (g_1, g_2) \in X$. *Then there exists a constant* $\rho_2 = \rho_2(L, \nu, \gamma, \|g\|_2)$ *such that for any* $R > 0$, *there is* $T_2(R) > 0$ *such that for any* $(\varepsilon_0, n_0) \in X$, $\|(\varepsilon_0, n_0)\|_2^2 \le R^2$, *we have*

$$\|S(t)(\varepsilon_0, n_0)\|_2^2 \le \rho_2^2, \quad \forall t \ge T_2(R). \tag{1.11.44}$$

In other words, in the closed ball in X_2,

$$B_2 = \{\xi = (\xi_1, \xi_2) \in X \mid \|\xi_1\|_2^2 + \|\xi_2\|_2^2 \le \rho_2^2\} \tag{1.11.45}$$

is the bounded set of the semigroup $S(t)$.

Proof. By Proposition 1.11.2, we only need to estimate the uniform upper bound of $|\varepsilon_{xx}|_0$ and $|n_{xx}|_0$. Denote $A = -\partial_{xx}$. Taking the inner product of (1.11.1) with $A\varepsilon_t + \gamma A\varepsilon$, and then its real part, we obtain

$$\frac{\mathrm{d}}{\mathrm{d}t}\left[\frac{1}{2}|A\varepsilon|_0^2 + \operatorname{Re}(g_1, A\varepsilon)\right] + b\gamma \operatorname{Re}\int_0^L (n\varepsilon)_x\overline{\varepsilon_x}\mathrm{d}x + \gamma|A\varepsilon|_0^2$$
$$+ \operatorname{Re}(g_1, A\varepsilon) + b\operatorname{Re}(A\varepsilon, A\varepsilon_t) = 0. \tag{1.11.46}$$

On the other hand, by equation (1.11.2) we get

$$\frac{\mathrm{d}}{\mathrm{d}t}\left(3\beta|An|_0^2 - 5\int_0^L f'(n)n_x^2\mathrm{d}x\right) = h_1 - 6(|\varepsilon|_{xx}^2, n_t) - 6\nu\beta|An|_0^2, \tag{1.11.47}$$

where

$$h_1 = 6\beta(g_{2xx}, n_{xx}) + 2\int_0^L f'(n)n_x|\varepsilon|_{xx}^2\mathrm{d}x$$
$$+ 10\nu\int_0^L f'(n)n_x^2\mathrm{d}x - 10(f'(n)n_{2x}, g_{2x})$$
$$+ \frac{2}{5}\int_0^L f'(n)n_x^2|\varepsilon|_{xx}^2\mathrm{d}x + 5\nu\int_0^L f''(n)nn_x^2\mathrm{d}x$$
$$- 5(f''(n)n_x^2, g_x) + \frac{5}{2}\int_0^L g''(n)f'(n)n_x^3\mathrm{d}x$$
$$+ \frac{\beta}{8}\int_0^L f^{(4)}(n)n_x^5\mathrm{d}x - 6(|\varepsilon|_{xx}^2, n - g_2). \tag{1.11.48}$$

Since

$$(|\varepsilon|_{xx}^2, n_t) = \frac{\mathrm{d}}{\mathrm{d}t}(|\varepsilon|_{xx}^2, n_t) - (|\varepsilon|_t^2, n_{xx})$$
$$= \frac{\mathrm{d}}{\mathrm{d}t}(|\varepsilon|_{xx}^2, n) - 2\operatorname{Re}\int_0^L \varepsilon\overline{\varepsilon_t}n_{xx}\mathrm{d}x - \operatorname{Re}(n\varepsilon, A\varepsilon_t)$$
$$= 2\operatorname{Re}(n_x\varepsilon_x, \varepsilon_t) + \operatorname{Re}(n_{xx}\varepsilon, \varepsilon_t) + \operatorname{Re}(n\varepsilon_{xx}, \varepsilon_t), \tag{1.11.49}$$

from equations (1.11.46), (1.11.48) and (1.11.49) we have

$$\frac{\mathrm{d}}{\mathrm{d}t}\{6|A\varepsilon|_0^2 + 3|b\beta||An|_0^2 + h_2\} + 12\gamma|A\varepsilon|_0^2 + 6\nu|b\beta||An|_0^2 + h_3 + h_4 + bh_1 = 0, \tag{1.11.50}$$

where

$$h_2 = 12\,\mathrm{Re}(g_1, A\varepsilon) + 5b\int_0^L f(n)n_x^2\mathrm{d}x + 6b(|\varepsilon|_x^2, n_x), \tag{1.11.51}$$

$$h_3 = 12\,\mathrm{Re}(g_1, A\varepsilon) + 12b\gamma\,\mathrm{Re}\int_0^L (\varepsilon n)_x\overline{\varepsilon_x}\mathrm{d}x - 24b\,\mathrm{Re}(n_x\varepsilon_x, \varepsilon_t), \tag{1.11.52}$$

$$h_4 = 12b\,\mathrm{Im}((n^2\varepsilon)_x, \varepsilon_x) - 12b\,\mathrm{Im}((i\gamma\varepsilon n - g_1 n)_x, \varepsilon_x). \tag{1.11.53}$$

By Proposition 1.11.2 we know, when $t \geq T_1(R)$,

$$\|\varepsilon(t)\|_1^2 + \|n(t)\|_1^2 \leq \rho_1^2,$$
$$\|\varepsilon(t)\|_{L^\infty} + \|n(t)\|_{L^\infty}^2 \leq C\rho_1^2.$$

Hence $|f^{(k)}(n)|_\infty \leq C(\rho_1)$. Then we know that

$$|h_2| \leq |A\varepsilon|_0^2 + C(\rho_1, \|g\|_1^2). \tag{1.11.54}$$

Since $|\mathrm{Re}(n_x, \varepsilon_x, \varepsilon_t)| \leq \theta|A\varepsilon|_0^2 + C(\theta, \rho, \|g\|_1^2)$, $\forall\theta > 0$, we have

$$|h_3| \leq \gamma|A\varepsilon|_0^2 + C(\rho_1, \|g\|_1^2), \tag{1.11.55}$$
$$|h_4| \leq C(\rho_1, \|g\|_1^2), \tag{1.11.56}$$
$$|bh_1| \leq |\beta b|\nu|n_{xx}|_0^2 + C(\rho_1, \|g\|_2^2). \tag{1.11.57}$$

From equations (1.11.50) and (1.11.54)–(1.11.57), when $t \leq T_1(R)$, we get

$$\frac{\mathrm{d}}{\mathrm{d}t}\{6|A\varepsilon|_0^2 + 3|b\beta|An|_0^2 + h_2\} + \gamma\{6|A\varepsilon|_0^2 + 3|b\beta||An|_0^2 + h_2\} \leq C(\rho_1, \|g\|_2^2). \tag{1.11.58}$$

Notice that H_2 satisfies equation (1.11.54). By Gronwall inequality, we obtain the claim. □

Define

$$\mathscr{A} = \omega^\omega(B_2) = \bigcap_{s\geq 0}\bigcup_{t\geq s}\overline{S(t)B_2}$$

where the closure is taken with respect to the weak topology of X_2 and B_2 is the absorbing set defined by equation (1.11.45). Using Proposition 1.11.3 and Lemma 1.11.1, we can prove

Theorem 1.11.2. *The set $\mathscr{A} = \omega^\omega(B_2)$ has the following properties:*
(1) *$\mathscr{A}$ is weakly compact in X, $S(t)\mathscr{A} = \mathscr{A}$, $\forall t \in \mathbf{R}^+$;*

(2) *for any bounded set* $B \subset X_2$,

$$\lim_{t\to\infty} d_H^{\omega}(S(t)B, \mathscr{A}) = 0;$$

(3) $\mathscr{A}$ *is connected in the weak sense in* X_2.

In other words, $\mathscr{A}$ *is the weak global attractor of* $S(t)$ *in* X_2.

Corollary 1.11.1. *For any bounded set* B *in* X_2, $S(t)B$ *converges to* $\mathscr{A}$ *when* $t \to \infty$ *with respect to the norm of* X_1.

Now we estimate the dimension of $\mathscr{A}$.

Let $\zeta_0 = (\varepsilon_0, n_0) \in X_2$ and let $S(t)\zeta_0 = (\varepsilon(t), n(t))$ be the semigroup defined by problem (1.11.1)–(1.11.4) with the initial value (ε_0, n_0). Consider

$$iU_t + U_{xx} - bV\varepsilon - bnU + i\gamma U = 0, \tag{1.11.59}$$

$$V_t + \frac{1}{2}(f'(n)V)_x + \frac{\beta}{2}V_{xxx} + \nu V + \frac{1}{2}(\varepsilon\overline{U} + U\overline{\varepsilon})_x = 0, \tag{1.11.60}$$

$$U(x,t) = U(x+L,t), \quad V(x,t) = V(x+L,t), \quad \forall x \in \mathbf{R}, \quad t \geq 0, \tag{1.11.61}$$

$$U(0) = u_0 \in IH_L^1, \quad V(0) = v_0 \in IH_L^1. \tag{1.11.62}$$

Since $S(t)\zeta_0 \in C(\mathbf{R}^+; X_2)$, it is easy to see that the linear problem (1.11.59)–(1.11.62) has a unique solution

$$(U(t), V(t)) \in C(\mathbf{R}^+; X_1).$$

It is easy to prove that the linear mapping

$$(DS(t)\zeta_0)(u_0, v_0) = (U(t), V(t)) \tag{1.11.63}$$

is the "uniform differential" of $S(t)$. Then we have

Proposition 1.11.4. *For any* $R, T, 0 < R, T < \infty$, *there exists a constant* $C = C(R,T)$ *such that for any* $\zeta_0 = (\varepsilon_0, n_0)$, $\eta_0 = (h_0, k_0)$, $\|\xi_0\|_1 \leq R_1$, $\|\zeta_0 + \eta_0\|_1 \leq R$, $t < T$, *we have*

$$\|S(t)(\zeta_0 + \eta_0) - S(t)\zeta_0 - (DS(t)\zeta_0)\eta_0\|_1 \leq C\|\eta_0\|_1^2, \tag{1.11.64}$$

where $(DS(t)\zeta_0)\eta_0 = (U(t), V(t))$ *is the solution of problem* (1.11.59)–(1.11.62) *with the initial value* $U(0) = h_0$, $V(0) = k_0$.

In order to study the variation of the m-dimensional volumes of $DS(t)$, we introduce the energy equation of problem (1.11.59)–(1.11.62). Let $(\varepsilon_0, n_0) \in X_2$, $(\varepsilon(t), n(t))$ be the solution of problem (1.11.1)–(1.11.4) with the initial value (ε_0, n_0). Then $(U(t), V(t))$ is the solution of problem (1.11.59)–(1.11.62) with the initial value $\eta_0 = (u_0, v_0)$. Set $u = e^{\gamma t}U$, $v = e^{\gamma t}V$. Then (u, v) satisfies

$$iu_t + v_{xx} - bv\varepsilon - bnu = 0, \tag{1.11.65}$$

$$v_t + \frac{1}{2}(f'(n)v)_x + \frac{\beta}{2}v_{xxx} + vu + \frac{1}{2}(\varepsilon\bar{u} + u\bar{\varepsilon})_x = 0. \tag{1.11.66}$$

Similar to equations (1.11.1) and (1.11.14)–(1.11.16), we have

$$\frac{1}{2}\frac{\mathrm{d}}{\mathrm{d}t}|u|_0^2 = \mathrm{Im}\ b(v\varepsilon, u), \tag{1.11.67}$$

$$\frac{\mathrm{d}}{\mathrm{d}t}\left(\frac{1}{2}|u_x|_0^2 + k_1\right) = k_2 + \mathrm{Re}\ b\int_0^L \varepsilon\bar{u}v_t\mathrm{d}x, \tag{1.11.68}$$

$$\frac{\mathrm{d}|v|_0^2}{\mathrm{d}t} = k_3, \tag{1.11.69}$$

$$\frac{\mathrm{d}}{\mathrm{d}t}\left(\frac{\beta}{2}|v_x|_0^2 - \frac{1}{2}\int_0^L f'(n)|v|^2\mathrm{d}x\right) = -(v-\gamma)\beta|v_x|_0^2 + 2(\mathrm{Re}\ \varepsilon\bar{u}, v_t) + k_4, \tag{1.11.70}$$

where

$$k_1 = \frac{b}{2}\int_0^L n|u|^2\mathrm{d}x + b\,\mathrm{Re}\int_0^L v\varepsilon\bar{u}\mathrm{d}x, \tag{1.11.71}$$

$$k_2 = -\frac{b}{2}\int_0^L\left(\frac{1}{2}f(n)_x + vn + \frac{1}{2}|\varepsilon|_x^2 - g_2\right)|u|^2\mathrm{d}x + \frac{b\beta}{2}\int_0^L n_{xx}\,\mathrm{Re}(u, \overline{u_x})\mathrm{d}x, \tag{1.11.72}$$

$$k_3 = -\int_0^L (f'(n)v)_x v\mathrm{d}x - 2(v-\gamma)|v|_0^2 - 2(\mathrm{Re}(\varepsilon\bar{u})_x, v), \tag{1.11.73}$$

$$\begin{aligned} k_4 = {} & \left(2\,\mathrm{Re}(\varepsilon\bar{u}), \frac{1}{2}(f'(n)v)_x + (v-\gamma)v\right) \\ & + \frac{1}{2}\left(f'(n)\left[\frac{1}{2}f(n)_x + vn + \frac{1}{2}|\varepsilon|_x^2 - g_2\right] - \frac{\beta}{2}f'''(n)n_x n_{xx}, v^2\right) \\ & (v-\gamma)(v, f'(n)v) - ((f'(n)v)_x, 2\,\mathrm{Re}(\varepsilon\bar{u})). \end{aligned} \tag{1.11.74}$$

Define

$$J_\mu(u,v) = |u_x|_0^2 + \frac{|b\beta|}{2}|v_x|_0^2 + 2k_1 + \frac{b}{2}\int_0^L f'(n)|v|^2\mathrm{d}x + \mu|u|_0^2 + \mu|v|_0^2, \tag{1.11.75}$$

$$I_\mu(u,v) = 2k_2 + 2\mu\,\mathrm{Im}(v\varepsilon, u) - \mu k_3 - bk_4 + (v-\gamma)b\beta|v_x|_0^2. \tag{1.11.76}$$

Then from equations (1.11.67)–(1.11.74), we have

$$\frac{\mathrm{d}J_\mu(u,v)}{\mathrm{d}t} = I_\mu(u,v), \tag{1.11.77}$$

where μ is a large enough positive constant and $J_\mu(u,v)$ is an equivalent norm in X_1.

Assume that X is a bounded invariant set in X_2, that is,

$$S(t)X = X, \quad \forall t \geq 0.$$

Set $\zeta_0 \in X$. Then $S(t)\zeta_0 = S(t)(\varepsilon_0, n_0) = (\varepsilon(t), n(t)) \in X$. Hence

$$|X|_\infty = \sup_{\zeta_0 \in X} \sup_{t \geq 0} \{|\varepsilon(t)|_{L^\infty} + |n(t)|_{L^\infty} + |n_x(t)|_{L^\infty}\} < \infty. \tag{1.11.78}$$

By equation (1.11.75) and from the definition of J_u, we know that there exist $\mu > 0$ and M_0, M_1 such that

$$M_0(\|u\|_1^2 + \|v\|_1^2) \leq J_u(u, v) \leq M_1(\|u\|_1^2 + \|v\|_1^2), \quad \forall (u, v) \in X_1. \tag{1.11.79}$$

So $J_\mu(u, v)$ is an equivalent norm in X_1. Now we introduce an **R**-linear inner product in X_1.

Let $\eta = (\eta_1, \eta_2)$, $\zeta = (\zeta_1, \zeta_2) \in X_1$. Define

$$\begin{aligned}\overline{\Psi}(\eta, \xi) = \operatorname{Re} \int_0^L \Big[\eta_{1x}\overline{\xi_{1x}} + \frac{|b\beta|}{2}\eta_{2x}\xi_{2x} + \frac{b}{2} f'(n)\eta_2\xi_2 + \mu\eta_1\overline{\xi_1} + \mu\eta_2\xi_2 + bn\eta_1\overline{\xi}_1 \Big] dx \\ + b \operatorname{Re} \int_0^L (\xi_2\varepsilon\overline{\eta_1} + \eta_1\varepsilon\overline{\xi_1}) dx.\end{aligned} \tag{1.11.80}$$

It is easy to prove that $\overline{\Psi}(\eta, \xi)$ is an **R**-linear symmetric form in X_1, and $\overline{\Psi}(\eta, \eta) = J_\mu(\eta, \eta)$ is coercive by (1.11.79). So $\overline{\Psi}(\eta, \eta)^{\frac{1}{2}}$ is an equivalent norm in X_1.

Now let $\xi_{0j} = (u_{0j}, v_0^j)$, $j = 1, 2, \ldots, m$ be m elements in X_1, and $\xi^j(t) = (DS(t)\xi_0)\xi^0$ be the corresponding solutions of (1.11.59)–(1.11.62). Let $\eta^j(t) = e^{\gamma t}\zeta^j(t)$, $j = 1, 2, \ldots, m$. Then $\eta^j(t) = (u^j(t), v^j(t))$ satisfies equation (1.11.66) and has initial value $\eta^j(0) = (u_{0j}, v_{0j})$. We study the following evolution of volume quantities:

$$|\xi^1(t) \wedge \xi^2(t) \wedge \cdots \wedge \xi^m(t)|_{\wedge^m(X_1)} = \det_{1 \leq i,j \leq m} (\xi^i, \xi^j). \tag{1.11.81}$$

Theorem 1.11.3. *Let X be the bounded invariant set in X_2. Then there exist constants C_1 and C_2 such that for any $\zeta_0(\varepsilon_0, n_0) \in X_2, m \geq 1, t \geq 0$, we have*

$$\begin{aligned}&|\xi^1(t) \wedge \xi^2(t) \wedge \cdots \wedge \xi^m(t)|_{\wedge^m(X_1)} \\ &\leq |\xi_0^1(t) \wedge \xi_0^2(t) \wedge \cdots \wedge \xi_0^m(t)|_{\wedge^m(X_1)} C_1 e^{(C_2\sqrt{m} - \gamma m)t}, \quad \forall \xi_0^j \in X_1.\end{aligned} \tag{1.11.82}$$

Proof. Firstly we note that

$$\begin{aligned}&|\xi^1(t) \wedge \xi^2(t) \wedge \cdots \wedge \xi^m(t)|^2_{\wedge^m(X_1)} \\ &= e^{-2\gamma mt}|\eta^1 \wedge \eta^2 \wedge \cdots \wedge \eta^m|_{\wedge^m(X_1)} \\ &= e^{-2\gamma mt} \det_{1 \leq i,j \leq m} \Psi(\eta^j(t)).\end{aligned} \tag{1.11.83}$$

So we only need to estimate $H_m(t) = \det_{1\le i,j\le m} \Psi(\eta^i(t), \eta^j(t))$. We have

$$\frac{dH_m(t)}{dt} = H_m(t)\sum_{l=1}^{m} \max_{F\subset \mathbf{R}^m, \dim F=l,}\ \min_{x\in F, x\neq 0} \frac{I_\mu(\sum_{j=1}^m x_j\eta^j(t))}{J_\mu(\sum_{j=1}^m x_j\eta^j(t))} \tag{1.11.84}$$

by equation (1.11.78). The norms $|n|_{L^\infty}$, $|n_x|_{L^\infty}$, $|\varepsilon|_{L^\infty}$, $|\varepsilon_x|_{L^\infty}$ are consistently bounded and, for $\zeta_0 = (\varepsilon_0, n_0) \in X$, we have

$$\begin{aligned}
|k_2| &\le C|u|_0^2 + C\int_0^L |n_{xx}||u||u_x| dx,\\
C|u|_0^2 &+ C\|n\|_2|u|_{L^\infty}|n_x|_0 \le C|u|_1^{\frac{3}{2}}|u|_0^{\frac{1}{2}},\\
\mu|k_3| &= |\mu| - \int_0^L (f'(n)v)_x v dx - 2(\nu-\gamma)|v|_0^2 + 2(\mathrm{Re}(\varepsilon\bar{u}), v_x)\\
&\le \frac{1}{8}(\nu-\gamma)|b\beta||v_x|_0^2 + C(|v|_0^2 + |v_x|_0|u|_0)\\
&\le \frac{1}{8}(\nu-\gamma)|b\beta||v_x|_0^2 + C(\|v\|_1^{\frac{3}{2}}|v|_0^{\frac{1}{2}} + \|u\|_1^{\frac{3}{2}}|u|_0^{\frac{1}{2}}),\\
|bk_4| &\le C(|v_x|_0|u|_0 + |u|_0^2 + |v|_0^2 + \|v\|_{L^4}^2)\\
&\le \frac{1}{4}(\nu-\gamma)|b\beta||v_x|_0^2 + C(\|u\|_1^{\frac{3}{2}}|u|_0^{\frac{1}{2}} + \|v\|_1^{\frac{3}{2}}|v|_0^{\frac{1}{2}}).
\end{aligned}$$

Hence, we note that

$$\begin{aligned}
I_\mu(u,v) &= 2k_2 + 2\mu\,\mathrm{Im}(v\varepsilon, u) - \mu k_3 - bk_4 + (\nu-\gamma)b\beta|v_x|_0^2\\
&\le \frac{1}{2}(\nu-\gamma)b\beta|v_x|_0^2 + C(\|u\|_1^{\frac{3}{2}}|u|_0^{\frac{1}{2}} + \|v\|_1^{\frac{3}{2}}|v|_0^{\frac{1}{2}})\\
&\le M(\|u\|_1^{\frac{3}{2}}|u|_0^{\frac{1}{2}} + \|v\|_1^{\frac{3}{2}}|v|_0^{\frac{1}{2}}),
\end{aligned} \tag{1.11.85}$$

since $\nu > \gamma$, $b\beta < 0$, where M is a positive constant.

On the other hand, $J_\mu = J_\mu(u,v) \ge M_0(\|u\|_1^2 + \|v\|_1^2)$. From (1.11.83) and (1.11.84), we have

$$\begin{aligned}
\frac{dH_m(t)}{dt} &\le \frac{M}{M_0}H_m(t)\sum_{l=1}^{m} \max_{F\subset \mathbf{R}^m, \dim F=l,}\ \min_{x\in F, x\neq 0} \frac{\|\sum_{j=1}^m x_ju_j(t)\|_1^{\frac{3}{2}}|\sum_{j=1}^m x_ju_j(t)|_0^{\frac{1}{2}}}{\|\sum_{j=1}^m x_ju_j(t)\|_1^2 + \|\sum_{j=1}^m x_jv_j(t)\|_1^2}\\
&\quad + \frac{\|\sum_{j=1}^m x_jv_j(t)\|_1^{\frac{3}{2}}|\sum_{j=1}^m x_jv_j(t)|_0^{\frac{1}{2}}}{\|\sum_{j=1}^m x_ju_j(t)\|_1^2 + \|\sum_{j=1}^m x_jv_j(t)\|_1^2}\\
&\le \frac{M}{M_0}H_m(t)\sum_{l=1}^{m} \max_{F\subset \mathbf{R}^m, \dim F=l,}\ \min_{x\in F, x\neq 0} \left\{\frac{|\sum_{j=1}^m x_ju_j(t)|_0^{\frac{1}{2}}}{\|\sum_{j=1}^m x_ju_j(t)\|_1^{\frac{1}{2}}} + \frac{|\sum_{j=1}^m x_jv_j(t)|_0^{\frac{1}{2}}}{\|\sum_{j=1}^m x_jv_j(t)\|_1^{\frac{1}{2}}}\right\}\\
&\le \frac{2M}{M_0}H_m(t)\sum_{l=1}^{m}\frac{1}{\lambda_l^{\frac{1}{4}}},
\end{aligned}$$

where we use Max–Min theorem and λ_j is the jth eigenvalue of $A = -\partial_{xx}$. We know that $\lambda_j \sim cj^2$ as $j \to \infty$, so there exists a constant C_2 such that

$$\frac{\mathrm{d}H_m(t)}{\mathrm{d}t} \le 2C_2\sqrt{m}H_m(t). \tag{1.11.86}$$

Hence, $H_m(t) \le e^{2C_2\sqrt{m}t}H_m(0)$, $t > 0$. Combining with equation (1.11.83), we get equation (1.11.82). □

As a corollary of Theorem 1.11.3, we have

Theorem 1.11.4. *The global attractor $\mathscr{A}$ defined by Theorem* 1.11.2 *in X_1 possesses bounded fractal and Hausdorff dimensions.*

Proof. By Theorem 1.11.3 we know that, for $\chi_0 \in \mathscr{A}$,

$$\omega_m(DS(t)\chi_0) \le C^m \exp(C_2\sqrt{m} - \gamma m)t,$$

where ω_m is the Lyapunov index. When m is large enough, $\overline{\omega_m}(\mathscr{A}) < 1$, where $\overline{\omega_m(\mathscr{A})}$ is the uniform Lyapunov index of $\mathscr{A}$. By Theorem V.3.1 in [197], we know that $\mathscr{A}$ in X_1 has bounded fractal and Hausdorff dimensions. □

1.12 The Landau–Lifshitz equation on a Riemannian manifold

Assume that (M,γ) and (N,g) are two Riemann manifolds, M is boundless and N is S^2. We will prove the existence of an attractor for the Landau–Lifshitz (LL) [25] equation in a Riemann manifold and will give upper and lower bounds on the attractor dimension. For this purpose, we must establish an a priori estimate to prove the existence of a global solution and of the attractor.

Consider the following Landau–Lifshitz equation on a Riemann manifold:

$$\partial_t u = -\alpha_1 u \times (u \times \Delta_M u) + \alpha_2 u \times \Delta_M u \tag{1.12.1}$$

with the initial value

$$u|_{t=0} = u_0(x), \quad |u_0(x)|^2 = 1, \quad x = (x_1, \dots, x_n) \in M, \tag{1.12.2}$$

where $u : M \to S^2$ and ΔM is the Laplace–Beltrami operator,

$$\Delta_M = \frac{1}{\sqrt{\gamma}}\frac{\partial}{\partial x^\beta}\left(\gamma^{\alpha\beta}\sqrt{\gamma}\frac{\partial}{\partial x^\alpha}\right) = \gamma^{\alpha\beta}\frac{\partial^2}{\partial x^\beta \partial x^\beta} - \Gamma^k_{\alpha\beta}\frac{\partial}{\partial x^k}.$$

In 1993, Guo and Hong [100] proved in the classical sense that u is the solution of problem (1.12.1)–(1.12.2) if and only if u is the solution of equation

$$\partial_t u = \alpha_1 \frac{1}{\sqrt{\gamma}}\frac{\partial}{\partial x^\alpha}\left(\gamma^{\alpha\beta}\sqrt{\gamma}\frac{\partial u}{\partial x^\beta}\right) + \alpha_1|\nabla u|^2 u + \alpha_2 u \times \frac{1}{\sqrt{\gamma}}\frac{\partial}{\partial x^\beta}\left(\gamma^{\alpha\beta}\sqrt{\gamma}\frac{\partial u}{\partial x^\alpha}\right) \tag{1.12.3}$$

with the same initial value as in (1.12.2). Meanwhile, they proved, also in the classical sense, that $u : M \to S^2$ is a harmonic mapping if and only if u satisfies (1.12.2), $\partial_t u(x,t) = 0$, $\forall t \geq 0$, and heat flow equation for a harmonic mapping $M \to S^2$ is

$$\partial_t u = \Delta_M u + |\nabla u|^2 u. \tag{1.12.4}$$

Now we establish a uniform a priori estimate.

Let (M, γ) be a bounded or unbounded compact Riemannian manifold and let ∇ denote the contact (covariant derivative) of γ,

$$\left|\nabla u(x)\right|^2 = \sum_{\alpha\beta}\sum_{i} \gamma^{\alpha\beta}\frac{\partial u^i}{\partial x^\alpha}\frac{\partial u^i}{\partial x^\beta}. \tag{1.12.5}$$

For the real function $\varphi \in C^k(M)$ ($k \geq 0$ is an integer), define

$$\left|\nabla^k\varphi\right|^2 = \nabla^{\sigma_1}\cdot\nabla^{\sigma_2}\cdots\nabla^{\sigma_k}\cdot\nabla_{\sigma_1}\cdot\nabla_{\sigma_2}\cdots\nabla_{\sigma_k}\varphi$$

In particular, $|\nabla^1\varphi| = |\nabla\varphi|$, $|\nabla^1\varphi|^2 = |\nabla\varphi|^2 = \nabla^\nu\varphi\nabla_\nu\varphi$, $\nabla^k\varphi$ is the kth covariant derivative of φ.

We consider the vector space L_k^p of C^∞ functions φ, $|\nabla^l\varphi| \in L^p(M)$, $\forall 0 \leq l \leq k$, where k and l are integers, and $p \geq 1$ is a real number. Sobolev space $W_p^k(M)$ is the completion of space $L_k^p(M)$ with respect to the norm

$$\|\varphi\|_{W_p^k(M)} = \sum_{i_0}^{k}\|\nabla^i\varphi\|_p.$$

In particular, $W_2^k(M) = H^k(M)$, $\|\cdot\|_2 = \|\cdot\|$.

Lemma 1.12.1. *Let $|u_0(x)|^2 = 1$. Then for a smooth solution of the initial value problem (1.12.1)–(1.12.2) we have*

$$\left|u(x,t)\right|^2 = 1, \quad \forall (x,t) \in M \times [0,\infty). \tag{1.12.6}$$

Proof. Multiplying equation (1.12.2) by u, we get

$$u \cdot \partial_t u = 0, \quad \forall (x,t) \in M \times [0,\infty).$$

Then through (1.12.2) we get (1.12.6). □

Lemma 1.12.2. *Suppose the condition of Lemma 1.12.1 is satisfied, and $\|\nabla u_0\| < \infty$. Then we have*

$$\left|\nabla u(\cdot,t)\right|^2 \leq \|\nabla u_0\|^2, \tag{1.12.7}$$

$$2\alpha_1\int_0^t \|u \times \Delta_M u\|^2 dt \leq \|\nabla u_0\|^2, \quad \forall 0 \leq t < \infty. \tag{1.12.8}$$

Proof. Multiplying equation (1.12.1) by $\Delta_M u$, we get

$$\Delta_M u \cdot u_t = -\alpha_1 \Delta_M u \cdot (u \times (u \times \Delta_M u))$$
$$= -\alpha_1 (u \times \Delta_M u) \cdot (\Delta_M u \times u) = \alpha_1 |u \times \Delta_M u|^2,$$
$$\int_M \Delta_M u \cdot u_t \mathrm{d}M = -\int_M \gamma^{\alpha\beta} \sqrt{\gamma} \frac{\partial u}{\partial x^\alpha} \frac{\partial u_t}{\partial x^\beta} \mathrm{d}x + \int_M u_t \gamma^{\alpha\beta} \sqrt{\gamma} \frac{\partial u}{\partial x^\alpha} \cos(\nu, x^\beta) \mathrm{d}s$$
$$= -\frac{1}{2} \frac{\mathrm{d}}{\mathrm{d}t} \int_M \gamma^{\alpha\beta} \frac{\partial u^i}{\partial x^\alpha} \frac{\partial u^i}{\partial x^\beta} \mathrm{d}M$$
$$= -\frac{1}{2} \frac{\mathrm{d}}{\mathrm{d}t} \|\nabla u(\cdot, t)\|^2.$$

Hence

$$\frac{\mathrm{d}}{\mathrm{d}t} \|\nabla u(\cdot, t)\|^2 + 2\alpha_1 |u \times \Delta_M u|^2 \mathrm{d}M = 0,$$

which yields

$$\frac{\mathrm{d}}{\mathrm{d}t} \|\nabla u(\cdot, t)\|^2 \le 0,$$

proving the lemma. □

Lemma 1.12.3 (Sobolev interpolation inequality on a compact Riemann manifold). *Let M be a Riemannian manifold with smooth boundary. Assume q, r are real numbers, $1 \le q, r \le \infty$, while j, m are integers, $0 \le j < m$. Then there exists a constant C, depending on n, m, j, q, r and a, such that for any $f \in W_r^m(M) \cap L_q(M)$, we have*

$$\|\nabla^j f\|_p \le C \|f\|^a_{W_r^m(M)} \|f\|_q^{1-a}, \tag{1.12.9}$$

where

$$\frac{1}{p} = \frac{j}{n} + a\left(\frac{1}{r} - \frac{m}{n}\right) + (1-a)\frac{1}{q}$$

for any a such that $\frac{j}{m} \le a \le 1$; p is a non-negative integer.

Proof. From Theorem 3.70 in [6] we have

$$\|\nabla^i F\|_p \le C \|\nabla^m F\|_r^a \|F\|_q^{1-a}, \tag{1.12.10}$$

where

$$F = f - \bar{f}, \quad \bar{f} = \frac{1}{\operatorname{vol} M} \int_M f \mathrm{d}M,$$
$$\frac{1}{p} = \frac{i}{n} + a\left(\frac{1}{r} - \frac{m}{n}\right) + (1-a)\frac{1}{q}.$$

Now consider several cases:

(1) When $i > 0$, by equation (1.12.10) we get

$$\begin{aligned}\|\nabla^i f\|_p &\le C\|\nabla^m f\|_r^a (\|f\|_q + \|\bar{f}\|_q)^{1-a} \\ &\le C'\|\nabla^m f\|_r^a \|f\|_q^{1-a} \le C'\|f\|_{W_r^m(M)}^a \|f\|_q^{1-a}.\end{aligned}$$

(2) When $i = 0$, using Hölder inequality, we have

$$\int |f|^p \mathrm{d}x = \int |f|^\alpha |f|^\beta \mathrm{d}x \le \left(\int |f|^{\alpha l}\mathrm{d}x\right)^{\frac{1}{l}} \left(\int |f|^{\beta l'}\mathrm{d}x\right)^{\frac{1}{l'}},$$

where

$$\alpha + \beta = p, \quad \alpha l = r, \quad \beta l' = q, \quad \frac{1}{l} + \frac{1}{l'} = 1.$$

Then

$$\frac{1}{l} = \left(\frac{(p-q)}{(q-r)}\right)\frac{r}{p}, \quad \frac{1}{l'} = 1 - \frac{1}{l}$$

and from equations (1.12.1) and (1.12.2) we get (1.12.9). □

Remark 1.12.1. Instead of the space $W_k^p(M)$, Aubin introduced the space $V_k^p(M)$ in [6], which is the completion of space $S_k^p(M)$ with respect to the norm

$$\|\varphi\|_{V_k^p(M)} = \sum_{0\le t\le \frac{k}{2}} \|\Delta_M^l \varphi\|_p + \sum_{0\le l\le \frac{k-1}{2}} \|\nabla \Delta_M^l \varphi\|_p,$$

where the vector space S_k^p comprises functions $\varphi \in C^\infty(M)$, $\Delta_M^l \varphi \in L^p(M)$, $0 \le l \le \frac{k}{2}$, such that $|\nabla \Delta_M^l \varphi| \in L^p(M)$, $0 \le l \le \frac{k-1}{2}$.

Lemma 1.12.4. *Suppose the condition of Lemma 1.12.3 is satisfied. Let*

$$\|\nabla u_0(x)\| \le \lambda, \quad n = 2, \tag{1.12.11}$$

where the constant λ is small enough. Then we have

$$\|\Delta_M u(\cdot, t)\| \le \frac{E_1}{t}, \quad \forall x = (x_1, \dots, x_n),\ t > 0,\ n \le 2, \tag{1.12.12}$$

where the constant E_1 depends only on $\|\nabla u_0(x)\|_{H^1(M)}$, $0 < T \le T$.

Proof. Acting with Δ_M on equation (1.12.2), then taking the inner product with $t\Delta_M u$, we get

$$(t\Delta_M u, t\Delta_M u_t - \alpha_1 \Delta_M^2 u_t - \alpha_1 \Delta_M(|\nabla u|^2 u) - \alpha_2 \Delta_M(u \times \Delta_M u)) = 0, \tag{1.12.13}$$

where

$$t(\Delta_M u, \Delta_M u_t) = \frac{1}{2}\frac{\mathrm{d}}{\mathrm{d}t}(t\|\Delta_M u\|^2) - \frac{1}{2}\|\Delta_M u\|^2,$$

$$\begin{aligned}
(\Delta_M u, \alpha_1 \Delta_M^2 u) &= \alpha_1 \int \Delta_M u \cdot \frac{1}{\sqrt{\gamma}} \frac{\partial}{\partial x^\beta}\left(\gamma^{\alpha\beta}\sqrt{\gamma}\frac{\partial \Delta_M u}{\partial x^\alpha}\right)\sqrt{\gamma}\mathrm{d}x \\
&= -\alpha_1 \int \gamma^{\alpha\beta}\sqrt{\gamma}\frac{\partial \Delta_M u}{\partial x^\alpha}\frac{\partial \Delta_M u}{\partial x^\beta}\mathrm{d}x \\
&= -\alpha_1 \|\nabla \Delta_M u\|^2,
\end{aligned}$$

$$\begin{aligned}
\left|\int \Delta_M u \cdot \Delta_M (u \times \Delta_M u)\sqrt{\gamma}\mathrm{d}x\right| &= \left|\int \Delta_M u \cdot \frac{\partial}{\partial x^\beta}\left(\gamma^{\alpha\beta}\sqrt{\gamma}\frac{\partial}{\partial x^\alpha}(u \times \Delta_M u)\right)\mathrm{d}x\right| \\
&= \left|\int \gamma^{\alpha\beta}\frac{\partial \Delta_M u}{\partial x^\beta}\frac{\partial (u \times \Delta_M u)}{\partial x^\alpha}\sqrt{\gamma}\mathrm{d}x\right| \\
&= \left|\int \gamma^{\alpha\beta}\frac{\partial \Delta_M u}{\partial x^\beta}\left[\frac{\partial u}{\partial x^\alpha} \times \Delta_m u + u \times \frac{\partial \Delta_M u}{\partial x^\alpha}\right]\sqrt{\gamma}\mathrm{d}x\right| \\
&\le C_1 \|\nabla \Delta_M u\| \|\nabla u\|_\infty \|\Delta_M u\|. \qquad (1.12.14)
\end{aligned}$$

In the above equation, we used

$$\int \gamma^{\alpha\beta}\frac{\partial \Delta_M u}{\partial x^\beta}\left(u \times \frac{\partial \Delta_M u}{\partial x^\alpha}\right)\sqrt{\gamma}\mathrm{d}x = 0. \qquad (1.12.15)$$

By Lemma 1.12.2, we get

$$\begin{aligned}
&\|\nabla u\|_\infty \le C_2 \|\nabla^3 u\|^{\frac{1}{2}} \|\nabla u\|^{\frac{1}{2}} + C_3, \quad n = 2, \\
&\|\Delta_M u\| \le C_4 \|\nabla^3 u\|^{\frac{1}{2}} \|\nabla u\|^{\frac{1}{2}} + C_5, \qquad (1.12.16)
\end{aligned}$$

where constants C_3 and C_5 depend on $\|\nabla u_0\|$.

Substituting equation (1.12.16) into equation (1.12.14), we get

$$\left|\int_M \Delta_M \cdot \Delta_M (u \times \Delta_M)\mathrm{d}M\right| \le 2C_1 C_2 C_4 C_6 \|\nabla u\| \|\nabla^3 u\|^2 + C_6', \qquad (1.12.17)$$

where

$$\|\nabla \Delta_M u\| \le C_6' \|\nabla^3 u\|$$

and constant C_6' depends on $\|\nabla u_0\|$.

Now we estimate the term $(\Delta_M(|\nabla u|u), \Delta_M^2 u)$ of equation (1.12.13) as follows:

$$\begin{aligned}
|(\Delta_M u, \Delta_M(|\nabla u|^2 u))| &= \left|\int_M \Delta_M \mu \cdot \frac{\partial}{\partial x^\beta}\gamma^{\alpha\beta}\sqrt{\gamma}\frac{\partial |\nabla u|^2 u}{\partial x^\alpha}\mathrm{d}x\right| \\
&= \left|\int_M \gamma^{\alpha\beta}\sqrt{\gamma}\frac{\partial}{\partial x^\alpha}|\nabla u|^2 u \cdot \frac{\partial}{\partial x^\beta}\Delta_M u\mathrm{d}x\right| \\
&= \left|\int_M \gamma^{\alpha\beta}\sqrt{\gamma}\left(\left(\frac{\partial}{\partial x^\alpha}|\nabla u|^2 u\right)u + |\nabla u|^2 \frac{\partial u}{\partial x^\alpha}\right)\frac{\partial}{\partial x^\beta}\Delta_M u\mathrm{d}x\right|
\end{aligned}$$

$$
\begin{aligned}
&= \left| \int_M \gamma^{\alpha\beta} \left[2\gamma^{l\delta}(x) \frac{\partial^2 u}{\partial x^\alpha \partial x^l} \frac{\partial u}{\partial x^\delta} u + \gamma^{l\delta}(x)' \frac{\partial u}{\partial x^l} \frac{\partial u}{\partial x^\delta} \right.\right. \\
&\quad \left.\left. + |\nabla u|^2 \frac{\partial u}{\partial x^\alpha} \right] \frac{\partial \Delta_M u}{\partial x^\beta} \mathrm{d}M \right| \\
&\le C_7[\|\nabla u\|_\infty \|u\|_\infty \|\nabla^2 u\| \|\nabla \Delta_M u\| + \|\nabla u\|^2 \|u\|_\infty \|\nabla \Delta_M u\| \\
&\quad + \|\nabla u\|_\infty^2 \|\nabla u\| \|\nabla \Delta_M u\|] \\
&\le C_7[2C_2C_4\|\nabla u\| + (2C_2^2+1)\|\nabla u\|^2] \|\nabla^3 u\| + C_8,
\end{aligned} \tag{1.12.18}
$$

where constants C_7, C_8 depend on $\|\nabla u_0\|$ and $\sup_{x\in M}(|\gamma^{\alpha\beta}(x)|, |\gamma^{\alpha\beta'}(x)|)$.

Hence, from equations (1.12.8), (1.12.12) and (1.12.13) we get

$$
\begin{aligned}
&\frac{1}{2}\frac{\mathrm{d}}{\mathrm{d}t} t\|\Delta_M u\|^2 - \frac{1}{2}\|\Delta_M u\|^2 + \alpha_1 t\|\Delta_M u\|^2 \\
&\quad \le 2t[|\alpha_2| C_1C_2C_4C_6 + \alpha_1 C_7C_2C_4 + \alpha_1(C_2^2+1)C_7\|\nabla u\|]\|\nabla u\|\|\nabla^3 u\| + C_9.
\end{aligned} \tag{1.12.19}
$$

Now we estimate a lower bound of $\|\nabla\Delta_M u\|^2$.

By Ricci formula,

$$
\Delta(\nabla^k f) = \nabla^k(\Delta f) + \sum_{i=0}^{k-1} S_{ki}(\nabla^{k-i} f), \tag{1.12.20}
$$

where S_{ki} is a linear functional which depends on the tensor covariant derivative $\nabla^i R$ curve. Moreover, we get

$$
\|\nabla \Delta_M u\|_2 = \|\Delta_M(\nabla u) - S_{10}(\nabla u)\| \ge \|\Delta_M(\nabla u)\| - S_{10}\|(\nabla u)\|. \tag{1.12.21}
$$

Since

$$
\begin{aligned}
\|\Delta_M(\nabla u)\|^2 &= \int_M \sum_j \Big(\sum_i u_{jii}\Big)\Big(\sum_k u_{jkk}\Big)\mathrm{d}M \\
&= \sum_j \int_M \Big(\sum_i u_{jii}\Big)\Big(\sum_k u_{jkk}\Big)\mathrm{d}M = -\sum_j \int_M \Big(\sum_i u_{ji}\Big)\Big(\sum_k u_{jkki}\Big)\mathrm{d}M \\
&= -\sum_j \int_M \Big(\sum_i u_{ji}\Big) \sum_k (u_{jkik} + u_{jl}R^l_{kki} + u_{lk}R^l_{jki})\mathrm{d}M \\
&= -\sum_j \int_M \Big(\sum_i u_{ji}\Big) \sum_k (u_{jikk} + u_l R^l_{jkik} + u_{lk}R^l_{jki} + u_{jl}R^l_{kki} + u_{lk}R^i_{jki})\mathrm{d}M \\
&\ge \int_M |\nabla^3 u|^2 \mathrm{d}M - C_{11}\int_M |\nabla u|^2 \mathrm{d}M,
\end{aligned} \tag{1.12.22}
$$

equations (1.12.19), (1.12.21) and (1.12.22) yield

$$
\begin{aligned}
&\frac{\mathrm{d}}{\mathrm{d}t} t\|\Delta_M u\|^2 - \|\Delta_M u\|^2 + 2t[\alpha_1 - (2\alpha_2C_1C_2C_4C_6 + 2\alpha_1C_7C_2C_4 \\
&\quad + 2\alpha_1(C_2^2+1)C_7\|\nabla u\| + C_{12}\|\nabla u\|)]\|\nabla u\|\|\nabla^3 u\|^2 \le C_{13}.
\end{aligned} \tag{1.12.23}
$$

Choose $\|\nabla u_0\|$ small enough so that

$$\alpha_1 - (2|\alpha_2|C_1C_2C_4C_6 - 2\alpha_1C_7C_2C_4 + 2\alpha_1(C_2^2+1)C_7\|\nabla u_0\| + C_{12}\|\nabla u_0\|)\|\nabla u_0\| > \frac{\alpha}{2} > 0.$$

From equation (1.12.23), we get

$$t\|\Delta_M u\|^2 - \int_0^t \|\Delta_M u\|^2 dt + \frac{1}{2}\alpha\int_0^t t\|\Delta_M^3 u\|^2 dt \le C_{13}t. \tag{1.12.24}$$

In order to estimate the middle term $\int_0^t \|\Delta_M u\|^2 dt$ on the left of equation (1.12.24), we need the following lemma:

Lemma 1.12.5. *Under the conditions of Lemma* 1.12.2, *we get*

$$\int_0^t \left\|\frac{\partial u}{\partial t}\right\|^2 dt \le \frac{(\alpha_1^2+\alpha_2^2)}{\alpha_1}\|\nabla u_0\|^2, \quad \forall t \in \mathbf{R}^+, \tag{1.12.25}$$

$$\int_0^t \|\Delta_M u\|^2 dt \le C_{14}, \quad 0 \le t \le T, \tag{1.12.26}$$

where the constant C_{14} depends on $\|\nabla u_0\|$ and T.

Proof. Multiplying equation (1.12.3) by $\partial_t u$, and integrating with respect to $(x,t) \in M \times [0,t)$, we get

$$\int_0^t\int_M |\partial u|^2 dM dt + \frac{\alpha_1}{2}\int_0^t \frac{d}{dt}\|\nabla u\|^2 dt - \alpha_2\int_0^t\int_M \partial_t u \cdot (u \times \Delta_M u) dM dt = 0. \tag{1.12.27}$$

Taking the cross-product of (1.12.3), we get

$$\begin{aligned} u \times \partial_t u &= \alpha_1 u \times (\Delta_M u) + \alpha_2(u \times (u \times \Delta_M u)) \\ &= \alpha_1(u \times \Delta_M u) - \alpha_2\Delta_M u - \alpha_2|\nabla u|^2 u. \end{aligned}$$

Since

$$-\Delta_M u - |\nabla u|^2 u = -\frac{1}{\alpha_1}\partial_t u + \frac{\alpha_2}{\alpha_1}(u \times \Delta_M u),$$

the latter equation yields

$$u \times \partial_t u + \frac{\alpha_2}{\alpha_1}\partial_t u = \left(\alpha_1 + \frac{\alpha_2^2}{\alpha_1}\right)(u \times \Delta_M u). \tag{1.12.28}$$

Multiplying equation (1.12.28) by $\partial_t u$, we have

$$\partial_t u \cdot (u \times \Delta_M u) = \alpha_2(\alpha_1^2+\alpha_2^2)^{-1}|\partial_t u|^2,$$

$$\alpha_2 \int_0^t \int_M \partial_t u \cdot (u \times \Delta_M u) \mathrm{d}M \mathrm{d}t = \frac{\alpha_2^2}{\alpha_1^2 + \alpha_2^2} \int_0^t \int_M |\partial_t u|^2 \mathrm{d}M \mathrm{d}t.$$

By equation (1.12.27) we have

$$\frac{\alpha_1^2}{\alpha_1^2 + \alpha_2^2} \int_0^t \int_M |\partial_t u|^2 \mathrm{d}M \mathrm{d}t + \frac{\alpha_1}{2} \left(\|\nabla u(\cdot, t)\|^2 - \|\nabla u_0\|^2 \right) = 0,$$

i. e.,

$$\int_0^t \int_M |\partial_t u|^2 \mathrm{d}M \mathrm{d}t \leq \frac{\alpha_1^2 + \alpha_2^2}{2\alpha_1} \|\nabla u_0\|^2, \quad \forall t \in \mathbf{R}^+. \tag{1.12.29}$$

From equation (1.12.3), we have

$$\int_0^t \|\Delta_M u\|^2 \mathrm{d}t \leq C_{15} \left(\int_0^t \|u_t\|^2 \mathrm{d}t + \int_0^t \|u \times \Delta_M u\|^2 \mathrm{d}t + \int_0^t \int_M |\nabla u|^4 \mathrm{d}M \mathrm{d}t \right). \tag{1.12.30}$$

With the aid of Sobolev inequality (1.12.9), we have

$$\int_M |\nabla u|^4 \mathrm{d}M \leq C_{16} \|\nabla^2 u\|^2 \|\nabla u\|^2 + C_{17}, \quad \forall 0 \leq t \leq T, \tag{1.12.31}$$

where the constant C_{17} depends on $\|\nabla u_0\|$ and $0 \leq t \leq T$.

By the definition of Laplace–Beltrami operator, we get

$$C_{18} \|\nabla^2 u\|^2 + C_{19} \|\nabla u\|^2 \geq \|\nabla_M u\|^2 \geq C_{18} \|\nabla^2 u\|^2 - C_{19} \|\nabla u\|^2, \tag{1.12.32}$$

where the constants C_{18}, C_{19} depend on $\sup_{x \in M} |\gamma^{\alpha\beta}(x)|$ and $\sup_{x \in M} |\Gamma_{\alpha\beta}^k|$. Hence, from equation (1.12.9), we get

$$\begin{aligned} & C_{18} \int_0^t \|\nabla^2 u\|^2 \mathrm{d}t - C_{15} C_{16} \int_0^t \|\nabla^2 u\|^2 \mathrm{d}t \cdot \|\nabla u_0\|^2 \\ & \leq C_{15} \left(\int_0^t \left\| \frac{\partial u}{\partial t} \right\|^2 \mathrm{d}t + \int_0^t \|u \times \Delta_M u\|^2 \mathrm{d}t + C_{17} \right) + C_{19} \|\nabla u_0\|^2 t. \end{aligned} \tag{1.12.33}$$

Choose $\|\nabla u_0\|$ small enough so that

$$C_{18} - C_{15} C_{16} \|\nabla u_0\|^2 \geq \frac{C_{18}}{2}. \tag{1.12.34}$$

From equations (1.12.7), (1.12.8), (1.12.31) and (1.12.32), we have

$$\int_0^t \|\Delta_M u\|^2 \mathrm{d}t \leq C_{14}, \quad 0 \leq t \leq T. \tag{1.12.35}$$

This completes the proof. □

Using equations (1.12.34) and (1.12.24), we get

$$\|\Delta_M u\|^2 \leq \frac{E_l}{t}, \quad \forall x = (x_1, x_2, \dots, x_n),\ n \leq 2,\ t > 0, \tag{1.12.36}$$

where the constant E_l only depends on $\|\nabla u_0\|$ and T. This completes the proof of the lemma. □

For the inequality (1.12.12) when $n = 1$, we do not need the restriction $\|\nabla u_0\| \leq \lambda$. In fact, by Lemma 1.12.2, we have

$$\begin{aligned}
\|\nabla u\|_\infty &\leq C_2 \|\nabla^3 u\|^{\frac{1}{4}} \|\nabla u\|^{\frac{3}{4}} + C_3, \\
\|\Delta_M u\| &\leq C_4 \|\nabla^3 u\|^{\frac{1}{2}} \|\nabla u\|^{\frac{1}{2}} + C_5 \quad (n = 1), \\
\int_M |\nabla u|^4 dM &\leq C_{16} \|\nabla^2 u\| \|\nabla u\|^3 + C_{17},
\end{aligned}$$

which yield inequality (1.12.12).

Now we prove the existence of a global unique smooth solution of problem (1.12.1)–(1.12.2)

Theorem 1.12.1 ([100]). *Let $u_0 : M^\varepsilon \equiv M \to S^2$ be a smooth mapping. Then there exists a constant $\varepsilon \geq 0$ and a unique smooth mapping $u : M \times [0, \varepsilon] \to S^2$, $u \in L_2^p(M^\varepsilon)$, such that*

$$\partial_t u = \alpha_1 \Delta_M u + \alpha_1 |\nabla u|^2 u + \alpha_2 u \times \Delta_M u, \quad (x, t) \in M \times [0, \varepsilon], \quad u = u_0, \quad x \in M \times \{0\}.$$

Lemma 1.12.6. *Let $u_0(x) \in H^1(M)$, $|u_0(x)|^2 = 1$. Then for a smooth solution of problem (1.12.1)–(1.12.2) we have*

$$\sup_{0 \leq t \leq T} \|u(\cdot, t)\|_{H^1(M)} \leq C_1, \tag{1.12.37}$$

where the constant C_1 depends on $\|u_0(x)\|_{H^1(M)}$.

Proof. Similar to that of Lemmas 1.12.1 and 1.12.2. □

Lemma 1.12.7. *Suppose the condition of Lemma 1.12.6 is satisfied, and*

$$\|\nabla u_0\| \leq \lambda, \quad n = 2, \tag{1.12.38}$$

where constant λ is small enough. Then we have

$$\sup_{0 \leq t \leq T} \|\nabla^2 u(\cdot, t)\| \leq C_2, \tag{1.12.39}$$

where constant C_2 depends on $\|u_0(x)\|_{H^2(M)}$.

Proof. Acting with Δ_M on equation (1.12.3), multiplying by $\Delta_M u$, and integrating with respect to $x \in M$, we get

$$(\Delta_M u, \Delta_M u_t - \alpha_1 \Delta_M^2 u - \alpha_1 \Delta_M(|\nabla u|^2 u) - \alpha_2 \Delta_M(u \times \Delta_M u)) = 0, \tag{1.12.40}$$

where

$$(\Delta_M u, \Delta_M u_t) = \frac{1}{2}\frac{\mathrm{d}}{\mathrm{d}t}\|\Delta_M u\|^2,$$
$$(\Delta_M u, \alpha_1 \Delta_M^2 u) = -\alpha_1 \|\nabla \Delta_M u\|^2.$$

Similarly as in the proof of Lemma 1.12.3, we get

$$\frac{1}{2}\frac{\mathrm{d}}{\mathrm{d}t}\|\Delta_M u\|^2 + [\alpha_1 - (C_3 + C_4\|\nabla u\|) \times \|\nabla u\|]\left\|\nabla^3 u\right\|^2 \le C_3, \tag{1.12.41}$$

where constant C_3 depends on $\|\nabla u_0\|$. Select $\|\nabla u_0\|$ small enough so that

$$\alpha_1 - (C_3 + C_4\|\nabla u_0\|)\|\nabla u_0\| \ge \frac{\alpha_1}{2}.$$

By equation (1.12.41), this implies

$$\sup_{0\le t\le T} \|\Delta_M u\|^2 \le C_4'. \tag{1.12.42}$$

□

Lemma 1.12.8. *Suppose the condition of Lemma* 1.12.7 *is satisfied. Set* $u_0(x) \in H^3(M)$. *Then we have*

$$\sup_{0\le t\le T} \left\|\nabla \Delta_M u(\cdot, t)\right\|^2 \le C_5, \tag{1.12.43}$$

where constant C_5 *depends on* $\|u_0(x)\|_{H^3(M)}$.

Proof. Acting with $\nabla\Delta_M$ on equation (1.12.3), and integrating with $\nabla\Delta_M u$, we get

$$(\nabla\Delta_M u, \nabla\Delta_M u_t - \alpha_1 \nabla\Delta_M^2 u - \alpha_1 \nabla\Delta_M(|\nabla u|^2 u) - \alpha_2 \nabla\Delta_M(u \times \Delta_M u)) = 0, \tag{1.12.44}$$

where

$$(\nabla\Delta_M u, \nabla\Delta_M u_t) = \frac{1}{2}\frac{\mathrm{d}}{\mathrm{d}t}\|\nabla\Delta_M u\|^2,$$
$$\begin{aligned}(\nabla\Delta_M^2 u, \nabla\Delta_M u) &= \int_M \gamma^{\delta l}\frac{\partial}{\partial x^\delta}\Delta_M^2 u \cdot \frac{\partial}{\partial x^l}\Delta_M u \mathrm{d}M \\ &= -\int_M \Delta_M^2 u \cdot \frac{\partial}{\partial x^\delta}\left(\gamma^{\delta l}\sqrt{\gamma}\frac{\partial}{\partial x^l}\Delta_M u\right)\mathrm{d}x \\ &= -\int_M (\Delta_M^2 u)^2 \mathrm{d}M = -\|\Delta_M^2 u\|^2,\end{aligned}$$

$$
\begin{aligned}
(\nabla\Delta_M(|\nabla u|^2u),\nabla\Delta_M u) &= \int\limits_M \gamma^{\delta l}\frac{\partial}{\partial x^\delta}\Delta_M(|\nabla u|^2u)\frac{\partial}{\partial x^l}\Delta_M u\cdot\sqrt{\gamma}dx\\
&= -\int\limits_M \Delta_M(|\nabla u|^2u)\frac{\partial}{\partial x^\delta}\sqrt{\gamma}\gamma^{\alpha\beta}\frac{\partial}{\partial x^l}\Delta_M udx\\
&= -\int\limits_M \Delta_M(|\nabla u|^2u)\Delta_M^2 udM.
\end{aligned}
\tag{1.12.45}
$$

Noting the following equations:

$$
\begin{aligned}
\Delta_M(|\nabla u|^2u) &= \sum_{i=1}^n(|\nabla u|^2u)_{ii}\\
&= \sum_{i=1}^n(|\nabla u|_i^2u+|\nabla u|^2u_i)_i\\
&= \sum_{i=1}^n(|\nabla u|_{ii}^2+2|\nabla u|_i^2u_i+|\nabla u|^2u_{ii})\\
&= \Delta_M(|\nabla u|^2)+2(|\nabla u|^2)\nabla u+|\nabla u|^2\Delta_M u,\\
\Delta_M(|\nabla u|^2) &= 2\sum u_{ij}^2+2\sum u_iu_{ijj}\\
&= 2\sum u_{ji}^2+2\sum u_i(\Delta_M u)_i+2\,\mathrm{Ric}(\nabla u,\nabla u)
\end{aligned}
$$

(see [190, p. 129]), from the equation (1.12.45) we get

$$
\begin{aligned}
&\int\limits_M \Delta_M(|\nabla u|^2u)\Delta_M^2 udM\\
&\quad= \int\limits_M [2\nabla^2u+2\nabla u\cdot\nabla\Delta_M u+2\,\mathrm{Ric}(\nabla u,\nabla u)]\Delta_M^2 udM,
\end{aligned}
\tag{1.12.46}
$$

where

$$
\left|2\int\limits_M \nabla^2u\Delta_M^2 udM\right| \le \frac{\alpha_1}{8}\|\Delta_M^2 u\|^2+C_7,
$$

$$
\begin{aligned}
\|\nabla u\|_\infty &\le C\|\nabla^4u\|^{\frac{1}{3}}\|\nabla u\|^{\frac{2}{3}},\\
\|\nabla\Delta_M u\| &= \|\Delta_M(\nabla u)-S_{10}(\nabla u)\| \le \|\Delta_M(\nabla u)\|+C\|\nabla u\|\\
&\le \|\nabla^3u\|+C\|\nabla^2u\|+C\|\nabla u\|\\
&\le \|\nabla^3u\|+C,
\end{aligned}
$$

$$
\begin{aligned}
\left|2\int\limits_M \nabla u\cdot\nabla\Delta_M u\cdot\Delta_M^2 udM\right| &\le 2\|\nabla u\|_\infty\|\nabla\Delta_M u\|\|\Delta_M^2 u\|\\
&\le \frac{\alpha_1}{8}\|\Delta^4u\|+C_8,\\
\left|\int\limits_M 2\,\mathrm{Ric}(\nabla u,\nabla u)\Delta_M^2 udM\right| &\le 2\|\nabla u\|_\infty\|\nabla u\|\|\Delta_M^2 u\|
\end{aligned}
$$

$$\leq \frac{\alpha_1}{8}\|\nabla^4 u\|^2 + C,$$
$$\|\nabla^4 u\|^2 - C(\|\nabla^3 u\|^2 + 1) \leq \|\Delta_M^2 u\|^2$$
$$\leq \|\nabla^4 u\|^2 + C[\|\nabla^3 u\|^2 + 1],$$

where we use Ricci formula. Then we have

$$\left|\int_M \Delta_M(|\nabla u|^2 u)\Delta_M^2 u \mathrm{d}M\right| \leq \frac{3\alpha_1}{8}\|\nabla^4 u\|^2 + C_{10}\|\nabla^3 u\|^2 + C_{11}. \tag{1.12.47}$$

Then, we estimate the terms in equation (1.12.44), namely

$$\alpha_2 \int_M \nabla\Delta_M u \cdot \nabla\Delta_M(u \times \Delta_M u)\mathrm{d}M$$

and

$$\int_M \nabla\Delta_M u \cdot \nabla\Delta_M(u \times \Delta_M u)\mathrm{d}M = -\int_M \Delta_M^2 u \cdot \Delta_M(u \times \Delta_M u)\mathrm{d}M,$$

where

$$\Delta_M(u \times \Delta_M u) = \sum_i \Big(u \times \sum_j u_{jj}\Big)_{ii}$$
$$= \sum_i \Big[\Big(u_i \times \sum_j u_{jj}\Big) + \Big(u \times \sum_j u_{jji}\Big)\Big]_i$$
$$= \sum_i \Big(u_{ii} \times \sum_{j=1} u_{jj}\Big) + \sum_i u_i \times \sum_{j=1} v_{jji} + \sum_i \Big(u_i \times \sum_{j=1} u_{jji}\Big) + \sum_j u_j \times \sum_{j=1} u_{jjii}$$
$$= 2\sum_{i=1} u_i \times \sum_{j=1} u_{jji} + u \times \Delta_M^2 u,$$

which implies

$$\left|\int_M \nabla\Delta_M u \cdot \nabla\Delta_M(u \times \nabla\Delta_M u)\mathrm{d}M\right|$$
$$\leq 2\int_M |\Delta_M^2 u| \cdot |\nabla u| \Big|\nabla\Delta_M u\Big| \mathrm{d}M \leq 2\|\nabla u\|_\infty \|\Delta_M^2 u\| \|\nabla\Delta_M u\|$$
$$\leq C\|\nabla^4 u\|^{\frac{1}{3}} \|\nabla u\|^{\frac{2}{3}} \|\Delta_M^2 u\| \|\nabla\Delta_M u\| + C$$
$$\leq \|\nabla^4 u\|^{\frac{1}{3}} \|\nabla^3 u\| \|\nabla^4 u\| + C\|\nabla^3 u\| + C$$
$$\leq \frac{\alpha_1}{4|\alpha_2|}\|\nabla^4 u\|^2 + C_{12}\|\nabla^3 u\|^2 + C_{13}, \tag{1.12.48}$$

where we used the following Sobolev inequality:

$$\|\nabla^3 u\| \le C\|\nabla^4 u\|^{\frac{1}{2}}\|\nabla^2 u\|^{\frac{1}{2}} + C.$$

Due to equations (1.12.47) and (1.12.48), we get

$$\frac{\mathrm{d}}{\mathrm{d}t}\|\nabla\Delta_M u\|^2 + \frac{3}{4}\alpha_1\|\nabla^4 u\| \le C_{14}\|\nabla^3 u\|^2 + C_{15}. \tag{1.12.49}$$

To the above inequalities, we use the inequality

$$\|\nabla^4 u\|^2 - C(\|\nabla^3 u\|^2 + 1) \le \|\Delta_M^2 u\|^2 \le \|\nabla^4 u\|^2 + C(\|\nabla^3 u\|^2 + 1).$$

Integrating equation (1.12.49) with respect to $t \in [0, T]$, we get

$$\|\nabla\Delta_M u(\cdot,t)\|^2 + \frac{3}{4}\alpha_1\int_0^t \|\nabla^4 u\|\mathrm{d}t \le C_{14}\int_0^t \|\nabla^3 u\|^2\mathrm{d}t + C_{15}t + \|\nabla\Delta_M u_0\|^2. \tag{1.12.50}$$

Using the inequality

$$\|\nabla^3 u\|^2 - C \le \|\nabla\Delta_M u\|^2 \le \|\nabla^3 u\|^2 + C,$$

and also Gronwall inequality, from equation (1.12.50) we have

$$\|\nabla\Delta_M u\|^2 \le C_{16},$$

where constant C_{16} depends on $\|u_0(x)\|_{H^3}$.

Using Lemmas 1.12.6, 1.12.2 and 1.12.8, we get

$$\sup_{0\le t\le T}\|u(\cdot,t)\|_{H^3(M)} \le C_{17},$$

where constant C_{17} depends on $\|u_0(x)\|_{H^3(M)}$. □

Lemma 1.12.9. *Suppose the condition of Lemma 1.12.8 is satisfied, and $u_0(x) \in H^m(M)$, $m \ge 4$. Then we have*

$$\sup_{0\le t\le T}\|u(\cdot,t)\|_{H^m(M)} \le C_{18} \tag{1.12.51}$$

where constant C_{18} depends on $\|u_0(x)\|_{H^m(M)}$.

Proof. Firstly, letting $m = 2n$, acting with Δ_M^{2n} on equation (1.12.3), and taking the inner product with $\Delta_M^{2n}u$, we get

$$(\Delta_M^{2n}u, \Delta_M^{2n}u_t - \alpha_1\Delta_M^{2n+1}u - \alpha_1\Delta_M^{2n}(|\nabla u|^2 u) - \alpha_2\Delta_M^{2n}(u \times \Delta_M u)) = 0.$$

By induction, Sobolev interpolation inequality, Ricci formula, and

$$C_3'\|\nabla^{4n}u\| - C_4'\|\nabla^{4n-1}u\| - C_5' \le \|\Delta_M^{2n}u\| \le \|\nabla^{4n}u\| + C_1'\|\nabla^{4n-1}u\| + C_2',$$
$$C_3''\|\nabla^{4n+1}u\| - C_4''\|\nabla^{4n}u\| - C_5'' \le \|\nabla\Delta_M^{2n}u\| \le \|\nabla^{4n+1}u\| + C_1''\|\nabla^{4n}u\| + C_2'',$$

we get

$$\frac{\mathrm{d}}{\mathrm{d}t}\|\nabla\Delta_M^{2n}u\|^2 + C_{19}\alpha_1\|\nabla^{4n-1}u\|^2 \le C_{20}. \tag{1.12.52}$$

Secondly, for $m = 2n+1$ we have

$$\frac{\mathrm{d}}{\mathrm{d}t}\|\Delta_M^{2n+1}u\|^2 + C_{21}\alpha_1\|\nabla^{4n+2}u\|^2 \le C_{22}. \tag{1.12.53}$$

From equations (1.12.52) and (1.12.53), we get inequality (1.12.51). □

From Lemmas 1.12.6–1.12.10 and Theorem 1.12.7, we get the existence of a smooth solution for problem (1.12.1)–(1.12.2). As for the uniqueness of a smooth solution, it is easy to obtain.

Theorem 1.12.2. *Let M be a boundless Riemannian manifold, which satisfies the following conditions:*

(1) $u_0(x) \in H^m(M)$, $m \ge 2$, $|u_0(x)|^2 = 1$,

$$x = (x_1, \ldots, x_n) \in M, \quad 1 \le n \le 2;$$

(2) *when* $n = 2$,

$$\|\nabla u_0(x)\| \le \lambda,$$

where the constant λ *is small enough. Then there exists a unique global smooth solution* $u(x,t) : M \times [0,\infty) \to S^2$ *for the initial value problem* (1.12.1)–(1.12.2),

$$u(x,t) \in L^\infty(0,\infty; H^m(M)).$$

Using Theorem 1.12.2 and Lemmas 1.12.1–1.12.2, we will prove that problem (1.12.1)–(1.12.2) possess an attractor.

Theorem 1.12.3. *Let M be a boundless Riemannian manifold* ($n \le 2$)*, which satisfies the following conditions:*

(1) $\alpha_1 > 0$, $|u_0(x)| = 1$, $u_0(x) \in H^1(M)$, $x = (x_1, \ldots, x_n) \in M$, $n \le 2$;

(2) $\|\nabla u_0(x)\| \le \lambda$, $x \in M$, $n = 2$, *where constant* λ *is small enough.*

Then on the manifold M the initial value problem (1.12.1)–(1.12.2) *of Landau–Lifshitz equation has a unique attractor* $\mathscr{A}$*. It is compact in* $H^1(M)$*, and*

$$\mathscr{A} = \omega(B_1) = \bigcap_{s\ge 0}\bigcap_{t\ge s}\overline{S(t)B_1}$$

where

$$B_1 = \{u \in H^1(M),\ |u(\cdot,t)| = 1,\ \|u(\cdot,t)\|_{H^1(M)} \le \rho_1\}$$

is a bounded absorbing subset of $E = \{u \in H^1(M) \mid |u(\cdot,t)| = 1\}$ *for* $S(t)$ *in* $H^1(M)$, ρ_1 *is a positive constant, and* $S(t)u_0$ *is the semigroup operator generated by problem* (1.12.1)–(1.12.2).

Proof. By Theorem 1.12.2, we can get the existence of a global smooth solution for problem (1.12.1)–(1.12.2). Moreover, it forms a semigroup. From Lemmas 1.12.1–1.12.2, we know that

$$B_1 = \{u \in H^1(M),\ |u| = 1,\ \|u\|_{H^1(M)} \le \|\nabla u_0(x)\| + \operatorname{vol} M = \rho_1\}$$

is a bounded subset of the set E in $H^1(M)$ for $S(t)$. By Lemma 1.12.4, we have

$$\|u(\cdot,t)\|_{H^2(M)} \le \frac{E_1}{t}, \quad t > 0,$$

where constant E_1 depends on $\|\nabla u_0(x)\|_{H^1(M)}$. This implies that the semigroup operator $S(t)$ is completely continuous in $H^1(M)$, $t > 0$. Then by a theorem in [197], we know that the semigroup $S(t)$ produces a compact attractor in $H^1(M)$,

$$\mathscr{A} = \bigcap_{s\ge 0}\bigcup_{t\ge s}\overline{S(t)B_1} = \omega(B_1).$$

Now we estimate the upper and lower Hausdorff and fractal dimensions of the attractor $\mathscr{A}$.

We consider the linear variational problem of (1.12.1)–(1.12.2):

$$v(t) + L(u(t))v = 0, \tag{1.12.54}$$

$$v(0) = v_0(x), \tag{1.12.55}$$

where

$$L(u(t))v = -\alpha_1\Delta_M v - \alpha_1|\nabla u|^2 v - 2\alpha_1\nabla u u\nabla v - \alpha_2 A'(u)\Delta_M u v - \alpha_2 A(u)\Delta_M v, \tag{1.12.56}$$

$$A(u) = \begin{pmatrix} 0 & -u_3 & u_2 \\ u_2 & 0 & -u_1 \\ -u_2 & u_1 & 0 \end{pmatrix}, \quad u = \begin{pmatrix} u_1 \\ u_2 \\ u_3 \end{pmatrix}.$$

Since the solution of problem (1.12.1)–(1.12.2) is sufficiently smooth, it is easy to prove that the linear problem (1.12.54)–(1.12.55) has a unique global solution $v(x,t) \in L^\infty(0,\infty; H^2(M))$, provided that the initial value $v_0(x)$ is sufficiently smooth. In fact, for the linear equation (1.12.54), its main part is

$$\sum_{\alpha,\beta=1}^{m} D_\alpha(a_{\alpha\beta}(x,t,u)D_\beta u) = \sum_{\alpha,\beta=1}^{m} D_\alpha\sqrt{\gamma}(\gamma^{\alpha\beta}D_\beta u + u \times \gamma^{\alpha\beta}D_\beta u), \tag{1.12.57}$$

the corresponding system $a^{ij}_{\alpha,\beta}$ is

$$a^{ij}_{\alpha,\beta} = \sqrt{\gamma}\gamma^{\alpha\beta}g_{ij}A\#,$$

where

$$g = (g_{ij}) = \begin{pmatrix} \alpha_1 & -\alpha_2 u_3 & \alpha_2 u_2 \\ \alpha_2 u_3 & \alpha_1 & -\alpha_2 u_1 \\ -\alpha_2 u_2 & \alpha_2 u_1 & \alpha_1 \end{pmatrix}. \tag{1.12.58}$$

Thus we have

$$\sum_{i,j=1}^{s}\sum_{\alpha,\beta=1}^{m} a^{ij}_{\alpha,\beta}(x,t,\eta)\xi^\alpha\xi^\beta\zeta_i\zeta_j = \alpha_1|\zeta|^2\sum_{\alpha,\beta=1}^{m}\gamma^{\alpha\beta}\sqrt{\gamma}\xi^\alpha\xi^\beta > 0 \tag{1.12.59}$$

for any $(x,t,\eta) \in M \times [0,T] \times \mathbf{R}^3$, $\xi = (\xi^1,\dots,\xi^m) \in \mathbf{R}^m \setminus \{0\}$, $\zeta = (\zeta_1,\zeta_2,\zeta_3) \in \mathbf{R}^3 \setminus \{0\}$. Thus the problem (1.12.54)–(1.12.55) is solvable.

Let G_t be the solution operator, $v(t) = G_t v_0$.

It is easy to prove that the semigroup $S_t v_0$ is differentiable in $L_2(M)$. The Fréchet derivative of $S_t v_0$ exists. And $G_t v_0 = S_t' v_0$. □

In order to estimate the dimension of $\mathscr{A}$, we need the following theorems.

Theorem 1.12.4 ([77]). *Let (M,g) be an n-dimensional Riemannian manifold. For any p such that*

$$\max\left\{1, \frac{n}{2m}\right\} < p \le 1 + \frac{n}{2m}$$

there exist two positive constants $k(M)$ and $\chi(M)$ such that for any finite set $\{\varphi_1,\dots,\varphi_N\} \in H^m(M)$, functions of which are orthogonal in $L^2(M)$,

$$\left(\int_M \varphi^{\frac{p}{p-1}}\mathrm{d}M\right)^{\frac{2m(p-1)}{n}} \le k(M)\sum_{j=1}^{N}\int_M |\nabla^m\varphi_j|^2\mathrm{d}M + \chi(M)\int_M \rho\,\mathrm{d}M, \tag{1.12.60}$$

where $\rho = \sum_{j=1}^N |\varphi_j(x)|^2$, and constants $k(M)$ and $\chi(M)$ depend on m, n, p and (M,g).

Lemma 1.12.10 ([190]). *Let M be an n-dimensional bounded or unbounded Riemannian manifold. Assume $\{\lambda_j\}$ is the spectrum of Laplace–Beltrami operator on M. Then the following inequality holds:*

$$\lambda_k \ge \frac{\delta}{e}\left(\frac{k}{\mathrm{vol}(M)}\right)^{\frac{2}{n}}, \tag{1.12.61}$$

where $n = \dim M$ and δ is the Sobolev constant for M. That is, for any $u \in C^\infty(M)$, we have

$$\int_k |\nabla u|^2\mathrm{d}M \ge \delta\left(\int_M |u|^{\frac{2n}{n-2}}\right)^{\frac{n-2}{n}}.$$

Theorem 1.12.5. *Suppose condition* (1.12.2) *is satisfied. Then the Hausdorff and fractal dimensions of the attractor* $\mathscr{A}$ *for the problem* (1.12.1)–(1.12.2) *are finite, that is,*

$$d_H(\mathscr{A}) \le J_0, \quad d_F(\mathscr{A}) \le 2J_0, \tag{1.12.62}$$

where J_0 *is the smallest integer such that*

$$J_0 \ge C_0 \alpha_1^{\frac{-4n}{(4-n)(2+n)}}, \quad 1 \le n \le 2, \tag{1.12.63}$$

where constant C_0 *depends on* M, $\|u\|_\infty$, $\|\nabla u\|_\infty$ *and* $\|\nabla^2 u\|_2$.

Proof. Based on Theorem 7.1 in [197], we only need to estimate the lower bound of $\operatorname{tr}(L(u(t)) \cdot Q_J(t))$. Assume that $\{\varphi_1(x), \dots, \varphi_J(x)\}$ is the standard orthogonal basis of $Q_J L_2$, then we have

$$\begin{aligned}
\operatorname{tr}(L(u(t)) \cdot Q_J) &= \sum_{j=1}^{J} (L(u(t)) \cdot Q_J(t)\varphi_j, \varphi_j) \\
&= \sum_{j=1}^{J} (L(u(t)) \cdot Q_J(t)\varphi_j(\tau), \varphi_j(\tau)) \\
&= \sum_{j=1}^{J} \{-\alpha_1(\Delta_M \varphi_j, \varphi_j) - \alpha_1(|\nabla u|^2 \varphi_j, \varphi_j) \\
&\quad - 2\alpha_1(\nabla u \cdot u \cdot \nabla\varphi_j, \varphi_j) - \alpha_2(A'(u)\nabla u \varphi_j, \varphi_j) - \alpha_2(A(u) \cdot \Delta_M \varphi_j, \varphi_j)\} \\
&\ge \alpha_1 \sum_{j=1}^{J} \lambda_j - \alpha_1 \big\||\nabla u|^2\big\|_2 \|\rho(x)\|_2 - 2\alpha_1 \|\nabla(\nabla u \cdot u)\|_2 \|\rho(x)\|_2 \\
&\quad - \alpha_2 \|A'(u)\|_\infty \|\nabla u\|_2 \|\rho(x)\|_2 \\
&\ge \alpha_1 \sum_{j=1}^{J} \lambda_j - 2\alpha_1 \|\nabla u\|_\infty \|\nabla u\|_2 \|\rho(x)\|_2 \\
&\quad - 2\alpha_1 \|u\|_\infty \|\nabla^2 u\|_2 \|\rho(x)\|_2 - \alpha_2 \|A'(u)\|_\infty \|\nabla u\|_2 \|\rho(x)\|_2 \\
&\ge \alpha_1 \sum_{j=1}^{J} \lambda_j - (2\alpha_1 \|\nabla u\|_\infty \|\nabla u\|_2 + 2\alpha_1 \|u\|_\infty \|\nabla^2 u\|_2 \\
&\quad + \alpha_2 \|A'(u)\|_\infty \|\nabla u\|_2) \|\rho(x)\|_2 \\
&\ge \alpha_1 \sum_{j=1}^{J} \lambda_j - H\bigg(k(M) \sum_{j=1}^{J} \lambda_j + \chi(M)\bigg)^{\frac{m}{4}} \\
&\ge \alpha_1 \bigg(\sum_{j=1}^{J} \lambda_j\bigg)^{\frac{m}{4}} \cdot \bigg[\bigg(\sum_{j=1}^{J} \lambda_j\bigg)^{1-\frac{m}{4}} - \frac{C_1}{\alpha_1} H\bigg] - C_2 H \\
&\ge \alpha_1 \bigg(\sum_{j=1}^{J} \lambda_j^{\frac{m}{4}} - C_2 H\bigg) > 0,
\end{aligned}$$

where

$$H = 2\alpha_1\|\nabla u\|_\infty\|\nabla u\|_2 + 2\alpha_1\|u\|_\infty\|\nabla^2 u\|_2 + \alpha_2\|A'(u)\|_\infty\|\nabla u\|_2$$

$$J \geq \max\left\{\left(C_3\left(\frac{2}{n}\right)\right)^{\frac{n}{2-n}} \cdot \left[\frac{C_1}{\alpha_1}H + 1\right]^{\frac{4n}{(4-n)(2-n)}}, \left(C_3\left(\frac{2}{n}\right)\right)^{\frac{n}{2+n}}\left[\frac{C_2}{\alpha_1}H\right]^{2-\frac{4}{n}}\right\},$$

and where we used the following inequality [197]:

$$\sum_{j=1}^{J} j^{\frac{2}{m}} \geq C_3\left(\frac{2}{m}\right)J^{\frac{2}{m}+1},$$

with C_1, C_2 depending on the constant of manifold M.

Now we estimate a lower bound of dimension for the attractor $\mathscr{A}$.

Consider a Banach space E and a continuous operator semigroup $S(t) : E \to E$,

$$S(t+s) = S(t)S(s), \quad \forall s, t > 0, \tag{1.12.64}$$

$$S(0) = I \quad \text{(the identity operator on } E). \tag{1.12.65}$$

Assume the semigroup $(t, u_0) \to S(t)u_0 : \mathbf{R}^+ \times E \to E$ is continuous.

Let Z be a fixed point of $S(t)$, that is,

$$S(t)Z = Z, \quad \forall t \in \mathbf{R}^+. \tag{1.12.66}$$

Assume the mapping $u \to S(t)u$ is Fréchet differentiable in a neighborhood of Z, say Θ, and its differential $S'(t)$ satisfies the Hölder condition

$$\|S'(t)u_1 - S'(t)u_2\| \leq C_3(T)\|u_1 - u_2\|^\alpha, \quad 0 < \alpha \leq 1, \ \forall u_1, u_2 \in \Theta, \ \forall t \in [0, T], \tag{1.12.67}$$

where constant C_3 depends on T, but not on u_1, u_2. □

Definition 1.12.1. We say that the fixed point Z is hyperbolic, provided that the following conditions are satisfied:

(1) The spectrum $\sigma(S'(t))$ of $S(t)$ does not intersect the circle $\{\lambda \in C, |\lambda| = 1\}$.

(2) E_+ has finite dimension, where $E_+ = E_+(Z)$ and $E_- = E_-(Z)$ are the linear invariant subspaces of E, which correspond to the subsets of $S'(t)$ included in $\{\lambda \in C, |\lambda| > 1\}$ and $\{\lambda \in C, |\lambda| < 1\}$, respectively. E_+ and E_- are independent of time t.

Definition 1.12.2. The unstable manifold $\mu_+(Z)$ of Z is the set of points $u_* \in H$ (may be empty), which totally belongs to the complete path $\{u(t), t \in \mathbf{R}\}$ and satisfies

$$u(t) \to u_0 \quad \text{as } t \to -\infty,$$

$$\mu_+(Z) = \{u_0 \in E, \forall t \leq 0, \exists u(t) \in S(-t)^{-1}u_0, \text{ and } u(t) \to Z, \text{ when } t \to -\infty\}.$$

The stable manifold $\mu_-(Z)$ of Z is the set of points $u_* \in H$ (may be empty), which belongs to the complete path $\{u(t), t \in \mathbf{R}\}$, $u_* = u(t_0)$ and satisfies

$$u(t) = S(t - t_0)u_* \to u_0, \quad t \to \infty,$$
$$\mu_-(Z) = \{u_0 \in E, \quad \forall t \leq 0, \exists u(t) \in S(-t)^{-1}u_0, \text{ and } S(t)u_0 \to Z, \text{ when } t \to \infty\}.$$

From the definition, we directly deduce that

$$S(t)\mu_+(Z) = \mu_+(Z), \quad S(t)\mu_-(Z) = \mu_-(Z), \quad \forall t \geq 0.$$

Definition 1.12.3. A heteroclinic orbit is an orbit which connects the unstable manifold of a steady point u_* to another stable manifold of a steady solution u_{**}, $u_{**} \neq u_*$; while if $u_{**} = u_*$, such an orbit is called a homoclinic orbit. The points belonging to a heteroclinic (homoclinic) orbit are called heteroclinic (homoclinic) points.

Consider the ball with radius $R > 0$,

$$\Theta_R(Z) = \{y \in E, \|y - Z\|_0 \leq R\},$$

and let

$$\mu_-^R(Z) = \Big\{u_0 \in \Theta(Z), \forall n \in N, \exists u_n \in S^n(u_0) \cap \Theta_R(Z), \text{and } u(t) \to Z \text{ when } n \to \infty\Big\}. \tag{1.12.68}$$

Theorem 1.12.6 ([51]). *Let E be a Banach space, and suppose $S(t)$ is a semigroup operator, $t \in \mathbf{R}^+$, which satisfies assumptions* (1.12.64), (1.12.65) *and* (1.12.67). *Suppose $S(t)$ possesses a unique global attractor $\mathscr{A}$, and $Z \in \mathscr{A}$ is the hyperbolic fixed point of $S(t)$. Then we have*

$$A \supset \mu_-(Z) \supset \mu_+^R(Z), \tag{1.12.69}$$

for $R > 0$ small enough.

Now we consider the following initial value problem for the Landau–Lifshitz equation:

$$Z_t = -\alpha_1(Z \times (Z \times \Delta_M Z)) + Z \times JZ, \tag{1.12.70}$$

$$Z|_t = 0 = Z_0(x), \quad x \in M, \quad \frac{\partial Z}{\partial \nu}\Big|_{\partial M} = 0, \tag{1.12.71}$$

where M is an unbounded compact Riemannian manifold and $J = \operatorname{diag}(J_1, J_2, 0)$, $\alpha_1 > 0$. By Theorem 1.12.6, we get a lower bound of the dimension for the attractor of problem (1.12.70)–(1.12.71).

Theorem 1.12.7. *Assume that $\mathscr{A}$ is the global attractor of problem* (1.12.70)–(1.12.71), *$J_1, J_2 < 0$, then we have*

$$\dim \mathscr{A} \geq C\alpha^{-\frac{n}{2}}, \tag{1.12.72}$$

where $\dim \mathscr{A}$ is Hausdorff or fractal dimension of A and C is a positive constant.

Proof. It is easy to see that $Z = (0,0,1)$ is the fixed point of $S(t)$ generated by problem (1.12.70)–(1.12.71), that is, $Z = (0,0,1)$ satisfies the equation

$$-A_1(Z) = -\alpha_1(Z \times (Z \times \Delta_M Z)) + Z \times JZ = 0. \tag{1.12.73}$$

The variational problem for equation (1.12.73) is

$$-A_1'(Z)v = \alpha_1 \Delta_M v + Z \times Jv = 0. \tag{1.12.74}$$

Let ζ be the eigenvalue of matrix $B(Z)v = Z \times Jv = \zeta v$, where

$$B(Z) = \begin{pmatrix} 0 & -J_2 & 0 \\ J_1 & 0 & 0 \\ 0 & 0 & 0 \end{pmatrix}.$$

Hence

$$\begin{aligned} &\zeta^2 + J_1 J_2 = 0, \\ &\zeta_1 = \sqrt{-J_1 J_2} > 0, \quad \zeta^2 = -\sqrt{-J_1 J_2}. \end{aligned}$$

Let λ_k, $k \in \mathbf{N}$ be the eigenvalues of the operator $-\Delta_M$ of M, that is,

$$\begin{aligned} &-\Delta_M \psi_k = \lambda_k \psi_k, \quad \frac{\partial \psi_k}{\partial v}\bigg|_{\partial M} = 0, \\ &0 < \lambda_1 \le \lambda_2 \le \cdots, \quad \lambda_k \to \infty, \quad k \to \infty. \end{aligned} \tag{1.12.75}$$

Let $\mu_k, k \in \mathbf{N}$ stand for the characteristic value sequence of the linear operator

$$-A_1'(Z)\omega_k = \alpha_1 \Delta_M \omega_k + B(Z)\omega_k = \mu_k \omega_k, \tag{1.12.76}$$

where $\omega_k(x) = \psi_k(x) p_k$, $p_k \in C^3$. Then we have

$$(\alpha_1 \lambda_k + B(Z)) p_k = \mu_k p_k. \tag{1.12.77}$$

If μ_k is the root of equation

$$\det(\alpha_1 \lambda_k + B(Z) - \mu_k I) = 0, \quad \operatorname{Re} \mu_k > 0, \tag{1.12.78}$$

then there exists a nonzero solution p_k. By the assumption of the theorem, when $\alpha_1 \ne 0$, there exists at least one root $\zeta_1 = \sqrt{-J_1 J_2} > 0$. Hence when $\alpha_1 \lambda_k < \delta$, there exists a root of equation (1.12.78) $\mu_k, \operatorname{Re} \mu_k > 0$, where δ is a small enough constant. Then due to the asymptotic behavior of characteristic values λ_k, $\lambda_k \sim Ck^{-\frac{n}{2}}$, we get the inequality

$$1 \le k \le C_1 \delta^{\frac{n}{2}} \alpha_1^{-\frac{n}{2}} = C_2 \alpha_1^{-\frac{n}{2}}.$$

By Theorem 1.12.6, we get

$$\dim \mathscr{A} \ge C\alpha_1^{-\frac{n}{2}}.$$ □

1.13 The dissipation Klein–Gordon–Schrödinger equations on $\mathbf{R}^3$

We consider the following Klein–Gordon–Schrödinger (KGS) equations [104, 136]

$$i\psi_t + \Delta\psi = -\phi\psi, \tag{1.13.1}$$

$$\phi_{tt} - \Delta\phi + \mu^2\phi = |\psi|^2, \tag{1.13.2}$$

where $\psi(x,t)$ is a complex nuclear field; ϕ is a real meson field; and μ is the quality of a meson. The Cauchy and initial–boundary value problems of KGS equations have been studied by many authors, e. g., in [9, 70, 106, 139], and so on. In [9], the existence of a global solution is obtained by use of the L^p–L^q estimates for Schrödinger equations. In [70], the asymptotic behavior of multidimensional KGS equations has been studied. The authors considered the initial–boundary value problem and the existence of a three-dimensional strong solution for KGS equations in [106], which was improved in [139].

When we consider the effect of damping, we have the following dissipation KGS equations:

$$i\psi_t + \Delta\psi + i\alpha\psi + \phi\psi = f, \tag{1.13.3}$$

$$\phi_{tt} + (I - \Delta)\phi + \beta\phi_t = |\psi|^2 + g, \tag{1.13.4}$$

where α, β are positive; $f(x)$, $g(x)$ are known functions; f is complex; and g is real.

The long time asymptotic behavior for equations (1.13.3)–(1.13.4) in a bounded domain Ω was obtained in [18, 161]. In [18], Biler proved the existence of a global attractor in the weak topology of $H_0^1 \times H_0^1(\Omega)$ and the finiteness of Hausdorff dimension. In [161], the authors proved the existence of a finite dimensional global attractor in $H^2 \cap H_0^1(\Omega) \times H^2 \cap H_0^1(\Omega)$. Here we consider the KGS equations (1.13.3)–(1.13.4) as in [8] in the whole space $\mathbf{R}^3$ and the initial conditions

$$\phi(0,x) = \psi_0(x), \quad \phi(0,x) = \phi_0(x), \quad \phi_t(0,x) = \psi_1(x), \quad x \in \mathbf{R}^3. \tag{1.13.5}$$

We prove that there exists a global attractor in $H^2 \times H^2 \times H^1(\mathbf{R}^3)$ for problem (1.13.3)–(1.13.5), which attracts bounded sets in $H^3 \times H^3 \times H^2(\mathbf{R}^3)$ with respect to $H^2 \times H^2 \times H^1(\mathbf{R}^3)$.

Because the domain is unbounded, the embedding of $H^s(\mathbf{R}^3)$ into $H^{s'}(\mathbf{R}^3)$ for $s > s'$ is not compact. In order to overcome this difficulty, we use the noncompact Kuratowski α measure to prove the asymptotic smoothness of the semigroup $S(t)$, and then reuse the theory in [137] to prove the existence of the attractor.

Let $\theta = \phi_t + \delta\phi$, with δ being an unknown positive constant. Equations (1.13.3)–(1.13.5) are equivalent to

$$i\psi_t + \Delta\psi + i\alpha\psi + \phi\psi = f, \tag{1.13.6}$$

$$\phi_t + \beta\phi = \theta, \tag{1.13.7}$$

$$\theta_t + (\beta - \delta)\theta + (1 - \delta(\beta - \delta) - \Delta)\phi = |\psi|^2 + g, \tag{1.13.8}$$

with the initial condition

$$(\psi, \phi, \theta)(0, x) = (\psi_0, \phi_0, \theta_0)(x), \quad x \in \mathbf{R}^3, \tag{1.13.9}$$

where $\theta_0 = \delta\phi_0 + \phi_1$. If $\delta \le \min\{\frac{\beta}{2}, \frac{1}{2\beta}\}$, then $A = 1 - \delta(\beta - \delta) - \Delta$ is a positive, self-adjoint second-order elliptic operator. Denote

$$\begin{aligned} H &= L^2 \times H^{\frac{1}{2}} \times H^{-\frac{1}{2}}(\mathbf{R}^3), \\ V &= H^1 \times H^1 \times L^2(\mathbf{R}^3), \\ X &= H^2 \times H^2 \times H^1(\mathbf{R}^3), \\ Y &= H^3 \times H^3 \times H^2(\mathbf{R}^3). \end{aligned}$$

Then $Y \hookrightarrow X \hookrightarrow V$ is a continuous embedding.

Lemma 1.13.1. *Let $f \in L^\infty(\mathbf{R}^+; L^2(\mathbf{R}^3))$. Then $\psi \in L^\infty(\mathbf{R}^+; L^2(\mathbf{R}^3))$, and it satisfies*

$$\|\psi(t)\|^2 \le \|\psi_0\|^2 \exp(-\alpha t) + \frac{|||f|||^2}{\alpha^2}(1 - \exp(-\alpha t)).$$

Hence, there exists $t_1(R) > 0$ such that

$$\|\psi(t)\|^2 \le 1 + \frac{|||f|||^2}{\alpha^2}, \quad \forall t \ge t_1(R),$$

provided $\|\psi_0\| < R$, where $|||f|||$ means the norm of f in $L^\infty(\mathbf{R}^-; L^2(\mathbf{R}^3))$.

Proof. Taking the inner product of (1.13.6) and ψ in $L^2(\mathbf{R}^3)$, then its imaginary part, we obtain

$$\frac{1}{2}\frac{\mathrm{d}}{\mathrm{d}t}\|\psi\|^2 + \alpha\|\psi\|^2 = \operatorname{Im}(f, \psi) \le \|f\|\|\psi\| \le \frac{\alpha}{2}\|\psi\|^2 + \frac{\|f\|^2}{2\alpha}.$$

By Gronwall inequality, we get the claim of the lemma. □

Lemma 1.13.2. *Let $f, g \in L^\infty(\mathbf{R}^+; L^2(\mathbf{R}^3))$. Then for $(\psi_0, \phi_0, \theta_0) \in V$, solution $(\psi, \phi, \theta) \in L^\infty(\mathbf{R}^+, V)$. Furthermore, there exists $t_2(R) > 0$ such that*

$$\|(\psi, \phi, \theta)\|_V \le C, \quad \forall t \ge t_2(R),$$

whenever $\|(\psi_0, \phi_0, \theta_0)\|_V \le R$.

Proof. Taking the inner product of (1.13.6) and $-(\psi_t + \alpha\psi)$ in L^2, then its real part, we obtain

$$\begin{aligned} &\frac{1}{2}\frac{\mathrm{d}}{\mathrm{d}t}\|\nabla\psi\|^2 + \alpha\|\nabla\psi\|^2 - \operatorname{Re}(\phi\psi, \psi_t) - \alpha\operatorname{Re}(\phi\psi, \psi) \\ &= -\frac{\mathrm{d}}{\mathrm{d}t}\operatorname{Re}(f, \psi) - \operatorname{Re}(f_t, \psi) - \alpha\operatorname{Re}(f, \psi). \end{aligned} \tag{1.13.10}$$

Noting that

$$-\operatorname{Re}(\phi\psi, \psi_t) = -\frac{1}{2}\frac{d}{dt}(\phi, |\psi|^2) + \frac{1}{2}(\phi_t, |\psi|^2),$$

from equation (1.13.10) we get

$$\begin{aligned}&\frac{1}{2}\frac{d}{dt}\left(\|\nabla\psi\|^2 - \int \phi|\psi|^2 dx + 2\operatorname{Re}\int f\overline{\psi} dx\right) + \alpha\|\nabla\psi\|^2 \\ &\quad - \alpha\int \phi|\psi|^2 dx + \alpha\int f\overline{\psi} dx - \frac{1}{2}\int \phi_t|\psi|^2 dx = 0. \end{aligned} \tag{1.13.11}$$

Taking the inner product of (1.13.8) and θ in L^2, then using equation (1.13.7), we get

$$\begin{aligned}&\frac{1}{2}\frac{d}{dt}(\|\theta\|^2 + (1-\delta(\beta-\delta))\|\phi\|^2 + \|\nabla\phi\|^2) \\ &\quad + (\beta-\delta)\|\theta\|^2 + \delta(1-\delta(\beta-\delta))\|\phi\|^2 + \delta\|\nabla\phi\|^2 \\ &= \int \phi_t|\psi|^2 dx + \delta\int \phi|\psi|^2 dx + \int g\theta dx. \end{aligned} \tag{1.13.12}$$

Then multiplying (1.13.11) by 2 and adding (1.13.12), we get

$$\frac{d}{dt}H_1(t) + I_1(t) = 0, \tag{1.13.13}$$

where

$$\begin{aligned} H_1(t) &= 2\|\nabla\psi\|^2 - 2\int \phi|\psi|^2 dx + 2\operatorname{Re}\int f\overline{\psi} dx + \|\theta\|^2 \\ &\quad + (1-\delta(\beta-\delta))\|\phi\|^2 + \|\nabla\phi\|^2, \\ I_1(t) &= 4\|\nabla\psi\|^2 - 2(2\alpha+\delta)\int \phi|\psi|^2 dx + 4\operatorname{Re}\int f\overline{\psi} dx + 2(\beta-\delta)\|\theta\|^2 \\ &\quad + 2\delta(1-\delta(\beta-\delta))\|\phi\|^2 + 2\delta\|\nabla\phi\|^2 - 2\int g\theta dx. \end{aligned} \tag{1.13.14}$$

Since $H^1(\mathbf{R}^3) \hookrightarrow L^6(\mathbf{R}^3)$

$$\|\psi\|_3 \le C\|\psi\|^{\frac{1}{2}}\|\nabla\psi\|^{\frac{1}{2}},$$

for any $\varepsilon_1, \varepsilon_2 > 0$ we have

$$\begin{aligned} \left|\int \phi|\psi|^2 dx\right| &\le C\|\psi\|_6\|\psi\|_3\|\psi\| \\ &\le C\|\nabla\phi\|\|\psi\|^{\frac{3}{2}}\|\nabla\psi\|^{\frac{1}{2}} \le \varepsilon_1\|\nabla\psi\|^2 + \varepsilon_2\|\nabla\phi\|^2 + C(\varepsilon_1,\varepsilon_2)\|\psi\|^6, \quad (1.13.15) \\ \left|\int f\overline{\psi} dx\right| &\le \|f\|\|\psi\|. \end{aligned}$$

In equation (1.13.15), we take $\varepsilon_1 = \frac{1}{2}$, $\varepsilon_2 = \frac{1}{4}$, and then obtain

$$\begin{aligned} H_1(t) &\geq \|\nabla\psi\|^2 + \|\theta\|^2 + (1 - \delta(\beta - \delta))\|\phi\|^2 \\ &\quad + \frac{1}{2}\|\nabla\phi\|^2 - C\|\psi\|^6 - 2\|f\|\|\psi\|, \\ H_1(t) &\leq 3\|\nabla\psi\|^2 + \|\theta\|^2 + (1 - \delta(\beta - \delta))\|\phi\|^2 \\ &\quad + \frac{3}{2}\|\nabla\phi\|^2 + C\|\psi\|^6 + 2\|f\|\|\psi\|. \end{aligned} \tag{1.13.16}$$

Taking $\varepsilon_1 = \frac{\alpha}{2\alpha+\delta}$, $\varepsilon_2 = \frac{\delta}{2(2\alpha+\delta)}$ in equation (1.13.15), together with

$$\left|\int g\theta \mathrm{d}x\right| \leq \|g\|\|\theta\| \leq \frac{\beta - \delta}{2}\|\theta\|^2 + \frac{1}{2(\beta - \delta)}\|g\|^2,$$

we know that

$$\begin{aligned} I_1(t) &\geq 2\alpha\|\psi\|^2 + (\beta - \delta)\|\theta\|^2 + \delta(1 - \delta(\beta - \delta))\|\phi\|^2 \\ &\quad + \delta\|\nabla\psi\|^2 - C\|\psi\|^6 - 4\|f\|\|\psi\| - \frac{1}{\beta - \delta}\|g\|^2. \end{aligned} \tag{1.13.17}$$

From equations (1.13.16) and (1.13.17), we can find $\beta_1 > 0$ such that

$$\beta_1 H_1(t) \leq I_1(t) + C\|\psi\|^6 + C\|f\|\|\psi\| + C\|g\|^2. \tag{1.13.18}$$

Hence, from equations (1.13.13) and (1.13.18), we get

$$\frac{\mathrm{d}}{\mathrm{d}t}H_1(t) + \beta_1 H_1(t) \leq C\|\psi\|^6 + C\|f\|\|\psi\|^2 + C\|g\|^2 \triangleq K_1.$$

By Gronwall inequality, we get

$$H_1(t) \leq H_1(0)e^{-\beta_1 t} + \frac{K_1}{\beta_1}(1 - e^{-\beta_1 t}), \tag{1.13.19}$$

and then from equations (1.13.16) and (1.13.19) obtain the claim of the lemma. □

Lemma 1.13.3. *Let $f, g \in L^\infty(\mathbf{R}^+; H^1(\mathbf{R}^3))$. Then for $(\psi_0, \phi_0, \theta_0) \in X$, solution $(\psi, \phi, \theta) \in L^\infty(\mathbf{R}^+, X)$. Furthermore, there exists $t_3(R) > 0$ such that*

$$\|(\psi, \phi, \theta)\|_X \leq C, \quad \forall t \geq t_2(R),$$

whenever $\|(\psi_0, \phi_0, \theta_0)\|_X \leq R$.

Proof. Taking the inner product of (1.13.6) and $\Delta\psi_t + \alpha\Delta\psi$, then its real part, we get

$$\begin{aligned} &\frac{1}{2}\frac{\mathrm{d}}{\mathrm{d}t}\|\Delta\psi\|^2 + \alpha\|\Delta\psi\|^2 + \operatorname{Re}\int \phi\psi\Delta\overline{\psi}_t \mathrm{d}x + \alpha \operatorname{Re}\int \phi\psi\Delta\overline{\psi}\mathrm{d}x \\ &= -\frac{\mathrm{d}}{\mathrm{d}t}\operatorname{Re}\int \nabla f\nabla\overline{\psi}\mathrm{d}x - \alpha \operatorname{Re}\int \nabla f\nabla\overline{\psi}\mathrm{d}x. \end{aligned} \tag{1.13.20}$$

Noting that

$$\operatorname{Re}\int \phi\psi\Delta\overline{\psi}_t dx = \frac{d}{dt}\int \operatorname{Re}\phi\psi\Delta\overline{\psi}dx - \operatorname{Re}\int \phi_t\psi\Delta\overline{\psi}dx - \operatorname{Re}\int \phi\psi_t\Delta\overline{\psi}dx,$$

from equation (1.13.7) we get

$$-\operatorname{Re}\int \phi_t\psi\Delta\overline{\psi}dx = -\operatorname{Re}\int \theta\psi\Delta\overline{\psi}dx + \delta\operatorname{Re}\int \phi\psi\Delta\overline{\psi}dx.$$

Through equation (1.13.6), we have

$$\begin{aligned}
\psi_t &= -i(f - \Delta\psi - i\alpha\psi - \phi\psi),\\
-\operatorname{Re}\int \phi\psi_t\Delta\overline{\psi}dx &= \operatorname{Re}\int i\phi[f - \Delta\psi - i\alpha\psi - \phi\psi]\Delta\overline{\psi}dx\\
&= -\operatorname{Im}\int \phi f\Delta\overline{\psi}dx + \alpha\operatorname{Re}\int \phi\psi\Delta\overline{\psi}dx + \operatorname{Im}\int \phi^2\psi\Delta\overline{\psi}dx.
\end{aligned}$$

From equation (1.13.20) we get

$$\begin{aligned}
&\frac{1}{2}\frac{d}{dt}\left(\|\Delta\psi\|^2 + 2\operatorname{Re}\int \phi\psi\Delta\overline{\psi}dx + 2\operatorname{Re}\int \nabla f\nabla\overline{\psi}dx\right)\\
&\quad + \alpha\|\Delta\psi\|^2 + (2\alpha + \delta)\operatorname{Re}\int \phi\psi\Delta\overline{\psi}dx + \alpha\operatorname{Re}\int \nabla f\nabla\overline{\psi}dx\\
&\quad - \operatorname{Re}\int \theta\psi\Delta\overline{\psi}dx - \operatorname{Im}\int \phi f\Delta\overline{\psi}dx + \operatorname{Im}\int \phi^2\psi\Delta\overline{\psi}dx = 0.
\end{aligned} \tag{1.13.21}$$

Taking the inner product of (1.13.8) and $-\Delta\theta$, we get

$$\begin{aligned}
&\frac{1}{2}\frac{d}{dt}(\|\nabla\theta\|^2 + (1 - \delta(\beta - \delta))\|\nabla\phi\|^2 + \|\Delta\phi\|^2)\\
&\quad + (\beta - \delta)\|\nabla\theta\|^2 + \delta(1 - \delta(\beta - \delta))\|\nabla\phi\|^2 + \delta\|\Delta\phi\|^2\\
&= -\int \theta\Delta|\psi|^2 dx + \int \nabla g\nabla\theta dx\\
&= -2\operatorname{Re}\int \theta\psi\Delta\overline{\psi}dx - 2\int \theta|\Delta\psi|^2 dx + \int \nabla g\nabla\theta dx.
\end{aligned} \tag{1.13.22}$$

Let

$$\begin{aligned}
H_2(t) &= \|\Delta\psi\|^2 + 2\operatorname{Re}\int \phi\psi\Delta\overline{\psi}dx + 2\operatorname{Re}\int \nabla f\nabla\overline{\psi}dx\\
&\quad + \frac{1}{2}\|\nabla\theta\|^2 + \frac{1}{2}(1 - \delta(\beta - \delta))\|\nabla\phi\|^2 + \frac{1}{2}\|\Delta\phi\|^2
\end{aligned} \tag{1.13.23}$$

$$\begin{aligned}
I_2(t) &= 2\alpha\|\Delta\psi\|^2 + 2(\alpha + \delta)\operatorname{Re}\int \phi\psi\Delta\overline{\psi}dx\\
&\quad + 2\alpha\operatorname{Re}\int \nabla f\nabla\overline{\psi}dx - 2\operatorname{Im}\int \phi f\Delta\overline{\psi}dx\\
&\quad + 2\operatorname{Im}\int \phi^2\psi\Delta\overline{\psi}dx + (\beta - \delta)\|\nabla\theta\|^2
\end{aligned}$$

$$+\delta(1-\delta(\beta-\delta))\|\nabla\phi\|^2+\delta\|\Delta\phi\|^2$$
$$+2\int\theta|\nabla\psi|^2\mathrm{d}x-\int\nabla g\nabla\theta\mathrm{d}x. \tag{1.13.24}$$

Then multiplying (1.13.21) by 2 and adding (1.13.22), we get

$$\frac{\mathrm{d}}{\mathrm{d}t}H_2(t)+I_2(t)=0. \tag{1.13.25}$$

We estimate the terms of $H_2(t)$ and $I_2(t)$ with undetermined signs as follows:

$$\left|\mathrm{Re}\int\phi\psi\Delta\bar{\psi}\mathrm{d}x\right|\le\|\phi\|_4\|\psi\|_4\|\Delta\psi\|\le\varepsilon_3\|\Delta\psi\|^2+C(\varepsilon_3)\|\nabla\phi\|^2\|\nabla\psi\|^2,$$
$$\left|\mathrm{Re}\int\nabla f\nabla\bar{\psi}\mathrm{d}x\right|\le\|\nabla f\|\|\nabla\psi\|,$$
$$\left|\mathrm{Im}\int\phi f\Delta\bar{\psi}\mathrm{d}x\right|\le\|\phi\|_4\|f\|_4\|\Delta\psi\|\le\frac{1}{4}\|\Delta\psi\|^2+C\|\nabla f\|^2\|\nabla\phi\|^2,$$
$$\left|\mathrm{Im}\int\phi^2\psi\Delta\bar{\psi}\mathrm{d}x\right|\le\|\phi\|_6^2\|\psi\|_6\|\Delta\psi\|\le\frac{1}{4}\|\Delta\psi\|^2+C\|\nabla\phi\|^4\|\nabla\phi\|^2,$$
$$\left|\int\theta|\nabla\psi|^2\mathrm{d}x\right|\le\|\theta\|\|\nabla\phi\|_4^2\le C\|\theta\|\|\nabla\psi\|^{\frac{1}{2}}\|\Delta\psi\|^{\frac{3}{2}}$$
$$\le\frac{1}{4}\|\Delta\psi\|^2+C\|\theta\|^4\|\nabla\psi\|^2,$$
$$\left|\int\nabla g\nabla\theta\mathrm{d}x\right|\le\|\nabla\theta\|\|\nabla g\|\le\frac{\beta-\delta}{2}\|\nabla\theta\|^2+\frac{1}{2(\beta-\delta)}\|\nabla g\|^2.$$

When estimating $H_2(t)$, we take $\varepsilon_3=\frac{1}{4}$; and when estimating $I_2(t)$, we take $\varepsilon_3=\frac{\alpha}{2(2\alpha+\delta)}$. Then from the above inequalities, we have

$$H_2(t)\ge\frac{1}{2}\|\Delta\psi\|^2+\frac{1}{4}[\|\nabla\theta\|^2+\|\Delta\phi\|^2$$
$$+(1-\delta(\beta-\delta))\|\nabla\phi\|^2-C\|\nabla\phi\|^2\|\nabla\psi\|^2-2\|\nabla f\|\|\nabla\psi\|],$$
$$H_2(t)\le\frac{3}{2}\|\Delta\psi\|^2+\|\nabla\theta\|^2+\|\Delta\phi\|^2$$
$$+(1-\delta(\beta-\delta))\|\nabla\phi\|^2+C\|\nabla\phi\|^2\|\nabla\psi\|^2+2\|\nabla f\|\|\nabla\psi\|, \tag{1.13.26}$$
$$I_2(t)\ge\alpha\|\Delta\psi\|^2+\frac{1}{2}(\beta-\delta)\|\nabla\theta\|^2+\frac{\delta}{2}\|\Delta\phi\|^2$$
$$+\delta(1-\delta(\beta-\delta))\|\nabla\phi\|^2-C(\varepsilon_3)\|\nabla\phi\|^2\|\nabla\psi\|^2$$
$$-C\|\nabla f\|\|\nabla\psi\|-C\|\nabla f\|^2\|\nabla\phi\|^2$$
$$-C\|\nabla\phi\|^4\|\nabla\psi\|^2-C\|\theta\|^4\|\nabla\psi\|^2-C\|g\|^2. \tag{1.13.27}$$

Hence, there exists a constant $\beta_2>0$ such that

$$\beta_2H_2(t)\le I_2(t)+C(\|\phi\|_{H^1},\|\psi\|_{H^1},\|\theta\|,\|f\|_{H^1},\|g\|_{H^1})$$

and

$$\frac{\mathrm{d}}{\mathrm{d}t}H_2(t) + \beta_2 H_2(t) \le K_2,$$

where $K_2 \triangleq C(\|\phi\|_{H^1}, \|\psi\|_{H^1}, \|\theta\|, \|f\|_{H^1}, \|g\|_{H^1})$. Hence by Gronwall inequality, we have

$$H_2(t) \le H_2(0)e^{-\beta_2 t} + \frac{K_2}{\beta_2}(1 - e^{-\beta t}). \tag{1.13.28}$$

Now from equations (1.13.26) and (1.13.28), we get the claim. □

Corollary 1.13.1. *Let $f, g \in L^\infty(\mathbf{R}^+; H^1(\mathbf{R}^3))$. Then for $(\psi_0, \phi_0, \theta_0) \in X$, solution $(\psi, \phi, \theta) \in L^\infty(\mathbf{R}^+ \times \mathbf{R}^3)$.*

Lemma 1.13.4. *Let $f, g \in L^\infty(\mathbf{R}^+; H^2(\mathbf{R}^3))$. Then for $(\psi_0, \phi_0, \theta_0) \in Y$, solution $(\psi, \phi, \theta) \in L^\infty(\mathbf{R}^+, Y)$. Furthermore, there exists $t_4(R) > 0$ such that for any $t \ge t_4(R)$,*

$$\|(\psi, \phi, \theta)\|_Y \le C,$$

whenever $\|(\psi_0, \phi_0, \theta_0)\|_Y \le R$.

Proof. The claim follows using similar arguments as in the proof of Lemma 1.13.3. □

From the estimates above, we get the following results:

Theorem 1.13.1. *Let $f, g \in L^\infty(\mathbf{R}^+; H^1(\mathbf{R}^3))$. Then for $(\psi_0, \phi_0, \theta_0) \in X$, there exist a unique solution $(\psi, \phi, \theta) \in L^\infty(\mathbf{R}^+, X)$ for problem (1.13.6)–(1.13.9). Furthermore, the solution operator $S(t)$ from X to X is continuous, and it has a bounded absorbing set $B_1 \subset X$.*

Proof. We firstly prove the existence of a local solution by the standard iterative procedure. Then with the aid of the a priori estimate, by Theorems 1.13.1–1.13.3, we may extent the local solution to the global one. The uniqueness of the solution can be deduced from the continuity of $S(t) : X \to X$.

In fact, let $(\psi_k, \phi_k, \theta_k)$, $k = 1, 2$, be two solutions of problem (1.13.6)–(1.13.9) with the initial value $(\psi_{0k}, \phi_{0k}, \theta_{0k})$. Let $(\psi, \phi, \theta) = (\psi_1 - \psi_2, \phi_1 - \phi_2, \theta_1 - \theta_2)$, $(\psi_0, \phi_0, \theta_0) = (\psi_{01} - \psi_{02}, \phi_{01} - \phi_{02}, \theta_{01} - \theta_{02})$. Then (ψ, ϕ, θ) satisfies

$$\begin{aligned}
i\psi_t + \Delta\psi + i\alpha\psi &= -\phi_1\psi - \phi\psi_2, \\
\phi_t + \delta\phi &= \theta, \\
\theta_t + (\beta - \delta)\theta + (1 - \delta(\beta - \delta) - \Delta)\phi &= \psi_1\overline{\psi} + \psi\overline{\psi_2}, \\
(\psi, \phi, \theta)|_{t=0} &= (\psi_0, \phi_0, \theta_0)(x), \quad x \in \mathbf{R}^3.
\end{aligned}$$

It is easily proved that

$$\begin{aligned}
&\|\psi\|^2 + \|\Delta\psi\|^2 + \|\theta\|^2 + \|\nabla\theta\|^2 + (1 - \delta(\beta - \delta))\|\phi\|^2 + (2 - \delta(\beta - \delta))\|\nabla\theta\|^2 + \|\Delta\theta\|^2 \\
&\quad \le C(\|\psi_0\|^2 + \|\Delta\psi_0\|^2 + \|\theta_0\|^2 + \|\nabla\theta_0\|^2 + \|\phi_0\|^2 + \|\Delta\phi_0\|^2)e^{Ct}.
\end{aligned}$$

From this we get the continuity of $S(t)$. The existence of a bounded absorbing set has been proved in Lemma 1.13.3. □

Remark 1.13.1. We can prove using the approximation method that the problem (1.13.6)–(1.13.9) has the solution $(\psi, \phi, \theta) \in V$ when $(\psi_0, \phi_0, \theta_0) \in V$. But the continuity of $S(t)$ in V is unknown. However, $S(t)$ is continuous from V to H. Hence, the solution is unique in V.

Theorem 1.13.2. *Let $f, g \in L^\infty(\mathbf{R}^+; H^2(\mathbf{R}^3))$. Then for any $(\psi_0, \phi_0, \theta_0) \in Y$, there exists a unique solution $(\psi, \phi, \theta) \in L^\infty(\mathbf{R}^+, Y)$ for problem* (1.13.6)–(1.13.9). *Furthermore, the solution operator $S(t)$ from Y to Y is continuous, and has a bounded absorbing set $\mathscr{B}_2 \subset Y$.*

Proof. The argument is similar to the proof of Theorem 1.13.1, hence omitted. □

In the following, assume that $f, g \in H^2(\mathbf{R}^3)$ are independent of t. Then $S(t)$ generates a semigroup. Let $B \subset Y$ be a bounded set, then $S(t)B \subset Y$ is bounded, too. We decompose $S(t)$, so that we can use the Kuratowski noncompact α measure to prove the asymptotic smoothness of $S(t)$. In other words, we decompose $S(t)$ into two parts, $S_1(t)$ and $S_2(t)$, where, when $t \to \infty$, $\alpha(S_1(t)B) \to 0$ and $S_2(t)$ is relatively compact in X. For a set $A \subset X$, its noncompact α measure is defined as

$$\alpha(A) = \inf\{d \mid \text{there exists a cover of } A \text{ with finitely many balls of radius } < d\}.$$

Hence

$$\alpha(S(t)B) \le \alpha(S_1(t)B) + \alpha(S_2(t)B) = \alpha(S_1(t)B) \to 0, \quad t \to \infty.$$

Let $B \subset Y, \sup_{\xi \in B} \|\xi\|_Y \le R$, and assume $(\psi, \phi, \theta) = S(t)(\psi_0, \phi_0, \theta_0) \in B$ is the solution of problem (1.13.6)–(1.13.9) with the initial value $(\psi_0, \phi_0, \theta_0) \in B$. We have shown that (ψ, ϕ, θ) is uniformly bounded in Y.

Let $\chi L(x) \in C_0^\infty(\mathbf{R}^3)$, $0 \le \chi L \le 1$ satisfy

$$\chi L(x) = \begin{cases} 1, & |x| \le L, \\ 0, & |x| \ge 1 + L. \end{cases}$$

Then for any $\eta \in (0, 1)$, there exists $L(\eta) > 0$ (large enough) such that

$$\begin{aligned} \|f - f_\eta\|_{H^2(\mathbf{R}^3)}^2 &\le \eta, \quad f_\eta = f\chi_{L(\eta)}, \\ \|g - g_\eta\|_{H^2(\mathbf{R}^3)}^2 &\le \eta, \quad g_\eta = g\chi_{L(\eta)}, \\ \||\psi|^2 - |\psi|^2\chi_{L(\eta)}\|_{H^2(\mathbf{R}^3)}^2 &\le \eta. \end{aligned}$$

Suppose $(\psi_\eta, \phi_\eta, \theta_\eta)$ is the solution of the following problem:

$$i\psi_{\eta t} + \Delta\psi_\eta + i\alpha\psi_\eta - i\eta\Delta\psi_\eta + \phi\psi_\eta = f - f_\eta - i\eta\Delta\psi, \tag{1.13.29}$$

$$\psi_{\eta t} + (I - \Delta)\phi_\eta + \beta\psi_{\eta t} = (|\psi|^2 + g)(I - \chi_{L(\eta)}), \tag{1.13.30}$$

$$\psi_\eta(0, x) = \psi_0(x), \quad \phi_\eta(0, x) = \phi_0(x),$$

$$\phi_{\eta t}(0, x) = \phi_1(x) = \theta_0 - \delta\phi_0, \quad x \in \mathbf{R}^3, \tag{1.13.31}$$

$$\theta_\eta = \phi_{\eta t} + \delta\phi_\eta.$$

Let $S_{1\eta}(t)(\psi_0, \phi_0, \theta_0) = (\psi_\eta, \phi_\eta, \theta_\eta)$. Then

$$\begin{aligned}(u_\eta, v_\eta, \omega_\eta) &= S_{2\eta}(t)(\psi_0, \phi_0, \theta_0)\\ &= S(t)(\psi_0, \phi_0, \theta_0) - S_{1\eta}(t)(\psi_0, \phi_0, \theta_0)\\ &= (\psi - \psi_\eta, \phi - \phi_\eta, \theta - \theta_\eta)\end{aligned}$$

is the solution of problem

$$iu_{\eta t} + \Delta u_\eta + i\alpha u_\eta - i\eta\Delta u_\eta + \phi\psi_\eta = f_\eta(x), \tag{1.13.32}$$

$$v_{\eta tt} + (I - \Delta)v_\eta + \beta v_{\eta t} = |\psi|^2\chi_{L(\eta)} + g_\eta(x), \tag{1.13.33}$$

$$u_\eta(0,x) = 0, \quad v_\eta(0,x) = v_{\eta t}(0,x) = 0, \quad x \in \mathbf{R}^3 \tag{1.13.34}$$

and $\omega_\eta = v_{\eta t} + \delta v_\eta$. We need the following lemma:

Lemma 1.13.5. *There exist a constant $C > 0$ and an increasing function $\omega(\eta)$ ($\omega(0) = 0$) such that the solution of (1.13.29)–(1.13.31) satisfies*

$$\begin{aligned}&\|\psi_\eta\|_{H^2}, \|\phi_\eta\|_{H^2}, \|\phi_{\eta t}\|_{H^1} \le C, \quad \forall 0 < \eta \le 1,\ t \ge 0,\\ &\|\psi_\eta\|_{H^2}, \|\phi_\eta\|_{H^2}, \|\phi_{\eta t}\|_{H^1} \le \omega(\eta), \quad \forall 0 < \eta \le 1,\ t \ge t_*\ (\exists t_* > 0).\end{aligned}$$

Proof. Taking the inner product of (1.13.29) and $2\psi_\eta$, and taking its imaginary part, we obtain

$$\begin{aligned}&\frac{\mathrm{d}}{\mathrm{d}t}\|\psi_\eta\|^2 + 2\alpha\|\psi_\eta\| + 2\eta\|\nabla\psi_\eta\|^2\\ &\quad = 2\,\mathrm{Im}(f - f_\eta, \psi_\eta) + 2\eta\,\mathrm{Im}(\nabla\psi, \nabla\psi_\eta)\\ &\quad \le C\|f - f_\eta\|^2 + \alpha\|\psi_\eta\|^2 + \eta\|\nabla\psi_\eta\|^2 + \eta\|\nabla\psi\|^2.\end{aligned}$$

Hence

$$\frac{\mathrm{d}}{\mathrm{d}t}\|\psi_\eta\|^2 + \alpha\|\psi_\eta\|^2 + \eta\|\nabla\psi_\eta\|^2 \le C\eta,$$

and by Gronwall inequality we have

$$\|\psi_\eta\| \le \|\psi_0\|^2 e^{-\alpha t} + \frac{C\eta}{\alpha}(1 - e^{-\alpha t}).$$

Therefore

$$\|\psi_\eta\|^2 \le C, \quad \forall t \ge 0,\ 0 < \eta \le 1. \tag{1.13.35}$$

Taking $t_1(R) > 0$ such that $e^{-\alpha t_1}\|\psi_0\|^2 \le \mathrm{e}^{-\alpha t_1}R^2 < \eta$, we deduce that

$$\|\psi_\eta\|^2 \le C\eta, \quad \forall t \ge t_1. \tag{1.13.36}$$

Thus $\|\psi_\eta\| \le C\sqrt{\eta}$, $t \ge t_1$.

Taking the inner product of (1.13.29) and $\Delta^2\psi_\eta$, and taking its real part, we have

$$\begin{aligned}
&\frac{1}{2}\frac{\mathrm{d}}{\mathrm{d}t}\|\Delta\psi_\eta\|^2 + \alpha\|\Delta\psi_\eta\|^2 + \eta\|\nabla\Delta\psi_\eta\|^2 \\
&\quad= -\operatorname{Im}(\Delta\phi\psi_\eta, \Delta\psi_\eta) - 2\operatorname{Im}(\nabla\phi\nabla\psi_\eta, \Delta\psi_\eta) \\
&\qquad+ \operatorname{Im}(\Delta(f - f_\eta), \Delta\psi_\eta) + \eta\operatorname{Im}(\nabla\Delta\psi, \nabla\Delta\psi_\eta) \\
&\quad\le C\|\Delta\phi\|\|\psi_\eta\|_\infty\|\Delta\psi_\eta\| + 2\|\nabla\phi\|_4\|\nabla\psi_\eta\|_4\|\Delta\psi_\eta\| \\
&\qquad+ \|\Delta(f - f_\eta)\|\|\Delta\psi_\eta\| + \eta\|\nabla\Delta\psi\|\|\nabla\Delta\psi_\eta\| \\
&\quad\le C\|\Delta\phi\|\|\psi_\eta\|^{\frac{1}{4}}\|\Delta\psi_\eta\|^{\frac{7}{4}} + C\|\phi\|_{H^2}\|\psi_\eta\|^{\frac{1}{8}}\|\Delta\psi_\eta\|^{\frac{15}{8}} \\
&\qquad+ \frac{\alpha}{6}\|\Delta\psi_\eta\|^2 + \frac{3}{2\alpha}\|\Delta(f - f_\eta)\|^2 + \frac{1}{4}\eta\|\nabla\Delta\psi_\eta\|^2 + \eta\|\nabla\Delta\psi\|^2 \\
&\quad\le \frac{\alpha}{2}\|\Delta\psi_\eta\|^2 + \frac{\eta}{2}\|\nabla\Delta\psi_\eta\|^2 + C\|\Delta\phi\|^8\|\psi_\eta\|^2 \\
&\qquad+ C\|\phi\|_{H^2}^{16}\|\psi_\eta\|^2 + C\|f - f_\eta\|_{H^2}^2 + \eta\|\nabla\Delta\psi\|^2.
\end{aligned}$$

Hence

$$\frac{\mathrm{d}}{\mathrm{d}t}\|\Delta\psi_\eta\|^2 + \alpha\|\Delta\psi_\eta\|^2 + \eta\|\nabla\Delta\psi_\eta\|^2 \le C\eta + C(\|\phi\|_{H^2}^{16} + 1)\|\psi_\eta\|^2.$$

By Gronwall inequality, we have

$$\|\Delta\psi_\eta\| \le \|\Delta\psi_0\|^2 e^{-\alpha t} + \frac{C(\|\phi\|_{H^2}^{16} + 1)}{2}\|\psi_\eta\|^2(1 - e^{-\alpha t}). \tag{1.13.37}$$

From equations (1.13.35), (1.13.36) and (1.13.37),

$$\|\Delta\psi_\eta\|^2 \le C, \quad \forall t \ge 0,\ 0 < \eta \le 1.$$

Take $t_2 = t_2(R) \ge t_1$ such that $e^{-\alpha t_2}\|\Delta\psi_0\|^2 \le e^{-\alpha t_2}R^2 < \eta$. Then by equations (1.13.37) and (1.13.36), we have

$$\|\Delta\psi_\eta\|^2 \le C\eta, \quad t \ge t_2.$$

Hence $\|\Delta\psi_\eta\| \le C$, $t \ge t_2$.

Now we estimate ϕ_η. Taking the inner product of (1.13.30) and $2\phi_{\eta t}$, we arrive at

$$\begin{aligned}
&\frac{1}{2}\frac{\mathrm{d}}{\mathrm{d}t}(\|\phi_{\eta t}\|^2 + \|\phi_\eta\| + \|\nabla\phi_\eta\|^2) + 2\beta\|\phi_{\eta t}\|^2 \\
&\quad= 2((|\psi|^2 + g)(1 - \chi_{L(\eta)}), \phi_{\eta t}) \\
&\quad\le \beta\|\phi_{\eta t}\|^2 + C(\big\||\phi|^2(1 - \chi_{L(\eta)})\big\|^2 + \|g - g_\eta\|^2),
\end{aligned}$$

which yields

$$\begin{aligned}
&\frac{\mathrm{d}}{\mathrm{d}t}(\|\phi_{\eta t}\|^2 + \|\phi_\eta\| + \|\nabla\phi_\eta\|^2) + \beta\|\phi_{\eta t}\|^2 \\
&\quad\le C\big\||\psi|^2(1 - \chi_{L(\eta)})\big\|^2 + C\|g - g_\eta\|^2 \le C\eta.
\end{aligned} \tag{1.13.38}$$

Taking the inner product of (1.13.30) and ϕ_η in L^2, we get

$$\begin{aligned}&\frac{\mathrm{d}}{\mathrm{d}t}\int \phi_\eta\phi_{\eta t}\mathrm{d}x - \|\phi_{\eta t}\|^2 + \|\phi_\eta\|^2 + \frac{1}{2}\beta\frac{\mathrm{d}}{\mathrm{d}t}\|\phi_\eta\|^2\\ &\quad = ((|\phi|^2 + g)(1 - \chi_{L(\eta)}), \phi_\eta) \le \frac{1}{2}\|\phi_\eta\|^2 + C\eta.\end{aligned} \tag{1.13.39}$$

Then multiplying (1.13.39) by δ and adding (1.13.38), we obtain

$$\frac{\mathrm{d}}{\mathrm{d}t}H_\eta + I_\eta \le C\eta, \tag{1.13.40}$$

where

$$\begin{aligned}H_\eta &= \|\phi_{\eta t}\|^2 + \left(1 + \frac{1}{2}\delta\beta\right)\|\phi_\eta\|^2 + \|\nabla\phi_\eta\|^2 + \delta\int \phi_\eta\phi_{\eta t}\mathrm{d}x,\\ I_\eta &= (\beta - \delta)\|\phi_{\eta t}\|^2 + \frac{\delta}{2}\|\phi_\eta\|^2 + \delta\|\nabla\phi_\eta\|^2.\end{aligned}$$

Note that

$$\left|\delta\int \phi_\eta\phi_{\eta t}\mathrm{d}x\right| \le \delta\|\phi_\eta\|\delta\|\phi_{\eta t}\| \le \frac{1}{2}\delta\beta\|\phi_\eta\|^2 + \frac{\delta}{2\beta}\|\phi_{\eta t}\|^2.$$

If $\delta \le \frac{1}{2}\beta$, then for large enough $L_1 > 0$, we have

$$\begin{aligned}&\frac{3}{4}\|\phi_{\eta t}\|^2 + \left(1 + \frac{1}{4}\delta\beta\right)\|\phi_\eta\|^2 + \|\nabla\phi_\eta\|^2 \le H_\eta(t)\\ &\quad \le \frac{5}{4}\|\phi_{\eta t}\|^2 + \left(1 + \frac{3}{4}\delta\beta\right)\|\phi_\eta\|^2 + \|\nabla\phi_\eta\|^2 \le L_1 I_\eta(t).\end{aligned} \tag{1.13.41}$$

Taking $\beta_3 = L_1^{-1}$, then from equations (1.13.40) and (1.13.41) we have

$$\frac{\mathrm{d}}{\mathrm{d}t}H_\eta(t) + \beta_3 H_3(t) \le C\eta.$$

Gronwall inequality now gives

$$H_\eta(t) \le H_\eta(0)e^{-\beta_3 t} + \frac{C\eta}{\beta_3}(1 - e^{-\beta_3 t}). \tag{1.13.42}$$

From equations (1.13.41) and (1.13.42), we have

$$\frac{3}{4}\|\phi_{\eta t}\|^2 + \left(1 + \frac{1}{4}\delta\beta\right)\|\phi_\eta\|^2 + \|\nabla\phi_\eta\|^2 \le C, \quad \forall t \ge 0,\ 0 < \eta \le 1.$$

Take $t_3 = t_3(R)$ such that $H_\eta(0)e^{-\beta_3 t_3} \le CR^2 e^{-\beta_3 t_3} \le \eta$. Then from equations (1.13.41)–(1.13.42) we get

$$\|\phi_{\eta t}\|^2 + \|\phi_\eta\|^2 + \|\nabla\phi_\eta\|^2 \le C\eta.$$

Multiplying (1.13.30) by $-2\Delta\phi_{\eta t}$, and integrate it by parts over $\mathbf{R}^3$, we get

$$\begin{aligned}&\frac{\mathrm{d}}{\mathrm{d}t}(\|\nabla\phi_{\eta t}\|^2+\|\nabla\phi_\eta\|^2+\|\Delta\phi_\eta\|^2)+2\beta\|\nabla\phi_{\eta t}\|^2\\&\quad=2(\nabla((|\psi|^2+g)(1-\chi_{L(\eta)})),\nabla\phi_{\eta t})\\&\quad\le\beta\|\nabla\phi_{\eta t}\|^2+C\big\|\nabla((|\psi|^2+g)(1-\chi_{L(\eta)}))\big\|^2.\end{aligned}$$

Hence

$$\frac{\mathrm{d}}{\mathrm{d}t}(\|\nabla\phi_{\eta t}\|^2+\|\nabla\phi_\eta\|^2+\|\Delta\phi_\eta\|^2)+\beta\|\nabla\phi_{\eta t}\|^2\le C\eta. \tag{1.13.43}$$

Taking the inner product of (1.13.30) and $-\Delta\phi_\eta$, we get

$$\begin{aligned}&-\frac{\mathrm{d}}{\mathrm{d}t}\int\psi_{\eta t}\Delta\phi_\eta\mathrm{d}x-\|\nabla\phi_{\eta t}\|^2+\|\nabla\phi_\eta\|^2+\|\Delta\phi_\eta\|^2+\frac{1}{2}\beta\frac{\mathrm{d}}{\mathrm{d}t}\|\nabla\phi_{\eta t}\|^2\\&=((|\psi|^2+g)(1-\chi_{L(\eta)}),\Delta\phi_\eta)\\&\le\frac{1}{2}\|\Delta\phi_\eta\|^2+C\eta.\end{aligned} \tag{1.13.44}$$

Adding equation (1.13.43) to equation (1.13.44) multiplied by δ, we deduce

$$\begin{aligned}&\frac{\mathrm{d}}{\mathrm{d}t}\left(\|\nabla\phi_{\eta t}\|^2+\left(1+\frac{1}{2}\delta\beta\right)\|\nabla\phi_\eta\|^2+\|\Delta\phi_\eta\|^2-\delta\int\phi_{\eta t}\Delta\phi_\eta\mathrm{d}x\right)\\&\quad+(\beta-\delta)\|\nabla\phi_{\eta t}\|^2+\delta\|\nabla\phi_\eta\|^2+\delta\|\Delta\phi_\eta\|^2\le C\eta.\end{aligned}$$

Let

$$J_\eta(t)=\|\nabla\phi_{\eta t}\|^2+\left(1+\frac{1}{2}\delta\beta\right)\|\nabla\phi_\eta\|^2+\|\Delta\phi_\eta\|^2-\delta\int\phi_{\eta t}\Delta\phi_\eta\mathrm{d}x.$$

Then

$$\begin{aligned}&J_\eta(t)\le J_\eta(0)e^{-\beta_4t}+\frac{C\eta}{\beta_4}(1-e^{-\beta_4t}),\\&\|\nabla\phi_{\eta t}\|^2+\|\Delta\phi_\eta\|^2\le C,\quad\forall t\ge0,\ 0<\eta\le1.\end{aligned}$$

Take $t_4=t_4(R)\ge t_3$ such that

$$J_\eta(0)e^{-\beta_4t}\le CRe^{-\beta_1t}\le\eta.$$

Then we have

$$\|\nabla\phi_{\eta t}\|^2+\|\Delta\phi_\eta\|^2\le C\eta,\quad t\ge t_4,$$

which proves the lemma. □

Lemma 1.13.6. *There exist constants $C_1(\eta)$, $C_2(\eta)$, $C_3(\eta)$, $C_4(\eta)$ such that*

$$\begin{aligned} \||x|u_\eta\| &\le C_1(\eta),\\ \||x|\nabla u_\eta\|, \||x|D_{jk}^2 u_\eta\| &\le C_2(\eta),\\ \||x|v_\eta\| + \||x|\nabla v_\eta\| + \||x|\nabla v_{\eta t}\| &\le C_3(\eta),\\ \||x|D_{jk}^2 v_\eta\| + \||x|\nabla v_\eta\| + \||x|\nabla v_{\eta t}\| &\le C_4(\eta),\end{aligned}$$

where u_η, v_η form the solution of problem (1.13.32), (1.13.33) *and* (1.13.34).

Proof. First we consider u_η. Taking the inner product of (1.13.32) and $2|x|^2 u_\eta$, and then taking its imaginary part, we get

$$\begin{aligned} &\frac{d}{dt}\int |x|^2|u_\eta|^2 dx + 2\alpha\int |x|^2|u_\eta|^2 dx + 2\eta\int |x|^2|\nabla u_\eta|^2 dx\\ &\quad = -2\,\mathrm{Im}\int f_\eta |x|^2\overline{u_\eta}dx + 4\eta\,\mathrm{Re}\int x\nabla u_\eta\overline{u_\eta}dx + 4\,\mathrm{Im}\int x\nabla u_\eta\overline{u_\eta}dx. \end{aligned} \tag{1.13.45}$$

Then the right-hand side of equation (1.13.45) is bounded as

$$\begin{aligned} &(4+4\eta)\||x|\nabla u_\eta\|\|u_\eta\| + 2\||x|f_\eta\|\||x|u_\eta\|\\ &\quad \le \eta\||x|\nabla u_\eta\|^2 + \left(4+\frac{4}{\eta}\right)\|u_\eta\|^2 + \alpha\||x|u_\eta\|^2 + \frac{1}{\alpha}\||x|f_\eta\|^2. \end{aligned}$$

Hence

$$\frac{d}{dt}\||x|u_\eta\|^2 + \alpha\||x|u_\eta\|^2 + \eta\||x|\nabla u_\eta\|^2 \le \left(4+\frac{4}{\eta}\right)\|u_\eta\|^2 + \frac{1}{4\alpha}\||x|f_\eta\|^2. \tag{1.13.46}$$

Since f_η has compact support, $\||x|f_\eta\|$ is finite. By Gronwall inequality, we get

$$\||x|u_\eta\|^2 \le C_1(\eta), \quad \forall t \ge 0.$$

Letting $D_{jk}^2 = D_{x_j x_k}^2$ act on the two sides of equation (1.13.32), we get

$$iD_{jk}^2 u_{\eta t} + \Delta D_{jk}^2 u_\eta + i\alpha D_{jk}^2 u_\eta - i\eta\Delta D_{jk}^2 u_\eta = F_\eta(x), \tag{1.13.47}$$

where

$$F_\eta = D_{jk}^2 f_\eta - D_{jk}^2\phi u_\eta - D_j\phi D_k u_\eta - D_k\phi D_j u_\eta.$$

Then we know that equation (1.13.45) still holds. If F_η is replaced by f_η, through integration by parts we get

$$\begin{aligned} &\int D_{jk}^2\phi u_\eta |x|^2 D_{jk}^2\overline{u_\eta}dx\\ &\quad = -\int D_j\phi[D_k u_\eta |x|^2 D_{jk}^2\overline{u_\eta} + 2u_\eta x_k D_{jk}^2\overline{u_\eta} + u_\eta|x|^2 D_k D_{jk}^2\overline{u_\eta}]dx, \end{aligned}$$

$$\frac{d}{dt}\int |x|^2|D_{jk}^2u_\eta|^2dx + 2\alpha\int |x|^2|D_{jk}^2u_\eta|^2dx + 2\eta\int |x|^2|\nabla D_{jk}^2u_\eta|^2dx$$
$$= -2\,\mathrm{Im}\int [D_{jk}^2f_\eta - D_j\phi D_k u_\eta - D_k\phi D_j u_\eta]|x|^2 D_{jk}^2\overline{u_\eta}dx$$
$$- 2\,\mathrm{Im}\int D_j\phi[D_k u_\eta|x|^2 D_{jk}^2\overline{u_\eta} + 2u_\eta x_k D_{jk}^2\overline{u_\eta} + u_\eta|x|^2 D_k D_{jk}^2\overline{u_\eta}]dx$$
$$+ 4\eta\,\mathrm{Re}\int x\nabla D_{jk}^2u_\eta D_{jk}^2\overline{u_\eta}dx + 4\,\mathrm{Im}\int x\nabla D_{jk}^2u_\eta D_{jk}^2\overline{u_\eta}dx. \tag{1.13.48}$$

Since ψ, ϕ are bounded in $H^3(\mathbf{R}^3)$, and ψ_η, ϕ_η are bounded in $H^2(\mathbf{R}^3)$, $u_\eta = \psi - \psi_\eta$, $v_\eta = \phi - \phi_\eta$ are bounded in $H^2(\mathbf{R}^3)$. Thus $\|u_\eta\|_{H^2}, \||x|u_\eta\|, \|f_\eta\|, \||x|f_\eta\| \le C(\eta)$, $\|\nabla\phi\|_\infty \le C\|\phi\|_{H^3} \le C$. Then

the right-hand side of (1.13.48)
$$\le 2\||x|D_{jk}^2f_\eta\|\||x|D_{jk}^2u_\eta\| + 6\|\nabla\phi\|_\infty\||x|\nabla u_\eta\|\||x|D_{jk}^2u_\eta\|$$
$$+ 4\|D_j\phi\|_\infty\||x|u_\eta\|\|D_{jk}^2u_\eta\| + 2\|D_j\phi\|_\infty\||x|u_\eta\|\||x|\nabla D_{jk}^2u_\eta\|$$
$$+ (4+4\eta)\||x|\nabla D_{jk}^2u_\eta\|\|D_{jk}^2u_\eta\|$$
$$\le \eta\||x|\nabla D_{jk}^2u_\eta\| + \alpha\||x|D_{jk}^2u_\eta\| + C(\eta)(\||x|D_{jk}^2u_\eta f_\eta\|^2 + \|\nabla\phi\|_{H^2}^2\||x|\nabla u_\eta\|^2$$
$$+ \|\nabla\phi\|_{H^2}^2\||x|\nabla u_\eta\|^2 + \|u_\eta\|_{H^2}^2)$$
$$\le \eta\int |x|^2|\nabla D_{jk}^2u_\eta|^2dx + C(\eta) + C(\eta)\||x|\nabla u_\eta\|^2.$$

Hence, we get

$$\frac{d}{dt}\int |x|^2|D_{jk}^2u_\eta|^2dx + \alpha\int |x|^2|D_{jk}^2u_\eta|^2dx + \eta\int |x|^2|\nabla D_{jk}^2u_\eta|^2dx$$
$$\le C(\eta) + C(\eta)\||x|\nabla u_\eta\|^2. \tag{1.13.49}$$

From equation (1.13.46), we have

$$\||x|\nabla u_\eta\|^2 \le C(\eta) - \frac{1}{\eta}\frac{d}{dt}\int |x|^2|u_\eta|^2dx.$$

By Gronwall inequality, we obtain (note that $u_\eta(0,x) = 0$)

$$\int |x|^2|D_{jk}^2u_\eta|^2dx \le \int_0^t e^{-2\alpha(t-s)}(C(\eta) + C(\eta)\||x|\nabla u_\eta\|^2(s))ds$$
$$\le C(\eta)\int_0^t e^{-2\alpha(t-s)}\left(C(\eta) - \frac{d}{ds}\int |x|^2|u_\eta(s)|^2dx\right)ds$$
$$\le C(\eta) - C(\eta)\int_0^t e^{-2\alpha(t-s)}\frac{d}{ds}\||x|^2u_\eta(s)\|^2ds$$

$$
\begin{aligned}
&\le C(\eta) - C(\eta)e^{-2\alpha(t-s)}\||x|^2u_\eta\|^2\Big|_{s=0}^{s=t} + C(\eta)\int_0^t 2\alpha\||x|u_\eta\|^2 e^{-2\alpha(t-s)}\mathrm{d}s\\
&\le C(\eta) - C(\eta)\||x|^2u_\eta\|^2 + C(\eta)C_1(\eta)(1-e^{-2\alpha t}) \le C_2(\eta).
\end{aligned}
$$

Integrating by parts, we have

$$
\begin{aligned}
\int |x|^2|\nabla u_\eta|^2\mathrm{d}x &= -\int 2x\nabla u_\eta u_\eta \mathrm{d}x - \int |x|^2\Delta u_\eta u_\eta \mathrm{d}x\\
&\le 2\||x|u_\eta\|\|\nabla u_\eta\| + \||x|\Delta u_\eta\|\||x|u_\eta\| \le C(\eta).
\end{aligned}
$$

Now we establish an inequality for v_η. Taking the inner product of (1.13.33) and $2|x|^2v_{\eta t}$, we get

$$
\begin{aligned}
&\frac{\mathrm{d}}{\mathrm{d}t}\left(\int |x|^2|v_{\eta t}|^2\mathrm{d}x + \int |x|^2|v_\eta|^2\mathrm{d}x + \int |x|^2|\nabla v_\eta|^2\mathrm{d}x\right) + \beta\int |x|^2|v_{\eta t}|^2\mathrm{d}x\\
&\quad = \int (|\psi|^2+g)\chi_{L(\eta)}|x|^2v_{\eta t}\mathrm{d}x - 4\int x\nabla v_\eta v_{\eta t}\mathrm{d}x
\end{aligned}
\tag{1.13.50}
$$

Taking the inner product of (1.13.33) and $|x|^2v_\eta$, we get

$$
\begin{aligned}
&\frac{\mathrm{d}}{\mathrm{d}t}\int |x|^2v_{\eta t}v_\eta\mathrm{d}x - \||x|v_{\eta t}\|^2 + \||x|v_\eta\|^2 + \||x|\nabla v_\eta\|^2 + 2\int x\nabla v_\eta v_\eta\mathrm{d}x + \frac{1}{2}\beta\frac{\mathrm{d}}{\mathrm{d}t}\||x|v_\eta\|^2\\
&\quad = \int (|\psi|^2+g)\chi_{L(\eta)}|x|^2v_\eta\mathrm{d}x.
\end{aligned}
\tag{1.13.51}
$$

Then by equations (1.13.50) and (1.13.51), we deduce

$$
\begin{aligned}
&\frac{\mathrm{d}}{\mathrm{d}t}\left(\||x|v_{\eta t}\|^2 + \left(1+\frac{1}{2}\delta\beta\right)\||x|v_\eta\|^2 + \||x|\nabla v_\eta\|^2 + \delta\int xv_\eta v_{\eta t}\mathrm{d}x\right)\\
&\qquad + (\beta-\delta)\||x|v_{\eta t}\|^2 + \delta\||x|v_\eta\|^2 + \delta\||x|\nabla v_\eta\|^2\\
&\quad = \int (|\psi|^2+g)\chi_{L(\eta)}|x|^2v_{\eta t}\mathrm{d}x - 4\int x\nabla v_\eta v_{\eta t}\mathrm{d}x\\
&\qquad - 2\delta\int x\nabla v_\eta v_\eta\mathrm{d}x + \delta\int (|\psi|^2+g)\chi_{L(\eta)}|x|^2v_\eta\mathrm{d}x.
\end{aligned}
\tag{1.13.52}
$$

If $\delta \le \frac{1}{2}\beta$, $(\psi,\phi,\theta) \in L^\infty(\mathbf{R}^+;Y)$, and, by Lemma 1.13.5, $\|\phi_\eta\|_{H^2} \le C$, thus we have $v_\eta = \phi - \phi_\eta \in L^\infty(\mathbf{R}^+;H^2)$. Hence,

$$
\begin{aligned}
&\text{the right-hand side of (1.13.52)}\\
&\quad\le \||x|\chi_{L(\eta)}(|\psi|^2+g)\|\||x|v_{\eta t}\| + 4\|\nabla v_\eta\|\||x|v_{\eta t}\|\\
&\qquad + 2\delta\||x|\nabla v_\eta\|\|v_\eta\| + \||x|\chi_{L(\eta)}(|\psi|^2+g)\|\||x|v_\eta\|\\
&\quad\le \frac{1}{2}(\beta-\delta)\||x|v_{\eta t}\|^2 + \frac{1}{2}\||x|\nabla v_\eta\|^2 + \frac{1}{2}\delta\||x|v_\eta\|^2\\
&\qquad + C(\eta)(\||x|\chi_{L(\eta)}(|\psi|^2+g)\|^2 + \|\nabla v_\eta\|^2 + \|v_\eta\|^2),
\end{aligned}
$$

$$\left|\delta\int xv_\eta v_{\eta t}\mathrm{d}x\right| \le \delta\|xv_{\eta t}\|\|v_\eta\| \le \frac{1}{2}\|xv_{\eta t}\|^2 + C\|v_\eta\|^2.$$

Inserting these inequalities into equation (1.13.41), and using Gronwall inequality, we get

$$\||x|v_{\eta t}\|^2 + (1+\delta\beta)\||x|v_\eta\|^2 + \||x|\nabla v_\eta\|^2 \le C_3(\eta).$$

Differentiating (1.13.33) with respect to x_k, $k = 1, 2, 3$, multiplying by $2|x|^2 v_{\eta x_k t}$ and integrating over $\mathbf{R}^3$ by parts, we also obtain

$$\||x|v_{\eta x_k t}\|^2 + (1+\delta\beta)\||x|v_{\eta x_k}\|^2 + \||x|\nabla v_{\eta x_k}\|^2 \le C_4(\eta),$$

finishing the proof of the lemma. □

Now we prove the existence of an attractor.

Theorem 1.13.3. *Let $f, g \in H^2(\mathbf{R}^3)$, and let $S(t)$ be the semigroup generated by problem (1.13.6)–(1.13.9). Then there exists a set $\mathscr{A} \subset X$ which satisfies*
(1) $S(t)\mathscr{A} = \mathscr{A}$, $\forall t \ge 0$;
(2) $\lim_{t\to\infty} \mathrm{dist}_X(S(t)B, \mathscr{A}) = \lim_{t\to\infty} \sup_{y\in\mathscr{A}} \mathrm{dist}_X(S(t)y, \mathscr{A}) = 0$, *where $B \subset Y$ is a bounded set;*
(3) *$\mathscr{A}$ is compact in X.*

That is to say, $\mathscr{A}$ is a global attractor in X, it attracts all the bounded set of Y with respect to the topology of X.

To prove the theorem, we need the following compact embedding lemma.

Lemma 1.13.7. *Let $s > s_1$ be an integer. The embedding from $H^s(\mathbf{R}^n) \cap H^{s_1}(\mathbf{R}^n, (1 + |x|^2)\mathrm{d}x)$ to $H^{s_1}(\mathbf{R}^n)$ is compact.*

Proof. Let $B \subset H^s \cap H^{s_1}((1 + |x|^2)\mathrm{d}x)$ be a bounded set. We only need to prove that B has a finite ε-net (for any $\varepsilon > 0$).

Firstly, since

$$\int_{R^n} |x|^2 \sum_{l\le s_1} |D^l u|^2 \mathrm{d}x \le C, \quad \forall u \in B,$$

there exists $A > 0$ such that

$$\int_{|x|>A} \sum_{l\le s_1} |D^l u|^2 \mathrm{d}x \le \frac{1}{A^2} \int_{|x|>A} |x|^2 \sum_{l\le s_1} |D^l u|^2 \mathrm{d}x \le \frac{C}{A^2} \le \frac{\varepsilon^2}{2}.$$

Let $\Omega = \{x \mid |x| < A\}$. Then the embedding $H^s(\Omega) \hookrightarrow H^{s_1}(\Omega)$ is compact. Hence $B|_\Omega = \{u \mid u = v|_{\overline{\Omega}}, v \in B\} \subset H^s(\Omega)$ is relatively compact in $H^{s_1}(\Omega)$, and it has finite $\frac{\varepsilon}{\sqrt{2}}$-net

$\{B(\widetilde{u_k}, \frac{\varepsilon}{\sqrt{2}}), k = 1, 2, \ldots, m\}$, $\widetilde{u_k} \in B|_\Omega$, $\widetilde{u_k} = u_k|_\Omega$, $u_k \in B$. We need to prove that $\{B(u_k, \frac{\varepsilon}{\sqrt{2}})\}$ is the ε-net of B in $H^{s_1}(\Omega)$. In fact, for any $u \in B$, $\tilde{u} = u|_{\overline{\Omega}}$, there exists $\tilde{u}_k$ such that

$$\|\tilde{u}_k - \tilde{u}\|_{H^{s_1}(\Omega)} < \frac{\varepsilon}{\sqrt{2}}.$$

Hence

$$\|u - u_k\|^2_{H^s(\Omega)} = \|\tilde{u} - \tilde{u}_k\|^2_{H^s(\Omega)} + \|\tilde{u} - \tilde{u}_k\|^2_{H^n(\Omega)} \le \frac{\varepsilon^2}{2} + \int\limits_{|x|>A} \sum_{l\le s_1} |D^l u|^2 \mathrm{d}x < \varepsilon^2,$$

and the lemma is proved. □

Proof of Theorem 1.13.3. From Lemmas 1.13.6 and 1.13.7 we know that the operator $S_{2\eta}(t)$ defined by (1.13.32)–(1.13.34) is compact from Y to X. For any bounded set $B \subset Y$, we have

$$\alpha(S_{2\eta}(t)B) = 0, \quad \forall t \ge 0.$$

From Lemma 1.13.5 we know, for $\forall\varepsilon > 0$, that there exist η and $t_0 > 0$ such that

$$\|S_{1\eta}(\psi_0, \phi_0, \theta_0)\| \le \varepsilon, \quad \forall t \ge t_0,$$
and $(\psi_0, \phi_0, \theta_0) \in B$, $B \subset Y$ is a bounded set.

That is, for $\eta > 0$,

$$\alpha(S_{1\eta}(t)B) \le 2\varepsilon, \quad t \ge t_0.$$

Then we have

$$\alpha(S(t)B) \le \alpha(S_{1\eta}(t)B) + \alpha(S_{2\eta}(t)B) = \alpha(S_{1\eta}(t)B) \le 2\varepsilon, \quad t \ge t_0.$$

Hence

$$\lim_{t\to\infty} \alpha(S(t)B) = 0$$

and $S(t)$ is asymptotically smooth. By the theory in [137], Theorem 1.13.3 is proved. □

1.14 Two-dimensional unbounded region derivative Ginzburg–Landau equation

As mentioned before, Guo and Wang [114] considered the following two-dimensional Ginzburg–Landau equation with derivative term (DGL):

$$u_t = \rho u + (1 + iv)\Delta u - (1 + iu)|u|^{2\sigma} u + \alpha\lambda_1 \cdot \nabla(|u|^2 u) + \beta(\lambda_2 \cdot \nabla u)|u|^2, \tag{1.14.1}$$

where $\rho > 0$, α, β, v, u are real constants and λ_1, λ_2 are real constant vectors. The authors proved that problem (1.14.1) with periodic initial value possesses a global attractor of bounded dimension, where they supposed that there exists a positive δ_0 such that the following inequality is valid:

$$\frac{1}{\sqrt{1+(\frac{\mu-v\delta^2}{1+\delta^2})^2}-1} \geq \sigma \geq 3. \tag{1.14.2}$$

In 1997, Gao and Duan [71] considered the Cauchy problem of the following two-dimensional derivative GL equation:

$$u_t = \alpha_0 u + \alpha_1 \Delta u + \alpha_2 |u|^2 u_x + \alpha_3 |u|^3 u_y + \alpha_4 u^2 \overline{u_x} + \alpha_5 u^2 \overline{u}_y - \alpha_6 |u|^{2\sigma} u,$$

where $\alpha_0 > 0$, $\alpha_j = a_j + ib_j$, $1 \leq j \leq 6$, $a_1 > 0$, $a_6 > 0$ and $\sigma > 0$. They proved the existence of a global solution in H^2. Assume that $b_1 b_6 > 0$, $\sigma \geq \frac{1+\sqrt{10}}{2}$; and, when $b_6 = 0$ or $b_1 b_6 < 0$, then there exists a positive number $\delta > 0$ such that

$$\frac{1}{\sqrt{1+\frac{(b_1\delta-b_6)^2}{(1+\delta)(a_1\delta+a_6)}}-1} \geq \sigma \geq \frac{1+\sqrt{10}}{2}. \tag{1.14.3}$$

In 1997, Guo and Li [163] proved the existence of a global solution of the above problem in an unbounded domain, and improved condition (1.14.3).

Now we consider the Cauchy problem for the following two-dimensional derivative GL equation:

$$\begin{aligned} u_t &= \gamma u + (1+iv)\Delta u - (1+i\mu)|u|^{2\sigma} u \\ &\quad + \lambda_1 \cdot \nabla(|u|^2 u) + (\lambda \cdot \nabla u)|u^2|, \quad t > 0,\ x \in \mathbf{R}^2, \end{aligned} \tag{1.14.4}$$

$$u(0,x) = u_0(x), \quad x \in \mathbf{R}^2, \tag{1.14.5}$$

where $\gamma > 0$, v, u are real constants and λ_1, λ_2 are complex constant vectors. Let σ, v, u satisfy the following condition (A):

$$\sigma > 2 \quad \text{and} \quad -1 - v\mu < \frac{\sqrt{2\sigma+1}}{\sigma}|v-\mu|.$$

Then we get the following lemma:

Theorem 1.14.1. *Assume that σ, v, μ satisfy condition (A). Then the Cauchy problem for DGL equation (1.14.4)–(1.14.5) generates the semigroup $S(t)$, which has a global attractor $\mathscr{A} \subset H^1_{lu}$ and the following properties:*

(1) *invariance, $S(t)\mathscr{A} = \mathscr{A}$, $\forall t \geq 0$;*

(2) *compactness, $\mathscr{A}$ is bounded in H^2_{lu}, so it is compact in H^1_ρ;*

(3) *attractiveness, for any bounded set $B \subset H^1_{lu}$,*

$$\lim_{t\to\infty} \operatorname{dist}_\rho(S(t)B, \mathscr{A}) = \lim_{t\to\infty} \sup_{v\in B} \operatorname{dist}_\rho(S(t)v, \mathscr{A}) = 0.$$

The definition of the weighted spaces H_ρ^m, H_{lu}^m is given below. Suppose that $\rho > 0$ is an appropriate weight function, which possesses the property:

$$|\nabla\rho(x)|, |\Delta\rho(x)| \le \rho_0\rho(x), \quad \text{and} \quad \int \rho(x)dx = \rho_0 < +\infty, \tag{1.14.6}$$

such as $\rho = \frac{1}{\cosh|x|}$, $\rho = e^{-|x|}$, and so on. Let $T_y\rho(x) = \rho(x-y)$ denote the translation of the weight function. The weighted L^p with norm is $\|\mu\|_{p,p} = (\int \rho|\mu|^p dx)^{\frac{1}{p}}$, $1 < p < \infty$. The uniform local norm is

$$\|u\|_{p,lu} = \sup_{y\in\mathbf{R}^2} \|u\|_{p,T_y p}$$

Let L_p^p denote the all weighted L^p norm function space, $\|u\|_{p,p} < \infty$; L_{lu}^p denotes all uniform local space of u,

$$\|u\|_{p,lu} < +\infty \quad \text{and} \quad \|T_y u - u\|_{p,lu} \to 0 \quad \text{when } y \to 0.$$

Both $(L_\rho^p, \|\cdot\|_{p,\rho})$ and $(L_{lu}^p, \|\cdot\|_{p,lu})$ are Banach spaces. We also define the weighted Sobolev space $W_\rho^{m,p}$, $\|u\|_{W_\rho^{m,p}} = (\sum_{k\le m} \|D^k u\|_{p,\rho}^p)^{\frac{1}{p}}$; the uniform local Sobolev space $W_{lu}^{m,p}$ = completion of function space C_b^∞ with respect to the norm $\|u\|_{W_{lu}^{m,p}} = \sup_{y\in\mathbf{R}^2} \|u\|_{W_{T_y p}^{m,p}}$. In particular, $H_p^m = W_\rho^{m,2}$, $H_{lu}^m = W_{lu}^{m,2}$.

Using the semigroup theory and contraction mapping principle, we can get the following existence theorem:

Theorem 1.14.2. *Let $u_0 \in H_{lu}^1$. Then there exists a unique solution of the Cauchy problem (1.14.4)–(1.14.5),*

$$u(t) \in C([0,T_*), H_{lu}^1) \cap C((0,T_*), H_{lu}^2).$$

If $T_* < +\infty$, then

$$\lim_{t\to T_*} \|u(t)\|_{H_{lu}^1} = +\infty.$$

In order to get the existence of a global solution and global attractor in an unbounded domain, we must give a uniform a priori estimate on the weighted space.

Lemma 1.14.1. *Suppose condition (A) is valid and*

$$2 \le p < \frac{2\sqrt{1+v^2}}{\sqrt{1+v^2}-1}. \tag{1.14.7}$$

Then there exists a constant C, which is independent of R, and constant $t_0(R) > 0$ such that

$$\|u\|_{p,\rho}^p \le C, \quad t \ge t_0(R),$$

whenever $\|u_0\|_{p,\rho} \le R$.

Proof. By direct calculation, it follows that

$$\begin{aligned}\frac{1}{p}\frac{\mathrm{d}}{\mathrm{d}t}\|u\|_{p,\rho}^{p} &= \mathrm{Re}\int \rho|u|^{p-2}\overline{u}u_t\mathrm{d}x\\ &= \mathrm{Re}\int \rho|u|^{p-2}\overline{u}(\gamma u + (1+i\nu)\Delta u - (1+i\mu)|u|^{2\sigma}u\\ &\quad + (\lambda_1\cdot\nabla)(|u|^2u) + (\lambda_2\cdot\nabla u)|u|^2)\mathrm{d}x\\ &= \gamma\|u\|_{p,\rho}^{p} - \|u\|_{p+2\sigma,\rho}^{p+2\sigma} + I_1 + I_2,\end{aligned} \tag{1.14.8}$$

where

$$\begin{aligned}I_1 &= \mathrm{Re}\int(1+i\nu)\rho|u|^{p-2}\overline{u}\Delta u\mathrm{d}x,\\ I_2 &= \int \rho|u|^{p-2}\overline{u}((\lambda_1\cdot\nabla)(|u|^2u) - (\lambda_2\cdot\Delta u)|u|^2)\mathrm{d}x.\end{aligned}$$

Integration by parts yields

$$\begin{aligned}I_1 &= -\mathrm{Re}\int(1+i\nu)\nabla\rho|u|^{p-2}\overline{u}\Delta u\mathrm{d}x - \mathrm{Re}\int(1+i\nu)\rho\nabla(|u|^{p-2}\overline{u})\Delta u\mathrm{d}x\\ &= -\mathrm{Re}\int(1+i\nu)\nabla\rho|u|^{p-2}\overline{u}\Delta u\mathrm{d}x\\ &\quad - \frac{1}{2}\mathrm{Re}\int\rho|u|^{p-4}(p|u|^2|\nabla u|^2 + (1+i\nu)(p-2)\overline{u}^2(\nabla u)^2)\mathrm{d}x\\ &= -\mathrm{Re}\int(1+i\nu)\nabla\rho|u|^{p-2}\overline{u}\Delta u\mathrm{d}x\\ &\quad - \frac{1}{4}\int\rho|u|^{p-4}\sum_{j=1}^{2}(\overline{u}\partial_j u, u\partial_j\overline{u})M(\nu,p)\begin{pmatrix}u\partial_j\overline{u}\\ \overline{u}\partial_j u\end{pmatrix}\mathrm{d}x,\end{aligned} \tag{1.14.9}$$

where

$$M(\nu,p) = \overline{M(\nu,p)}^{\mathrm{tr}} = \begin{pmatrix}p & (1-i\nu)(p-2)\\ * & p\end{pmatrix}. \tag{1.14.10}$$

When equation (1.14.7) is satisfied, the smallest eigenvalue of $M(\nu,p)$ is

$$\lambda_M(\nu,p) = p - |p-2|\sqrt{1+\nu^2} > 0. \tag{1.14.11}$$

Hence $M(\nu,p)$ is positive, and we have

$$\begin{aligned}I_1 &\le \rho_\nu\int\rho|u|^{p-1}|\nabla u|\mathrm{d}x - \frac{1}{4}\lambda_M(\nu,p)\int\rho|u|^{p-2}|\nabla u|^2\mathrm{d}x\\ &\le \frac{2\rho_\nu^2}{\lambda_M(\nu,p)}\int\rho|u|^p\mathrm{d}x - \frac{1}{8}\lambda_M(\nu,p)\int\rho|u|^{p-2}|\nabla u|^2\mathrm{d}x,\end{aligned}$$

where $\rho_\nu = \rho_0\sqrt{1+\nu^2}$. By Cauchy inequality, we get

$$\begin{aligned}|I_2| &\le (3|\lambda_1| + |\lambda_2|)\int\rho|u|^{p+1}|\nabla u|\mathrm{d}x\\ &\le \frac{1}{8}\lambda_M(\nu,p)\int\rho|u|^{p-2}|\nabla u|^2\mathrm{d}x + \frac{2(3|\lambda_1| + |\lambda_2|)^2}{\lambda_M(\nu,p)}\int\rho|u|^{p+4}\mathrm{d}x.\end{aligned}$$

Since $\sigma > 2$, by Hölder inequality we have

$$\begin{aligned}
&\frac{2(3|\lambda_1| + |\lambda_2|)^2}{\lambda_M(v,p)} \int \rho|u|^{p+1}\mathrm{d}x \\
&\leq \frac{2(3|\lambda_1| + |\lambda_2|)^2}{\lambda_M(v,p)} \Big(\int \rho|u|^p \mathrm{d}x\Big)^{1-2\sigma} \Big(\int \rho|u|^{p+2\sigma}\mathrm{d}x\Big)^{2\sigma} \\
&\leq \frac{1}{2}\|u\|_{p+2\sigma,\rho}^{p+2\sigma} + C_1\|u\|_{p,\rho}^p,
\end{aligned}$$

where $C_1 = \frac{\sigma-2}{\sigma}(\frac{\sigma}{4})^{\frac{2(\sigma-2)}{\sigma^2}}(\frac{2(3|\lambda_1|-|\lambda_2|)^2}{\lambda_M(v,p)})^{\frac{2\sigma}{\sigma-2}}$. Then we get

$$\frac{1}{p}\frac{\mathrm{d}}{\mathrm{d}t}\|u\|_{p,\rho}^p \leq C_2\|u\|_{p,\rho}^p - \frac{1}{2}\|u\|_{p+2\sigma,\rho}^{p+2\sigma} \leq -\frac{C_2}{p}\|u\|_{p,\rho}^p + \frac{C_3}{p},$$

where $C_2 = \gamma + \frac{2p^2}{\lambda_M(v,p)} + C_1$ and $C_3 = \sigma\rho_0(\frac{2C_2(p+1)}{p+2\sigma})^{\frac{p+2\sigma}{2\sigma}}$. Gronwall inequality gives

$$\|u\|_{p,\rho}^p \leq \|u_0\|_{p,\rho}^p\, e^{-C_2 t} + \frac{C_3}{C_2}, \quad t \geq 0,$$

proving the lemma. □

Remark 1.14.1. If $\sigma = 2$, we can suppose $6|\lambda_1| + 2|\lambda_2| < \lambda_M(v,p)$, then Lemma 1.14.1 is also valid.

Lemma 1.14.2. *Under condition (A), there exists a constant C, independent of R, and constant $t_1(R)$ such that*

$$\|\nabla u(t)\|_{2,\rho}^2 \leq C, \quad t \geq t_1(R),$$

whenever $\|u_0\|_{H_\rho^1} \leq R$.

Proof. By equation (1.14.4), and integrating by parts, we obtain

$$\begin{aligned}
\frac{1}{2}\frac{\mathrm{d}}{\mathrm{d}t}\int \rho|\nabla u|^2\mathrm{d}x &= \mathrm{Re}\int \rho\nabla\bar{u}\nabla u_t \mathrm{d}x \\
&= \mathrm{Re}\int \rho\nabla\bar{u}\nabla(\gamma u + (1+iv)\Delta u - (1+i\mu)|u|^{2\sigma}u \\
&\quad + (\lambda_1\cdot\nabla)(|u|^2u) + (\lambda_2\cdot\nabla u)|u|^2)\mathrm{d}x \\
&= \gamma\|\nabla u\|_{2,\rho}^2 - \|\Delta u\|_{2,\rho}^2 - \mathrm{Re}\int (1+iv)\nabla\rho\nabla\bar{u}\Delta u \mathrm{d}x \\
&\quad + \mathrm{Re}\int (1+i\mu)\rho|u|^{2\sigma}u\Delta\bar{u}\mathrm{d}x + \mathrm{Re}\int (1+i\mu)\nabla\rho\nabla\bar{u}|u|^{2\sigma}u\mathrm{d}x \\
&\quad - \mathrm{Re}\int \rho\Delta\bar{u}((\lambda_1\cdot\nabla)(|u|^2u) + (\lambda_2\cdot\nabla u)|u|^2)\mathrm{d}x \\
&\quad - \mathrm{Re}\int \nabla\rho\nabla\bar{u}((\lambda_1\cdot\nabla)(|u|^2u) + (\lambda_2\cdot\nabla u)|u|^2)\mathrm{d}x
\end{aligned}$$

$$= \gamma\|\nabla u\|_{2,\rho}^2 - \|\Delta u\|_{2,\rho}^2 + \sum_{k=3}^{7} I_k, \tag{1.14.12}$$

where

$$\begin{aligned}
|I_3| &= \left| -\operatorname{Re}\int (1+i\nu)\nabla\rho\nabla\bar{u}\Delta u \mathrm{d}x \right| \le \rho_\nu \int \rho|\nabla u||\Delta u|\mathrm{d}x,\\
I_4 &= \operatorname{Re}\int \rho(1-i\mu)|u|^{2\sigma}\bar{u}\Delta u \mathrm{d}x,\\
|I_5| &= \left| -\operatorname{Re}\int (1+i\mu)\nabla\rho\nabla\bar{u}|u|^{2\sigma}u \mathrm{d}x \right|\\
&\le \rho_u \int \rho|u|^{2\sigma+1}|\nabla u|\mathrm{d}x,\\
|I_6| &= \left| -\operatorname{Re}\int \nabla\rho\nabla\bar{u}((\lambda_1\cdot\nabla)(|u|^2u) + (\lambda_2\cdot\nabla u)|u|^2)\mathrm{d}x \right|\\
&\le (3|\lambda_1| + |\lambda_2|)\rho_0 \int \rho|u|^2|\nabla u||\nabla u|^2\mathrm{d}x,\\
|I_7| &= \left| -\operatorname{Re}\int \rho\Delta\bar{u}((\lambda\cdot\nabla)(|u|^2u) + (\lambda_2\cdot\nabla u)|u|^2\mathrm{d}x \right|\\
&\le (3|\lambda_1| + |\lambda_2|) \int \rho|u|^2|\nabla u||\nabla u||\Delta u|\mathrm{d}x.
\end{aligned}$$

Let $\delta > 0$ (choosing properly) and define

$$V_\delta(u(t)) = \int \rho\left(\frac{1}{2}|\nabla u|^2 + \frac{\delta}{2\sigma+2}|u|^{2\sigma+2}\right)\mathrm{d}x. \tag{1.14.13}$$

From equations (1.14.12) and (1.14.18), for $p = 2\sigma + 2$, we get

$$\begin{aligned}
\frac{\mathrm{d}}{\mathrm{d}t}V_\delta(u(t)) &= \gamma(\|\nabla u\|_{2,\rho}^2 - \delta\|u\|_{2\sigma+2,\rho}^{2\sigma+2}) - (\|\Delta u\|_{2,\rho}^2 + \delta\|u\|_{4\sigma+2,\rho}^{4\sigma+2})\\
&\quad + (\delta I_1 + I_4) + I_3 + I_5 + I_6 + \delta I_2 + I_7\\
&\le \gamma(\|\nabla u\|_{2,\rho}^2 + \delta\|u\|_{2\sigma+2,\rho}^{2\sigma+2}) - (\|\Delta u\|_{2,\rho}^2 + \delta\|u\|_{4\sigma+2,\rho}^{4\sigma+2})\\
&\quad + \frac{1}{2}\operatorname{Re}\int \rho(|u|^{2\sigma}u, \Delta u)\cdot N_0 \cdot \begin{pmatrix} |u|^{2\sigma}\bar{u} \\ \Delta u \end{pmatrix} \mathrm{d}x\\
&\quad + \rho_\nu \int \rho|\nabla u||\Delta u|\mathrm{d}x + \rho_\mu \int \rho|u|^{2\sigma+1}|\nabla u|\mathrm{d}x\\
&\quad + (3|\lambda_1| + |\lambda_2|)\rho_0 \int \rho|u|^2|\nabla u|^2\mathrm{d}x\\
&\quad + \delta(3|\lambda_1| + |\lambda_2|) \int \rho|u|^{2\sigma+3}|\nabla u|\mathrm{d}x\\
&\quad + (3|\lambda_1| + |\lambda_2|) \int \rho|u|^2|\nabla u||\Delta u|\mathrm{d}x,
\end{aligned} \tag{1.14.14}$$

where $N_0 = \overline{N_0^T} = \begin{pmatrix} 0 & 1+\delta-i(\delta\nu-\mu) \\ * & 0 \end{pmatrix}$. Arguing similarly as with equations (1.14.9)–(1.14.10), for any α such that

$$|\alpha| < \frac{\sqrt{2\sigma+1}}{\sigma},$$

the smallest eigenvalue $\lambda_M(\alpha, 2\sigma+2)$ of $M(\alpha, 2\sigma+2)$ is positive. Hence $M(\alpha, 2\sigma+2)$ is positive. It follows that

$$\begin{aligned}\operatorname{Re}\int(1+i\alpha)\rho|u|^{2\sigma}\overline{u}\nabla u\mathrm{d}x &= -\operatorname{Re}\int(1+i\alpha)\nabla\rho|u|^{2\sigma}\overline{u}\Delta u\mathrm{d}x\\ &\quad -\int\rho|u|^{2\sigma-2}\sum_{j=1}^{2}(\overline{u}\partial_j u, u\partial_j\overline{u})M(\alpha, 2\sigma+2)\begin{pmatrix}u\partial_j\overline{u}\\ \overline{u}\partial_j u\end{pmatrix}\mathrm{d}x\\ &\le -\lambda_M(\alpha, p)\int\rho|u|^{2\sigma}|\nabla u|^2\mathrm{d}x+\rho_\alpha\int\rho|u|^{2\sigma-1}|\nabla u|\mathrm{d}x,\end{aligned}$$

that is,

$$\operatorname{Re}\int(1+i\alpha)\rho|u|^{2\sigma}\overline{u}\Delta u\mathrm{d}x+\lambda_M(\alpha, 2\sigma)\int\rho|u|^{2\sigma}|\nabla u|^2\mathrm{d}x-\rho_\alpha\int\rho|u|^{2\sigma+1}|\nabla u|\mathrm{d}x\le 0. \tag{1.14.15}$$

Multiplying equation (1.14.15) by $-\eta$ ($\eta > 0$ is unknown), and adding to equation (1.14.14), we arrive at

$$\begin{aligned}\frac{\mathrm{d}}{\mathrm{d}t}V_\delta(u(t)) &\le \gamma(\|\nabla u\|_2^2+\delta\|u\|_{2\sigma+2}^{2\sigma+2})-(1-\kappa)(\|\Delta u\|_2^2+\delta\|u\|_{4\sigma+2}^{4\sigma+2})\\ &\quad -\eta\lambda_M(\alpha, 2\sigma+2)\int\rho|u|^{2\sigma}|\nabla u|^2\mathrm{d}x\\ &\quad +\frac{1}{2}\operatorname{Re}\int\rho(|u|^{2\sigma}u, \Delta u)\cdot N\cdot\begin{pmatrix}|u|^{2\sigma}\overline{u}\\ \Delta\overline{u}\end{pmatrix}\mathrm{d}x\\ &\quad +(\rho_\mu+\eta\rho_\alpha)\int\rho|u|^{2\sigma+1}|\nabla u|\mathrm{d}x+\rho_\nu\int\rho|\nabla u||\Delta u|\mathrm{d}x\\ &\quad +(3|\lambda_1|+|\lambda_2|)\rho_0\int\rho|u|^2|\nabla u|^2\mathrm{d}x\\ &\quad +\delta(3|\lambda_1|+|\lambda_2|)\int\rho|u|^{2\sigma+3}|\nabla u|\mathrm{d}x\\ &\quad +(3|\lambda_1|+|\lambda_2|)\int\rho|u|^2|\nabla u||\Delta u|\mathrm{d}x,\end{aligned}\tag{1.14.16}$$

where $0\le\kappa<1$ is undetermined and

$$N=\overline{N^{\mathrm{tr}}}=\begin{pmatrix}-2\delta\kappa & 1+\delta-\eta-i(\delta\nu-\mu-\alpha\eta)\\ * & -2\kappa\end{pmatrix}.$$

For this matrix N, we have

Proposition 1.14.1. *When σ, ν and μ satisfy condition (A), we can choose proper δ, η which are positive, $\kappa\in(0,1)$ and $|\alpha|<\frac{\sqrt{2\sigma+1}}{\sigma}$ such that N is non-positive. Hence*

$$\operatorname{Re}\int\rho(|u|^{2\sigma}u, \Delta u)\cdot N\cdot\begin{pmatrix}|u|^{2\sigma}\overline{u}\\ \Delta u\end{pmatrix}\mathrm{d}x\le 0.$$

For the last five integrals of equation (1.14.16), we perform the following estimation:

$$(\rho_\mu + \eta\rho_\alpha)\int \rho|u|^{2\sigma+1}|\nabla u|dx \le \frac{1}{8}(1-\kappa)\int \rho|u|^{4\sigma-2}dx + \frac{2(\rho_\mu+\eta\rho_\alpha)^2}{(1-\kappa)}\int \rho|\nabla u|^2dx,$$

$$\rho_v\int \rho|\nabla u||\Delta u|dx \le \frac{1}{8}(1-\kappa)\int \rho|\Delta u|^2dx + \frac{2\rho_\mu^2}{(1-\kappa)}\int \rho|\nabla u|^2dx,$$

$$\begin{aligned}(3|\lambda_1|+|\lambda_2|)\rho_0\int \rho|u|^2|\nabla u|^2dx &= (3|\lambda_1|+|\lambda_2|)\rho_0\int \rho^{\frac{1}{\sigma}}|u|^2|\nabla u|^{\frac{2}{\sigma}}\cdot\rho^{1-\frac{1}{\sigma}}|\nabla u|^{2-2/\sigma}dx\\
&\le (3|\lambda_1|+|\lambda_2|)\rho_0\Big(\int \rho|u|^{2\sigma}|\nabla u|^2dx\Big)^{\frac{1}{\sigma}}\Big(\int \rho|\nabla u|^2dx\Big)^{1-\frac{1}{\sigma}}\\
&\le \frac{1}{2}\eta\lambda_M(\alpha,2\sigma+2)\int \rho|u|^{2\sigma}|\nabla u|^2dx\\
&\quad + \frac{(\sigma-1)[(3|\lambda_1|+|\lambda_2|)\rho_0]^{\frac{\sigma}{\sigma-1}}}{\sigma}\Big(\frac{2}{\sigma}\Big)^{\frac{1}{\sigma}}\int \rho|\nabla u|^2dx,\end{aligned}$$

$$\int \rho|u|^2|\nabla u||\Delta u|dx \le \varepsilon_1\int \rho|\Delta u|^2dx + \frac{1}{4\varepsilon_1}\int \rho|u|^4|\nabla u|^2dx,$$

$$\int \rho|u|^{2\sigma+3}|\nabla u|dx \le \varepsilon_2\int \rho|u|^{4\sigma+2}|\nabla u|^2dx + \frac{1}{4\varepsilon_2}\int \rho|u|^4|\nabla u|^2dx,$$

where $\varepsilon_1 > 0, \varepsilon_2 > 0$ are arbitrary. For $\sigma > 2$, by Young inequality, we have

$$\begin{aligned}\int \rho|u|^4|\nabla u|^2dx &= \int \rho|u|^4|\nabla u|^{\frac{4}{\sigma}}|\nabla u|^{2(1-2/\sigma)}dx\\
&\le \varepsilon_3\int \rho|u|^{2\sigma}|\nabla u|^2dx + \frac{\sigma-2}{\sigma}\Big(\frac{2}{\sigma\varepsilon_3}\Big)^{\frac{2}{\sigma-2}}\int \rho|\nabla u|^2dx, \quad \forall\varepsilon_3 > 0.\end{aligned}$$

Choosing ε_1, ε_2 and ε_3 small enough so that

$$\begin{aligned}\varepsilon_1(3|\lambda_1|+|\lambda_2|) &\le 2(1-\kappa),\\
\varepsilon_1\delta(3|\lambda_1|+|\lambda_2|) &\le \frac{1}{2}(1-\kappa),\\
\varepsilon_3\Big(\frac{1}{4\varepsilon_1}+\frac{\delta}{4\varepsilon_2}\Big)(3|\lambda_1|+|\lambda_2|) &\le \frac{1}{2}\eta\lambda_M(\alpha,2\sigma+2),\end{aligned}$$

we obtain

$$\begin{aligned}\frac{d}{dt}V_\delta(u(t)) &\le (\gamma+C_4)\|\nabla u\|_{2,\rho}^2 + \delta\gamma\|u\|_{2\sigma+2,\rho}^{2\sigma+2}\\
&\quad - \frac{1}{2}(1-\kappa)(\|\Delta u\|_2^2 + \delta\|u\|_{4\sigma+2,\rho}^{2\sigma+2}),\end{aligned} \tag{1.14.17}$$

where C_4 is the sum of integral coefficients of $\int \rho|\nabla u|^2 dx$ in the above inequalities. Note that there exists $C_5 = C_5(\gamma, \sigma, \kappa) > 0$ such that

$$\delta\gamma\|u\|_{2\sigma+2,\rho}^{2\sigma+2} - \frac{1}{2}\delta(1-\kappa)\|u\|_{4\sigma+2,\rho}^{2\sigma+2}$$
$$\le \delta\int \rho\left(C_5 - \frac{2\gamma}{2\sigma+2}|u|^{2\sigma+2}\right)dx = -\frac{2\delta\gamma}{2\sigma+2}\|u\|_{2\sigma+2,\rho}^{2\sigma+2} + \delta C_5\rho_0.$$

Using integration by parts and Cauchy inequality, we arrive at

$$\begin{aligned}\|\nabla u\|_{2,\rho} &= \int \rho\nabla u\nabla\bar{u}dx = -\int \nabla\rho\nabla u\bar{u}dx - \int \rho\Delta u\bar{u}dx\\ &\le \rho_0\|u\|_{2,\rho}\|\nabla u\|_{2,\rho} + \|u\|_{2,\rho}\|\Delta u\|_{2,\rho}\\ &\le \frac{1}{2}\|\nabla u\|_{2,\rho}^2 + \frac{1-\kappa}{4(2\gamma+C_4)}\|\Delta u\|_{2,\rho}^2 + \left(\frac{\rho_0^2}{2} + \frac{(2\gamma+C_4)}{1-\kappa}\right)\|u\|_{2,\rho}^2.\end{aligned}$$

By Lemma 1.14.1, $\|u\|_{2,\rho} \le C_0$, and

$$(\gamma + C_4)\|\nabla u\|_{2,\rho} \le C_6 + \frac{1}{2}(1-\kappa)\|\Delta u\|_{2,\rho}^2 - \gamma\|\nabla u\|_{2,\rho}^2,$$

where $C_6 = (\rho_0^2(2\gamma + C_4) + \frac{2(2\gamma+C_4)^2}{1-\kappa})C_0$, hence we have

$$\frac{d}{dt}V_\delta(u(t)) \le -2\gamma V_\delta(u(t)) + \delta C_5\rho_0 + C_6. \tag{1.14.18}$$

Using Gronwall inequality, we obtain

$$V_\delta(u(t)) \le V(u_0)e^{-2\gamma t} + \frac{\delta C_5\rho_0 + C_6}{2\gamma}, \quad t \ge 0, \tag{1.14.19}$$

proving the lemma. □

Corollary 1.14.1. *Under the assumptions of Lemma* 1.14.2, *we have*

$$\|u(t)\|_{H^1_{lu}} \le C, \quad t \ge t_1(R),$$

whenever $\|u_0\|_{H^1_{lu}} \le R$, *with* C *independent of* R.

Remark 1.14.2. When $\sigma = 2$ and $3|\lambda_1| + |\lambda_2|$ is small enough, Lemma 1.14.2 and Corollary 1.14.1 are valid.

Based on local existence theorem and above a priori estimates, we have

Theorem 1.14.3 (Global existence). *Suppose condition* (*A*) *is valid. Then for any* $u_0 \in H^1_{l\mu}$, *the problem* (1.14.4)–(1.14.5) *has a unique solution*

$$u(t) \in C([0,+\infty); H^1_{lu}) \cap C((0,+\infty); H^2_{lu}).$$

The semigroup $S(t)$ generated by the DGL equation is continuous in H^1_{lu} $(t > 0)$ and there exist constants L_1 and $t_1(R) > 0$ such that

$$\|u(t)\|_{H^1_{lu}} \le L_1, \quad t \ge t_1(R),$$

whenever $\|u_0\|_{H^1_{lu}} \le R$. This time, $B(0, L_1)$ is an absorbing set in H^1_{lu}. For any $q > 2$, there exist constants $L_2 > 0$ and $t_*(R) > 0$ such that

$$\|u(t)\|_{W^{l,q}_{lu}} \le L_2, \quad t \ge t_*(R).$$

Proof. We first prove the regularity of a solution in H^2_{lu}. Since $M_1 = \sup_{t\ge0} \|u(t)\|_{H^1_{lu}} < \infty$, when $1 < p < 2$,

$$\|F(u(t))\|_{L^p_{lu}} \le C(p)(M_1 + M_1^{2\sigma+1} + M_1^3) \triangleq M_2, \quad t \ge 0,$$

and then equation (1.14.4) can be written as

$$\begin{aligned} u_t &= B_p u + F(u), \\ B_p &= A_p - (R_1 + 1), \\ F(u) &= (\gamma + R_1 + 1)u + (1 + i\mu)|u|^{2\sigma} u \\ &\quad + (\lambda_1 \cdot \nabla)(|u|^2 u) + (\lambda_2 \cdot u)|u|^2, \\ A_p u &= (1 + i\nu)\Delta u. \end{aligned}$$

Hence, for $q > 2$, by the interpolation inequality (setting $p = \frac{2q}{1+q} \in (1,2)$, $\theta = \frac{1}{2} - \frac{1}{q} + \frac{1}{p} = 1 - \frac{1}{2q} \in (0,1)$) and

$$\|e^{B_q t} u\|_{W^{s+l,q}_{lu}} \le M t^{-(\frac{1}{p} + \frac{1}{2} - \frac{1}{q})} e^{-\omega t} \|u\|_{W^{s,p}_{lu}}, \quad \forall u \in W^{s,p}_{lu},\ 1 < p \le q,\ t > 0, \tag{1.14.20}$$

we arrive at

$$\begin{aligned} \|e^{B_q(t-s)} F(u(s))\|_{W^{l,q}_{lu}} &\le \left(\|e^{B_q(t-s)} F(u(s))\|_{L^p_{lu}}\right)^{1-\theta} \left(\|e^{B_q(t-s)} F(u(s))\|_{W^{2,p}_{lu}}\right)^{\theta} \\ &\le M(t-s)^{-\theta} e^{-\omega(t-s)} \|F(u(s))\|_{L^p_{lu}} \\ &\le M M_2 (t-s)^{-\theta} e^{-\omega(t-s)}, \quad t > s \ge 0. \end{aligned}$$

Through

$$u(t) = e^{B_q t} u(0) + \int_0^t e^{B_q(t-s)} F(u(s)) ds,$$

we arrive at

$$\|u(t)\|_{W^{1,q}_{lu}} \le Mt^{-\frac{1}{2}}e^{-\omega t}\|u(0)\|_{L^q_{lu}} + \int_0^t \|e^{B_q(t-s)}F(u(s))\|_{W^{1,q}_{lu}}\mathrm{d}s,$$

$$CMt^{-\frac{1}{2}}e^{-\omega t}\|u_0\|_{H^1_{lu}} + \int_0^t MM_2(t-s)^{-\theta}e^{-\omega(t-s)}\mathrm{d}s$$

$$\le MM_1t^{-\frac{1}{2}}e^{-\omega t} + \frac{MM_2\Gamma(\frac{1}{2q})}{\omega}, \quad t > 0.$$

In particular, we have $\|u(t)\|_{W^{1,4}_{lu}} \le M_3(t_0)$, $t \ge t_0$, for fixed $t_0 > 0$.

It follows that

$$\begin{aligned}\|F(u(t))\|_{H^1_{lu}} &\le C\|u(t)\|_{H^1_{lu}} + \|u(t)\|^{2\sigma+1}_{W^{1,4}_{lu}} + \|u(t)\|^2_{W^{1,4}_{lu}}\|u(t)\|_{H^2_{lu}} \\ &\le C(M_3+1)^{2\sigma+1}(1+\|u(t)\|_{H^2_{lu}}) \\ &\le M_4(1+\|u(t)\|_{H^2_{lu}}), \quad t \ge t_0,\end{aligned}$$

which yields

$$\begin{aligned}\|u(t)\|_{H^2_{lu}} &\le \|e^{B_2(t-t_0)}u(t_0)\|_{H^2_{lu}} + \int_{t_0}^t \|e^{B_2(t-s)}F(u(t))\|_{H^2_{lu}}\mathrm{d}s \\ &\le M(t-t_0)^{-\frac{1}{2}}e^{-\omega(t-t_0)}\|u(t_0)\|_{H^1_{lu}} \\ &\quad + \int_{t_0}^t M(t-s)^{-\frac{1}{2}}e^{-\omega(t-s)}\|F(u(t))\|_{H^1_{lu}}\mathrm{d}s \\ &\le M(t-t_0)^{-\frac{1}{2}}e^{-\omega(t-t_0)}\|u(t_0)\|_{H^1_{lu}} \\ &\quad + \int_{t_0}^t MM_4(t-s)^{-\frac{1}{2}}e^{-\omega(t-s)}(1+\|u(s)\|_{H^2_{lu}})\mathrm{d}s, \quad t > t_0.\end{aligned}$$

From this, by Gronwall inequality, we get

$$\|u(s)\|_{H^2_{lu}} \le M_5(t-t_0)^{-\frac{1}{2}}, \quad t_0 < t \le T.$$

Since $t_0 > 0$ is arbitrary, $u(t) \in H^2_{lu}$, and it is continuous for $t > 0$.

Secondly, we prove that there exists $t_* > 0$ such that $u(t)$ is uniformly bounded in $W^{1,q}_{lu}$, where $\|u_0\|_{H^2_{lu}} \le R$. Form this we will know that $B(0, L_2)$ (that is, the ball with radius L_2 in $W^{1,q}_{lu}$) is an absorbing set in H^1_ρ.

Since $\sup_{t \ge t_1(R)} \|u(t)\|_{H^1_{lu}} \le L_1$, when $1 < p < 2$,

$$\|F(u(t))\|_{L^p_{lu}} \le C(p)(L_1 + L_1^{2\sigma+1} + L_1^3) \triangleq C_1(p), \quad t \ge t_1(R),$$

where $C_1(p)$ is independent of R, similarly to the proof above. For $q > 2$, by interpolation formula (setting $p = \frac{2q}{1+q}$, $\theta = \frac{1}{2} - \frac{1}{q} + \frac{1}{p} = 1 - \frac{1}{2q}$) and equation (1.14.20), we have

$$\begin{aligned}\|e^{B_q(t-s)}F(u(s))\|_{W^{1,q}_{lu}} &\le (\|e^{B_q(t-s)}F(u(s))\|_{L^p_{lu}})^{1-\theta}(\|e^{B_q(t-s)}F(u(s))\|_{W^{2,p}_{lu}})^{\theta}\\ &\le M(t-s)^{-\theta}e^{-\omega(t-s)}\|F(u(s))\|_{L^p_{lu}}\\ &\le MC_1(p)t^{-\theta}e^{-\omega t}, \quad t > s \ge t_1.\end{aligned}$$

Since

$$u(t) = e^{B_q(t-t_1)}u(t_1) + \int_{t_1}^{t} e^{B_q(t-s)}F(u(s))\mathrm{d}s,$$

we arrive at

$$\begin{aligned}\|u(t)\|_{W^{1,q}_{lu}} &\le M(t-t_1)^{-\frac{1}{2}}e^{-\omega(t-t_1)}\|u(t_1)\|_{L^q_{lu}} + \int_{t_1}^{t} MC_1(p)(t-s)^{-\theta}e^{-\omega(t-s)}\mathrm{d}s\\ &\le M(t-t_1)^{-\frac{1}{2}}e^{-\omega(t-t_1)}\|u(t_1)\|_{H^1_{lu}} + \frac{MC_1(p)\Gamma(\frac{1}{2q})}{\omega}, \quad t > t_1.\end{aligned}$$

Taking $t_* = t_1 + 1 + \frac{1}{\omega}\lg(ML_1)$, we obtain

$$\|u(t)\|_{W^{1,q}_{lu}} \le L_2 = 1 + \frac{MC_1(p)\Gamma(\frac{1}{2q})}{\omega}, \quad t \ge t_*(R). \qquad \square$$

Employing the theory of Hale [137] and Temam [197], together with the compact embedding from H^2_{lu} to H^1_{ρ}, Theorem 1.14.1 can be deduced. The global attractor $\mathscr{A}$ can be represented as the ω-limit set which is generated by the semigroup $S(t)$ for DGL equation (1.14.4) with initial value (1.14.5). That is,

$$\mathscr{A} = \omega(B(0,L_1)) = \bigcap_{s\ge 0}\overline{\bigcup_{t\ge s} S(t)B(0,L_1)},$$

where the closure is taken with respect to H^1_{ρ} topology, and this set possesses the following properties:

(1) $\mathscr{A}$ is translation invariant;

(2) $\mathscr{A}$ is rotation invariant;

(3) If σ is an integer and ρ is smooth, $|D^m\rho(x)| \le \rho_m\rho(x)$, $\forall m \ge 1$, then

$$\mathscr{A} \subset \bigcap_{m=1}^{\infty} H^m_{lu}.$$

Properties (1) and (2) can be obtained from the translation invariance and rotation invariance of DGL equation (1.14.4). Property (3)can be deduced by induction through a similar argument as the proof of Theorem 1.14.1.

1.15 The relation between attractor and turbulence

With the in-depth research on dynamical system properties of Navier–Stokes equations, these results can reveal certain analogies with the turbulence theory of Kolmogorov. Now we indicate them briefly as follows:

1.15.1 The algebraic decay of characteristic norm

Suppose that the solution u of Navier–Stokes equation can be expanded into a series:

$$u(x,t) = \sum_{j=1}^{\infty} \hat{u}_j(t)\omega_j(x) \tag{1.15.1}$$

where $\{\omega_j(x)\}$ is an orthogonal complete set of functions having eigenvalues $\{\lambda_j\}$,

$$\begin{cases} -\Delta\omega_j + \operatorname{grad} r_j = \lambda_j\omega_j, \\ \operatorname{div}\omega_j = 0, \end{cases} \tag{1.15.2}$$

ω_j satisfies the boundary conditions as for u, and r_j corresponds to the pressure term.

Let $|\cdot|_0$ denote the L^2-norm of a vector function ϕ in Ω,

$$|\phi|_0 = \left\{ \int_\Omega |\phi(x)|^2 \mathrm{d}x \right\}^{\frac{1}{2}}.$$

Use $\|\cdot\|$ to denote the L^2-norm of the gradient of ϕ:

$$\|\phi\| = |\operatorname{grad}\phi|_0 = \left\{ \int_\Omega |\operatorname{grad}\phi(x)|^2 \mathrm{d}x \right\}^{\frac{1}{2}}.$$

Use R_0 to denote an upper of the norm in L^2 for u with initial value $u(x,0)$,

$$|u(\cdot,0)|_0 \le R_0. \tag{1.15.3}$$

Foias and coauthors got the following results in [68]:

Theorem 1.15.1. *Suppose that the space dimension is 2. Then there exists a constant k_1, which only depends on μ, $|f|_0$, Ω, and constant t_1, depending on the above parameters and R_0, such that*

$$|\hat{u}_m(t)| \le k_1\left(\frac{\lambda_1}{\lambda_m}\right)\left(1 + \ln\left(\frac{\lambda_1}{\lambda_m}\right)\right)^{\frac{1}{2}}, \quad \forall m,\ \forall t \ge t_1, \tag{1.15.4}$$

where ν is viscosity coefficient and f is external force. Here $\lambda_m = k_m^2$, k_m is wave number. Denote $\hat{u}_m(t) = \hat{u}(k_m, t)$. From equation (1.15.4) we get

$$|\hat{u}(k_m, t)| \le \left(\frac{\kappa_1}{k_m^2}\right)\left(1 + \ln\left(\frac{k_1}{k_m}\right)\right)^{\frac{1}{2}}.$$

The decay rate of energy spectrum E is

$$E(k, t) \approx k|\hat{u}(k, t)|^2 \le \left(\frac{\kappa}{k^3}\right)\ln k, \quad t \ge t_1.$$

Up to the factor $\ln k$*, it is consistent with the two-dimensional Kraichnan decay rate (see* [152]*).*

Similar results are valid in dimension 3. Suppose that no singularity occurs, and $\|u(\cdot, t)\|$ stays uniformly bounded

$$M_1 = \sup_{t\ge 0} \|u(\cdot, t)\| < +\infty. \tag{1.15.5}$$

Similar to equation (1.15.4), we have

$$|\hat{u}_m(t)| \le \kappa_1'\left(\frac{\lambda_1}{\lambda_m}\right)^{\frac{2}{3}}, \quad \forall m, \ \forall t \ge t_1'.$$

Hence

$$|\hat{u}(k_m, t)| \le \frac{\chi}{k_m^{\frac{4}{3}}}.$$

Energy spectrum is

$$E(k, t) \approx k^2|\hat{u}(k, t)|^2 \le \frac{\kappa}{k^{\frac{2}{3}}}, \quad t \ge t_1.$$

If Grashof or Reynolds number is large enough, then it is consistent with Kolmogorov decay rate.

1.15.2 The Fourier coefficients of exponential decay

Suppose that equation (1.15.1) can be expanded into standard Fourier series,

$$u(x, t) = \sum_{j\in \mathbf{Z}^n} \hat{u}_j(t)e^{ijx}, \tag{1.15.6}$$

where $j = \{j_1, j_2\}$ or $\{j_1, j_2, j_3\}$. Suppose that $\int_\Omega u(x, t)dx = 0$, hence $\hat{u}_0 = 0$, and

$$\hat{u}_j = \overline{\hat{u}_j}.$$

Due to incompressibility condition $\operatorname{div} u = 0$,

$$j \cdot \hat{u}_j = 0, \quad \forall j. \tag{1.15.7}$$

Assume that $f(x)$ and pressure p also have Fourier series expansions,

$$f(x) = \sum_{j\in \mathbf{Z}^n} \hat{f}_j e^{ij\cdot x}, \quad p(x,t) = \sum_{j\in \mathbf{Z}^n} \hat{p}_j(t) e^{ij\cdot x}.$$

Then Navier–Stokes equation is equivalent to the algebraic differential equations

$$\frac{d\hat{u}_j}{dt} + 4\pi^2 \nu \hat{u}_j + i \sum_{k,l\in \mathbf{Z}^n, k+l=j} (l\cdot \hat{u}_k)\hat{u}_l + ij\hat{p}_j = f_j, \quad \forall j \neq 0.$$

From equation (1.15.7) we get

$$\frac{d\hat{u}_j}{dt} + 4\pi^2 \nu \hat{u}_j + i \sum_{k\in \mathbf{Z}^n} (j\cdot \hat{u}_k)\hat{u}_{j-k} + ij\hat{p}_k = f_j, \quad \forall j \neq 0. \tag{1.15.8}$$

Taking the inner product of (1.15.8) and j, and using equation (1.15.7), we obtain the expression of $\hat{p}_j$ as

$$\hat{p}_j = -i\frac{j_0 \cdot \hat{f}_j}{|j|^2} - \sum_{k\in \mathbf{Z}^n} (j\cdot \hat{u}_k)(j\cdot \hat{u}_{j-k}).$$

Substituting $\hat{p}_j$ into equation (1.15.8), we have

$$\frac{d\hat{u}_j}{dt} + 4\pi^2 \nu \hat{u}_j + i \sum_{k\in \mathbf{Z}^n} (j\cdot \hat{u}_k)\left\{ j\cdot \hat{u}_{j-k} - \frac{j}{|j|}\left(\frac{j}{|j|}\cdot \hat{u}_{j-k}\right)\right\} = \hat{f}_j - \frac{j}{|j|}\left(\frac{j}{|j|}\cdot \hat{f}_j\right). \tag{1.15.9}$$

If $|j|$ is large, $\hat{f}_j$ disappears, so that

$$\hat{f}_j = 0, \quad |j| > j_0.$$

Suppose $\hat{f}_j$ is exponentially decaying in $|j|$, that is, there exist $\sigma_1, \sigma_2 > 0$ such that

$$|\hat{f}_j| \leq \sigma_1 e^{-\sigma_2 |j|}, \quad \forall j.$$

Foias et al. [64] proved the following:

Theorem 1.15.2. *Suppose that the space dimension is 2. Then there exist constants k_2, k_3, which only depend on ν, $|f|_0$, σ_1, σ_2, and t_2 depending on ν and κ_0 such that*

$$|\hat{u}_j(t)| \leq \left(\frac{k_2}{|j|}\right) e^{-k_3 |j|}, \quad \forall j \neq 0, \ \forall t \geq t_2. \tag{1.15.10}$$

When the space dimension is 3, as long as smoothness is kept, similar results are also valid. These results are consistent with the Kolmogorov turbulence. Equation (1.15.10) is completely consistent with the dissipation expression of Kolmogorov turbulence. The results are more detailed than Kolmogorov's given in 1941, because they concretely apply to the decision flow and allow obtaining the obvious exponential decay rate. Meanwhile, their precision is inferior to the results of Kolmogorov, because the constants k_2, k_3 are too large.

1.15.3 The dimension of the attractor

Navier–Stokes equation can be written in functional form as

$$\frac{du}{dt} + \nu Au + B(u,u) = f, \tag{1.15.11}$$

where A is Stokes operator, $B(\cdot,\cdot)$ is a nonlinear term, f is external force, $u = u(t)$ $(= u(\cdot,t))$. Considering the solution $u(t)$ of the linear operator for equation (1.15.11),

$$\Phi \to \nu A\Phi + B(u(t),\Phi) + B(\Phi,u(t)),$$

we need to estimate the time average of the trace of the finite dimensional projection operator for any possible orbit $u(t)$. For the proper function family Φ_j, $j = 1,2,\ldots,m$ (these functions are assumed to satisfy $\Phi_j \in D(A^{\frac{1}{2}})$, and are orthogonal in $L^2(\Omega)^m$, $D(A^{\frac{1}{2}}) = V$, $\phi \in L^2$, $\|\phi\| = |\nabla\phi|_0 < +\infty$). We consider the following sum:

$$\begin{aligned}&\sum_{j=1}^{m}\langle(\nu A\Phi_j + B(u(t),\Phi_j) + B(\Phi_j,u(t)),\Phi)\rangle\\&\quad= \sum_{j=1}^{m}\{\nu\|\Phi_j\|^2 + B(\Phi_j,u(t),\Phi_j)\}.\end{aligned}$$

Let $q_m(t)$ be a lower bound of the above sum, where for any orbit and any Φ_j on the attractor, let

$$q_m = \liminf_{T\to\infty}\frac{1}{T}\int_0^T q_m(t)\mathrm{d}t.$$

We can prove that if m is large enough, $q_m > 0$, and then the dimension of the attractor is m. Suppose that the standard length (i. e., for the diameter or $|\Omega|^{\frac{1}{n}}$) of field Ω is $l_0 = k_0^{-1}$. Kolmogorov dissipation length $l_d = k_d^{-1}$ is defined as

$$l_d = \left(\frac{\Delta^3}{\varepsilon}\right)^{\frac{1}{4}},$$

where ε is the energy when $|\nabla u(x,t)|^2 >$ average dissipation rate. The average rate of the attractor $\mathscr{A}$ is

$$\varepsilon = \nu \limsup_{t\to\infty} \left\{ \sup_{u(0)\in\mathscr{A}} \frac{1}{t} \int_0^T \sup_{x\in\Omega} |\nabla u(\tau,x)|^2 d\tau \right\}.$$

Constantin et al. [32] gave the following result:

An upper bound of Hausdorff and fractal dimensions of attractor $\mathscr{A}$ is

$$C\left(\frac{k_d}{k_0}\right)^n, \quad n = 3, \tag{1.15.12}$$

where C is an absolute constant.

The following remark is compelling. Let $\overline{|\nabla v|}$ denote an upper bound of $|\nabla v(x)|$, $x \in \Omega$, $v \in \mathscr{A}$. Obviously,

$$\varepsilon \le \bar{\varepsilon} = \nu(\overline{|\nabla v|})^2,$$
$$k_d \le \overline{k_d} = \left(\frac{\bar{\varepsilon}}{\nu^3}\right)^{\frac{1}{4}} = \left(\frac{\overline{|\nabla v|}}{\nu}\right)^{\frac{1}{2}}.$$

By equation (1.15.12) we deduce that

$$\dim \mathscr{A} \le C l_0^n \left(\frac{\overline{|\nabla v|}}{\nu}\right)^{\frac{1}{2}n}, \quad n = 3. \tag{1.15.13}$$

The length of $(\frac{\nu}{\overline{|\nabla v|}})^{\frac{1}{2}}$ is called the smallest scale by Henshow et al. (1991) and Bartucelli et al. (1990) [10]. Constantin, Foias, Temam and coauthors have proved in [32] that

$$\left(\frac{l_0^2}{l_d}\right) \le G.$$

Hence

$$\dim \mathscr{A} \le CG, \tag{1.15.14}$$

where G is Grashof number

$$G = \frac{L^{\frac{1}{2}n-3}}{\nu^2} \left\{ \int_\Omega |f(x)|^2 dx \right\}^{\frac{1}{2}},$$

with ν denoting dynamic viscosity, L the diameter of Ω or $|\Omega|^{\frac{1}{n}}$, and $Re = G^{\frac{1}{2}}$. This result improves those of Babin and Vishik (1983) [7], Ladyzhenskaya (1982, 1985), Foias and Temam (1983) [61].

For the periodic case, the result has been further improved. Constantin et al. [87] have proved

$$\dim \mathscr{A} \leq C\left(\frac{l_0}{l_\eta}\right)^2 \left(1 + \ln\left(\frac{l_0}{l_\eta}\right)\right)^{\frac{1}{3}}, \tag{1.15.15}$$

where Kraichnan dissipation length l_η replaces with the Kolmogorov dissipative length l_d,

$$l_\eta = \left(\frac{\nu_3}{\eta}\right)^{\frac{1}{6}},$$

and η is the space and time average of $\nu|\operatorname{curl}\operatorname{curl} u(x,t)^2|$ on the attractor $\mathscr{A}$, that is,

$$\eta = \nu \limsup_{t\to\infty} \sup_{u(0)\in\mathscr{A}} \int_0^t \frac{1}{|\Omega|} \int_\Omega |\Delta u(x,s)|^2 dx ds.$$

Meanwhile, this proves $\frac{l_0}{l_\eta} \leq G^{\frac{1}{3}}$. Hence by equation (1.15.15) one obtains

$$\dim \mathscr{A} \leq CG^{\frac{2}{3}}(1 + \ln G)^{\frac{1}{3}}. \tag{1.15.16}$$

A simplified proof of equations (1.15.15) and (1.15.16) appeared in Doering and Gillon (1991) and Ghidagila, Temam (1990) [77].

1.15.4 The best dimensional estimate of attractor

Babin and Vishik [36] considered the situation of a spacial periodic flow in a narrow and long area $\Omega = (0, \frac{2\pi L}{\alpha}) \times (0, 2\pi L)$. Firstly, Ghidaglia and Temam (1990) proved:

$$\dim \mathscr{A} \leq C\left(1 + \frac{G^{\frac{2}{3}}}{\alpha}\right)\left(\ln\left(1 + \frac{g}{\alpha^{\frac{1}{4}}}\right)\right)^{\frac{1}{3}}, \quad \alpha \neq 1, \tag{1.15.17}$$

where C is an absolute constant. When $\alpha \to 0$, we have the estimate

$$\dim \mathscr{A} \leq C\left(1 + \frac{G}{\alpha^{\frac{1}{2}}}\right). \tag{1.15.18}$$

Babin and Vishik have given an estimate of a lower bound of $\dim \mathscr{A}$. Suppose that the pressure $f(x_1, x_2)$ possesses the form

$$f(x_1, x_2) = (g(x_2), 0), \tag{1.15.19}$$

$$\int_0^{2\pi L} g(x_2)\mathrm{d}x_2 = 0,$$

and Grashof number G is

$$G = \frac{\overline{G}}{\alpha^{\frac{1}{2}}}, \tag{1.15.20}$$

$$\overline{G} = \frac{(2\pi)^{\frac{1}{2}}}{\nu^2}\left(\int_0^{2\pi L} g(y)\mathrm{d}y\right)^{\frac{1}{2}}.$$

It is easy to see that Navier–Stokes equation possesses a stationary solution u_s, p_s, $p_s = 0$, $u_s = (U(x_2), 0)$,

$$-\nu U''(x_2) = g(x_2),$$

$$U(x_2) = -\frac{1}{\nu}\int_0^{x_2}(x_2 \cdot s)g(s)\mathrm{d}s - \frac{x_2}{2\pi L\nu}\int_0^{2\pi L} sg(s)\mathrm{d}s + \text{const.}$$

Choosing the constant such that

$$\int_0^{2\pi L} U(x_2)\mathrm{d}x_2 = 0$$

for small α, the region Ω becomes slender and the stationary solution becomes unstable. For the time being, we know that the global attractor contains $\mathscr{A}$ steady solution u_s and its unstable manifold of $\mathscr{U}_{\text{uns}}$. Thus

$$\dim \mathscr{A} \geq \dim \mathscr{U}_{\text{uns}}.$$

By using the properties of Orr–Somerfeld equation, one shows that there exists at least $\frac{\kappa}{\alpha}$ unstable norm, where the parameter κ depends on $\overline{G}$, and

$$\dim \mathscr{A} \geq \frac{\kappa(\overline{G})}{\alpha}. \tag{1.15.21}$$

Ghidaglia and Temam [76] generalized these results in dimension 3.

Let $\Omega = (0, \frac{2\pi L}{\alpha}) \times (0, 2\pi L) \times (0, \frac{2\pi L}{\beta})$, $0 < \beta \leq \alpha$, and

$$f(x_1, x_2, x_3) = (g(x_2), 0, 0).$$

When α is small enough, the stationary solution u_s belongs to unstable manifold and so

$$\dim \mathscr{A} \geq \dim \mathscr{U}_{\text{uns}} \geq \frac{\kappa(\overline{G})}{\alpha\beta}.$$

On the other hand, we can estimate

$$\dim \mathscr{U} \leq C\frac{(1+Re^3)}{\alpha\beta},$$

where

$$Re = \frac{L}{2}\sup_u \limsup_{t\to\infty}\left(\frac{1}{t}\int_0^t \sup_{x\in\Omega}|u(x,s)|^2 ds\right)^{\frac{1}{2}} < \infty.$$

Hence

$$\dim \mathscr{U} \simeq \frac{\kappa}{\alpha\beta}.$$

This is the best dimensional estimate of the invariant set $\mathscr{U}$.

2 Inertial manifold

The concept of inertial manifold was first put forward by Foias, Sell and Temam in 1985. The inertial manifold is a finite-dimensional manifold which is at least Lipschitz continuous, is invariant in the phase space, approaches the trajectory exponentially, and contains the global attractor [35].

In [65], the authors studied the general initial value problems of nonlinear evolution equations

$$\begin{aligned} \frac{\mathrm{d}u}{\mathrm{d}t} + Au &= f(u), \\ u(0) &= u_0, \end{aligned}$$

where the nonlinear semigroup $S(t)$ defined on a Hilbert space is a self-adjoint operator. In 1988, Chow and Lu [28] took investigated a general equation with the bounded nonlinear term $f \in C^1$ in a Banach space, but the index attracting to manifold was not proved to be on the phase space of bounded subsets uniformly. Mallet-Paret and Sell [171] introduced the space average principle in 1988; for the case, when the spectral gap condition is not completely satisfied, they proved the existence of inertial manifolds for diffusion equations. In 1988, Constantin et al. [34] tried to portray the spectral gap condition by the concept of spectral barrier in a Hilbert space. In 1990, Bernal [17] considered the situation in a Banach space, but the proof was more complex, and the spectral gap condition was demanded more critically. In 1991, Demengel, Ghidaglia [47] first demonstrated the existence of an inertial manifold in a Hilbert space when A is self-adjoint and f is unbounded. In 1993, Debussche and Temam [43] gave another proof when f was essentially unbounded. In 1990, Debussche [40] by using Sacker's equation gave another proof of the existence of the manifold for a general Banach space with $f \in C^1$. In 1991, Fabes, Luskin and Sell [54] used the elliptic regularization method to construct the inertial manifold. For a Hilbert space, when A is a self-adjoint operator, the proof of existence of inertial manifolds could be found in [42] and [191]. For the construction of an inertial manifold in a Hilbert space, the strong extrusion and cone conditions, we refer to [188]. Conway, Hoff and Smoller [37] in 1978, Mane [172] in 1981, and Mora [177] in 1983 studied the reaction diffusion equations and parabolic equations which yielded the proof of existence of inertial manifolds.

There were lots of works which developed general theory to be applicable to some specific equations. Especially, for the estimation of the minimum dimension of inertial manifolds; for instance, for the KS equation, see Foias et al. [69] and Temam and Wang [199]; for Cahn–Hilliard equation, nonlocal Burgers equation and some reaction diffusion equations, see Constantin et al. [34]; for compressible gas dynamics model, see Nicolaenko [180]; for one dimensional reaction–diffusion equations (including explicit structure of inertial manifolds), see Jolly [145]; for Swift–Hohenberg

https://doi.org/10.1515/9783110549652-002

convection model, see Jaboada [196]; for Ginzburg–Landau equation, see Demengel and Ghidaglia [47]; for phase flow equation, see Bates and Zheng [11], and so on.

Many existence results depend on the spectral gap condition, which is very restrictive. For instance, for the Navier–Stokes equations this condition is not satisfied. In 1992, Kwak [158] pointed out that the partial periodic boundary conditions for 2D Navier–Stokes equation, and the ratio of the periods in the two directions being a rational number, could overcome this difficulty. His key idea is converting the original equation into a set of reaction diffusion equations by a nonlinear transformation of dependent variable. And then these equations satisfy the spectral gap condition and possess the asymptotic properties of the original NS equation. Of course, the inertial manifold of reaction diffusion equations is not proved cross-sectionally in manifolds. Therefore, whether any trajectory of NS equation is attracted to the finite-dimensional manifold exponentially is still an open question.

New progress on inertial manifolds is mainly visible when discussing:

(1) Generalized properties of equations, the accuracy of the spectral gap condition, the completely asymptotic behavior of inertial manifolds;
(2) The existence of continuous invariant foliation, the growth features of a complete trajectory, $u(t) = O(e^{-\sigma t})$, $\sigma > 0$, $t \to -\infty$;
(3) C^1 regularity and normal hyperbolicity;
(4) $C^{m,\alpha}$ regularity.

For these aspects we can refer to Rasa and Temam [189]. In 1995, Guo [26, 89] proved the existence of an inertial manifold of generalized Kuramoto–Sivashinsky type; in 1996, Gao and Guo [91] proved the existence of a finite-dimensional inertial manifold of the one-dimensional generalized Ginzburg–Landau equation. For other related work, see [66, 146, 173, 174, 198, 41, 63, 52, 44, 47, 205].

2.1 The inertial manifold for a class of nonlinear evolution equations

In a Hilbert space H, let an inner product $(\cdot,\cdot)$ be given. The nonlinear evolution equation has the form

$$\frac{du}{dt} + Au + R(u) = 0, \tag{2.1.1}$$

where

$$R(u) = B(u,u) + Cu - f, \tag{2.1.2}$$

A is a linear unbounded self-adjoint operator on H, and $D(A)$ is dense in H. Let A be positive, that is,

$$(Av, v) > 0, \quad \forall v \in D(A),\ v \neq 0.$$

Assume A^{-1} is compact and the mapping $u \to Au$ is an isomorphism from $D(A)$ to H; A^s means the sth power of A for $s \in \mathbf{R}$. The space $V_{2s} = D(A^s)$ is a Hilbert space, which is endowed with the inner product

$$(u, v)_{2s} = (A^s u, A^s v), \quad \forall u, v \in D(A^s),$$

$u \in V_s$, and $|u|_s = (u, u)_s^{\frac{1}{2}}$.

Because A^{-1} is compact and self-adjoint, there exists an orthogonal basis $\{\omega_j\}$ of H which consists of the eigenvectors of A,

$$A\omega_j = \lambda_j \omega_j, \tag{2.1.3}$$

and the eigenvalues satisfy

$$0 < \lambda_1 \le \lambda_2 \le \cdots, \quad \lambda_j \to +\infty, \quad j \to \infty. \tag{2.1.4}$$

From equations (2.1.3)–(2.1.4) we readily get

$$|A^{\frac{1}{2}} u| \ge \lambda_1^{\frac{1}{2}} |u|, \quad u \in D(A^{\frac{1}{2}}), \tag{2.1.5}$$

$$|A^{p+\frac{1}{2}} u| \ge \lambda_1^{\frac{1}{2}} |A^p u|, \quad \forall u \in D(A^{p+\frac{1}{2}}),\ \forall p. \tag{2.1.6}$$

Let P_N be the orthogonal projection of H in the subspace spanned by $\{\omega_1, \omega_2, \dots, \omega_N\}$, $N = 1, 2, \dots$, $Q_N = I - P_N$. The nonlinear terms $R(u)$, $B(u, u)$ in (2.1.2) are bilinear operators, $D(A) \times D(A) \to H$, C is a linear operator from $D(A)$ to H, and $f \in D(A^{\frac{1}{2}})$. Furthermore, set

$$(B(u, v), v) = 0, \quad \forall u, v \in D(A), \tag{2.1.7}$$

$$|B(u, v)| \le C_1 |u|^{\frac{1}{2}} |A^{\frac{1}{2}} u|^{\frac{1}{2}} |A^{\frac{1}{2}} v|^{\frac{1}{2}} |Av|^{\frac{1}{2}}, \quad \forall u, v \in D(A), \tag{2.1.8}$$

$$|Cu| \le C_2 |A^{\frac{1}{2}} u|^{\frac{1}{2}} |Av|^{\frac{1}{2}}, \quad \forall u \in D(A), \tag{2.1.9}$$

where C_1, C_2 and the following C_i, $i = 3, 4$, are positive constants. For operators B and C, the following continuity properties are assumed:

$$|A^{\frac{1}{2}} B(u, v)| \le C_3 |Au||Av|, \quad \forall u, v \in D(A), \tag{2.1.10}$$

$$|A^{\frac{1}{2}} Cu| \le C_4 |Au|, \quad \forall u \in D(A). \tag{2.1.11}$$

Finally, if $A + C$ is positive, then

$$((A + C)u, u) \ge \alpha |A^{\frac{1}{2}} u|^2, \quad \forall u \in D(A),\ \alpha > 0. \tag{2.1.12}$$

Consider the initial value problem of equation (2.1.1), i. e., equation (2.1.1) with the initial condition

$$u(0) = u_0 \in H. \tag{2.1.13}$$

Assume that problem (2.1.1), (2.1.13) has a unique solution $S(t)u_0$, $\forall t \in \mathbf{R}_+$, $S(t)u_0 \in D(A)$, $\forall t \in \mathbf{R}_+$. The mapping $S(t)$ has the property of an ordinary semigroup.

Now we establish an a priori estimate for equation (2.1.1). To proceed, we need the following inequality and lemma: for $\beta > 0$, $\frac{1}{p} + \frac{1}{q} = 1$, $1 < p, q < +\infty$, we have

$$\Sigma|x_i y_i| = \Sigma|\beta x_i||\beta^{-1} y_i| \le \frac{\beta^p}{p}(\Sigma|x_i|^p) + \frac{\beta - q}{q}(\Sigma|y_i|^q). \tag{2.1.14}$$

Lemma 2.1.1. *Assume that $g(t)$, $h(t)$, $y(t)$ are three positive integrable functions, $t_0 \le t < \infty$, which satisfy*

$$\frac{dy}{dt} \le gy + h, \quad \forall t \ge t_0, \tag{2.1.15}$$

and

$$\int_t^{t-1} g(s)ds \le \alpha_1, \quad \int_t^{t+1} h(s)ds \le \alpha_2,$$
$$\int_t^{t-1} y(s)ds \le \alpha_3, \quad \forall t \ge t_0, \tag{2.1.16}$$

where α_1, α_2, α_3 are positive constants. Then we have

$$y(1+t) \le (\alpha_3 + \alpha_2)\exp(\alpha_1), \quad \forall t \ge t_0. \tag{2.1.17}$$

Taking the inner product of (2.1.1) and u, and using equations (2.1.7) and (2.1.12), we get

$$\frac{1}{2}\frac{d}{dt}|u|^2 + \alpha|A^{\frac{1}{2}}u|^2 \le |(f,u)| \le \lambda_1^{-\frac{1}{2}}|f||A^{\frac{1}{2}}u| \le \frac{\alpha}{2}|A^{\frac{1}{2}}u|^2 + \frac{1}{2\lambda_1\alpha}|f|^2,$$

which yields

$$\frac{d}{dt}|u|^2 + \alpha\lambda_1|u|^2 \le \frac{d}{dt}|u|^2 + \alpha|A^{\frac{1}{2}}u|^2 \le \frac{1}{\alpha\lambda_1}|f|^2. \tag{2.1.18}$$

Taking the inner product of equation (2.1.1) and Δu, from equations (2.1.8), (2.1.9) and (2.1.14), we have

$$\begin{aligned}|B(u,u) + (u, Au)| &\le C|u|^{\frac{1}{2}}|A^{\frac{1}{2}}u||Au|^{\frac{3}{2}} + C_2|A^{\frac{1}{2}}u|^{\frac{1}{2}}|Au|^{\frac{3}{2}}\\ &\le 54(C_1^4|u|^2|A^{\frac{1}{2}}u|^4 + C_2^4|A^{\frac{1}{2}}u|^2) + \frac{1}{4}|Au|^2.\end{aligned}$$

Similarly, we have $|(f, Au)| \le |f||Au| \le |f|^2 + \frac{1}{4}|Au|^2$. From this we get

$$\begin{aligned}\frac{1}{2}\frac{d}{dt}|A^{\frac{1}{2}}u|^2 + \lambda_1|A^{\frac{1}{2}}u|^2 &\le \frac{d}{dt}|A^{\frac{1}{2}}u|^2 + |Au|^2\\ &\le C_6|u|^2|A^{\frac{1}{2}}u|^4 + C_7|A^{\frac{1}{2}}u|^2 + 2|f|^2.\end{aligned} \tag{2.1.19}$$

Taking the inner product of equation (2.1.1) and A^2u, we have

$$\begin{aligned}\frac{1}{2}\frac{\mathrm{d}}{\mathrm{d}t}|Au|^2 + |A^{\frac{3}{2}}u|^2 &\le |B(u,u) + (Cu, A^2u)| + |(f, A^2u)|\\ &\le |(A^{\frac{1}{2}}(B(u,u) + Cu), A^{\frac{3}{2}}u)| + |(A^{\frac{1}{2}}f, A^{\frac{3}{2}}u)|\\ &\le C_3|Au|^2|A^{\frac{3}{2}}u| + C_4|Au||A^{\frac{3}{2}}u| + |A^{\frac{1}{2}}f||A^{\frac{3}{2}}u|\\ &\le \frac{1}{2}C_8|Au|^4 + \frac{1}{2}C_9|Au|^2 + \frac{3}{2}|A^{\frac{1}{2}}f|^2 + \frac{1}{2}|A^{\frac{3}{2}}u|^2,\end{aligned}$$

which yields

$$\begin{aligned}\frac{\mathrm{d}}{\mathrm{d}t}|Au|^2 + \lambda_1|Au|^2 &\le \frac{\mathrm{d}}{\mathrm{d}t}|Au|^2 + |A^{\frac{3}{2}}u|^2\\ &\le C_8|Au|^4 + C_9|Au|^2 + 3|A^{\frac{1}{2}}f|^2.\end{aligned} \tag{2.1.20}$$

Applying equation (2.1.15) to (2.1.18), we have $u(t) = S(t)u_0$,

$$|u(t)|^2 \le |u(0)|^2\exp(-\alpha\lambda_1 t) + \rho_0^2(1 - \exp(-\alpha\lambda_1 t)), \tag{2.1.21}$$

where $\rho_0 = \frac{1}{2\lambda_1}|f|$. Hence $|u(t)|$ is uniformly bounded in t, and we have

$$\lim_{t\to\infty}\sup|u(t)|^2 \le \rho_0^2. \tag{2.1.22}$$

From equation (2.1.18), we have that $\int_t^{t+1}|A^{\frac{1}{2}}u|^2\mathrm{d}s$ is uniformly bounded. By equation (2.1.19), we get

$$\frac{\mathrm{d}}{\mathrm{d}t}|A^{\frac{1}{2}}u|^2 \le C_{10}|A^{\frac{1}{2}}u|^4 + (C_7 - \lambda_1)|A^{\frac{1}{2}}u|^2 + 2|f|^2,$$

where $C_{10} = C_6b_0^2$, $|u(t)|^2 \le b_0^2$, $t \ge 0$. Then by Lemma 2.1.1, setting

$$\begin{aligned}&g = C_{10}|A^{\frac{1}{2}}u|^2, \quad y = |A^{\frac{1}{2}}u|^2,\\ &h = (C_7 - \lambda_1)|A^{\frac{1}{2}}u|^2 + 2|f|^2,\end{aligned}$$

we get that $|A^{\frac{1}{2}}u|^2$ is uniformly bounded in H. By equation (2.1.19), we know that $\int_t^{t+1}|Au(s)|^2\mathrm{d}s$ is uniformly bounded. Moreover, by equation (2.1.20), we get $|Au(t)|^2$ and $\int_t^{t+1}|A^{\frac{3}{2}}u(s)|^2\mathrm{d}s$ are uniformly bounded in t. Hence

$$\lim_{t\to+\infty}\sup|A^{\frac{1}{2}}u(t)|^2 \le \rho_1^2, \tag{2.1.23}$$

$$\lim_{t\to+\infty}\sup|Au(t)|^2 \le \rho_2^2. \tag{2.1.24}$$

From equations (2.1.22), (2.1.23) and (2.1.24) we know that any solution of equation (2.1.1) enters into the following balls after some time($t \ge t_0 > 0$),

$$B_0 = \{x \in H,\ |x| \le 2\rho_0\},$$
$$B_1 = \{x \in D(A^{\frac{1}{2}}),\ |A^{\frac{1}{2}}x| \le 2\rho_1\},$$
$$B_2 = \{x \in D(A),\ |Ax| \le 2\rho_2\},$$

respectively, where the ω-limit set of B_2 given by

$$\mathscr{A} = \omega(B_2) = \bigcap_{s\ge 0} \mathrm{Cl}\Big(\bigcap_{t\ge s} S(t)B_2\Big)$$

is the global attractor of equation (2.1.1); the closure Cl is taken on the set H, and

$$\mathscr{A} \subseteq B_2 \cap B_1 \cap B_0.$$

We consider the inertial manifold of a truncated equation of (2.1.1). Let $\theta(s)$ be a smooth function from $\mathbf{R}_+$ to $[0,1]$: $\theta(s) = 1,\ 0 \le s \le 1$; $\theta(s) = 0,\ s \ge 2$, $|\theta'(s)| \le 2,\ s \ge 0$. Fix $\rho = 2\rho_2$ and define

$$\theta_\rho(s) = \theta\Big(\frac{s}{\rho}\Big), \quad s \ge 0.$$

Then the inertial manifold equation of (2.1.1) is

$$\frac{du}{dt} + Au + \theta_\rho(|Au|)R(u) = 0, \tag{2.1.25}$$

for which one can prove directly the existence and uniqueness of solution to (2.1.25) with initial value $u(0) = u_0 \in H$. Obviously, when $|Au| \le \rho$, $\theta_s(|Au|) = 1$, equation (2.1.24) and equation (2.1.1) are consistent. When $|Au| \ge 2\rho$, $\theta_\rho(|Au|) = 0$, taking the inner product of equation (2.1.24) and A^2u, we get

$$\frac{1}{2}\frac{d}{dt}|Au|^2 + \lambda_1|Au|^2 \le \frac{1}{2}\frac{d}{dt}|Au|^2 + \big|A^{\frac{3}{2}}u\big|^2 \le 0.$$

Thus the trajectory $u(t)$ will exponentially converge to the ball with radius $\rho_3 \ge 2\rho$ in $D(A)$. In addition, $R(u)$ is locally Lipschitz continuous in u. And $F(u) = \theta_\rho(|u|)R(u)$ is globally Lipschitz continuous. That is, there exists K, such that

$$\big|F(u) - F(v)\big| \le K|u - v|, \quad \forall u, v \in H. \tag{2.1.26}$$

Definition 2.1.1. The inertial manifold for a semigroup $\{S(t)\}_{t\ge 0}$ is a finite-dimensional smooth manifold $\mu \in H$ (at least Lipschitz), which satisfies the following properties:
(1) μ is invariant, that is,

$$S(t)\mu \subset \mu; \tag{2.1.27}$$

(2) μ attracts all the solutions of equation (2.1.25) exponentially. That is, there exist constants $k_1 > 0$, $k_2 > 0$ such that for $u_0 \in H$, we have

$$\text{dist}(S(t)u_0, \mu) \leq k_1 e^{-k_2 t}, \quad \forall t \geq 0; \tag{2.1.28}$$

(3) The attractor $\mathscr{A}$ belongs to μ.

Now we will construct the inertial manifold, and in doing so we will prove its existence. Suppose P_N is an N-dimensional orthogonal projection of H, $Q_N = I - P_N$. And denote $P = P_N$, $Q = Q_N$. Set $u(t)$ to be the solution of equation (2.1.25). Let $p(t) = Pu$, $q(t) = Qu$. Then $p(t)$, $q(t)$ belong to PH and QH and satisfy

$$\frac{dp}{dt} + Ap + PF(u) = 0, \tag{2.1.29}$$

$$\frac{dq}{dt} + Aq + QF(u) = 0, \tag{2.1.30}$$

where $F(u) = \theta_\rho(|Au|)R(u)$, and $u = p + q$. We are looking for the inertial manifold μ, which can be constructed as the graph of the Lipschitz function $\Phi : PD(A) \to QD(A)$. That is, $\mu = \text{graph}\,\Phi$. The function Φ is an operator, which is obtained by the fixed point of a function $\mathscr{F}_{b,l}$, which is a function $\Phi : PD(A) \to QD(A)$, satisfying

$$|A\Phi(p)| \leq b, \quad \forall p \in PD(A), \tag{2.1.31}$$

$$|A\Phi(p_1) - A\Phi(p_2)| \leq l|Ap_1 - Ap_2|, \quad \forall p_1, p_2 \in PD(A), \tag{2.1.32}$$

$$\text{supp}\ \Phi \subset \{p \in PD(A) \mid |Ap| \leq 4\rho\},$$

where $b > 0$, $l > 0$. When $p = p(t)$, $q = \Phi(p(t))$ satisfy equations (2.1.29)–(2.1.30), $u = p(t) + \Phi(p(t))$ is the solution of (2.1.25). Suppose that Φ is given by $\mathscr{F}_{b,l}$, $p_0 \in PD(A)$, then the solution $p = p(t; p_0, \Phi)$ for equation

$$\frac{dp}{dt} + Ap + PF(p + \Phi(p)) = 0, \quad p(0) = p_0, \tag{2.1.33}$$

exists and is unique. This is because $\sigma \to \theta_\rho R(\sigma + \Phi(\sigma))$ is Lipschitz continuous. Since $p = p(t; p_0, \Phi)$, $\forall t \in \mathbf{R}$, similar to equation (2.1.30), we have

$$\frac{dq}{dt} + Aq + QF(p + \Phi(p)) = 0, \quad p(0) = p_0. \tag{2.1.34}$$

Since $QF(p + \Phi(p))$ is bounded, $\mathbf{R} \to H$, a solution of equation (2.1.34) exists and is unique; it is a bounded solution $q(t)$ when $t \to \infty$. From this, we get

$$q(0) = -\int_{-\infty}^{0} e^{\tau AQ} QF(p + \Phi(p))d\tau, \tag{2.1.35}$$

where $p = p(\tau) = p(\tau, \Phi, p_0)$. This $q(0)$ depends on $\Phi \in \mathscr{F}_{b,l}$ and $p \in PD(A)$; $q(0) = q(0, p_0, \Phi)$. Function

$$p_0 \in PD(A) \to q(0; p_0, \Phi) \in QD(A)$$

maps $PD(A)$ into $QD(A)$, and is denoted by $\mathscr{F}\Phi$. Hence

$$\mathscr{F}\Phi(p_0) = -\int_{-\infty}^{0} e^{\tau AQ} QF(u) d\tau, \tag{2.1.36}$$

where $u = u(\tau) = p(\tau, \Phi, p_0) + \Phi(p(\tau; \Phi, p_0))$. Since $q(0) = \Phi(p_0)$, we seek conditions on N, b and l such that

(1) $\mathscr{F}$ maps $\mathscr{F}_{b,l}$ into itself;

(2) $\mathscr{F}$ is dense in $\mathscr{F}_{b,l}$.

Now we consider the function $\Phi : PD(A) \to QD(A)$, $P = P_N$, $Q = Q_N = I - P_N$. Then $\Phi \in \mathscr{F}_{b,l}$, i. e., equations (2.1.31)–(2.1.33) are satisfied.

With the distance

$$\|\Phi - \Psi\| = \sup_{p \in D(A)} |A\Phi(p) - A\Psi(p)|, \tag{2.1.37}$$

$\mathscr{F}_{b,l}$ is a complete metric space. For $\Phi \in \mathscr{F}_{b,l}$, the mapping $\mathscr{F}$ is defined on $PD(A)$ as

$$\mathscr{F}\Phi(p_0) = -\int_{-\infty}^{0} e^{\tau AQ} QF(u) d\tau, \quad p_0 \in PD(A), \tag{2.1.38}$$

where $u(\tau) = p(\tau; \Phi, p_0) + \Phi(p(\tau; \Phi, p_0))$, $p(\tau; \Phi, p_0)$ is the solution of equation (2.1.29) such that $p(0; \Phi, p_0) = p_0$. In the following we study the properties of operator $\mathscr{F}$.

Lemma 2.1.2. *Let $\alpha > 0$, and $\tau < 0$. The operator $(AQ)^\alpha e^{\tau AQ}$ is linear and continuous on QH, its norm on $\mathscr{F}$ (that is, $|(AQ)^\alpha e^{\tau AQ}|_{op}$) is bounded by*

$$\begin{aligned} &K_3|\tau|^{-\alpha}, \quad \text{when } -\alpha\lambda_{N+1}^{-1} \le \tau < 0; \\ &\lambda_{N+1}^{\alpha} e^{\tau\lambda_{N+1}}, \quad \text{when } -\infty < \tau \le -\alpha\lambda_{N-1}^{-1}. \end{aligned} \tag{2.1.39}$$

Proof. Let $v = \sum_{j=N+1}^{\infty} b_j \omega_j$ be an element of QH. Then

$$\begin{aligned} |(AQ)^\alpha e^{\tau AQ} v|^2 &= \sum_{j=N+1}^{\infty} (\lambda_j^\alpha e^{\tau\lambda_j}) b_j^2 \\ &\le \sup_{\lambda \ge \lambda_{N+1}} (\lambda^\alpha e^{\tau\lambda}) \Sigma b_j^2 \\ &= \sup_{\lambda \ge \lambda_{N+1}} (\lambda^\alpha e^{\tau\lambda})^2 |v|^2. \end{aligned}$$

Hence

$$|(AQ)^{\alpha}e^{\tau AQ}|_{op} \le \sup_{\lambda \ge \lambda_{N-1}} \lambda^{\alpha}e^{\tau\lambda}.$$

Through elementary calculation, this shows that

$$\sup_{\lambda\ge\lambda_{N+1}} (\lambda^{\alpha}e^{\tau A\lambda}) = \begin{cases} |\tau|^{-\alpha}(\alpha e^{-1})^{\alpha}, & \text{when } -\alpha\lambda_{N+1}^{-1} \le \tau < 0, \\ \lambda_{N+1}^{\alpha}e^{\tau\lambda N+1}, & \text{when } \tau \le -\alpha\lambda_{N+1}^{-1}. \end{cases}$$

Hence we get the equation (2.1.39), where $K_3 = K_3(\alpha) = (\alpha e^{-1})^{\alpha}$.

As the direct corollary of equation (2.1.39), we have

$$\int_{-\infty}^{0} |(AQ)^{\alpha}e^{\tau AQ}|_{op}\mathrm{d}\tau \le (1-\alpha)^{-1}e^{\alpha}\lambda_{N+1}^{-\alpha-1}, \quad 0<\alpha<1. \tag{2.1.40}$$

From equations (2.1.10) and (2.1.11) we get

$$\begin{aligned} \left|(AQ)^{\frac{1}{2}}R(u)\right| &\le \left|A^{\frac{1}{2}}B(u,u)\right| + \left|A^{\frac{1}{2}}Cu\right| + \left|A^{\frac{1}{2}}f\right| \\ &\le C_3|Au|^2 - C_4|Au| + \left|A^{\frac{1}{2}}f\right|. \end{aligned}$$

When $|Au| > 2\rho$ and $\theta_{\rho}(|Au|) = 0$, we have

$$\left|(AQ)^{\frac{1}{2}}F(u)\right| \le K_4, \tag{2.1.41}$$

where $K_4 = 4C_3\rho^2 + 2C_4\rho + |A^{\frac{1}{2}}f|$ and $F(u)$ is determined by equation (2.1.30). □

Lemma 2.1.3. *Let $p_0 \in PD(A)$, $\mathscr{F}\Phi(p_0) \in QD(A)$. Then*

$$\left|A\mathscr{F}\Phi(p_0)\right| \le K_5\lambda_{N+1}^{-\frac{1}{2}}, \tag{2.1.42}$$

$$\left|A^{\frac{5}{4}}\mathscr{F}\Phi(p_0)\right| \le K_6\lambda_{N+1}^{-\frac{1}{4}}, \tag{2.1.43}$$

where K_5, K_6 are absolute constants, independent of p_0, Φ.

Proof. Since $Qe^{\tau AQ} = e^{\tau AQ}$, it is easy to see that $\mathscr{F}\Phi(p_0) \in QD(A)$ and

$$\begin{aligned} \left|A\mathscr{F}\Phi(p_0)\right| &\le \int_{-\infty}^{0} \left|AQe^{\tau AQ}F(u)\right|\mathrm{d}\tau \\ &\le \int_{-\infty}^{0} \left|(AQ)^{\frac{1}{2}}e^{\tau AQ}\right|_{op}\left|(AQ)^{\frac{1}{2}}F(u)\right|\mathrm{d}\tau, \\ \left|A^{\frac{5}{4}}\mathscr{F}\Phi(p_0)\right| &\le \int_{-\infty}^{0} \left|(AQ)^{\frac{5}{4}}e^{\tau AQ}F(u)\right|\mathrm{d}\tau \\ &\le \int_{-\infty}^{0} \left|(AQ)^{\frac{3}{4}}e^{\tau AQ}\right|_{op}\left|(AQ)^{\frac{1}{2}}F(u)\right|\mathrm{d}\tau. \end{aligned}$$

Inequalities (2.1.43)–(2.1.42) can be obtained from equations (2.1.40)–(2.1.41). Now we take

$$b = K_5\lambda_{N+1}^{-\frac{1}{2}}. \tag{2.1.44}$$

Then for $\Phi \in \mathscr{F}_{b,l}$, similar to equation (2.1.31), $\mathscr{F}\Phi$ satisfies

$$\left|A\mathscr{F}\Phi(p_0)\right| \le b, \quad \forall p_0 \in PD(A). \tag{2.1.45}$$

By equation (2.1.43), $\mathscr{F}\Phi$ is a bounded set in $D(A^{\frac{5}{4}})$. Since $A^{-\frac{1}{4}}$ is compact, the range of $\mathscr{F}\Phi$ is a compact subset of $QD(A)$, which does not depend on Φ. □

Now we consider the properties of the support and continuity properties of $\mathscr{F}\Phi$.

Lemma 2.1.4. *For every $\Phi \in \mathscr{F}_{b,l}$, the support of $\mathscr{F}\Phi$ is included in $\{p \in PD(A);\ |Ap| \le 4\rho\}$.*

Proof. Set $u = p + \Phi(p)$. If $|Ap| > 2\rho$, then

$$|Au| = \left(|Ap|^2 + \left|A\Phi(p)\right|^2\right)^{\frac{1}{2}} \ge |Ap| > 2\rho.$$

Hence, $\theta_\rho(|Au|) = 0$.

Now if we set $|Ap_0| > 4\rho$, then in some interval of t, $|Ap(t)| > 2\rho$. Then equation (2.1.28) becomes

$$\frac{\mathrm{d}p}{\mathrm{d}t} + Ap = 0,$$

from which we infer that

$$\frac{1}{2}\frac{\mathrm{d}}{\mathrm{d}t}|Ap|^2 + \lambda_1|Ap|^2 \le \frac{1}{2}\frac{\mathrm{d}|Ap|^2}{\mathrm{d}t} + \left|A^{\frac{3}{2}}p\right|^2 = 0.$$

Hence, for $\tau > 0$, we have

$$2\rho < \left|Ap(0)\right| \le \left|Ap(\tau)\right|\exp(\lambda_1\tau) \le \left|Ap(\tau)\right|.$$

Thus $\theta_\rho(|Au(\tau)|) = 0$, $\forall \tau < 0$. By virtue of equation (2.1.38), we have

$$\mathscr{F}\Phi(p_0) = 0, \quad \forall \Phi \in \mathscr{F}_{b,l}.$$ □

To proceed, we verify the Lipschitz property of nonlinear $F(u)$.

Lemma 2.1.5. *If $p_1, p_2 \in PD(A)$, $\Phi_1, \Phi_2 \in \mathscr{F}_{b,l}$, $u_i = p_i + \Phi_i(p_i)$, then*

$$\left|A^{\frac{1}{2}}F(u_1) - A^{\frac{1}{2}}F(u_2)\right| \le K_7\left[(1+l)|Ap_1 - Ap_2| + \|\Phi_1 - \Phi_2\|\right], \tag{2.1.46}$$

where constant K_7 does not depend on p_i or Φ_i, $i = 1, 2$.

Proof. First, noting equations (2.1.10) and (2.1.10), we deduce

$$\begin{aligned}
&|A^{\frac{1}{2}}R(u_1) - A^{\frac{1}{2}}R(u_2)| \\
&\quad \le |A^{\frac{1}{2}}[B(u_1,u_1) - B(u_1,u_2) + B(u_1,u_2) - B(u_2,u_2)]| + |A^{\frac{1}{2}}C(u_1 - u_2)| \\
&\quad \le C_3(|Au_1| + |Au_2|)|Au_1 - Au_2| + C_4|Au_1 - Au_2|
\end{aligned}$$

and

$$|A^{\frac{1}{2}}R(u_1)| \le C_3|Au_1|^2 + C_4|Au_1|^2 + |A^{\frac{1}{2}}f|.$$

Define G as follows:

$$\begin{aligned}
G &= A^{\frac{1}{2}}F(u_1) - A^{\frac{1}{2}}F(u_2) \\
&= \theta_\rho(|Au_1|)A^{\frac{1}{2}}R(u_1) - \theta_\rho(|Au_2|)A^{\frac{1}{2}}R(u_2).
\end{aligned}$$

We discuss three different cases:

(1) $2\rho \le |Au_1|$, $2\rho \le |Au_2|$;

(2) $|Au_1| < 2\rho \le |Au_2|$ or $|Au_2| < 2\rho \le |Au_1|$;

(3) $|Au_1| \le 2\rho$, $|Au_2| \le 2\rho$.

For case (1), when $|Au| \ge 2\rho$, $\theta_\rho(|Au|) = 0$ and $|\theta'| \le 2\rho^{-1}$, we get $G = 0$. In case (2), we have

$$\begin{aligned}
|G| &= |\theta_\rho(|Au|)A^{\frac{1}{2}}R(u)| \\
&= |\theta_\rho(|Au_1|)A^{\frac{1}{2}}R(u_1) - \theta_\rho(|Au_2|)A^{\frac{1}{2}}R(u_1)| \\
&\le |\theta_\rho(|Au_1|) - \theta_\rho(|Au_2|)||A^{\frac{1}{2}}R(u_1)| \\
&\le 2\rho^{-1}\|Au_1\| - \|Au_2\|(C_3|Au_1|^2 + C_4|Au_2|^2 + |A^{\frac{1}{2}}f|) \\
&\quad + [C_3(|Au_1| + |Au_2|) + C_4]|Au_1 - Au_2|.
\end{aligned}$$

Hence

$$|A^{\frac{1}{2}}F(u_1) - A^{\frac{1}{2}}F(u_2)| \le K_7|Au_1 - Au_2|, \tag{2.1.47}$$

where $K_7 = 2\rho^{-1}(C_3 4\rho^2 + C_4 2\rho + |A^{\frac{1}{2}}f|) + C_3 4\rho + C_4$. Since $u_1 - u_2 = p_1 - p_2 + (\Phi_1(p_1) - \Phi_1(p_2)) + (\Phi_1(p_2) - \Phi_2(p_2))$, we have $|Au_1 - Au_2| \le (1+l)|Ap_1 - Ap_2| + \|\Phi_1 - \Phi_2\|$. Combining with equation (2.1.47), we get equation (2.1.46). □

Now we prove that under appropriate assumptions, the map $\mathscr{F}$ is Lipschitz from $\mathscr{F}_{b,l}$ to $\mathscr{F}_{b,l}$, and this is a strict restriction.

First, we set Φ to be fixed, take $p_{01}, p_{02} \in PD(A)$, and let $p = p_1(t)$, $p = p_2(t)$ be the solutions of equation (2.1.28) which satisfy the respective initial conditions $p_i(0) = p_{0i}$, $i = 1, 2$.

Let $\Delta = p_1 - p_2$, then Δ satisfies the equation

$$\frac{d\Delta}{dt} + A\Delta + PF(u_1) - PF(u_2) = 0, \tag{2.1.48}$$

where $u_i = p_i + \Phi(p_i)$, $i = 1, 2$. Taking the inner product of (2.1.48) with $A^2\Delta$, we get

$$\frac{1}{2}\frac{\mathrm{d}}{\mathrm{d}t}|A\Delta|^2 + \left|A^{\frac{3}{2}}\Delta\right|^2 = -(A^{\frac{1}{2}}P(F(u_1) - F(u_2)), A^{\frac{3}{2}}\Delta). \tag{2.1.49}$$

By equation (2.1.46), we get

$$\frac{1}{2}\frac{\mathrm{d}}{\mathrm{d}t}|A\Delta|^2 + \left|A^{\frac{3}{2}}\Delta\right|^2 \le K_7(1+l)|A\Delta||A^{\frac{3}{2}\Delta}|.$$

Hence $|A\Delta|\frac{\mathrm{d}}{\mathrm{d}t}|A\Delta| \ge -|A^{\frac{3}{2}}\Delta|^2 - K_7(1+l)|A\Delta||A^{\frac{3}{2}}\Delta|$. Since $\Delta \in PD(A)$, we have

$$\left|A^{\frac{3}{2}}\Delta\right| \le \lambda_N^{\frac{1}{2}}|A\Delta|,$$

which yields

$$|A\Delta|\frac{\mathrm{d}}{\mathrm{d}t}|A\Delta| \ge -\lambda_N|A\Delta|^2 - K_7(1+l)\lambda_N^{\frac{1}{2}}|A\Delta|^2,$$

or equivalently,

$$\frac{\mathrm{d}}{\mathrm{d}t}|A\Delta| + (\lambda_N + K_7(1+l)\lambda_N^{\frac{1}{2}})|A\Delta| \ge 0. \tag{2.1.50}$$

From equation (2.1.50) we get

$$\left|A\Delta(\tau)\right| \le \left|A\Delta(0)\right| \exp(-\tau(\lambda_N + K_7(1+l)\lambda_N^{\frac{1}{2}})),\ \ \tau \le 0. \tag{2.1.51}$$

Lemma 2.1.6. *Assume that* $\gamma_N = \lambda_{N+1} - \lambda_N - K_7(1+l)\lambda_N^{\frac{1}{2}} > 0$. *Then for* $\Phi \in \mathscr{F}_{b,l}$ *and* $p_{01}, p_{02} \in PD(A)$, *we have*

$$|A\mathscr{F}\Phi(p_{01}) - A\mathscr{F}\Phi(p_{02})| \le L|Ap_{01} - Ap_{02}|, \tag{2.1.52}$$

where

$$L = K_7(1+l)\lambda_{N+1}^{-\frac{1}{2}}[1 + (1-\gamma_N\alpha_N)^{-1}]e^{-\frac{1}{2}}\exp\left(\frac{\gamma_N\alpha_N}{2}\right),$$

$$\gamma_N = \frac{\lambda_N}{\lambda_{N+1}},\quad \alpha_N = 1 + K_7(1+l)\lambda_{N+1}^{-\frac{1}{2}},$$

which implies that $\Phi \in \mathscr{F}_{b,l}$.

Proof. By equations (2.1.38) and (2.1.46), we have

$$\begin{aligned}|A\mathcal{F}\Phi(p_{01}) - A\mathcal{F}\Phi(p_{02})| &\le \int_{-\infty}^{0} |AQe^{\tau AQ}(F(u_1) - F(u_2))| d\tau \\ &\le K_7(1+l)\int_{-\infty}^{0} |(AQ)^{\frac{1}{2}}e^{\tau AQ}|_{0p} |A\Delta(\tau)| d\tau,\end{aligned}$$

where $\Delta = p_1 - p_2$. By Lemma 2.1.2 and equation (2.1.51), we have

$$\begin{aligned}&\int_{-\infty}^{0} |(AQ)^{\frac{1}{2}}e^{\tau AQ}|_{0p} |A\Delta(\tau)| d\tau \\ &\le \int_{-\infty}^{-\frac{1}{2}\lambda_{N+1}^{-1}} \lambda_{N+1}^{-\frac{1}{2}} \exp[\tau(\lambda_{N+1} - \lambda_N - K_7(1+l)\lambda_N^{-\frac{1}{2}})] d\tau \\ &\quad + \int_{-\frac{1}{2}\lambda_{N+1}^{-1}}^{0} K_3\Big(\frac{1}{2}\Big)|\tau|^{-\frac{1}{2}} \exp[-\tau(\lambda_N + K_7(1+l)\lambda_N^{\frac{1}{2}})] d\tau |Ap_{01} - Ap_{02}|.\end{aligned}$$

Through elementary calculations, we show that the right-hand expression is bounded by

$$\lambda_{N+1}^{-\frac{1}{2}} e^{-\frac{1}{2}} [1 + (1 - \gamma_N \alpha_N)^{-1}] \exp\Big(\frac{\gamma_N \alpha_N}{2}\Big)|Ap_{01} - Ap_{02}|,$$

which proves equation (2.1.52). By equations (2.1.44), (2.1.52) and Lemma 2.1.4, we get

$$\mathcal{F}\Phi \in \mathcal{F}_{b,l}.$$ □

Up to now, we have proved that $\mathcal{F} : \mathcal{F}_{b,l} \to \mathcal{F}_{b,l}$. Now we prove that the map $\mathcal{F}$ is Lipschitz. To this end, we consider two functions Φ_1 and Φ_2 with the same initial value. Let

$$p_i = p(t; \Phi_i, p_0), \quad u_i = p_i + \Phi_i(p_i), \quad \text{where } i = 1, 2.$$

We estimate $|A\mathcal{F}\Phi_1(p_0) - A\mathcal{F}\Phi_2(p_0)|$. Using a similar method as before, we get

$$\frac{d}{dt}|A\Delta| + \lambda_N \alpha_N |A\Delta| \ge -K_7 \lambda_N^{\frac{1}{2}} \|\Phi_1 - \Phi_2\|, \tag{2.1.53}$$

where $\Delta = p_1 - p_2$, $\alpha_N = (1 + K_7(1+l)\lambda_{N+1}^{\frac{1}{2}})$. Since $\Delta(0) = 0$, from equation (2.1.53) we get

$$\begin{aligned}|A\Delta(\tau)| &\le \frac{K_7 \lambda_N^{\frac{1}{2}} \|\Phi_1 - \Phi_2\|}{\alpha_N \lambda_N}(\exp(-\alpha_N \lambda_N \tau)) - 1 \\ &\le K_7 \lambda_N^{\frac{1}{2}} \|\Phi_1 - \Phi_2\| \exp(-\alpha_N \lambda_N \tau), \quad \tau \le 0.\end{aligned} \tag{2.1.54}$$

As in Lemma 2.1.6, from equations (2.1.38),(2.1.46) and (2.1.54), we get

$$\begin{aligned}
&|A\mathscr{F}\Phi_1(p_0) - A\mathscr{F}\Phi_2(p_0)| \\
&\quad \le \int_{-\infty}^{0} |AQe^{\tau AQ}(F(u_1) - F(u_2))| d\tau \\
&\quad \le K_7 \int_{-\infty}^{0} |(AQ)^{\frac{1}{2}} e^{\tau AQ}|_{op} [(1+l)|A\Delta| + \|\Phi_1 - \Phi_2\|] d\tau \\
&\quad \le K_7 \|\Phi_1 - \Phi_2\| \int_{-\infty}^{0} |(AQ)^{\frac{1}{2}} e^{\tau AQ}|_{op} (1 + K_7 \lambda_N^{-\frac{1}{2}} (1-l) e^{-\lambda_N \alpha_N \tau}) d\tau. \qquad (2.1.55)
\end{aligned}$$

From Lemma 2.1.2, the integral of the right-hand side of (2.1.55) is bounded by

$$\begin{aligned}
&2e^{-\frac{1}{2}} \lambda_{N+1}^{-\frac{1}{2}} + \int_{-\alpha}^{0} K_3\left(\frac{1}{2}\right) |\tau|^{-\frac{1}{2}} \exp(-\lambda_N \alpha_N \tau) d\tau \\
&\quad + K_7(1+l)\lambda_N^{-\frac{1}{2}} \left(\int_{-\infty}^{-\alpha} \lambda_{N+1}^{-\frac{1}{2}} \exp[\tau(\lambda_{N+1} - \lambda_N \alpha_N)] d\tau \right) \\
&\le 2e^{-\frac{1}{2}} \lambda_{N+1}^{-\frac{1}{2}} + K_7(1+l)\lambda_N^{-\frac{1}{2}} \left[\lambda_{N+1}^{-\frac{1}{2}} e^{-\frac{1}{2}} (1 + (1 - \gamma_N \alpha_N)^{-1}) \exp\left(\frac{\gamma_N \alpha_N}{2} \right) \right] \\
&\le 2e^{-\frac{1}{2}} \lambda_{N+1}^{-\frac{1}{2}} + \lambda_N^{-\frac{1}{2}} L.
\end{aligned}$$

Then, we have

$$|A\mathscr{F}\Phi_1(p_0) - A\mathscr{F}\Phi_2(p_0)| \le L' \|\Phi_1 - \Phi_2\|, \quad \forall p_0 \in PD(A), \qquad (2.1.56)$$

where $L' = K_7(2e^{-\frac{1}{2}} \lambda_{N+1}^{-\frac{1}{2}} + \lambda_N^{-\frac{1}{2}} L)$.

As mentioned above, we seek conditions to ensure that the mapping $\mathscr{F}$ is from $\mathscr{F}_{b,l}$ to itself, as well as a strict contractive compression on $\mathscr{F}_{b,l}$. This requires us to find sufficient conditions (for λ_N and λ_{N-1}) to guarantee

$$L \le l; \quad L' < 1.$$

Firstly, we note that $\gamma_N = \lambda_{N+1} - \lambda_N - k_l(1+l)\lambda_N^{-\frac{1}{2}} > 0$ is equivalent to

$$1 - \gamma_N \alpha_N > 0, \qquad (2.1.57)$$

or

$$1 > \gamma_N \alpha_N > 0. \qquad (2.1.58)$$

Then from equation (2.1.57) we deduce that

$$L \le K_7(1+l)\lambda_{N+1}^{-\frac{1}{2}} [1 + (1 - \gamma_N \alpha_N)^{-1}].$$

In order to have $L \leq l$, we need to select N appropriately so that the following two inequalities are valid:

$$K_7(1+l)\lambda_{N+1}^{\frac{1}{2}} \leq \frac{l}{2}, \tag{2.1.59}$$

$$K_7(1+l)\lambda_{N+1}^{-\frac{1}{2}}(1-\gamma_N\alpha_N)^{-1} \leq \frac{l}{2}. \tag{2.1.60}$$

Inequality (2.1.59) can be rewritten as

$$K_{10} \leq \lambda_{N+1}^{\frac{1}{2}}, \tag{2.1.61}$$

where $K_{10} = 2K_7(1+l)L^{-1}$. Now we select N, such that equation (2.1.61) holds. Inequality (2.1.60) can be written as

$$K_{10}\lambda_{N+1}^{-\frac{1}{2}} \leq 1-\gamma_N\alpha_N, \tag{2.1.62}$$

which is equivalent to

$$K_{10}\lambda_{N+1}^{-\frac{1}{2}} - 1 + \gamma_N + K_7(1+l)\lambda_{N+1}^{-\frac{1}{2}}\gamma_N^{\frac{1}{2}} \leq 0, \tag{2.1.63}$$

where $\gamma_N = \frac{\lambda_N}{\lambda_{N+1}}$. Let

$$\gamma_N^{\frac{1}{2}} + K_{10}\lambda_{N+1}^{-\frac{1}{2}} = (\gamma_N\lambda_{N+1}^{-1})^{-\frac{1}{2}} + K_{10}\gamma_{N+1}^{-\frac{1}{2}} \leq 1, \tag{2.1.64}$$

apply to equation (2.1.64) twice, to obtain

$$K_{10}\lambda_{N+1}^{-\frac{1}{2}} - 1 + \gamma_N + K_{10}\lambda_{N+1}^{-\frac{1}{2}}\gamma_N^{\frac{1}{2}} \leq K_{10}\lambda_{N+1}^{-\frac{1}{2}} - 1 + \gamma_N^{\frac{1}{2}} \leq 0. \tag{2.1.65}$$

Since $l \leq \frac{1}{8}$, we have $K_7(1+l) \leq K_{10}$. By equation (2.1.64), we deduce equation (2.1.63). Moreover, we obtain that (2.1.62) holds.

Therefore, in order to ensure the map $\mathscr{F}$ takes $\mathscr{F}_{b,l}$ into itself, we need to suppose $\gamma_N > 0$ or $1-\gamma_N\alpha_N > 0$. This assumption is assured by equation (2.1.62). Sufficient conditions for which the mapping $\mathscr{F}$ takes $\mathscr{F}_{b,l}$ into itself are (2.1.59) and (2.1.62), this is assured by equations (2.1.61), (2.1.64). It is easy to see that these two inequalities are corollaries of

$$K_{10} \leq \lambda_{N+1}^{\frac{1}{2}} - \lambda_N^{\frac{1}{2}}. \tag{2.1.66}$$

In order to let the mapping $\mathscr{F}$ be a contraction on $\mathscr{F}_{b,l}$, we must have $L' < 1$. Set $L' \leq \frac{1}{2}$, then we can deduce

$$K_{11} \leq \lambda_{N+1}^{\frac{1}{2}}, \tag{2.1.67}$$

where $K_{11} = 2K_7(2e^{-\frac{1}{2}}+L)$. Hence, under conditions in (2.1.66) and (2.1.67), the mapping $\mathcal{F}$ takes $\mathcal{F}_{b,l}$ into itself and is a contraction. Therefore, there exists a fixed point of $\mathcal{F}$.

Now we prove $M = \text{graph}(\Phi)$ is fixed under the action of $S(t)$. That is,

$$S(t)M \subset M, \quad \forall t \geq 0, \tag{2.1.68}$$

and prove that it attracts all the orbits and approaches M exponentially.

We firstly prove the invariance of M. In fact, from the expression

$$\Phi(p_0) = -\int_{-\infty}^{0} e^{\tau AQ} QF(u(\tau, p_0))d\tau, \tag{2.1.69}$$

where $u(\tau, p_0) = p(\tau, \Phi, p_0) + \Phi(p(\tau), \Phi, p_0)$, inserting $p(t) = p(t; \Phi, p_0)$ with p_0 into equation (2.1.69), and noting that

$$p(\tau, \Phi, p(t; \Phi, p_0)) = p(\tau + t; \Phi, p_0),$$

we deduce that

$$\begin{aligned}\Phi(p(t)) &= -\int_{-\infty}^{0} e^{\tau AQ} QF(u(\tau, p(t)))d\tau \\ &= -\int_{-\infty}^{t} e^{-(t-\tau)AQ} QF(u(\tau), p_0)d\tau, \quad \forall t \in \mathbf{R}.\end{aligned} \tag{2.1.70}$$

Differentiating with respect to t, it is readily seen that $(p(t), q(t))$ is the solution of problem (2.1.29)–(2.1.30), and $u(t) = p(t) + q(t)$ is the solution of problem (2.1.25), where $q(t) = \Phi(p(t))$. This means that $S(t)M \subseteq M$, $\forall t \geq 0$.

In order to prove that the set M attracts all solutions of (2.1.25) exponentially, we first describe excursion properties of equation (2.1.20):

Excursion properties

For every $T > 0$, $\gamma > 0$, $r > 0$, there exist constants K_2, K_3 (they depend on T, γ, r and constant $C_1 = C_4$, but not on $S(t)$ or N), such that for every $N \geq 1$, one of the following inequalities is valid:

$$\left|Q_N(S(t)u_0 - S(t)v_0)\right| \leq \gamma \left|P_N(S(t)u_0 - S(t)v_0)\right|, \tag{2.1.71}$$

or

$$\left|S(t)u_0 - S(t)v_0\right| \leq K_2 \exp(-K_3 \alpha \lambda_{N+1} t)|u_0 - v_0|. \tag{2.1.72}$$

The results are used for all t, satisfying $t_0 \leq t \leq 2t_0$, where $t_0 = (\frac{1}{2K_1}) \lg 2$ and K_1 are constants. When $|Au_0| \leq r$, $|Au_0| \leq r$, we have

$$\left|S(t)u_0 - S(t)v_0\right| \leq \exp(K_1 t)|u_0 - v_0|, \quad \forall t \geq t_0,$$

$\gamma = \frac{1}{8}$, $N \geq N_0$, where N_0 satisfies

$$\lambda_{N_0+1} \geq (2K_3 a t_0)^{-1} \lg(2K_2). \tag{2.1.73}$$

Now equations (2.1.71)–(2.1.72) become

$$\big|Q_N(S(t)u_0 - S(t)v_0)\big| \leq \frac{1}{8}\big|P_N(S(t)u_0 - S(t)v_0)\big|, \tag{2.1.74}$$

$$\big|S(t)u_0 - S(t)v_0\big| \leq \frac{1}{2}|u_0 - v_0|, \tag{2.1.75}$$

where $u_0, v_0 \in D(A)$, $|Au_0| \leq r$, $|Av_0| \leq r$, $t_0 \leq t \leq 2t_0$.

For a fixed $r = 4\rho + b$, the orbit of equations (2.1.24)–(2.1.25) will enter into the ball with origin 0 and radius $4\rho = 8\rho_2$ in $D(A)$. Let

$$|Au_0| \leq 4\rho, \quad \big|AS(t)u_0\big| \leq 4\rho, \quad \forall t \geq 0.$$

We first prove that for any t_1, $t_0 \leq t_1 \leq 2t_0$, we have

$$\operatorname{dist}(S(t_1)u_0, \mu) \leq \frac{1}{2}\operatorname{dist}(u_0, M),$$

where $\operatorname{dist}(\Phi, M) = \inf\{|\phi - v| : v \in M\}$. To achieve this, we select v_0 such that $|u_0 - v_0| = \operatorname{dist}(u_0, M)$. Then $v_0 = Pv_0 + \Phi(P(v_0))$. We request $|APv_0| \leq 4\rho$. Otherwise, if $|APv_0| > 4\rho \geq |APu_0|$, then $\Phi(Pv_0) = 0$, $v_0 = Pv_0$. In addition, there exists $\beta, 0 < \beta \leq 1$, such that $|APv_\beta| = 4\rho$, where $v_\beta = \beta Pu_0 + (1-\beta)v_0 \in PD(A)$. Then we have $\Phi(v_\beta) = 0$, hence $v_\beta \in u$, and

$$\begin{aligned}
|v_\beta - u_0|^2 &= |v_\beta - Pu_0|^2 + |Qu_0|^2 \\
&= \big|(1-\beta)(v_0 - Pu_0)\big|^2 + |Qu_0|^2 \\
&< |v_0 - Pu_0|^2 + |Qu_0|^2 = |v_0 - u_0|^2.
\end{aligned}$$

This is contradiction to $|v_0 - u_0| = \operatorname{dist}(v_0, M)$. Since $|A\Phi(Pv_0)| \leq b$, we have

$$|Av_0| \leq |APv_0| + \big|A\Phi(Pv_0)\big| \leq 4\rho + b = r.$$

Secondly, for $S(t)u_0$ and $S(t)v_0$, using the excursion property (2.1.74)–(2.1.75), if equation (2.1.75) is established, then

$$\begin{aligned}
\operatorname{dist}(S(t_1)u_0, \mu) &\leq \big|S(t_1)u_0 - S(t_1)v_0\big| \\
&\leq \frac{1}{2}|u_0 - v_0| \leq \frac{1}{2}\operatorname{dist}(u_0, \mu).
\end{aligned}$$

If (2.1.74) is established, then

$$\begin{aligned}
\operatorname{dist}(S(t_1)u_0, \mu) &\leq \big|S(t_1)u_0 - (P(S(t_1)u_0) + \Phi(PS(t_1)u_0))\big| \\
&\leq \big|QS(t_1)u_0 - QS(t_1)v_0\big| + \big|\Phi(PS(t_1)v_0) - \Phi(PS(t_1)u_0)\big| \\
&\leq \left(\frac{1}{8} + l\frac{\lambda_N}{\lambda_{N+1}}\right)\big|PS(t_1)v_0 - PS(t_1)u_0\big|.
\end{aligned}$$

From equations (2.1.74)–(2.1.32), and

$$|q| \le \lambda_{N+1}^{-1}|Aq|, \quad q \in QD(A),$$
$$|Ap| \le \lambda_N|p|, \quad p \in PD(A),$$

we have

$$\operatorname{dist}(S(t_1)u_0, M) \le \frac{1}{4}|S(t_1)v_0 - S(t_1)u_0| \le \frac{1}{2}|v_0 - u_0| \le \frac{1}{2}\operatorname{dist}(u_0, M).$$

Then, for any $t \ge t_0$, $t = nt_1$, $t_0 \le t_1 \le 2t_0$, we have

$$\begin{aligned} \operatorname{dist}(S(t)u_0, \mu) &\le \left(\frac{1}{2}\right)^n \operatorname{dist}(u_0, M) \\ &= \exp\left(-\frac{t}{t_1}\ln 2\right)\operatorname{dist}(u_0, M) \\ &\le \exp\left(-\frac{t}{2t_0}\ln 2\right)\operatorname{dist}(u_0, M), \end{aligned} \tag{2.1.76}$$

which implies that the exponential decay rate is $\frac{1}{2t_0}\ln 2$.

We prove the global attractor $\mathscr{A} \subset M$. In fact, if $u \in \mathscr{A}$, then the solution $S(t)u$ is defined for all t. By equations (2.1.23)–(2.1.24), we have

$$\operatorname{dist}(S(t)u, M) \le 2\rho_2, \quad \forall t \in \mathbf{R}.$$

Letting $v = S(-t)u$, where $t \ge t_0$, from equation (2.1.76), we have

$$\operatorname{dist}(u, M) = \operatorname{dist}(S(t)v, M) \le \exp\left(-\frac{t}{t_0}\right) \cdot 2\rho_2.$$

From this we get $\operatorname{dist}_{u\in\mathscr{A}}(u, M) = 0$. Hence $\mathscr{A} \subset M$.

All in all, we get the following theorem:

Theorem 2.1.1. *Assume that equations* (2.1.1)–(2.1.11) *are satisfied, and* $0 < l < \frac{1}{8}$. *Let* N_0 *be determined by equation* (2.1.73). *Then if there are constants* K_{10}, K_{11} *(depending on* l *and the initial value) such that*

$$N \ge N_0, \quad \lambda_{N+1}^{\frac{1}{2}} \ge K_{11}, \quad \lambda_{N+1}^{\frac{1}{2}} - \lambda_N^{\frac{1}{2}} \ge K_{10}, \tag{2.1.77}$$

then there exists $b > 0$ *such that*

(i) *The mapping* $\mathscr{F}$ *maps* $\mathscr{F}_{b,l}$ *into itself;*
(ii) $\mathscr{F}_{b,l}$ *has one fixed point;*
(ii) $M = \operatorname{graph}(\Phi)$ *is the inertial manifold of* (2.1.25);
(iv) M *includes the global attractor of* (2.1.1).

Theorem 2.1.2. *Assume that equations* (2.1.1) *and* (2.1.25) *are given in* H, *where the nonlinear term* $F(u) = \theta_\rho(|Au|)R(u)$ *satisfies equation* (2.1.26). *Suppose* l *is given*, $0 < l < \frac{1}{8}$. *Assume there exists* ρ_0 *such that for every solution of equation* (2.1.1) *inequality* (2.1.21) *is satisfied. If there exist constants* N_0, K_{12}, K_{13} *(they depend on* l *and the initial value) such that*

$$N \geq N_0, \quad \lambda_{N+1} \geq K_{12}, \quad \lambda_{N+1} - \lambda_N \geq K_{13}, \tag{2.1.78}$$

then the conclusion of Theorem 2.1.1 *remains true.*

Proof. The nonlinear term $F(u) = \theta_\rho(|Au|)R(u)$ satisfies the Lipschitz condition

$$\left|F(u) - F(v)\right| \leq K|u - v|, \quad \forall u, v \in H,$$

where the parameter is taken as $\rho = 2\rho_0$. The space $\mathscr{F}_{b,l}$ is comprises functions $\Phi : PH \to QH$, satisfying:

$$\begin{gathered}\left|\Phi(p)\right| \leq b, \quad \forall p \in PD(A), \\ \left|\Phi(p_1) - \Phi(p_2)\right| \leq l|p_1 - p_2|, \quad \forall p_1, p_2 \in D(A), \\ \operatorname{supp}\Phi \subseteq \left\{p \in PD(A)\|p\| \leq 4\rho\right\}.\end{gathered}$$

The operator $\mathscr{F}$ is defined as

$$\mathscr{F}\Phi(p_0) = -\int_{-\infty}^{0} e^{\tau AQ}QF(u)\mathrm{d}\tau,$$

where $u = u(\tau) = p(\tau; \Phi, p_0) + \Phi(p(\tau; \Phi, p_0))$. The inequality (2.1.41) becomes

$$\left|F(u)\right| \leq K_4',$$

where K_4' depends on $R(u)$, θ and ρ. Lemma 2.1.2 and inequality (2.1.40) is both valid when $\alpha = 0$. Equation (2.1.42) becomes

$$\left|\mathscr{F}\Phi(p_0)\right| \leq K_5'\lambda_{N+1}^{-1},$$

where $K_5' = K'$. Hence, we can take $b = K_5'\lambda_{N+1}^{-1}$. Lemma 2.1.4 allows changing the norm $|Av|$ into norm $|v|$, then inequality (2.1.46) turns into

$$\left|F(u_1) - F(u_2)\right| \leq K_7'[(1 + l)|p_1 - p_2| + \|\Phi_1 - \Phi_2\|], \tag{2.1.79}$$

where $\|\Phi\| = \sup\{\Phi(p)|;\ p \in PD(A)\}$.

Let $\Delta = p_1 - p_2$, then equation (2.1.48) stays the same. Taking the inner product of equation (2.1.48) and Δ, we obtain

$$\frac{1}{2}\frac{\mathrm{d}}{\mathrm{d}t}|\Delta|^2 + \left|A^{\frac{1}{2}}\Delta\right|^2 = -(P(F(u_1) - F(u_2)), \Delta).$$

From equation (2.1.79), we get

$$\left|\frac{1}{2}\frac{\mathrm{d}}{\mathrm{d}t}|\Delta|^2 + \left|A^{\frac{1}{2}}\Delta\right|^2\right| \le K_7'(1+l)|\Delta|^2,$$

from which we deduce

$$\begin{aligned}|\Delta|\frac{\mathrm{d}}{\mathrm{d}t}|\Delta| &\ge -\left|A^{\frac{1}{2}}\Delta\right|^2 - K_7'(1+l)|\Delta|^2\\ &\ge -|\lambda_N\Delta|^2 - K_7'(1+l)|\Delta|^2.\end{aligned}$$

Then we have

$$|\Delta(\tau)| \le |\Delta(0)|\exp(-\tau(\lambda_N + K_7'(1+l))), \quad \tau < 0,$$

which can be replaced with inequality (2.1.51). In Lemma 2.1.6 the assumptions for y_N can be replaced with

$$|\mathcal{F}\Phi(p_{01}) - \mathcal{F}\Phi(p_{02})| \le L|p_{01} - p_{02}|,$$

where $L = K_7'(1+l)\lambda_N^{-1}$. In fact,

$$\begin{aligned}&|\mathcal{F}\Phi(p_{01}) - \mathcal{F}\Phi(p_{02})|\\ &\le \int_{-\infty}^{0} |e^{\tau AQ}|_{op}|F(u_1) - F(u_2)|\mathrm{d}\tau\\ &\le K_7'(1+l)|p_{01} - p_{02}|\\ &\quad\cdot \int_{-\infty}^{0} \exp(\tau(\lambda_{N+1} - \lambda_N - K_7'(1+l))\mathrm{d}\tau\\ &\le K_7'(1+l)(\lambda_{N+1} - \lambda_N - K_7'(1+l)))^{-1}|p_{01} - p_{02}|.\end{aligned}$$

Similarly, we have

$$|\mathcal{F}\Phi_1(p_0) - \mathcal{F}\Phi_2(p_0)| \le L'\|\Phi_1 - \Phi_2\|,$$

where $L' = K_7'\lambda_{N+1}^{-1} + K_7(\lambda_N + K_7'(1+l))^{-1}L$.

Set $L \le l$, $L' \le \frac{1}{2}$. Then, when $l < \frac{1}{8}$ and

$$K_{12} \le \lambda_{N+1}, \quad K_{13} \le \lambda_{N+1} - \lambda_N, \tag{2.1.80}$$

the mapping $\mathcal{F}$ maps $\mathcal{F}_{b,l}$ into $\mathcal{F}_{b,l}$, and has a fixed point. The lemma is proved. □

Now we consider the Galerkin approximation equation of (2.1.25),

$$\frac{du_M}{\mathrm{d}t} + Au_M + P_M F(u_M) = 0, \tag{2.1.81}$$

where $F(u) = \theta_p(|Au|)R(u)$, u_M is defined on $P_M D(A)$.

Regarding the Galerkin approximation equation (2.1.81), we have the following theorem:

Theorem 2.1.3. *Assume that the assumptions of Theorem 2.1.1 hold, $l > 0$, and N satisfies the condition of Theorem 2.1.1. then for every $M > N$, equation (2.1.81) possesses an inertial manifold M_M. It is made of Lipschitz graph of function Φ_M, where*

$$\Phi_M : P_M D(A) \to QP_M D(A) \subset QD(A),$$

the Lipschitz constant L of Φ_M is the same as that of $\Phi : PD(A) \to QD(A)$ in Theorem 2.1.1. Finally,

$$\|\Phi_M - \Phi\| \le 2K_6 \lambda_{N+1}^{-\frac{1}{4}} \lambda_{M+1}^{\frac{1}{4}},$$

where

$$\|\Phi_M - \Phi\| = \sup_{p \in PD(A)} \left| A\Phi_M(p) - A\Phi(p) \right|.$$

2.2 Inertial manifold and normal hyperbolicity property

We consider the following abstract evolution equation in a Banach space $\mathscr{E}$:

$$\frac{\mathrm{d}u}{\mathrm{d}t} + Au = f(u), \quad u(0) = u_0, \tag{2.2.1}$$

where f is globally Lipschitz continuous, mapping a Banach space E into another Banach space F, and we assume $E \subset F \subset \mathscr{E}$, the inclusion mappings are continuous, E in F and F in $\mathscr{E}$ are both are dense.

Suppose there exists $M_1 > 0$, which is the Lipschitz constant of f, such that

$$|f(u) - f(v)|_F \le M_1 |u - v|_E, \quad u, v \in E. \tag{2.2.2}$$

Suppose that $-A$ generates a strongly continuous semigroup $\{e^{-tA}\}_{t\ge 0}$ on ε, $e^{-tA} \subset \mathscr{E}$, $\forall t > 0$, and there exist two sequences $\{\lambda_n\}_{n\in\mathbf{N}}$, $\{\Lambda_n\}_{n\in\mathbf{N}}$, $0 < \lambda_n < \Lambda_n$, $\forall n \in \mathbf{N}$. Assume that there exists a sequence of finite-dimensional projections $\{P_n\}_{n\in\mathbf{N}}$, such that $Q_n = I - P_n$, and the following exponential dichotomy was established:

$P_n\mathscr{E}$ is invariant under the action of $\{e^{-tA}\}_{t\ge 0}$, which can be extended on $P_n\mathscr{E}$ to a strongly continuous semigroup $\{e^{-tA}P_n\}_{t\in\mathbf{R}}$ such that

$$\begin{cases} \|e^{-tA}P_n\|_{\mathscr{L}(E)} \le K_1 e^{-\lambda n t}, & \forall t \le 0, \\ \|e^{-tA}P_n\|_{\mathscr{L}(F,E)} \le K_1 \lambda_n^{\alpha} e^{-\lambda n t}, & \forall t \le 0, \end{cases} \tag{2.2.3}$$

and $Q_n\mathscr{E}$ is positively invariant with respect to e^{-tA}, $\forall t \ge 0$. We have

$$\begin{cases} \|e^{-tA}Q_n\|_{\mathscr{L}(E)} \le K_2 e^{-\Lambda_n t}, & \forall t \ge 0, \\ \|e^{-tA}Q_n\|_{\mathscr{L}(F,E)} \le K_2 (t^{-\alpha} + \Lambda_n^{\alpha}) e^{-\Lambda_n t}, & \forall t > 0, \end{cases} \tag{2.2.4}$$

where $K_1, K_2 \ge 1$, $0 \le \alpha < 1$.

Assume equation (2.2.1) determines a continuous flow $\{S(t)\}_{t\geq 0}$, $S(t)u_0 = u(t)$ in E, where $u(t)$ is the solution for the following integral equation (2.2.1):

$$u(t) = e^{-tA}u_0 + \int_0^t e^{-(t-s)A}f(u(s))ds, \quad \forall t > 0.$$

Setting $(t, u_0) \to S(t)u_0$ to be continuous, $[0, \infty) \times E \to E$, and

$$\gamma_\alpha = \begin{cases} \int_0^{+\infty} e^{-\tau}y^{-\alpha}dy, & 0 < \alpha < 1, \\ 0, & \alpha = 0, \end{cases}$$

we can get the following result:

Theorem 2.2.1. *Under the above assumptions, if for some $n \in \mathcal{N}$, the following spectral gap condition is satisfied:*

$$\Lambda_n - \lambda_n > 3M_1K_1K_3\lambda_n^\alpha + 3M_1K_1K_2(1+\gamma_\alpha)\Lambda_n^\alpha, \tag{2.2.5}$$

then the semiflow $\{S(t)\}_{t\geq 0}$ generated by equation (2.2.1) possesses the inertial manifold $M = \text{graph}\,\Phi$, where $\Phi : P_nE \to Q_nE$ is Lipschitz continuous and has Lipschitz constant < 1. Manifold M has the positive and negative constants. Furthermore, M is asymptotically complete, that is, for any $u_0 \in E$, there exists $v_0 \in E$ such that

$$|S(t)u_0 - S(t)v_0|_E \leq K_\eta(|u_0|_E)e^{-\eta t}, \quad \forall t \geq 0,$$

where η is an arbitrary number satisfying the following inequality:

$$0 < \eta < \Lambda_n - 2M_1K_2(1+\gamma_\alpha)\Lambda_n^\alpha.$$

Here K_η depends on η and $|u_0|_E$.

Remark 2.2.1. Theorem 2.2.1 only includes the main results. Actually, under the same assumptions as in Theorem 2.2.1, we can get more results, namely, the inertial manifold is the union of all complete orbits, the orbits are confined by $e^{-\sigma t}$, $t \to -\infty$. For some $\sigma > 0$, the existence of a continuous foliation $E = \bigcup_{v_0 \in M} N_{v_0}$ in the space E also can be obtained, where every leaf is a Lipschitz function from Q_nE to P_nE, and N_{v_0} can be expressed as

$$N_{v_0} = \{u_0 \in E,\ |S(t)u_0 - S(t)v_0|_E = o(e^{-\eta t}),\ t \to +\infty\}, \quad \eta > 0.$$

We can also get a continuous contraction mapping $\pi : E \to M$, there is $\pi^{-1}\pi u_0 = N_{u_0}$ such that $S(t)\pi u_0$ is the unique orbit in M, which possesses $|S(t)u_0 - S(t)\pi u_0|_E = \mathcal{O}(e^{-\eta t})$, $t \to \infty$, $\eta > 0$. This gives the asymptotic completeness of M.

Usually, the nonlinear term f in (2.2.1) is Lipschitz on a bounded set. Set

$$|f(u) - f(v)|_F \le d_1(r)|u - v|_E, \quad |f(u)|_F \le d_0(r),$$

$\forall u, v \in B(E)(r) = \{u \in E; |u|_E \le r\}$. Now we can get for $\{S(t)\}_{t\ge 0}$ in E that there exists an absorbing set $\mathscr{B}$, that is, $\mathscr{B} \subset E$ is bounded, and for any bounded set $B \subset E$, there exists $t_0(B) > 0$, such that $S(t)B \subset \mathscr{B}$, $\forall t \ge t_0$.

Take a function $\theta \in C^1([0, +\infty))$, $\theta(s) = 1$, $s \in [0, 1]$, $\theta(s) = 0$, $4s \in [2, -\infty)$, $|\theta'(s)| \le 2$, $\forall s \ge 0$. Using the truncation function $f_\theta(\mu)$ instead of f, we have

$$f_\theta(u) = \theta\left(\frac{|u|_E^2}{\rho^2}\right) f(u), \quad \forall u \in E,$$

where $\rho > 0$, such that $\overline{\mathscr{B}} \subset B_E(\rho)$; here $\overline{\mathscr{B}}$ means the closure in E of $\mathscr{B}$. We consider the truncation function in the equation

$$\frac{\mathrm{d}u}{\mathrm{d}t} + Au = f_\theta(u), \quad u(0) = u_0,$$

which determines the continuous semiflow $\{S_\theta(t)\}_{t\ge 0}$ in E, and possesses the same initial value as equation (2.2.1).

Theorem 2.2.2. *Under the above assumptions, if for some $n \in \mathscr{N}$, the following spectral gap condition is met:*

$$\Lambda_n - \lambda_n > 3M_1K_1K_2\lambda_n^\alpha + 3M_1K_1K_2(1 + \gamma_\alpha)\Lambda_n^\alpha,$$

where $M_1 := d_1(\sqrt{2}\rho) + 4\sqrt{2}\frac{d_0(\sqrt{2}\rho)}{\rho}$, then equation (2.2.1) has inertial manifold M, which is the graph of a Lipschitz function $\Phi : O \subset P_nE \to Q_nE$, where O is an open subset of P_nE. Furthermore, M is asymptotically complete, that is, for any $u_0 \in E$, there exists $v_0 \in E$ such that

$$|S(t + t_0)u_0 + S(t)v_0|_E \le e^{-\eta t}, \quad \forall t \ge 0,$$

where η is an arbitrary number satisfying the following inequality:

$$0 < \eta < \Lambda_n - 2M_1K_2(1 + \gamma_\alpha)\Lambda_n^\alpha,$$

where t_0 depends on η and $|u_0|_E$.

Proof of Theorem 2.2.1. For simplicity, fix $n \in \mathscr{N}$, suppose equation (2.2.5) is valid, and denote $P = P_n$, $Q = Q_n$, $\lambda = \lambda_n$, $\Lambda = \Lambda_n$, $0 < \alpha < 1$. When $\alpha = 0$, the proof is almost the same. Take $\gamma_\alpha = 0$. Introduce the space

$$\mathscr{F}_\sigma = \left\{\varphi \in C((-\infty, 0], E) : \|\varphi\|_{\mathscr{F}_\sigma} = \sup_{t\le 0}(e^{\sigma t}|\varphi(t)|_E) < +\infty\right\},$$

which is a Banach space with the norm $\|\cdot\|_{\mathscr{F}_\sigma}$. For $\varphi \in \mathscr{F}_\sigma$, $y \in PE$, we consider the formal mapping

$$\mathscr{J}(\varphi, y)(t) = e^{-tA}Py - \int_t^0 e^{-(t-s)A}Pf(\varphi(s))ds + \int_{-\infty}^t e^{-(t-s)A}Qf(\varphi(s))ds, \quad t \le 0. \tag{2.2.6}$$

The construction of an invariant manifold is based on the fact that for appropriate σ, a function $\varphi \in \mathscr{F}_\sigma$ is the solution of equation (2.2.1) if and only if φ is the fixed point of $\mathscr{J}$, which requires proving that for an appropriate σ, the mapping $\mathscr{J} : \mathscr{F}_\sigma \times PE \to \mathscr{F}_\sigma$ defined by equation (2.2.6) is a strict contraction in $\mathscr{F}_\sigma$ and uniform for PE. Therefore, there exists a mapping $\varphi : PE \to \mathscr{F}_\sigma$ such that $\mathscr{J}(\varphi(y_0), y_0) = \varphi(y_0)$, $\forall y_0 \in PE, \varphi(y_0)$ is a solution of equation (2.2.1). We define the mapping: $\Phi : PE \to QE$

$$\Phi(y_0) = Q\varphi(y_0)(0) = \int_{-\infty}^0 e^{sA}Qf(\varphi(y_0)(s))ds.$$

Hence

$$\varphi(y_0)(0) = y_0 + \Phi(y_0).$$

Let M = graph Φ, and prove M is Lipschitz, invariant and possesses asymptotic completeness property. Thus M is our sought inertial manifold. □

Lemma 2.2.1. *When $\lambda < \sigma < \Lambda$, we have $\mathscr{J} : \mathscr{F}_\sigma \times PE \to \mathscr{F}_\sigma$.*

Proof. Choose $\varphi \in \mathscr{F}_\sigma$, $y \in PE$. Since f is globally Lipschitz, there exists a constant $M_0 > 0$ such that

$$|f(u)|_F \le M_0 + M_1|u|_E, \quad \forall u \in E. \tag{2.2.7}$$

It follows that

$$\begin{aligned}|Q\mathscr{J}(\varphi, y)|_E &\le \int_{-\infty}^t |e^{-(t-s)A}Qf(\varphi(s))|_E ds \\ &\le K_2 \int_{-\infty}^t (e^{-\Lambda(t-s)}((t-s)^{-\alpha} + \Lambda^\alpha)(M_0 + M_1)|\varphi(s)|_E)ds.\end{aligned}$$

Hence, for $\lambda < \sigma < \Lambda$, we have

$$e^{\sigma t}|Q\mathscr{J}(\varphi, y)(t)|_E \le M_0K_2\Lambda^\alpha e^{(\sigma-\Lambda)t} \int_{-\infty}^t e^{\Lambda s}ds$$

$$
\begin{aligned}
&+ M_1 K_2 e^{\sigma t} \int_{-\infty}^{t} e^{-\Lambda(t-s)}(t-s)^{-\alpha}\mathrm{d}s \\
&+ M_1 K_2 \Lambda^{\sigma} \|\varphi\|_{\mathcal{F}_\sigma} \int_{-\infty}^{t} e^{(\sigma-\Lambda)(t-s)}\mathrm{d}s \\
&+ M_1 K_2 \|\varphi\|_{\mathcal{F}_\sigma} \int_{-\infty}^{t} e^{(\sigma-\Lambda)(t-s)}(t-s)^{-\alpha}\mathrm{d}s \\
\le{}& \frac{M_0 K_2}{\Lambda^{1-\alpha}} e^{\sigma t} + \frac{\gamma_\alpha M_0 K_2}{\Lambda^{1-\alpha}} e^{\sigma t} + \frac{M_1 K_2 \Lambda^{\alpha}}{\Lambda-\sigma}\|\varphi\|_{\mathcal{F}_\sigma} + \frac{M_1 K_2 \gamma_\alpha}{(\Lambda-\sigma)^{1-\sigma}}\|\varphi\|_{\mathcal{F}_\sigma},
\end{aligned}
$$

where we used

$$
\begin{aligned}
\int_{-\infty}^{t} e^{-r(t-s)}(t-s)^{-\alpha}\mathrm{d}s &= \frac{\gamma_\alpha}{\gamma^{1-\alpha}}, \\
\int_{-\infty}^{t} e^{-r(t-s)}\mathrm{d}s &= \frac{1}{r}, \quad \forall r > 0.
\end{aligned}
\tag{2.2.8}
$$

Hence, $Q\mathcal{J}(\varphi, y)(t) \in QE$, $\forall t \le 0$, and

$$
\sup_{t\le 0}(e^{\sigma t}|Q\mathcal{J}(\varphi, y)(t)|_E) \le M_0 K_2(1+\gamma_\alpha) + \frac{M_1 K_2(1+\gamma_\alpha)\Lambda^{\alpha}}{\Lambda-\sigma}\|\varphi\|_{\mathcal{F}_\sigma}. \tag{2.2.9}
$$

Similarly,

$$
\begin{aligned}
|P\mathcal{J}(\varphi, y)(t)|_E &\le |e^{-tA}Py|_E + \int_t^0 |e^{-\Lambda(t-s)}Pf(\varphi(s))|_E \mathrm{d}s \\
&\le K_1 e^{-\lambda t}|y|_E + K_1\lambda^{\alpha}\int_t^0 e^{-\lambda(t-s)}(M_0 + M_1|\varphi(s)|_E)\mathrm{d}s.
\end{aligned}
$$

Hence

$$
\begin{aligned}
&e^{\sigma t}|P\mathcal{J}(\varphi, y)(t)|_E \\
&\le K_1 e^{(\sigma-\lambda)t}|y|_E + M_0 K_1 \lambda^{\alpha} e^{(\sigma-\lambda)t}\int_t^0 e^{\lambda s}\mathrm{d}s + M_1 K_1 \lambda^{\alpha}\|\varphi\|_{\mathcal{F}_\sigma}\int_t^0 e^{(\sigma-\lambda)(t-s)}\mathrm{d}s \\
&\le K_1 e^{(\sigma-\lambda)t}|y|_E + \frac{M_0 K_1}{\lambda^{1-\alpha}} e^{(\sigma-\lambda)t} + \frac{M_1 K_1 \lambda^{\alpha}}{\sigma-\lambda}\|\varphi\|_{\mathcal{F}_\sigma}.
\end{aligned}
$$

Thus, $P\mathcal{J}(\varphi, y)(t) \in PE$, $\forall t \le 0$, and

$$
\sup_{t\le 0}(e^{\sigma t}|P\mathcal{J}(\varphi, y)(t)|_E) \le K_1|y|_E + \frac{M_0 K_1}{\lambda^{1-\alpha}} + \frac{M_1 K_1 \lambda^{\alpha}}{\sigma-\lambda}\|\varphi\|_{\mathcal{F}_\sigma}. \tag{2.2.10}
$$

Hence, $\mathcal{J}(\varphi,y)(t) \in E$, $\forall t \le 0$. The continuity of the mapping $t \to \mathcal{J}(\varphi,\varphi)$ from $(-\infty,0]$ to E is also proved. From equations (2.2.9)–(2.2.10), we know $\mathcal{J}(\varphi,y) \in \mathcal{F}_\sigma$, and

$$\|\mathcal{J}(\varphi,y)(t)\|_{\mathcal{F}_\sigma} \le K_1|y|_E + \frac{M_0K_1}{\lambda^{1-\alpha}} + (1+\gamma_\alpha)\frac{M_0K_2}{\Lambda^{1-\alpha}} + \left\{\frac{M_1K_2\lambda^\alpha}{\sigma-\lambda} + \frac{M_1K_2(1+\gamma_\alpha)\Lambda^\alpha}{\Lambda-\sigma}\right\}\|\varphi\|_{\mathcal{F}_\sigma}.$$

This proves that $\mathcal{J}$ is a well-defined mapping, $\mathcal{F}_\sigma \times PE \to \mathcal{F}_\sigma$. □

Lemma 2.2.2. *For any σ, such that $\lambda < \sigma < \Lambda$, we have*

$$\|\mathcal{J}(\varphi_1,y) - \mathcal{J}(\varphi_2,y)\|_{\mathcal{F}_\sigma} \le \theta_\sigma\|\varphi_1-\varphi_2\|_{\mathcal{F}_\sigma}, \quad \forall \varphi_1,\varphi_2 \in \mathcal{F}_\sigma,\ y \in PE,$$

where

$$\theta_\sigma = \frac{M_1K_1\lambda^\alpha}{\sigma-\lambda} + \frac{M_1K_2(1+\gamma_\alpha)\Lambda^\alpha}{\Lambda-\sigma}.$$

Proof. Taking $\varphi_1,\varphi_2 \in \mathcal{F}_\sigma$, $y \in PE$, for $t \le 0$, we have

$$\begin{aligned}
&|Q\mathcal{J}(\varphi_1,y)(t) - Q\mathcal{J}(\varphi_2,y)(t)|_E \\
&\le \int_{-\infty}^{t} |e^{-(t-s)A}Q[f(\varphi_1(s)) - f(\varphi_2(s))]|_E \mathrm{d}s \\
&\le K_2 \int_{-\infty}^{t} ((t-s)^{-\alpha} + \Lambda^\alpha)e^{-\Lambda(t-s)}|f(\varphi_1(s)) - f(\varphi_2(s))|_F \mathrm{d}s \\
&\le M_1K_2 \int_{-\infty}^{t} ((t-s)^{-\alpha} + \Lambda^\alpha)e^{-\Lambda(t-s)}|\varphi_1(s) - \varphi_2(s)|_E \mathrm{d}s \\
&\le M_1K_2\|\varphi_1-\varphi_2\|_{\mathcal{F}_\sigma} \int_{-\infty}^{t} ((t-s)^{-\alpha} + \Lambda^\alpha)e^{-\Lambda(t-s)}e^{-\sigma s}\mathrm{d}s.
\end{aligned}$$

Hence, using equations (2.2.7) and (2.2.8), we get

$$\begin{aligned}
&e^{\sigma t}|Q\mathcal{J}(\varphi_1,y)(t) - Q\mathcal{J}(\varphi_2,y)(t)|_E \\
&\le M_1K_2\|\varphi_1-\varphi_2\|_{\mathcal{F}_\sigma} \int_{-\infty}^{t} e^{(\sigma-\Lambda)(t-s)}(t-s)^{-\alpha}\mathrm{d}s \\
&\quad + M_1K_2\Lambda^\alpha\|\varphi_1-\varphi_2\|_{\mathcal{F}_\sigma} \int_{-\infty}^{t} e^{(\sigma-\Lambda)(t-s)}\mathrm{d}s \\
&\le \frac{M_1K_2\gamma^\alpha}{(\Lambda-\sigma)^{1-\alpha}}2\|\varphi_1-\varphi_2\|_{\mathcal{F}_\sigma} + \frac{M_1K_2\Lambda^\alpha}{(\Lambda-\sigma)}\|\varphi_1-\varphi_2\|_{\mathcal{F}_\sigma}
\end{aligned}$$

$$\leq \frac{M_1 K_2(1+\gamma_\alpha)}{(\Lambda-\sigma)}\|\varphi_1-\varphi_2\|_{\mathcal{F}_\sigma}. \tag{2.2.11}$$

Similarly we have

$$\begin{aligned}|P\mathcal{J}(\varphi_1,y)(t)-P\mathcal{J}(\varphi_2,y)(t)|_E &\leq \int_t^0 |e^{-(t-s)A}P[f(\varphi_1(s))-f(\varphi_2(s))]|_E \mathrm{d}s\\ &\leq M_1K_1\lambda^\alpha\int_t^0 e^{-\lambda(t-s)}|\varphi_1(s)-\varphi_2(s)|_E\mathrm{d}s.\end{aligned}$$

Hence

$$\begin{aligned}e^{\sigma t}|P\mathcal{J}(\varphi_1,y)(t)-P\mathcal{J}(\varphi_2,y)(t)|_E &\leq M_1K_1\|\varphi_1-\varphi_2\|_{\mathcal{F}_\sigma}\int_t^0 e^{(\sigma-\Lambda)(t-s)}\mathrm{d}s\\ &\leq \frac{M_1K_1\Lambda^\alpha}{\sigma-\lambda}\|\varphi_1-\varphi_2\|_{\mathcal{F}_\sigma}.\end{aligned} \tag{2.2.12}$$

Then by equations (2.2.11) and (2.2.12) we get

$$\|\mathcal{J}(\varphi_1,y)-\mathcal{J}(\varphi_2,y)\|_{\mathcal{F}_\sigma}\leq\theta_\sigma\|\varphi_1-\varphi_2\|_{\mathcal{F}_\sigma},$$

where

$$\theta_\sigma=\frac{M_1K_1\lambda^\alpha}{\sigma-\lambda}+\frac{M_1K_2(1+\gamma_\alpha)\Lambda^\alpha}{\Lambda-\sigma}.$$

Now by the spectral gap condition (2.2.5), we have

$$\begin{aligned}\lambda+2M_1K_1\lambda^\alpha &< \lambda+3M_1K_1K_2\lambda^\alpha\\ &< \Lambda-3M_1K_1K_2(1+\gamma_\alpha)\Lambda^\alpha\\ &< \Lambda-2M_1K_2(1+\gamma_\alpha)\Lambda^\alpha.\end{aligned}$$

Thus, we can select σ which satisfies

$$\lambda-2M_1K_1\lambda^\alpha<\sigma<\Lambda-2M_1K_2(1+\gamma_\alpha)\Lambda^\alpha, \tag{2.2.13}$$

such that

$$\frac{M_1K_1\lambda^\alpha}{\sigma-\lambda}<\frac{1}{2},\quad \frac{M_1K_2(1+\gamma_\alpha)\Lambda^\alpha}{\Lambda-\sigma}<\frac{1}{2}.$$

Therefore, $\theta_\sigma<1$. That is, $\mathcal{J}$ is a strict contraction in $\mathcal{J}_\sigma$, uniform for PE, where σ satisfies equation (2.2.13), which allows us to define the mapping, $PE\to QE$. □

Lemma 2.2.3. *Assume that for any σ satisfying equation* (2.2.13), $\theta_\sigma < 1$, $\mathcal{J}$ *is a strict contraction in* $\mathcal{J}_\sigma$, *for PE being uniform. Then there exists a mapping* $\varphi : PE \to \mathcal{F}_\sigma$ *such that* $\mathcal{J}(\varphi(y_0), y_0) = \varphi(y_0)$, $\forall \varphi(y_0) \in PE$. *Hence we can define mapping, PE* $\to$ *QE, by*

$$\Phi(y_0) = Q\varphi(y_0)(0) = \int_{-\infty}^{0} e^{sA} Qf(\varphi(y_0)(s))\mathrm{d}s. \tag{2.2.14}$$

Proposition 2.2.1. *Consider the manifold* $M = \operatorname{graph}\Phi$, *where* Φ *is defined by equation* (2.2.14). *Then M is invariant under the equation* (2.2.1), *that is,* $S(t)M = M$, $\forall t \geq 0$, *of finite dimension, Lipschitz continuous, and has Lipschitz constant less than 1; M can be characterized as*

$$M = \{u_0 \in E,\ u_0 \text{ belongs to a complete orbit } \{u(t;u_0)\}_{t\in\mathbf{R}},$$
$$\text{the solution of equation (2.2.1) satisfies } |u(t;u_0)|_E = O(e^{-\sigma t}),\ t \to -\infty\}, \tag{2.2.15}$$

where σ is an arbitrary number satisfying equation (2.2.13). *Furthermore, for any* $y_0 \in PE$, Φ *satisfies*

$$\Phi(y_0) = \int_{-\infty}^{0} e^{sA} Qf(y(s) + \Phi(y(s)))\mathrm{d}s, \tag{2.2.16}$$

where $y = y(t)$, $t \leq 0$, *is the solution of*

$$\frac{\mathrm{d}y}{\mathrm{d}t} + Ay = Pf(y + \Phi(y)), \quad y(0) = y_0. \tag{2.2.17}$$

Meanwhile, $\Phi(y_0) = z(0)$, *where* $z = z(t) \in QE$ *is the solution of*

$$\frac{\mathrm{d}z}{\mathrm{d}t} + Az = Qf(y + \Phi(y)), \quad t \leq 0, \tag{2.2.18}$$

$|z(t)|_E = O(e^{-\sigma t})$, $t \to -\infty$, *where σ satisfies equation* (2.2.13), $y = y(t) \in PE$, $t \leq 0$ *is the solution of equation* (2.2.17).

Proof. To see that M is characterized by equation (2.2.15), we fix σ to satisfy equation (2.2.13), and use M_σ to denote the right-hand side of equation (2.2.15). By definition, $u_0 \in M_\sigma$ if and only if there exists a continuous function $(-\infty, 0] \ni t \to u(t;u_0) \in E$, $u(0;u_0) = u_0$, $u(\cdot, u_0) \in \mathcal{F}_\sigma$ is the solution of equation (2.2.1) and

$$u(t;u_0) = e^{-(t-\tau)A}u(\tau;u_0) + \int_{\tau}^{t} e^{-(t-\tau)A} f(u(s;u_0))\mathrm{d}s, \tag{2.2.19}$$

$\forall \tau \leq t \leq 0$. Then we have

$$\begin{aligned} |e^{-(t-\tau)A}Qu(\tau;u_0)|_E &\leq K_2 |e^{-\Lambda(t-\tau)}u(\tau;u_0)|_E \\ &\leq K_2 e^{-\Lambda t}\|u(\cdot;u_0)\|_{\mathcal{F}_\sigma} e^{(\Lambda-\sigma)\tau} \to 0, \quad \tau \to -\infty. \end{aligned} \tag{2.2.20}$$

Acting with Q on the integral equation (2.2.19), and setting $\tau \to -\infty$, we deduce

$$Qu(t;u_0) = \int_{-\infty}^{t} e^{-(t-s)}Qf(u(s;u_0))\mathrm{d}s, \quad \forall t \le 0. \tag{2.2.21}$$

On the other hand, acting with P on equation (2.2.19), setting $t = 0$, we get

$$Pu_0 = e^{\tau A}Pu(\tau;u_0) + \int_{-\infty}^{t} e^{sA}Pf(u(s;u_0))\mathrm{d}s. \tag{2.2.22}$$

Since $\{e^{tA}P\}_{t\in\mathbf{R}}$ generates a group, multiplying equation (2.2.22) by $e^{-\tau A}P$, we get

$$Pu(\tau;u_0) = e^{-\tau A}Pu_0 - \int_{\tau}^{0} e^{-(\tau-s)A}Pf(u(s;u_0))\mathrm{d}s, \quad \tau \le 0. \tag{2.2.23}$$

Therefore, based on equations (2.2.21) and (2.2.23), by (2.2.19) we deduce

$$\begin{aligned} u(t;u_0) = {} & e^{-\tau A}Pu_0 - \int_{t}^{0} e^{-(\tau-s)A}Pf(u(s;u_0))\mathrm{d}s \\ & + \int_{-\infty}^{t} e^{-(\tau-s)A}Qf(u(s;u_0))\mathrm{d}s, \quad \forall t \le 0. \end{aligned} \tag{2.2.24}$$

This means that $u(0;u_0) \in \mathscr{F}_\sigma$ is the fixed point of $\mathscr{J}(\cdot, Pu_0)$. By Lemma 2.2.3, we have

$$u(\cdot, u_0) = \varphi(Pu_0).$$

Hence

$$u_0 = \varphi(Pu_0)(0) = Pu_0 + \Phi(Pu_0) \in \operatorname{graph}\Phi = M,$$

which proves $M_\sigma \subset M$. On the contrary, if $u_0 \in \operatorname{graph}\Phi$, then we have $u_0 = Pu_0 + \Phi(Pu_0) = \varphi(Pu_0)(0)$. Since $\varphi(Pu_0)$ is the fixed point of $\mathscr{J}(\cdot, Pu_0)$, we get

$$\begin{aligned} & \varphi(Pu_0)(t) - e^{-(t-\tau)A}\varphi(Pu_0)(\tau) \\ & = e^{-tA}Pu_0 - e^{-(t-s)A-\tau A}Pu_0 - \int_{\tau}^{0} e^{-(t-s)A}Pf(\varphi(Pu_0))(s)\mathrm{d}s \\ & + e^{-(t-\tau)A}\int_{\tau}^{0} e^{-(\tau-s)A}Pf(\varphi(Pu_0))(s)\mathrm{d}s \\ & + \int_{-\infty}^{t} e^{-(t-s)A}Qf(\varphi(Pu_0)(s))\mathrm{d}s - e^{-(t-\tau)A}\int_{-\infty}^{\tau} e^{-(\tau-s)A}Qf(\varphi(Pu_0)(s))\mathrm{d}s \end{aligned}$$

$$= \int_{\tau}^{t} e^{-(t-s)A} Pf(\varphi(Pu_0)(s))ds + \int_{\tau}^{t} e^{-(t-s)A} Qf(\varphi(Pu_0)(s))ds, \quad \tau \le t \le 0.$$

Hence

$$\varphi(Pu_0)(t) = e^{-(t-\tau)A}\varphi(Pu_0)(\tau) + \int_{\tau}^{t} e^{-(t-s)A} f(\varphi(Pu_0)(s))ds,$$

which means $\varphi(Pu_0)$ is the solution of equation (2.2.1), $u(Pu_0) \in \mathcal{F}_\sigma$ is the fixed point of $\mathcal{J}(\cdot, Pu_0)$. Hence, $u_0 \in M_\sigma$, $M \subset M_\sigma$, thus $M = M_\sigma$.

It can be seen from the characterizing equation (2.2.15) that M is invariant. Hence equations (2.2.16)–(2.2.18) are valid. For the Lipschitz continuity of Φ, taking $y_1, y_2 \in PE$, and noting that σ satisfies equation (2.2.13), we get

$$\begin{aligned}
|\Phi(y_1) - \Phi(y_2)|_E &\le \int_{-\infty}^{0} |e^{\tau A} Q[f(\varphi(y_1)(s)) - f(\varphi(y_2)(s))]|_E ds \\
&\le M_1 K_2 \int_{-\infty}^{0} e^{\Lambda s}(|s|^{-\alpha} + \Lambda^\alpha)|\varphi(y_1)(s) - \varphi(y_2)(s)|_E ds \\
&\le M_1 K_2 \|\varphi(y_1) - \varphi(y_2)\|_{\mathcal{F}_\sigma} \int_{-\infty}^{0} (|s|^{-\alpha} + \Lambda^\alpha) e^{(\Lambda-\sigma)s} ds \\
&\le \frac{M_1 K_2 (1+\gamma_\alpha)\Lambda^\alpha}{\Lambda - \sigma} \|\varphi(y_1) - \varphi(y_2)\|_{\mathcal{F}_\sigma},
\end{aligned} \tag{2.2.25}$$

where

$$\begin{aligned}
\|\varphi(y_1) - \varphi(y_2)\|_{\mathcal{F}_\sigma} &= \|\mathcal{J}(\varphi(y_1), y_1) - \mathcal{J}(\varphi(y_2), y_2)\|_{\mathcal{F}_\sigma} \\
&\le \|\mathcal{J}(\varphi(y_1), y_1) - \mathcal{J}(\varphi(y_2), y_1)\|_{\mathcal{F}_\sigma} \\
&\quad + \|\mathcal{J}(\varphi(y_2), y_1) - \mathcal{J}(\varphi(y_2), y_2)\|_{\mathcal{F}_\sigma} \\
&\le \theta_\sigma \|\varphi(y_1) - \varphi(y_2)\|_{\mathcal{F}_\sigma} + \|e^{-tA} P(y_1 - y_2)\|_{\mathcal{F}_\sigma} \\
&\le \theta_\sigma \|\varphi(y_1) - \varphi(y_2)\|_{\mathcal{F}_\sigma} + K_1 |y_1 - y_2|_E
\end{aligned} \tag{2.2.26}$$

and $\theta_\sigma < 1$. Hence by equation (2.2.26), we obtain

$$\|\varphi(y_1) - \varphi(y_2)\|_{\mathcal{F}_\sigma} \le \frac{K_1}{1 - \theta_\sigma} |y_1 - y_2|_E. \tag{2.2.27}$$

Inserting equation (2.2.27) into equation (2.2.25), we get

$$\|\Phi(y_1) - \Phi(y_2)\|_E \le \frac{M_1 K_1 K_2 (1+\gamma_\alpha)\Lambda^\alpha}{(1-\theta_\sigma)(\Lambda - \alpha)} |y_1 - y_2|_E. \tag{2.2.28}$$

By the spectral gap condition (2.2.25), we select

$$\sigma = \Lambda - 3M_1K_1K_2(1+\gamma_\alpha)\Lambda^\alpha,$$

where σ satisfies equation (2.2.13). For the σ, we have

$$\begin{aligned}\theta_\sigma &= \frac{M_1K_1\Lambda_\alpha}{\sigma-\lambda} + \frac{M_1K_2(1+\gamma_\alpha)}{\Lambda-\alpha}\\ &= \frac{M_1K_1\Lambda_\alpha}{\Lambda-\lambda-3M_1K_1K_2(1+\gamma_\alpha)\Lambda^\alpha} + \frac{M_1K_2(1+\gamma_\alpha)\Lambda^\alpha}{3M_1K_1K_2(1+\gamma_\alpha)\Lambda^\alpha}.\end{aligned}$$

Since

$$\Lambda - \lambda - 3M_1K_1K_2(1+\gamma_\alpha)\Lambda^\alpha > 3M_1K_1K_2\lambda^\alpha,$$

we obtain

$$\theta_\sigma < \frac{1}{3K_2} + \frac{1}{3K_2} \le \frac{2}{3},$$

that is,

$$\frac{1}{1-\theta_\sigma} < 3.$$

By equation (2.2.28), we obtain

$$|\Phi(y_1) - \Phi(y_2)|_E \le l|y_1 - y_2|_E, \tag{2.2.29}$$

where

$$l \le \frac{M_1K_1K_2(1+\gamma_\alpha)\Lambda^\alpha}{(1-\theta_\sigma)(\Lambda-\sigma)} < \frac{3M_1K_1K_2(1+\gamma_\alpha)\Lambda^\alpha}{3M_1K_1K_2(1+\gamma_\alpha)\Lambda^\alpha} = 1.$$

This proves that Φ is Lipschitz continuous, and has Lipschitz constant less than 1. We clearly know that the set M possesses the PE dimension, which is finite. Since P is a finite-dimensional projection, we finish the proof of Proposition 2.2.1. □

In what follows, we consider the asymptotic completeness property. Firstly we compare two different trajectories of (2.2.1). Fix $u_0 \in E$, and set $v_0 = y_0 + z_0$, $y_0 \in PE$, $z_0 \in QE$. We study the difference

$$\psi(t) = S(t)v_0 - S(t)u_0, \quad t \ge 0, \tag{2.2.30}$$

for any $0 \le t \le \tau$, for $Q\psi \in [0,t]$ and $P\psi \in [t,\tau]$. We use the constant variation method to obtain

$$\begin{aligned}\psi(t) &= e^{-tA}(z_0 - Qu_0) + \int_0^t e^{-(t-s)A}Q[f(S(s)v_0) - f(S(s)u_0)]ds\\ &\quad + e^{-(t-\tau)A}P(S(\tau)v_0 - S(\tau)u_0) - \int_t^\tau e^{-(t-s)A}P[f(S(s)v_0) - f(S(s)u_0)]ds.\end{aligned} \tag{2.2.31}$$

Let $|\psi(t)|_E = O(e^{-\sigma t})$, $t \to +\infty$. For $\sigma > \lambda$, we have

$$|e^{-(t-\tau)A}P[S(\tau)v_0 - S(\tau)u_0]|_E \le K_1 e^{-\lambda(t-\tau)}|\psi(t)|_E = O(e^{(\lambda-\sigma)t}), \quad \tau \to +\infty.$$

Hence, if we let $\tau \to +\infty$ in equation (2.2.31), using equation (2.2.30), we get

$$\psi(t) = e^{-tA}(z_0 - Qu_0) + \int_0^t e^{-(t-s)A}Q[f(S(s)u_0 + \psi(s)) - f(S(s)u_0)]ds$$
$$- \int_t^\infty e^{-(t-s)A}P[f(S(s)v_0 + \psi(s)) - f(S(s)u_0)]ds. \tag{2.2.32}$$

On the contrary, let the function $\psi \in C([0,+\infty), E)$ be such that $\psi(t) = O(e^{-\sigma t})$, $t \to +\infty$. And for $z_0 \in QE$, assume $u_0 \in E$ satisfies equation (2.2.32). Then $y_0 = Pu_0 + P\psi(0)$, $v_0 = y_0 + z_0$. Set

$$v(t) = \psi(t) + S(t)u_0, \quad t \ge 0. \tag{2.2.33}$$

First, we note that

$$v(0) = \psi(0) + u_0 = P\psi(0) + z_0 - Qu_0 + u_0$$
$$= y_0 - Pu_0 + z_0 - Qu_0 + u_0 = y_0 + z_0 = v_0.$$

Second,

$$y_0 = Pu_0 + P\psi(0)$$
$$= Pu_0 - \int_0^{+\infty} e^{sA}P[f(S(s)u_0 + \psi(s)) - f(S(s)u_0)]ds.$$

Thus, equation (2.2.32) can be rewritten as

$$\psi(t) = e^{-tA}(z_0 - Qu_0) + e^{-tA}(y_0 - Pu_0)$$
$$+ \int_0^t e^{-(t-s)A}Q[f(S(s)u_0 + \psi(s)) - f(S(s)u_0)]ds$$
$$+ \int_0^t e^{-(t-s)A}P[f(S(s)u_0 + \psi(s)) - f(S(s)u_0)]ds$$
$$= e^{-tA}(v_0 - u_0) + \int_0^t e^{-(t-s)A}[f(S(s)u_0 + \psi(s)) - f(S(s)u_0)]ds.$$

Then

$$v(t) = \psi(t) + S(t)u_0 = e^{-tA}v_0 + \int_0^t e^{-(t-s)A}f(v(s))ds$$

and $v(t)$ is the solution having $v(0) = v_0 = y_0 + z_0$ and satisfying equation (2.2.1), $v(t) = S(t)v_0$. Finally, we have

$$|S(t)v_0 - S(t)u_0|_E = |\psi(t)|_E = O(e^{-\sigma t}), \quad t \to +\infty.$$

Hence, we have the following lemma:

Lemma 2.2.4. *Let $\psi \in C([0,+\infty), E)$, $|\psi(t)|_E = O(e^{-\sigma t})$, $t \to +\infty$, where $\sigma > \lambda$. Then ψ satisfies equation (2.2.32), $u_0 \in E$, $z_0 \in QE$, if and only if ψ possesses the form*

$$\psi(t) = S(t)v_0 - S(t)u_0, \quad \forall t \geq 0,\ v_0 \in E.$$

Furthermore, v_0, u_0, z_0 and ψ have the following relation:

$$v_0 = Pu_0 + P\psi(0) + z_0.$$

The above lemma enlightens us to make a similar definition of $\mathscr{F}_\sigma$. Define the space

$$\mathscr{G}_\sigma = \left\{\psi \in C([0,+\infty), E) : \|\psi\|_{\mathscr{G}_\sigma} = \sup_{t\geq 0}(e^{\sigma t}|\psi(t)|_E) < +\infty\right\},$$

which is a Banach space with norm $\|\cdot\|_{\mathscr{G}_\sigma}$. For $u_0 \in E$, $z_0 \in QE$, $\psi \in \mathscr{G}_\sigma$, $t \geq 0$. We define a formal mapping

$$\begin{aligned} W(\psi, u_0, z_0)(t) = e^{-tA}(z_0 - Qu_0) + \int_0^t e^{-(t-s)A} Q[f(S(s)u_0 + \psi(s)) - f(S(s)u_0)]\mathrm{d}s \\ - \int_t^{+\infty} e^{-(t-s)A} P[f(S(s)u_0 + \psi(s)) - f(S(s)u_0)]\mathrm{d}s. \end{aligned} \tag{2.2.34}$$

Note that for $u_0 \in E$, $z_0 \in QE$, ψ satisfies (2.2.32) if and only if $\psi = W(\psi, u_0, z_0)$.

Lemma 2.2.5. *Let σ satisfy (2.2.32). Equation (2.2.34) defines a mapping $W : \mathscr{G}_\sigma \times E \times QE \to \mathscr{G}_\sigma$ which satisfies*

$$\|W(\psi, u_0, z_0)\|_{\mathscr{G}_\sigma} \leq K_2|z_0 - Qu_0|_E + \theta_\sigma\|\psi\|_{\mathscr{G}_\sigma}, \tag{2.2.35}$$

$$\|W(\psi_1, u_0, z_0) - W(\psi_2, u_0, z_0)\|_{\mathscr{G}_\sigma} \leq \theta_\sigma\|\psi_1 - \psi_2\|_{\mathscr{G}_\sigma}, \tag{2.2.36}$$

where $u_0 \in E$, $z_0 \in QE$, $\psi, \psi_1, \psi_2 \in \mathscr{G}_\sigma$; θ_σ is defined in Lemma 2.2.2, in particular, $\theta_\sigma < 1$.

Proof. Fixing σ, which satisfies (2.2.13), taking $u_0 \in E$, $z_0 \in QE$, $\psi \in \mathscr{G}_\sigma$, we have

$$\begin{aligned} |W(\psi, u_0, z_0)(t)|_E \leq K_2 e^{-\Lambda t}|z_0 - Qu_0|_E + M_1K_2\int_0^t ((t-s)^{-\alpha} + \Lambda^\alpha)e^{-\Lambda(t-s)}|\psi(s)|_E\mathrm{d}s \\ + M_1K_1\lambda^\alpha \int_t^{+\infty} e^{-\lambda(t-s)}|\psi(s)|_E\mathrm{d}s. \end{aligned}$$

Hence,

$$\begin{aligned}e^{\sigma t}\big|W(\psi,u_0,z_0)(t)\big|_E &\le K_2 e^{(\sigma-\Lambda)t}|z_0-Qu_0|_E\\&\quad+M_1K_2\|\psi\|_{\mathcal{G}_\sigma}\int_0^t((t-s)^{-\alpha}+\Lambda^\alpha)e^{-(\Lambda-\sigma)(t-s)}\mathrm{d}s\\&\quad+M_1K_1\lambda^\alpha\|\psi\|_{\mathcal{G}_\sigma}\int_t^{+\infty}e^{(\sigma-\Lambda)(t-s)}\mathrm{d}s\\&\le K_2|z_0-Qu_0|_E+\left\{\frac{M_1K_2(1+\gamma_\alpha)\Lambda^\alpha}{\Lambda-\alpha}+\frac{M_1K_1\lambda^\alpha}{\sigma-\lambda}\right\}\|\psi\|_{\mathcal{G}_\sigma},\end{aligned}$$

and then we have

$$\big\|W(\psi,u_0,z_0)(t)\big\|_{\mathcal{G}_\sigma}\le K_2|z_0-Qu_0|_E+\theta_\sigma\|\psi\|_{\mathcal{G}_\sigma}.$$

This proves equation (2.2.35). Now for $\psi_1,\psi_2\in\mathcal{G}_\sigma$,

$$\begin{aligned}&\big|W(\psi_1,u_0,z_0)(t)-W(\psi_2,u_0,z_0)(t)\big|_E\\&\quad\le\int_0^t K_2((t-s)^{-\alpha}+\Lambda^\alpha)e^{-\Lambda(t-s)}M_1|\psi_1(s)-\psi_2(s)|_E\mathrm{d}s\\&\quad\quad+\int_t^{+\infty}K_1\lambda^\alpha e^{-\lambda(t-s)}M_1|\psi_1(s)-\psi_2(s)|_E\mathrm{d}s.\end{aligned}$$

Hence

$$\begin{aligned}&e^{\sigma t}\big|W(\psi_1,u_0,z_0)(t)-W(\psi_2,u_0,z_0)(t)\big|_E\\&\quad\le M_1K_2\|\psi_1-\psi_2\|_{\mathcal{G}_\sigma}\int_0^t((t-s)^{-\alpha}+\Lambda^\alpha)e^{(\sigma-\Lambda)(t-s)}\mathrm{d}s\\&\quad\quad+M_1K_1\lambda^\alpha\|\psi_1-\psi_2\|_{\mathcal{G}_\sigma}\int_t^{+\infty}e^{(\sigma-\lambda)(t-s)}\mathrm{d}s,\end{aligned}$$

and then we have

$$\big\|W(\psi_1,u_0,z_0)(t)-W(\psi_2,u_0,z_0)(t)\big\|_{\mathcal{G}_\sigma}\le\theta_\sigma\|\psi_1-\psi_2\|_{\mathcal{G}_\sigma}.$$

This proves equation (2.2.36). Since $\theta_\sigma<1$, we have proved the lemma. □

Definition 2.2.1. Using Lemma 2.2.5, we define the mapping $W:\mathcal{G}_\sigma\times E\times QE\to\mathcal{G}_\sigma$, which is a strict contraction in $\mathcal{F}_\sigma$, uniformly in $E\times QE$, hence there exists a mapping $\psi:E\times QE\to\mathcal{G}_\sigma$, such that $W(\psi(u_0,z_0),u_0,z_0)=\psi(u_0,z_0)$, $\forall u_0\in E$, $\forall z_0\in QE$. We can define a mapping $\Psi_{u_0}:QE\to PE$, for any $u_0\in E$, by

$$\Psi_{u_0}(z_0)=Pu_0+P\psi(u_0,z_0)(0),\quad z_0\in QE.\tag{2.2.37}$$

Defining $N_{u_0} = \operatorname{graph}\Psi_{u_0}$, $\forall u_0 \in E$, we have

Proposition 2.2.2. *For any $u_0 \in E$, we have*

$$|\Psi_{u_0}(z_1) - \Psi_{u_0}(z_2)|_E \le l'|z_1 - z_2|_E, \quad \forall z_1, z_2 \in QE, \tag{2.2.38}$$

where $0 < l' < 1$. Furthermore, Ψ_{u_0} satisfies

$$\Psi_{u_0}(z_0) = Pu_0 - \int_0^{+\infty} e^{sA}P[f(S(s)(\Psi_{u_0}(z_0) + z_0)) - f(S(s)(u_0))]\mathrm{d}s, \tag{2.2.39}$$

and N_{u_0} can be characterized as

$$N_{u_0} = \{v_0 \in E;\ |S(t)v_0 - S(t)u_0|_E = O(e^{-\sigma t}),\ t \to +\infty\}, \tag{2.2.40}$$

where σ is an arbitrary number, which satisfies equation (2.2.13).

Proof. Fix $u_0 \in E$, and consider $z_1, z_2 \in QE$, $\psi_i = \psi(u_0, z_i)$, $i = 1, 2$. By definition,

$$\begin{aligned}
\Psi_{u_0}(z_1) - \Psi_{u_0}(z_2) &= Pu_0 + P\psi(u_0, z_1)(0) - Pu_0 - P\psi(u_0, z_0)(0) \\
&\quad + PW(\psi_1, u_0, z_1)(0) - PW(\psi_2, u_0, z_2)(0) \\
&= -\int_0^{+\infty} e^{sA}P[f(S(s)u_0 + \psi_1(s)) - f(S(s)u_0 + \psi_2(s))]\mathrm{d}s.
\end{aligned}$$

Since σ satisfies equation (2.2.13), we have

$$\begin{aligned}
|\Psi_{u_0}(z_1) - \Psi_{u_0}(z_2)|_E &\le M_1K_1\lambda^\alpha \int_0^{+\infty} e^{-\lambda s}|\psi_1(s) - \psi_2(s)|_E\mathrm{d}s \\
&\le M_1K_1\lambda^\alpha \|\psi_1 - \psi_2\|_{\mathcal{G}_\sigma} \int_0^{+\infty} e^{(\sigma-\lambda)s}\mathrm{d}s \\
&= \frac{M_1K_1\lambda^\alpha}{\sigma - \lambda}\|\psi_1 - \psi_2\|_{\mathcal{G}_\sigma}.
\end{aligned}$$

Since ψ_i is the fixed point of $W(\cdot, u_0, z_i)$, $i = 1, 2$, we have

$$\begin{aligned}
\|\psi_1 - \psi_2\|_{\mathcal{G}_\sigma} &= \|W(\psi_1, u_0, z_1) - W(\psi_2, u_0, z_2)\|_{\mathcal{G}_\sigma} \\
&\le \|W(\psi_1, u_0, z_1) - W(\psi_2, u_0, z_1)\|_{\mathcal{G}_\sigma} + \|W(\psi_2, u_0, z_1) - W(\psi_2, u_0, z_2)\|_{\mathcal{G}_\sigma} \\
&\le \theta_\sigma\|\psi_1 - \psi_2\|_{\mathcal{G}_\sigma} + \|e^{-tA}Q(z_1 - z_2)\|_{\mathcal{G}_\sigma} \\
&\le \theta_\sigma\|\psi_1 - \psi_2\|_{\mathcal{G}_\sigma} + K_2|z_1 - z_2|_E.
\end{aligned}$$

Since $0 < \theta_\sigma < 1$, σ satisfies equation (2.2.13), and then we get

$$\|\psi_1 - \psi_2\|_{\mathcal{G}_\sigma} \le \frac{K_2}{1 - \theta_\sigma}|z_1 - z_2|_E.$$

Therefore

$$|\Psi_{u_0}(z_1) - \Psi_{u_0}(z_2)|_E \le l'|z_1 - z_2|_E,$$

where

$$l' \le \frac{M_1K_1K_2\lambda^\alpha}{(1-\theta_\sigma)(\sigma-\lambda)}.$$

By the spectral gap condition (2.2.5), we can take $\sigma = \lambda + 3M_1K_1K_2\lambda^\alpha$ to satisfy equation (2.2.13). Then by Proposition 2.2.1, $\theta_\sigma < \frac{2}{3}$, and

$$\frac{1}{1-\theta_\sigma} < 3.$$

Thus

$$l' < \frac{3M_1K_1K_2\lambda^\alpha}{3M_1K_1K_2\lambda^\alpha} = 1.$$

This proves that (2.2.38) is true. Now fix $u_0 \in E$, let N_σ denote the right-hand side of (2.2.40) and note that, by Lemma 2.2.4, we have $v_0 \in N_\sigma$ if and only if

$$\begin{aligned} v_0 &= Pu_0 + P\psi(u_0, Qu_0) + Qv_0 \\ &= \Psi_{u_0}(Qv_0) + Qv_0 \in \operatorname{graph}\Psi_{u_0} = N_{u_0}. \end{aligned}$$

Hence, $N_{u_0} = N_\sigma$, and equation (2.2.40) is proved. In order to prove equation (2.2.39), by the definition of Ψ_{u_0} we know that

$$\Psi_{u_0}(z_0) = Pu_0 - \int_0^{+\infty} e^{sA}P[f(S(s)u_0 + \psi(u_0,z_0)(s)) - f(S(s)u_0)]ds.$$

By Lemma 2.2.4, we get

$$S(s)u_0 + \psi(u_0,z_0)(s) = S(s)(\Psi_{u_0}(z_0) + z_0), \quad s \ge 0,\ u_0 \in E,\ z_0 \in QE.$$

And so the lemma is proved. □

Lemma 2.2.6. *Given any $u_0 \in E$, there exists a unique $v_0 \in M \cap N_{u_0}$, which defines a mapping $\pi : E \to M$, $\pi u_0 = v_0$. Furthermore, π maps bounded subsets of E into bounded subsets of M.*

Proof. Fixing $u_0 \in E$, we prove that there exists a unique $v_0 \in M \cap N_{u_0}$. To achieve this, we note that $v_0 \in M \cap N_{u_0}$ if and only if

$$v_0 = Pv_0 + \Phi(Pv_0) \in \operatorname{graph}\Phi$$

and

$$v_0 = \Psi_{u_0}(Qv_0) + Qv_0 \in \operatorname{graph} \Psi_{u_0},$$

which is equivalent to

$$Pv_0 = \Psi_{u_0}(Qv_0), \quad Qv_0 = \Phi(Pv_0). \tag{2.2.41}$$

But equation (2.2.41) is also equivalent to

$$Pv_0 = \Psi_{u_0}(\Phi(Pv_0)), \quad Qv_0 = \Phi(Pv_0). \tag{2.2.42}$$

Since Φ, Ψ_{u_0} both are Lipschitz continuous, and their Lipschitz constant is strictly less than 1, equation (2.2.42) defines a strict contraction:

$$PE \ni y_0 \to \Psi_{u_0}(\Phi(y_0)) \in PE.$$

Hence equation (2.2.42) has a unique solution $v_0 \in E$, and v_0 is the unique element in $M \cap N_{u_0}$. Let $\pi u_0 = v_0$. For boundedness preservation property of π, fix $\widetilde{v_0} \in M$, $u_0 \in E$ arbitrarily. Since $\pi u_0 = P\pi u_0 + \Phi(\pi u_0) \in \operatorname{graph} \Phi$, $\widetilde{v_0} = P\widetilde{v_0} - \Phi(P\widetilde{v_0})$, we have

$$\begin{aligned} |Q\pi u_0 - Q\widetilde{v_0}|_E &\le \left|\Phi(P\pi u_0) - \Phi(P\widetilde{v_0})\right|_E \\ &\le l|P\pi u_0 - P\widetilde{v_0}|_E, \end{aligned} \tag{2.2.43}$$

where $l < 1$ is the Lipschitz constant of Φ. By equation (2.2.43) and

$$\begin{aligned} u_0 &= \Psi_{u_0}(Qu_0) + Qu_0 \in \operatorname{graph} \Psi_{u_0}, \\ \pi u_0 &= \Psi_{u_0}(Q\pi u_0) + Q\pi u_0 \in \operatorname{graph} \Psi_{u_0}, \end{aligned}$$

we have

$$\begin{aligned} |P\pi u_0 - P\widetilde{v_0}|_E &\le |P\pi u_0 - Pu_0|_E + |Pu_0 - P\widetilde{v_0}|_E \\ &= \left|\Psi_{u_0}(Q\pi u_0) - \Psi_{u_0}(Qu_0)\right|_E + |Pu_0 - P\widetilde{v_0}|_E \\ &\le l'|Q\pi u_0 - Qu_0|_E + |Pu_0 - P\widetilde{v_0}|_E \\ &\le l'|Q\pi u_0 - Q\widetilde{v_0}|_E + l'|Q\widetilde{v_0} - Qu_0|_E + |Pu_0 - P\widetilde{v_0}|_E \\ &\le ll'|P\pi u_0 - P\widetilde{v_0}|_E + l'|Q\widetilde{v_0} - Qu_0|_E + |Pu_0 - P\widetilde{v_0}|_E, \end{aligned}$$

where $l' < 1$. Thus

$$\begin{aligned} |P\pi u_0 - P\widetilde{v_0}|_E &\le \frac{1}{1 - ll'}\{l'|Q\widetilde{v_0} - Qu_0|_E + |Pu_0 - P\widetilde{v_0}|_E\} \\ &\le \frac{K_1 + l'K_2}{1 - ll'}|u_0 - \widetilde{v_0}|_E. \end{aligned} \tag{2.2.44}$$

By virtue of equations (2.2.43)–(2.2.44), we obtain

$$
\begin{aligned}
|\pi u_0|_E \leq& |\pi u_0 - \widetilde{v_0}|_E + |\widetilde{v_0}|_E \\
&\leq |\widetilde{v_0}|_E + |P\pi u_0 - P\widetilde{v_0}|_E + |Q\pi u_0 - Q\widetilde{v_0}|_E \\
&\leq |\widetilde{v_0}|_E + \frac{(1+l)(K_1 + l'K_2)}{1 - ll'}|u_0 - \widetilde{v_0}|_E.
\end{aligned} \tag{2.2.45}
$$

Hence, we have

$$
|\pi u_0|_E \leq m_0 + m_1 |u_0|_E,
$$

where

$$
\begin{aligned}
m_1 &= \frac{(1+l)(K_1 + l'K_2)}{1 - ll'}, \\
m_0 &= (1 + m_1)|\widetilde{v_0}|_E.
\end{aligned}
$$

Since $\widetilde{v_0}$ is fixed, according to equation (2.2.45), we can deduce that π maps bounded subsets of E into bounded subsets of M. This completes the proof of the lemma. □

Proposition 2.2.3 (Asymptotic completeness). *For any $u_0 \in E$,*

$$
\left|S(t)u_0 - S(t)\pi u_0\right|_E \leq K_\eta(|u_0|_E)e^{-\eta t}, \quad t \geq 0,
$$

where $\eta < \Lambda - 2M_1K_2(1 + \gamma_\alpha)\Lambda^\alpha$, K_η depends on η and $|u_0|_E$.

Proof. Let $u_0 \in E$. Then by definition, $\pi u_0 \in N_{u_0}$, and by Lemma 2.2.4,

$$
S(t)u_0 - S(t)\pi u_0 = \psi(u_0, Q\pi u_0)(t), \quad t \geq 0.
$$

From equation (2.2.35), since σ satisfies equation (2.2.13), we have

$$
\begin{aligned}
\|\psi(u_0, Q\pi u_0)\|_{\mathcal{G}_\sigma} &\leq \|W(\psi(u_0, Q\pi u_0), u_0, Q\pi u_0)\|_{\mathcal{G}_\sigma} \\
&\leq K_2|Qu_0 - Q\pi u_0|_E + \theta_\sigma\|\psi(u_0, Q\pi u_0)\|_{\mathcal{G}_\sigma}.
\end{aligned}
$$

Hence

$$
\|\psi(u_0, Q\pi u_0)\|_{\mathcal{G}_\sigma} \leq \frac{K_2}{1 - \theta_\sigma}|Qu_0 - Q\pi u_0|_E,
$$

and for $t \geq 0$,

$$
\begin{aligned}
\left|S(t)u_0 - S(t)\pi u_0\right|_E &\leq \left|\psi(u_0, Q\pi u_0)\right|_E \\
&\leq \|\psi(u_0, Q\pi u_0)\|_{\mathcal{G}_\sigma} e^{-\sigma t} \\
&\leq \frac{K_2}{1 - \theta_\sigma}|Qu_0 - Q\pi u_0|_E e^{-\sigma t}.
\end{aligned}
$$

By Lemma 2.2.6, when u_0 is in a bounded subset, πu_0 is uniformly bounded in E_0, hence we have

$$|S(t)u_0 - S(t)\pi u_0|_E \le K_\sigma(|u_0|_E)e^{-\sigma t}, \quad t \ge 0,$$

where $K_\sigma = K_\sigma(|u_0|_E)$ depends on the bounded σ and $|u_0|_E$. Obviously in the above relations, replacing σ with $\eta < \Lambda - 2M_1K_2(1+\gamma_\alpha)\Lambda^\alpha$, $K_\eta = K_{\sigma_0}, \eta \le \sigma_0 = \lambda + 2M_1K_1\lambda^\alpha$ is also allowed. Then we finish the proof. □

We want to get further properties of mapping π and manifold N_{u_0} and present

Proposition 2.2.4. *The mapping $\pi : E \to M$ is a continuous shrinkage, that is, $\pi : E \to M$ is continuous and its range on $M \subset E$ is equivalent to M.*

Proof. From the definition of πu_0 and $u_0 \in N_{u_0}$, we directly deduce that $\pi|M$ is equivalent to M. For the continuity of π, fixing a point $u_0 \in E$, selecting any sequence $\{u_j\}_{j\in\mathbf{N}}$, $u_j \to u_0, j \to +\infty$, if πu_j does not converge to πu_0, then we have

$$|\pi u_{jk} - \pi u_0| > \varepsilon, \tag{2.2.46}$$

where $\varepsilon > 0$, and $\{u_{jk}\}_{j\in\mathbf{N}}$ is a subsequence. By Lemma 2.2.6, $\{\pi u_{jk}\}$ is bounded in M. Since M is locally compact (its dimension is finite), a subsequence of $\{\pi u_{jk}\}$ still denoted by πu_{jk} converges to $v_0 \in M$, $k \to \infty$:

$$\pi u_{jk} \to v_0, \quad k \to +\infty.$$

By Proposition 2.2.3, it follows that

$$|S(t)u_{jk} - S(t)\pi u_{jk}|_E \le Ce^{-\sigma t}, \quad \forall t \ge 0, \forall k \in \mathbf{N}, \tag{2.2.47}$$

where σ satisfies equation (2.2.13), C is independent of k, t. For each fixed point $t \ge 0$, in equation (2.2.47) by letting $k \to +\infty$, we obtain

$$|S(t)u_0 - S(t)v_0|_E \le Ce^{-\sigma t}, \quad \forall t \ge 0.$$

Hence, by the characterization of N_{u_0}, we get $v_0 \in N_{u_0}$. But $v_0 \in M$, by the definition, $\pi u_0 = u_0$. This contradicts equation (2.2.46), hence we know $\pi u_j \to \pi u_0$, and π is continuous. □

Lemma 2.2.7. *The mapping*

$$(u_0, z_0) \to \Psi_{u_0}(z_0) \tag{2.2.48}$$

is continuous, $E \times QE \to PE$, and the mapping

$$(u_0, z_0, t) \to \psi(u_0, z_0)(t) \tag{2.2.49}$$

is continuous, $E \times QE \times [0, \infty) \to E$.

Proof. Fix $u_0 \in E$, $z_0 \in QE$, and select any sequences $\{u_j\}_{j\in\mathbf{N}}$ and $\{z_j\}_{j\in\mathbf{N}}$ such that

$$u_j \to u_0, \quad z_j \to z_0, \quad j \to +\infty.$$

From equation (2.2.35), if $\psi(u_j, z_j)$ is the fixed point of $W(\cdot, u_j, z_j)$ we know

$$\|\psi(u_j, z_j)\|_{\mathcal{G}_\sigma} \leq \frac{K_2}{1-\theta_\sigma}|z_j - Qu_j|_E.$$

Hence

$$\begin{aligned} |\psi_{u_j}(z_j)|_E &= |Pu_j + P\psi(u_j, z_j)(0)|_E \\ &\leq K_1|u_j|_E + K_1\|\psi(u_j, z_j)\|_{\mathcal{G}_\sigma} \\ &\leq K_1|u_j|_E + \frac{K_1K_2}{1-\theta_\sigma}|z_j - Qu_j|_E, \end{aligned}$$

which means $\{\Psi_{u_j}(z_j)\}$ is bounded in PE, for the PE's dimension is finite. If $\{\Psi_{u_j}(z_j)\}$ does not converge to $\Psi_{u_0}(z_0)$, then as in the previous proof, we can select subsequences $\{u_{jk}\}_{k\in\mathbf{N}}$ and $\{z_{jk}\}_{k\in\mathbf{N}}$ such that

$$\begin{aligned} &|\Psi_{u_{jk}}(z_{jk}) - \Psi_{u_0}(z_0)|_E \geq \varepsilon, \quad \varepsilon > 0, \\ &\Psi_{u_{jk}}(z_{jk}) \to \widetilde{y_0}, \quad k \to +\infty, \quad \widetilde{y_0} \in PE. \end{aligned} \tag{2.2.50}$$

Noting that $z_j + \Psi_{u_j}(z_j) \in N_{u_j}$, and it is bounded, by Propositions 2.2.2 and 2.2.3, it follows that

$$|S(t)(z_{jk} + \Psi_{u_{jk}}(z_{jk})) - S(t)\pi u_{jk}|_E \leq Ce^{-\sigma t}, \quad \forall t \geq 0,$$

where σ satisfies equation (2.2.13), C is independent of k, t. Letting $k \to \infty$, we obtain

$$|S(t)(z_0 + \widetilde{y_0}) - S(t)\pi u_0|_E \leq Ce^{-\sigma t}, \quad \forall t \geq 0.$$

Thus $z_0 + \widetilde{y_0} \in N_{u_0} = \text{graph}\,\Psi_{u_0}$. Then we have $\widetilde{y_0} = \Psi_{u_0}(z_0)$. Since

$$\Psi_{u_{jk}}(z_{jk}) \to \Psi_{u_0}(z_0), \quad k \to +\infty,$$

we obtain a contradiction to equation (2.2.50). Hence $\Psi_{u_j}(z_j) \to \pi u_0(z_0)$, and equation (2.2.48) is proved.

For the equation (2.2.49), by Lemma 2.2.4 we have

$$\psi(u_0, z_0)(t) = S(t)(\Psi_{u_0}(z_0) + z_0) - S(t)u_0.$$

Hence, from equation (2.2.48) and $(t, u_0) \to S(t)u_0$, the continuity of the mapping follows by equation (2.2.49). □

Finally, we have

Proposition 2.2.5. *A continuous foliation structure of E is given by* $E = \bigcup_{u_0 \in M} N_{u_0}$*; furthermore, the foliation along the semiflow possesses the "translation" property:* $S(t)N_{u_0} = N_{S(t)u_0}$, $t \geq 0$, $\forall u_0 \in E$. $S(t) \cdot \pi = \pi \cdot S(t)$, $\forall t \geq 0$.

Proof. Firstly, we note that $E = \bigcup_{v_0 \in M} N_{u_0}$, by Propositions 2.2.2 and 2.2.3, $u_0 \in N_{\pi u_0}$, $\forall u_0 \in E$.

Let $h : PE \times QE \to E$ be defined as

$$h(y, z) = \Psi_{y+\Phi(y)}(\Phi(y) + z) + \Phi(y) + z.$$

From equation (2.2.48), we know h is continuous. Furthermore, $h(y, QE) = \text{graph}\,\Psi_{y+\Phi(y)} =$ foliation across $y + \Phi(y) =$ foliation through $\Psi_{y+\Phi(y)}(\Phi(y)) + \Phi(y) =$ foliation through $h(y, 0)$, where we use $y = \Psi_{y+\Phi(y)}(\Phi(y))$ (Lemma 2.2.6). In a similar way, we have

$$h(y, 0) = \Psi_{y+\Phi(y)}(\Phi(y)) + \Phi(y) = y + \Phi(y),$$

whence it follows that

$$h(PE, 0) = \text{graph}\,\Phi = M$$

and then $E = \bigcup_{u_0 \in M} N_{u_0}$ is a continuous leaf structure.

From the characterization of N_{u_0}, it is easy to get $S(t)N_{u_0} = N_{S(t)u_0}$, the same characterization and the definition of π give $S(t) \cdot \pi = \pi \cdot S(t)$. □

Proof of Theorem 2.2.2. We know that

$$|f_\theta(u) - f_\theta(v)|_F \leq M_1 |u - v|_E, \quad \forall u, v \in E, \tag{2.2.51}$$

where

$$M_1 = d_1(\sqrt{2}\rho) + \frac{4\sqrt{2}d_0(\sqrt{2}\rho)}{\rho}.$$

Based on equation (2.2.51) and the spectral gap condition, using Theorem 2.2.2 for truncation equation, we know that $\{S_\theta(t)\}_{t\geq 0}$ has an inertial manifold $M_\theta = \text{graph}\,\Phi_\theta$, where $\Phi_{\theta_\sigma} : PE \to QE$ has Lipschitz constant less than 1. Now we start from M_θ to get the inertial manifold of equation (2.2.1).

Denote $B_0 = B_E(\rho)$ and let $\mathscr{B} \subset B_0$ be an absorbing set of the original equation. Since $\mathscr{B}$ is absorbing, there exists $t_0 = t_0(B_0)$ such that

$$S(t)B_0 \subset \mathscr{B}, \quad \forall t \geq t_0. \tag{2.2.52}$$

Let

$$\mathscr{B}_1 = \bigcup_{t \geq t_0} S(t)B_0.$$

then $\mathscr{B}_1 \subset \mathscr{B}$, $\overline{\mathscr{B}_1} \subset B_0$, $S(t)\mathscr{B}_1 \subset \mathscr{B}$, $\forall t \geq 0$. Hence $S(t)|_{\mathscr{B}_1} = S_\theta(t)|_{\mathscr{B}_1}$, and $\mathscr{B}_1$ is an absorbing set of the original equation. Let $M_1 = M_\theta \cap \mathscr{B}_1$. Then

$$\begin{aligned} S(t)M_1 &= S_\theta(t)M_1 = S_\theta(t)M_\theta \cap S_\theta(t)\mathscr{B}_1 \\ &\subset M_\theta \cap \mathscr{B}_1 = M_1, \quad \forall t \geq 0. \end{aligned} \tag{2.2.53}$$

Thus, M_1 is positive invariant for the original and truncated equations.

Since $\overline{\mathscr{B}_1} \subset B_0$, we can select ε, $0 < \varepsilon < 1$, such that $v_\varepsilon(\mathscr{B}_1) \subset B_0$, where

$$v_\varepsilon(\mathscr{B}_1) = \{v \in E,\ |v - u|_E < \varepsilon,\ u \in \mathscr{B}_1\}.$$

Let $M_\varepsilon = M_\theta \cap v_\varepsilon(\mathscr{B}_1)$. Then M_ε is an open neighborhood of M_1 in M_θ. Now we consider the inertial form $\{S_\theta(t)|_{M_\theta}\}_{t\geq 0}$ in M_θ. Since M_θ is Lipschitz continuous, of finite dimension and invariant, we get that $S_\theta(\cdot)$ is continuous, $[0, t_0] \times M_\theta \to M_\theta$. Hence $S_\theta(t)|_{M_\theta}$ is a homeomorphism of M_θ, $t \geq 0$. Since $S_\theta(t)M_1 \subset M_1$, $t \geq 0$, M_ε is a domain of $M_\theta(t)$ in M_1, and from the continuity of $S_\theta(\cdot)$ as a map $[0, t_0] \times M_\theta \to M_\theta$, we deduce that there exists M_θ in a domain M_δ of M_1 such that

$$S_\theta(t)M_\delta \subset M_\varepsilon, \quad \forall t \in [0, t_0]. \tag{2.2.54}$$

Without loss of generality, we can set M_δ to have the form $M_\delta = M_\theta \cap v_\delta(\mathscr{B}_1)$, $0 < \delta < \varepsilon$.

By equations (2.2.52) and (2.2.54), we get

$$S_\theta(t)M_\delta \subset M_\varepsilon, \quad S(t)|_{M_\delta} = S_\theta(t)|_{M_\delta}, \quad \forall t \geq 0.$$

Denote

$$M = \bigcup_{t\geq 0} S(t)M_\delta.$$

We show that M is the required inertial manifold. In fact, by definition $S(t)M \subset M$, $\forall t \geq 0$. Now we know that $S_\theta(t)|_{M_\theta}$ is a homeomorphism of M_θ, $t \geq 0$. Thus it maps an open set of M_θ into an open set of M_θ. Since $S(t)M_\delta(= S_\theta(t)M_\delta)$ is open in M_θ, $t \geq 0$, the set M itself is open in M_θ and $I + \Phi_\theta : PE \to M_\theta$ is continuous. The set

$$O = (I - \Phi_\theta)^{-1}(M)$$

is open in PE, $M = \operatorname{graph}\Phi$, where Φ is a Lipschitz function given by the following map:

$$\Phi = \Phi_\theta|_O : O \subset PE \to QE,$$

and has Lipschitz constant less than 1.

From the asymptotic completeness property of M, if $\mathscr{B} \subset E$ is bounded, because $\mathscr{B}_1$ for the original equations is absorbing, then there exists $t_1 \geq 0$, $t_1 = t_1(\mathscr{B})$, such that $S(t)\mathscr{B} \subset \mathscr{B}_1$, $\forall t \geq t_1$.

Given $u_0 \in \mathscr{B}$, set $\tilde{u}_0 = S(t)u_0$, $\widetilde{u_0} \in \mathscr{B}_1$. By the asymptotic completeness of truncated equation M_θ, there exists $\widetilde{u_0} \subset M_\theta$, such that

$$|S_\theta(t)\widetilde{u_0} - S_\theta(t)\widetilde{v_0}|_E \le K_\eta e^{-\eta t}, \quad \forall t \ge 0, \tag{2.2.55}$$

where

$$0 < \eta < \Lambda - 2M_1K_2(1+\gamma_\alpha)\lambda^\alpha$$

and K_η only depends on η, since $\widetilde{u_0} \in \mathscr{B}_1$.

Taking $t_2 = t_2(\eta)$ such that

$$K_\eta e^{-\eta t_2} < \delta \tag{2.2.56}$$

and setting $v_0 = S_\theta(t_2)\widetilde{v_0}$, we then deduce that $v_0 \in M$. Furthermore, from equations (2.2.55)–(2.2.56), we get

$$\begin{aligned}
&|S(t+t_1+t_2)u_0 - S(t)v_0|_E \\
&= |S(t+t_2)S(t_1)u_0 - S_\theta(t)v_0|_E \\
&= |S(t+t_2)\widetilde{u_0} - S_\theta(t)S_\theta(t_2)\widetilde{v_0}|_E \\
&= |S_\theta(t+t_2)\widetilde{u_0} - S_\theta(t+t_2)\widetilde{v_0}|_E \le K_\eta e^{-\eta(t+t_2)} \\
&= K_\eta e^{-\eta t_2} e^{-\eta t} \le \delta e^{-\eta t}.
\end{aligned}$$

Hence, setting $t_3 = t_1 + t_2$, $t_3 = t_3(\mathscr{B}, \eta)$, and $\delta < \varepsilon < 1$, we obtain

$$|S(t+t_2)u_0 - S(t)v_0|_E \le e^{-\eta t}, \quad \forall t \ge 0,$$

which proves the asymptotic completeness of M. Finally, if the original equation has a global absorbing attractor $\mathscr{A}$, then from the asymptotic completeness and invariance property, $S(t)\mathscr{A} = \mathscr{A}$, $t \ge 0$, we get

$$\mathrm{dist}_E(\mathscr{A}, M) = \mathrm{dist}_E(S(t)\mathscr{A}, M) = O(e^{-\eta t}), \quad t \to +\infty,\ \eta > 0.$$

Hence $\mathscr{A} \subset \overline{M}$. But since $\mathscr{A} \subset \mathscr{B}_1$, $M \supset M_\theta \cap v_\delta(\mathscr{B}_1)$, thus $\mathscr{A} \subset M$. This finishes the proof. □

In the following, we consider the normal and normal hyperbolicity property of inertial manifold.

Theorem 2.2.3. *If $f \in C^1(E,F)$, then Theorem 2.2.1 defines inertial manifold M = graph $\Phi \in C^1$, where Φ satisfies Sack equation*

$$D\Phi(y)(-Ay + P_nf(y+\Phi(y)) + A\Phi(y)) = Q_nf(y+\Phi(y)), \tag{2.2.57}$$

where y is defined in the domain of Φ. If we consider that it satisfies the assumption of Theorem 2.2.2, $f_\eta \in C^1(E,F)$, then the same conclusion is valid.

Proof. Because the inertial manifold in Theorem 2.2.2 is seen as a restriction of the inertial manifold for the truncated equation, this fully proves the result when f is globally Lipschitz continuous. Assume that all the conditions of Theorem 2.2.1 are satisfied and $f \in C^1(E,F)$.

The rest of the proof can be divided into three steps. For convenience, set $\varphi(y)(t) = \varphi(y,t)$, $y \in E$, $t \le 0$, where $\varphi : PE \to \mathcal{F}_\sigma$ is given by Lemma 2.2.3.

Step 1: The selection of differential.

In Lemma 2.2.3, we have seen that $\Phi(y) = P\varphi(y,0)$ and

$$\varphi(y,t) = e^{-tA}Py - \int_t^0 e^{-(t-s)A}Pf(\varphi(y,s))ds + \int_{-\infty}^t e^{-(t-s)A}Qf(\varphi(y,s))ds. \tag{2.2.58}$$

We consider the differential of Φ. First, we seek the differential of φ. Since $D\Phi(y) = Q\partial_y\varphi(y,0)$, by formally differentiating equation (2.2.58) with respect to y, we see that $\partial_y\varphi(y)$ is a fixed point of $\mathcal{L}_1(\cdot,y)$, where $\mathcal{L}_1$ is given by

$$\mathcal{L}_1(\Delta,y)(t) = e^{-tA}P - \int_t^0 e^{-(t-s)A}PDf(\varphi(y,s))\Delta(s)ds + \int_{-\infty}^t e^{-(t-s)A}QDf(\varphi(y,s))\Delta(s)ds.$$

Similar as for the map $\mathcal{L}$, we must verify that the mapping $\mathcal{L}_1$ is completely determined, which is strictly contracting in some appropriate space of Δ and is consistent with y.

When σ satisfies equation (2.2.13), we consider the following space:

$$\mathcal{F}_{1,\sigma} = \left\{\Delta \in C((-\infty,0],\ \mathcal{L}(PE,E));\ \|\Delta\|_{\mathcal{F}_{1,\sigma}} = \sup_{t\le 0}\left(e^{\sigma t}\|\Delta(t)\|_{\mathcal{L}(PE,F)}\right) < +\infty\right\},$$

which is a Banach space with norm $\|\cdot\|_{\mathcal{F}_{1,\sigma}}$. By equation (2.2.2), we have

$$\|Df(u)\|_{\mathcal{L}(E,F)} \le M_1, \quad \forall u \in E. \tag{2.2.59}$$

From this, we clearly see that $\mathcal{L}_1$ is entirely determined as a function, $\mathcal{F}_{1,\sigma}\times PE \to \mathcal{F}_{1,\sigma}$, which is Lipschitz continuous in Δ and has Lipschitz constant θ_σ. When σ satisfies equation (2.2.13), $\theta_\sigma < 1$. From this we deduce that there exists a function $\Delta : PE \to \mathcal{F}_{1,\sigma}$ such that

$$\mathcal{L}_1(\Delta(y),y) = \Delta(y), \quad \forall y \in PE.$$

For convenience, set $\Delta(y)(t) = \Delta(y,t)$; Δ is related with the choice of differential φ.

Step 2: Δ is continuous.

For fixed $y_0 \in PE$, we consider $y \in PE$ which approaches y_0. Similar as equation (2.2.26), we get

$$\|\Delta(y) - \Delta(y_0)\|_{\mathscr{F}_{1,\sigma}} \leq \frac{1}{1-\theta_\sigma}\|\mathscr{L}_1(\Delta(y_0), y) - \mathscr{L}_1(\Delta(y_0), y_0)\|_{\mathscr{F}_{1,\sigma}}.$$

Hence, in order to prove the continuity of Δ, we only need to prove that

$$\|\mathscr{L}_1(\Delta(y_0), y) - \mathscr{L}_1(\Delta(y_0), y_0)\|_{\mathscr{F}_{1,\sigma}} \to 0, \quad y \to y_0.$$

Taking $\mu < \sigma$, it satisfies equation (2.2.13). Then by step 1, we know $\Delta(y_0) \in \mathscr{F}_{1,\mu}$. Since

$$\begin{aligned}
&|\mathscr{L}_1(\Delta(y_0), y)(t) - \mathscr{L}_1(\Delta(y_0), y_0)(t)|_E \\
&\leq \|\Delta(y_0)\|_{\mathscr{F}_{1,u}} K_2 \int_{-\infty}^{t} e^{-\Lambda(t-s)}((t-s)^{-\alpha} + \Lambda^\alpha) N(s,y) e^{-\mu s} ds \\
&\quad + \|\Delta(y_0)\|_{\mathscr{F}_{1,u}} K_1 \lambda^\alpha \int_{t}^{0} e^{-\lambda(t-s)} N(s,y) e^{-\mu s} ds,
\end{aligned}$$

where

$$N(s,y) = \|Df(\varphi(y_0,s)) - Df(\varphi(y,s))\|_{\mathscr{J}(E,F)},$$

we get

$$\|\mathscr{L}_1(\Delta(y_0), y) - \mathscr{L}_1(\Delta(y_0), y_0)\|_{\mathscr{F}_{1,\sigma}} \leq \|\Delta(y_0)\|_{\mathscr{F}_{1,u}} (K_1\lambda^\alpha + K_2)\widetilde{N}(y),$$

where

$$\widetilde{N}(y) = \sup_{t \leq 0}\left[e^{(\sigma-\Lambda)t}\int_{-\infty}^{t}((t-s)^{-\alpha} + \Lambda^\alpha)e^{(\Lambda-\mu)s}N(s,y)ds + e^{(\sigma-\lambda)s}\int_{t}^{0} e^{(\lambda-\mu)t}N(s,y)ds\right].$$

In order to prove that, when $y \to y_0$, $\widetilde{N}(y) \to 0$, we use proof by contradiction. Assume $\widetilde{N}(y_j) > \varepsilon$, where $\varepsilon > 0$, for a sequence $\{y_j\}_{j\in\mathbf{N}}$ in PE which satisfies $|y_j - y_0|_E \to 0$, $j \to +\infty$. Then there exists a non-positive sequence $\{t_j\}_{j\in\mathbf{N}}$ such that

$$e^{(\sigma-\Lambda)t_j}\int_{-\infty}^{t_j}((t_j-s)^{-\alpha} + \Lambda^\alpha)e^{(\lambda-u)t}N(s,y_j)ds + e^{(\sigma-\lambda)t_j}\int_{t_j}^{0} e^{(\lambda-u)s}N(s,y_j)ds \geq \varepsilon, \quad \forall j. \tag{2.2.60}$$

But by equation (2.2.59), $N = N(s,y)$ is consistently bounded in $2M_1$, hence

$$\begin{aligned}
&|\text{left-hand side of equation (2.2.60)}| \\
&\leq 2M_1 e^{(\sigma-\Lambda)t_j}\int_{-\infty}^{t_j}((t_j-s)^{-\alpha} + \Lambda^\alpha)e^{(\Lambda-\mu)s}ds + 2M_1 e^{(\sigma-\Lambda)t_j}\int_{t_j}^{0} e^{(\lambda-\mu)s}ds
\end{aligned}$$

$$\leq 2M_1 e^{(\sigma-\mu)t_j}\left[\frac{(1+\gamma_\alpha)\Lambda^\alpha}{\Lambda-\mu}+\frac{1}{\mu-\lambda}\right].$$

Therefore, based on equation(2.2.60), t_j must have a lower bound. Let $-\infty < T \leq t_j \leq 0$, $\forall j$, $T \leq 0$, Then we have

$$\begin{aligned}&|\text{left-hand side of equation(2.2.60)}|\\&\leq e^{(\sigma-\Lambda)t_j}\int_{-\infty}^{t_j}((t_j-s)^{-\alpha}e^{(\Lambda-u)(t_j-s)}N(s,y_j)ds+\Lambda^\alpha e^{(\sigma-\Lambda)T}\int_{-\infty}^{0}e^{(\Lambda-\mu)s}N(s,y_j)ds\\&+e^{(\sigma-\Lambda)t_j}\int_T^0 e^{(\lambda-\mu)s}N(s,y_j)ds.\end{aligned}$$

Then, by changing variables in the first integral on the right-hand side,

$$\begin{aligned}&|\text{left-hand side of equation (2.2.60)}|\\&\leq\int_0^{+\infty}s^{-\alpha}e^{-(\Lambda-\mu)s}N(t_j-sy_j)ds+\Lambda^\alpha e^{(\sigma-\Lambda)T}\int_{-\infty}^{0}e^{(\Lambda-\mu)s}N(s,y_j)ds\\&+\int_T^0 e^{(\lambda-\mu)s}N(s,y_j)ds.\end{aligned}\tag{2.2.61}$$

But from equation (2.2.27), we have

$$\begin{aligned}|\varphi(y_0,t_j-s)-\varphi(y,t_j-s)|_E&\leq\|\varphi(y_0)-\varphi(y)\|_{\mathscr{I}_\sigma}e^{-\sigma(t_j-s)}\\&\leq\frac{K_1}{1-\theta_\sigma}e^{-\sigma(T-s)}|y_j-y_0|_E\to 0,\quad j\to\infty.\end{aligned}$$

Thus $N(t_j-s,y_j)\to 0, j\to+\infty$. For a fixed $s\geq 0$, $N(s,y_j)\to 0$. By applying Lebesgue dominated convergence theorem on the right-hand side of equation (2.2.61), we obtain that |left-hand side of equation (2.2.60)| $\to 0, j\to+\infty$. This contracts (2.2.60). Therefore, when $|y-y_0|_E\to 0$, we have $\widetilde{N}(y)\to 0$, hence $\Delta=\Delta(y)$ is a continuous function from PE to $\mathscr{F}_{1,\sigma}$.

Step 3: $\partial_y\varphi(y)=\Delta(y)$.

Considering $y,h\in PE$, we have

$$\begin{aligned}&\varphi(y+h,t)-\varphi(y,t)-\Delta(y,t)h\\&=\int_{-\infty}^{t}e^{-(t-s)A}Q[f(\varphi(y+h,s))-f(\varphi(y,s))-Df(\varphi(y,s))\Delta(y,s)h]ds\\&+\int_t^0 e^{-(t-s)A}P[f(\varphi(y+h,s))-f(\varphi(y,s))\\&+Df(\varphi(y,s))\Delta(y,s)h]ds.\end{aligned}\tag{2.2.62}$$

Let

$$\rho(y,h,t)=\frac{|\varphi(y+h,t)-\varphi(y,t)-\Delta(y,t)h|_E}{|h|_E},\quad \forall y,h\in PE,\ \forall t\le 0,$$

$$r(u,\omega)=\frac{|f(u+\omega)-f(u)-Df(u)\omega|_F}{|\omega|_E},\quad \forall u,\omega\in E,$$

$$R(y,h,t)=r(\varphi(y,t),\varphi(y+h,t)-\varphi(y,t)),\quad \forall y,h\in PE,\ \forall t\le 0.$$

Hence, adding and subtracting $Df(\varphi(y,s)(\varphi(y+h,s)-(\varphi(y,s))))$ in the brackets of equation (2.2.62), we can estimate $\rho=\rho(y,h,t)$ by

$$\begin{aligned}\rho(y,h,t)\le{}& K_2\int_{-\infty}^{t} e^{-\Lambda(t-s)}((t-s)^{-\alpha}+\Lambda^{\alpha})R(y,h,s)\frac{|\varphi(y+h,t)-\varphi(y,t)|_E}{|h|_E}\mathrm{d}s\\&+K_1\lambda^{\alpha}\int_{t}^{0} e^{-\lambda(t-s)}R(y,h,s)\frac{|\varphi(y+h,t)-\varphi(y,t)|_E}{|h|_E}\mathrm{d}s\\&+M_1K_2\int_{-\infty}^{t} e^{-\Lambda(t-s)}((t-s)^{-\alpha}+\Lambda^{\alpha})\rho(y,h,s)\mathrm{d}s\\&+M_1K_1\int_{t}^{0} e^{-\lambda(t-s)}\rho(y,h,s)\mathrm{d}s.\end{aligned}$$

Let

$$\widetilde{\rho}(y,h)=\sup_{t\le 0}(e^{\sigma t}\rho(y,h,t))=\frac{\|\varphi(y+h,\cdot)-\varphi(y,\cdot)-\Delta(y,\cdot)h\|_{\mathscr{F}_\sigma}}{|h|_E},\quad \forall y,h\in PE.$$

Thus, by the above inequalities we have

$$\begin{aligned}\widetilde{\rho}(y,h)&\le \widetilde{R}(y,h)+\widetilde{\rho}(y,h)\left\{\int_{-\infty}^{t} e^{(\sigma-\lambda)(t-s)}((t-s)^{-\alpha}+\Lambda^{\alpha})\mathrm{d}s+M_1K_1\lambda^{\alpha}\int_{t}^{0} e^{(\sigma-\lambda)(t-s)}\mathrm{d}s\right\}\\&\le \widetilde{R}(y,h)+\theta_\sigma\widetilde{\rho}(y,h),\end{aligned}$$

where

$$\begin{aligned}\widetilde{R}(y,h)=\sup_{t\le 0}\Bigg\{&K_2e^{\sigma t}\int_{-\infty}^{t}((t-s)^{-\alpha}+\Lambda^{\alpha})e^{-\Lambda(t-s)}R(y,h,s)\frac{|\varphi(y+h,s)-\varphi(y,s)|_E}{|h|_E}\mathrm{d}s\\&+K_1\lambda^{\alpha}e^{\sigma t}\int_{t}^{0} e^{-\lambda(t-s)}R(y,h,s)\frac{|\varphi(y+h,t)-\varphi(y,t)|_E}{|h|_E}\mathrm{d}s\Bigg\}.\end{aligned}$$

Then, since $\theta_\sigma<1$, we obtain

$$\widetilde{\rho}(y,h)\le\frac{1}{1-\theta_\sigma}\widetilde{R}(y,h).$$

As in step 2 for $\widetilde{N} = \widetilde{N}(y)$, we can prove $\widetilde{R}(y,h) \to 0$, when $|h|_E \to 0$. Here we use the following inequality:

$$\frac{|\varphi(y+h,s) - \varphi(y,s)|_E}{|h|_E} \le \frac{K_1}{1-\theta_\mu} e^{-\mu s}, \quad \mu \text{ satisfies equation (2.2.13)}, \ \mu < \sigma.$$

Then $\widetilde{\rho}(y,h) \to 0$, when $|h|_E \to 0$. This proves $\partial_y \varphi(y) = \Delta(y)$.

Step 4: $\Phi \in C^1(PE, QE)$.

From steps 2 and 3, $\varphi(y) = \partial_y \varphi(y, 0)$, and we have

$$\begin{aligned}\frac{|\Phi(y+h) - \Phi(y) - D\Phi(y)|_E}{|h|_E} &= \frac{|Q\Phi(y+h,0) - Q\Phi(y,0) - Q\partial_y\Phi(y,0)h|_E}{|h|_E} \\ &= \rho(y,h,0) \\ &\le \widetilde{\rho}(y,h) \to 0, \quad h \to 0, \quad \forall y \in PE,\end{aligned}$$

where $\widetilde{\rho}(y,h,t)$ and $\widetilde{\rho}(y,h)$ have been defined in step 3. □

Proposition 2.2.6. *If $f \in C^1(E,F)$, then each leaf $N_{u_0} \in C^1$. More specifically, for all $u_0 \in E$, we have $\Psi_{u_0}(z_0) \in C^1(QE, PE)$, $(u_0, z_0) \to D\Psi_{u_0}(z_0)$ is continuous, $E \times QE \to \mathscr{L}(QE, PE)$.*

Proof. Similar to the proof of Theorem 2.2.3 on the regularity of Φ (omitted). □

Since Φ and $\Psi_{u_0} \in C^1$, we consider the tangent bundles

$$\begin{aligned}T_{u_0}M &= \{\eta + D\Phi(Pu_0)\eta;\ \eta \in PE\}, \\ T_{u_0}N_{u_0} &= \{\xi - D\Phi_{u_0}(Qu_0)\xi;\ \xi \in QE\}, \quad \forall u_0 \in M.\end{aligned}$$

The next lemma states that $E = T_{u_0}M \oplus T_{u_0}N_{u_0}$. Hence $T_{u_0}N_{u_0}$ is the normal vector bundle of M at u_0.

Lemma 2.2.8. *For any $u_0 \in M$, there is the decomposition $E = T_{u_0}M \oplus T_{u_0}N_{u_0}$, and the decomposition is continuous in u_0.*

Proof. For a fixed $u_0 \in M$ and given $\mu \in E$, the decomposition $\mu = \eta + \xi$ is unique, $\eta \in T_{u_0}M$, $\xi \in T_{u_0}N_{u_0}$. This is equivalent to

$$\eta = \eta + D\Phi(Pu_0)\eta + D\Psi_{u_0}(Qu_0)\xi + \xi, \tag{2.2.63}$$

where $\eta \in PE$, $\xi \in QE$. But equation (2.2.36) is equivalent to

$$P\mu = \eta + D\Psi_{u_0}(Qu_0)\xi + \xi, \quad Qu = \xi + D\Phi(Pu_0)\eta,$$

that is,

$$\begin{cases}\eta = P\mu - D\Psi_{u_0}(Qu_0)Q\mu + D\Psi_{u_0}(Qu_0)D\Phi(Pu_0)\eta, \\ \xi = Q\mu - D\Phi(Pu_0)P\mu + D\Phi(Pu_0)D\Psi_{u_0}(Qu_0).\end{cases} \tag{2.2.64}$$

Since Φ and Ψ_{u_0} have Lipschitz constant less than 1, their differentials $D\Phi(Pu_0)$ and $D\Psi_{u_0}(Qu_0)$ have norm less than 1, so $I - D\Psi_{u_0}(Qu_0)D\Phi_{u_0}(Pu_0)$ and $I - D\Phi_{u_0}(Pu_0) \times D\Psi_{u_0}(Qu_0)$ are invertible operators in PE and QE, respectively. Then, equation (2.2.64) is equivalent to

$$\begin{cases} \eta = (I - D\Psi_{u_0}(Qu_0)D\Phi(Pu_0))^{-1}(P - D\Psi_{u_0}(Qu_0)Q)\mu, \\ \xi = (I - D\Phi(Pu_0)D\Psi_{u_0}(Qu_0))^{-1}(Q - D\Phi(Pu_0)P)\mu. \end{cases}$$

Hence, it can be written uniquely as

$$\mu = P(u_0)\mu + Q(u_0)\mu,$$

where $P(u_0)$ is the projection of E to $T_{u_0}N_{u_0}$ along $T_{u_0}M$, given as

$$P(u_0) = (I + D\Phi(Pu_0))(I - D\Psi_{u_0}(Qu_0)D\Phi(Pu_0))^{-1}(P - D\Psi_{u_0}(Qu_0)Q),$$

and $Q(u_0)$ is the projection of E to $T_{u_0}N_{u_0}$ along $T_{u_0}M$, given as

$$Q(u_0) = (I + D\Psi_{u_0}(u_0))(I - D\Phi(u_0)D\Psi_{u_0}(Qu_0))^{-1}(Q - D\Phi(Pu_0)P).$$

By the regularity of Φ and Proposition 2.2.6, $P(u_0)$ and $Q(u_0)$ are continuous in u_0, so the decomposition $E = T_{u_0}M \oplus T_{u_0}N_{u_0}$ is continuous. □

Now we define the tangent and normal bundles

$$\begin{aligned} TM &= \{(u,\mu) \in E \times E;\ u \in M,\ \mu \in T_uM\}, \\ NM &= \{(u,\mu) \in E \times E;\ u \in M,\ \mu \in N_uM\}, \end{aligned}$$

where $N_uM = T_uN_u$.

We consider equation (2.2.1) and its first variation

$$\begin{gathered} \frac{\mathrm{d}u}{\mathrm{d}t} + Au = f(u), \quad \frac{\mathrm{d}\mu}{\mathrm{d}t} + A\mu = f(u)\mu, \\ u(0) = u_0, \quad \mu(0) = \mu_0. \end{gathered} \tag{2.2.65}$$

We need to prove M is normal hyperbolic, namely, the tangent bundle TM and normal bundle NM are invariant under equation (2.2.65), and the exponential dichotomy of equation (2.2.65) applies to these bundles. More precisely, we have the following results:

Lemma 2.2.9. *The tangent bundle is invariant under equation* (2.2.65), *and*

$$|\mu(t)|_E \le 2K_1^2|\mu_0|_E e^{-(\lambda+2M_1K_1\lambda^a)t}, \quad \forall t \le 0,\ (u_0,\mu_0) \in TM.$$

Proof. The invariance of TM follows from the invariance of M. In fact, set $(u_0, \mu_0) \in TM$. Then

$$u_0 = y_0 + \Phi(y_0), \quad \mu_0 = \eta_0 + D\Phi(y_0)\eta_0, \quad y_0, \eta_0 \in PE.$$

Let $y = y(t, y_0)$, $t \in \mathbf{R}$ be a global solution with the following inertial form:

$$\frac{dy}{dt} + Ay = Pf(y + \Phi(y)), \quad y(0, y_0) = y_0, \tag{2.2.66}$$

and let $\eta = \eta(t)$ be the first variational global solution, that is,

$$\frac{d\eta}{dt} + A\eta = PDf(y + \Phi(y))(\eta + D\Phi(y)\eta), \quad \eta(0) = \eta_0. \tag{2.2.67}$$

Since M = graph Φ is invariant under equation (2.2.1), $\Phi \in C^1$. We deduce that $\Phi(y)$ satisfies

$$\begin{aligned} &\frac{d\Phi(y)}{dt} + A\Phi(y) = Qf(y + \Phi(y)), \\ &\Phi(y(0, y_0)) = \Phi(y_0). \end{aligned} \tag{2.2.68}$$

Differentiating equation (2.2.68) with respect to y_0 and acting on η_0, we arrive at

$$\frac{dD\Phi(y)\eta}{dt} + AD\Phi(y)\eta = QDf(y + \Phi(y))(\eta - D\Phi(y)\eta), \quad \forall t \in \mathbf{R}. \tag{2.2.69}$$

Adding equations (2.2.69) and (2.2.67), we get

$$\frac{d(\eta + D\Phi(y)\eta)}{dt} + A(\eta + D\Phi(y)\eta) = Df(y + \Phi(y))(\eta + D\Phi(y)\eta).$$

Thus, $\mu(t) = \eta(t) + D\Phi(y(t; y_0))\eta(t)$ is the solution for first variational equation (2.2.1), $\forall t \in \mathbf{R}$. And it belongs to the tangent space to Φ in $y(t; y_0) \in \Phi(y(t; y_0)) \in$ graph $\Phi = M$. Hence, if we set $u(t) = y(t; y_0) + \Phi(y(t; y_0))$, then we find that (u, μ) is the solution of equation (2.2.65) with $(u(0), \mu(0)) = (u_0, \mu_0)$. It belongs to TM, $\forall t \in \mathbf{R}$. This proves that TM is invariant under equation (2.2.65). □

Regarding the estimation of $\mu(t)$, from equation (2.2.67) we note that

$$\eta(t) = e^{-tA}\eta_0 - \int_t^0 e^{-(t-s)A} PDf(y(s) + \Phi(y(s))(\eta(s) + D\Phi(y(s))\eta(s))ds.$$

Thus for $t \le 0$, we arrive at

$$|\eta(t)|_E \le K_1 e^{-\lambda t} |\eta_0|_E + M_1 K_1 (1 + l)\lambda^\alpha \int_t^0 e^{-\lambda(t-s)} |\eta(s)|_E ds,$$

where $l < 1$ is the Lipschitz constant of Φ. Then

$$e^{\lambda t}|\eta(t)|_E \le K_1|\eta_0|_E + 2M_1K_1\lambda^\alpha \int_t^0 e^{\lambda s}|\eta(s)|_E ds,$$

and, by Gronwall lemma, we obtain

$$|\eta(t)|_E \le K_1|\eta_0|_E e^{-(\lambda+2M_1K_1\lambda^\alpha)t}, \quad \forall t \le 0.$$

Therefore

$$\begin{aligned}|\mu(t)|_E &= |\eta(t) + D\Phi(y(t))\eta(t)|_E \\ &\le (1+l)K_1|y_0|_E e^{-(\lambda+2M_1K_1\lambda^\alpha)t} \\ &\le 2K_1|P\mu_0|_E e^{-(\lambda+2M_1K_1\lambda^\alpha)t}.\end{aligned}$$

Since $\|P\|_{\mathscr{L}(E)} \le K_1$, we finally get

$$|\mu(t)|_E \le 2K_1^2|\mu_0|_E e^{-(\lambda+2M_1K_1\lambda^\alpha)t}, \quad \forall t \le 0.$$

Lemma 2.2.10. *The normal bundle NM is invariant under equation* (2.2.65), *and*

$$|\mu(t)|_E \le 2K_2^2\frac{1+l'}{1-l'}|\mu_0|_E e^{-(\Lambda-2M_1K_2(1+\gamma_\alpha)\Lambda^\alpha)t}, \quad \forall t \ge 0,$$

where $(u_0, \mu_0) \in NM$, $l' < 1$.

Proof. The invariance of NM comes from the invariance of M and translation invariance of the fiber. In fact, set $(u_0, \mu_0) \in NM$. For each $v_0 \in N_{u_0} = \operatorname{graph}\Psi_{u_0}$, it can be written as

$$v_0 = \Psi_{u_0}(z_0) + z_0, \quad z_0 \in QE.$$

Noticing that

$$S(t)N_{u_0} = N_{S(t)u_0} = \operatorname{graph}\Psi_{S(t)u_0},$$

we obtain

$$S(t)u_0 = \Psi_{S(t)u_0}(z(t)) + z(t), \quad z(t) \in QE,\ t \ge 0, \tag{2.2.70}$$

that is,

$$z(t) = QS(t)v_0, \quad \Psi_{S(t)u_0}(z(t)) = PS(t)v_0.$$

Hence, for $t \geq 0$, we have

$$z(t) = e^{-tA}z_0 + \int_0^t e^{-(t-s)A}Qf(S(s)(\Psi_{u_0}z_0) + z_0)ds$$
$$= e^{-tA}z_0 + \int_0^t e^{-(t-s)A}Qf(\Psi_{u_0}(z_0) + z_0)ds \tag{2.2.71}$$

and

$$\Psi_{S(t)u_0}(z(t)) = e^{-tA}\Psi_{u_0}(z_0) + \int_0^t e^{-(t-s)A}Pf(\Psi_{S(t)u_0}(z(s)) + z(s))ds. \tag{2.2.72}$$

Let $\xi = \xi(t)$, $t \geq 0$ be the solution for the first variational equation (2.2.71), that is,

$$\xi(t) = e^{-tA}\xi_0 + \int_0^t e^{-(t-s)A}QDf(\Psi_{S(t)u_0}(z(s)) + z(s))(D\Psi_{S(t)u_0}(z(s))\xi(s) + \xi(s))ds, \tag{2.2.73}$$

where we take $\xi_0 = Qu_0$, $\mu_0 = D\Psi_{u_0}(Pu_0)\xi_0 + \xi_0$.

Now differentiating equation (2.2.72) with respect to z_0, and acting on ξ_0, we get

$$D\Psi_{S(t)u_0}(z(t))\xi(t)$$
$$= e^{-tA}D\Psi_{u_0}(z_0)\xi_0 + \int_0^t e^{-(t-s)A}PDf(\Phi_{S(t)u_0}(z(s)) + z(s))(D\Psi_{S(s)u_0}(z(s))\xi(s) + \xi(s))ds. \tag{2.2.74}$$

Thus, from equations (2.2.73) and (2.2.74) we obtain that

$$\mu(t) = D\Psi_{S(t)u_0}(z(t))\xi(t) + \xi(t) \in T_{S(t)v_0}N_{S(t)u_0}$$

is the solution for the first variational equation (2.2.1):

$$\mu'(t) + A\mu = PDf(S(t)v_0)\mu, \quad \mu(0) = D\Psi_{u_0}(z_0)\xi_0 + \xi_0.$$

If we set $z_0 = Pu_0$, then we have $v_0 = u_0$, $\mu(0) = \mu_0$. Hence

$$\mu(t) \in T_{S(t)u_0}N_{S(t)u_0} = N_{S(t)u_0}M, \quad \forall t \geq 0.$$

Therefore $(u(t), \mu(t)) \in NM$, $\forall t \geq 0$, where $u(t) = S(t)u_0$. This means that NM is positive invariant under equation (2.2.65).

For the estimation of $\xi(t)$, from equation (2.2.73) when $t \geq 0$ we have

$$|\xi(t)|_E \leq K_2e^{-\Lambda t}|\xi(0)|_E + M_1K_2(1 + l')\int_0^t ((t-s)^{-\alpha} + \Lambda^\alpha)e^{-\Lambda(t-s)}|S(s)|_E ds,$$

where $l' < 1$ is a bound on the Lipschitz constant for $\Psi_{S(t)u_0}$, and $b = 2M_1K_2(1+\gamma_\alpha)\Lambda^\alpha$. Then

$$e^{(\Lambda-b)t}|\xi(t)|_E \le K_2e^{-bt}|\xi_0|_E + M_1K_2(1+l')\int_0^t ((t-s)^{-\alpha}+\Lambda^\alpha)e^{-b(t-s)}e^{(\Lambda-b)s)}|\xi(s)|_E ds.$$

Let

$$G(t) = \max_{0\le s\le t} e^{(\Lambda-b)s)}|\xi(s)|_E.$$

Hence, for $0 \le t \le T$,

$$\begin{aligned}e^{(\Lambda-b)t}|\xi(t)|_E &\le K_2|\xi_0|_E + M_1K_2(1+l')G(t)\int_0^t ((t-s)^{-\alpha}+\Lambda^\alpha)e^{-b(t-s)}ds\\ &\le K_2|\xi_0|_E + \frac{M_1K_2(1+l')(1+\gamma_\alpha)}{b}G(t).\end{aligned}$$

Taking the maximum of the above equation over $t \in [0,T]$ and inserting it into the expression of b, we obtain

$$G(T) \le K_2|\xi_0|_E + \frac{1+l'}{2}G(T),$$

which yields $G(t) \le \frac{2K_2}{1-l'}|\xi_0|_E$, $\forall t \ge 0$. Then

$$\begin{aligned}|\xi(t)|_E &\le \frac{2K_2}{1-l'}|\xi_0|_E e^{-(\Lambda-2M_1K_2(1+\gamma_\alpha)\Lambda^\alpha)t}, \quad \forall t\ge 0,\\ |\mu(t)|_E &= |D\Psi_{S(t)u_0}(z(t))\xi(t)+\xi(t)|_E\\ &\le (1+l')|\xi(t)|_E\\ &\le 2K_2\frac{1+l'}{1-l'}|\xi_0|_E e^{-(\Lambda-2M_1K_2(1+\gamma_\alpha)\Lambda^\alpha)t}, \quad \forall t\ge 0.\end{aligned}$$

Since

$$|\xi_0|_E = |Q\mu_0|_E \le K_2|\mu_0|_E,$$

we finally obtain

$$|\mu(t)|_E \le 2K_2^2\frac{1+l'}{1-l'}|\mu_0|_E e^{-(\Lambda-2M_1K_2(1+\gamma_\alpha)\Lambda^\alpha)t}, \quad \forall t\ge 0. \qquad \square$$

Theorem 2.2.4. *Under the assumptions of Theorem 2.2.1, if $f \in C^1(E,F)$, then the inertial manifold is normal hyperbolic.*

Proof. From Lemmas 2.2.9, 2.2.10 and 2.2.11, together with the spectral gap condition, we arrive at

$$\lambda + 2M_1K_1\lambda^\alpha < \Lambda - 2M_1K_2(1+\gamma_\alpha)\Lambda^\alpha.$$

Hence the contraction along the normal direction of M is stronger than along the tangential. The theorem is proved. □

Now we consider the high-order regularity of inertial manifold.

Theorem 2.2.5. *Suppose the assumptions of Theorem 2.2.1 are satisfied, and* $f \in C^{j,\nu}(E,F)$, $j = 1,2,\ldots,k$; $0 \le \nu \le 1$, $k \in \mathbf{N}$. *Then* $D^j f$ *is uniformly bounded and possesses global Hölder continuity with the index* ν. *Furthermore, if the following strong spectral gap condition is valid:*

$$\Lambda_n - 2M_1K_2(1+\gamma_\alpha)\Lambda_n^\alpha \ge (k+\nu)(\lambda_n + 2M_1K_1\lambda_n^\alpha), \tag{2.2.75}$$

then the inertial manifold $M = \operatorname{graph}\Phi$ *is satisfied* $\Phi \in C^{j,\nu}(P_nE, Q_nE)$, $j = 1,2,\ldots,k$.

In order to prove Theorem 2.2.5, we introduce the space:

$$\mathscr{F}_{k,\sigma} = \Big\{\Delta \in C((-\infty,0], M_k^s(PE,E));\ \|\Delta\|_{\mathscr{F}_{k,\sigma}} \equiv \sup_{t\le 0} e^{k\sigma t}\|\Delta(t)\|_{M_k^s(PE,E)} < +\infty\Big\},$$

which is a Banach space with norm $\|\cdot\|_{\mathscr{F}_{k,\sigma}}$; $M_k^s(PE,E)$ means the space of all the continuous symmetric k-linear mappings $PE \to E$, with the norm $\|\cdot\|_{(PE,E)}$. For convenience, set $\lambda = \lambda_n$, $\Lambda = \Lambda_n$, $P = P_n$, $Q = Q_n$, and assume condition (2.2.75).

Lemma 2.2.11. *Under the assumptions of Theorem 2.2.5, there exist* $\partial_y^j\varphi$, $j = 1,\ldots,k$, *and* $\partial_y^j\varphi \in \mathscr{F}_{j,\sigma}$, *where* $\varphi(y)$ *is given by Lemma 2.2.2, and* σ *satisfies*

$$\Lambda - 2M_1K_2(1+\gamma_\alpha)\Lambda^\alpha > k\sigma > \sigma > \lambda + 2M_1K_1\lambda^\alpha. \tag{2.2.76}$$

Proof. We prove by induction on k. When $k = 1$, the claim has been proved in Theorem 2.2.3. Now suppose the claim is valid when j ranges from 1 to $k-1$, $k \ge 2$. Next we prove that the theorem is true when $j = k$. The proof is divided into three steps.

Step 1: Preparations.

Since $\varphi(y)$ is the fixed point of $\mathscr{L}(\cdot,y)$, differentiating equation $\varphi(y) = \mathscr{L}(\varphi(y),y)$ with respect to y, we know that $\partial_y^k\varphi(y)$ is the fixed point of the formal mapping $\mathscr{L}_k(\cdot,y)$,

$$\begin{aligned}\mathscr{L}_k(\Delta,y)(t) = &-\int_t^0 e^{-(t-s)A}P[G(y,s) + Df(\varphi(y,s))\Delta(s)]ds\\ &+\int_{-\infty}^t e^{-(t-s)A}Q[G(y,s) + Df(\varphi(y,s))\Delta(s)]ds,\end{aligned}$$

where $\Delta \in \mathscr{F}_{k,\sigma}$, $G = G(y,s)$ involves derivatives of φ of order from the first to $(k-1)$th. From the induction hypothesis, they exist. Now we set $\partial_y^j \varphi(y) \in \mathscr{F}_{j,\sigma}$, $j = 1,\dots,k$, and since the derivative of f is bounded, we have

$$\|G(y,t)\|_{M_k^s(PE,E)} \le m_\sigma e^{-k\sigma t}, \quad \forall t \le 0, \tag{2.2.77}$$

where σ satisfies equation (2.2.77), $\mathscr{L}_k$ is a well-defined mapping, $\mathscr{F}_{k,\sigma} \times PE \to \mathscr{F}_{k,\sigma}$, which satisfies

$$\|\mathscr{L}_k(\Delta_1,y) - \mathscr{L}_k(\Delta_2,y)\|_{\mathscr{F}_{k,\sigma}} \le \theta_{k\sigma}\|\Delta_1 - \Delta_2\|_{\mathscr{F}_{k,\sigma}},$$

where $y \in PE$, $\Delta_1, \Delta_2 \in \mathscr{F}_{k,\sigma}$,

$$\theta_{k\sigma} = \frac{M_1 K_2 \lambda^\alpha}{k\sigma - \lambda} + \frac{M_1 K_2 (1+\lambda_\alpha)\Lambda^\alpha}{\Lambda - k\sigma},$$

and σ satisfies equation (2.2.76). Then $\theta_{k\sigma} < 1$. Hence, $\mathscr{L}_k$ is a strictly contracting mapping in $\mathscr{F}_{k,\sigma}$, uniform in PE. Then there exists a function $\Delta : PE \to \mathscr{F}_{k,\sigma}$, which satisfies $\mathscr{L}_k(\Delta(y),y) = \Delta(y)$, $\forall y \in PE$.

Step 2: The continuity of Δ.

Fix $y_0 \in PE$, and consider $y \in PE$. Let σ satisfy equation (2.2.76), and assume μ satisfies equation (2.2.76), too, $\mu < \sigma$. Similar to equation (2.2.26), we get

$$\|\Delta(y) - \Delta(y_0)\|_{\mathscr{F}_{k,\sigma}} \le \frac{1}{1-\theta_{k,\sigma}} \|\mathscr{L}_k(\Delta(y_0),y) - \mathscr{L}_k(\Delta(y_0),y_0)\|_{\mathscr{F}_{k,\sigma}}, \tag{2.2.78}$$

$\Delta \in \mathscr{F}_{k,\sigma}$, which yields

$$\|\mathscr{L}_k(\Delta(y_0),y) - \mathscr{L}_k(\Delta(y_0),y_0)\|_{\mathscr{F}_{k,\sigma}} \le M(y) + \|\Delta(y_0)\|_{\mathscr{F}_{k,\sigma}} N(y), \tag{2.2.79}$$

where

$$\begin{aligned} M(y) = \sup_{t\le 0} \Bigg\{ & K_1 \lambda^\alpha e^{k\sigma t} \int_t^0 e^{-\lambda(t-s)} \|G(y,s) - G(y_0,s)\|_{M_k^s(PE,F)} \mathrm{d}s \\ & + K_2 e^{k\sigma t} \int_{-\infty}^t ((t-s)^{-\alpha} + \Lambda^\alpha) e^{-\Lambda(t-s)} \|G(y,s) - G(y_0,s)\|_{M_k^s(PE,F)} \mathrm{d}s \Bigg\}, \\ N(y) = \sup_{t\le 0} \Bigg\{ & K_1 \lambda^\alpha e^{k\sigma t} \int_t^0 e^{-\lambda(t-s)} \|Df(\varphi(y,s)) - Df(\varphi(y_0,s))\|_{\mathscr{L}(E,F)} e^{-k\mu s} \mathrm{d}s \\ & + K_2 e^{k\sigma t} \int_{-\infty}^t e^{-\Lambda(t-s)} \|Df(\varphi(y,s)) - Df(\varphi(y_0,s))\|_{\mathscr{L}(E,F)} e^{-k\mu s} \mathrm{d}s \Bigg\}. \end{aligned}$$

As in the proof of Theorem 2.2.3, when $y \to y_0$, $M(y), N(y) \to 0$. Then from equations (2.2.78)–(2.2.79), we deduce that when $y \to y_0$, $\|\Delta(y) - \Delta(y_0)\|_{\mathscr{F}_{k,\sigma}} \to 0$. This proves the continuity of Δ.

Step 3: $\partial_y^k \varphi(y) = \Delta(y)$.

For $y, h \in PE$, set

$$\rho(y,h,t) = \frac{1}{|h|_E^k}\Big[\varphi(y+h,t) - \varphi(y,t) - \partial_y\varphi(y,t)h - \cdots - \frac{1}{(k-1)!}\partial_y^{k-1}\varphi(y)(h,\dots,h) - \frac{1}{k!}[\Delta(y)(h,\dots,h)|_E]\Big].$$

By induction, we prove that μ and σ satisfy equation (2.2.76), $\mu < \sigma$. We have

$$\begin{aligned} e^{k\sigma t}\rho(y,h,t) \le{}& K_1\lambda^\alpha e^{k\sigma t}\int_t^0 e^{-\lambda(t-s)}R(y,h,s)e^{-k\mu s}\mathrm{d}s \\ &+ K_2 e^{k\sigma t}\int_{-\infty}^t ((t-s)^{-\alpha} + \Lambda^\alpha)e^{-\Lambda(t-s)}R(y,h,s)e^{-k\mu\sigma}\mathrm{d}s \\ &+ \theta_{k\sigma}\sup_{s\le 0}(e^{k\sigma s}\rho(y,h,s)), \end{aligned}$$

where $R(y,h,s)$ is uniformly bounded and, when $|h|_E \to 0$, $R(y,h,s) \to 0$. For each point s and $t \le 0$, taking an upper bound, we have

$$\sup_{t\le 0}(e^{k\sigma t}\rho(y,h,t)) \le \frac{1}{1-\theta_{k\sigma}}\widetilde{R}(y,h),$$

where by Lebesgue dominated convergence theorem, when $|h|_E \to 0$, $\widetilde{R}(y,h,s) \to 0$. This gives the existence of $\partial_y^k\varphi(y)$ and $\partial_y^k\varphi(y) = \Delta(y) \in \mathscr{F}_{k,\sigma}$. □

Proof of Theorem 2.2.5. By the spectral gap condition (2.2.75), we can select σ which satisfies equation (2.2.76), $\partial_y^j\varphi \in \mathscr{F}_{j,\sigma}$, $j = 1,2,\dots,k$. Since $\Phi(y) = Q\varphi(y,0)$, then $\Phi \in C^k(PE, QE)$.

We show Hölder continuity of $D\Phi$. Taking $y_1, y_2 \in PE$, we have

$$\begin{aligned} &\|\partial_y\varphi(y_1,t) - \partial_y\varphi(y_2,t)\|_{\mathscr{L}(E,F)} \\ &\le K_1\lambda^\alpha\int_t^0 e^{-\lambda(t-s)}\|Df(\varphi(y_1,s)) - Df(\varphi(y_2,s))\|_{\mathscr{L}(E,F)}\|\partial_y\varphi(y_1,s)\|_{\mathscr{L}(E,F)}\mathrm{d}s \\ &\quad + K_1\lambda^\alpha\int_t^0 e^{-\lambda(t-s)}\|Df(\varphi(y_2,s))\|_{\mathscr{L}(E,F)}\|\partial_y\varphi(y_1,s) - \partial_y\varphi(y_2,t)\|_{\mathscr{L}(E,F)}\mathrm{d}s \\ &\quad + K_1\int_{-\infty}^t ((t-s)^{-\alpha}\Lambda\alpha)e^{-\Lambda(t-s)} \\ &\quad \times \|Df(\varphi(y_1,s)) - Df(\varphi(y_2,s))\|_{\mathscr{L}(E,F)}\|\partial_y\varphi(y_1,s)\|_{\mathscr{L}(E,F)}\mathrm{d}s \end{aligned}$$

$$+ K_2 \int_{-\infty}^{t} ((t-s)^{-\alpha} \Lambda\alpha) e^{-\Lambda(t-s)}$$

$$\times \|Df(\varphi(y_2, s))\|_{\mathcal{L}(E,F)} \|\partial_y \varphi(y_1, s) - \partial_y \varphi(y_2, s)\|_{\mathcal{L}(E,F)} ds$$

$$\leq K_1 \lambda^\alpha \|\partial_y \varphi(y_1)\|_{\mathcal{B}_{1,\mu}} \int_{t}^{0} e^{-\lambda(t-s)} M_2 \|\varphi(y_1, s) - \varphi(y_2, s)\|_E^\nu e^{-\mu s} ds$$

$$+ K_2 \|\partial_y \varphi(y_1)\|_{\mathcal{B}_{1,\mu}} \int_{-\infty}^{t} ((t-s)^{-\alpha} \Lambda\alpha) e^{-\Lambda(t-s)} M_2 |\varphi(y_1, s) - \varphi(y_2, s)|_E^\nu e^{-\mu s} ds$$

$$+ \theta_\sigma e^{-\sigma t} \|\partial_y \varphi(y_1) - \partial_y \varphi(y_2)\|_{\mathcal{B}_{1,\sigma}},$$

where σ, μ satisfy equation (2.2.13), $(1+\nu)\mu < \sigma$, and M_2 is the Hölder constant of M_2. If we replace μ with σ in equation (2.2.27), we get

$$\|\partial_y \varphi(y_1) - \partial_y \varphi(y_2)\|_{\mathcal{B}_{1,\sigma}} \leq \frac{M_2}{1-\theta_\sigma} \frac{K_1^\nu}{(1-\theta_\sigma)^\nu} \|\partial_y \varphi(y_1)\|_{\mathcal{B}_{1,\mu}}$$

$$|y_1 - y_2|_E^\nu \left\{ \frac{K_1 \lambda^\alpha}{\sigma - \lambda} + \frac{K_1(1+\lambda_\alpha)\Lambda^\alpha}{\Lambda - \sigma} \right\}.$$

Since $\partial_y \varphi(y_1) = \mathcal{L}_1(\partial_y \varphi(y_1), y_1)$, we have

$$\|\partial_y \varphi(y_1)\|_{\mathcal{B}_{1,\sigma}} \leq \frac{K_1}{(1-\theta_\mu)}.$$

Therefore

$$\|\partial_y \varphi(y_1, s) - \partial_y \varphi(y_2, s)\|_{\mathcal{B}_{1,\sigma}} \leq C |y_1 - y_2|_E^\nu,$$

where

$$C = \frac{M_2}{1-\theta_\sigma} \frac{K_1^{1+\nu}}{(1-\theta_\sigma)^{1+\nu}} \left\{ \frac{K_1 y^\alpha}{\sigma - \lambda} + \frac{K_1(1+\lambda_\alpha)\Lambda^\alpha}{\Lambda - \sigma} \right\}.$$

Thus for $D\Phi(y) = Q\partial_y \varphi(y, 0)$, it follows that

$$\|D\Phi(y_1) - D\Phi(y_2)\|_{\mathcal{L}(PE,QE)} \leq C |y_1 - y_2|_E^\nu,$$

which proves the Hölder continuity of $D\Phi$. The Hölder continuity of $D^j\Phi$, $j = 1, \ldots, k$, can be proved by induction. □

As a consequence, we get

Theorem 2.2.6. *Suppose the assumptions of Theorem 2.2.5 are met; in particular, the spectral gap condition (2.2.75) is satisfied. Then the leaf N_{v_0} = graph Ψ_{v_0} is such that $\Psi_{v_0} \in C^{j\nu}(QE, PE)$, $j = 1, 2, \ldots, k$, $\forall v_0 \in E$.*

Proof. Similar to the proof of Theorem 2.2.5. □

2.3 The finite-dimensional inertial form for the one-dimensional generalized Ginzburg–Landau equation

In [97], the authors consider the following equation:

$$\begin{aligned}\partial_t u + \nu u_x = {} & \chi u + (\gamma_r + i\gamma_i)u_{xx} - (\beta_r + i\beta_i)|u|^2 u \\ & - (\delta_r + i\delta_i)|u|^4 u - (\lambda_r + i\lambda_i)|u|^2 u_x \\ & - (\mu_r + i\mu_i)u^2 \overline{u}_x, \quad x \in \mathbf{R},\ t > 0,\end{aligned} \tag{2.3.1}$$

which was put forward in [46]. As shown in [209], when $\gamma_r > 0$, $\nu > 0$, $\beta_i = 1$, $\beta_r < 0$, $\chi < 0$, $\delta_r = \lambda_r = \mu_r = \mu_i = 0$, and when the initial value of u is large enough, the spacial periodic solution of the equation will blow up. But the corresponding physical motion did not provide evidence of the blow-up phenomenon [79]. So the above equation was proposed in [46] to describe the objective reality more accurately. Therefore studies of the well-posedness of the solution of this equation and its long time behavior are necessary.

In [97], the authors got the following conclusion:

Theorem 2.3.1. *When $\chi > 0$, $\delta_r > 0$, $\gamma_r > 0$, $4\delta_r\gamma_r > (\lambda_i - \mu_i)^2$, a spacial periodic solution for equation (2.3.1) exists and is unique in $H^1_{\text{per}}[0, L]$. And for the semigroup $S(t)$ determined by the solution (namely, $S(t)u(t) = S(t)u_0$, $t \geq 0$, where u_0 is the initial value) there exists a finite-dimensional attractor in $H^1_{\text{per}}[0, L]$.*

In the following, we will prove that the condition $4\delta_r\gamma_r > (\lambda_i - \mu_i)^2$ cannot be improved. If $4\delta_r\gamma_r \leq (\lambda_i - u_i)^2$, we consider a spacial periodic solution $u = Re^{ikx}$. Substituting it into equation (2.3.1), taking the real part and solving for R^2, we get

$$\begin{aligned}R^2 &= \frac{1}{2\delta_r}(-(\beta_r - (\lambda_i - \mu_i)k)) \pm \sqrt{(\beta_r - (\lambda_i - u_i)k)^2 - (4\delta_r\gamma_r k^2 - 4\delta_r\chi)} \\ &= \frac{k}{2\delta_r}\left(-\left(\frac{\beta_r}{k} - (\lambda_i - \mu_i)\right)\right) \pm \sqrt{\left(\frac{\beta_r}{k} - (\lambda_i - \mu_i)\right)^2 - 4\delta_r\gamma_r + \frac{4\delta_r\chi}{k^2}}.\end{aligned}$$

We can see that, when $k \to \pm\infty$, $R^2 \to +\infty$. This illustrates that the solution blows up. Thus $4\delta_r\gamma_r > (\lambda_i - \mu_i)$ cannot be improved.

Because the semigroup $S(t)$ of equation (2.3.1) has a finite-dimensional attractor, in order to effectively consider the long time behavior of solutions, we expect to use finite-dimensional ordinary differential equations to describe the attractor. Therefore, we must take into account the inertial manifold; for the definitions of attractor and inertial manifold, refer to [197]. An important condition for the existence of inertial manifolds is the spectral gap condition, $\lambda_{m+1} - \lambda_m > M_0^2 \frac{1+l}{l}(\lambda^\alpha_{m+1} + \lambda^\alpha_m)$, here $\{\lambda_m\}_{m=1}^\infty$ is a sequence of eigenvalues for $-\gamma_r\partial_{xx}$ ($\lambda_m = \gamma_r(\frac{2\pi m}{L})^2$, L is the length of spacial period) with periodic boundary conditions, $l \leq \frac{1}{8}$ is given, M_0 is a constant of estimation. The spectral gap conditions are met if

$$\gamma_r\left(\frac{2\pi}{L}\right)^2(2m+1) > M_0\frac{1+l}{l}\frac{2\pi}{L}\gamma_r^{\frac{1}{2}}(2m+1),$$

or

$$\frac{1+l}{l}\gamma_r^{-\frac{1}{2}}L \le \frac{2\pi}{M_0}.$$

From the above equation we can see that, when L is small enough or γ_r is large enough, this is indeed the case. By the conclusions in [197], now we get the existence of an inertial manifold of equation (2.3.1); here we do not discuss this in detail.

The purpose of this section is to try to improve the results. Consider a simplified form of equation (2.3.1):

$$u_t = \chi\mu + \gamma u_{xx} - (\beta_r + i\beta_i)|u|^2u - (\delta_r + i\delta_i)|u|^4u - (\lambda_r + i\lambda_i)|(|u|^2u)_x, \tag{2.3.2}$$

$x \in \mathbf{R}$, $t > 0$, $\gamma, \delta_r > 0$. Obviously, equation (2.3.3) is a special case of equation (2.3.2). Equation (2.3.2) is a generalized form for the following Ginzburg–Landau equation studied in [53]:

$$u_t = u_{xx} + (1 - |u|^2)u,$$

where u is a complex function.

Here using a nonlinear term linearized method, the long time dynamic behavior of equation (2.3.2) can be described by a group of finite-dimensional ordinary differential equations.

Firstly, we introduce some notations:

$$H = L^2_{\text{per}}[0,L] = \{u \in L^2[0,1],\ u(x) = u(x+L)\},$$

its inner product and norm are respectively defined by

$$(u,v) = \int_0^L u\bar{v}\,\mathrm{d}x, \quad |u|_0^2 = (u,u), \quad u,v \in H,$$

$$H^n_{\text{per}}[0,L] = \{u : u \in H,\ u_x \in H, \dots, u_{x\dots x} \in H\}, \quad n \ge 1.$$

Suppose $A = -(\mu + \gamma\partial_{xx})$, for $\gamma > 0$ and appropriate μ. Then the characterization problem with periodic boundary condition, $-(\mu + \gamma\partial_{xx})g = \lambda g$, has no zero eigenvalue. Hence A is a linear self-adjoint unbounded positive operator. It follows that we can define the power of A as A^α, $\alpha \in [0,1]$, $V_{2\alpha} = D(A^\alpha)$ (domain of A^α). By [141], $V_1 = D(A^{\frac{1}{2}}) = H^1_{\text{per}}[0,L]$, $V_2 = D(A) = H^2_{\text{per}}[0,L]$; the norms of V_1 and V_2 can be defined as $|\cdot|_1$ and $|\cdot|_2$.

Since equation (2.3.2) is a special case of equation (2.3.1), then $\gamma, \delta_r > 0$ and, whenever $4\delta_r\gamma > \lambda_i^2$, the global solution of equation (2.3.2) exists uniquely in V_1, and there exists a finite-dimensional attractor $\mathscr{A}_{\text{GGL}}$ in V_1. Further, we have

Proposition 2.3.1. *If $u \in \mathscr{A}_{\mathrm{GGL}}$, and if there exist $\rho_0, \rho_1, \rho_2 > 0$ such that*

$$|u|_0 \le \frac{\rho_0}{2}, \quad |u|_1 \le \frac{\rho_1}{2}, \quad |u|_2 \le \frac{\rho_2}{2},$$

then we have $\mathscr{A}_{\mathrm{GGL}} \subset H^2_{\mathrm{per}}[0, L]$.

Proof. By [97] we know that, if $u \in \mathscr{A}_{\mathrm{GGL}}$, then it is obvious that $|u|_0 \le \frac{\rho_0}{2}$, $|u|_1 \le \frac{\rho_1}{2}$. We only need to prove $|u|_2 \le \frac{\rho_2}{2}$.

By [97] we know, when $t > 0$, $u(t)$ is smooth enough, then taking the inner product of (2.3.2) and u_{xxxx} and then the real component, we obtain

$$\begin{aligned}\frac{1}{2}\frac{\mathrm{d}}{\mathrm{d}t}|u_{xx}|_0^2 + \gamma|u_{xxx}|_0^2 = \chi|u_{xx}|_0^2 &- \mathrm{Re}\left((\beta_r + i\beta_i)\int_0^L |u|^2 u\bar{u}_{xxxx}\mathrm{d}x\right)\\ &- \mathrm{Re}\left((\delta_r + i\delta_i)\int_0^L |u|^4 u\bar{u}_{xxxx}\mathrm{d}x\right)\\ &- \mathrm{Re}\left((\lambda_r + i\lambda_i)\int_0^L (|u|^2 u)_x\bar{u}_{xxxx}\mathrm{d}x\right).\end{aligned} \tag{2.3.3}$$

By [97] we know that, there exists $T > 0$ such that when $t \ge T$ we have

$$|u|_0 \le \rho_0, \quad |u|_1 \le \rho_1.$$

In the following estimation, we use the embedding inequality $|u|_{L^\infty} \le C|u|_1$ and Gagliardo–Nirenberg inequality. The constant C in the estimate is a positive number only depending on the coefficients ρ_0, ρ_1 in equation (2.3.2). Here we do not distinguish them. For $t \ge T$, we have

$$\begin{aligned}&\frac{1}{2}\frac{\mathrm{d}}{\mathrm{d}t}|u_{xx}|_0^2 + \gamma|u_{xxx}|_0^2\\ &\quad\le |\chi||u_{xx}|_0^2 + \sqrt{\beta_r^2 + \beta_i^2}\left|\int_0^L (|u|^2 u)_x\bar{u}_{xxx}\mathrm{d}x\right| + \sqrt{\delta_r^2 + \delta_i^2}\left|\int_0^L (|u|^4 u)_x\bar{u}_{xxx}\mathrm{d}x\right|\\ &\quad= \sqrt{\lambda_r^2 + \lambda_i^2}\left|\int_0^L (|u|^2 u)_{xx}\bar{u}_{xxx}\mathrm{d}x\right|.\end{aligned} \tag{2.3.4}$$

Obviously, we only need to deal with the last term:

$$\begin{aligned}\left|\int_0^L (|u|^2 u)_{xx}\bar{u}_{xxx}\mathrm{d}x\right| &\le \frac{\gamma}{4}|u_{xxx}|_0^2 + \frac{1}{\gamma}\int_0^L (6|u||u_x|^2 + 3|u|^2|u_{xx}|)^2\mathrm{d}x\\ &\le \frac{\gamma}{4}|u_{xxx}|_0^2 + C(\rho_0, \rho_1)\int_0^L |u_x|^4\mathrm{d}x + C(\rho_0, \rho_1)|u_{xx}|_0^2.\end{aligned}$$

For $\int_0^L |u_x|^4 \mathrm{d}x$, we use Gagliardo–Nirenberg inequality, and equation (2.3.4) becomes

$$\frac{\mathrm{d}}{\mathrm{d}t}|u_{xx}|_0^2 \le C + C|u_{xx}|_0^2.$$

By Lemma 2.4 in [97], we have

$$\int_t^{t+1} |u_{xx}(\cdot,\tau)|_0^2 \mathrm{d}\tau \le C(\rho_0,\rho_1), \quad t \ge T.$$

Using the uniform Gronwall inequality, we obtain that $|u_{xx}|_0^2 \le \rho^2$, $t \ge T+1$, ρ only depends on the coefficients ρ_0, ρ_1 of the equation. Hence

$$|u|_2 \le \frac{\rho_2}{2}, \quad u \in \mathscr{A}_{\mathrm{GGL}},$$

and Proposition 2.2.3 has been proved. This generalized the results of [97]. □

In order to keep the dissipation of equations after transformation, we need to alter equation (2.3.2) appropriately, namely

$$\begin{aligned} u_t &= \gamma u_{xx} - (\lambda_r + i\lambda_i)(|u|^2 u)_x - \varphi_\rho \cdot g(u) \\ &\quad - (1-\varphi_\rho)\left(\delta_r + 9\gamma + 9\frac{\lambda_r^2+\lambda_i^2}{\gamma}\right) \times |u|^4 u, \end{aligned} \tag{2.3.5}$$

where $g(u) = (\beta_r + i\beta_i)|u|^2 u + (\delta_r + i\delta_i)|u|^4 u - \chi u$,

$$\varphi_\rho = \varphi\left(\frac{|u|_0^2 + 2|u_x|_0^2 + |u|_{L^6}^6}{\rho^2}\right), \quad 0 < \rho \le \infty,$$

$\varphi : \mathbf{R}_+ \to [0,1]$ is a smooth monotonic function such that $\varphi(s) = 1$, $0 \le s \le 1$, $\varphi(s) = 0$, $s \ge 2|\varphi'(s)| \le 2$. We notice that φ_ρ is independent of the spacial variables. If $\rho = \infty$, then $\varphi_\rho = 1$, and equation (2.2.5) is (2.2.2). Given the initial value $u_0 \in V_1$, similar to the discussion in [97], we have

Proposition 2.3.2. *Given the initial value $u_0 \in V_1$, $\gamma, \delta_r > 0$, $4\delta_r\gamma > \lambda_i^2$, equation (2.2.5) has a unique solution $u \in C([0,\infty); V_1) \cap L^2(0,T;V_2) \cap C(0,+\infty; H^n_{\mathrm{per}}[0,L])$ ($n \ge 2$ is arbitrary). Furthermore, there exist constants r_0, r_1, r_2 (independent of u_0 and ρ) such that*

$$\overline{\lim_{t\to\infty}}\, |u|_0 \le r_0, \quad \overline{\lim_{t\to\infty}}\, |u|_1 \le r_1, \quad \overline{\lim_{t\to\infty}}\, |u|_2 \le r_2.$$

Thus it can be seen that equation (2.2.5) possesses the global attractor $\mathscr{A}_\rho$ (its finite dimension is shown similarly as in [97]), and when $\rho^2 \ge 4r^2 = 4(r_0^2 + 2r_1^2 + r_3^2)$ (here r_3 is a bound of $|u|_{L^6}^3$), $\mathscr{A}_\rho = \mathscr{A}_{\mathrm{GGL}}$.

The proof of Proposition 2.3.2 is similar to the proof of existence of an absorbing set in [97] and the proof of Proposition 2.3.2; here we omit it.

By Proposition 2.3.2, we can introduce the following function transform:

$$J(u) = (u, u_x, |u|^2 u) = (u, v, \omega).$$

Then u, v, ω respectively satisfy the following equations (we are only using equation (2.3.5) and calculating directly):

$$\begin{cases} u_t = \gamma u_{xx} - (\lambda_r + i\lambda_i)\omega_x + \eta_1, \\ v_t = \gamma v_{xx} - (\lambda_r + i\lambda_i)\omega_{xx} + \eta_2, \\ \omega_t = \gamma\omega_{xx} + \eta_3, \end{cases} \tag{2.3.6}$$

where

$$\begin{aligned} \eta_1 &= -\varphi_\rho g(u) - (1-\varphi_\rho)\Big(9\gamma + \delta_r + 9\frac{\lambda_r^2 + \lambda_i^2}{\gamma}\Big)|u|^4 u, \\ \eta_2 &= -\varphi_\rho h(u, v) - (1-\varphi_\rho)\Big(9\gamma + \delta_r + 9\frac{\lambda_r^2 + \lambda_i^2}{\gamma}\Big)(3|u|^4 v + 2|u|^2 u^2 \overline{v}), \\ h(u, v) &= (\beta_r + i\beta_i)(2|u|^2 v + u^2\overline{v}) + (\delta_r + i\delta_i)(2|u|^4 v + 2|u|^2 u^2 \overline{v}), \\ \eta_3 &= -(4\gamma|v|^2 u^2 + 2\gamma v^2 \overline{u}) + 2|u|^2\eta_1 + u^2\overline{\eta}_1 \\ &\quad - 2(\lambda_r + i\lambda_i)(2|u|^4 v + |u|^2 u^2 \overline{v}) - (\lambda_r - i\lambda_i)(2|u|^2\overline{v} + |u|^4 v). \end{aligned}$$

For equations (2.3.6), coupled with some additional terms below (and to keep the dissipative property),

$$\begin{cases} u_t = \gamma u_{xx} - (\lambda_r + i\lambda_i)\omega_x + \eta_1 + \xi_1, \\ v_t = \gamma v_{xx} - (\lambda_r + i\lambda_i)\omega_{xx} + \eta_2 - k_1(v - u_x) + \xi_2, \\ \omega_t = \gamma\omega_{xx} - (4\gamma|v|^2 u + 2\gamma v^2\overline{u}) \\ \qquad + 2|u|^2(\eta_1 + \xi_1) + u^2(\overline{\eta}_1 + \overline{\xi}_1) \\ \qquad - (k_2 - 16(\lambda_r^2 + \lambda_i^2))\Big(1 + \frac{1}{\gamma}|u|^2\Big)(\omega - f(u)), \end{cases} \tag{2.3.7}$$

where $f(u) = |u|^2 u$; k_1, k_2 are unknown constants; ξ_1, ξ_2 can be taken as

$$\begin{aligned} \xi_1 &= -2\gamma|u|^2(\omega - f(u)) + \gamma u^2\overline{(\omega - f(u))}, \\ \xi_2 &= 2\gamma\overline{u}v(\omega - f(u)) + 2\gamma uv\overline{(\omega - f(u))} + 2\gamma u\overline{v}(\omega - f(u)). \end{aligned}$$

Note that if $v = u_x$, $\omega = f(u)$, the additional terms disappear. Setting $J(u) = (u, u_x, f(u))$, $u \in V_1$, these additional terms will lead to the solutions of (2.3.7), which are exponentially converging to $J(u)$.

So far we have seen that if $u(t)$ is the solution of equation (2.3.5) with initial value $u_0 \in V_1$, then $J(u(t))$ is the solution of equation (2.3.7). By the uniqueness of solution of equation (2.3.7), the reversed conclusion is also valid. The uniqueness and existence of solution of equation (2.3.7) will be given in the following. Hence, the solutions of equations (2.3.5) and (2.3.7) have the same dynamics in the set $J(V_1)$.

Proposition 2.3.3. *If $u(t)$ is the solution of equation (2.3.5) with the initial value $u_0 \in V_1$, then $J(u(t))$ is the solution of equation (2.3.7). Conversely, if (u, v, ω) is the solution of equation (2.3.7) with the initial value $J(u_0)$, $u_0 \in V_1$, then $u(t)$ is the solution of equation (2.3.5).*

In the following we study the uniqueness and existence of solution for equation (2.3.7). Letting $U = (u, v, \omega)^t$ (t denotes the transpose), in the space $D(A) \times D(A) \times D(A)$ we define the operator $\mathbf{A}$ by

$$\mathbf{A}U = \begin{pmatrix} Au + (\lambda_r + i\lambda_i)\omega_x \\ Av + k_1 u_x + (\lambda_r + i\lambda_i)\omega_{xx} \\ A\omega \end{pmatrix},$$

where A is defined as before. Similarly, define

$$\tilde{F} = (F_1, F_2, F_3)^t$$

where $F_1 = \eta_1 + \xi_1 + \mu u$, $F_2 = \eta_2 + \xi_2 - k_1 v + \mu v$, F_3 is the right-hand side of the third equation in (2.3.7) with the term $\mu\omega$ added. Then equation (2.3.7) can be written as

$$\frac{\mathrm{d}}{\mathrm{d}t}U = -AU + \overline{F}(U). \tag{2.3.8}$$

It is obvious that the operator A is not self-adjoint, but we can prove A is a fan-shaped operator, that is, $-A$ generates an analytic semigroup in $\mathscr{H} = H \times H \times H$.

Lemma 2.3.1. *The operator* $\mathbf{A}$ *is a fan-shaped operator in* $\mathscr{H}$.

Proof. First, we note that A is a fan-shaped operator in H, hence there exist $0 < \theta < \frac{\pi}{2}$ and $M \geq 1$ such that

$$\begin{aligned} &\rho(A) \supset \Sigma = \{\lambda \mid \theta < |\arg \lambda| \leq \pi,\ \lambda \neq 0\}, \\ &\|(\lambda - A)^{-1}\|_{\mathrm{op}} \leq \frac{M}{|\lambda|}, \quad \lambda \in \Sigma. \end{aligned} \tag{2.3.9}$$

□

Define

$$(A - \lambda)U = \overline{F}$$

where $\lambda \in \Sigma$, $\overline{F} = (f_1, f_2, f_3)^t \in \mathscr{H}$. Then the equation for ω can be written as

$$A\omega - \lambda\omega = f_3. \tag{2.3.10}$$

From equation (2.3.9) we have

$$|\omega|_0 \leq \frac{M}{|\lambda|}|f_3|_0. \tag{2.3.11}$$

Taking the inner product of equation (2.3.10) and ω, and then taking its real component, and using equation (2.3.11), we get

$$|A^{\frac{1}{2}}\omega|_0^2 \le |\lambda||\omega|_0^2 + |f_3|_0|\omega|_0 \le \frac{M^2+M}{|\lambda|}|f_3|_0^2. \tag{2.3.12}$$

From equation (2.3.10), we get

$$|\omega_{xx}|_0 \le (|\lambda| + |u|)|\omega|_0 + |f_3|_0. \tag{2.3.13}$$

The equation about u is

$$Au + (\lambda_r + i\lambda_i)\omega_x - \lambda_u = f_1.$$

Using equation (2.3.9) and (2.3.12), we arrive at

$$\begin{aligned} |u|_0 &\le \frac{M}{|\lambda|}\left(|f_1|_0 + \sqrt{\lambda_r^2+\lambda_i^2}|\omega_x|_0\right) \\ &\le \frac{M}{|\lambda|}\left(|f_1|_0 + \left(\frac{M^2+M}{|\lambda|}\right)^{\frac{1}{2}}\sqrt{\lambda_r^2+\lambda_i^2}|f_3|_0\right) \\ &\le \frac{M_1}{|\lambda|}(|f_1|_0 + |f_3|_0), \quad M_1 \ge 1, \end{aligned} \tag{2.3.14}$$

where we used $|u_x|_0 \le |A^{\frac{1}{2}}u|_0$, $u \in D(A^{\frac{1}{2}})$. Similarly, we have

$$\begin{aligned} |A^{\frac{1}{2}}u|_0^2 &\le \lambda|u|_0^2 + \sqrt{\lambda_r^2+\lambda_i^2}|\omega_x|_0|u|_0 + |f_1|_0|u|_0 \\ &\le (|\lambda|+1)|u|_0^2 + \frac{1}{2}(\lambda_r^2+\lambda_i^2)|\omega_x|_0^2 + \frac{1}{2}|f_1|_0^2. \end{aligned} \tag{2.3.15}$$

Finally, we consider the equation for v. From equation (2.3.9) we have

$$|v|_0 \le \frac{M}{|\lambda|}\left(k_1|u_x|_0 + \sqrt{\lambda_r^2+\lambda_i^2}|\omega_{xx}|_0 + |f_2|_0\right).$$

By equations (2.3.12), (2.3.13) and (2.3.15), we have

$$|v|_0 \le \frac{M_2}{|\lambda|}(|f_1|_0 + |f_2|_0 + |f_3|_0), \quad M_2 \ge 1. \tag{2.3.16}$$

From equations (2.3.11), (2.3.14) and (2.3.16), we obtain that for $\lambda \in \Sigma$, $(A-\lambda I)^{-1}$ exists, and

$$\|(-\lambda I + A)^{-1}\|_{op} \le \frac{M_3}{|\lambda|}, \quad M_3 \ge 1.$$

This proves that **A** is a fan-shaped operator in $\mathscr{H}$. By [185, Th. 5.2], we get that $-A$ generates an analytic semigroup in $\mathscr{H}$.

Since $-A$ is a sectoral operator, we can define the fractional power of A, and it is easy to see that $\widetilde{F} : D(A^{\frac{1}{2}}) = V_1 \times V_1 \times V_1 \to \mathscr{H}$ is locally Lipschitz continuous. Hence if $U(0) = U_0 \in D(A^{\frac{1}{2}})$, then there exists a unique strong solution of equation (2.3.7) such that

$$U \in C([0,T); D(A^{\frac{1}{2}})), \quad 0 < T \le \infty.$$

In the following, we will prove equations (2.3.7) possess the global attractor $\mathscr{A}$. In fact, $\mathscr{A}$ is the image of $\mathscr{A}_\rho$ under embedding J, that is, $J(\mathscr{A}_\rho) = \mathscr{A}$.

If U is a smooth enough solution of equation (2.3.7), then the equation of u is

$$u_t = \gamma u_{xx} - (\lambda_r + i\lambda_i)\omega_x + \eta_1 + \xi_1. \tag{2.3.17}$$

Differentiating equation (2.3.17) with respect to x, we get

$$(u_x)_t = \gamma u_{xxx} - (\lambda_r + i\lambda_i)\omega_{xx} + \eta_{1x} + \xi_{1x}.$$

Then subtracting the above equation from that of v in (2.3.7), we obtain

$$(v - u_x)_t = \gamma(v - u_{xx})_x + \eta_2 - \eta_{1x} + \xi_2 - \xi_{1x} - k_1(v - u_x). \tag{2.3.18}$$

Letting $\widetilde{\omega} = f(|u|^2 u)$, it is easy to verify that $\widetilde{\omega}$ satisfies

$$\begin{aligned}\widetilde{\omega}_t = {} & \gamma\widetilde{\omega}_{xx} - (4\gamma|u_x|^2 u + 2\gamma u_x^2 \overline{u}) \\ & + 2|u|^2(\eta_1 + \xi_1) + u^2(\overline{\eta}_1 + \overline{\xi}_1) \\ & - 2|u|^2(\lambda_r + i\lambda_i)\omega_x - u^2(\lambda_r - i\lambda_i)\overline{\omega}_x,\end{aligned}$$

and then we have

$$\begin{aligned}(\omega - \widetilde{\omega})_t = {} & \gamma(\omega - \widetilde{\omega})_{xx} - 4\gamma u(|v|^2 - |u_x|^2) - 2\gamma\overline{u}(v^2 - u^2) \\ & - 2(\lambda_r + i\lambda_i)|u|^2(2|u|^2 v + u^2 v - \omega_x) \\ & - (\lambda_r - i\lambda_i)u^2(2|u|^2\overline{v} + \overline{u}^2 v - \overline{\omega}_x) \\ & - \left(k_2 - 16(\lambda_r^2 + \lambda_i^2)\left(1 + \frac{1}{\gamma}\right)|u|^4\right)(\omega - \widetilde{\omega}).\end{aligned} \tag{2.3.19}$$

Firstly, we estimate a uniform bound of

$$|u|_0^2 + |u_x|_0^2 + |v - u_x|_0^2 + |\omega - \widetilde{\omega}|_0^2.$$

Then by Minkowski inequality, we get a uniform bound of $|U|_{\mathscr{K}}$. That is, we need to prove the following proposition:

Proposition 2.3.4. *Let $k_1 > 0$, $k_2 > \gamma + \frac{\lambda_r^2 + \lambda_i^2}{2\alpha}$, $\gamma > 2\sqrt{\lambda_r^2 + \lambda_i^2}$ (α is a constant). If U is the solution of equation (2.3.7) with the initial value $U_0 \in D(A^{\frac{1}{2}})$, then, when $\rho_4 \ge \sqrt{2}\rho$, there exists*

$$\varlimsup_{t\to\infty} |U|_{\mathscr{K}}^2 \le \rho_4^2.$$

Proof. Let $U = (u, v, \omega)^t$ be the smooth solution of equation (2.3.7). Then taking the inner product of equation (2.3.17) and u, and then taking its real part, we obtain

$$\frac{1}{2}\frac{\mathrm{d}}{\mathrm{d}t}|u|_0^2 = -\gamma|u_x|_0^2 - \mathrm{Re}\left((\lambda_r + i\lambda_i)\int_0^L \omega_x\bar{u}\mathrm{d}x\right) + \mathrm{Re}(\eta_1 + \xi_1, u).$$

Since

$$\begin{aligned}
&-\mathrm{Re}\left((\lambda_r + i\lambda_i)\int_0^L \omega_x\bar{u}\mathrm{d}x\right)\\
&\quad= \mathrm{Re}\left((\lambda_r + i\lambda_i)\int_0^L \omega\bar{u}_x\mathrm{d}x\right)\\
&\quad= \mathrm{Re}\left((\lambda_r + i\lambda_i)\int_0^L (\omega - f(u))\bar{u}_x\mathrm{d}x\right) + \mathrm{Re}\left((\lambda_r + i\lambda_i)\int_0^L f(u)\bar{u}_x\mathrm{d}x\right)\\
&\quad= \mathrm{Re}((\lambda_r + i\lambda_i)(\omega - \tilde{\omega}, u_x)) - \lambda_i\,\mathrm{Im}\int_0^L |u|^2u\bar{u}_x\mathrm{d}x,\\
&\mathrm{Re}(\eta_1 + \xi_1, u)\\
&\quad= -\mathrm{Re}\left(\varphi_\rho\int_0^L ((\beta_r + i\beta_i)|u|^4 + (\delta_r + i\delta_i)|u|^6 - \chi|u|^2)\mathrm{d}x\right)\\
&\qquad- \mathrm{Re}\left((1-\varphi_\rho)\left(\delta_r + 9\gamma + 9\frac{\lambda_r^2 + \lambda_i^2}{\gamma}\right)\int_0^L |u|^6\mathrm{d}x\right) + \mathrm{Re}(\xi_1, u)\\
&\quad\le -\delta_r\int_0^L |u|^6\mathrm{d}x + |\beta_r|\int_0^L |u|^4\mathrm{d}x + |\chi|\int_0^L |u|^2\mathrm{d}x - \mathrm{Re}\left((1-\varphi_\rho)9\gamma\int_0^L |u|^6\mathrm{d}x\right) + \mathrm{Re}(\xi_1, u).
\end{aligned}$$

According to the choice of ξ_1, we have

$$\mathrm{Re}(\xi_1, u) \le 3\gamma\int_0^L |u|^3|\omega - \tilde{\omega}|\mathrm{d}x.$$

Hence

$$\begin{aligned}
\frac{1}{2}\frac{\mathrm{d}}{\mathrm{d}t}|u|_0^2 \le{}& -\gamma|u_x|_0^2 - \mathrm{Re}((\lambda_r + i\lambda_i)(\omega - \tilde{\omega}, u_x))\\
&+ |\lambda_i|\int_0^L |u|^3|u_x|\mathrm{d}x - \delta_r\int_0^L |u|^6\mathrm{d}x + |\beta_r|\int_0^L |u|^4\mathrm{d}x + |\chi|\int_0^L |u|^2\mathrm{d}x\\
&- (1-\varphi_\rho)9\gamma\int_0^L |u|^6\mathrm{d}x + 3\gamma\int_0^L |u|^3|\omega - \tilde{\omega}|\mathrm{d}x.
\end{aligned}$$

By virtue of

$$3\gamma\int_0^L |u|^3|\omega-\tilde{\omega}|\mathrm{d}x \le 9\gamma\int_0^L |u|^6\mathrm{d}x + \frac{\gamma}{4}|\omega-\tilde{\omega}|_0^2$$

and the assumption $4\delta_r\gamma > \lambda_i^2$, we can choose

$$2\alpha = 2\gamma - b^2, \quad \beta = 2\delta_r - a^2, \quad |a\cdot b| = |\lambda|.$$

The rest of estimation follows [97], and we have

$$\begin{aligned}\frac{1}{2}\frac{\mathrm{d}}{\mathrm{d}t}|u|_0^2 &\le -2\alpha|u_x|_0^2 - |u|_0^2 + P + 18\varphi_\rho\gamma\int_0^L |u|^6\mathrm{d}x \\ &\quad + \frac{\gamma}{2}|\omega-\tilde{\omega}|_0^2 - 2\operatorname{Re}((\lambda_r+i\lambda_i)(\omega-\tilde{\omega},u_x)) \\ &\le -\alpha|u_x|_0^2 - |u|_0^2 + P + 18\varphi_\rho\gamma\int_0^L |u|^6\mathrm{d}x \\ &\quad + \left(\frac{\gamma}{2} + \frac{\lambda_r^2+\lambda_i^2}{\alpha}\right)|\omega-\tilde{\omega}|_0^2,\end{aligned} \tag{2.3.20}$$

where $P = \frac{1}{8}(\frac{(|\beta_r|+1)^2}{\beta} + 2|\chi| + 1)^2$.

Taking the inner product of equation (2.3.17) and u_{xx}, and then taking its real part, we obtain

$$\frac{1}{2}\frac{\mathrm{d}}{\mathrm{d}t}|u|_0^2 = -\gamma|u_{xx}|_0^2 + \operatorname{Re}\left((\lambda_r+i\lambda_i)\int_0^L \omega_x\bar{u}_{xx}\mathrm{d}x\right) - \operatorname{Re}(\eta_1+\xi_1,u_{xx}).$$

By virtue of

$$\begin{aligned}\operatorname{Re}\left((\lambda_r+i\lambda_i)\int_0^L \omega_x\bar{u}_{xx}\mathrm{d}x\right) &= \operatorname{Re}\left((\lambda_r+i\lambda_i)\int_0^L (\omega_x-\tilde{\omega})_x\bar{u}_{xx}\mathrm{d}x\right) \\ &\quad + \operatorname{Re}\left((\lambda_r+i\lambda_i)\int_0^L (|u|^2u)_x\bar{u}_{xx}\mathrm{d}x\right) \\ &\le \frac{\gamma}{2}|u_{xx}|_0^2 + \frac{\lambda_r^2+\lambda_i^2}{\gamma}|(\omega-\tilde{\omega})_x|_0^2 + \frac{9}{\gamma}(\lambda_r^2+\lambda_i^2)\int_0^L |u|^4|u_x|^2\mathrm{d}x,\end{aligned}$$

$$\begin{aligned}\operatorname{Re}(\eta_1+\xi_1,u_{xx}) &= -\varphi_\rho\operatorname{Re}(g(u),u_{xx}) + \operatorname{Re}(\xi_1,u_{xx}) \\ &\quad - (1-\varphi_\rho)\left(9\gamma+\delta_r+9\frac{\lambda_r+i\lambda_i}{\gamma}\right)\operatorname{Re}(|u|^4u,u_{xx}),\end{aligned}$$

$$\begin{aligned}\operatorname{Re}(|u|^4u, u_{xx}) &= 3\int_0^L |u|^4|u_x|^2\mathrm{d}x + 2\int_0^L |u|^2u^2\bar{u}_x^2\mathrm{d}x\\ &\geq \int_0^L |u|^4|u_x|^2\mathrm{d}x,\end{aligned}$$

$$\begin{aligned}\operatorname{Re}(\xi_1, u_{xx}) &\leq 3\gamma\int_0^L |u|^2|\omega - \widetilde{\omega}|\mathrm{d}x\\ &\leq 9\gamma\int_0^L |u|^4|\omega - \widetilde{\omega}|^2\mathrm{d}x + \frac{\gamma}{4}|u_{xx}|_0^2.\end{aligned}$$

Together with the above calculation, we obtain

$$\begin{aligned}\frac{1}{2}\frac{\mathrm{d}}{\mathrm{d}t}|u_x|_0^2 \leq & -\frac{\gamma}{2}|u_{xx}|_0^2 + \varphi_\rho\cdot\frac{18}{\gamma}(\lambda_r^2+\lambda_i^2)\int_0^L |u|^4|u_x|^2\mathrm{d}x\\ & - 2\varphi_\rho\operatorname{Re}(g(u), u_{xx}) + \frac{2(\lambda_r^2+\lambda_i^2)}{\gamma}|(\omega_x - \overline{\omega})_x|_0^2\\ & + 9\gamma\int_0^L |u|^4|\omega - \widetilde{\omega}|^2\mathrm{d}x.\end{aligned} \tag{2.3.21}$$

Taking the inner product of equation (2.3.18) and $v - u_x$, and then taking its real part, we obtain

$$\begin{aligned}\frac{1}{2}\frac{\mathrm{d}}{\mathrm{d}t}|v - u_x|_0^2 = & -\gamma|(v-u_x)_x|_0^2 + \operatorname{Re}(\eta_2 - \eta_{1x}, v - u_x)\\ & + \operatorname{Re}(\xi_2 - \xi_{1x}, v - u_x) - k_1|v - u_x|_0^2.\end{aligned}$$

Let

$$\begin{aligned}g_1(u) = & (\beta_r + i\beta_i)(2|u|^2u_x + u^2\overline{u}_x)\\ & + (\delta_r + i\delta_i)(3|u|^4u_x + 2|u|^2u^2\overline{u}_x) - \chi u_x.\end{aligned}$$

It follows that

$$\begin{aligned}&\operatorname{Re}(\eta_2 - \eta_{1x}, v - u_x)\\ &= \varphi_\rho\operatorname{Re}((g_1(u) - h(u,v), v - u_x)) - (1-\varphi_\rho)\Big(9\gamma + \delta_r + 9\frac{\lambda_r^2+\lambda_i^2}{\gamma}\Big)\\ &\times \operatorname{Re}\int_0^L (3|u|^4|v-u_x|^2 + 2|u|^2u^2(\bar{v} - \bar{u}_x)^2)\mathrm{d}x\end{aligned}$$

$$
\begin{aligned}
&\leq \varphi_\rho \operatorname{Re}((g_1(u) - h(u,v), v - u_x)) \\
&\quad - (1-\varphi_\rho)\left(9\gamma + \delta_r + 9\frac{\lambda_r^2 + \lambda_i^2}{\gamma}\right)\int_0^L |u|^4 |v - u_x|^2 \mathrm{d}x,
\end{aligned}
$$

$$
\begin{aligned}
&\operatorname{Re}(\xi_2 - \xi_{1x}, v - u_x) \\
&\quad = \gamma\left[\operatorname{Re}\int_0^L (2\overline{u}(v+u_x)(\omega - \widetilde{\omega})(\overline{v} - \overline{u}_x) + 2u(\overline{v} + \overline{u}_x)\overline{(\omega - \widetilde{\omega})}(\overline{v} - \overline{u}_x) \right. \\
&\quad\quad \left. + 2u(\overline{v} + \overline{u}_x)(\omega - \widetilde{\omega})(\overline{v} - \overline{u}_x))\mathrm{d}x\right] \\
&\quad\quad + \gamma \operatorname{Re}\left[\int_0^L (2|u|^2(\omega - \widetilde{\omega})_x + u^2\overline{(\omega - \widetilde{\omega})_x})(\overline{v} - \overline{u}_x)\mathrm{d}x\right],
\end{aligned}
$$

$$
\begin{aligned}
&\operatorname{Re}[2\overline{u}(v+u_x)(\omega - \widetilde{\omega})(\overline{v} - \overline{u})_x + 2u(v+u_x)\overline{(\omega - \widetilde{\omega})}(\overline{v} - \overline{u}_x)] \\
&\quad = 4\operatorname{Re}(\operatorname{Re}(\overline{u}(\omega - \widetilde{\omega})(|v|^2 - |u_x|^2 - v\overline{u}_x + \overline{v}u_x) \\
&\quad = 4\operatorname{Re}(\overline{u}(\omega - \widetilde{\omega}))(|v|^2 - |u_x|^2).
\end{aligned}
$$

Therefore, we have

$$
\begin{aligned}
\operatorname{Re}(\xi_2 - \xi_{1x}, v - u_x) &\leq 4\gamma \operatorname{Re}\int_0^L \overline{u}(\omega - \widetilde{\omega})(|v|^2 - |u_x|^2)\mathrm{d}x \\
&\quad + 2\gamma \operatorname{Re}\int_0^L u(\omega - \widetilde{\omega})(\overline{v}^2 - \overline{u}_x)^2\mathrm{d}x + 3\gamma\int_0^L |u|^2(\omega - \widetilde{\omega})_x\|v - u_x\|\mathrm{d}x \\
&\leq I + \frac{\gamma}{2}|(\omega - \widetilde{\omega})_x|_0^2 + \frac{9}{2}\gamma\int_0^L |u|^4|v - u_x|^2\mathrm{d}x,
\end{aligned}
$$

where

$$
I = 4\gamma \operatorname{Re}\int_0^L \overline{u}(\omega - \widetilde{\omega})(|v|^2 - |u_x|^2)\mathrm{d}x + 2\gamma \operatorname{Re}\int_0^L u(\omega - \widetilde{\omega})(\overline{v}^2 - \overline{u}_x^2)\mathrm{d}x.
$$

Thus, we arrive at

$$
\begin{aligned}
&\frac{\mathrm{d}}{\mathrm{d}t}|v - u_x|_0^2 \leq -2\gamma|(v-u_x)_x|_0^2 - 2k_1|v - u_x|_0^2 + 2I + \gamma|(\omega - \widetilde{\omega})_x|_0^2 \\
&\quad + \varphi_\rho\left(9\gamma\int_0^L |u|^4|v - u_x|^2\mathrm{d}x - (h(u,v) - g_1(u), v - u_x)\right). \qquad (2.3.22)
\end{aligned}
$$

Taking inner product of equation (2.3.19) and $\omega-\widetilde{\omega}$, and taking its real part, we obtain

$$\begin{aligned}\frac{1}{2}\frac{\mathrm{d}}{\mathrm{d}t}|\omega-\widetilde{\omega}|_0^2 &= -\gamma|(\omega-\widetilde{\omega})_x|_0^2-k_2|\omega-\widetilde{\omega}|_0^2\\ &\quad-\mathrm{Re}\left(\int_0^L(4\gamma u(|v|^2-|u_x|^2)-2\gamma\bar{u}(v^2-u_x^2)\overline{(\omega-\widetilde{\omega})})\mathrm{d}x\right)\\ &\quad-\mathrm{Re}\left(\int_0^L(2|u|^2(\lambda_r+i\lambda_i)(2|u|^2v+u^2v-\omega_x)\overline{(\omega-\widetilde{\omega})})\mathrm{d}x\right)\\ &\quad-\mathrm{Re}\left(\int_0^L u^2(\lambda_r-i\lambda_i)(2|u|^2\bar{v}+\bar{u}^2v-\overline{\omega}_x)\overline{(\omega-\widetilde{\omega})}\mathrm{d}x\right)\\ &\quad-\left(9\gamma+16(\lambda_r^2+\lambda_i^2)\left(1+\frac{1}{\gamma}\right)\right)\int_0^L|u|^4|\omega-\widetilde{\omega}|^2\mathrm{d}x.\end{aligned}$$

For the terms $2|u|^2v+u^2\bar{v}-\omega_x$ and $2|u|^2\bar{v}+\bar{u}^2v-\overline{\omega}_x$, we handle them by inserting $\widetilde{\omega}_x$ and $\overline{\widetilde{\omega}_x}$ ($\widetilde{\omega}=f(u)=|u|^2u$). We omit the specific steps. Finally, we arrive at

$$\begin{aligned}\frac{1}{2}\frac{\mathrm{d}}{\mathrm{d}t}|\omega-\widetilde{\omega}|_0^2 &\le -\gamma|(\omega-\widetilde{\omega})_x|_0^2-k_2|(\omega-\widetilde{\omega})|_0^2-I\\ &\quad+12\sqrt{\lambda_r^2+\lambda_i^2}\int_0^L|u|^4|v-u_x||\omega-\widetilde{\omega}|\mathrm{d}x\\ &\quad+4\sqrt{\lambda_r^2+\lambda_i^2}\int_0^L|u|^2|\omega-\widetilde{\omega}||(\omega-\widetilde{\omega})_x|\mathrm{d}x\\ &\quad-\left(9\gamma+16(\lambda_r^2+\lambda_i^2)\left(1+\frac{1}{\gamma}\right)\right)\int_0^L|u|^4|\omega-\widetilde{\omega}|^2\mathrm{d}x\\ &\le -\frac{3}{4}\gamma|(\omega-\widetilde{\omega})_x|_0^2-k_2|\omega-\widetilde{\omega}|_0^2-I\\ &\quad+9\int_0^L|u|^4|v-u_x|^2\mathrm{d}x-9\gamma\int_0^L|u|^4|\omega-\widetilde{\omega}|^2\mathrm{d}x.\end{aligned}\tag{2.3.23}$$

Adding equation (2.3.20) to (2.3.23), we get

$$\begin{aligned}&\frac{\mathrm{d}}{\mathrm{d}t}(|u|_0^2+|u_x|_0^2+|v-u_x|_0^2+|\omega-\widetilde{\omega}|_0^2)\\ &\le -|u|_0^2+P-\alpha|u_x|_0^2+\left(\frac{\gamma}{2}+\frac{\lambda_r^2+\lambda_i^2}{\alpha}-2k_2\right)|\omega-\widetilde{\omega}|_0^2\\ &\quad+18\varphi_p\gamma\int_0^L|u|^6\mathrm{d}x-\frac{\gamma}{2}|u_{xx}|_0^2+18\varphi_p\frac{\lambda_r^2+\lambda_i^2}{\gamma}\int_0^L|u|^4|u_x|^2\mathrm{d}x\end{aligned}$$

$$- 2\varphi_\rho \operatorname{Re}(g(u), u_{xx}) + 18\gamma\varphi_\rho \int_0^L |u|^4 |v - u_x|^2 \mathrm{d}x$$
$$+ \left(\frac{\lambda_r^2 + \lambda_i^2}{\gamma} - \frac{\gamma}{2}\right) |(\omega - \widetilde{\omega})_x|_0^2 - 2k_1 |v - u_x|_0^2 - 2\gamma |(v - u_x)_x|_0^2. \qquad (2.3.24)$$

For a given $\rho > 0$, without loss of generality, suppose $|u|_0^2 + |u_x|_0^2 + |v|_0^2 + |\omega|_0^2 \geq 2\rho^2$, otherwise Proposition 2.3.4 is proved. At this time, $\varphi_\rho = 0$, where $\varphi_\rho = \varphi(\frac{|u|_0^2+|u_x|_0^2+|v|_0^2+|\omega|_0^2}{\rho^2})$. Hence, when $k_1 > 0$, $k_2 > \gamma + \frac{\lambda_r^2+\lambda_i^2}{2\alpha}$, $\gamma > 2\sqrt{\lambda_r^2 + \lambda_i^2}$, from equation (2.3.24) we deduce that $|u|_0^2 + |u_x|_0^2 + |v - u_x|_0^2 + |\omega - \widetilde{\omega}|_0^2$ exponentially decays, and is uniformly bounded. Also

$$|u|_0^2 + |u_x|_0^2 + |v|_0^2 + |\omega|_0^2 \leq |u|_0^2 + 2|u_x|_0^2 + |v - u_x|_0^2 + |\omega - \widetilde{\omega}|_0^2 + |\widetilde{\omega}|_0^2,$$
$$|\widetilde{\omega}| = \int_0^L |u|^6 \mathrm{d}x.$$

By using Sobolev embedding theorem,

$$\int_0^L |u|^6 \mathrm{d}x \leq C(|u_x|_0^2 + |u|_0^2)^3$$

and there exists $\rho_4 \geq 2\rho$ such that

$$\overline{\lim_{t\to\infty}} \{|u|_0^2 + |u_x|_0^2 + |v|_0^2 + |\omega|_0^2\} \leq \rho_4^2.$$

Thus Proposition 2.3.4 has been proved. □

As corollaries of Proposition 2.3.4, we get that the solution of equation (2.3.7) exponentially converges to the set $J(V_1)$.

Corollary 2.3.1. *Under the assumptions of Proposition* 2.3.4, *if* $U(t)$ *is the solution of equation* (2.3.7), *then there exist* $T_1, K_1 > 0$ *such that, when* $t \geq T_1$, *we have*

$$\frac{\mathrm{d}}{\mathrm{d}t}(|v - u_x|_0^2 + |\omega - \widetilde{\omega}|_0^2) \leq (K_1 - 2k_1)|v - u_x|_0^2 - k_2|\omega - \widetilde{\omega}|_0^2.$$

Proof. Adding equation (2.3.22) to (2.3.23), we get

$$\frac{\mathrm{d}}{\mathrm{d}t}(|v - u_x|_0^2 + |\omega - \widetilde{\omega}|_0^2) \leq -2\gamma|(v - u_x)_x|_0^2 - 2k_1|v - u_x|_0^2 - \frac{3}{2}\gamma|(\omega - \widetilde{\omega})_x|_0^2$$
$$- 2k_2|\omega - \widetilde{\omega}|_0^2 + 18\gamma\varphi_\rho \int_0^L |u|^4 |v - u_x|^2 \mathrm{d}x.$$

By virtue of

$$\int_0^L |u|^4|v-u_x|^2 dx \le |u|_{L^\infty}^4 |v-u_x|_0^2,$$

due to Proposition 2.3.4 and inclusion $H^1_{\text{per}}(I) \hookrightarrow L^\infty(I)$, we know that there exists $T_1 > 0$ such that

$$|u|_{L^\infty} \le C_1, \quad t \ge T_1.$$

Hence $k_1 = 18\gamma C_1^4$, and Corollary 2.3.1 has been proved. □

In the following, we will prove that the solution of equation (2.3.7) is uniformly bounded in $D(A^{\frac{1}{2}})$ and has the global attractor in $D(A^{\frac{1}{2}})$.

Proposition 2.3.5. *When $k_1 > \frac{K_1}{2}$, $k_2 > \gamma + \frac{\lambda_r^2+\lambda_i^2}{2\alpha}$, $\gamma > 2\sqrt{\lambda_r^2+\lambda_i^2}$, the solution of equation (2.3.7) is uniformly bounded in $D(A^{\frac{1}{2}})$ and has the global attractor $\mathscr{A}$ in $D(A^{\frac{1}{2}})$.*

Proof. Taking the inner product of equation (2.3.7) and v, u, ω, respectively, then taking its real parts, and using Hölder inequality and Proposition 2.3.4, we obtain

$$\frac{1}{2}\frac{d}{dt}|u|_0^2 \le -\gamma|u_x|_0^2 + \sqrt{\lambda_r^2+\lambda_i^2}|\omega_x|_0|u|_0 + C_1,$$

$$\frac{1}{2}\frac{d}{dt}|v|_0^2 \le -\gamma|v_x|_0^2 + \sqrt{\lambda_r^2+\lambda_i^2}|\omega_x|_0|v|_0 + C_2,$$

$$\frac{1}{2}\frac{d}{dt}|\omega|_0^2 \le -\gamma|\omega_x|_0^2 + C_3(|v|_{L^4}^2 + 1),$$

where C_i, $i = 1,2,3$, only depend on the coefficients of the equation and the uniform bound of Proposition 2.3.4. By Sobolev embedding theorem and Young inequality, we have

$$|v|_{L^4}^2 \le C|v|_0|v|_1 \le C_1(|v|_0^2 + |v|_0|v_x|_0) \le C_4(1+|v_x|_0), \tag{2.3.25}$$

which yields

$$\frac{1}{2}\frac{d}{dt}(|u|_0^2+|v|_0^2+|\omega|_0^2) \le -\left(\gamma|u_x|_0^2 + \frac{\gamma}{4}|v_x|_0^2 + \left(\gamma - \frac{\lambda_r^2+\lambda_i^2}{\gamma}\right)|\omega_x|_0^2\right) + C_5. \tag{2.3.26}$$

Taking the inner product of the three equation of equations (2.3.7) and $-u_{xx}$, $-v_{xx}$, $-\omega_{xx}$, respectively, then taking its real parts, and using Hölder inequality, we obtain

$$\frac{1}{2}\frac{d}{dt}|u_x|_0^2 \le -\gamma|u_{xx}|_0^2 + \sqrt{\lambda_r^2+\lambda_i^2}|\omega_x|_0|u_{xx}|_0 + C_6|u_{xx}|_0,$$

$$\frac{1}{2}\frac{\mathrm{d}}{dt}|v_x|_0^2 \le -\gamma|v_{xx}|_0^2 + \sqrt{\lambda_r^2+\lambda_i^2}|\omega_{xx}|_0|v_{xx}|_0 + C_7|v_{xx}|_0,$$

$$\frac{1}{2}\frac{\mathrm{d}}{dt}|\omega_x|_0^2 \le -\gamma|\omega_{xx}|_0^2 + C_8(|v|_{L^4}^2+1)|\omega_{xx}|_0.$$

Using Young inequality and equation (2.3.25), we get

$$\frac{1}{2}\frac{\mathrm{d}}{dt}(|u_x|_0^2+|v_x|_0^2+|\omega_x|_0^2) \le -\frac{1}{2}\left(\gamma - \frac{\lambda_r^2+\lambda_i^2}{\gamma}\right)|\omega_{xx}|_0^2 + \frac{\lambda_r^2+\lambda_i^2}{2\gamma}|\omega_x|_0^2 + C_9|v_x|_0^2 + C_{10}. \tag{2.3.27}$$

From equations (2.3.26)–(2.3.27), assumption $\gamma > 2\sqrt{\lambda_r^2+\lambda_i^2}$ and uniform Gronwall inequality, we get that U is uniformly bounded in $D(A^{\frac{1}{2}})$. As for the existence of the global attractor, which can be established as in [197, Th. 1.1] with operator A being sectoral operator (it generates the semigroup which is compact in $D(A^{\frac{1}{2}})$ when $t>0$). Similarly as C_1, C_2, C_3, the above C_i, $i=4,\dots,10$, only depend on the coefficients of the equation and the uniform bound of Proposition 2.3.4. □

Using Corollary 2.3.1, we can prove the following lemma:

Lemma 2.3.2. *Assume that* $k_1 > \frac{K_1}{2}$, $k_2 > \gamma + \frac{\lambda_r^2+\lambda_i^2}{2\alpha}$, $\gamma > 2\sqrt{\lambda_r^2+\lambda_i^2}$. *Then* $\mathscr{A}$ *is included into the set* $J(V_1)$.

Proof. First we note that there exists a constant ρ_5 such that

$$|v-u_x|_0^2 + |\omega - f(u)|_0^2 \le \rho_5^2, \quad \forall (u,v,\omega)\in\mathscr{A}.$$

Now set $U=(U_0,t)=(u(t),v(t),\omega(t))^t$ to be the solution of equation (2.3.7) with the initial value $U_0\in\mathscr{A}$. By Corollary 2.3.1, we get

$$\begin{aligned}
&|v(t)-u_x(t)|_0^2 + |\omega(t)-\widetilde{\omega}(t)|_0^2 \\
&\le e^{-\overline{\mu}t}(|v(T_1)-u_x(T_1)|_0^2 + |\omega(T_1)-\widetilde{\omega}(T_1)|_0^2) \\
&\le \rho_5^2 e^{-\overline{\mu}t} \quad \text{(due to the invariance of } \mathscr{A}\text{)}.
\end{aligned} \tag{2.3.28}$$

For $t\ge T_1$, here $\overline{\mu}=\min(2k_1-K_1, 2k_2)$. Furthermore, by the invariance of $\mathscr{A}$, for any $U^*\in\mathscr{A}$ and $t\ge T_1$, there exists U_0 such that $U(U_0,t)=U^*$. Hence by equation (2.3.28), we obtain

$$\begin{aligned}
&\operatorname{dist}(U^*, J(V_1)) \le \rho_5^2 e^{-\overline{\mu}t}, \quad t\ge T_1, \\
&\operatorname{dist}(\mathscr{A}, J(V_1)) = \sup_{U\in\mathscr{A}} \operatorname{dist}(U, J(V_1)) \le \rho_5^2 e^{-\overline{\mu}t}, \quad t\ge T_1, \\
&\operatorname{dist}(\mathscr{A}, J(V_1)) = 0.
\end{aligned}$$

Next we only need to prove that $J(V_1)$ is closed in $\mathscr{H}$, which can be deduced from $\mathscr{A} \subset J(V_1)$. In fact, if $(u_n, (u_n)_x, f(u_n)) \in \mathscr{H}$ converges to $(u, v, \omega) \in \mathscr{H}$, then v is the weak derivative of $u, v \in H$, and it follows that

$$|v|_0 \le \left|v - (u_n)_x\right|_0 + \left|(u_n)_x\right|_0 < \infty,$$

and

$$\left|\omega - f(u)\right|_0 \le \left|\omega - f(u_n)\right|_0 + \left|f(u_n) - f(u)\right|_0,$$

$$\begin{aligned}\left|f(u_n) - f(u)\right|_0 &= \left(\int_0^L (|u|^6 - |u|^6)^2 \mathrm{d}x\right)^{\frac{1}{2}}\\ &\le (|u|_{L^\infty}^2 + |u_n|_{L^\infty}|u|_{L^\infty} + |u_n|_{L^\infty}^2)\\ &\quad \times (|u_n|_{L^\infty}^3 + |u|_{L^\infty}^3)|u_n - u_0|_0.\end{aligned}$$

Since $|u_n|_{L^\infty}$ is uniformly bounded, then so is $|u|_{L^\infty}$. Hence $\omega = f(u)$. The lemma has been proved. □

In the following, we give the main result:

Theorem 2.3.2. *Assume that* $k_1 > \frac{K_1}{2}$, $k_2 > \gamma + \frac{\lambda_r^2 + \lambda_i^2}{2\alpha}$, $\gamma > 2\sqrt{\lambda_r^2 + \lambda_i^2}$. *Then* $\mathscr{A} = J(\mathscr{A}_\rho)$.

Proof. It is obvious that $J(\mathscr{A}_\rho) \subset \mathscr{A}$, and $\mathscr{A}$ is the invariant set in $J(V_1)$, but since $\mathscr{A}_\rho$ is the global attractor of equation (2.3.5), $J(\mathscr{A}_\rho)$ is the maximal invariant set in $J(V_1)$. Hence, $\mathscr{A} \subset J(\mathscr{A}_\rho)$, and it follows that $\mathscr{A} = J(\mathscr{A}_\rho)$. □

Next, we consider the existence of inertial manifold and inertial form.

According to the previous discussion, we know that equation (2.3.7) keeps the long time dynamic behavior of equation (2.3.2). In particular, by virtue of Propositions 2.3.4 and 2.3.5, there exists a constant $\rho_6 > 0$ such that

$$\left|\mathbf{A}^{\frac{1}{2}} U\right|_0 \le \rho_6, \quad \forall U \in \mathscr{A}.$$

In order to prove the existence of an inertial manifold for equation (2.3.7), we truncate the nonlinear term $\widetilde{F}$. That is, considering

$$\frac{\mathrm{d}U}{\mathrm{d}t} + \mathbf{A}U = F(U) \tag{2.3.29}$$

where $F(U) = \varphi(\frac{|A^{\frac{1}{2}}U|_0^2}{\rho_6})\widetilde{F}(U)$, we can see that equations (2.3.29) and (2.3.7) for the global attractor $\mathscr{A}$ have the same long time behavior. For simplicity, here we just illustrate the conclusion.

Proposition 2.3.6. *Under the above assumptions, equation* (2.3.29) *has the inertial manifold* $\mu = \operatorname{graph}\Phi$, *where* $\Phi : P\mathscr{H} \to Q\mathscr{H} \cap D(A^{\frac{1}{2}})$ *is a Lipschitz mapping; P is the* *projection onto the subspace spanned by the first $N+1$ eigenvectors of* $\mathbf{A}$ *in* $\mathscr{H}$*;* $Q = I - P$*; and* $\mathbf{A}$ *is the differential operator having the following form:*

$$\mathbf{A} = \begin{pmatrix} A & 0 & (\lambda_r + i\lambda_i)\partial_x \\ -k_1\partial_x & A & -(\lambda_r + i\lambda_i)\partial_{xx} \\ 0 & 0 & A \end{pmatrix}.$$

Similar to the discussion of Lemma 2.3.1, $\mathbf{A}$ has discrete spectrum. By the classical functional analysis, $P\mathscr{H}$ and $Q\mathscr{H}$ are invariant under the action of A. Because $\mathbf{A}$ is not a self-adjoint operator, we can prove Proposition 2.3.6 using the results of [191]. At this time, the spectrum of $\mathbf{A}$ is $\{(\frac{2\pi n}{L})^2\}_{n\geq 1}$, so the gap condition is satisfied.

Corollary 2.3.2. *The essential long time behavior dynamics for the semigroup of equation* (2.3.2) *can be described by the following ordinary differential equation:*

$$\frac{\mathrm{d}}{\mathrm{d}t}PU = -APU + PF(PU + \Phi(PU)). \tag{2.3.30}$$

From the exponential attractivity property of inertial manifold, we have

Corollary 2.3.3. *If $u(t)$ is a solution of equation* (2.3.2)*, then there exists a solution $PU(t)$ of ordinary differential equation* (2.3.30) *which satisfies (C, α are positive constants)*

$$\left|u(t) - (P_1(t) + \Phi_1(PU))\right| \leq Ce^{-\alpha t}, \quad \forall t > 0,$$

where

$$PU(t) = (P_1(t), P_2(t), P_3(t)), \quad \Phi = (\Phi_1, \Phi_2, \Phi_3).$$

In the following, we give the corresponding conclusion of Kwak. Define

$$S = \mathscr{U} \cap J(V_1).$$

We denote S^u as the first component of S.

Proposition 2.3.7. *The set S^u is invariant under the action of solution semigroup of equation* (2.3.2) *and it attracts all the orbits of equation* (2.3.2).

We cannot prove that S^u is a finite-dimensional manifold, but can prove $\mathscr{U}^u$ (the first component of $\mathscr{U}$) is a finite-dimensional manifold which attracts all bounded set in $D(A^{\frac{1}{2}})$, and it has no invariance under the action of the semigroup.

Finally, we provide a lower dimensional ordinary differential equation than equation (2.3.30) to describe the dynamics of S^u.

Theorem 2.3.3. *The essential long time dynamics of the solution semigroup of equation* (2.3.2) *can be completely described by the following ordinary differential equation:*

$$\frac{\mathrm{d}}{\mathrm{d}t}PU + APU = PF(J(u))$$

where $PU = (P_1, P_{1x}, P_3)$, $u(t) = P_1(t) - \Phi_1(P_1(t), P_{1x}(t), P_3(t))$.

2.4 The existence of inertial manifolds for the generalized KS equation

Consider the existence of inertial manifold and its dimension estimation of the periodic initial value problem for the following generalized Kuramoto–Sivashinsky equation (GKS) [59]:

$$u_t + \alpha u_{xx} + \gamma u_{xxx} + f(u)_x + \varphi(u)_{xx} = g(u) + h(x), \quad (x,t) \in \mathbf{R} \times \mathbf{R}_+, \tag{2.4.1}$$

$$u|_{t=0} = u_0(x), \quad x \in \mathbf{R}, \tag{2.4.2}$$

$$u(x+D,t) = u(x-D,t), \quad \forall x \in \mathbf{R},\ t \in \mathbf{R}_+,\ D > 0, \tag{2.4.3}$$

where $\alpha \geq 0, \gamma > 0$. Obviously, when $\alpha = 0, f(u) = 0$, equation (2.4.1) is the well-known generalized Cahn–Hilliard equation. For the problem (2.4.1)–(2.4.3) we will give a time t a priori estimate and will then prove the invariant cone property (ICP) and strong squeezing property (SSP) of the GKS equation, giving the existence of inertial manifolds for the prepared GKS equation (PGKS). Finally, we will prove the existence of an inertial manifold for the periodic initial value problem of GKS equation.

Lemma 2.4.1. *If the following conditions are satisfied:*

(1) $\varphi'(u) \leq \varphi_0, \varphi_0 > 0$;

(2) $g(0) = 0, g'(u) \leq g_0, g_0 < -\frac{\alpha+\varphi_0+1}{2}, \gamma > \frac{1}{2}(\alpha + \varphi_0)$;

(3) $h(x) \in L_2(\Omega), \Omega = (-D, D)$;

(4) $u_0(x) \in L_2(\Omega)$,

then for any smooth solution of problem (2.4.1)–(2.4.3), *we have the following estimate:*

$$\|u(\cdot,t)\|^2_{L_2(\Omega)} \leq e^{2(g_0+\frac{\alpha+\varphi_0+1}{2})t}\|u_0(x))\|^2_{L_2(\Omega)} + \frac{\|h(x)\|^2_{L_2(\Omega)}}{|2g_0 + \alpha + \varphi_0 + 1|}(1 - e^{2(g_0+\frac{\alpha+\varphi_0+1}{2})t})$$

$$0 \leq t < \infty.$$

Furthermore, we have

$$\overline{\lim_{t\to\infty}}\ \|u(\cdot,t)\|^2_{L_2(\Omega)} + \overline{\lim_{t\to\infty}}\ \frac{1}{t}\int_0^1 \|u_{xx}(\cdot,s)\|^2_{L_2(\Omega)}ds \leq E_0, \tag{2.4.4}$$

where the constant E_0 *depends on* $\|u_0(x)\|^2_{L_2(\Omega)}$ *and* $\|h_0(x)\|_{L_2(\Omega)}$.

Proof. Taking the inner product of equation (2.4.1) and u, we obtain

$$(u, u_t + \alpha u_x + \gamma u_{xxxx} + f(u)_x + \varphi(u)_{xx} - g(u) - h(x)) = 0, \tag{2.4.5}$$

where

$$(f(u)_x, u) = -(f(u), u_x) = 0,$$

$$(\varphi(u)_{xx}, u) = -(\varphi'(u)u_x, u_x) \geq -\varphi_0\|u_x\|_{L_2}^2,$$
$$(u, g(u)) \leq g_0\|u\|_{L_2}^2,$$
$$(u, h(x)) \leq \frac{1}{2}(\|u\|_{L_2}^2 + \|h\|_{L_2}^2).$$

For the periodic function $u(\cdot, t)$, we have

$$\|u_x\|_{L_2}^2 \leq \|u\|_{L_2}\|u_{xx}\|_{L_2} \leq \frac{1}{2}(\|u\|_{L_2}^2 + \|u_{xx}\|_{L_2}^2).$$

Therefore, from (2.4.5) we have

$$\frac{1}{2}\frac{\mathrm{d}}{\mathrm{d}t}\|u\|_{L_2}^2 + \left(\gamma - \frac{\alpha + \varphi_0}{2}\right)\|u_{xx}\|_{L_2}^2 \leq \left(g_0 + \frac{\alpha + \varphi_0 + 1}{2}\right)\|u\|_{L_2}^2 + \frac{1}{2}\|h(x)\|_{L_2}^2. \tag{2.4.6}$$

Since $\gamma - \frac{\alpha+\varphi_0}{2} > 0$, using

$$\frac{\mathrm{d}}{\mathrm{d}t}\|u(\cdot, t)\|_{L_2}^2 \leq 2\left(g_0 + \frac{\alpha + \varphi_0 + 1}{2}\right)\|u(\cdot, t)\|_{L_2}^2 + \|h(x)\|_{L_2}^2,$$

it is then easy to get

$$\|u(\cdot, t)\|_{L_2}^2 \leq e^{(2g_0+\alpha+\varphi_0+1)t}\|u_0(x)\|_{L_2}^2 + \frac{\|h(x)\|_{L_2(\Omega)}^2}{|2g_0 + \alpha + \varphi_0 + 1|}(1 - e^{(2g_0+\alpha+\varphi_0+1)t})$$

Integrating equation (2.4.6) with respect to time, we get the estimation

$$\overline{\lim_{t\to\infty}} \frac{1}{t}\int_0^t \|u_{xx}(\cdot, s)\|_{L_2}^2 \mathrm{d}s \leq E_0. \qquad \square$$

Lemma 2.4.2. *Suppose the conditions of Lemma 2.4.1 are met, and let*

(1) $|f(u)| \leq A|u|^p, 1 \leq p < 7, A > 0,$

$$|\varphi'(u)| \leq B|u|^q, \quad 0 \leq q < 4,\ B > 0;$$

(2) $u_0(x) \in H^1(\Omega)$, $h(x) \in L_2(\Omega)$.

Then we have the following estimate of a solution for problem (2.4.1)–(2.4.3)*:*

$$\|u_x(\cdot, t)\|_{L_2(\Omega)}^2 \leq e^{g_0 t}\|u_{0x}(x)\|_{L_2(\Omega)}^2 + \frac{2}{|g_0|}\max_{t\in[0,\infty)}|C_5(D, p,\ \ \|u(\cdot, t)\|_{L_2}, \|h(x)\|_{L_2})|, \quad \forall t \geq 0, \tag{2.4.7}$$

where the function $C_5(\cdot, \cdot, \xi, \eta)$ is continuous and increasing in the variables ξ, η. Furthermore, we have

$$\overline{\lim_{t\to\infty}}\|u_x(\cdot, t)\|_{L_2(\Omega)}^2 + \overline{\lim_{t\to\infty}}\frac{1}{t}\int_0^t \|u_{xxx}(\cdot, s)\|_{L_2}^2 \mathrm{d}s \leq E_1, \tag{2.4.8}$$

where the constant E_1 depends on $\|u_0(x)\|_{H^1(\Omega)}$, $\|h(x)\|_{L^2(\Omega)}$ and E_0.

Proof. Taking the inner product of equation (2.4.1) and u_{xx}, this gives

$$(u_{xx}, u_t + \alpha u_{xx} + \gamma u_{xxxx} + f(u)_x + \varphi(u)_{xx} - g(u) - h(u)) = 0, \tag{2.4.9}$$

where

$$\begin{aligned}|(f(u)_x, u_{xx})| &= |(f(u), u_{xxx})| \le \|f(u)\|_{L_2}\|u_{xxx}\|_{L_2}\\ &\le \frac{\gamma}{6}\|u_{xxx}\|^2_{L_2} + \frac{3}{2\gamma}\|f(u)\|^2_{L_2}.\end{aligned} \tag{2.4.10}$$

Using the assumptions of the lemma and Sobolev estimate for a smooth periodic function [88], we have

$$\begin{aligned}&\|u(x)\|_{L_\infty(\Omega)} \le \frac{1}{\sqrt{D}}\|u\|_{L_2} + \sqrt{2}\|u\|^{\frac{5}{6}}_{L_2}\|u_{xx}\|^{\frac{1}{6}}_{L_2},\\ &\|u(x)\|_{L_\infty(\Omega)} \le (2D)^{\frac{1}{2}}\|u_{xx}\|_{L_2},\\ &\|D^j_x u\|_{L_\infty(\Omega)} \le \|D^{j-1}_x u\|^{\frac{1}{2}}_{L_\infty(\Omega)}\|D^{j+1}_x\|^{\frac{1}{2}}_{L_\infty(\Omega)}, \quad j = 1, 2, \dots .\end{aligned}$$

Then we use the generalized Young inequality

$$ab \le \varepsilon^p \frac{a^p}{p} + \varepsilon^{-q}\frac{b^q}{q}, \quad \frac{1}{p} + \frac{1}{q} = 1, \quad p, q > 1,\ \varepsilon, a, b > 0,$$

which yields

$$\begin{aligned}\frac{3}{2\gamma}|f(u)|^2_{L_2} &\le \frac{3}{2\gamma}A_2\|u\|^{2p}_{L_{2p}} \le \frac{3}{2\gamma}A_2\|u\|^{2p-2}_{L_2}\|u\|^2_{L_2}\\ &\le \frac{\gamma}{6}\|u_{xxx}\|^2_{L_2} + c_1\|u\|^{2p}_{L_2} + c_2\|u\|^{\frac{2(5p+1)}{7-p}}_{L_2},\end{aligned} \tag{2.4.11}$$

where

$$\begin{aligned}&c_1 = \frac{3A^2(p-1)}{\gamma D^{p-1}},\\ &c_2 = (7-p)\cdot 2^{\frac{(7p-13)}{7-p}}(p-1)^{\frac{5+p}{7-p}}\cdot 3^{\frac{p-1}{7-p}}\cdot A^{\frac{12}{7-p}}\cdot \gamma^{-\frac{5+p}{7-p}}\\ &|(\varphi(u)_{xx}, u_{xx}| = |(\varphi'(u)u_x, u_{xxx})| \le \frac{\gamma}{6}\|u_{xxx}\|^2 + \frac{3}{2\gamma}\|\varphi'(u)u_x\|^2_{L_2},\end{aligned}$$

$$\begin{aligned}\frac{3}{2\gamma}\|\varphi'(u)u_x\|^2_{L_2} &\le \frac{3}{2\gamma}\|\varphi'(u)\|^2_{L_\infty}\cdot\|u_x\|^2_{L_2}\\ &\le \frac{\gamma}{6}\|u_{xxx}\|^2_{L_2} + c_3\|u\|^{3p-2}_{L_2} + c_4\|u\|^{\frac{2(5p+4)}{4-q}}_{L_2},\end{aligned} \tag{2.4.12}$$

where

$$\begin{aligned}&c_3 = 4\sqrt{3}\gamma^{-2}q^{\frac{3}{2}}B^3D^{-\frac{4}{3}q},\\ &c_4 = 2^{\frac{7p+2}{4-q}}3^{\frac{6}{4-q}}\gamma^{-\frac{q+8}{4-q}}B^{\frac{12}{7-p}}q^{\frac{6}{4-q}}(q+2)^{\frac{2+q}{4-q}},\end{aligned}$$

$$-(g(u), u_{xx}) = (g'(u)u_x, u_x) \le g_0\|u_x\|_{L_2}^2, \tag{2.4.13}$$

$$\begin{aligned}|(h(x), u_{xx})| &\le \|h(x)\|_{L_2}\|u_{xx}\|_{L_2}\\ &\le \|h(x)\|_{L_2}\|u\|_{L_2}^{\frac{1}{2}}\|u_{xxx}\|_{L_2}^{\frac{2}{3}}\\ &\le \frac{\gamma}{6}\|u_{xxx}\|_{L_2}^2 + \frac{2}{3}\left(\frac{\gamma}{2}\right)^{-\frac{q}{2}}\|u\|_{L_2}^{\frac{1}{2}}\|h(x)\|_{L_2}^{\frac{3}{2}}.\end{aligned} \tag{2.4.14}$$

Using Young inequality,

$$\begin{aligned}\alpha\|u_{xx}\|_{L_2}^2 &\le \alpha\|u_{xx}\|_{L_2}^{\frac{2}{3}}\|u_{xxx}\|_{L_2}^{\frac{4}{3}}\\ &\le \frac{\gamma}{6}\|u_{xxx}\|_{L_2}^2 + \frac{\alpha^3}{3}\left(\frac{\gamma}{4}\right)^{-2}\|u\|_{L_2}^2,\end{aligned} \tag{2.4.15}$$

and then from equations (2.4.9)–(2.4.15) we deduce

$$\frac{1}{2}\frac{\mathrm{d}}{\mathrm{d}t}\|u_x(\cdot,t)\|_{L_2}^2 + \frac{\gamma}{6}\|u_{xxx}\|_{L_2}^2 \le g_0\|u_x\|_{L_2}^2 + 2c_5(p, D, q, \|u\|_{L_2}, \|h\|_{L_2}), \tag{2.4.16}$$

where

$$\begin{aligned}&c_5(p, q, D, \|u\|_{L_2}, \|h\|_{L_2})\\ &= c_1\|u\|_{L_2}^{2p} + c_2\|u\|_{L_2}^{\frac{2(5p+1)}{7-p}} + c_3\|u\|_{L_2}^{3q+2} + c_4\|u\|_{L_2}^{\frac{2(5p+4)}{4-p}}\\ &\quad + \frac{2}{3}\left(\frac{\gamma}{2}\right)^{-\frac{q}{2}}\|u\|_{L_2}^{\frac{1}{2}}\|h\|_{L_2}^{\frac{3}{2}} + \frac{\alpha^3}{3}\left(\frac{\gamma}{4}\right)^{-2}\|u\|_{L_2}^2.\end{aligned}$$

From equation (2.4.16) we get

$$\|u_x(\cdot,t)\|_{L_2(\Omega)}^2 \le e^{2g_0t}\|u_{0x}(x)\|_{L_2(\Omega)}^2 + \frac{2}{|g_0|}\max_{t\in[0,\infty)}|C_5(p, D, q, \|u\|_{L_2}, \|h\|_{L_2})|.$$

By Lemma 2.4.1 and equation (2.4.16), we get

$$\overline{\lim_{t\to\infty}}\|u_x(\cdot,t)\|_{L_2(\Omega)}^2 + \overline{\lim_{t\to\infty}}\frac{1}{t}\int_0^1\|u_{xxx}(\cdot,s)\|_{L_2(\Omega)}^2\mathrm{d}s \le E_1. \tag{2.4.17}$$

□

Lemma 2.4.3. *Assume that the conditions of Lemma 2.4.2 are met, and set $u_0(x) \in H^2(\Omega)$. Then for a smooth solution of problem (2.4.1)–(2.4.3), we have*

$$\|u_{xx}(\cdot,t)\|_{L_2(\Omega)}^2 \le e^{2g_0t}\|u_{0xx}(x)\|_{L_2(\Omega)}^2 + \frac{1}{|g_0|}(1 - e^{2g_0t})\left(\frac{3}{\gamma}\|h\|_{L_2(\Omega)}^2 + C_6\right), \tag{2.4.18}$$

where the continuous function C_6 depends on $\|u(\cdot,t)\|_{H^1(\Omega)}$. Further, we have

$$\overline{\lim_{t\to\infty}}\|u_{xx}(\cdot,t)\|_{L_2(\Omega)}^2 + \overline{\lim_{t\to\infty}}\frac{1}{t}\int_0^t\|u_{xxxx}(\cdot,s)\|_{L_2(\Omega)}^2\mathrm{d}s \le E_2, \tag{2.4.19}$$

where the constant E_2 depends on $\|u_{0xx}(x)\|_{L_2}$, $\|h(x)\|_{L_2}$ and E_1.

Proof. Similar to the proof of Lemmas 2.4.1 and 2.4.2. □

Lemma 2.4.4. *Suppose the conditions of Lemma 2.4.3 are met, and set $u_0(x) \in H^3(\Omega)$. Then for a smooth solution of problem* (2.4.1)–(2.4.3), *we have*

$$\|u_{xxx}(\cdot,t)\|^2_{L_2(\Omega)} \le e^{2g_0 t}\|u_0(x)\|_{H^3(\Omega)} + \frac{1}{|g_0|}\max_{0\le t<\infty}\big|C_7(\|u\|_{H^2(\Omega)}, \|h\|_{H^1(\Omega)})\big|. \tag{2.4.20}$$

Furthermore, we have

$$\overline{\lim_{t\to\infty}}\|u_{xxx}(\cdot,t)\|^2_{L_2(\Omega)} + \overline{\lim_{t\to\infty}}\frac{1}{t}\int_0^t \|u_{x5}(\cdot,s)\|^2_{L_2(\Omega)}ds \le E_3, \tag{2.4.21}$$

$$\overline{\lim_{t\to\infty}}\big(\|u(\cdot,t)\|^2_{L_\infty(\Omega)} + \|u_x(\cdot,t)\|_{L_\infty(\Omega)} + \|u_{xx}(\cdot,t)\|_{L_\infty(\Omega)}\big) \le E_3', \tag{2.4.22}$$

where constants E_3 and E_3' depend on $\|u_0(x)\|_{H^3(\Omega)}$, $\|h(x)\|_{H^1(\Omega)}$.

Proof. Inequalities (2.4.20)–(2.4.21) can be proved similarly as those of Lemmas 2.4.2 and 2.4.3. Inequality (2.4.22) is obtained similarly as in the proof of Lemmas 2.4.1–2.4.3, and (2.4.21) follows from Sobolev inequality. □

We consider ICK and SSP properties for the GKS equation (2.4.1).

Now we select an eigenfunction $W_k(x) = e^{i\frac{2\pi k}{D}x}$, which satisfies the characteristic equation

$$W_{kxxxx} = \lambda_k W_k,$$

where $\lambda_k = (\frac{2\pi k}{D})^4$, let P_N denote the orthogonal projection of the Hilbert space H to the subspace spanned by $\{W_1, W_2, \dots, W_N\}$, let Q_N be the infinite-dimensional projection, $Q_N = I - P_N$. Let $P_N H \oplus Q_N H$ denote the direct sum of the orthogonal projection P_N and Q_N; P_N's rank is $N < \infty$, $S(t)u$ denotes the orbit of equation (2.4.2) passing through $u_0(x)$ when $t = 0$.

Definition 2.4.1 (Invariant cone property). If for some invariant absorbing set B, the "cone" with index $N \in \{1, 2, \dots\}$, $\gamma \in (0, \infty)$,

$$C_N(\gamma) = \{(u,v) \in B \times B,\ \|Q_N(u-v)\|_H \le \gamma\|P_N(u-v)\|_H\} \tag{2.4.23}$$

is strictly invariant, that is, if

$$\|Q_N(u-v)\|_H = \gamma\|P_N(u-v)\|_H, \quad (u,v) \in B\times B,\ u \ne v, \tag{2.4.24}$$

then we have

$$\|Q_N(S(t)u - S(t)v)\|_H = \gamma\|P_N(S(t)u - S(t)v)\|_H, \quad \forall t > 0, \tag{2.4.25}$$

and say that it possesses the invariant cone property.

Definition 2.4.2 (Squeezing property). If there exists a constant $\beta > 0$ such that

$$\|(S(t)u - S(t)v\|_H \le e^{-\beta t}\|u - v\|_H, \quad \forall t \ge 0, \tag{2.4.26}$$

where $(u, v) \in B{\times}B$, $(S(t)u, S(t)v) \notin C_N(\gamma)$, then we call it the squeezing property, which means $B \times B \setminus C_N(\gamma)$ is exponentially squeezed.

Definition 2.4.3 (Strong squeezing property). If ICP and the squeezing property hold simultaneously, we call it the strong squeezing property, SSP for short.

For the GKS equation (2.4.1), setting the operator $A = \frac{d^4u}{dx^4}$ be defined on subspace

$$D_A = \{u \in V : \frac{d^4u}{dx^4} \in L_2(-D, D), \frac{d^3u}{dx^3} \in V\},$$

A is a self-adjoint operator in $L_2(\Omega)$, which satisfies $AW_k = \lambda_k W_k$, $k = 1, 2, \ldots,$ where $\lambda_k = (\frac{2\pi k}{D})^4$. It is easy to verify that

$$V = D_{A^{\frac{1}{4}}}, \quad \|u\|_1 = \|A^{\frac{1}{4}}u\|_{L_2} = \|A^{\frac{1}{4}}u\|, \quad u \in V,$$
$$A^{\frac{1}{2}}u = \frac{d^2u}{dx^2}, \quad u \in D_{A^{\frac{1}{2}}}, \quad \|u\|_3 = \|A^{\frac{3}{4}}u\|, \quad u \in D_{A^{\frac{3}{4}}}.$$

Then GKS equation (2.4.1) can be rewritten in the operator form:

$$\frac{du}{dt} + \gamma Au - \alpha A^{\frac{1}{2}}u + f(u)_x + \varphi(u)_{xx} = g(u) + h(x). \tag{2.4.27}$$

Lemma 2.4.5. *Let $u_1^0, u_2^0 \in B$ and assume $u_i^0(x)$, $i = 1, 2$, and $h(x)$ are odd periodic functions, $u_1(t) = S(t)u_1^0$, $u_2(t) = S(t)u_2^0$. Then we have the following estimate*

$$\|u_1(t) - u_2(t)\| \le \|u_1^0 - u_2^0\|e^{2Kt}, \quad \forall t \ge 0, \tag{2.4.28}$$

where $S(t)u_i^0$ means the semigroup generated by equation (2.4.27) with the initial value $u_i^0(x)$, $i = 1, 2$,

$$B = \bigcup_{t\ge 0} S(t)B_{2E},$$
$$B_{2E} = \{u \in D_{A^{\frac{3}{4}}}, \|u\|_3 \le 2(E_0 + E_1 + E_2 + E_3)\}, \tag{2.4.29}$$

and the constants are determined by equations (2.4.4), (2.4.8), (2.4.19), and (2.4.21), as well as

$$\begin{cases} K = \frac{1}{2}\sup_{0\le t<\infty}\left(\|R_{2x}\|_{L_\infty(\Omega)} + 2\|R_1\|_{L_\infty(\Omega)} + \frac{3}{4\gamma}\|R_3\|^2_{L_\infty(\Omega)}\right) + \frac{3\alpha^2}{4\gamma} - \frac{2\gamma}{3}\lambda_1, \\ \lambda_1 = \left(\frac{2\pi}{D}\right)^4, \\ R_1 = \int_0^1 f'_u(\tau u_1 + (1-\tau)u_2)\mathrm{d}\tau + \int_0^1 f''_{ux}(\tau u_1 + (1-\tau)u_2)\mathrm{d}\tau \\ \qquad + \int_0^1 g'_u(\tau u_1 + (1-\tau)u_2)\mathrm{d}\tau + \int_0^1 \varphi''_{ux}(\tau u_1 + (1-\tau)u_2)\mathrm{d}\tau, \\ R_2 = \int_0^1 f'_u(\tau u_1 + (1-\tau)u_2)\mathrm{d}\tau + 2\int_0^1 \varphi''_{ux}(\tau u_1 + (1-\tau)u_2)\mathrm{d}\tau, \\ R_3 = \int_0^1 \varphi'_u(\tau u_1 + (1-\tau)u_2)\mathrm{d}\tau. \end{cases} \tag{2.4.30}$$

Proof. Let $W = u_1 - u_2$. Then by equation (2.4.27) we get

$$\begin{aligned} &\frac{\mathrm{d}W}{\mathrm{d}t} + \gamma AW - \alpha A^{\frac{1}{2}}W + f(u_1)_x \\ &\quad = f(u_2)_x + \varphi(u_1)_{xx} - \varphi(u_2)_{xx} - g(u_1) + g(u_2) = 0, \end{aligned} \tag{2.4.31}$$

where

$$\begin{aligned} f(u_1) - f(u_2) &= W\int_0^1 f'_u(\tau u_1 + (1-\tau)u_2)\mathrm{d}\tau \\ (f(u_1) - f(u_2))_x &= W_x\int_0^1 f'_u\mathrm{d}\tau + W\int_0^1 f''_{ux}\mathrm{d}\tau, \\ f''_{ux} &= f''_{uu}(\tau u_1 + (1-\tau)u_2)(\tau u_{1x} + (1-\tau)u_{2x}) \\ (\varphi(u_1) - \varphi(u_2))_{xx} &= W_{xx}\int_0^1 \varphi'_u(\tau u_1 + (1-\tau)u_2)\mathrm{d}\tau \\ &\quad + 2\int_0^1 \varphi''_{u_x}(\tau u_1 + (1-\tau)u_2)\mathrm{d}\tau W + W\int_0^1 \varphi'''_{uxx}\mathrm{d}\tau, \\ \varphi'''_{uxx} &= \varphi'''_{uuu}(\tau u_{1x} + (1-\tau)u_{2x}) + \varphi''_{ux}(\tau u_{1xx} + (1-\tau)u_{2xx}) \\ g(u_1) - g(u_2) &= W\int_0^1 g'_u(\tau u_1 + (1-\tau)u_2)\mathrm{d}\tau. \end{aligned}$$

Thus equation (2.4.31) can be rewritten as

$$\frac{\mathrm{d}W}{\mathrm{d}t} + \gamma AW - \alpha A^{\frac{1}{2}}W + R_1 W + R_2 W_x + R_3 W_{xx} = 0, \tag{2.4.32}$$

W has the initial value

$$W(0) = u_1^0 - u_2^0, \tag{2.4.33}$$

where

$$\begin{cases} R_1 = \int_0^1 f'_u \mathrm{d}\tau + \int_0^1 f''_{ux} \mathrm{d}\tau + \int_0^1 \varphi'''_{uxx} \mathrm{d}\tau + \int_0^1 g'_u \mathrm{d}\tau, \\ R_2 = \int_0^1 f'_u \mathrm{d}\tau + 2\int_0^1 \varphi''_{ux} \mathrm{d}\tau, \\ R_3 = \int_0^1 \varphi'_u \mathrm{d}\tau. \end{cases} \tag{2.4.34}$$

Taking the inner product of equation (2.4.33) and W, we obtain

$$\begin{aligned} &\frac{1}{2}\frac{\mathrm{d}}{\mathrm{d}t}\|W\|^2 + \gamma\|A^{\frac{1}{2}} W\|^2 - \alpha\|A^{\frac{1}{4}} W\|^2 \\ &\quad + (R_1 W, W) + (R_2 W_x, W) + (R_3 W_{xx}, W) = 0, \end{aligned} \tag{2.4.35}$$

where

$$\begin{aligned} |(R_1 W, W)| &\le \|R_1\|_{L_\infty} \|W\|^2, \\ |(R_2 W_x, W)| &\le \frac{1}{2}\|R_{2x}\|_{L_\infty} \|W\|^2, \\ |(R_3 W_{xx}, W)| &\le \|R_2\|_{L_\infty} \|W_{xx}\| \|W\| \le \frac{\gamma}{3}\|A^{\frac{1}{2}} W\|^2 + \frac{3}{4\gamma}\|R_3\|^2_{L_\infty} \|W\|^2, \\ \alpha\|A^{\frac{1}{4}} W\|^2 &\le \alpha\|A^{\frac{1}{2}} W\|\|W\| \le \frac{\gamma}{3}\|A^{\frac{1}{2}} W\|^2 + \frac{3\alpha^2}{4\gamma}\|W\|^2. \end{aligned}$$

Hence, by equation (2.4.35) we get

$$\begin{aligned} &\frac{1}{2}\frac{\mathrm{d}}{\mathrm{d}t}\|W\|^2 + \frac{\gamma}{3}\|A^{\frac{1}{2}} W\|^2 \\ &\quad \le \frac{1}{2}\left(\|R_{2x}\|_{L_\infty} + \|R_1\|_{L_\infty} + \frac{3}{4\gamma}\|R_3\|^2_{L_\infty} + \frac{3\alpha^2}{4\gamma} \right)\|W\|^2_{L_\infty}. \end{aligned} \tag{2.4.36}$$

By virtue of

$$\begin{aligned} &\|W_x\| \le \|W\|^{\frac{1}{2}} \|W_{xx}\|^{\frac{1}{2}}, \\ &\|W_{xx}\|^2 \ge \frac{\|W_x\|^4}{\|W\|^2} = \frac{\|W_x\|^4}{\|W\|^4}\|W\|^2 \ge \lambda_1\|W\|^2, \end{aligned}$$

from equation (2.4.36), we get

$$\begin{aligned} &\frac{\mathrm{d}}{\mathrm{d}t}\|W\|^2 + \frac{2\gamma}{3}\lambda_1\|W\|^2 \\ &\quad \le \left(\|R_{2x}\|_{L_\infty} + 2\|R_1\|_{L_\infty} + \frac{3}{4\gamma}\|R_3\|^2_{L_\infty} + \frac{3\alpha^2}{2\gamma} \right)\|W\|^2, \end{aligned}$$

which yields

$$\|W(\cdot,t)\|2 = \|u_1(t) - u_2(t)\|^2 \le \|W(0)\|^2 e^{2Kt}, \quad t \ge 0. \qquad \square$$

Lemma 2.4.6. *Let $u_1, u_2 \in B$, and $(S(t)u_1, S(t)u_2) \notin C_N(\gamma_1)$, $\forall t \ge 0$. Then for any N, we only need to require that*

$$N + 1 \ge N_0(\gamma, \widetilde{D}), \tag{2.4.37}$$

where

$$\begin{aligned} N_0(\gamma,\widetilde{D}) = (3\widetilde{D}^4\gamma^{-1})^{\frac{1}{4}}\Bigg[&\frac{\sqrt{2}}{\gamma_1}\sup_{0\le t<\infty}\big(\|R_1\|_{L_\infty(\Omega)} + \|R_{2x}\|_{L_\infty(\Omega)} + \|R_{3x}\|_{L_\infty(\Omega)} + \|R_{3xx}\|_{L_\infty(\Omega)}\big) \\ &+ \frac{4\alpha^2}{\gamma} + \frac{3}{4}\gamma^{-\frac{1}{3}}\left(\frac{\sqrt{2}}{\gamma_1}\right)^{\frac{4}{3}}\Big(\sup_{0\le t<\infty}\|R_2\|_{L_\infty}\Big)^{\frac{4}{3}} + \frac{8}{\gamma\gamma_1^2}\sup_{0\le t<\infty}\|R_{3x}\|_{L_\infty}^2 \\ &+ \frac{8}{\gamma\gamma_1^2}\Big(\sup_{0\le t<\infty}\|R_{3x}\|_{L_\infty}\Big)^2\Bigg]^{\frac{1}{4}}, \quad \widetilde{D} = \frac{D}{2\pi}, \end{aligned}$$

to have

$$\|S(t)u_1 - S(t)u_2\| \le \frac{\sqrt{2}}{\gamma_1}\|u_1 - u_2\| \exp\left[-\frac{\gamma}{2}\frac{(N-1)^4}{\widetilde{D}^4}t\right], \quad t \ge 0. \tag{2.4.38}$$

Proof. Let

$$p_N = P_N(S(t)u_1 - S(t)u_2), \quad q_N = Q_N(S(t)u_1 - S(t)u_2).$$

Since

$$\begin{aligned} &B = \bigcup_{t\ge 0} S(t)B_{2E}, \\ &B_{2E} = \{u \in D_{A^{\frac{3}{4}}}, \ \|u\|_3 \le 2(E_0 + E_1 + E_2 + E_3)\} \quad (u_1, u_2) \in B \times B \setminus C_N(\gamma_1), \end{aligned}$$

where

$$C_N(\gamma_1) = \{0 < \gamma_1 \le 1, \ \|Q_N(u_1 - u_2)\| \le \gamma\|P_N(u_1 - u_2)\|\},$$

for the problem (2.4.32)–(2.4.33), setting $W = p_N \oplus q_N$, taking the inner product of equation (2.4.32) and q_N, we obtain

$$\left(q_N, \frac{dW}{dt} + \gamma AW - \alpha A^{\frac{1}{2}}W + R_1W + R_2W_x + R_3W_{xx}\right) = 0, \tag{2.4.39}$$

where

$$\left(\frac{dW}{dt}, q_N\right) = \left(\frac{d}{dt}(p_N + q_N), q_N\right) = \frac{1}{2}\frac{d}{dt}\|q_N\|^2,$$

$$
\begin{aligned}
(\gamma AW, q_N) &= \gamma(A(p_N+q_N), q_N) = \gamma\|A^{\frac{1}{2}}q_N\|^2,\\
\alpha(-A^{\frac{1}{2}}W, q_N) &= -\alpha(A^{\frac{1}{2}}q_N, q_N) = -\alpha\|A^{\frac{1}{4}}q_N\|^2,\\
|(R_1W, q_N)| &\le \|R_1\|_{L_\infty}\|W\|\|q_N\| \le \frac{\sqrt{2}}{\gamma_1}\|R_1\|_{L_\infty}\|q_N\|^2,\\
|(R_2W_x, q_N)| &= |-(R_{2x}W, q_N) - (R_2W, q_{Nx})|\\
&\le \|R_{2x}\|_{L_\infty}\|W\|\|q_N\| + \|R_2\|_{L_\infty}\|W\|\|q_{Nx}\|\\
&\le \frac{\sqrt{2}}{\gamma_1}\|R_{2x}\|_{L_\infty}\|q_N\|^2 + \frac{\sqrt{2}}{\gamma_1}\|R_2\|_{L_\infty}\|q_N\|\|q_{Nx}\|, \qquad (2.4.40)\\
|(R_3W_{xx}, q_N)| &= |-(R_{3x}W_x, q_N) - ((R_3W_x, q_{Nx}))|\\
&= |(R_{3xx}W, q_N) + 2(R_{3x}W, q_{Nx}) + (R_3W, q_{Nxx})|\\
&\le (\|R_{3xx}\|_{L_\infty}\|q_N\| - 2\|R_{3x}\|_{L_\infty}\|q_{Nx}\| + \|R_3\|_{L_\infty}\|q_{Nxx}\|)\frac{\sqrt{2}}{\gamma_1}\|q_N\|. \qquad (2.4.41)
\end{aligned}
$$

Using the inequality

$$\|q_{Nx}\|^2 \le \|q_N\|\|q_{Nxx}\|,$$

equation (2.4.40) can be estimated as follows:

$$
\begin{aligned}
|(R_2W, q_N)| &\le \frac{\sqrt{2}}{\gamma_1}\|R_{2x}\|_{L_\infty}\|q_N\|^2 + \frac{\sqrt{2}}{\gamma_1}\|R_2\|_{L_\infty}\|q_N\|^{\frac{3}{2}}\|q_{Nxx}\|^{\frac{1}{2}}\\
&\le \left[\frac{\sqrt{2}}{\gamma_1}\|R_{2x}\|_{L_\infty} + \frac{3}{4}\left(\frac{\sqrt{2}}{\gamma_1}\right)^{\frac{4}{3}}\gamma^{-\frac{1}{3}}\|R_2\|_{L_\infty}^{\frac{4}{3}}\right]\|q_N\|^2 + \frac{\gamma}{4}\|q_{Nxx}\|^2. \qquad (2.4.42)
\end{aligned}
$$

Equation (2.4.41) can be estimated as follows:

$$
\begin{aligned}
|(R_3W_{xx}, q_N)| &\le \frac{\sqrt{2}}{\gamma_1}\|R_{2xx}\|_{L_\infty}\|q_N\|^2 + \frac{\sqrt{2}}{\gamma_1}\|R_{3x}\|_{L_\infty}(\|q_N\|^2 + \|q_{Nxx}\|^2)\\
&\quad + \frac{\sqrt{2}}{\gamma_1}\|R_3\|_{L_\infty}\|q_N\|\|q_{Nxx}\|\\
&\le \frac{\sqrt{2}}{\gamma_1}(\|R_{3xx}\|_{L_\infty} + \|R_{3x}\|_{L_\infty})\|q_N\|^2 + \frac{\gamma}{16}\|q_{Nxx}\|^2\\
&\quad + \frac{8}{\gamma\gamma_1^2}\|R_{3x}\|_{L_\infty}^2\|q_N\|^2 + \frac{\gamma}{16}\|q_{Nxx}\|^2 + \frac{8}{\gamma\gamma_1^2}\|R_3\|_{L_\infty}^2\|q_N\|^2\\
&\le \left[\frac{\sqrt{2}}{\gamma_1}(\|R_{3xx}\|_{L_\infty} + \|R_{3x}\|_{L_\infty}) + \frac{8}{\gamma\gamma_1^2}\|R_{3x}\|_{L_\infty}^2 + \frac{8}{\gamma\gamma_1^2}\|R_3\|_{L_\infty}^2\right]\|q_N\|^2\\
&\quad + \frac{\gamma}{8}\|q_{Nxx}\|^2, \qquad (2.4.43)\\
\alpha\|q_{Nx}\|^2 &\le \frac{\gamma}{16}\|q_{Nxx}\|^2 + \frac{4\alpha^2}{\gamma}\|q_N\|_{L_\infty}^2. \qquad (2.4.44)
\end{aligned}
$$

From equations (2.4.39), (2.4.42), (2.4.43) and (2.4.44), we obtain

$$\frac{1}{2}\frac{\mathrm{d}}{\mathrm{d}t}\|q_N\|^2 + \frac{9}{16}\gamma\|A^{\frac{1}{4}}q_N\|^2 \le R\|q_N\|^2, \qquad (2.4.45)$$

where

$$R = \frac{\sqrt{2}}{\gamma_1}(\|R_1\|_{L_\infty} + \|R_{2x}\|_{L_\infty} + \|R_{3x}\|_{L_\infty} + \|R_{3xx}\|_{L_\infty})$$
$$+ \frac{3}{4}\left(\frac{\sqrt{2}}{\gamma_1}\right)^{\frac{4}{3}} \gamma^{-\frac{1}{3}} \|R_2\|_{L_\infty}^{\frac{4}{3}} + \frac{8}{\gamma\gamma_1^2}\|R_{3x}\|_{L_\infty}^2 + \frac{8}{\gamma\gamma_1^2}\|R_3\|_{L_\infty}^2 + \frac{4\alpha^2}{\gamma}. \tag{2.4.46}$$

Choosing N large enough, that is, $N + 1 \geq N_6(\gamma, \widetilde{D})$, such that

$$\gamma\lambda_{N+1} - \sup_{0\leq t<\infty} R \geq 0,$$

where

$$\lambda_{N+1} = \left(\frac{2\pi(N+1)}{D}\right)^4 = \frac{(N-1)^4}{\widetilde{D}^4}, \quad \widetilde{D} \equiv \frac{D}{2\pi},$$

from equation (2.4.45), we then get

$$\frac{\mathrm{d}}{\mathrm{d}t}\|q_N\|^2 + \frac{\gamma}{8}\lambda_{N+1}\|q_N\|^2 \leq 0. \tag{2.4.47}$$

Since

$$\|W\|^2 \leq \|p_N\|^2 + \|q_N\|^2 \leq \frac{2}{\gamma_1^2}\|q_N\|^2,$$

by equation (2.4.47) we get the estimate

$$\|S(t)u_1 - S(t)u_2\| \leq \frac{\sqrt{2}}{\gamma_1}\|u_1 - u_2\| \exp\left[-\frac{\gamma}{16}\frac{(N+1)^4)}{\widetilde{D}^4}t\right], \quad \forall t \geq 0. \tag{2.4.48}$$

□

Lemma 2.4.7. *If $N + 1 \geq \widetilde{N}_0(\gamma, \widetilde{D})$ is chosen such that*

$$\frac{19}{48}\gamma\lambda_{N+1} - \frac{7\gamma}{6\gamma_1^2}\lambda_N - \overline{R} > 0, \tag{2.4.49}$$

where

$$\begin{cases} \lambda_N = \dfrac{(N+1)^4}{\widetilde{D}^4}, \\ \overline{R}(x,0) = \sup\limits_{0\leq t<\infty} R + \sup\limits_{0\leq t<\infty}\left[\dfrac{\sqrt{2}}{\gamma_1}\|R_1\|_{L_\infty} \right. \\ \qquad \left. + \dfrac{1}{2\gamma_1^2}\|R_{2x}\|_{L_\infty} + \dfrac{3}{\gamma\gamma_1^2} + \|R_3\|_{L_\infty} + \dfrac{1}{2\gamma_1^2}\|R_2\|_{L_\infty}^2\right] + \dfrac{3}{8\gamma}, \end{cases} \tag{2.4.50}$$

then $C_N(\gamma_1) = \{0 < \gamma_1 \leq 1, \|Q_N(u_1 - u_2)\| \leq \gamma_1\|P_N(u_1 - u_2)\|\}$ is strictly positive invariant, that is, if $(u_1, u_2) \in C_N(\gamma_1)$, then

$$(S(t)u_1, S(t)u_2) \in C_N(\gamma_1) - \partial C_N(\gamma_1), \ \forall t \geq 0.$$

Proof. Let $p_N = P_N(S(t)u_1 - S(t)u_2)$, $q_N = Q_N(S(t)u_1 - S(t)u_2)$, $W = p_N \oplus q_N$. We know that the estimate (2.4.48) holds on $\|q_N\|^2 = \gamma_1^2 \|p_N\|^2$ as in the proof of Lemma 2.4.6. Now we take the inner product of equation (2.4.32) and p_N, from which it follows that

$$\left(p_N, \frac{\mathrm{d}W}{\mathrm{d}t} + \gamma AW - \alpha A^{\frac{1}{2}} W + R_1 W + R_2 W_x + R_3 W_{xx}\right) = 0. \tag{2.4.51}$$

Since

$$\begin{aligned}
&\left(\frac{\mathrm{d}}{\mathrm{d}t}(p_N \oplus q_N), p_N\right) = \frac{1}{2}\frac{\mathrm{d}}{\mathrm{d}t}\|p_N\|^2,\\
&\left(\gamma A(p_N \oplus q_N), p_N\right) = \gamma \left\|A^{\frac{1}{2}} p_N\right\|^2,\\
&\alpha\left(A^{\frac{1}{2}}(p_N \oplus p_N), p_N\right) = \alpha(A^{\frac{1}{2}} p_N, p_N) + \alpha(q_N, A^{\frac{1}{2}} p_N) = \alpha\left\|A^{\frac{1}{2}} p_N\right\|^2,\\
&\left|(R_1 W, p_N)\right| \le \|R_1\|_{L_\infty} \|W\| \|p_N\| \le \frac{\sqrt{2}}{\gamma_1} \|R_1\|_{L_\infty} \|q_N\|^2,\\
&\left|(R_2 W_x, p_N)\right| = \left|(R_2(p_{Nx} + q_{Nx}), p_N)\right|\\
&\qquad \le \frac{1}{2}\|R_{2x}\|_{L_\infty} \|p_N\|^2 + \frac{1}{\gamma_1}\|R_2\|_{L_\infty} \|q_N\| \|q_{Nx}\|,\\
&\left|(R_3 W_{xx}, p_N)\right| = \left|(R_3(p_{Nxx} + q_{Nxx}), p_N)\right|\\
&\qquad \le \left|(R_3(p_{Nxx}, p_N)\right| + \left|(R_3 q_{Nxx}, p_N)\right|\\
&\qquad \le \frac{\gamma}{6}\|p_{Nxx}\|^2 + \frac{3}{2\gamma}\|R_3\|_{L_\infty}^2 \|p_N\|^2 + \frac{\gamma}{6}\|p_{Nxx}\|^2 + \frac{3}{2\gamma}\|R_3\|_{L_\infty}^2 \|p_N\|^2\\
&\qquad \le \frac{\gamma}{6}\|p_{Nxx}\|^2 + \frac{\gamma}{6}\|p_{Nxx}\|^2 + \frac{3}{\gamma\gamma_1^2}\|R_3\|_{L_\infty}^2 \|q_N\|^2,\\
&\frac{1}{2}\frac{\mathrm{d}}{\mathrm{d}t} p_N + \gamma\left\|A^{\frac{1}{2}} p_N\right\|^2 \ge \alpha\left\|A^{\frac{1}{4}} p_N\right\|^2 - \frac{\sqrt{2}}{\gamma_1^2}\|R_1\|_{L_\infty}\|q_N\|^2\\
&\qquad - \frac{1}{2}\|R_{2x}\|_{L_\infty}\|q_N\|^2 - \frac{1}{\gamma_1}\|R_2\|_{L_\infty}\|q_N\|\|q_{Nxx}\|\\
&\qquad - \frac{\gamma}{6}\|p_{Nxx}\|^2 - \frac{\gamma}{6}\|q_{Nxx}\|^2 - \frac{3}{2\gamma\gamma_1^2}\|R_3\|_{L_\infty}^2 \|q_N\|^2.
\end{aligned}$$

Hence we have$(0 < \gamma_1 \le 1)$

$$\begin{aligned}
-\frac{\gamma_1}{2}\frac{\mathrm{d}}{\mathrm{d}t}\|p_N\|^2 &\le \gamma\left\|A^{\frac{1}{2}} p_N\right\|^2 - \gamma_1\alpha\left\|A^{\frac{1}{4}} p_N\right\|^2 + \frac{\sqrt{2}}{\gamma_1^2}\|R_1\|_{L_\infty}\|q_N\|^2\\
&\quad + \frac{1}{2}\|R_{2x}\|_{L_\infty} \cdot \frac{1}{\gamma_1^2}\|q_N\|^2 + \frac{1}{\gamma_1}\|R_2\|_{L_\infty}\|q_N\|\|q_{Nx}\|\\
&\quad + \frac{\gamma}{6}\|p_{Nxx}\|^2 + \frac{\gamma}{6}\|q_{Nxx}\|^2 + \frac{3}{\gamma\gamma_1^2}\|R_3\|_{L_\infty}^2\|q_N\|^2 = S.
\end{aligned} \tag{2.4.52}$$

By equations (2.4.45) and (2.4.52), we obtain

$$\frac{1}{2}\frac{\mathrm{d}}{\mathrm{d}t}[\|q_N\|^2 - \gamma_1\|p_N\|^2] + \frac{9}{16}\gamma\left\|A^{\frac{1}{2}} q_N\right\|^2 \le R\|q_N\|^2 + S, \tag{2.4.53}$$

where

$$\|A^{\frac{1}{4}}p_N\|^2 \le \lambda_N\|p_N\|^2 \le \frac{\lambda_N}{\gamma_1^2}\|q_N\|^2,$$

$$\begin{aligned}\frac{1}{\gamma_1}\|R_2\|_{L_\infty}\|q_N\|\|q_{Nx}\| &\le \frac{1}{2}\Big[\|p_{Nx}\|^2 + \frac{1}{\gamma_1^2}\|R_2\|_{L_\infty}^2\|q_N\|^2\Big]\\ &\le \frac{1}{2}\Big[\frac{\gamma}{3}\|q_{Nxx}\|^2 + \frac{3}{4\gamma}\|q_N\|^2 + \frac{1}{\gamma_1^2}\|R_2\|_{L_\infty}^2\|q_N\|^2\Big]\\ &\le \frac{\gamma}{6}\|A^{\frac{1}{2}}q_N\|^2 + \frac{1}{4}\Big(\frac{1}{\gamma_1^2}\|R_2\|_{L_\infty}^2 + \frac{3}{4\gamma}\|q_N\|^2\Big).\end{aligned}$$

Then by equation (2.4.53), we have

$$\begin{aligned}&\frac{1}{2}\frac{\mathrm{d}}{\mathrm{d}t}[\|q_N\|^2 - \gamma_1\|p_N\|^2] + \frac{9\gamma}{16}\|A^{\frac{1}{2}}q_N\|^2\\ &\quad\le \Big[R + \frac{\sqrt{2}}{\gamma_1^2}\|R_1\|_{L_\infty} + \frac{1}{2\gamma_1^2}\|R_{2x}\|^2 + \frac{1}{2}\Big(\frac{\|R_2\|_{L_\infty}^2}{\gamma_1^2} + \frac{3}{4\gamma}\Big) + \frac{3}{\gamma\gamma_1^2}\|R_3\|_{L_\infty}^2 + \frac{7}{6}\frac{\gamma}{\gamma_1^2}\lambda_N\Big]\|q_N\|^2\\ &\quad\le \Big[\overline{R} + \frac{7\gamma}{6\gamma_1^2}\lambda_N\Big]\|q_N\|^2.\end{aligned}$$

Hence when $N+1 \ge \widetilde{N}_0(\gamma, \gamma_1, \widetilde{D})$ is such that

$$\frac{19}{48}\gamma\lambda_{N+1} - \frac{7\gamma}{6\gamma_1^2}\lambda_N - \overline{R} > 0, \quad \lambda_{N+1} = \frac{(N+1)^4}{\widetilde{D}^4}, \tag{2.4.54}$$

we have

$$\frac{\mathrm{d}}{\mathrm{d}t}[\|q_N\|^2 - \gamma_1\|p_N\|^2] < 0.$$

The lemma has been proved. □

Through Lemmas 2.4.6 and 2.4.7, we deduce the following theorem:

Theorem 2.4.1. *Suppose the conditions of Lemmas 2.4.6 and 2.4.7 are satisfied. Then for the GKS equation there exists* $N_0 = N_0(\gamma, \gamma_1, \widetilde{D})$ *such that when* $N+1 > N_0$*:*

(1) *The cone* $C_N(\gamma_1)$ *is strictly invariant, that is, if for* $(u_1, u_2) \in C_N(\gamma_1)$*, then* $(S(t)u_1, S(t)u_2) \in C_N(\gamma_1) \setminus \partial C_N(\gamma_1)$, $\forall t \ge 0$*;*

(2) $B \times B \setminus C_N(\gamma_1)$ *is exponentially squeezing, that is, for* $u_1, u_2 \in B$ *and* $(S(t)u_1, S(t)u_2) \notin C_N(\gamma_1)$*, we have*

$$\|S(t)u_1 - S(t)u_2\| \le \frac{\sqrt{2}}{\gamma_1}\|u_1 - u_2\| \exp\Big[-\frac{\gamma}{16}\frac{(N+1)^4}{\widetilde{D}^4}t\Big], \quad \forall t \ge 0.$$

Now we prove the existence of inertial manifolds for PGKS equation.

Suppose that $\theta(s) : \mathbf{R}^+ \to [0,1]$ is a smooth increasing function such that $\theta(s) = 1$, $0 \le s \le 1$, $\theta(s) = 0$, $s \ge 2$, $\sup_{s\ge 0} |\theta'(s)| \le 2$.

We consider the following PGKS equation:

$$\frac{du}{dt} + \gamma AW + \theta\Big(\frac{\|u\|_1}{E}\Big)(-\alpha A^{\frac{1}{2}}u) + F(u)_x + \Phi(u)_{xx} + \theta\Big(\frac{\|u\|_1}{E}\Big)(-g(u) - h(x)) = 0. \tag{2.4.55}$$

where $E = 2(E_0 + E_1)$, $B_E = \{u : \|u\|_1 \le E\}$ is the absorbing set of PGKS equation, $F(u) = f(\theta(\frac{\|u\|_1}{E})u)$, $\Phi(u) = \varphi(\theta(\frac{\|u\|_1}{E})u)$. Obviously, when $u \in B_E$, then PGKS equation and GKS equation are the same. For some positive integer N, the operator $P = P_N$ means the orthogonal projection from the Hilbert space H to the subspace spanned by N eigenfunctions of operator A, $Q = Q_N = I - P_N$, obviously operators P, Q and operators A^β, $\forall \beta \in \mathbf{R}$, are commutative.

For given $b, l > 0$, $b, l \in (0,1]$, we define the Lipschitz function family $\mathscr{F} = \mathscr{F}^\alpha_{b,l} : PD(A^\alpha) \to QD(A^\alpha)$ such that

$$\begin{cases} (1) & \operatorname{supp} \Phi \subset \{p \in PD(A^\alpha),\ \|A^\alpha p\| < 2E\}; \\ (2) & \|A^\alpha \Phi(p)\| \le b, \quad \forall p \in PD(A^\alpha); \\ (3) & \|A^\alpha \Phi(p_1) - A^\alpha \Phi(p_2)\| \le l\|A^\alpha(p_1 - p_2)\|, \qquad \forall p_1, p_2 \in PD(A^\alpha). \end{cases} \tag{2.4.56}$$

If we define the following distance in $\mathscr{F} = \mathscr{F}^\alpha_{b,l}$:

$$d(\Phi_1, \Phi_2) = \sup_{p\in PD(A^\alpha)} \|A^\alpha(\Phi_1(p) - \Phi_2(p))\|, \tag{2.4.57}$$

then $\mathscr{F}$ is complete.

Let

$$u = p + \overline{\varphi}(p), \tag{2.4.58}$$

where $\overline{\varphi}(p) \in \mathscr{F}^{\frac{1}{4}}_{b,l}$. Then $(p, \overline{\varphi}(p))$ satisfies the equations

$$\frac{dp}{dt} + \gamma Ap + \theta\Big(\frac{\|u\|_1}{E}\Big)(-\alpha A^{\frac{1}{2}}p) + p_N B(u,v) = 0, \tag{2.4.59}$$

$$\frac{d\overline{\varphi}(p)}{dt} + \gamma A\overline{\varphi}(p) + \theta\Big(\frac{\|u\|_1}{E}\Big)(-\alpha A^{\frac{1}{2}}\overline{\varphi}(p)) + Q_N B(u,v) = 0, \tag{2.4.60}$$

$$B(u,v) \triangleq F(u)_x + \Phi(u)_{xx} + \theta\Big(\frac{\|u\|_1}{E}\Big)(-g(u) - h(x)). \tag{2.4.61}$$

Obviously, if $p(t)$ is a bounded and unique solution of equation (2.4.59), then $\overline{\varphi}(p)$ is a bounded solution of equation (2.4.60), and there is an expression

$$\overline{\varphi}(p_0) = -\int_{-\infty}^{0} e^{\tau\gamma AQ_N}\Big\{-\alpha A^{\frac{1}{2}}\bar{\varphi}(p(\tau))\theta\Big(\frac{\|p(\tau) + \overline{\varphi}(p(\tau))\|_1}{E}\Big) + Q_N B(p(\tau) + \overline{\varphi}(p(\tau)),\ p(\tau) + \overline{\varphi}(p(\tau)))\Big\}d\tau, \tag{2.4.62}$$

where

$$p_0 = p(0), \quad p(\tau) = p(\tau; p_0, \overline{\varphi}), \quad p(0; p_0, \overline{\varphi}) = p_0. \tag{2.4.63}$$

Now define the functional operator $\mathscr{F}_0 : \mathscr{F}^{\frac{1}{4}}_{b,l} \to \mathscr{F}$ by

$$\mathscr{F}_0\overline{\varphi} = -\int_{-\infty}^{0} e^{\tau\gamma A Q_N}\left\{-\alpha A^{\frac{1}{2}}\overline{\varphi}(p)\theta\left(\frac{\|u\|_1}{E}\right) + Q_N(u,u)\right\}d\tau, \tag{2.4.64}$$

where $u = p + \overline{\varphi}(p)$, and

$$p = p(\tau; p_0, \overline{\varphi}) \tag{2.4.65}$$

is the solution of equation (2.4.59) satisfying the initial condition

$$p(0; p_0, \overline{\varphi}) = p_0. \tag{2.4.66}$$

So to show the existence of an inertial manifold, that is, to establish the existence of the pair $(p, \overline{\varphi})$, amounts to establishing the existence of a fixed point to the functional equation

$$\mathscr{F}_0\Phi = \Phi, \quad \Phi \in \mathscr{F}^{\frac{1}{4}}_{b,l}. \tag{2.4.67}$$

Therefore, in order to consider the existence of bounded solutions for equation (2.4.59) on the interval $(-\infty, +\infty)$, we need the following lemma:

Lemma 2.4.8. *Given $\overline{\varphi} \in \mathscr{F}$, $p_0 \in PD(A^{\frac{1}{4}})$ and setting $f(0) = \varphi(0) = 0$, there exists a unique solution $p(t; p_0, \overline{\varphi}) \in L^\infty(\mathbf{R}; D(A^{\frac{1}{4}}))$ of problem (2.4.59), (2.4.63) for $t \in (-\infty, +\infty)$.*

Proof. First we prove that the mapping $\sigma \to \alpha\theta(\frac{\|\sigma+\overline{\varphi}(\sigma)\|_1}{E})A^{\frac{1}{2}}\sigma - p_N B(\sigma + \overline{\varphi}(\sigma), \sigma + \overline{\varphi}(\sigma))$ is Lipschitz continuous, $D(A^{\frac{1}{4}}) \to D(A^{\frac{-1}{4}})$. In fact, setting $u_1, u_2 \in D(A^{\frac{1}{4}})$, $\theta_i = \theta_i(\frac{\|u_i\|_1}{E})$, $i = 1, 2$, we prove

$$\begin{aligned} L &= \alpha\|\theta_1 A^{\frac{1}{4}} p_1 - \theta_2 A^{\frac{1}{4}} p_2\| + \|A^{-\frac{1}{4}}(B(u_1, u_1) - B(u_2, u_2))\| \\ &\le C\|A^{\frac{1}{4}}(u_1 - u_2)\|. \end{aligned}$$

Therefore, we consider three different cases:

(1) $\|A^{\frac{1}{4}}u_1\| = \|u_1\|_1$, $\|A^{\frac{1}{4}}u_2\| = \|u_2\|_1 \ge 2E$. Then $\theta_i = 0$, $i = 1, 2$, $L = 0$, and the claim is obvious.

(2) $\|u_1\|_1 \le 2E \le \|u_2\|_1$, then $\theta_2 = 0$, and

$$\begin{aligned}
L &= \alpha\left|\theta\left(\frac{\|u_1\|_1}{E}\right) - \theta\left(\frac{\|u_2\|_1}{E}\right)\right|\|A^{\frac{1}{4}}p_1\| \\
&\quad + \|F(u_1) + A^{\frac{1}{4}}\Phi(u_1) + \theta_1 A^{-\frac{1}{4}}(-g(u) - h)\| \\
&= \alpha|\theta_1 - \theta_2|\|p_1\|_1 + \|f((\theta_1 - \theta_2)u_1) + \varphi((\theta_1 - \theta_2)u_1)_x + (\theta_1 - \theta_2)A^{-\frac{1}{4}}(-g(u) - h)\| \\
&\le \frac{2\alpha}{E}\|u_1 - u_2\|_1\|p_1\|_1 + 2/E\|f'_u\|_{L_\infty}\|u_1\|_{L_\infty}\|u_1 - u_2\|_1 \\
&\quad + \frac{2}{E}\|\varphi'_u\|_{L_\infty}\|u_1\|_{L_\infty}\|u_1 - u_2\|_1 + \frac{2}{E}\|A^{-\frac{1}{4}}(-g(u) - h)\|\|u_1 - u_2\|_{L_\infty} \\
&\le \left(\frac{2\alpha}{E}\|p_1\|_1 + \|f'_u\|_{L_\infty}2/E\|u_1\|_{L_\infty} + 2/E\|\varphi'_u\|_{L_\infty} + 2/E\|A^{-\frac{1}{4}}(-g(u) + h)\|\right)\|u_1 - u_2\|_1 \\
&\le C\|A^{\frac{1}{4}}(u_1 - u_2)\|,
\end{aligned}$$

where

$$\begin{aligned}
|(g(u) + h, v)| &\le (\|g(u)\| + \|h(x)\|)\|v\| \\
&\le C\|v_1\|\|A^{-\frac{1}{4}}(g(u) + h)\| \le C.
\end{aligned}$$

(3) $\|u_1\|_1, \|u_2\|_1 < 2E$, then we have

$$\begin{aligned}
L &= \alpha\|\theta_1 A^{\frac{1}{4}}p_1 - \theta_2 A^{\frac{1}{4}}p_2\| + \|A^{-\frac{1}{4}}(B(u_1, u_1) - B(u_2, u_2))\| \\
&\le \alpha\|(\theta_1 - \theta_2)A^{\frac{1}{4}}p_1 + \theta_2(A^{\frac{1}{4}}p_1 - A^{\frac{1}{4}}p_2)\| \\
&\quad + \|A^{-\frac{1}{4}}[F(u_1)_x + \Phi(u_1)_{xx} + \theta_1(-g(u_1) - h(x)) \\
&\quad - F(u_2)_x - \Phi(u_2)_{xx} - \theta_2(-g(u_2) - h(x))]\| \\
&\le \frac{2\alpha}{E}\|u_1 - u_2\|_1\|A^{\frac{1}{4}}p_1\|_1 + \alpha\|u_1 - u_2\|_1 + \|F'(u)\|_{L_\infty}\|u_1 - u_2\|_1 \\
&\quad + (4E\|\Phi''(u)\|_{L_\infty} + \|\Phi'(u)\|_{L_\infty})\|u_1 - u_2\|_1 \\
&\quad + (2/E\|g(u)\|_{L_\infty} + \|g'(u)\|_{L_\infty})\|u_1 - u_2\|_1 \\
&\le C\|u_1 - u_2\|_1.
\end{aligned}$$

So we prove that the mapping $\sigma \to \alpha\theta(\frac{\|\sigma+\overline{\varphi}(\sigma)\|_1}{E})A^{\frac{1}{2}}\sigma - p_N B(\sigma+\overline{\varphi}(\sigma), \sigma+\overline{\varphi}(\sigma))$ is Lipschitz continuous and $PAp = APp = Ap$. Moreover, $PD(A^{\frac{1}{4}})$ has finite dimension. Thus from the existence of solution theorem for ordinary differential equations, we know that the initial problem (2.4.59), (2.4.63) possesses a unique solution $p = p(t; p_0, \overline{\varphi})$, $\forall t \in \mathbf{R}$. □

Lemma 2.4.9. *For any $\alpha \in \mathbf{R}$ and $\sigma \in L^\infty(\mathbf{R}; D(A^{\alpha-\frac{1}{2}}))$, there exists a unique continuous bounded function $\xi : \mathbf{R} \to D(A^\alpha)$, which satisfies the equation*

$$\frac{d\xi}{dt} + A\xi = \sigma. \tag{2.4.68}$$

Lemma 2.4.10. *If $p_0 \in PD(A^{\frac{1}{2}}), \overline{\varphi} \in \mathscr{F}_{b,l}^{\frac{1}{4}}$, then there exists a unique continuous bounded solution $\overline{\varphi}(p) : \mathbf{R} \to QD(A^{\frac{1}{4}})$.*

Proof. We only need to verify the conditions of Lemma 2.4.9, namely to prove $\sigma = \alpha A^{\frac{1}{2}}\overline{\varphi}(p) + Q_N B(p + \overline{\varphi}(p), p + \overline{\varphi}(p)) \in L^\infty(\mathbf{R}; D(A^{-\frac{1}{4}}))$. In fact,

$$\begin{aligned}
&|(\alpha A^{\frac{1}{2}}\overline{\varphi}(p), v)| = |(\alpha A^{\frac{1}{4}}\overline{\varphi}(p), A^{\frac{1}{4}}v)| \le C\|A^{\frac{1}{4}}v\| = C\|v\|_1,\\
&\left|(F(u)_x + \Phi(u)_{xx} + \theta\left(\frac{\|u\|_1}{E}\right)(-g(u) - h(x), v)\right|\\
&\quad \le \|F(u)\|\|u_x\| + \|\Phi'(u)\|_{L_\infty}\|u_x\|\|v_x\| + (\|g(u)\| + \|h(x)\|)\|v\|\\
&\quad \le C\|v\|_1.
\end{aligned}$$

Hence $\sigma \in L^\infty(\mathbf{R}; D(A^{-\frac{1}{4}}))$, and the lemma has been proved. □

In the following, we establish certain properties of the functional operator $\mathscr{F}_0$.

Lemma 2.4.11. *Let $\overline{\varphi} \in \mathscr{F}_{b,l}^{\frac{1}{4}}$. Then* supp $\mathscr{F}_0\overline{\varphi} \subset B_{4E} = \{u : \|u\|_1 \le 4E\}$.

Proof. If not, set $\|p_0\|_1 > 4E$, and then $\|U(0)\|_1 = (\|p_0\|_1^2 + \|\overline{\varphi}(p_0)\|_1^2)^{\frac{1}{4}} > 4E$. Due to the continuity of the solution of equation (2.4.59) in t, there must exist a neighborhood D such that the solution of equation (2.4.59) satisfies $\|p(t)\|_1 > 2E + \rho, \rho > 0, \forall t \in D$. Hence

$$\|p(t) + \overline{\varphi}(p(t))\|_1 > 2E,$$

and then we have

$$\begin{aligned}
&F(p(t) + \overline{\varphi}(p(t))) = 0, \quad \Phi(p(t) + \Phi(p(t))) = 0,\\
&\theta\left(\frac{\|p + \overline{\varphi}(p)\|_1}{E}\right) = 0.
\end{aligned}$$

Equation (2.4.59) boils down to

$$A^{\frac{1}{2}}p(\tau) = e^{-\gamma AP_N\tau}A^{\frac{1}{4}}p_0.$$

The above equation is well-defined at least for small $|\tau|$, for $\tau < 0$ we have

$$\|p(\tau)\|_1 = \|A^{\frac{1}{4}}p(\tau)\| \ge e^{\gamma\lambda_1|\tau|}\|A^{\frac{1}{4}}p_0\| = e^{\gamma\lambda_1|\tau|}\|p_0\|_1.$$

Thus

$$\|p(\tau) + \overline{\varphi}(p(\tau))\|_1 \ge \|p(\tau)\|_1 > 4E, \quad \forall \tau \le 0,$$

which yields that

$$\mathscr{F}_0\overline{\varphi}(p_0) = 0.$$

The lemma has been proved. □

Lemma 2.4.12 ([69]). *For $\alpha > 0$ and $\tau < 0$, the operator $(AQ)^\alpha e^{\tau AQ}$ is linear and continuous in QH. Furthermore, its norm $\|(AQ)^\alpha e^{\tau AQ}\|_{Op}$ is bounded in $\mathscr{L}(QH)$ by*

$$k_2(\alpha)|\tau|^{-\alpha}, \quad \text{when } -\alpha\lambda_{N+1}^{-1} \le \tau < 0, \tag{2.4.69}$$

$$\lambda_{N+1}^{\alpha} e^{\tau\lambda_{N+1}}, \quad \text{when } -\infty < \tau \le -\alpha\lambda_{N+1}^{-1}. \tag{2.4.70}$$

If $\alpha < 1$, then

$$\int_{-\infty}^{0} \|(AQ)^\alpha e^{\tau AQ}\|_{\mathscr{L}(QH)} \mathrm{d}\tau \le k_3(\alpha)\lambda_{N+1}^{\alpha-1}, \tag{2.4.71}$$

where λ_{N+1} is the smallest eigenvalue of $A|_{QH}$, $k_2(\alpha)$, $k_3(\alpha)$ are all some known constants, depending on α.

Lemma 2.4.13. *Let $\overline{\varphi} \in \mathscr{F}_{b,l}^{\frac{1}{4}}$, $b, l \in (0,1]$. Then for $N+1 \ge N_0'(E)$, we have*

$$\|\mathscr{F}_0(\overline{\varphi})(p)\|_1 = \|A^{\frac{1}{4}}\mathscr{F}_0\overline{\varphi}(p)\| \le b. \tag{2.4.72}$$

Proof. From the expression (2.4.64) of $\mathscr{F}_0\overline{\varphi}(p)$, we get

$$\begin{aligned} \|A^{\frac{1}{4}}\mathscr{F}_0\overline{\varphi}(p)\| \le \alpha &\int_{-\infty}^{0} \|(AQ_N)^{\frac{1}{2}} e^{\tau\gamma AQ_N}\| \|A^{\frac{1}{4}}\overline{\varphi}(p)\| \mathrm{d}\tau \\ &+ \int_{-\infty}^{0} \|(AQ_N)^{\frac{1}{4}} e^{\tau\gamma AQ_N}\| \|Q_N B_1(u,u)\| \mathrm{d}\tau \\ &+ \int_{-\infty}^{0} \|(AQ_N)^{\frac{1}{2}} e^{\tau\gamma AQ_N}\| \|A^{\frac{1}{4}}\Phi(u)\| \mathrm{d}\tau, \end{aligned} \tag{2.4.73}$$

where

$$B_1(u,u) = f(u)_x + \theta\left(\frac{\|u\|_1}{E}\right)(-g(u) - h(x))$$

$$\|Q_N B_1(u,v)\| \le C\|f'(u)\|_{L_\infty}\|u\|_1 + \|g(u)\| + \|h(x)\|_2 \le C_1(E)$$

$$\|A^{\frac{1}{4}}\Phi(u)\| \le \|\varphi'(u)\|_{L_\infty}\|u\|_1 \le C_2(E).$$

Then by Lemma 2.4.12 and equation (2.4.73) we get

$$\begin{aligned} \|\mathscr{F}_0\overline{\varphi}(p)\|_1 &= \|A^{\frac{1}{4}}\mathscr{F}_0\overline{\varphi}(p)\| \\ &\le abk_3\left(\frac{1}{2}\right)\lambda_{N+1}^{\frac{1}{2}} + C_1(E)k_3\left(\frac{1}{4}\right)\lambda_{N+1}^{-\frac{3}{4}} + C_2(E)k_3\left(\frac{1}{2}\right)\lambda_{N+1}^{-\frac{1}{2}} \\ &= k_3\left(\frac{1}{2}\right)\lambda_{N+1}^{-\frac{1}{2}}(ab + C_2(E)) + C_1(E)k_3\left(\frac{1}{4}\right)\lambda_{N+1}^{-\frac{3}{4}} \\ &\le C_3(E)[\lambda_{N+1}^{-\frac{1}{2}} + \lambda_{N+1}^{-\frac{3}{4}}], \end{aligned}$$

where $\lambda_{N+1} = \frac{(N+1)^4}{\widetilde{D}^4}$. Hence for $N+1 \geq \widetilde{N}_0(E)$, we have

$$\|\mathcal{F}_0\overline{\varphi}(p)\|_1 \leq b. \qquad \square$$

Suppose there are two groups of differential inequality:

$$\frac{\mathrm{d}y}{\mathrm{d}t} + ay + bz \geq 0, \tag{2.4.74}$$

$$\frac{\mathrm{d}z}{\mathrm{d}t} - cy + dz \leq 0, \quad t \in I \subset \mathbf{R}. \tag{2.4.75}$$

Consider the set C_γ given by

$$C_\gamma = \{(y,e) \in \mathbf{R}^+ \times \mathbf{R}^+ \mid e \geq \gamma_y\}, \quad \gamma > 0. \tag{2.4.76}$$

Lemma 2.4.14. *Let $a, b, c, d, \gamma > 0$ and*

$$d - \gamma^{-1}c > 0, \quad d - a - \gamma b - \gamma^{-1}c > 0. \tag{2.4.77}$$

Then for the orbit $(y(t), z(t))$ of (2.4.74)–(2.4.75), when $(y(t_0), z(t_0)) \in C_\gamma$ we have

$$z(t) \leq z(t_0)\exp\{-(d - \gamma^{-1}C)(t - t_0)\}, \quad \forall t \geq t_0,\ (y(t), z(t)) \in C_\gamma. \tag{2.4.78}$$

Theorem 2.4.2. *For any $b, l \in (0,1]$, there exists $N_1 = N_1(\gamma, \widetilde{D})$ such that when $N+1 > N_1$ we have*

$$\mathcal{F}_0\mathcal{F}^{\frac{1}{4}}_{b,l} \subset \mathcal{F}^{\frac{1}{4}}_{b,l}. \tag{2.4.79}$$

Proof. By Lemmas 2.4.11 and 2.4.14, we only need to prove

$$\begin{aligned} &\|A^{\frac{1}{4}}(\mathcal{F}_0\overline{\varphi}(p_{01}) - \mathcal{F}_0\overline{\varphi}(p_{02}))\| \leq l\|A^{\frac{1}{4}}(p_{01} - p_{02})\|, \\ &l > 0,\ \forall p_{01}, p_{02} \in PD(A^{\frac{1}{4}}). \end{aligned} \tag{2.4.80}$$

Now we consider the functional operator $\mathcal{F}_0\overline{\varphi}$ with two different initial values p_{01}, p_{02}. Let $U_j = p_j + \overline{\varphi}(p_j)$, $j = 1, 2$, where $p_j = p_j(\tau; p_{0j}; \overline{\varphi})$ is the solution of the equation

$$\dot{p}_j + \gamma A p_j + \theta\left(\frac{\|U_j\|_1}{E}\right)(1 - \alpha A^{\frac{1}{2}} p_j) + p_N B(U_j, U_j) = 0, \tag{2.4.81}$$

which has the initial value

$$p_j(0; p_{0j}, \overline{\varphi}) = p_{0j}. \tag{2.4.82}$$

Then

$$\mathcal{F}_0\overline{\varphi}(p_{0j}) = q_j(0), \tag{2.4.83}$$

where $q_j(\tau)$ is the solution of equation

$$\dot{q}_j + \gamma A Q_N q_j + \theta\Big(\frac{\|U_j\|_1}{E}\Big)(-\alpha A^{\frac{1}{2}} Q_N U_j) + Q_N B(U_j, U_j) = 0. \tag{2.4.84}$$

Let

$$\begin{cases} \delta(\tau) = p_1(\tau; p_{01}; \overline{\varphi}) - p_2(\tau; p_{02}; \overline{\varphi}), \\ \Delta(\tau) = q_1(\tau) - q_2(\tau). \end{cases} \tag{2.4.85}$$

Then $\delta(\tau)$, $\Delta(\tau)$ satisfy the equations

$$\dot{\delta} + \gamma A \delta + P_N D(U_1, U_2) = 0, \quad \delta(0) = p_{01} - p_{02}, \tag{2.4.86}$$

$$\dot{\Delta} + \gamma A \Delta + Q_N D(U_1, U_2) = 0, \tag{2.4.87}$$

where

$$\begin{aligned} D(U_1, U_2) = \theta\Big(\frac{\|U_1\|_1}{E}\Big)(-\alpha A^{\frac{1}{2}} U_1) + B(U_1, U_1) \\ - \theta\Big(\frac{\|U_2\|_1}{E}\Big)(-\alpha A^{\frac{1}{2}} U_2) - B(U_2, U_2). \end{aligned} \tag{2.4.88}$$

$$\begin{aligned} B(U, U) &= F(U)_x + \Phi(U)_{xx} + \theta\Big(\frac{\|U_1\|_1}{E}\Big)(-g(U) - h(x)) \\ &= B_1(U, U) + \Phi(U)_{xx}. \end{aligned} \tag{2.4.89}$$

For simplicity, introduce the shorthand

$$\theta_1 = \theta\Big(\frac{\|U_1\|_1}{E}\Big), \quad \theta_2 = \theta\Big(\frac{\|U_2\|_1}{E}\Big).$$

Then $D(U_1, U_2)$ can be written as

$$\begin{aligned} D(U_1, U_2) &= \frac{1}{2}(\theta_1 + \theta_2)(-\alpha A^{\frac{1}{2}}(U_1 - U_2)) + \frac{1}{2}(\theta_1 - \theta_2)(-\alpha A^{\frac{1}{2}}(U_1 + U_2)) \\ &\quad + (F'(U_1) - F'(U_2))U_{1x} + F'(U_2)(U_{1x} - U_{2x}) \\ &\quad + (\Phi(U_1) - \Phi(U_2))_{xx} - \frac{1}{2}(\theta_1 + \theta_2)(g(U_1) - g(U_2)) \\ &\quad - \frac{1}{2}(\theta_1 - \theta_2)(g(U_1) + g(U_2)) \\ &= D_1(U_1, U_2) + (\Phi(U_1) - \Phi(U_2))_{xx}. \end{aligned} \tag{2.4.90}$$

Taking the inner product of equation (2.4.86) and $A^{\frac{1}{2}}\delta$, we get

$$\frac{1}{2}\frac{\mathrm{d}}{\mathrm{d}t}\|A^{\frac{1}{4}}\delta\|^2 + \gamma\|A^{\frac{3}{4}}\delta\|^2 + (P_N D(U_1, U_2), A^{\frac{1}{2}}\delta) = 0, \tag{2.4.91}$$

where

$$
\begin{aligned}
(D(U_1,U_2),A^{\frac{1}{2}}\delta) &= (D_1(U_1,U_2)+(\Phi(U_1)-\Phi(U_2))_{xx},A^{\frac{1}{2}}\delta)\\
&= (D_1,A^{\frac{1}{2}}\delta)-((\Phi(U_1)-\Phi(U_2))_x,\mathscr{A}^{\frac{3}{4}}\delta)\\
&= (D_1,A^{\frac{1}{2}}\delta)-((\Phi'(U_1)-\Phi'(U_2))U_{1x}+\Phi'(U_2)(U_{1x}-U_{2x}),A^{\frac{3}{4}}\delta).
\end{aligned}
$$

Therefore

$$
\begin{aligned}
|(D(U_1,U_2),A^{\frac{1}{2}}\delta)| \le{}& \frac{\alpha}{2}(\theta_1+\theta_2)\|A^{\frac{3}{4}}\delta\|\|A^{\frac{1}{4}}(U_1-U_2)\|\\
&+\frac{\alpha}{2}(\theta_1-\theta_2)\|A^{\frac{1}{4}}(U_1+U_2)\|\|A^{\frac{3}{4}}\delta\|\\
&+\|F''(U_2)\|_{L_\infty}\|U_{1x}\|\|U_1-U_2\|_{L_\infty}\|A^{\frac{1}{2}}\delta\|\\
&+\|F'(U_2)\|_{L_\infty}\|U_1-U_2\|_1\|A^{\frac{1}{2}}\delta\|\\
&+\frac{1}{2}(\theta_1+\theta_2)\|g'(U)\|_{L_\infty}\|U_1-U_2\|\|A^{\frac{1}{2}}\delta\|\\
&+\frac{1}{2}|\theta_1-\theta_2|\|g(U_1)+g(U_2)\|\|A^{\frac{1}{2}}\delta\|\\
&+\|\Phi''(U)\|_{L_\infty}\|U_{1x}\|\|U_1-U_2\|_{L_\infty}\|A^{\frac{3}{4}}\delta\|\\
&+\|\Phi'(U_2)\|_{L_\infty}\|U_1-U_2\|_1\|A^{\frac{3}{4}}\delta\|,
\end{aligned}
$$

where

$$
\begin{aligned}
\|U_1-U_2\|_1 &\le \|p_1-p_2\|_1+\|\overline{\varphi}(p_1)-\overline{\varphi}(p_2)\|_1\\
&\le (1+l)\|A^{\frac{1}{4}}\delta\| \le 2\|A^{\frac{1}{4}}\delta\|,\\
|\theta_1-\theta_2| &= \left|\theta\left(\frac{\|U_1\|_1}{E}\right)-\theta\left(\frac{\|U_2\|_1}{E}\right)\right|\\
&\le \frac{2}{E}\|U_1-U_2\|_1\\
&\le \frac{4}{E}\|A^{\frac{1}{4}}\delta\|.
\end{aligned}
$$

Then

$$
\begin{aligned}
|(D(U_1,U_2),A^{\frac{1}{2}}\delta| \le{}& 2\alpha\|A^{\frac{3}{4}}\delta\|\|A^{\frac{1}{2}}\delta\|+2\alpha\|U_1+U_2\|_1\|A^{\frac{1}{4}}\delta\|\|A^{\frac{3}{4}}\delta\|\\
&+C\|F''(U)\|_{L_\infty}\|U_{1x}\|\|A^{\frac{1}{4}}\delta\|\|A^{\frac{1}{2}}\delta\|\\
&+2\|F'(U_2)\|_{L_\infty}\|A^{\frac{1}{4}}\delta\|\|A^{\frac{1}{2}}\delta\|\\
&+2\|g'(U)\|_{L_\infty}\|A^{\frac{1}{4}}\delta\|\|A^{\frac{1}{2}}\delta\|\\
&+2\|g(U_1)+g(U_2)\|\|A^{\frac{1}{4}}\delta\|\|A^{\frac{1}{2}}\delta\|\\
&+2\|\Phi''(U)\|_{L_\infty}\|U_{1x}\|\|A^{\frac{1}{4}}\delta\|\|A^{\frac{3}{4}}\delta\|
\end{aligned}
$$

$$
\begin{aligned}
&+ 2\|\Phi'(U_2)\|_{L_\infty}\|A^{\frac{1}{4}}\delta\|\|A^{\frac{3}{4}}\delta\| \\
&\leq \frac{\gamma}{2}\|A^{\frac{3}{4}}\delta\|^2 + C_1(\|A^{\frac{3}{4}}\delta\| + \|A^{\frac{1}{4}}\delta\|)\|A^{\frac{1}{4}}\delta\|.
\end{aligned}
$$

Thus by equation (2.4.91) we get

$$
\begin{aligned}
&\left|\frac{1}{2}\frac{\mathrm{d}}{\mathrm{d}t}\|A^{\frac{1}{4}}\delta\|^2 + \gamma\|A^{\frac{3}{4}}\delta\|^2\right| \\
&\quad\leq C_1(\|A^{\frac{3}{4}}\delta\| + \|A^{\frac{1}{4}}\delta\|)\|A^{\frac{1}{4}}\delta\| + \frac{\gamma}{2}\|A^{\frac{3}{4}}\delta\|^2.
\end{aligned}
\tag{2.4.92}
$$

In a similar way, we get

$$
\begin{aligned}
\frac{1}{2}\frac{\mathrm{d}}{\mathrm{d}t}\|A^{\frac{1}{4}}\Delta\|^2 + \frac{\gamma}{2}\|A^{\frac{3}{4}}\Delta\|^2 &\leq C_1'(\|A^{\frac{3}{4}}\Delta\| + \|A^{\frac{1}{4}}\Delta\|)\|U_1 - U_2\|_1 \\
&\leq C_2(\|A^{\frac{3}{4}}\Delta\| + \|A^{\frac{1}{4}}\Delta\|)\|A^{\frac{1}{4}}\delta\|.
\end{aligned}
\tag{2.4.93}
$$

Let $y = A^{\frac{1}{4}}\delta$, $z = A^{\frac{1}{4}}\Delta$. Then from equations (2.4.92)–(2.4.93) we get

$$
\begin{aligned}
\frac{\mathrm{d}}{\mathrm{d}t}\|y\|^2 &\geq -\frac{2}{3\gamma}\lambda_N\|y\|^2 - C_1(\lambda_N^{\frac{1}{2}}\|y\| + \|y\|)\|y\| \\
\frac{\mathrm{d}}{\mathrm{d}t}\|z\|^2 + \gamma\|A^{\frac{3}{4}}\Delta\|^2 &\leq 2C_2(\|A^{\frac{3}{4}}\Delta\| + \|z\|)\frac{\|z\|}{\gamma_1}.
\end{aligned}
$$

By virtue of

$$
\begin{aligned}
\gamma\|A^{\frac{3}{4}}\Delta\|^2 &\geq \gamma\lambda_{N+1}\|z\|^2, \\
\frac{2C_2}{\gamma_1}\|A^{\frac{3}{4}}\Delta\|\|z\| &\leq \gamma/2\|A^{\frac{3}{4}}\Delta\|^2 + \frac{1}{2\gamma}\left(\frac{2C_2}{\gamma_1}\right)^2\|z\|^2,
\end{aligned}
$$

we thus have

$$
\frac{\mathrm{d}}{\mathrm{d}t}\|y\| + \left(\frac{3\gamma}{4}\lambda_N + C_1\lambda_N^{\frac{1}{2}} + 1\right)\|y\| \geq 0, \tag{2.4.94}
$$

$$
\frac{\mathrm{d}}{\mathrm{d}t}\|z\| + \left(\frac{\gamma}{4}\lambda_{N+1} - C_3(\gamma,\gamma_1)\right)\|z\| \leq 0, \tag{2.4.95}
$$

where

$$
C_3(\gamma,\gamma_1) = \frac{C_2}{\gamma\gamma_1^2} + \frac{2C_2}{\gamma_1}.
$$

By Lemma 2.4.14, where

$$
\begin{aligned}
&d = \gamma/4\lambda_{N+1} - C_3(\gamma,\gamma_1) > 0, \quad N \geq N_0, \\
&C = 0, \quad b = 0, \quad a = 3\gamma/4\lambda_N + C_1\lambda_N^{\frac{1}{2}} + 1,
\end{aligned}
$$

$$d - a - \gamma_1 b - \gamma_1^{-1} C = \frac{\gamma}{4}\lambda_{N+1} - \frac{3\gamma}{4}\lambda_N - C_1\lambda_N^{\frac{1}{2}} - 1 - C_3 > 0, \quad N \geq N_0,$$

we arrive at

$$\|z(t)\| \leq \|z(t_0)\| \exp\left\{-\left(\frac{\gamma}{4}\lambda_{N+1} - C_3\right)(t - t_0)\right\} \tag{2.4.96}$$

$$\|z(0)\| \leq \|z(t_0)\| \exp\left\{-\left(\frac{\gamma}{4}\lambda_{N+1} - C_3\right)t_0\right\}.$$

In the above equation, letting $t_0 \to -\infty$, by Lemma 2.4.10, since

$$\sup\{\|A^{\frac{1}{4}}\Delta(t)\| = \|z(t)\| : -\infty < t \leq 0\} < \infty,$$

we have

$$\|z(0)\| = 0.$$

But $\|z(0)\| > l\|y(0)\|(\neq 0)$ is not possible, so it follows that

$$\|z(0)\| \leq l\|y(0)\|,$$

namely

$$\|A^{\frac{1}{4}}(\mathscr{F}_0\overline{\varphi}(p_{01}) - \mathscr{F}_0\overline{\varphi}(p_{02}))\| \leq l\|A^{\frac{1}{4}}(p_{01} - p_{02})\|. \tag{2.4.97}$$

If N is large enough, $N \geq \max(N_0, N_2') = N_1(\gamma, \gamma_1, \widetilde{D})$. Theorem 2.4.2 is proved. □

Now we prove the contractivity of functional operator $\mathscr{F}_0$.

Theorem 2.4.3. *For any $b, l \in (0, 1]$, there exists $N_2 = N_2(\gamma, \gamma_1, b, l, \widetilde{D})$ such that when $N + 1 \geq N_2$ we have*

$$d(\mathscr{F}_0\overline{\varphi_1}, \mathscr{F}_0\overline{\varphi_2}) \leq \frac{1}{2}d(\overline{\varphi_1}, \overline{\varphi_2}), \quad \forall \overline{\varphi_1}, \overline{\varphi_2} \in \mathscr{F}_{b,l}^{\frac{1}{4}}. \tag{2.4.98}$$

Proof. Let $\overline{\varphi_1}, \overline{\varphi_2} \in \mathscr{F}_{b,l}^{\frac{1}{4}}$ and set

$$d_0 = d(\overline{\varphi_1}, \overline{\varphi_2}) = \max\{\|\overline{\varphi_1}(p) - \overline{\varphi_2}(p)\|_1, \ p \in PD(A^{\frac{1}{4}})\}. \tag{2.4.99}$$

In the proof of Theorem 2.4.2, set

$$U_j = p_j + \overline{\varphi_j}(p_j), \quad j = 1, 2, \tag{2.4.100}$$

where $p_j = p_j(\tau; p_0; \overline{\varphi_j})$ is the solution of equation (2.4.81), namely

$$p_j' + \gamma A p_j + \theta\left(\frac{\|U_j\|_1}{E}\right)(-\alpha A^{\frac{1}{2}} p_j) + P_N B(U_j, U_j) = 0, \tag{2.4.101}$$

satisfying the same initial value condition

$$p_j(0; p_0; \overline{\varphi_j}) = p_0. \tag{2.4.102}$$

Let $q_j(\tau)$ be the bounded solution of equation (2.4.84) in $(-\infty, 0]$, that is,

$$q_j' + \gamma A Q_N q_j + \theta\Big(\frac{\|U_j\|_1}{E}\Big)(-\alpha A^{\frac{1}{2}} Q_N U_j) + Q_N B(U_j, U_j) = 0. \tag{2.4.103}$$

Let

$$\begin{aligned} \delta(\tau) &= p_1(\tau; p_0, \overline{\varphi_1}) - p_2(\tau; p_0, \overline{\varphi_2}), \\ \Delta(\tau) &= q_1(\tau) - q_2(\tau). \end{aligned} \tag{2.4.104}$$

Then by equations (2.4.101)–(2.4.102), we have

$$\dot{\delta} + \gamma A\delta + P_N D(U_1, U_2) = 0, \quad \delta(0) = 0, \tag{2.4.105}$$

$$\dot{\Delta} + \gamma A\Delta + Q_N D(U_1, U_2) = 0. \tag{2.4.106}$$

Similar to inequality (2.4.92) and equation (2.4.93), we have

$$\left|\frac{1}{2}\frac{\mathrm{d}}{\mathrm{d}t}\|A^{\frac{1}{4}}\delta\|^2 + \frac{\gamma}{2}\|A^{\frac{3}{4}}\delta\|^2\right| \le C_1(\|A^{\frac{3}{4}}\delta\| + \|A^{\frac{1}{4}}\delta\|)\|U_1 - U_2\|_1, \tag{2.4.107}$$

$$\frac{1}{2}\frac{\mathrm{d}}{\mathrm{d}t}\|A^{\frac{1}{4}}\delta\|^2 + \frac{\gamma}{2}\|A^{\frac{3}{4}}\delta\|^2 \le C_2(\|A^{\frac{3}{4}}\Delta\| + \|A^{\frac{1}{4}}\Delta\|)\|U_1 - U_2\|_1. \tag{2.4.108}$$

Since

$$\begin{aligned} \|U_1 - U_2\|_1 &\le \|p_1 - p_2\|_1 + \|\overline{\varphi_1}(p_1) - \overline{\varphi_1}(p_2)\|_1 \\ &\le \|p_1 - p_2\|_1 + \|\overline{\varphi_1}(p_1) - \overline{\varphi_1}(p_2)\|_1 + \|\overline{\varphi_1}(p_1) - \overline{\varphi_1}(p_2)\|_1 \\ &\le 2\|A^{\frac{1}{4}}\Delta\| + d_0, \end{aligned}$$

setting $y = A^{\frac{1}{4}}\delta$, $z = A^{\frac{1}{4}}\Delta$, we then get

$$\left|\frac{1}{2}\frac{\mathrm{d}}{\mathrm{d}t}\|y\|^2 + \frac{\gamma}{2}\|A^{\frac{1}{2}}y\|^2\right| \le C_1(\|A^{\frac{1}{2}}y\| + \|y\|)(2\|y\| + d_0), \tag{2.4.109}$$

$$\begin{aligned} \frac{\mathrm{d}}{\mathrm{d}t}\|z\|^2 + \frac{\gamma}{2}\|A^{\frac{1}{2}}z\|^2 &\le C_2(\|A^{\frac{1}{2}}z\| + \|z\|)(2\|y\| + d_0) \\ &\le \frac{C_2}{\gamma_1}(\|A^{\frac{1}{2}}z\| + \|z\|)(2\|z\| + \gamma_1 d_0), \quad \text{for } \|z\| > \gamma_1\|y\|, \end{aligned} \tag{2.4.110}$$

where

$$\|y(0)\| = 0, \quad \sup\{\|A^{\frac{1}{4}}\Delta(t)\| = \|z(t)\| : -\infty < t \le 0\} < \infty. \tag{2.4.111}$$

Consider the open subset $J = \{t \in (-\infty, 0] : \|z\| > \frac{1}{2}d_0\}$ on $(-\infty, 0]$. If $0 \in J$, set $J_0 = (-T, 0]$ to the part of J including 0. On J_0, by inequalities (2.4.110)–(2.4.111) we get

$$\begin{aligned}
\left|\frac{1}{2}\frac{\mathrm{d}}{dt}\|y\|^2 + \frac{\gamma}{2}\|A^{\frac{1}{2}}y\|^2\right| &\le 2C_1(\|A^{\frac{1}{2}}y\| + \|y\|)^2(\|y\| + \|z\|)\\
\frac{1}{2}\frac{\mathrm{d}}{dt}\|z\|^2 + \frac{\gamma}{2}\|A^{\frac{1}{2}}z\|^2 &\le \frac{2C_2}{\gamma_1}(\|A^{\frac{1}{2}}z\| + \|z\|)(\|z\| + \gamma_1\|z\|)\\
&= \frac{2C_2}{\gamma_1}(\gamma_1 + 1)[\|A^{\frac{1}{2}}z\|\|z\| + \|z\|^2]\\
&\le \frac{\gamma}{4}\|A^{\frac{1}{2}}z\|^2 + \frac{2C_1(\gamma_1 + 1)}{\gamma_1}\left(\frac{2C_1(\gamma_1 + 1)}{\gamma\gamma_1} + 1\right)\|z\|^2,
\end{aligned}$$

hence we have

$$\frac{\mathrm{d}}{dt}\|y\| + \frac{\gamma}{2}\lambda_N\|y\| + 2C_1(\lambda_N^{\frac{1}{2}} + 1)(\|y\| + \|z\|) \ge 0, \tag{2.4.112}$$

$$\frac{\mathrm{d}}{dt}\|z\| + \frac{\gamma}{4}\lambda_{N+1}\|z\| - \frac{2C_1(\gamma_1 + 1)}{\gamma_1}\left(\frac{2C_1(\gamma_1 + 1)}{\gamma\gamma_1} + 1\right)\|z\| \le 0. \tag{2.4.113}$$

Since $\|z(0)\| \ge \frac{1}{2}d_0$, $\|y(0)\| = 0$, and

$$\begin{aligned}
d &= \frac{\gamma}{4}\lambda_{N+1} - \frac{2C_1(\gamma_1 + 1)}{\gamma_1}\left(\frac{2C_1(\gamma_1 + 1)}{\gamma\gamma_1} + 1\right) > 0, \quad N + 1 \ge N_2^0,\\
c &= 0, \quad a = \frac{\gamma}{2}\lambda_N + 2C_1(\lambda_{N+1}^{\frac{1}{2}} + 1) > 0,\\
b &= 2C_1(\lambda_N^{\frac{1}{2}} + 1) > 0,\\
d - a - \gamma_1 b - \gamma_1^{-1}c &= \frac{\gamma}{4}\lambda_{N+1} - \frac{\gamma}{2}\lambda_N - 2C_1(\lambda_{N+1}^{\frac{1}{2}} + 1)\\
&\quad - 2C_1\gamma_1(\lambda_{N+1}^{\frac{1}{2}} + 1) - \frac{2C_1(\gamma_1 + 1)}{\gamma_1}\left(\frac{2C_1(\gamma_1 + 1)}{\gamma\gamma_1} - 1\right)\\
&= \frac{\gamma}{4}\lambda_{N+1} - \frac{\gamma}{2}\lambda_N - 2C_1(\gamma_1 + 1)\lambda_N^{\frac{1}{2}} - 2C_1(\gamma_1 + 1)\\
&\quad - \frac{2C_1(\gamma_1 + 1)}{\gamma_1}\left(\frac{2C_1(\gamma_1 + 1)}{\gamma\gamma_1} + 1\right) > 0, \quad N \ge N_2^1.
\end{aligned}$$

By Lemma 2.4.14, when $N + 1 \ge \max(N_2^0, N_2^1) = N_2$, we have

$$\|z(0)\| \le \|z(t)\|\exp(dt), \quad t \in J_0. \tag{2.4.114}$$

Since $\sup\{\|z(t)\| : -\infty \le t \le 0\} < \infty$, if $T = \infty$ then we have

$$\|z(0)\| = 0 \le \frac{1}{2}d_0.$$

Otherwise by equation (2.4.114) and for $-T \notin J$, we have

$$\|z(0)\| \le \|z(-T)\| \le \frac{1}{2}d_0.$$

Thus we prove

$$\|\mathscr{F}_0\overline{\varphi_1}(p_0) - \mathscr{F}_0\overline{\varphi_2}(p_0)\| = \|z(0)\| \le \frac{1}{2}d_0 = \frac{1}{2}d(\overline{\varphi_1}, \overline{\varphi_2}).$$

Since $p_0 \in PD(A^{\frac{1}{2}})$ is arbitrary, then we have

$$d(\mathscr{F}_0\overline{\varphi_1}, \mathscr{F}_0\overline{\varphi_2}) \le \frac{1}{2}d(\overline{\varphi_1}, \overline{\varphi_2}), \quad N+1 \ge N_2.$$

□

Now we reveal the structure of the inertial manifold M of PGKS equation, namely:

(1) $M = \text{graph}(\Phi)$, where Φ is a Lipschitz continuous mapping from $PD(A^{\frac{1}{4}})$ to $QD(A^{\frac{1}{4}})$;
(2) Φ has compact support, $\text{supp}\,\Phi \subset \{p \in PD(A^{\frac{1}{4}}) : \|A^{\frac{1}{4}}p\| < 2E\}$;
(3) $S_p(t)M \subset M, t \ge 0$;
(4) There exists a constant $\lambda > 0$, such that for any $u_0 \in D(A^{\frac{1}{4}})$, there exists $\mu_0 > 0$ (u_0 is established uniformly for bounded set) such that

$$\text{dist}(S_p(t)u_0, M) \le \mu_0 \exp(-\lambda t), \quad t \ge 0;$$

(5) The attractor $\mathscr{A}_p$ of PGKS equation is in M.

Since $\Phi \in \mathscr{A}_{b,l}^{\frac{1}{4}}$ is known, claims (1) and (2) are true. Now we prove that (3), (4) and (5) are also true.

Theorem 2.4.4. *Let $b, l \in (0,1]$ and*

$$N+1 > N_2(\gamma, \gamma_1, b, l, \widetilde{D}).$$

Then $\mathscr{F}_0$ is the unique fixed point $\Phi \in \mathscr{F}_{b,l}^{\frac{1}{4}}$ such that

$$d(\Phi, \mathscr{F}_0\overline{\varphi}) \le \frac{1}{2}d(\Phi, \overline{\varphi}), \quad \forall \overline{\varphi} \in \mathscr{F}_{b,l}^{\frac{1}{4}}, \tag{2.4.115}$$

and the Lipschitz manifold

$$M = \text{graph}(\Phi) = \{p + \overline{\varphi}(p) : p \in PD(A^{\frac{1}{4}})\} \tag{2.4.116}$$

is invariant for $\{S_p(t)\}_{t\ge 0}$.

Proof. The first part of the claim follows from Theorem 2.4.2. Since Φ is the fixed point of $\mathscr{F}_0$, namely $\mathscr{F}_0\Phi = \Phi$, by Theorem 2.4.3 we have

$$d(\Phi, \mathscr{F}_0\overline{\Phi}) = d(\mathscr{F}_0\phi, \mathscr{F}_0\overline{\varphi}) \le \frac{1}{2}d(\phi, \overline{\varphi}).$$

By the definition of $\mathscr{F}_0$ and from $\mathscr{F}_0\Phi = \Phi$, $\Phi \in \mathscr{F}_{b,l}^{\frac{1}{4}}$, we have

$$S_p(t)M \subset M.$$

□

Theorem 2.4.5. *For any $u \in B_{2E}$ and $\tau \geq 0$, we have*

$$\begin{aligned} \operatorname{dist}(S_p(\tau)u, M) &= \inf\{\|A^{\frac{1}{4}}(S_p(\tau)u - v)\|, \ v \in M\} \\ &\leq 4E\exp\{-d\tau\}, \end{aligned} \tag{2.4.117}$$

where M is defined in (2.4.116) and

$$d = \frac{\gamma}{4}\frac{(N+1)^4}{\widetilde{D}^4} - \frac{2C_1(\gamma_1+1)}{\gamma_1}\left(\frac{2C_1(\gamma_1+1)}{\gamma\gamma_1}+1\right) > 0.$$

Proof. For $u \in B_{2E}$ and $\tau > 0$, let

$$p(t) = p(t; p_N S_p(t)u; \Phi)$$

be the solution of equation

$$\frac{dp}{dt} + \gamma A p + \theta\left(\frac{\|U\|_1}{E}\right)(-\alpha A^{\frac{1}{2}} p) + P_N B(U,U) = 0, \tag{2.4.118}$$

where $U = p + \Phi(p)$. Define $v = p(-\tau) + \Phi(p(-\tau))$, and set

$$\begin{aligned} \widetilde{\delta}(t) &= A^{\frac{1}{4}} P_N[S_p(t)u - S_p(t)v], \\ \widetilde{\Delta}(t) &= A^{\frac{1}{4}} Q_N[S_p(t)u - S_p(t)v], \quad t \in [0,\tau]. \end{aligned}$$

Then as in the proof of Theorem 2.4.2, we have

$$\begin{aligned} &\left|\frac{1}{2}\frac{d}{dt}\|\widetilde{\delta}\|^2 + \frac{\gamma}{2}\|A^{\frac{1}{2}}\widetilde{\delta}\|^2\right| \leq C_1(\|A^{\frac{1}{2}}\widetilde{\delta}\| + \|\widetilde{\delta}\|)\|\widetilde{\delta}\|, \\ &\frac{1}{2}\frac{d}{dt}\|\widetilde{\Delta}\|^2 + \frac{\gamma}{2}\|A^{\frac{1}{2}}\widetilde{\Delta}\|^2 \leq C_2(\|A^{\frac{1}{2}}\widetilde{\Delta}\| + \|\widetilde{\Delta}\|)\|\widetilde{\delta}\|. \end{aligned}$$

Since $|\widetilde{\Delta}(0)| \geq |\widetilde{\delta}(0)| = 0$, by Lemma 2.4.14 we then have

$$\begin{aligned} &\|S_p(\tau)u - \{P_N S_p(\tau)u + \Phi[P_N S_p(t)u]\}\|_1 \\ &= \|\widetilde{\Delta}(\tau)\| \leq \|\widetilde{\Delta}(0)\|e^{-d\tau}. \end{aligned} \tag{2.4.119}$$

If $\|p(-\tau)\|_1 \geq 2E$, then $Q_N v = 0$ and

$$\begin{aligned} \|\widetilde{\Delta}(0)\| &= \|A^{\frac{1}{4}} Q_N[S_p(t)u - S_p(0)v]\| \\ &= \|A^{\frac{1}{4}} Q_N u\|_1 = \|Q_N u\|_1 \leq 2E. \end{aligned}$$

If $\|p(-\tau)\|_1 < 2E$, then

$$\begin{aligned} \|Q_N v\|_1 &= \|\Phi[p(-\tau)]\|_1 = \|\Phi[p(-\tau)] - \Phi(2E)\|_1 \\ &\leq l(2E - \|p(-\tau)\|_1) \leq 2E. \end{aligned}$$

Hence $\|\widetilde{\Delta}(0)\| \leq \|Q_N(u - v)\|_1 \leq 4E$, which yields

$$\|\widetilde{\Delta}(0)\| \leq \max(2E, 4E) = 4E.$$

The theorem has been proved. □

Corollary 2.4.1. *The attractor $\mathscr{A}_p$ of PGKS equation is in the set M.*

Proof. Since $\mathscr{A}_p = S_p(t)\mathscr{A}_p \subset S(t)B_{2E}$, $\forall t \geq 0$, where B_{2E} is an absorbing set, for $u \in \mathscr{A}_p$, by inequality (2.4.119) we have

$$\operatorname{dist}(u, M) \leq 4Ee^{-dt}, \quad d > 0.$$

When $t \to \infty$, we can get that $\mathscr{A}_p$ is included in the set M. □

By the above results, we conclude that

Theorem 2.4.6. *For any $N \geq \widetilde{N}(\gamma, \gamma_1, b, l, \widetilde{D})$, there exists an inertial manifold M of PGKS equation, which is the graph of the fixed point of equation* (2.4.64) *defined by the operator $\mathscr{F}_0$ on $\mathscr{F}_{b,l}^{\frac{1}{4}}$.*

Now we will consider the relations between the inertial manifold of PGKS equation and that of GKS equation (2.4.1). Moreover, we will consider the Galerkin approximation of inertial manifold M for PGKS equation.

Lemma 2.4.15. *The attractor $\mathscr{A}$ and absorbing set of GKS equation* (2.4.1) *given by*

$$B_{\frac{3}{2}E} = \left\{u \in H^1,\ \|u\|_1 \leq \frac{3}{2}(E_0 + E_1)\right\}$$

have the following relation:

$$\mathscr{A} = \bigcap_{t\geq 0} S(t)B_{\frac{3}{2}E}. \tag{2.4.120}$$

Proof. By the definition of the global attractor $\mathscr{A}$, we know

$$\mathscr{A} = \bigcap_{t\geq 0}\bigcup_{t\geq \tau} \overline{S(t)B_{\frac{3}{2}E}} \tag{2.4.121}$$

where the closure is taken in L_2. Denoting $\mathscr{A}' = \bigcap_{t\geq 0} S(t)B_{\frac{3}{2}E}$, from equation (2.4.121) we know $\mathscr{A}' \subset \mathscr{A}$. On the other hand, since $B_{\frac{3}{2}E}$ is absorbing, there exists $t_1 > 0$ such that

$$S(t)B_{\frac{3}{2}E} \subset B_{\frac{3}{2}E}, \quad t \geq t_1.$$

Hence

$$\mathscr{A} \subset \bigcup_{t\geq\tau} S(t)B_{\frac{3}{2}E} \subset S(\tau - t_1)B_{\frac{3}{2}E}, \quad \tau \geq t_1,$$

and it follows that $\mathscr{A} \subset \mathscr{A}'$, thus we have

$$\mathscr{A} = \mathscr{A}'.$$ □

Theorem 2.4.7. *Let M be the inertial manifold of the PGKS equation, which satisfies:*
(1) $M = \operatorname{graph}(\Phi)$, *where Φ is the Lipschitz continuous mapping, $P_N D(\mathscr{A}^{\frac{1}{4}}) \to Q_N D(\mathscr{A}^{\frac{1}{4}})$;*
(2) *Φ has compact support, when Φ is outside of B_{2E}, $\Phi \geq 0$.*

Then in an open neighborhood of M, M_2 is an inertial manifold of GKS equation (2.4.1).

Proof. Let $B_{\frac{1}{2}E} = \{u \in H^1 : \|u\|_1 \leq \frac{1}{2}E\}$. Then there exists $t_0 \geq 0$ such that

$$B_1 = \bigcup_{t \geq t_0} S(t) B_{\frac{1}{2}E} \subset B_E \{u \in H^1 : \|u\|_1 \leq E\}.$$

Noting that $S(t)B_1 \subset B_1$, $S_p(t)B_1 = S(t)B$, $\forall t \geq 0$, and setting

$$M_1 = M \cap B_1,$$

we then have

$$\begin{aligned} S(t)M_1 &= S_p(t)M_1 \subset S_p(t)M \cap S_p(t)B_1 \\ &= S_p(t)M \cap S(t)B_1 \subset M \cap B_1 = M_1, \quad \forall t \geq 0. \end{aligned}$$

Let

$$M_1 = \partial M + M_2,$$

that is, M_2 is the set of inner points of M_1 in M. Because $S_p(t)M$ is a homeomorphism on M, then $S(t)M_2$ is also in the interior of $S_p(t)M_1 = S_p(t)M_1 \subset M_1$. Therefore M_2 satisfies properties (1) and (2), namely:
(1) $M_2 = \operatorname{graph}(\Phi)$, where Φ is a Lipschitz continuous mapping, $P_N D(A^{\frac{1}{4}}) \to Q_N D(A^{\frac{1}{4}})$;
(2) $S(t)M_2 \subset M_2$, $t \geq 0$.

As in [182], we can show that the attractor $\mathscr{A}$ of GKS equation $\subset M_2$. And there exists

$$\operatorname{dist}(S(t)u, M_2) \leq \mu_1 e^{-\delta t}, \quad \mu_1 > 0,\ \delta > 0,\ t \geq 0,\ u \in B,$$

where B is any bounded set in B_1. Then M_2 is an inertial manifold of GKS equation. □

Now consider the mth order approximation of PGKS equation: Acting with P_m on PGKS equation, we get

$$\begin{aligned} &\frac{\mathrm{d}}{\mathrm{d}t} u^{(m)} + \gamma A u^{(m)} + \theta\Big(\frac{\|u^{(m)}\|_1}{E}\Big)(-\alpha A^{\frac{1}{2}} u^{(m)}) + P_m B(u^{(m)}, u^{(m)}) = 0, \\ &\quad u^{(m)} \in P_m D(A^{\frac{1}{4}}). \end{aligned} \tag{2.4.122}$$

When $m \geq N \geq \widetilde{N}$, the Galerkin approximate equation (2.4.122) has m-dimensional inertial manifold $M^{(m)}$, and we have the following theorem:

Theorem 2.4.8. *Let $m \geq N \geq \widetilde{N}$, then the Galerkin approximate equation* (2.4.122) *has inertial manifold $M^{(m)} = \{p + \Phi_m(p), p \in P_N D(A^{\frac{1}{4}})\}$, where Φ_m is a fixed point of operator $P_m \mathscr{F}$,*

$$P_m \mathscr{F}_0 : P_m \mathscr{F}^{\frac{1}{4}}_{b,l} \to P_m \mathscr{F}^{\frac{1}{4}}_{b,l}, \quad P_m \mathscr{F}^{\frac{1}{4}}_{b,l} = \{P_m \varphi : \varphi \in P_m \mathscr{F}^{\frac{1}{4}}_{b,l}\}.$$

Suppose m is large enough, and we can set up an approximate estimation of Φ_m.

Theorem 2.4.9. *Suppose $m \geq \widetilde{N}$ is fixed. Then we have*

$$d(\Phi, \Phi_m) \leq \frac{C(\widetilde{D})}{(m+1)^3} \tag{2.4.123}$$

where

$$d(\varphi_1, \varphi_2) = \max\{\|\varphi_1(p) - \varphi_2(p)\|_1, \ p \in P_N D(A^{\frac{1}{4}})\},$$

and N is the dimension of inertial manifolds graph Φ *and* graph Φ_m.

Proof. Let $m > N \geq \widetilde{N}$, and set

$$U^{(m)} = p^{(m)} + \Phi_m(p^{(m)}), \quad P_m U^{(m)} = U^{(m)}, \tag{2.4.124}$$

where $p^{(m)}(\tau, p_0; \Phi_m)$ is the solution of equation

$$\frac{\mathrm{d}}{\mathrm{d}t} u^{(m)} + \gamma A u^{(m)} + \theta\Big(\frac{\|U^{(m)}\|_1}{E}\Big)(-\alpha A^{\frac{1}{2}} p^{(m)}) + P_N B(U^{(m)}, U^{(m)}) = 0, \tag{2.4.125}$$

$$p^{(m)}(0; p_0; \Phi_m) = p_0, \tag{2.4.126}$$

which yields

$$\begin{aligned}\Phi_m(p_0) &= P_m \mathscr{F}_0 \Phi_m(p_0) \\ &= -\int_{-\infty}^{0} e^{\tau\gamma A Q_N} \cdot \theta\Big(\frac{\|U^{(m)}\|_1}{E}\Big) Q_N[-\alpha A^{\frac{1}{2}} U^{(m)} \\ &\quad + P_N B(U^{(m)}, U^{(m)})]\mathrm{d}\tau.\end{aligned} \tag{2.4.127}$$

Now considering the definition of $\mathscr{F}_0 \Phi_m$, equations (2.4.58)–(2.4.59) are consistent with equations (2.4.124)–(2.4.125). Since $P_N D(A^{\frac{1}{2}}) \subset P_m D(A^{\frac{1}{2}})$, $P_m U^{(m)} = U^{(m)}$, then

$$\mathscr{F}_0 \Phi_m(p_0) = -\int_{-\infty}^{0} e^{\tau\gamma A Q_N} \cdot \theta\Big(\frac{\|U^{(m)}\|_1}{E}\Big) Q_N[-\alpha A^{\frac{1}{2}} U^{(m)} + P_N B(U^{(m)}, U^{(m)})]\mathrm{d}\tau. \tag{2.4.128}$$

From equations (2.4.127)–(2.4.128), we arrive at

$$\begin{aligned}
&\|\Phi_m(p_0) - \mathscr{F}_0\Phi_m(p_0)\|_1 \\
&\quad = \|P_m\mathscr{F}_0\Phi_m(p_0) - \mathscr{F}_0\Phi_m(p_0)\|_1 \\
&\quad = \left|\int_{-\infty}^{0} (AQ_N)^{\frac{1}{4}} e^{\tau\gamma AQ_N} \cdot Q_N(1-P_m)B(U^{(m)}, U^{(m)})\mathrm{d}\tau\right| \\
&\quad \le \int_{-\infty}^{0} (AQ_m)^{\frac{1}{4}} e^{\tau\gamma AQ_m} \cdot Q_mB(U^{(m)}, U^{(m)})\mathrm{d}\tau \\
&\quad \le \frac{Ce^{\frac{-\gamma}{4}}}{(\lambda_{m+1})^{\frac{3}{4}}}.
\end{aligned}$$

□

3 The approximate inertial manifold

In the previous discussions on the global attractor and inertial manifold, we saw that the global attractor may not be smooth. Although the inertial manifold is smooth, we must solve the integral equations on an infinite interval to seek them, which brings much trouble in the calculation. Therefore it is natural that we would like to use the approximate and smooth manifold which can be obtained more easily to approximate the global attractor and inertial manifold. This chapter describes recent developments [213, 214] on approximate inertial manifold. In the following, we would like to use the two-dimensional Navier–Stokes equations as an example to illustrate the variety of approximate inertial manifolds and their error estimation.

3.1 Two-dimensional Navier–Stokes equation

Let Navier–Stokes equation have the following form [31, 60, 61]:

$$\frac{du}{dt} + \nu A\mu + B(u,u) = f, \tag{3.1.1}$$

$$u(0) = u_0, \tag{3.1.2}$$

where $Au = -P\Delta u$, $\forall u \in D(A)$, $B(u,v) = P[(u\cdot\nabla)\omega]$, $\forall u, \omega \in D(A)$, here P means the orthogonal projection from $(L^2(\Omega))^2$ to H, while H means the closure of V with respect to $(L^2(\Omega))^2$. When $u|_{\partial\Omega} = 0$, $V = \{v \in \{C_0(\Omega)\}^2, \operatorname{div} v = 0\}$, here A is a linear unbounded self-adjoint operator, A^{-1} is compact, $D(A)$ is dense in H. Hence, H has the orthogonal basis $\{\omega_j\}_{j=1}^{\infty}$, $A\omega_j = \lambda_j\omega_j$, $j = 1, 2, \ldots$, which are the eigenvectors of operator A, and

$$0 < \lambda_1 \le \lambda_2 \le \cdots,$$

eigenvalues λ_m satisfy

$$C_0\lambda_1 m \le \lambda_m \le C_1\lambda_1 m, \quad m = 1, 2, \ldots, \tag{3.1.3}$$

where C_0, C_1 are some known constants. In the following $C_0, C_1, \ldots$ represent positive constants.

It is easy to verify that $B(u,v)$ satisfies

$$|(B(u,v),\omega)| \le C_2|u|^{\frac{1}{2}}\|u\|^{\frac{1}{2}}\|v\|^{\frac{1}{2}}|\omega|^{\frac{1}{2}}\|\omega\|^{\frac{1}{2}}, \quad \forall u, v, \omega \in V, \tag{3.1.4}$$

$$|(B(u,v),\omega)| \le C_3\|u\|_{L^\infty(\Omega)}\|v\||\omega|, \quad \forall u \in D(A),\ \forall v \in V,\ \forall \omega \in H, \tag{3.1.5}$$

where $u \in H$, $|u|^2 = \int_\Omega |u(x)|^2 dx$; $u \in V$, $\|u\|^2 = \int_\Omega |\nabla v(x)|^2 dx$. By Brezis–Gallouet inequality, we have

$$\|u\|_{L^\infty(\Omega)} \le C_4\|u\|\left[1 + \lg\left[\frac{|Au|}{\lambda_1^{\frac{1}{2}}\|u\|}\right]\right]^{\frac{1}{2}}, \quad \forall u \in D(A). \tag{3.1.6}$$

https://doi.org/10.1515/9783110549652-003

From equations (3.1.5)–(3.1.6), we have

$$|(B(u,v),\omega)| \le C_5\|v\||\omega|\|u\|\left[1+2\lg\left[\frac{|Au|}{\lambda_1^{\frac{1}{2}}\|u\|}\right]\right]^{\frac{1}{2}}, \quad \forall u \in D(A),\ \forall v \in V,\ \forall \omega \in H, \tag{3.1.7}$$

$$|(B(u,v),\omega)| \le C_6\|v\||u|\|\omega\|\left[1+2\lg\left[\frac{|A\omega|}{\lambda_1^{\frac{1}{2}}\|\omega\|}\right]\right]^{\frac{1}{2}}, \quad \forall u \in H,\ \forall v \in V,\ \forall \omega \in D(A). \tag{3.1.8}$$

Furthermore, $B(u,v)$ also satisfies the following functional equation $(*)$:

$$(B(u,v),\omega) = -(B(u,\omega),v), \quad \forall u \in H,\ \forall v,\omega \in D(A).$$

For the solution $u(t)$ of problem (3.1.1)–(3.1.2), we have proved that there exists t_0, which depends on u_0, v, $|f|$, and λ_1, such that

$$|u(t)| \le M_0, \quad \|u(t)\| \le M_1, \quad \forall t \ge t_0, \tag{3.1.9}$$

where the constants M_0, M_1 depend on v, $|f|$, and λ_1.

Now we consider the approximate inertial manifold of problem (3.1.1)–(3.1.2).

Let P_m be the orthogonal projection of H onto $H_m = \operatorname{span}\{\omega_1,\dots,\omega_m\}$, $Q_m = I - P_m$. Let $p = P_m u$, $q = Q_m u$. Then equation (3.1.1) is equivalent to

$$\frac{\mathrm{d}p}{\mathrm{d}t} + vAp + P_m B(p+q,p+q) = P_m f, \tag{3.1.10}$$

$$\frac{\mathrm{d}q}{\mathrm{d}t} + vAq + Q_m B(p+q,p+q) = Q_m f. \tag{3.1.11}$$

The inertial manifold of equation (3.1.1) is the subset of $M \subset H$, possessing the following properties:

(i) M is a finite-dimensional Lipschitz manifold. (3.1.12)

(ii) The convection M is a positive invariant set. That is, if $u_0 \in M$, then the solution of equation (3.1.1)–(3.1.2) $u(t) \in M$, $\forall t > 0$. (3.1.13)

(iii) M attracts all orbits exponentially. For any solution $u(t)$ of equations (3.1.1)–(3.1.2) it follows that $\operatorname{dist}(u(t), M) \to 0$, exponentially as $t \to \infty$. From this we deduce that the global attractor $\mathscr{A} \subset M$. (3.1.14)

If we demand that M is the graph of a Lipschitz function $\Phi : H_m \to Q_m H$, then the invariance condition (3.1.13) is equivalent to $q(t) = \Phi(p(t))$ being valid for any solution $p(t)$, $q(t)$ of problem (3.1.10)–(3.1.11) with $q(0) = \Phi(p(0))$. Therefore, if the function Φ exists, then equations (3.1.10)–(3.1.11) for the inertial manifold M are equivalent to the following system of differential equations (called the inertia form):

$$\frac{\mathrm{d}p}{\mathrm{d}t} + vAp + P_m B(p+\Phi(p), p+\Phi(p)) = P_m f, \quad p \in H_m. \tag{3.1.15}$$

In order to use the smooth manifold to approximate the global attractor, we introduce the approximate inertial manifold.

Obviously, in equation (3.1.15), if $\Phi \triangleq 0$, then we can obtain the ordinary Galerkin approximation [140]:

$$\frac{\mathrm{d}u_m}{\mathrm{d}t} + vAu_m + P_mB(u_m, u_m) = P_mf, \quad u_m \in H_m. \tag{3.1.16}$$

Now we introduce a finite-dimensional analytic manifold, $\mu_0 = \text{graph}(\Phi_0)$,

$$\Phi_0(p) = (vA)^{-1}[Q_mf - Q_mB(p,p)], \quad p \in H_m, \tag{3.1.17}$$

which can be a better approximation of the global attractor.

Theorem 3.1.1. *Let m be large enough such that*

$$\lambda_{m+1} \geq \left(\frac{2C_2M}{v}\right)^2. \tag{3.1.18}$$

Then for any solution $u(t) = p(t) + q(t)$ *of equation* (3.1.10)–(3.1.11), *which satisfies*

$$|q(t)| \leq K_0\lambda_{m+1}^{-1}L^{\frac{1}{2}}, \tag{3.1.19}$$

$$|q(t)| \leq K_1\lambda_{m+1}^{-\frac{1}{2}}L^{\frac{1}{2}}, \tag{3.1.20}$$

$$|Aq(t)| \leq K_2L^{\frac{1}{2}}, \tag{3.1.21}$$

$$\left|\frac{\mathrm{d}q}{\mathrm{d}t}\right| \leq K_0'\lambda_{m+1}^{-1}L^{\frac{1}{2}}, \tag{3.1.22}$$

$$\|q(t) - \Phi_m(p(t))\| \leq K_1\lambda_{m+1}^{-1}L, \quad \forall t \geq T_*, \tag{3.1.23}$$

where $T_* > 0$ *only depends on* v, λ_1, $|f|$ *and* R_0, $|u(0)| \leq R_0$,

$$L = \left(\lg\frac{\lambda_m}{\lambda_1}\right) + 1,$$

and where K_0, K_0', K_1, K_2 *are positive constants depending on* v, λ_1 *and* $|f|$.

Let $\mathscr{B} = \{p \in H_m \mid \|p\| \leq 2M_1\}$, $\mathscr{B}^- = \{q \in Q_mV \mid \|p\| \leq 2M_1\}$, where M_1 satisfies equation (3.1.13). When m is large enough, there exists a mapping $\Phi^s : \mathscr{B} \to Q_mV$ which satisfies

$$\Phi^s(p) = (vA)^{-1}[Q_mf - Q_mB(p + \Phi^s(p), p + \Phi^s(p))], \quad \forall p \in \mathscr{B}. \tag{3.1.24}$$

Let $M^s = \text{graph}\,\Phi^s$, which is a C-analytic manifold, and contains all the stationary solutions of equation (3.1.1). Now to prove the existence of Φ^s, its upper bound is given.

Theorem 3.1.2. *Let m be large enough such that*

$$\lambda_{m+1} \geq \max\left\{4r_2^2, \frac{r_2^2}{4M_1^2}\right\}. \tag{3.1.25}$$

Then there exists a unique mapping $\Phi^s : \mathscr{B} \to Q_m V$, *which satisfies equation (3.1.24), and*

$$\|\Phi^s(p)\| \leq \lambda_{m+1}^{-\frac{1}{2}} r_1, \tag{3.1.26}$$

where

$$\begin{aligned}
r_1 &= \nu^{-1} C_5 8 M_1^2 L^{\frac{1}{2}} + \nu^{-1} C_2 8 M_1^2 + \nu^{-1} \lambda_{m+1}^{-\frac{1}{2}} |f|, \\
r_2 &= [\nu^{-1} C_5 2 M_1 L^{\frac{1}{2}} + \nu^{-1} C_2 6 M_1], \\
L &= 1 + \lg \frac{\lambda_m}{\lambda_1}.
\end{aligned}$$

Proof. Suppose $p \in \mathscr{B}$ is fixed, define $T_p : \mathscr{B}^\perp \to Q_m V$ such that

$$T_p(q) = (\nu A)^{-1}[Q_m f - Q_m B(p+q, p+q)].$$

To prove that T_p has a unique fixed point, first we need to prove $T_p : \mathscr{B}^\perp \to \mathscr{B}^\perp$. Let $q \in \mathscr{B}^\perp$, $\omega \in H$, $|\omega| = 1$, then

$$\begin{aligned}
\left|(A^{\frac{1}{2}} T_p(q), \omega)\right| &\leq \nu^{-1}\left[\left|(B(p+q, p+q), A^{-\frac{1}{2}} Q_m \omega)\right| + |A^{-1} Q_m f||\omega|\right] \\
&\leq \nu^{-1}\left[\left|(B(p, p+q), A^{-\frac{1}{2}} Q_m \omega)\right|\right. \\
&\quad \left. + \left|(B(q, p+q), A^{-\frac{1}{2}} Q_m \omega)\right|\right] + \nu^{-1} \lambda_{m+1}^{-1} |f|.
\end{aligned}$$

By equations (3.1.7) and (3.1.4) we get

$$\begin{aligned}
|(A^{\frac{1}{2}} T_p(q), \omega)| &\leq \nu^{-1} C_5 \|p+q\| |A^{-\frac{1}{2}} Q_m \omega| \|p\| \left(1 + \lg \frac{|Ap|}{\|p\| \lambda_1^{\frac{1}{2}}}\right)^{\frac{1}{2}} \\
&\quad + \nu^{-1} C_2 |q|^{\frac{1}{2}} \|q\|^{\frac{1}{2}} \|p+q\| |A^{-\frac{1}{2}} Q_m \omega|^{\frac{1}{2}} |\omega|^{\frac{1}{2}} + (\nu \lambda_{m+1})^{-1} |f| \\
&\leq \nu^{-1} C_5 8 M_1^2 \lambda_{m+1}^{-\frac{1}{2}} (1 + \lg \lambda_m / \lambda_1)^{\frac{1}{2}} \\
&\quad + \nu^{-1} C_2 \lambda_{m+1}^{-\frac{1}{2}} 8 M_1^2 + (\nu \lambda_{m+1})^{-1} |f|,
\end{aligned}$$

which yields

$$\|T_p(q)\| \leq \lambda_{m+1}^{-\frac{1}{2}} r. \tag{3.1.27}$$

From equation (3.1.25), we obtain $\|T_p(q)\| \leq 2M_1$.

Next we prove that T_p is contracting. We observe that

$$\frac{\partial}{\partial q}T_p(q)\eta = (\nu A)^{-1}Q_m[B(p+q,\eta)+B(\eta,p+q)], \quad \forall \eta \in Q_m V,$$

set $\omega \in H$, $|\omega| = 1$, and then arrive at

$$\begin{aligned}\left|\left(A^{\frac{1}{2}}\frac{\partial}{\partial q}T_p(q)\eta,\omega\right)\right| &\le \nu^{-1}|(B(p,\eta),A^{-\frac{1}{2}}Q_m\omega)| \\ &\quad + \nu^{-1}|(B(q,\eta),A^{-\frac{1}{2}}Q_m\omega)| + \nu^{-1}|(B(\eta,p+q),A^{-\frac{1}{2}}Q_m\omega)| \\ &\le \nu^{-1}C_5\|\eta\||A^{-\frac{1}{2}}Q_m\omega|\|p\|(1+\lg|Ap|/\lambda_1^{\frac{1}{2}}\|p\|)^{\frac{1}{2}} \\ &\quad + \nu^{-1}C_2|q|^{\frac{1}{2}}\|q\|^{\frac{1}{2}}\|\eta\||A^{-\frac{1}{2}}Q_m\omega|^{\frac{1}{2}}|\omega|^{\frac{1}{2}} \\ &\quad + \nu^{-1}C_2|\eta|^{\frac{1}{2}}\|\eta\|^{\frac{1}{2}}\|p+q\||A^{-\frac{1}{2}}Q_m\omega|^{\frac{1}{2}}|\omega|^{\frac{1}{2}} \\ &\le \left[\nu^{-1}C_5 2M_1\left(1+\lg\left(\frac{\lambda_m}{\lambda_1}\right)\right)^{\frac{1}{2}} + \nu^{-1}C_2 6M_1\right]\lambda_{m+1}^{-\frac{1}{2}}\|\eta\|,\end{aligned}$$

which yields

$$\left\|\frac{\partial}{\partial q}T_p(q)\right\|_{\mathscr{L}(Q_m V)} \le r_3\lambda_{m+1}^{-\frac{1}{2}}. \tag{3.1.28}$$

In virtue of equation (3.1.25), from (3.1.28) we deduce

$$\left\|\frac{\partial}{\partial q}T_p(q)\right\|_{\mathscr{L}(Q_m V)} \le \frac{1}{2},$$

completing the proof. □

From the contraction mapping principle, we deduce that there exists a unique $q(p) \in \mathscr{B}^{\perp}$ such that $q(p) = T_p(q)$. Let $\Phi^s(p) = q(p)$. From equation (3.1.27), we deduce that (3.1.26) is valid and $M^s = \operatorname{graph}\Phi^s$ is a C-analytic manifold. Every orbit $u(t) = p(t) + q(t)$ is located in a small neighborhood of manifold M, and the global attractor is also included in this small neighborhood.

Theorem 3.1.3. *Suppose that m is large enough, so that equation (3.1.25) is true, then for every solution $u(t) = p(t) + q(t)$ of equations (3.1.10)–(3.1.11), we have*

$$\|q(t) - \Phi^s(p(t))\| \le \frac{2K_0'}{\nu}\lambda_{m+1}^{-\frac{3}{2}}L^{\frac{1}{2}}, \quad \forall t \ge T_*, \tag{3.1.29}$$

where T_* and K_0' are the constants from Theorem 3.1.1.

Proof. Let $\Delta(t) = q(t) - \Phi^s(p(t))$. By equations (3.1.11) and (3.1.17), we have

$$\nu A\Delta + Q_m[B(\Delta, p+\Phi^s(p(t)) + B(p+q,\Delta)] + \frac{\mathrm{d}q}{\mathrm{d}t} = 0.$$

Taking the inner product of the equation above and Δ on H, by $(*)$ we have

$$\nu\|\Delta\|^2 \le |B(\Delta, p + \Phi^s(p(t)), \Delta)| + \left|\left(\frac{\mathrm{d}q}{\mathrm{d}t}, \Delta\right)\right|.$$

From equation (3.1.4) we have

$$\nu\|\Delta\|^2 \le C_2|\Delta|\|\Delta\|\|p + \Phi^s(p)\| + \left|\frac{\mathrm{d}q}{\mathrm{d}t}\right||\Delta|, \tag{3.1.30}$$

when $t > T_*$, $\|p(t)\| \le M_1$. By Theorem 3.1.2, we have $\Phi^s(p(t)) \le 2M$. Substituting equation (3.1.22) into equation (3.1.30), we get

$$\nu\|\Delta\|^2 \le C_2\lambda_{m+1}^{-\frac{1}{2}}\|\Delta\|^2(M_1 + 2M_1) + K_0'\lambda_{m+1}^{-\frac{1}{2}}L^{\frac{1}{2}}\|\Delta\|.$$

From equation (3.1.25), we get

$$\|\Delta\| \le \frac{2K_0'}{\nu}\lambda_{m+1}^{-\frac{3}{2}}L^{\frac{1}{2}},$$

proving the theorem. □

Now we consider another approximation Φ^s, which can be successively approximated and explicitly solved for. We have the following theorem:

Theorem 3.1.4. *Assume that m is large enough, so that equation* (3.1.25) *is true. As in Theorem* 3.1.2, *we define $T_p : \mathscr{B}^\perp \to \mathscr{B}^\perp$ by*

$$T_p(q) = (\nu A)^{-1}[Q_m f - Q_m B(p + q, p + q)], \quad \forall q \in \mathscr{B}^\perp.$$

Denote

$$\begin{cases} \Phi_0^s(p) = T_p(0), & \forall p \in \mathscr{B}, \\ \Phi_{n+1}^s(p) = T_p(\Phi_n^s(p)), & \forall p \in \mathscr{B};\ n = 0, 1, 2, \ldots. \end{cases} \tag{3.1.31}$$

Then

$$|\Phi^s(p) - \Phi_n^s(p)| \le \left(2r_2\lambda_{m+1}^{-\frac{1}{2}}\right)^{n+1}\lambda_{m+1}^{-\frac{1}{2}}\nu^{-1}\left[|f| + 4C_5M_1^2L^{\frac{1}{2}}\right], \tag{3.1.32}$$

where r_2 is defined in Theorem 3.1.2.

Proof. Firstly, we note that $\Phi_0^s(p) \triangleq \Phi_0(p)$, $\forall p \in \mathscr{B}$. By Theorem 3.1.2 and equations (3.1.28)–(3.1.35), it is easy to get

$$|\Phi^s(p) - \Phi_n^s(p)| \le 2\left(r_2\lambda_{m+1}^{-\frac{1}{2}}\right)^{n+1}\|\Phi_0^s(p)\|. \tag{3.1.33}$$

Hence, we only need to estimate $\|\Phi_0^s(p)\|$. From equation (3.1.31), we have

$$\Phi_0^s(p) = \Phi_0(p) = (\nu A)^{-1}[Q_m f - Q_m B(p, q)], \tag{3.1.34}$$

which yields

$$|A\Phi_0^s(p)| \le \nu^{-1}|f| + \nu^{-1}C_5\|p\|^2\left(1 + \lg\frac{|Ap|}{\|p\|\lambda_1^{\frac{1}{2}}}\right)^{\frac{1}{2}},$$

$$|A\Phi_0^s(p)| \le \nu^{-1}|f| + \nu^{-1}C_5 4M_1^2 L^{\frac{1}{2}}.$$

Hence

$$\|\Phi_0^s(p)\| \le \lambda_{m+1}^{-\frac{1}{2}}\nu^{-1}[|f| + 4C_5M_1^2L^{\frac{1}{2}}]. \tag{3.1.35}$$

Combining with equations (3.1.33)–(3.1.35), we deduce that equation (3.1.32) holds. □

Corollary 3.1.1. *Assume that m is large enough, so that equation* (3.1.25) *is true. Then for every solution* $u(t) = p(t) + q(t)$ *of equations* (3.1.10)–(3.1.11), *we have*

$$\|q(t) - \Phi_n^s(p(t))\| \le \lambda_{m+1}^{-\frac{3}{2}}\frac{2K_0}{\nu}L^{\frac{1}{2}} + 2(r_2\lambda_{m+1}^{-\frac{1}{2}})^{n+1}\lambda_{m+1}^{-\frac{1}{2}}\nu^{-1}[|f| + 4C_5M_1^2L^{\frac{1}{2}}],$$
$$\forall t \ge T_*,\ n = 0, 1, 2, \ldots, \tag{3.1.36}$$

where Φ_n^s *is determined by equation* (3.1.31); T_*, L *and* K_0' *are determined in Theorem* 3.1.1; r_2 *is determined in Theorem* 3.1.2.

Proof. This is a corollary of Theorems 3.1.3 and 3.1.4. □

In order to estimate the finite-dimensional approximation error of the Galerkin method, we need to prove the following lemma:

Lemma 3.1.1. *Assume that m is large enough, such that equation* (3.1.25) *is valid, then for any positive integer* $k \ge m + 1$, *we have*

$$\|Q_k\Phi^s(p)\| \le K_1\lambda_{k+1}^{-\frac{1}{2}}, \tag{3.1.37}$$

where $K_1 = \frac{16C_2M_1^2}{\nu}(1 + \lg\frac{\lambda_k}{\lambda_1})^{\frac{1}{2}} + 1$.

Proof. From equation (3.1.24), we have

$$\nu AQ_k\Phi^s(p) + Q_kB(p + \Phi^s(p), p + \Phi^s(p)) = Q_kf.$$

Taking the inner product of the equation above and $\Phi^s(p)$ on H, we have

$$\begin{aligned}\nu\|Q_k\Phi^s(p)\|^2 &\le |(B(p + \Phi^s(p), p + \Phi^s(p)), Q_k\Phi^s(p))| \\ &\quad + |f||Q_k\Phi^s(p)|,\\ \nu\|Q_k\Phi^s(p)\|^2 &\le |(B(p + P_k\Phi^s(p), p + \Phi^s(p)), Q_k\Phi^s(p))| \\ &\quad + |(B(Q_k\Phi^s(p), p + \Phi^s(p)), Q_k\Phi^s(p))| + |f|\|Q_k\Phi^s(p)\|.\end{aligned}$$

From equations (3.1.4) and (3.1.7), we have

$$\nu\|Q_k\Phi^s(p)\|^2 \le C_5\|p+\Phi^s(p)\|^2 + |Q_k\Phi^s(p)|\left(1+\lg\frac{\lambda_k}{\lambda_1}\right)^{\frac{1}{2}}$$
$$+ C_2|Q_k\Phi^s(p)|\|Q_k\Phi^s(p)\|\|p+\Phi^s(p)\|\| + |f|\|Q_k\Phi^s(p)\|.$$

Through equation (3.1.26) and by definition of $\mathscr{B}$, we have

$$\nu\|Q_k\Phi^s(p)\| \le C_5 8M_1^2\lambda_{k+1}^{-\frac{1}{2}}\left(1+\lg\frac{\lambda_k}{\lambda_1}\right)^{\frac{1}{2}}$$
$$+ C_2\sqrt{8}M_1\lambda_{k+1}^{-\frac{1}{2}}\|Q_k\Phi^s(p)\| + \lambda_{k+1}^{-\frac{1}{2}}|f|.$$

From equation (3.1.25), we can deduce that equation (3.1.37) is valid.

Let $k \ge m+1$, where m be large enough and satisfies equation (3.1.25). Consider the ordinary kth order Galerkin approximation:

$$\frac{du_k}{dt} + \nu Au_k + P_kB(u_k,u_k) = P_kf, \quad u_k \in H_k. \tag{3.1.38}$$

By using Theorem 3.1.2, it is easy to prove that equation (3.1.38) possesses a unique solution $\Phi^{s,k} : \mathscr{B} \to P_kQ_mV$, which satisfies

$$\nu A\Phi^{s,k}(p) + P_kQ_mB(p+\Phi^{s,k}(p), p+\Phi^{s,k}(p)) = P_kQ_mf, \quad \forall p \in \mathscr{B}. \tag{3.1.39}$$

We note that $\Phi^{s,k}$ possesses all the stationary solutions of equation (3.1.38). □

Lemma 3.1.2. *Assume that m is large enough, so that equation* (3.1.25) *is true. Then for any positive integer $k \ge m+1$, we have*

$$\|\Phi^s(p) - \Phi^{s,k}(p)\| \le K_3\lambda_{m+1}^{-\frac{1}{2}}, \quad \forall p \in \mathscr{B}, \tag{3.1.40}$$

where

$$K_3 = \left[1 + \frac{(2C_2+C_3)}{\nu}2\sqrt{8}M_1\lambda_{m+1}^{-\frac{1}{2}}\left(1+\lg\frac{\lambda_k}{\lambda_1}\right)^{\frac{1}{2}}\right]K_1$$

and K_1 is defined in Lemma 3.1.1.

Proof. For $p \in \mathscr{B}$, set $u = p + \Phi^s(p)$, $v = p + \Phi^{s,k}(p)$, $\Delta = P_k(u-v)$, $\eta = Q_k(u-v)$, $u - v = \Delta + \eta$. From equation (3.1.24) and (3.1.39), we have

$$\nu A\Delta + P_kQ_m[B(u-v,u) + B(v,u-v)] = 0,$$
$$\nu A\Delta + P_kQ_m[B(\Delta+\eta,u) + B(v,\Delta+\eta)] = 0.$$

Taking the inner product of above equation and Δ on H, and using equation $(*)$, we get

$$\nu\|\Delta\|^2 \le |(B(\Delta+\eta,u),\Delta)| + |(B(v,\eta),\Delta)|.$$

From equation (∗), we get

$$\nu\|\Delta\|^2 \le |(B(\Delta,u),\Delta)| + |(B(\eta,u),\Delta)| + |(B(P_k v,\Delta),\eta)| + |(B(Q_k v,\Delta),\eta)|.$$

Using equations (3.1.4), (3.1.8) and (3.1.7), we get

$$\begin{aligned}\nu\|\Delta\|^2 &\le C_2|\Delta|\|\Delta\|\|u\| \\ &\quad + C_5|\eta|\|u\|\|\Delta\|\left(1+\lg\frac{\lambda_k}{\lambda_1}\right)^{\frac{1}{2}} \\ &\quad + C_5\|v\|\|\Delta\||\eta|\left(1+\lg\frac{\lambda_k}{\lambda_1}\right)^{\frac{1}{2}} \\ &\quad + C_2|Q_k v|^{\frac{1}{2}}\|Q_k v\|^{\frac{1}{2}}\|\Delta\||\eta|^{\frac{1}{2}}\|\eta\|^{\frac{1}{2}}.\end{aligned} \tag{3.1.41}$$

By equation (3.1.26) we deduce that $\|u\| \le \sqrt{8}M_1$. Similarly, we have $\|v\| \le \sqrt{8}M_1$. From equation (3.1.41), we deduce

$$\begin{aligned}\nu\|\Delta\| &\le C_2\lambda_{m+1}^{-\frac{1}{2}}\|\Delta\|\sqrt{8}M_1 \\ &\quad + C_2 2\sqrt{8}M_1\|\eta\|\lambda_{m+1}^{-\frac{1}{2}}\left(1+\lg\frac{\lambda_k}{\lambda_1}\right)^{\frac{1}{2}} \\ &\quad + C_2\sqrt{8}M_1\|\eta\|\lambda_{m+1}^{-\frac{1}{2}}\left(1+\lg\frac{\lambda_k}{\lambda_1}\right)^{\frac{1}{2}}.\end{aligned}$$

From equation (3.1.25), we get

$$\|\Delta\| \le \frac{2(2C_5+C_2)}{\nu}\sqrt{8}M_1\|\eta\|\lambda_{m+1}^{-\frac{1}{2}}\left(1+\lg\frac{\lambda_k}{\lambda_1}\right)^{\frac{1}{2}}. \tag{3.1.42}$$

Finally, by equations (3.1.37) and (3.1.42), we get equation (3.1.40). □

Theorem 3.1.5. *Assume that m is large enough, so that equation* (3.1.25) *is true. Then for any positive integer* $k \ge m+1$, *for any* $p \in \mathscr{B}$, *we define* $T_p : \mathscr{B}^{\perp} \to \mathscr{B}^{\perp}$, *like in Theorem* 3.1.2,

$$T_p(q) = (\nu A)^{-1}[Q_m f - Q_m B(p+q,p+q)], \quad \forall q \in \mathscr{B}.$$

Let

$$\begin{aligned}\Phi_0^{s,k}(p) &= P_k T_p(0), \quad \forall p \in \mathscr{B}, \\ \Phi_{m+1}^{s,k}(p) &= P_k T_p(\Phi_0^{s,k}(p)), \quad \forall p \in \mathscr{B};\ n = 0,1,2,\ldots\end{aligned} \tag{3.1.43}$$

Then

$$\|\Phi^{s,k}(p) - \Phi_n^{s,k}(p)\| \le 2(r_2\lambda_{k+1}^{-\frac{1}{2}})^{n+1}\lambda_{m+1}^{-\frac{1}{2}}\nu^{-1}[|f| + 4C_5M_1^2L^{\frac{1}{2}}]. \tag{3.1.44}$$

Furthermore, for a solution $u(t) = p(t) + q(t)$ *of equations* (3.1.10)–(3.1.11), *we have*

$$
\begin{aligned}
\|q(t) - \Phi_n^{s,k}(p(t))\| \leq \; & \frac{2K_0'}{\nu}\lambda_{m+1}^{-\frac{3}{2}}L^{\frac{1}{2}} \\
& + (2r_2\lambda_{m+1}^{-\frac{1}{2}})^{n+1}\lambda_{m+1}^{-\frac{1}{2}}\nu^{-1}[|f| + 4C_5M_1^2L^{\frac{1}{2}}] \\
& + K_2\lambda_{k+1}^{-\frac{1}{2}}, \quad \forall t \geq T_*, \; n = 0, 1, 2, \ldots, \qquad (3.1.45)
\end{aligned}
$$

where T_*, L *and* K_0' *are determined in Theorem* 3.1.1, r_2 *is determined in Theorem* 3.1.2, *and* K_3 *is determined in Lemma* 3.1.2.

Proof. To get (3.1.44), we repeat the proof of Theorem 3.1.4, and only replace Φ^s with $\Phi^{s,k}$. The estimate (3.1.45) is a direct consequence of equations (3.1.29), (3.1.40) and (3.1.44). □

3.2 The Gevrey regularity of solutions

Consider Navier–Stokes equations with spatial periodic boundary conditions

$$\frac{\partial u}{\partial t} - \nu\Delta u + (u \cdot \nabla)u + \nabla p = f, \qquad (3.2.1)$$

$$\nabla \cdot u = 0, \qquad (3.2.2)$$

where $u = u(x,t)$, $p = p(x,t)$ are unknown functions, $u = \{u_1, u_2\}$ in dimension 2; $u = \{u_1, u_2, u_3\}$ in dimension 3; $x \in \mathbf{R}^n$, $n = 2$ or 3. The external force f is given, while $\nu > 0$ is dynamic viscosity. Let f, u, p along any spatial direction have period 2π. Denote Ω as the cube of period $(0, 2\pi)^n$ and assume that

$$u, p \text{ are periodic on } \Omega. \qquad (3.2.3)$$

For simplicity, suppose that the average of u over Ω is zero, namely

$$\int_\Omega u(x,t)\mathrm{d}x = 0, \quad \forall t \in \mathbf{R}^+. \qquad (3.2.4)$$

Usually problem (3.2.1)–(3.2.4) can be written in an abstract form

$$\frac{\mathrm{d}u}{\mathrm{d}t} + \nu Au + B(u) = f. \qquad (3.2.5)$$

In the Hilbert space H, operator A (corresponding to the Stokes operator with spatial periodic boundary conditions) is a linear self-adjoint unbounded positive operator, $D(A) \subset H$.

We expand u into Fourier series with respect to x,

$$u = \sum_{j \in Z^n} u_j e^{ijx}, \quad u_j \in C^n, \quad u_{-j} = \bar{u}_j, \quad u_0 = 0, \qquad (3.2.6)$$

$$j \cdot u_j = 0, \quad \forall j, \tag{3.2.7}$$

$$\sum_{j \in Z^n} |u_j|^2 = \frac{2}{(2\pi)^n} |u|^2 < \infty. \tag{3.2.8}$$

Then equation (3.2.7) is equivalent to equation (3.2.2). The condition $u_0 = 0$ is implied by equation (3.2.4).

Consider the domain for a positive power of operator of A, $D(A^\alpha)$, $\alpha > 0$, that is, the set of functions u which satisfy problem (3.2.6)–(3.2.8) and

$$(2\pi)^n \sum_{j \in Z^n} |j|^{2\alpha} |u_j|^2 = |A^\alpha u|^2 < \infty. \tag{3.2.9}$$

For the given $\tau, s > 0$, we consider the Gevrey category $D(e^{\tau A^s})$, which is the set of solutions u of equations (3.2.6)–(3.2.8) satisfying

$$(2\pi)^n \sum_{j \in \mathbf{Z}^n} e^{2\tau |j|^{2s}} |u_j|^2 = \left|e^{\tau A^s} u\right|^2 < \infty. \tag{3.2.10}$$

Finally, in equation (3.2.5), $B(u) = B(u, u)$, where $B(u, v)$ is defined as

$$(B(u, v), \omega) = b(u, v, \omega) = \sum_{j,k=1}^{n} \int_\Omega u_j \frac{\partial v_k}{\partial x_j} \omega_k \mathrm{d}x.$$

We supplement equation (3.2.5) with the initial condition

$$u(0) = u_0. \tag{3.2.11}$$

Then we consider the Gevrey regularity for the solution of the initial value problem (3.2.5) and (3.2.11).

Lemma 3.2.1. *Assume that u, v, ω are given in $D(e^{\tau A^{\frac{1}{2}}})$, $\tau > 0$. Then $B(u, v)$ belongs to $D(e^{\tau A^{\frac{1}{2}}})$, and for the two- and three-dimensional spaces, we have*

$$\begin{aligned} &\left|(e^{\tau A^{\frac{1}{2}}} B(u, v), e^{\tau A^{\frac{1}{2}}} A\omega)\right| \\ &\quad \le C_1 \left|e^{\tau A^{\frac{1}{2}}} A^{\frac{1}{2}} u\right|^{\frac{1}{2}} \left|e^{\tau A^{\frac{1}{2}}} Au\right|^{\frac{1}{2}} \left\|e^{\tau A^{\frac{1}{2}}} A^{\frac{1}{2}} v\right\| \left|e^{\tau A^{\frac{1}{2}}} A\omega\right|, \end{aligned} \tag{3.2.12}$$

where $C_1 > 0$ is an appropriate constant.

Proof. Let

$$u = \sum_{j \in \mathbf{Z}^n} u_j e^{ijx}, \quad u^* = \sum_{j \in \mathbf{Z}^n} u_j^* e^{ijx}, \quad u^* = e^{\tau |j|} u_j.$$

Using similar symbols for v, ω, we have

$$(B(u, v), \omega) = (2\pi)^n i \sum_{j-k=l} (u_j \cdot k)(v_k \cdot \overline{\omega}_l), \tag{3.2.13}$$

where $j, k, l \in \mathbf{Z}^n$. It follows that

$$\begin{aligned}(e^{\tau A^{\frac{1}{2}}}B(u,v), e^{\tau A^{\frac{1}{2}}}A\omega) &= (2\pi)^n i \sum_{j+k=l}(u_j\cdot k)(v_k\cdot\overline{\omega}_l)|l|^2 e^{2\tau|l|}\\ &= (2\pi)^n i \sum_{j+k=l}(u_j^*\cdot k)(v_k^*\cdot\overline{\omega}_l^*)|l|^2 e^{\tau(|l|-|j|-|k|)}.\end{aligned} \tag{3.2.14}$$

Noting that $|l| - |j| - |k| = |j+k| - |j| - |k| \le 0$, we have

$$(e^{\tau A^{\frac{1}{2}}}B(u,v), e^{\tau A^{\frac{1}{2}}}A\omega) \le (2\pi)^n \sum_{j+k=l}|u_j^*||k||v_k^*||\overline{\omega}_l^*||l|^2. \tag{3.2.15}$$

The right-hand side of equation (3.2.15) equals the integral

$$\int_\Omega \xi(x)\Phi(x)\theta(x)\mathrm{d}x,$$

where

$$\xi(x) = \sum_{j\in\mathbf{Z}^n}|u_j^*|e^{ijx}, \quad \Phi(x) = \sum_{j\in\mathbf{Z}^n}|k||v_k^*|e^{ikx},$$
$$\theta(x) = \sum_{j\in\mathbf{Z}^n}|l|^2|\omega_l^*|e^{ilx}.$$

This integral can be estimated by $(B(u^*, v^*), A\omega^*)$; an estimate involving B is standard in the NS equation. So we get equation (3.2.12). □

Theorem 3.2.1. *Assume that u_0 is given in $D(A^{\frac{1}{2}})$, and f is given in $D(e^{\sigma_1 A^{\frac{1}{2}}})$, $\sigma_1 > 0$. Then there exists T_*, which depends on the initial value u_0 and $|A^{\frac{1}{2}}u_0|$, such that the following claims are valid:*

(1) *For the two- and three-dimensional spaces, equations (3.2.5) and (3.2.11) on $(0, T_*)$ possess a unique regular solution (it is continuous from $[0, T_*]$ to $D(A^{\frac{1}{2}})$), such that $t \to e^{\Psi(t)A^{\frac{1}{2}}}A^{\frac{1}{2}}u(t)$ is analytic in $(0, T_*)$, $\Psi(t) = \min(t, \sigma_1, T_*)$;*

(2) *If solutions of equations (3.2.5) and (3.2.11) for any $t > 0$ exist and stay bounded in $D(A^{\frac{1}{2}})$, then any such u is analytic in the interval (T_*, ∞), its value belongs to $D(A^{\frac{1}{2}}e^{\sigma A^{\frac{1}{2}}})$, where $\sigma > 0$, and T_* is defined as before.*

Proof. The main idea is to establish an a priori estimate of a solution. The rest involves using the Galerkin method to establish the approximate solution. Suppose that we take the complex framework, $t = \zeta \in \mathbf{C}$, $\varphi(t) = \min(t, \sigma_1)$;

(i) Real case. At the time τ, taking the inner product of equation (3.2.5) and $Au(\tau)$ with respect to $D(e^{\varphi(\tau)A^{\frac{1}{2}}})$ gives

$$\begin{aligned}&(e^{\varphi(\tau)A^{\frac{1}{2}}}u'(\tau), Ae^{\varphi(\tau)A^{\frac{1}{2}}}u(\tau)) + \nu|e^{\varphi(\tau)A^{\frac{1}{2}}}Au(\tau)|^2\\ &\quad = (e^{\varphi(\tau)A^{\frac{1}{2}}}f, e^{\varphi(\tau)A^{\frac{1}{2}}}Au(\tau)) - (e^{\varphi(\tau)A^{\frac{1}{2}}}B(u(\tau)), e^{\varphi(\tau)A^{\frac{1}{2}}}Au(\tau)),\end{aligned} \tag{3.2.16}$$

where $(\cdot,\cdot)_2$ and $|\cdot|_2$ mean the inner product and norm in $D(e^{\tau A^{\frac{1}{2}}})$, respectively, but $((\cdot,\cdot))$ and $\|\cdot\|_2$ mean the inner product and norm in $D(e^{\tau A^{\frac{1}{2}}} A^{\frac{1}{2}})$, respectively. Then

$$\begin{aligned}
&(e^{\varphi(t)A^{\frac{1}{2}}} u'(t), e^{\varphi(t)A^{\frac{1}{2}}} Au(t)) \\
&= (A^{\frac{1}{2}}(e^{\varphi(t)A^{\frac{1}{2}}} u(t))' - \varphi'(t)Ae^{\varphi(t)A^{\frac{1}{2}}} u(t), e^{\varphi(t)A^{\frac{1}{2}}} A^{\frac{1}{2}} u(t)) \\
&= \frac{1}{2}\frac{\mathrm{d}}{\mathrm{d}t}\left|A^{\frac{1}{2}} e^{\varphi(t)A^{\frac{1}{2}}} u(t)\right|^2 - \varphi'(t)(Ae^{\varphi(t)A^{\frac{1}{2}}} u(t), e^{\varphi(t)A^{\frac{1}{2}}} A^{\frac{1}{2}} u(t)) \\
&= \frac{1}{2}\frac{\mathrm{d}}{\mathrm{d}t}\|u(t)\|_{\varphi(t)}^2 - \varphi'(t)(Au(t), A^{\frac{1}{2}} u(t))_{\varphi(t)} \\
&\geq \frac{1}{2}\frac{\mathrm{d}}{\mathrm{d}t}\|u(t)\|_{\varphi(t)}^2 - |Au(t)|_{\varphi(t)}\|u(t)\|_{\varphi(t)} \\
&\geq \frac{1}{2}\frac{\mathrm{d}}{\mathrm{d}t}\|u(t)\|_{\varphi(t)}^2 - \frac{\nu}{4}|Au(t)|_{\varphi(t)}^2 - \frac{1}{\nu}\|u(t)\|_{\varphi(t)}^2 .
\end{aligned} \tag{3.2.17}$$

The right-hand side of equation (3.2.16) can be written as

$$(f, Au)_\varphi - (B(u), Au)_\varphi . \tag{3.2.18}$$

By Lemma 3.2.1, (3.2.12) and Schwarz inequality, we get the bound of equation (3.2.18), namely

$$|f|_\varphi |Au|_\varphi + C_1\|u\|_\varphi^{\frac{3}{2}} |Au|_\varphi^{\frac{3}{2}} \leq \frac{\nu}{4}|Au|_\varphi^2 + \frac{2}{\nu}|f|_\varphi^2 + \frac{C_2}{\nu^3}\|u\|_\varphi^6,$$

where C_1, C_2 and C_i, C_i' are constants. Then by equation (3.2.16) we have

$$\begin{aligned}
\frac{\mathrm{d}}{\mathrm{d}t}\|u\|_\varphi^2 + \nu|Au|_\varphi^2 &\leq \frac{4}{\nu}|f|_\varphi^2 + \frac{2}{\nu}\|u\|_\varphi^2 + \frac{2C_2}{\nu^3}|u|_\varphi^6 \\
&\leq \frac{4}{\nu}|f|_\varphi^2 + C_3 + \frac{3C_2}{\nu^3}|u|_\varphi^2 .
\end{aligned} \tag{3.2.19}$$

Finally, we get

$$\begin{aligned}
&y' \leq K_1 y^3, \\
&y(t) = 1 + \|u(t)\|_\varphi^2, \quad K_1 = \frac{4}{\nu}|f|_{\sigma_1}^2 + C_3 + \frac{(3C_2)^{\frac{1}{3}}}{\nu} .
\end{aligned} \tag{3.2.20}$$

Hence

$$y(t) = 1 + \left|e^{\varphi(t)A^{\frac{1}{2}}} A^{\frac{1}{2}} u(t)\right|^2 \leq 2y(0) = 2 + 2\left|A^{\frac{1}{2}} u(0)\right|^2, \tag{3.2.21}$$

$$t \leq T_1(|A^{\frac{1}{2}} u(0)|) = \frac{2}{K_1}(1 + |A^{\frac{1}{2}} u(0)|^2)^{-2}, \tag{3.2.22}$$

and then $u(t)$ belongs to $D(e^{\varphi(t)A^{\frac{1}{2}}} A^{\frac{1}{2}})$, so equation (3.2.21) is established when $t \in (0, T_1)$, $u(0) \in D(A^{\frac{1}{2}})$. In particular,

$$\left|e^{\varphi(T_1)A^{\frac{1}{2}}} A^{\frac{1}{2}} u(T_1)\right|^2 \leq 2 + 2\left|A^{\frac{1}{2}} u(0)\right|^2 . \tag{3.2.23}$$

If we know that

$$|A^{\frac{1}{2}}u(t)| \le M, \quad \forall t \ge 0, \tag{3.2.24}$$

then equation (3.2.23) for two-dimensional space is always true, while in dimension 3 must give some assumptions. Next we repeat the above principle: for any $t_0 > 0$,

$$|e^{\sigma_2 A^{\frac{1}{2}}} A^{\frac{1}{2}} u(t)|^2 \le 2 + 2M_1^2, \quad \forall t \ge T_2, \tag{3.2.25}$$

where $\sigma_2 = \varphi(T_2) = \min(T_2, \sigma_1)$, and we have

$$T_2 = T_2(M_1) = \frac{2}{K_1}(1 + M_1^2)^{-2}. \tag{3.2.26}$$

(ii) Complex case. In order to get the analyticity of time t, we consider equation (3.2.5) with complex time $\zeta \in \mathbf{C}$. Then u is a complex function and $\mathbf{H}$ is the complexification of the original space H. The inner product, norm and operators A, B can be also extended accordingly.

Equation (3.2.5) can then be written as

$$\frac{du}{d\zeta} + \nu Au + Bu = f, \tag{3.2.27}$$

where $\zeta = se^{i\theta}$, $s > 0$, $\cos\theta > 0$, hence $\operatorname{Re}\zeta > 0$. Taking the inner product of equation (3.2.27) and $Au(se^{i\theta})$ with respect to $D(e^{\varphi(s\cos\theta)A^{\frac{1}{2}}})$, multiplying this equation by $e^{-i\theta}$, and taking its real part, we get

$$\begin{aligned}
&\operatorname{Re} e^{-i\theta}\Big(e^{\varphi(s\cos\theta)A^{\frac{1}{2}}}\frac{du}{d\zeta}(se^{i\theta}), e^{\varphi(s\cos\theta)A^{\frac{1}{2}}}Au(se^{i\theta})\Big)\\
&= \operatorname{Re}\Big(A^{\frac{1}{2}}\frac{d}{ds}(e^{\varphi(s\cos\theta)A^{\frac{1}{2}}}u(se^{i\theta}))\\
&\quad - (e^{-i\theta}\varphi'(s\cos\theta)\cos\theta Au(se^{i\theta}), e^{\varphi(s\cos\theta)A^{\frac{1}{2}}}A^{\frac{1}{2}}u(se^{i\theta}))\Big)\\
&= \frac{1}{2}\frac{d}{ds}|A^{\frac{1}{2}}u(se^{i\theta})|^2_{\varphi(s\cos\theta)}\\
&\quad - \cos^2\theta\varphi'(s\cos\theta)|Au(se^{i\theta})|_{\varphi(s\cos\theta)}|A^{\frac{1}{2}}u(se^{i\theta})|_{\varphi(s\cos\theta)}\\
&\ge \frac{1}{2}\frac{d}{ds}|A^{\frac{1}{2}}u(se^{i\theta})|^2_{\varphi(s\cos\theta)} - \frac{\nu\cos\theta}{4}|Au(se^{i\theta})|^2_{\varphi(s\cos\theta)}\\
&\quad - \frac{1}{\nu\cos\theta}\|u(se^{i\theta})\|^2_{\varphi(s\cos\theta)}\\
&\ge \operatorname{Re} e^{-i\theta}(e^{\varphi(s\cos\theta)A^{\frac{1}{2}}}Au(se^{i\theta}), e^{\varphi(s\cos\theta)A^{\frac{1}{2}}}Au(se^{i\theta}))\\
&= \cos\theta|Au(se^{i\theta})|^2_{\varphi(s\cos\theta)}.
\end{aligned}$$

Replacing φ with $\varphi(s\cos\theta)$ (s can be ignored if necessary), we get

$$\begin{aligned}\frac{1}{2}\frac{\mathrm{d}}{\mathrm{d}s}\|u\|_\varphi^2 + \frac{3}{4}\nu\cos\theta|Au|_\varphi^2 - \frac{1}{\nu\cos\theta}\|u\|_\varphi^2\\ \leq \operatorname{Re} e^{-i\theta}(f, Au)_\varphi - \operatorname{Re} e^{-i\theta}(B(u), Au)_\varphi.\end{aligned} \tag{3.2.28}$$

The right-hand side of inequality (3.2.28) can be estimated as follows:

$$|f|_\varphi|Au|_\varphi + C_1\|u\|_\varphi^{\frac{3}{2}}\|Au\|_\varphi^{\frac{3}{2}} \leq \frac{\nu\cos\theta}{4}\|Au\|_\varphi^2 + \frac{2}{\nu\cos\theta}\|f\|_\varphi^2 + \frac{C_4}{(\nu\cos\theta)^3}\|u\|_\varphi^6.$$

Similar to equation (3.2.19), we get

$$\begin{aligned}\frac{\mathrm{d}}{\mathrm{d}s}\|u\|_\varphi^2 + \nu\cos\theta|Au|_\varphi^2 &\leq \frac{4}{\nu\cos\theta}\|f\|_\varphi^2 + \frac{2}{\nu\cos\theta}\|u\|_\varphi^2 + \frac{2C_4}{(\nu\cos\theta)^3}\|u\|_\varphi^6\\ &\leq \frac{4}{\nu\cos\theta}\|f\|_\varphi^2 + C_5 + \frac{4C_4}{(\nu\cos\theta)^3}\|u\|_\varphi^6,\end{aligned} \tag{3.2.29}$$

where $u = u(se^{i\theta})$, $\varphi = \varphi(s\cos\theta)$. If we restrict $\cos\theta \geq \sqrt{2}/2$, then we obtain equations whose form is similar to equation (3.2.20), i. e.,

$$\begin{aligned}&\frac{\mathrm{d}y}{\mathrm{d}s} \leq K_2 y^3,\\ &y(s) = 1 + |e^{\varphi(s\cos\theta)A^{\frac{1}{2}}}A^{\frac{1}{2}}u(se^{i\theta})|^2,\\ &K_2 = \frac{8}{\nu}|f|_{\sigma_1}^2 + C_5 + \frac{2^5 C_4}{\nu^3},\end{aligned} \tag{3.2.30}$$

then also $y(s) \leq 2y(0)$. That is,

$$|e^{\varphi(s\cos\theta)A^{\frac{1}{2}}}A^{\frac{1}{2}}u(se^{i\theta})|^2 \leq 2 + 2|A^{\frac{1}{2}}u(0)|^2, \tag{3.2.31}$$

where

$$0 \leq s \leq T_3(|A^{\frac{1}{2}}u(0)|) = \frac{2}{K_2}[1 + |A^{\frac{1}{2}}u(0)|^2]^{-2}, \tag{3.2.32}$$

$$\frac{\sqrt{2}}{2} \leq \cos\theta \leq 1. \tag{3.2.33}$$

This implies that, if $u(0) \in D(A^{\frac{1}{2}})$, then the complex angle region which is determined by $u(se^{i\theta})$ in equations (3.2.32)–(3.2.33) belongs to $D(e^{\varphi(s\cos\theta)A^{\frac{1}{2}}}A^{\frac{1}{2}})$. In particular,

$$|e^{\sigma_3 A^{\frac{1}{2}}}A^{\frac{1}{2}}u(T_3e^{i\theta})|^2 \leq 2 + 2|A^{\frac{1}{2}}u(0)|^2. \tag{3.2.34}$$

From this we have $1 \geq \cos\theta \geq \sqrt{2}/2$, $\sigma_3 = \varphi(T_3, \sqrt{2}/2) = \min(T\sqrt{2}/2, \sigma_1)$.

If equation (3.2.24) is valid, then we can repeat the above process, such that for any $t_0 > 0$,

$$|e^{\sigma_4 A^{\frac{1}{2}}} A^{\frac{1}{2}} u(se^{i\theta} + t_0)|^2 \le 2 + 2M_1^2, \tag{3.2.35}$$

where $0 \le s \le T_4$, $\sqrt{2}/2 \le \cos\theta \le 1$,

$$T_4 = T_4(M_1) = \frac{2}{K_2}(1 + M_1^2)^{-2}, \tag{3.2.36}$$

$\sigma_4 = \varphi(T_4, \sqrt{2}/2) = \min(T_4\sqrt{2}/2, \sigma_1)$, in particular the estimate (3.2.25) holds for $\varphi = se^{i\theta} + t_0$. On the domain $\Delta(M_1)$,

$$\operatorname{Re}(s) \ge T_4(M_1), \quad |\operatorname{Im}(s)| \le \frac{\sqrt{2}}{2} T_4(M_1) \tag{3.2.37}$$

are established. Here

$$T_* = T_4(M_1), \quad \sigma = \min((\sqrt{2}/2)T_*, \sigma_1).$$

Thus the theorem has been proved. □

Remark 3.2.1. **(i)** If f depends on t, and f is analytic with respect to time t in $D(e^{\sigma_1 A^{\frac{1}{2}}})$, then the analyticity domain of u is the intersection of $\Delta(M_1)$ and the analyticity domain of f.

(ii) In the proof of Lemma 3.2.1, for the left-side estimate of equation (3.2.12), we can use another estimate for B. For instance, for two-dimensional space, from

$$|(B(u^*, v^*), \omega^*)| \le C|A^{\frac{1}{2}} u^*||A^{\frac{1}{2}} v^*||A\omega^*|\left(1 + \lg \frac{|Au^*|^2}{4\pi^2 |A^{\frac{1}{2}} u^*|^2}\right)^{\frac{1}{2}}$$

we have

$$\begin{aligned} &|(e^{\tau A^{\frac{1}{2}}} B(u, v), e^{\tau A^{\frac{1}{2}}} \omega)| \\ &\le C_1' |e^{\tau A^{\frac{1}{2}}} A^{\frac{1}{2}} u||e^{\tau A^{\frac{1}{2}}} A^{\frac{1}{2}} v||e^{\tau A^{\frac{1}{2}}} A\omega|\left(1 + \lg \frac{|e^{\tau A^{\frac{1}{2}}} Au|^2}{4\pi^2 |e^{\tau A^{\frac{1}{2}}} A^{\frac{1}{2}} u|^2}\right)^{\frac{1}{2}}. \end{aligned} \tag{3.2.38}$$

Replacing equation (3.2.38) with equation (3.2.12) could increase the size of $T_4(M_1)$ and area (3.2.37). If which replace with equation (3.2.39), we get

$$\frac{\mathrm{d}y}{\mathrm{d}s} \le C_7 (v\cos\theta)^{-1} y^2 \lg y. \tag{3.2.39}$$

Hence, if equation (3.2.24) is satisfied, we can get equation (3.2.36) for any $t_0 \ge 0$. For any s, θ satisfying $0 \le s \le T_5(M_1)$, $\sqrt{2}/2 \le \cos\theta \le 1$,

$$T_5(M_1) = \frac{C_8}{(\frac{1}{v})(|f|_{\sigma_1}^2 + M_1^2) - \lg((1/v^2)(|f|_{\sigma_1}^2 + M_1^2))}. \tag{3.2.40}$$

The analyticity domain of u contains $\Delta'(M)$, similar to $\Delta(M_1)$; T_5 is replaced with T_4. The analyticity domain is in $D(A^{\frac{1}{2}}e^{\sigma_5}A^{\frac{1}{2}})$, $\sigma_5 = \min((T_5\sqrt{2}/2), \sigma_1)$.

(iii) Let $u^*(t) = e^{\varphi(t)A^{\frac{1}{2}}}u(t)$. Then from equation (3.2.5) we get the following equation for u^*:

$$\frac{du^*}{dt} + \nu Au^* - \varphi' A^{\frac{1}{2}}u + e^{-\varphi}B(u^*) = f^*. \tag{3.2.41}$$

The related results in this aspect can refer to the reference [121, 122].

3.3 Time analyticity of solution for a class of dissipative nonlinear evolution equations

Suppose that H is an infinite-dimensional Hilbert space, which possesses inner product $(\cdot,\cdot)$ and norm $|\cdot|$, A is a given unbounded, positive self-adjoint linear operator, $D(A) \subset H$, A^{-1} is compact in H, a nonlinear operator $R : D(A) \to H$ on $D(A)$ is analytic in a finite-dimensional subspace, which can be extend in accordance with the complexity of the subspace.

Because A^{-1} is compact in H, there exists a basis of functions $\{\omega_j\}$ of H, which are the eigenvectors of A. Namely, there exists an orthogonal basis in H of eigenfunctions of A with corresponding set of eigenvalues $\lambda_j \in \mathbf{R}_+$,

$$A\omega_j = \lambda_j \omega_j,$$
$$|\omega_j| = 1,$$
$$0 < \lambda_1 \le \lambda_2 \le \cdots.$$

When $j \to \infty$, $\lambda_j \to +\infty$. Likewise we can define powers of A, namely A^s, $s \in \mathbf{R}$. Then A^s maps $D(A^s)$ to H. In particular, for $V = D(A^{\frac{1}{2}})$, its norm and inner product are denoted by $\|\cdot\|$ and $((\cdot))$, respectively, and $|\cdot|_1 = (\|\cdot\|^2 + |\cdot|^2)^2$. For the vector u, $Au = (Au_1, \dots, Au_k)$.

Consider the following abstract initial value problem:

$$u'(t) + DAu(t) + R(u(t)) = f(t), \quad t \in \mathbf{R}_+, \tag{3.3.1}$$

$$u(0) = u_0, \quad t = 0, \tag{3.3.2}$$

where u is a vector function $\mathbf{R}_+ \to D(A)$, $f : \mathbf{R}_+ \to H$ is analytic, $D = (d_{ij})_{k\times k}$ is a real positive-definite matrix.

Use H_c, V_c and $D(A)_c$ to denote the complexification of H, V and $D(A)$, respectively. For example, H_c is the subset of $H \oplus H$ which consists of the basic elements $h_1 + ih_2$, $h_1, h_2 \in \mathbf{R}$, $A(h_1 + ih_2) = Ah_1 + iAh_2$,

$$(h_1 + ih_2, g_1 + ig_2) = (h_1, g_1) + (h_2, g_2) + i[(h_2, g_1) - (h_1, g_2)],$$

A^{-1} is compact in H_c. In fact, using the orthogonal basis $\{\omega_j\}$ of space H, it is easy to construct the H_c's orthogonal basis $\{\omega_j\}$ which is formed from the eigenvectors of A.

Suppose that $1 \le \gamma < \infty$, $K > 0$, and the function $C \in C((0,\infty); \mathbf{R}_+)$ is such that for any $\varepsilon > 0$, R satisfies

$$\left|(R(u), Au + u)\right| \le \varepsilon |Au|^2 + C(\varepsilon)|u|_1^{2\gamma} + K, \quad \forall u \in D(A)_c. \tag{3.3.3}$$

When $v_m \to v$ from X to $D(A)$ converge in the weak topology,

$$R(v_m) \to R(v) \quad \text{weakly with respect to } L^2(X, H). \tag{3.3.4}$$

Suppose there exists $M > 0$ such that f satisfies

$$\|f(t)\|_* \le M, \quad \forall t \in \mathbf{R}_+. \tag{3.3.5}$$

Theorem 3.3.1. *Suppose u is a solution of equations (3.3.1)–(3.3.2). Under the above assumptions (3.3.3), (3.3.4) and (3.3.5), there exists θ_0, $|\theta_0| \le \frac{\pi}{4}$, and function $T_1 \in C(\mathbf{R}_+, \mathbf{R}_-)$ such that if $|u(0)|_1$ is finite, then u takes values in $D(A)_c$ and can be extended analytically to*

$$\Delta(|u(0)|_1) = \{z = se^{i\theta} \mid |\theta| \le \theta_0, 0 \le s \le T_1(|u_0|_1)\}. \tag{3.3.6}$$

Furthermore, if $|u(t)|_1$ is constrained by B for $t \in (a, b)$, then the above set can be extended to

$$\Delta = \bigcup_{t \in (a,b)} t + \Delta(B).$$

For any compact set K included in the area of analyticity, the following inequalities hold:

$$\sup_{z \in K} \left|\frac{d^k u}{dz^k}(z)\right|_1 \le 2^{\frac{1}{2}}(2/d)^k(k!)(1 + |u_0|_1^2)^{\frac{1}{2}}, \quad d = \operatorname{dist}(K, \partial\Delta), \tag{3.3.7}$$

$$\sup_{z \in K} |Au(z)| \le T_2(K) < \infty, \tag{3.3.8}$$

$$\sup_{z \in K} \left|A\left(\frac{d^k u}{dz^k}(z)\right)\right| \le 2^k (k!)[d(K, \partial\Delta(u_0))]^{-k} T_2(K'), \tag{3.3.9}$$

where $K' \triangleq \{z \in \Delta(u_0) \mid d(z, \partial\Delta(u_0)) \ge \frac{1}{2} d(K, \partial\Delta(u_0))\}$.

Proof. Considering the Galerkin approximation of complex time, for the complex differential equations on $H_m \triangleq \{C\omega_1 + \cdots + C\omega_m\}$. Let P_m be the projection of H_m. We seek solutions

$$u_m(z) = \sum_{i=1}^{m} g_i(z)\omega_i, \quad g_i : \mathbf{C} \to \mathbf{C}$$

$$\left(\frac{\partial u_m}{\partial z} + DAu_m + R(u_m) - f(t), v\right) = 0, \quad \forall v \in H_m, \tag{3.3.10}$$

$$u_m(0) = P_m u_0. \tag{3.3.11}$$

Because on H_m the operator A possesses a very simple form, namely $Au_m(z) = \sum \lambda_i g_i(z)\omega_i$, equations (3.3.10)–(3.3.11) boil down to m ordinary differential equations. Based on the well-known Cauchy–Kovalevskaya theorem, we obtain a unique analytic solution in a complex neighborhood of the origin. Moreover, u_m can also be used in the Galerkin approximation of the real time problem (3.3.1)–(3.3.2) (restriction on the real axis). Now we make an a priori estimate. Setting $v = Au_m + u_m$ in equation (3.3.10) and ignoring the subscript m, we have

$$\left(\frac{\partial u}{\partial z}, Au\right) + \left(\frac{\partial u}{\partial z}, u\right) + (DAu, Au) + (DAu, u)$$
$$+ (R(u), Au + u) - (f(t), Au + u) = 0. \tag{3.3.12}$$

Let $z = se^{i\theta}$, $-\frac{\pi}{2} \le \theta \le \frac{\pi}{2}$. Multiply equation (3.3.12) by $e^{i\theta}$ and take its real part. Since D is a real positive matrix, there exists $\alpha_0 \ge 0$, such that $\operatorname{Re}(Dz, z) \ge \alpha_0 |z|^2$, $\forall z \in \mathbf{C}^n$. Then we get

$$\frac{1}{2}\frac{\mathrm{d}}{\mathrm{d}s}(\|u\|^2 + |u|^2) + \alpha_0 \cos\theta(|Au|^2 + \|u\|^2)$$
$$\le \sin\theta \|D\|_*(|Au|^2 + \|u\|^2)$$
$$+ \left|(R(u), Au + u)\right| + \left|(f(t), Au + u)\right|.$$

Using condition (3.3.3), taking $\varepsilon = (\alpha_0 \cos\theta)/8$, and combining with condition (3.3.5), we get

$$\frac{1}{2}\frac{\mathrm{d}}{\mathrm{d}s}(\|u\|^2 + |u|^2) + \alpha_0 \cos\theta(|Au|^2 + \|u\|^2)$$
$$\le |\sin\theta| \|D\|_*(|Au|^2 + \|u\|^2) + (\alpha_0/8)\cos\theta |Au|^2$$
$$+ K + C(\theta)|u|_1^{2\gamma} + M(|Au| + |u|). \tag{3.3.13}$$

Further we restrict θ so that

$$\alpha_0 \cos\theta \ge \|D\|_* |\sin\theta|,$$

for example, we can select

$$|\theta| \le \min\left(\arctan(\alpha_0/4\|D\|_*) \cdot \frac{\pi}{4}\right). \tag{3.3.14}$$

Hence, using Young inequality in the last term of equation (3.3.13), we get

$$\frac{1}{2}\frac{\mathrm{d}}{\mathrm{d}s}|u|_1^2 + \alpha_0 \cos\theta(|Au|^2 + \|u\|^2)$$
$$\le (\alpha_0/4)\cos\theta[|Au|^2 + \|u\|^2]$$
$$+ (\alpha_0/8)\cos\theta|Au|^2 + K + C(\theta)|u|_1^{2\gamma}$$
$$+ 4M^2/(\alpha_0 \cos\theta) + (\alpha_0/8)\cos\theta|Au|^2 + M|u|^2,$$

which can be reduced to

$$\begin{aligned}\frac{d}{ds}|u|_1^2 + \alpha_0 \cos\theta(|Au|^2 + \|u\|^2)\\ \le C(\theta)|u|_1^{2\gamma} + 4M^2/(\alpha_0\cos\theta) + K.\end{aligned} \tag{3.3.15}$$

If θ satisfies equation (3.3.14), then the coefficient in the bound of equation (3.3.15) is independent of θ. Thus

$$\frac{d}{ds}|u(se^{i\theta})|_1^2 + C_1(|Au|^2 + \|u\|^2) \le C_2 + C_3|u(se^{i\theta})|_1^{2\gamma}. \tag{3.3.16}$$

Ignoring $C_1(|Au|^2 + \|u\|^2)$, and denoting $y(s) = 1 + |u(se^{i\theta})|_1^2$, we get the differential inequality

$$y'(s) \le C_4 y^\gamma(s), \quad s \ge 0,$$

where $C_4 = \max\{C_2, C_3\}$. Integrating the above inequality, we obtain

$$\begin{aligned}0 < y(s) \le \left(y(0)^{1-\gamma} - (\gamma-1)C_4 s\right)^{1/(1-\gamma)},\\ 0 \le s < y(0)^{1-\gamma}\left[(\gamma-1)C_4\right]^{-1}.\end{aligned}$$

This means that there exists a constant $T_1 = y(0)^{1-\gamma}(1 - (\frac{1}{2})^{\gamma-1})/((\gamma-1)C_4)$, which depends on $|u_0|_1$, but not on m, such that

$$\begin{aligned}|u_m(se^{i\theta})|_1 \le 2(1 + |u_m(0)_1^2|)^{\frac{1}{2}} \le 2(1 + |u_0|_1^2)^{\frac{1}{2}},\\ \forall\theta \text{ satisfies equation (3.3.14)}, \ 0 \le s \le T_1(|u_0|_1),\end{aligned} \tag{3.3.17}$$

where we have recovered the subscript m. So from the existence of solutions for ordinary differential equations, we can deduce that u_m can be extended to an analytic solution of equation (3.3.16) in the set $\Delta(|u(0)|_1) = \{z = se^{i\theta} \mid 0 < s < T_1(|u_0|_1), \ \theta \text{ satisfies the equation (3.3.14)}\}$, shown in Figure 3.1.

Namely, we have

$$\sup |u_m(z)|_1 \le 2(1 + |u_0|_1^2)^{\frac{1}{2}}, \quad z \in \Delta(|u_0|_1).$$

Furthermore, by Cauchy formula we have

$$\frac{d^k u_m}{dz^k} = (k!)/(2\pi i) \int_{|z-\eta| = \frac{d}{2}} u_m(\eta)(\eta - z)^{-(k+1)} d\eta,$$

where $d = d(z, \partial\Delta(u_0))$. Hence

$$\left|\frac{d^k u_m}{dz^k}\right|_1 \le (2/d)^k (k!) \sup_{z \in \Delta(|u_0|_1)} |u_m(z)|_1.$$

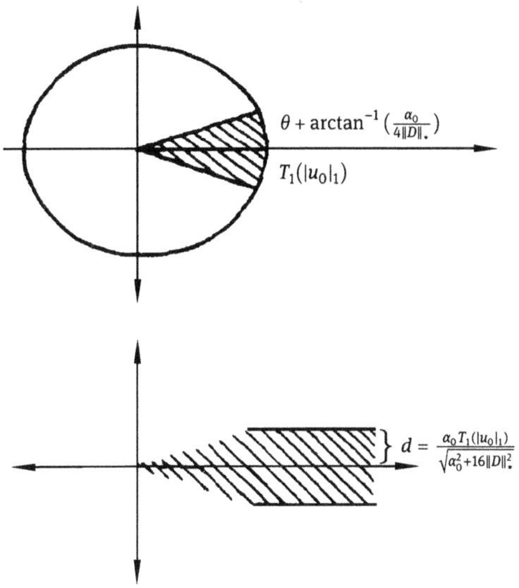

Figure 3.1

In particular, for any K which is a compact subset of $\Delta(|u_0|_1)$, we have

$$\sup_{z\in K}\left|\frac{\mathrm{d}^k u_m}{\mathrm{d}z^k}(z)\right|_1 \le 2^{\frac{1}{2}}(2/d)^k(k!)(1+|u_0|_1^2)^{\frac{1}{2}}, \tag{3.3.18}$$

where $d = d(K, \partial\Delta(|u_0|_1))$. In order to get a bound of u_m in $D(A)$, for any compact subset K of $\Delta(|u_0|_1)$, and any $z = se^{i\theta} \in K$, we have

$$\begin{aligned}\left|\frac{\mathrm{d}}{\mathrm{d}s}|u_m(se^{i\theta})|_1^2\right| &= \left|2\left(\frac{\mathrm{d}u_m}{\mathrm{d}z}, u_m\right)\right| \\ &\le 2\left|\frac{\mathrm{d}u_m}{\mathrm{d}z}\right|_1 |u_m|_1 \le 4(2/d)(1+|u_0|_1^2),\end{aligned} \tag{3.3.19}$$

where $d = d(K, \partial\Delta(|u_0|_1))$. Inserting this into equation (3.3.15), and ignoring $\|u_m\|^2$, we get

$$C_1|Au_m|^2 \le C_2 + C_3(2(1+|u_0|_1^2)^\gamma) + 4(2/d)(1+|u_0|_1^2). \tag{3.3.20}$$

Hence, for any compact subset $K \subset \Delta(|u_0|_1)$, u_m in $L^\infty(K, D(A))$ is uniformly bounded (is independent of m).

$$\sup_{z\in K}|Au_m(z)| \le T_2 < \infty, \quad T_2 = T_2(K).$$

By Cauchy formula, we get, for any compact subset K, that when $d = d(K, \partial\Delta(|u_0|_1)) > d > 0$,

$$A(d^k u_m/dz^k)(z) = (k!)/(2\pi i)\int_{|z-\eta|=\frac{d}{2}} Au_m(\eta)/(\eta - z)^{k+1}\mathrm{d}\eta, \quad \forall z \in K.$$

Letting K' be the compact subset $\{z \in \Delta(u_0) \mid d(z, \partial\Delta(|u_0|_1)) \geq \frac{1}{2}d(K, \partial\Delta(|u_0|_1))\} \supset K$, we have

$$|A(d^k u_m/dz^k)(z)| \leq 2^k(k!)[d(z, \partial\Delta(u_0))]^{-k} \sup_{z\in K'} |Au_m(z)|,$$

$$\sup_{z\in K} |A(d^k u_m/dz^k)(z)| \leq 2^k(k!)[d(z, \partial\Delta(u_0))]^{-k} T_2(K'). \tag{3.3.21}$$

Now we take the limit $m \to \infty$. The functions $u_m : \mathbf{C} \to D(A)$ on $\Delta(|u_0|_1)$ are uniformly bounded and analytic. On the basis of classical Montel theorem, for the sequence $\{u_m\}$, we can select a subsequence which converges uniformly to function a u^* with respect to the $D(A)$ norm on any compact set $\Delta(u_0)$, which is $D(A)$ analytic in $\Delta(u_0)$. Furthermore, because $u_m|_{0,T}$ corresponds to the real Galerkin approximation, $u_m|_{0,T} \to u$ is a solution of equation (3.3.31)–(3.3.32). Thus u^* is just the analytic continuation of u on $\Delta(u_0)$. So u^* is the unique analytical continuation of u on $\Delta(u_0)$, $u^* = u$. Therefore any subsequence of $\{u_m\}$ converges to u, and in fact the whole sequence converges to u. Then we have

$$\sup_{z\in K} |Au(z)| \leq T_2(K), \quad \forall \text{ compact set } K \subset \Delta(u_0).$$

Similarly, by equations (3.3.18)–(3.3.21) we deduce that

$$\frac{d^k u_m}{dz^k} \to \frac{d^k u}{dz^k}$$

is consistently convergent in $\Delta(u_0)$ with respect to $D(A)$. Since u satisfies equations (3.3.38)–(3.3.39), from the results on the uniform convergence, it is readily seen that u is the solution of equations (3.3.31)–(3.3.32).

If $|u|_1 \leq R$, $t \in (\alpha, \beta)$, then we can repeat the principle at $t = 0$ to get the same at any point $t \in (\alpha, \beta)$, and we have that $u : \mathbf{C} \to D(A)_c$ is a $D(A)$-valued analytic function on the open set

$$\bigcup_{t\geq 0}[t + \Delta(|u(t)|_1)] \supset \bigcup_{t\in(\alpha,\beta)} [t + \Delta(R)]. \qquad \square$$

Remark 3.3.1. Under the assumptions of Theorem 3.3.1, for $0 < t \leq (\frac{3}{4})T_1(|u_0|_1)$, we have

$$\left|\frac{d^k u}{dz^k}(t)\right|_1 \leq 2^k k! C^{-k}(2^{\frac{1}{2}})(1 + |u_0|_1^2)^{\frac{1}{2}}, \tag{3.3.22}$$

$$\left|A\frac{d^k u}{dz^k}(t)\right| \leq 2^k k! C^{-k}(C_2 + C_3(2(1 + |u_0|_1^2)^{\gamma}) + 4(4/C)(1 + |u_0|_1^2)^{\frac{1}{2}} C_1), \tag{3.3.23}$$

where $C = (t_m/(1 + m)^2)^{\frac{1}{2}}$ is the distance from t to $\partial\Delta$, $m = \alpha_0/4\|D\|_*$, C_1, C_2 and γ are constants.

In the following, we list some specific applications of Theorem 3.3.1.

Example: The reaction diffusion equation

Let $\Omega \subset \mathbf{R}^n$ be a bounded open set. The reaction diffusion equation is written as

$$\frac{\partial u}{\partial t} - D\Delta u + g(u) = 0, \tag{3.3.24}$$

$$u(0) = u_0, \tag{3.3.25}$$

where $u = (u_1, u_2, \dots, u_k)$ is the vector function defined on $\Omega \times \mathbf{R}_+$, D is a positive diagonal matrix, the diagonal elements of which are $d_1, d_2, \dots, d_k$. The function $g : \mathbf{R}^k \to \mathbf{R}^k$ has its components of rth order ($r \geq 2$) polynomial form:

$$g_i(x) = \sum_{|\alpha| \leq r} C^i_\alpha x_1^{\alpha_1} \cdots x_k^{\alpha_k}. \tag{3.3.26}$$

Supplement one of the following boundary conditions:

$$u(x,t) = 0, \quad x \in \partial\Omega,\ t > 0, \tag{3.3.27}$$

$$u(\cdot,t) \text{ is } \Omega \text{ periodic}, \quad t \geq 0,\ \Omega = (0,L)^n, \tag{3.3.28}$$

$$\frac{\partial u}{\partial n}(x,t) = 0, \quad \forall x \in \partial\Omega,\ \forall t \geq 0. \tag{3.3.29}$$

Corresponding to the above boundary conditions, we have

$$\begin{aligned} H &= L^2(\Omega), && L^2(\Omega), && L^2(\Omega), \\ V &= H_0^1(\Omega), && H^1_{\text{per}}(\Omega), && H^1(\Omega), \\ D(A) &= H^2(\Omega) \cap H_0^1(\Omega), && H^2_{\text{per}}(\Omega), && \left\{ v \in H^2(\Omega) \mid \frac{\partial v}{\partial n} = 0 \right\}. \end{aligned}$$

Write equation (3.3.24) in the form of equation (3.3.31), so that we have $A = -\Delta$, $R(u) = g(u)$, $u \in D(A)$, and consider that $R(u)$ satisfies condition (3.3.3) and inequality

$$\begin{aligned} |R(u)| &= \int \sum_{1 \leq i \leq k} |(C^i_\alpha x_1^{\alpha_1} \cdots u_k^{\alpha_k})|^2 \mathrm{d}x \\ &\leq C \int \sum_{1 \leq i \leq k} \sum_{|\alpha| \leq \gamma} |(C^i_\alpha x_1^{\alpha_1} \cdots u_k^{\alpha_k})|^2 \mathrm{d}x \leq C \int \sum_{1 \leq i \leq k} \sum_{|\alpha| \leq \gamma} C^i_\alpha \|u\|^{2|\alpha} \mathrm{d}x. \end{aligned}$$

We can find constants C_1 and C_2, which are independent of i, such that

$$\sum_{|\alpha| \leq \gamma} |C^i_\alpha| \|u\|^{2|\alpha} \leq C_1 |u|^{2\gamma} + C_2.$$

Hence

$$\begin{aligned} |R(u)| &\leq C(|u|_{L^{2\gamma}})^\gamma + C|\Omega|, \\ |R(u), Au + u| &\leq \left| \int \sum_{1 \leq i \leq k} \sum_{|\alpha| \leq \gamma} (C^i_\alpha u_1^{\alpha_1} \cdots u_k^{\alpha_k})(-\Delta u_i + u_i) \mathrm{d}x \right| \end{aligned}$$

$$\begin{aligned}
&\le \Big|\sum_{1\le i\le k}\Big\{\int\sum_{|\alpha|\le\gamma}|C_\alpha^i||u|^{|\alpha|}|\Delta u_i|\mathrm{d}x+\int\sum_{|\alpha|\le\gamma}|C_\alpha^i||u|^{|\alpha|+1}\mathrm{d}x\Big\}\Big|\\
&\le \sum_{1\le i\le k}\Big\{\int(C_1|u|^\gamma|\Delta u_i|+C_2|\Delta u_i|)\mathrm{d}x+\int(C_3|u|^{\gamma+1}+C_4)\mathrm{d}x\Big\}\\
&\le C_1(|u|_{L^{2\gamma}})^\gamma|Au|+C_2|\Omega|^{\frac12}|Au|+C_3(|u|_{L^{\gamma+1}})^{\gamma+1}+C_4|\Omega|. \qquad (3.3.30)
\end{aligned}$$

When $n=1$ or $n=2$, we have

$$|u|_{L^p}\le C_p|u|_{H^1},\quad p\ge 1.$$

Thus, using Young inequality again, we obtain

$$\begin{aligned}
|(R(u),Au+u)|&\le C_1(C_{2\gamma}|u|_1)^\gamma|Au|+C_2|\Omega|^{\frac12}|Au|+C_3(C_{\gamma+1}|u|_1)^{\gamma+1}+C_4|\Omega|\\
&\le \varepsilon|Au|^2+C\Big(\frac{1}{\varepsilon}\Big)(|u|_1)^{2\gamma}+|\Omega|+C(|u|_1)^{\gamma+1}+C_1|\Omega|.
\end{aligned}$$

Therefore the condition of equation (3.3.3) holds. At this point γ is an arbitrary positive integer.

When $n\ge 3$, we have the following lemma:

Lemma 3.3.1. *There exist σ and τ which satisfy $1>\tau,\sigma\ge 0$, $r\sigma<1$, $(r+1)\tau<2$ such that*

$$\begin{aligned}
(|u|_{L^{2r}})^r&\le C(r,n,\Omega)(|Au|^2+|u|^2)^{r\sigma/2}(|u|_1)^{r(1-\sigma)},\\
&\quad 1\le r<(n+2)/(n-2), \qquad (3.3.31)\\
(|u|_{L^{r+1}})^{r+1}&\le C(r,n,\Omega)(|Au|^2+|u|^2)^{(r+1)r/2}(|u|_1)^{(r+1)(1-r)},\\
&\quad 1\le r<(n+6)/(n-2). \qquad (3.3.32)
\end{aligned}$$

Proof. First, we have the embedding theorem: $H^{\sigma+1}$ is embedded into L^{2r}, when

$$\frac{1}{2r}\ge\frac12-(1+\sigma)/n,\quad 2r\le 2n/(n-2(1+\sigma)). \qquad (3.3.33)$$

By using $H^{1+\sigma}$ which can be obtained by interpolation of H^1 and H^2, and $(|Au|^2+|u|^2)^{\frac12}$ which is equivalent to the H^2-norm:

$$\begin{aligned}
|u|_{H^{1+\sigma}}&\le C(|u|_{H^2})^\sigma(|u|_{H^1})^{1+\sigma},\\
(|u|_{H^{1+\sigma}})^r&\le C(|Au|^2+|u|^2)^{r\sigma/2}(|u|_{H^1})^{r(1-\sigma)},
\end{aligned}$$

as long as $r\sigma<1$, we have

$$\begin{aligned}
&2r<2n/(n-2)\Big(1+\frac1r\Big),\\
&r<(n+2)/(n-2).
\end{aligned}$$

Equation (3.3.31) has been proved. For equation (3.3.32), by Sobolev embedding theorem, $H^{1+\tau}$ embeds into L^{r+1}, when

$$1/(r+1) \geq \frac{1}{2} - (1+\tau)/n,$$
$$r+1 \leq 2n/(n-2(1+\tau)). \tag{3.3.34}$$

Considering $H^{1+\tau}$, which can be obtained by interpolating H^1 and H^2, we arrive at

$$(|u|_{H^{1+\tau}})^{1+r} \leq C(|Au|^2 + |u|^2)^{(r+1)\tau/2}(|u|_{H^1})^{(r+1)(1-\tau)}.$$

Taking $(r+1)\tau/2 < 1$, that is, $(r+1)\tau < 2$, and substituting into equation (3.3.34), we have

$$r+1 < 2n/(n-2(1+\tau/(r+1))),$$
$$r < (n+6)/(n-2),$$

which yields equation (3.3.32). By use of Lemma 3.3.1, we get

$$\begin{aligned}|(R(u), Au+u)| &\leq C(|Au|^2 + |u|^2)^{rs/2}(|u|_1)^{r(1-s)}|Au| + C_2|\Omega|^{\frac{1}{2}}|Au| \\ &\quad + C(|Au|^2 + |u|^2)^{(r+1)\tau/2}(|u|_1)^{(r+1)(1-\tau)} + C_4|\Omega| \\ |(R(u), Au+u)| &\leq \frac{C}{2}(|Au|^{rs+1} + |u|^{rs})(|u|_1)^{r(1-s)} + C_2|\Omega|^{\frac{1}{2}}|Au| \\ &\quad + C(|Au|^{(r+1)\tau} + |u|^{(r+1)\tau})(|u|_1)^{(r+1)(1-\tau)} + C_4|\Omega|.\end{aligned}$$ □

From this we can obtain

Theorem 3.3.2. *In dimension 1 or 2, there exist θ_0, $|\theta_0| \leq \frac{\pi}{4}$, and function $T_1 \in C(\mathbf{R}_+, \mathbf{R}_+)$, which depends on g and spacial dimension, such that if for any t, $|u(t)|_1$ is finite, then u has $D(A)_c$ analytic extension to the complex domain*

$$\Delta = t + \Delta(|u(t)|_1),$$

and equations (3.3.37), (3.3.38) and (3.3.39) are valid. While in dimension 3, the same holds as long as the degree of polynomial of g(u) is 1, 2, 3 or 4. For space dimension 4 or 5, we restrict r to 1 or 2.

Example: Ginzburg–Landau equation

Let $\Omega \subset \mathbf{R}^n$ be an bounded open set, $n = 1, 2, 3$. The Ginzburg–Landau equation is

$$\frac{\partial u}{\partial t} - (\nu + i\alpha)\Delta u + (k + i\beta)|u|^2 u - \gamma u = 0, \tag{3.3.35}$$
$$u(0) = u_0, \tag{3.3.36}$$

where $u(x,t)$ is a complex function, $(x,t) \in \Omega \times \mathbf{R}_+$, the parameters ν, α, k, β, γ are real constants, and $\nu > 0$, $k > 0$. Suppose that one of the following boundary conditions holds:

$$u(x,t) = 0, \quad \forall x \in \partial\Omega,\ t \geq 0, \tag{3.3.37}$$

$$u(\cdot,t) \text{ is } \Omega \text{ periodic}, \quad \forall t \geq 0,\ \Omega = (0,L)^n, \tag{3.3.38}$$

$$\frac{\partial u}{\partial n}(x,t) = 0, \quad \forall x \in \partial\Omega,\ t \geq 0. \tag{3.3.39}$$

Write equation (3.3.35) in the form of equation (3.3.31), $u = u_1 + iu_2$, $A = -\Delta$,

$$D = \begin{pmatrix} \nu & -\alpha \\ \alpha & \nu \end{pmatrix}, \qquad \begin{aligned} R_1(u) &= (u_1^2 + u_2^2)(ku_1 - \beta u_2) - \gamma u_1, \\ R_2(u) &= (u_1^2 + u_2^2)(ku_2 - \beta u_1) - \gamma u_2. \end{aligned}$$

Now we check condition (3.3.3) which involves estimating

$$\begin{aligned} \big|(R(u), Au + u)\big| &\leq \big|(k + i\beta)\big| \left| \int \nabla(|u|^2 u + (|\gamma| + 1)u) \cdot \nabla u dx \right| \\ &\leq C|u|_{L^4}^2 |\nabla u|_{L^4}^2 + (|\gamma| + 1)\|u\|^2. \end{aligned}$$

When $n = 1,2,3$, $H^{\frac{3}{4}}$ can be continuously embedded into L^4, and $H^{\frac{3}{4}}$ can be constructed by the interpolation of L^2 and H^1. We have

$$|u|_{L^4} \leq C|u|^{\frac{1}{4}}(|u|_1)^{\frac{3}{4}}. \tag{3.3.40}$$

By the equivalence of the $(|Au|^2 + |u|^2)^{\frac{1}{2}}$ and H^2 norms, we get

$$|\nabla u|_{L^4} \leq C|u|^{\frac{1}{4}}(|Au|^2 + |u|^2)^{\frac{3}{8}}. \tag{3.3.41}$$

From equations (3.3.40) and (3.3.41), we get

$$\begin{aligned} \big|(R(u), Au + u)\big| &\leq C|u|^{\frac{1}{2}}(|u|_1)^{\frac{3}{2}}\|u\|^{\frac{1}{2}}(|u|^2 + |Au|^2)^{\frac{3}{4}} + (|\gamma| + 1)\|u\|^2 \\ &\leq \varepsilon|Au|^2 + \varepsilon|u|^2 + (\varepsilon)^{-3}C|u|^2\|u\|^2(|u|_1)^6 + (|\gamma| + 1)\|u\|^2 \\ &\leq \varepsilon|Au|^2 + C(\varepsilon)^{-3}(|u|_1)^{10} + K, \end{aligned}$$

where $C = \max((4\varepsilon)^{-1}C, |\gamma| + 1)$. Then condition (3.3.3) is satisfied, where $\gamma = 5$. Hence the conclusion of Theorem 3.3.2 is true. Then the two real components of the solution of Ginzburg–Landau equation are $D(A)$-analytic functions. Since $n = 1,2$, there exists an absorbing set in V. We have

Theorem 3.3.3. *In dimension 1 or 2, there exist $d, \tau > 0$ such that the real and imaginary components of the solution for the Ginzburg–Landau equation with the initial–boundary value problem* (3.3.35) *can be analytically continued on the set*

$$\Delta_d = \{z \in C \mid \mathrm{Re}(z) > \tau,\ |\mathrm{Im}(z)| < d\},$$

and the equations (3.3.37)–(3.3.39) are valid in Δ_d *(particularly for* $t > \tau$*). For spacial dimension 3, when* $|u(t)|_1$ *is finite, we can continue the real and imaginary parts of the solution of this problem analytically to* $\Delta(|u(t)|_1)$.

To introduce the following Gevrey regularity classes, suppose we are given a $(2p)$th order linear self-adjoint unbounded elliptic operator A, which possesses the form

$$A = \sum_{|\alpha|=p} a_\alpha D^{2\alpha},$$

where $\alpha = (\alpha_1, \dots, \alpha_n)$, $|\alpha| = \alpha_1 + \cdots + \alpha_n$. Let $\Omega = [0, 2\pi]^n$ be the set in the periodic boundary condition. The Fourier transform of A is $\hat{A}$, which is a homogeneous $(2p)$th order positive polynomial:

$$C^{-1}|\xi|^{2p} \le \hat{A}(\xi) \le C|\xi|^{2p}. \tag{3.3.42}$$

The eigenvectors of A are the exponential functions $\{e^{ijx}\}_{j\in\mathbf{Z}^n}$, the corresponding eigenvalues are $\{\lambda_j = \hat{A}(j)\}_{j\in\mathbf{Z}^n}$,

$$u = \sum u_j e^{ijx}, \quad u_j = (u_j^1, \dots, u_j^n) \in \mathbf{C}^n,$$
$$|u|^2 \triangleq \left(\frac{1}{2\pi}\right)^n \sum |u_j|^2 < \infty. \tag{3.3.43}$$

For simplicity, set

$$\int_\Omega u(x,t)\mathrm{d}x = 0, \quad \forall t \ge 0, \text{ for all the solutions } u. \tag{3.3.44}$$

Equation (3.3.44) is equivalent to condition $u_0 = 0$. On $D(A)$,

$$|Au|^2 = \left(\frac{1}{2\pi}\right)^n \sum |\lambda_j|^2 |u_j|^2.$$

Defining the differential operator $B = (-1)^p \Delta^p$, we have $\hat{B}(\xi) = |\xi|^{2p}$ and

$$|Bu|^2 = \left(\frac{1}{2\pi}\right)^n \sum |j|^{4p} |u_j|^2.$$

Hence, in equation (3.3.42), if we set $\xi = j$, then on $D(A)$ we have

$$C^{-1}|Bu|^2 \le |Au|^2 \le C|Bu|^2. \tag{3.3.45}$$

Now for $\tau > 0$, we define the Gevrey category, $D(\exp(2B^{\frac{1}{2p}}))$, as the set of functions having the form (3.3.43)–(3.3.44) such that

$$|u|_\tau^2 = \left|\exp(\tau B^{\frac{1}{2p}})u\right|^2 = \left(\frac{1}{2\pi}\right)^n \sum e^{2\tau|j|}|u_j|^2 < \infty. \tag{3.3.46}$$

Now we consider an abstract initial value problem (3.3.31)–(3.3.32) with periodic boundary conditions. And nonlinear term R is a polynomial of $\{D^\alpha\}_{|\alpha|\le d}$. Since Fourier transform preserves distances in L^2, there exists a function F, which only depends on Fourier coefficient u_j and multiindex α_j of R, such that

$$(R(u), Au) = F(u_j, j^{\alpha j}).$$

We require that γ, K and C are as in equation (3.3.33), and for arbitrary $\varepsilon > 0$, F satisfies

$$|F(u_j, |j^{\alpha j}|)| \le \varepsilon |Au|^2 + C(\varepsilon)\|u\|^{2\gamma} + K. \tag{3.3.47}$$

But for condition (3.3.34), there is a replacement: there are constants M, $\sigma > 0$ such that f satisfies

$$|f(t)|_\sigma \le M, \quad \forall t \in \mathbf{R}_+. \tag{3.3.48}$$

Theorem 3.3.4. *Suppose that a differential operator A satisfies the previous conditions, the initial value is $u_0 \in V$, f satisfies equation (3.3.48), and $R(u)$ satisfies equations (3.3.47) and (3.3.34). There exist T_*, which only depends on $\|u_0\|$, and region $\Delta \supseteq (0, T_*)$ in the complex plane such that the following statements are valid:*

(i) *There exists a unique regular solution u for the problem (3.3.31) and equation (3.3.32) such that the mapping $t \to [A^{\frac{1}{2}} \exp(\phi(t)B^{\frac{1}{2p}})]u$ takes values on H and is analytic in $(0, T_*) \subset \Delta$, $\phi(t) = \min(t, \sigma, T_*)$.*

(ii) *If a solution of equation (3.3.31)–(3.3.32) exists and is uniformly bounded in V ($t \in \mathbf{R}_+$), then u takes values on $D(A^{\frac{1}{2}} \exp(\sigma B^{\frac{1}{2p}}))$ and is analytic for $t \in (0, \infty)$, here Δ is included in $(0, \infty)$.*

In order to prove Theorem 3.3.4, we first need the following lemma:

Lemma 3.3.2. *Let $u \in D(A^{\frac{1}{2}} e^{\tau B^{\frac{1}{2p}}})$, $\tau > 0$. If R satisfies equation (3.3.47), then we have*

$$|(R(u), Au)_\tau| \le \varepsilon(|Au|_\tau^2) + C(\varepsilon)|A^{\frac{1}{2}} u|_\tau^{2\gamma} + K, \quad \forall \varepsilon > 0, \tag{3.3.49}$$

where γ, k, C are as in equation (3.3.33).

Proof. Let

$$u = \sum u_j e^{ijx},$$
$$u^* = \sum u_j^* e^{ijx} = \exp(\tau B^{\frac{1}{2p}} u),$$

where

$$u^* = e^{\tau |j|} u_j, \quad j \in \mathbf{Z}^n.$$

Let

$$R(u) = \sum \prod_{k=1}^{d} D^{\alpha_k} u^{i_k},$$

where $\alpha_1, \ldots, \alpha_d$ are arbitrary multiindices; $i_1, \ldots, i_d$ are integers between 1 and n; $u = (u^1, \ldots, u^n)$. Then we have

$$(R(u), Au)_\tau = (R(u), \exp(2\tau B^{1+(2p)})Au)_\tau,$$

$$(R(u), Au)_\tau = \int_\Omega \Big(\prod_{k=1}^{d} \sum_{j_k \in \mathbf{Z}^n} u_{j_k}^{i_k} j_k^{\alpha_k} e^{ij_k x} \Big) \Big(\sum_{j_0 \in \mathbf{Z}^n} \overline{u}_{j_0}^{j_0} \lambda_{j_0} e^{2\tau |j_0|} e^{-ij_0 x} \Big) \mathrm{d}x.$$

This integral is zero, due to the property of the exponential function, unless

$$-j_0 + j_1 + j_2 + \cdots + j_k = 0.$$

In this case, the integral value is $(2\pi)^n$. So

$$(R(u), Au)_\tau = (2\pi)^n \sum \sum_{j_1 + \cdots + j_d = j_0} u_{j_d}^{i_1} \cdots u_{j_d}^{i_d} \overline{u}_{j_d}^{i_d} j_1^{\alpha_d} \cdots j_d^{\alpha_d} \lambda_{j_0} e^{2\tau |j_0|},$$

$$|(R(u), Au)_\tau| \le (2\pi)^n \sum \sum_{j_1 + \cdots + j_d = j_0} |u_{j_0}^{*i_0}| \cdots |u_{j_d}^{*i_d}| |j_1^{\alpha_1}| \cdots |j_d^{\alpha_d}| \lambda_{j_0} e^{\tau(|j_0| - |j_1| - \cdots - |j_d|)},$$

but $|j_0| = |j_1 + \cdots + j_k| \le |j_1| + \cdots + |j_k|$, yielding $\exp(\tau(|j_0| - |j_1| - \cdots - |j_d|)) \le 1$. Therefore we get

$$|(R(u), Au)_\tau| \le (2\pi)^n \sum_{j_1 + \cdots + j_d = j_0} \sum |u_{j_0}^{*i_0}| \cdots |u_{j_d}^{*i_d}| \cdot |j_d^{\alpha_d}| \lambda_{j_0},$$

where $u^* = \exp(\tau B^{\frac{1}{2p}})u$. Then we have

$$(R(u), Au)_\tau \le |F(|u_j^*|, |j^{\alpha_j}|)| \le \varepsilon |Au^*|^2 + C(\varepsilon) |A^{\frac{1}{2}} u^*|^{2\gamma} + R,$$

completing the proof of the lemma. □

Now we use Lemma 3.3.2 to prove Theorem 3.3.4. As in the framework of Theorem 3.3.1, the key point is to get an a priori estimate. Firstly, complexify equation (3.3.31), and set $\phi(t) = \min(t, \sigma)$, $z = se^{i\theta}$, $s > 0$, $\cos\theta > 0$, $E_s = \exp(\phi(s\cos\theta)B^{\frac{1}{2p}})$, $\theta \in (-\frac{\pi}{2}, \frac{\pi}{2})$. Taking the inner product of equation (3.3.31) and $Au(se^{i\theta})$ in $D(E_s)$, multiplying by $e^{i\theta}$, and taking its real part, we get

$$\begin{aligned} &\operatorname{Re} e^{i\theta} \Big[\Big(E_s \frac{\partial u}{\partial t}(se^{i\theta}), E_s Au(se^{i\theta}) \Big) + (E_s DAu, E_s Au) + (E_s R(u), E_s Au) \Big] \\ &\quad = \operatorname{Re} e^{i\theta} (E_s f, E_s Au). \end{aligned} \tag{3.3.50}$$

Using the relation $\frac{\mathrm{d}}{\mathrm{d}s} = e^{-i\theta}\frac{\mathrm{d}}{\mathrm{d}z}$, we get

$$\begin{aligned}
&\operatorname{Re} e^{i\theta}\left\{\left(E_s\frac{\partial u}{\partial t}(se^{i\theta}), E_sAu(se^{i\theta})\right)\right\} \\
&= \operatorname{Re}\left(A^{\frac{1}{2}}\frac{\mathrm{d}}{\mathrm{d}s}(E_su(se^{i\theta})) - \phi'(s\cos\theta)(\cos\theta)E_sAu(se^{i\theta}), E_sA^{\frac{1}{2}}u(se^{i\theta})\right) \\
&= \frac{1}{2}\frac{\mathrm{d}}{\mathrm{d}s}|A^{\frac{1}{2}}u(se^{i\theta})|^2_{\phi(s\cos\theta)} - (\cos\theta)\phi'(s\cos\theta)|Au|_{\phi(s\cos\theta)}|A^{\frac{1}{2}}u|_{\phi(s\cos\theta)} \\
&\geq \frac{1}{2}\frac{\mathrm{d}}{\mathrm{d}s}|A^{\frac{1}{2}}u(se^{i\theta})|^2_{\phi(s\cos\theta)} - \alpha_0\cos\theta/4|Au|^2_{\phi(s\cos\theta)} \\
&\quad - (\cos\theta/\alpha_0)|A^{\frac{1}{2}}u|^2_{\phi(s\cos\theta)}.
\end{aligned} \tag{3.3.51}$$

As estimated before, we have

$$\operatorname{Re} e^{i\theta}(E_sDAu, E_sAu) \geq \cos\theta\alpha_0|Au|^2_{\phi(s\cos\theta)} - |\sin\theta|\|D\|_*|Au|^2_{\phi(s\cos\theta)}.$$

Restricting θ to satisfy equation (3.3.51), we get

$$\operatorname{Re} e^{i\theta}(E_sDAu, E_sAu) \geq \left(\frac{3}{4}\right)\cos\theta\alpha_0|Au|^2_{\phi(s\cos\theta)}. \tag{3.3.52}$$

Substituting equations (3.3.51) and (3.3.52) into (3.3.50), and replacing $\phi(s\cos\theta)$ with ϕ, we get

$$\frac{1}{2}\frac{\mathrm{d}}{\mathrm{d}s}\|u\|^2_\phi + (\alpha_0\cos\theta)/2|Au|^2_\phi \leq |(f, Au)_\phi| - |(R(u), Au)_\phi|. \tag{3.3.53}$$

Dealing with the term $|(f, Au)_\phi|$ as in Lemma 3.3.1, we get the inequality

$$\frac{\mathrm{d}}{\mathrm{d}s}\|u(se^{i\theta})\|^2_\phi + C_1|Au(se^{i\theta})|^2_\phi \leq C_2 + C_3|u(se^{i\theta})|^{2\sigma}_\phi, \tag{3.3.54}$$

where constants C_2, C_3 only depend on the initial value, and not on θ. Letting

$$y(s) = 1 + |E_sA^{\frac{1}{2}}u(se^{i\theta})|^2, \tag{3.3.55}$$

$$y'(s) \leq C_4y^\gamma(s), \tag{3.3.56}$$

we get

$$|E_sA^{\frac{1}{2}}u(se^{i\theta})|^2 \leq 2 + 2|A^{\frac{1}{2}}u_0|^2. \tag{3.3.57}$$

The region $\Delta(\|u_0\|)$ is given by

$$\begin{aligned}
&0 \leq s \leq T_1(\|u_0\|) = y(0)^{1-\gamma}\left(1 - \left(\frac{1}{2}\right)^{\gamma-1}\right)/((\gamma-1)C_4), \\
&|\theta| \leq \min\left(\arctan(\alpha_0/4\|D\|_*), \frac{\pi}{4}\right).
\end{aligned} \tag{3.3.58}$$

So, if $u_0 \in D(A^{\frac{1}{2}})$, then $u(se^{i\theta}) \in D(E_s A^{\frac{1}{2}})$ in the angular region $\Delta(\|u_0\|)$. If $\|u\| \leq M$, $t \in \mathbf{R}_+$, then $\Delta(\|u_0\|)$ can be extended to $\Delta = \bigcup_{t>0}(t + \Delta(M))$. Similarly, for any compact set $K \subset \Delta$, we have the following inequality;

$$\sup_{z\in K} \left\| \frac{d^k u}{dz^k}(se^{i\theta}) \right\|_\phi \leq 2^{\frac{1}{2}}(2/d)^k(k!)(1 + \|u_0\|^2)^{\frac{1}{2}}, \quad d = \operatorname{dist}(K, \partial\Delta), \tag{3.3.59}$$

$$\sup_{z\in K} |Au(se^{i\theta})|_\phi \leq T_2(K) < \infty, \tag{3.3.60}$$

$$\sup_{z\in K} \left\| A\left(\frac{d^k u}{dz^k}(se^{i\theta}) \right) \right\|_\phi \leq 2^k(k!)[d(K, \partial\Delta(u_0))]^{-k} T_2(K'), \tag{3.3.61}$$

where

$$K' \triangleq \left\{ z \in \Delta(u_0) \mid d(z, \partial\Delta(u_0)) \geq \frac{1}{2} d(K, \partial\Delta(u_0)) \right\}.$$

Remark 3.3.2. If solution u is uniformly bounded by M ($\forall t > 0$) in V, then from equation (3.3.57) we have

$$|E_t A^{\frac{1}{2}} u(t)|^2 \leq 2 + 2M^2. \tag{3.3.62}$$

In particular, by the definition of E_t we have

$$|u_j(t)|^2 \leq (2 + 2M^2)\lambda_j^{-1} e^{-2|j|\phi(t)}. \tag{3.3.63}$$

This suggests that the Fourier coefficients decay exponentially.

Remark 3.3.3. Suppose that α is a given multiindex. Then

$$|D^\alpha u| \leq \left(\frac{1}{2\pi}\right)^n \sum_{j\in Z^n} |j^\alpha|^2 |u_j|^2,$$

$$|u|_\tau^2 = \left(\frac{1}{2\pi}\right)^n \sum_j e^{2\tau|j|} |u_j|^2.$$

There exists a constant $M(\alpha, \tau) > 0$ such that

$$|j^\alpha| \leq e^{2|j|}, \quad j \in \mathbf{Z}^n, \ |j| \geq M.$$

From this we can see, if $|u|_\tau < \infty$ (for some $\tau > 0$), then $|D^\alpha u| < \infty$. Hence the Gevrey function category is included in $C^\infty(\Omega)$.

3.4 Two-dimensional Ginzburg–Landau equation

In 1996, Guo and Wang considered the time analyticity, Gevrey regularity and approximate inertial manifold for the two-dimensional Ginzburg–Landau equation [25, 117].

Suppose we are given the following Ginzburg–Landau equation:

$$\frac{\partial u}{\partial t} = \rho u + (1 + iv)\Delta u - (1 + iu)|u|^{2\sigma} u + \alpha\lambda_1 \cdot \nabla(|u|^2 u) + \beta(\lambda_2 \cdot \nabla u)|u|^2,$$
$$(x, t) \in \Omega \times \mathbf{R}^+, \tag{3.4.1}$$
$$u(x, 0) = u_0(x), \quad x \in \Omega, \tag{3.4.2}$$
$$u \text{ is } \Omega \text{ periodic}, \quad \Omega = (0, L_1) \times (0, L_2), \tag{3.4.3}$$

where u is an unknown complex function, $\sigma \in \mathbf{N}$, $\rho > 0$, u, α, β are constants, λ_1, λ_2 are real vectors.

In [114], we have proved that if $u_0 \in H^2(\Omega)$, and there exists $\delta > 0$ such that

$$2 < \sigma \le \frac{1}{\sqrt{1 + (\frac{\mu - v\delta^2}{1+\delta^2})^2} - 1}, \quad \sigma \in \mathbf{N}, \tag{3.4.4}$$

where **N** is set of the natural numbers, then there exists a unique global solution $u(x, t)$ of equation (3.4.1)–(3.4.3) such that

$$u(x, t) \in L^\infty(0, T; H^2(\Omega)) \cap L^2(0, T; H^3(\Omega)), \quad \forall T > 0, \tag{3.4.5}$$

and there exists a constant K, which depends the parameters $(\sigma, \rho, v, \mu, \alpha, \beta, \lambda_1, \lambda_2, \delta, \Omega)$, such that

$$\|u(t)\|_{H^1} \le K_1, \quad \forall t \ge t_1, \tag{3.4.6}$$

where t_1 depends on the parameters $(\sigma, \rho, v, \mu, \alpha, \beta, \lambda_1, \lambda_2, \delta, \Omega)$ and R, whenever $\|u_0\|_{H^1} \le R$.

Let $u(t) = u_1(t) + iu_2(t)$, $u_1(t)$ and $u_2(t)$ are real functions. Then taking the real and imaginary parts of equation (3.4.1), it follows that

$$\frac{\partial u_1}{\partial t} = \rho u_1 + \Delta u_1 - v\Delta u_2 - |u|^{2\sigma}(u_1 - \mu u_2)$$
$$+ \alpha\lambda_1 \cot \nabla(|u|^2 u_1) + \beta(\lambda_2 \cdot \nabla u_1)|u|^2, \tag{3.4.7}$$
$$\frac{\partial u_2}{\partial t} = \rho u_2 + \Delta u_2 + v\Delta u_1 - |u|^{2\sigma}(u_2 + \mu u_1)$$
$$+ \alpha\lambda_1 \cdot \nabla(|u|^2 u_2) + \beta(\lambda_2 \cdot \nabla u_2)|u|^2. \tag{3.4.8}$$

For simplicity, use $u(t)$ to denote the vector $(u_1(t), u_2(t))$. Then the formulas (3.4.7)–(3.4.8) can be written as

$$\frac{\partial u}{\partial t} = \rho u + D\Delta u - D_1|u|^{2\sigma} u + \alpha\lambda_1 \cdot \nabla(|u|^2 u) + \beta(\lambda_2 \cdot \nabla u)|u|^2,$$

where

$$D = \begin{pmatrix} 1 & -v \\ v & 1 \end{pmatrix}, \quad D_1 = \begin{pmatrix} 1 & -\mu \\ \mu & 1 \end{pmatrix}.$$

Then we have

$$\frac{\mathrm{d}u(t)}{\mathrm{d}t} + DAu(t) + R(u(t), u(t), u(t)) = 0, \tag{3.4.9}$$

where $A = -\Delta$ is an unbounded self-adjoint operator, $D(A) = \{u \in H^2(\Omega) \times H^2(\Omega) :$ u satisfies equation (3.4.3)$\}$, and

$$R(u, v, \omega) = -\rho\omega + D_1(u \cdot v)^\sigma \omega - \alpha\lambda_1 \cdot \nabla((u \cdot v)\omega) - \beta(\lambda_2 \cdot \nabla\omega)(u \cdot v). \tag{3.4.10}$$

The operator $R : D(A) \times D(A) \times D(A) \to \mathscr{H} = H \times H$, and for $R(u, v, \omega)$ we can establish the following estimate.

Lemma 3.4.1. *Suppose that $u, v, \omega \in D(A)$. Then $R(u, v, \omega) \in \mathscr{H}$, and*

$$\begin{aligned}\|R(u, v, \omega)\| \le{}& \rho\|\omega\| + C\|u\|_{H^1}^\sigma \|v\|_{H^1}^\sigma \|\omega\|_{H^1} \\ &+ C\|\omega\|_{H^1}^{\frac{1}{2}} \|A\omega\|^{\frac{1}{2}} \|u\|_{H^1} \|v\|_{H^1} \\ &+ C\|\omega\|_{H^1} \|u\|_{H^1} \|v\|_{H^1}.\end{aligned}$$

Hereafter, we use C and C_i, $i = 1, 2, \ldots$, to denote any constants which only depend on the parameters $(\sigma, \rho, \nu, \mu, \alpha, \beta, \lambda_1, \lambda_2, \delta, \Omega)$.

Proof. From equation (3.4.10) we get

$$\begin{aligned}\|R(u, v, \omega)\| \le{}& \rho\|\omega\| + \|D_1(u, v)^\sigma \omega\| \\ &+ \|\alpha\lambda_1 \cdot \nabla((u, v)\omega)\| + \|\beta(\lambda_2 \cdot \nabla\omega)(u \cdot v)\|,\end{aligned} \tag{3.4.11}$$

$$\begin{aligned}\|D_1(u, v)^\sigma \omega\| ={}& \sqrt{1 + \mu^2}\Big(\int_\Omega |u|^{2\sigma} |v|^{2\sigma} |\omega|^2 \mathrm{d}x\Big)^{\frac{1}{2}} \\ \le{}& \sqrt{1 + \mu^2}\|u\|_{8\sigma}^\sigma \|v\|_{8\sigma}^\sigma \|\omega\|_4 \le C_1\|u\|_{H^1}^\sigma \|v\|_{H^1}^\sigma \|\omega\|_{H^1},\end{aligned} \tag{3.4.12}$$

$$\begin{aligned}\|\beta(\lambda_2 \cdot \nabla\omega)(u \cdot v)\| \le{}& |\beta\lambda_2|\Big(\int_\Omega |\nabla\omega|^2 |u|^2 |v|^2 \mathrm{d}x\Big)^{\frac{1}{2}} \\ \le{}& \beta\lambda_2\|\nabla\omega\|_4 \|u\|_8 \|v\|_8 \le C\|\nabla\omega\|^{\frac{1}{2}} \|\nabla\omega\|_{H^1}^{\frac{1}{2}} \|u\|_{H^1} \|v\|_{H^1} \\ \le{}& C\|\omega\|_{H^1}^{\frac{1}{2}} \|\omega\|_{H^2}^{\frac{1}{2}} \|u\|_{H^1} \|v\|_{H^1} \\ \le{}& C\|\omega\|_{H^1}^{\frac{1}{2}} (\|\omega\| + \|A\omega\|)^{\frac{1}{2}} \|u\|_{H^1} \|v\|_{H^1} \\ \le{}& C\|\omega\|_{H^1}^{\frac{1}{2}} (\|\omega\|^{\frac{1}{2}} + \|A\omega\|^{\frac{1}{2}}) \|u\|_{H^1} \|v\|_{H^1} \\ \le{}& C_2\|\omega\|_{H^1}^{\frac{1}{2}} \|A\omega\|^{\frac{1}{2}} \|u\|_{H^1} \|v\|_{H^1} \\ &+ C_3\|\omega\|_{H^1} \|u\|_{H^1} \|v\|_{H^1}.\end{aligned} \tag{3.4.13}$$

Since $\alpha\lambda_1 \cdot \nabla((u \cdot v)\omega) = \alpha(\lambda_1 \cdot \nabla\omega)(u \cdot v) + \alpha(\lambda_1 \cdot \nabla u)(v \cdot \omega) + \alpha(\lambda_1 \cdot \nabla v)(u \cdot \omega)$, similar to equation (3.4.13), we get

$$\begin{aligned}\|\alpha\lambda_1 \cdot \nabla((u \cdot v)\omega)\| \le\ & C_4\|\omega\|_{H^1}^{\frac{1}{2}}\|A\omega\|^{\frac{1}{2}}\|u\|_{H^1}\|v\|_{H^1} \\ & + C_5\|u\|_{H^1}^{\frac{1}{2}}\|Au\|^{\frac{1}{2}}\|v\|_{H^1}\|\omega\|_{H^1} \\ & + C_6\|v\|_{H^1}^{\frac{1}{2}}\|Av\|^{\frac{1}{2}}\|u\|_{H^1}\|\omega\|_{H^1} \\ & + C_7\|\omega\|_{H^1}\|u\|_{H^1}\|v\|_{H^1}.\end{aligned} \tag{3.4.14}$$

Lemma 3.4.1 can be deduced from equations (3.4.11)–(3.4.14). □

As a corollary of Lemma 3.4.1, we have

$$\begin{aligned}\|R(u,u,u)\| \le\ & \rho\|u\|_{H^1} + C_1\|u\|_{H^1}^{2\sigma+1} \\ & + C_8\|u\|_{H^1}^{\frac{5}{2}}\|Au\|^{\frac{1}{2}} + C_9\|u\|_{H^1}^{3}.\end{aligned} \tag{3.4.15}$$

From this we get

$$\begin{aligned}|(R(u,u,u), Au + u)| &\le \|R(u)\|\|Au\| + \|R(u)\|\|u\| \\ &\le \|R(u)\|\|Au\| + \|R(u)\|\|u\|_{H^1} \\ &\le \rho\|u\|_{H^1}\|Au\| + C_1\|u\|_{H^1}^{2\sigma+1}\|Au\| + C_8\|u\|_{H^1}^{\frac{5}{2}}\|Au\|^{\frac{3}{2}} \\ &\quad + C_9\|u\|_{H^1}^{3}\|Au\| + \rho\|u\|_{H^1}^{2} + C_1\|u\|_{H^1}^{2\sigma+2} \\ &\quad + C_8\|u\|_{H^1}^{\frac{7}{2}}\|Au\|^{\frac{1}{2}} + C_9\|u\|_{H^1}^{4} \\ &\le \varepsilon\|Au\|^2 + C_{10}\|u\|_{H^1}^{4\sigma+2} + C_{11}, \quad \forall\varepsilon > 0.\end{aligned} \tag{3.4.16}$$

From equations (3.4.16) and (3.4.6), we get

Theorem 3.4.1. *Suppose that equation (3.4.4) holds and $u \in H^2(\Omega)$. Then there exist θ_0 and T_0 such that each component of the solution for the problem (3.4.1)–(3.4.2) possesses $D(A)$-valued analytic continuation to the following complex region:*

$$\Delta_1 = \{t + se^{i\theta} : t \ge t_1,\ |\theta| \le \theta_0,\ 0 \le s \le T_0\},$$

where t_1 is determined in equation (3.4.6); θ_0 and T_0 depend on the initial value and $|\theta_0| \le \frac{\pi}{4}$. Moreover, there exists a constant K, depending on the initial value, such that

$$\|u(z)\|, \|A^{\frac{1}{2}}u(z)\|, \|Au(z)\| \le K, \quad \forall z \in \Delta_2, \tag{3.4.17}$$

where $\Delta_2 = \{z : \operatorname{Re} z \ge a,\ |\operatorname{Im} z| \le b\}$, with a, b constants, which depend on initial value and R, whenever $|u_0|_{H^1} \le R$.

Proof. Noting equations (3.4.16) and (3.4.6), we deduce that the conditions of Theorem 1.1 in [186] are satisfied, and from it we get the implication of Theorem 3.4.1. □

By Theorem 3.4.1 and Cauchy formula, we get

Proposition 3.4.1. *Suppose that equation (3.4.3) is true and $u_0 \in H^2(\Omega)$. Then we have*

$$\left\|\frac{\mathrm{d}}{\mathrm{d}t}u(t)\right\|, \left\|A^{\frac{1}{2}}\frac{\mathrm{d}}{\mathrm{d}t}u(t)\right\|, \left\|A\frac{\mathrm{d}}{\mathrm{d}t}u(t)\right\| \leq K_2, \quad \forall t \geq t_2,$$

where constant K_2 depends on the initial data, $t_2 > t_1$ only depends on the initial value and R, whenever $\|u_0\|_{H^1} \leq R$.

In the following, we construct an approximate inertial manifold for problem (3.4.1)–(3.4.3). First of all, we know that the eigenvectors of $A = -\Delta$ form an orthogonal basis $\{\omega_j\}_{j=1}^{\infty}$ in H such that

$$A\omega_j = \lambda_j \omega_j, \quad 0 = \lambda_1 < \lambda_2 \leq \cdots \leq \lambda_j \to \infty,\ j \to \infty.$$

Given m, let $P = P_m$ be the orthogonal projection of H to the subspace spanned by $\{\omega_1, \ldots, \omega_m\}$. Letting $Q = Q_m = I - P_m$ and acting with P_m and Q_m on equation (3.4.9), we have

$$\frac{\mathrm{d}p}{\mathrm{d}t} + DAp + P_m R(p+q, p+q, p+q) = 0, \tag{3.4.18}$$

$$\frac{\mathrm{d}q}{\mathrm{d}t} + DAq + Q_m R(p+q, p+q, p+q) = 0, \tag{3.4.19}$$

where $p = P_m u$, $q = Q_m u$, and we have

$$\|A^\gamma p\| \leq \lambda_m^\gamma \|p\|, \quad \gamma > 0,\ p \in P_m D(A^\gamma), \tag{3.4.20}$$

$$\|A^\gamma q\| \leq \lambda_{m+1}^\gamma \|q\|, \quad \gamma > 0,\ q \in Q_m D(A^\gamma), \tag{3.4.21}$$

$$\|A^{\frac{1}{2}} u\| = \left\|\frac{\partial u}{\partial x}\right\|, \quad u \in H^1(\Omega), \tag{3.4.22}$$

$$\|P_m u\| \leq \|u\|, \quad \|Q_m u\| \leq \|u\|, \quad \forall u \in H. \tag{3.4.23}$$

By equation (3.4.7) and Proposition 3.4.1, we obtain

$$\|Au(t)\| \leq C, \quad \left\|\frac{\mathrm{d}}{\mathrm{d}t}u(t)\right\| \leq C, \quad \forall t \geq t_*. \tag{3.4.24}$$

For this C and t_*, as in Proposition 3.4.1, equations (3.4.21), (3.4.23) and (3.4.24) imply

$$\|q(t)\| \leq C\lambda_{m+1}^{-1}, \quad \|A^{\frac{1}{2}} q(t)\| \leq C_{m+1}^{-\frac{1}{2}}$$
$$\left\|\frac{\mathrm{d}}{\mathrm{d}t}q(t)\right\| \leq C\lambda_{m+1}^{-1}, \quad \forall t > t_*. \tag{3.4.25}$$

Now we construct an approximate inertial manifold of problem (3.4.1)–(3.4.3). To achieve this, we define the mapping $\Phi : P_m H \to Q_m H$ such that $\forall p \in P_m H$, $\Phi(p) = \Psi$ is given by the following equation:

$$DA\Psi + Q_m R(p, p, p) = 0. \tag{3.4.26}$$

Letting $\Sigma = \operatorname{graph}(\Phi)$, we prove that Σ is an approximate inertial manifold. Then we get

Theorem 3.4.2. *Let equation* (3.4.4) *be valid, and* $u_0 \in H^2(\Omega)$. *Then there exists constant* K*, depending on the initial value, such that*

$$\operatorname{dist}_H(u(t), \Sigma) \le K\lambda_{m+1}^{-\frac{3}{2}}, \quad t \ge t_*, \tag{3.4.27}$$

where $u(t)$ *is the solution of problem* (3.4.1)–(3.4.3), t_* *depends on the initial value and* R_**, whenever* $\|u_0\|_{H^1} \le R$.

Proof. From equations (3.4.19) and (3.4.26), we get

$$D(A\Psi - Aq) = \frac{dq}{dt} + Q_m R(u) - Q_m R(p), \tag{3.4.28}$$

$$\begin{aligned} R(u) - R(p) = {} & -\rho q + D_1 |u|^{2\sigma} u - D_1 |p|^{2\sigma} p \\ & - \alpha\lambda_1 \cdot \nabla(|u|^2 u) + \alpha\lambda_1 \cdot \nabla(|p|^2 p) \\ & - \beta(\lambda_2 \cdot \nabla u)|u|^2 + \beta(\lambda_2 \cdot \nabla p)|p|^2, \end{aligned} \tag{3.4.29}$$

where in every term of estimate (3.4.29), $f(s) = s^{2\sigma}$, and ξ is between $|u|$ and $|p|$. Moreover,

$$\begin{aligned} \|D_1|u|^{2\sigma}u - D_1|p|^{2\sigma}p\| & \le \|D_1(u^{2\sigma} - |p|^{2\sigma})u\| + \|D_1|p|^{2\sigma}(u-p)\| \\ & \le \sqrt{1+\mu^2}\|u\|_\infty \||u|^{2\sigma} - |p|^{2\sigma}\| + \sqrt{1+\mu^2}\|p\|_\infty^{2\sigma}\|q\| \\ & \le \sqrt{1+\mu^2}\|u\|_\infty \|f'(\xi)(|u| - |p|)\| + \sqrt{1+\mu^2}\|p\|_\infty^{2\sigma}\|q\| \\ & \le \sqrt{1+\mu^2}\|p\|_\infty \|f'(\xi)\|_\infty \|q\| \\ & \quad + \sqrt{1+\mu^2}\|p\|_\infty^{2\sigma}\|q\|. \end{aligned} \tag{3.4.30}$$

Equation (3.4.17) and the following inequalities

$$C_1\|u\|_{H^2(\Omega)} \le \|u\| + \|\Delta u\| \le C_2\|u\|_{H^2(\Omega)}, \quad \forall u \in H^2(\Omega),$$

yield

$$\|u\|_{H^2(\Omega)} \le C, \quad \forall t \ge t_*, \tag{3.4.31}$$

where t_* depends the initial value and R, whenever $\|u_0\|_{H^1} \le R$. From this, we deduce

$$\|u\|_\infty \le C\|u\|^{\frac{1}{2}}\|u\|_{H^2(\Omega)}^{\frac{1}{2}} \le C, \quad \forall t \ge t_*. \tag{3.4.32}$$

Similarly, we have

$$\begin{aligned} \|p\|_\infty & \le C_2\|p\|^{\frac{1}{2}}\|p\|_{H^2(\Omega)}^{\frac{1}{2}} \\ & \le C_2\|p\|^{\frac{1}{2}}(\|p\| + \|\Delta p\|)^{\frac{1}{2}} \\ & \le C_2\|u\|^{\frac{1}{2}}(\|u\| + \|Au\|)^{\frac{1}{2}} \le C_3. \end{aligned} \tag{3.4.33}$$

From equations (3.4.32) and (3.4.33) we get

$$\|\xi\|_\infty \le \|p\|_\infty + \|u\|_\infty \le C, \quad \forall t \ge t_*. \tag{3.4.34}$$

By virtue of equations (3.4.40), (3.4.32)–(3.4.34), we get

$$\begin{aligned}
&\big\|D_1|u|^{2\sigma}u - D_2|p|^{2\sigma}p\big\| \le C_5\|q\|, \qquad (3.4.35)\\
&\big\|\beta(\lambda_2\cdot\nabla u)|u|^2 - \beta(\lambda_2\cdot\nabla p)|p|^2\big\|\\
&\quad \le \big\|\beta(\lambda_2\cdot(\nabla u-\nabla p))|u|^2\big\| + \big\|\beta(\lambda_2\cdot\nabla p)(|u|^2-|p|^2)\big\|\\
&\quad \le |\beta\lambda_2|\|u\|_\infty^2\|\nabla q\| + \big\|\beta(\lambda_2\cdot\nabla p)(u+p)(u-p)\big\|\\
&\quad \le |\beta\lambda_2|\|u\|_\infty^2\|\nabla q\| + |\beta\lambda_2|\|u+p\|_\infty\Big(\int_\Omega(\nabla p)^2|q|^2\Big)^{\frac12}\\
&\quad \le C_6\|\nabla q\| + C_7\|\nabla p\|_4\|q\|_4\\
&\quad \le C_6\|\nabla q\| + C_8\|\nabla p\|^{\frac12}\|\nabla p\|_{H^1}^{\frac12}\|q\|_{H^1}\\
&\quad \le C_6\big\|A^{\frac12}q\big\| + C_8\big\|A^{\frac12}u\big\|^{\frac12}\big(\big\|A^{\frac12}p\big\| + \|Ap\|\big)^{\frac12}\big(\|q\| + \big\|A^{\frac12}q\big\|\big)\\
&\quad \le C_6\big\|A^{\frac12}q\big\| + C_8\big\|A^{\frac12}u\big\|^{\frac12}\big(\big\|A^{\frac12}u\big\| + \|Au|\big)^{\frac12}\big(\|q\| + \big\|A^{\frac12}q\big\|\big)\\
&\quad \le C_9\|q\| + C_{10}\big\|A^{\frac12}q\big\|, \qquad (3.4.36)\\
&\alpha\lambda_1\cdot\nabla(|u|^2u) = \alpha(\lambda_1\cdot\nabla u)|u|^2 + 2\alpha(\lambda_1\cdot\nabla u)uu. \qquad (3.4.37)
\end{aligned}$$

By using equation (3.4.37), similar to equation (3.4.36), we get

$$\begin{aligned}
&\big\|\alpha\lambda_1\cdot\nabla(|u|^2u) - \alpha\lambda_1\cdot\nabla(|p|^2p)\big\|\\
&\quad \le C_{11}\|q\| + C_{12}\big\|A^{\frac12}q\big\|.
\end{aligned} \tag{3.4.38}$$

From equations (3.4.29), (3.4.35), (3.4.36) and (3.4.38), we know that there exists a constant C such that

$$\big\|R(u) - R(p)\big\| \le C\|q\| + C\big\|A^{\frac12}q\big\|. \tag{3.4.39}$$

By equations (3.4.28) and (3.4.39) we get

$$\begin{aligned}
\big\|D(A\Psi - Aq)\big\| &\le \left\|\frac{dq}{dt}\right\| + \big\|R(u) - R(p)\big\|\\
&\le C\lambda_{m+1}^{-1} + C\lambda_{m+1}^{-\frac12} \le C\lambda_2^{-\frac12}\lambda_{m+1}^{-\frac12} + C\lambda_{m+1}^{-\frac12}.
\end{aligned} \tag{3.4.40}$$

From the formula

$$\begin{aligned}
\big\|D(A\Psi - Aq)\big\| &= \sqrt{1+\nu^2}\|A\Psi - Aq\|\\
&\ge \sqrt{1+\nu^2}\lambda_{m+1}\|\Psi - q\|,
\end{aligned} \tag{3.4.41}$$

together with equations (3.4.40) and (3.4.41), we get

$$\|\Psi - q\| \le C\lambda_{m+1}^{-\frac{3}{2}}, \quad \forall t \ge t_*, \tag{3.4.42}$$

where t_* is taken as in equation (3.4.31). Then we know

$$\begin{aligned} d_H(u(t), \Sigma) &\le \|u(t) - (p(t) + \Phi(p(t)))\| \\ &= \|\Psi(t) - q(t)\| \le C\lambda_{m+1}^{-\frac{3}{2}}, \end{aligned}$$

and the theorem has been proved. □

Now we consider the Gevrey regularity of the solution for problem (3.4.1)–(3.4.3), which can be used to improve the convergence rate of the approximate inertial manifold. Set $\Omega = (0, 2\pi)^2$, and

$$\int_\Omega u(x,t)\mathrm{d}x = 0, \quad \forall t > 0. \tag{3.4.43}$$

Lemma 3.4.2. *Let $u, v, \omega \in D(e^{\tau A^{\frac{1}{2}}}A)$, $\tau > 0$. Then for $R(u, v, \omega)$ we have the following estimate:*

$$\begin{aligned} (e^{\tau A^{\frac{1}{2}}}R(u,v,\omega), e^{\tau A^{\frac{1}{2}}}Ay) \le{}& \rho\|e^{\tau A^{\frac{1}{2}}}\omega\|\|e^{\tau A^{\frac{1}{2}}}Ay\| \\ &+ C\|e^{\tau A^{\frac{1}{2}}}A^{\frac{1}{2}}u\|^\sigma\|e^{\tau A^{\frac{1}{2}}}A^{\frac{1}{2}}v\|^\sigma\|e^{\tau A^{\frac{1}{2}}}A^{\frac{1}{2}}\omega\|\|e^{\tau A^{\frac{1}{2}}}Ay\| \\ &+ C\|e^{\tau A^{\frac{1}{2}}}A^{\frac{1}{2}}u\|\|e^{\tau A^{\frac{1}{2}}}A^{\frac{1}{2}}v\|\|e^{\tau A^{\frac{1}{2}}}A^{\frac{1}{2}}\omega\|^{\frac{1}{2}} \\ &\times\|e^{\tau A^{\frac{1}{2}}}A^{\frac{1}{2}}\omega\|^{\frac{1}{2}}\|e^{\tau A^{\frac{1}{2}}}Ay\| \\ &+ C\|e^{\tau A^{\frac{1}{2}}}A^{\frac{1}{2}}u\|^{\frac{1}{2}}\|e^{\tau A^{\frac{1}{2}}}Au\|^{\frac{1}{2}}\|e^{\tau A^{\frac{1}{2}}}A^{\frac{1}{2}}v\| \\ &\times\|e^{\tau A^{\frac{1}{2}}}A^{\frac{1}{2}}\omega\|\|e^{\tau A^{\frac{1}{2}}}Ay\| \\ &+ C\|e^{\tau A^{\frac{1}{2}}}A^{\frac{1}{2}}u\|\|e^{\tau A^{\frac{1}{2}}}A^{\frac{1}{2}}v\|\|e^{\tau A^{\frac{1}{2}}}A^{\frac{1}{2}}v\|^{\frac{1}{2}} \\ &\times\|e^{\tau A^{\frac{1}{2}}}A^{\frac{1}{2}}\omega\|\|e^{\tau A^{\frac{1}{2}}}Ay\|. \end{aligned}$$

Proof. Letting

$$u = \sum_{j\in\mathbf{Z}^2} u_j e^{ijx}, \quad u^* = e^{\tau A^{\frac{1}{2}}}u = \sum_{j\in\mathbf{Z}^2} u_j^* e^{ijx}, \quad u^* = e^{\tau|j|}u_j, \tag{3.4.44}$$

$$v = \sum_{j\in\mathbf{Z}^2} v_j e^{ijx}, \quad v^* = e^{\tau A^{\frac{1}{2}}}v = \sum_{j\in\mathbf{Z}^2} v_j^* e^{ijx}, \quad v^* = e^{\tau|j|}v_j, \tag{3.4.45}$$

$$\omega = \sum_{j\in\mathbf{Z}^2} \omega_j e^{ijx}, \quad \omega^* = e^{\tau A^{\frac{1}{2}}}\omega = \sum_{j\in\mathbf{Z}^2} \omega_j^* e^{ijx}, \quad \omega^* = e^{\tau|j|}\omega_j, \tag{3.4.46}$$

$$y = \sum_{j\in\mathbf{Z}^2} y_j e^{ijx}, \quad y^* = e^{\tau A^{\frac{1}{2}}} y = \sum_{j\in\mathbf{Z}^2} y_j^* e^{ijx}, \quad y^* = e^{\tau|j|} y_j, \tag{3.4.47}$$

we have

$$\begin{aligned}(R(u,v,\omega),y) = &-\rho(\omega,y) + ((u,v)^\sigma D_1\omega, y)\\ &- (\alpha\lambda_1\cdot\nabla(u,v)\omega, y)\\ &- (\beta(\lambda_2\cdot\nabla\omega)(u,v),y).\end{aligned} \tag{3.4.48}$$

All the terms of estimate (3.4.48) are as follows:

$$-\rho(\omega,y) = -\rho\int_\Omega \sum_l \omega_l e^{ilx}\cdot\sum_s \bar{y}_s e^{-isx}\mathrm{d}x = 4\pi^2\rho\sum_{l=s}\omega_l\bar{y}_s, \tag{3.4.49}$$

$$\begin{aligned}((u,v)^\sigma D_1\omega,y) &= \int_\Omega\Big(\sum_{j_1}u_{j_1}e^{ij_1x}\cdot\sum_{k_1}v_{k_1}e^{ik_1x}\Big)\cdots\Big(\sum_{j_\sigma}u_{j_\sigma}e^{ij_\sigma x}\cdot\sum_{k_\sigma}v_{k_\sigma}e^{ik_\sigma x}\Big)\\ &\quad\times\Big(D_1\sum_l\omega_l e^{ilx}\cdot\sum_s\bar{y}_s e^{-isx}\Big)\mathrm{d}x\\ &= 4\pi^2\sum_{j_1+k_1+\cdots+j_\sigma+k_\sigma+l=s}(u_{j_1}\cdot v_{k_1})\cdots(u_{j_\sigma}\cdot v_{k_\sigma})(D_1\omega_l\cdot\bar{y}_s),\end{aligned} \tag{3.4.50}$$

$$\begin{aligned}-(\beta(\lambda_2\cdot\nabla\omega)(u,v),y) &= -\beta\int_\Omega(u\cdot v)(\lambda_2\cdot\nabla\omega)\bar{y}\mathrm{d}x\\ &= -\beta\int_\Omega\Big(\sum_j u_j e^{ijx}\cdot\sum_k v_k e^{ikx}\Big)\\ &\quad\times\Big(\sum_l(\lambda_2\cdot il)\omega_l e^{ilx}\cdot\sum_s\bar{y}_s e^{-isx}\Big)\mathrm{d}x\\ &= -4\pi^2\beta\sum_{j+k+l=s}(u_j\cdot v_k)(\lambda_2\cdot il)(\omega_l\cdot\bar{y}_s),\end{aligned} \tag{3.4.51}$$

$$\begin{aligned}-\alpha(\lambda_1\cdot\nabla((u,v)\omega),y) &= -\alpha((\lambda_1\cdot\nabla\omega)(u\cdot v),y)\\ &\quad-\alpha((\lambda_1\cdot\nabla u)\omega,y)-\alpha(((\lambda_1\cdot\nabla v)u)\omega,y).\end{aligned} \tag{3.4.52}$$

Similar to equation (3.4.51) we get

$$-\alpha((\lambda_1\cdot\nabla\omega)(u,v),y) = -4\pi^2\alpha\sum_{j+k+l=s}(u_j\cdot v_k)(\lambda_1\cdot il)(\omega_l\cdot\bar{y}_s), \tag{3.4.53}$$

$$\begin{aligned}-\alpha(((\lambda_1\cdot\nabla u)v)\omega,y) &= -4\pi^2\alpha\sum_{j+k+l=s}(\lambda_1\cdot ij)(u_j\cdot v_k)(u_l\cdot\bar{y}_s)\\ &= -\alpha\int_\Omega\Big(\sum_j(\lambda_1\cdot ij)u_j e^{ijx}\cdot\sum_k v_k e^{ikx}\Big)\\ &\quad\times\Big(\sum_l\omega_l e^{ilx}\sum_s\bar{y}_s e^{-isx}\Big)\mathrm{d}x.\end{aligned} \tag{3.4.54}$$

Similarly, we arrive at

$$
\begin{aligned}
&-\alpha(((\lambda_1\cdot\nabla v)u)\omega,y)\\
&=-4\pi^2\alpha\sum_{j+k+l=s}(\lambda_1\cdot ik)(u_j\cdot v_k)(\omega_l\cdot\overline{y}_s).
\end{aligned}\tag{3.4.55}
$$

By equations (3.4.48)–(3.4.55) we get

$$
\begin{aligned}
(R(u,v,\omega),y)=&-4\pi^2\rho\sum_{l=s}\omega_l\cdot\overline{y}_s\\
&+4\pi^2\sum_{j_1+k_1+\cdots+j_\sigma+k_\sigma+l=s}(u_{j_1}\cdot v_{k_1})\cdot(u_{j_\sigma}\cdots v_{k_\sigma})(D_1\omega_l\cdot\overline{y}_s)\\
&-4\pi^2\alpha\sum_{j+k+l=s}(u_j\cdot v_k)(\lambda_1\cdot il)(\omega_l\cdot\overline{y}_s)\\
&-4\pi^2\alpha\sum_{j+k+l=s}(\lambda_1\cdot ij)(u_j\cdot v_k)(\omega_l\cdot\overline{y}_s)\\
&-4\pi^2\alpha\sum_{j+k+l=s}(\lambda_1\cdot ik)(u_j\cdot v_k)(\omega_l\cdot\overline{y}_s)\\
&-4\pi^2\beta\sum_{j+k+l=s}(u_j\cdot v_k)(\lambda_2\cdot il)(\omega_l\cdot\overline{y}_s),
\end{aligned}\tag{3.4.56}
$$

$$
\begin{aligned}
(e^{\tau A^{\frac12}}R(u,v,\omega),e^{\tau A^{\frac12}}Ay)=&(R(u,v,\omega),e^{2\tau A^{\frac12}}Ay)\\
=&-4\pi^2\rho\sum_{l=s}(\omega_l^*,\overline{y}_s^*)|s|^2\\
&+4\pi^2\sum_{j_1+k_1+\cdots+j_\sigma+k_\sigma+l=s}(u_{j_1}^*\cdot v_{k_1}^*)\cdot(u_{j_\sigma}^*\cdots v_{k_\sigma}^*)\\
&\times(D_1\omega_l^*\cdot\overline{y}_s^*)|s|^2e^{\tau(|s|-|j_1|-|k_1|-\cdots-|j_\sigma|-|k_\sigma|-|l|)}\\
&-4\pi^2\alpha\sum_{j+k+l=s}(u_j^*\cdot v_k^*)(\lambda_1\cdot il)(\omega_l^*\cdot\overline{y}_s^*)|s|^2e^{\tau(|s|-|j|-|k|-|l|)}\\
&-4\pi^2\alpha\sum_{j+k+l=s}(\lambda_1\cdot ij)(u_j^*\cdot v_k^*)(\omega_l^*\cdot\overline{y}_s^*)|s|^2e^{\tau(|s|-|j|-|k|-|l|)}\\
&-4\pi^2\alpha\sum_{j+k+l=s}(\lambda_1\cdot ik)(u_j^*\cdot v_k^*)(\omega_l^*\cdot\overline{y}_s^*)|s|^2e^{\tau(|s|-|j|-|k|-|l|)}\\
&-4\pi^2\beta\sum_{j+k+l=s}(u_j^*\cdot v_k^*)(\lambda_2\cdot il)(\omega_l^*\cdot\overline{y}_s^*)|s|^2e^{\tau(|s|-|j|-|k|-|l|)}.
\end{aligned}\tag{3.4.57}
$$

Note that

$$
\begin{aligned}
|s|&=|j_1+k_1+\cdots+j_\sigma+k_\sigma+l|\\
&\le|j_1|+|k_1|+\cdots+|j_\sigma|+|k_\sigma|+|l|,
\end{aligned}
$$

hence

$$
e^{\tau(|s|-|j_1|-|k_1|-\cdots-|j_\sigma|-|k_\sigma|-|l|)}\le1.\tag{3.4.58}
$$

Similarly, for $s = j + k + l$, we have

$$e^{\tau(|s|-|j|-|k|-|l|)} \le 1. \tag{3.4.59}$$

Then from equations (3.4.57)–(3.4.59) we have

$$\begin{aligned}
&|(e^{\tau A^{\frac{1}{2}}} R(u,v,\omega), e^{\tau A^{\frac{1}{2}}} Ay)| \\
&\quad \le 4\pi^2 \rho \sum_{l=s} |\omega_l^*||\bar{y}_s^*||s|^2 \\
&\qquad + 4\pi^2 \sqrt{1+\mu^2} \sum_{j_1+k_1+\cdots+j_\sigma+k_\sigma+l=s} |u_{j_1}^*||v_{k_1}^*| \cdots |u_{j_\sigma}^*||v_{k_\sigma}^*||\omega_l^*||\bar{y}_s^*||s|^2 \\
&\qquad + 4\pi^2 \alpha|\lambda_1| \sum_{j+k+l=s} |u_j^*||v_k^*||l||\omega_l^*||\bar{y}_s^*||s|^2 \\
&\qquad + 4\pi^2 \alpha|\lambda_1| \sum_{j+k+l=s} |j||u_j^*||v_k^*||\omega_l^*||\bar{y}_s^*||s|^2 \\
&\qquad + 4\pi^2 \alpha|\lambda_1| \sum_{j+k+l=s} |u_j^*||k||v_k^*||\omega_l^*||\bar{y}_s^*||s|^2 \\
&\qquad + 4\pi^2 \beta|\lambda_2| \sum_{j+k+l=s} |u_j^*||v_k^*||l||\omega_l^*||\bar{y}_s^*||s|^2.
\end{aligned} \tag{3.4.60}$$

Obviously,

$$4\pi^2 \rho \sum_{l=s} |\omega_l^*||\bar{y}_s^*||s|^2 = \rho \int_\Omega \xi(x)\theta(x)\mathrm{d}x, \tag{3.4.61}$$

where

$$\xi(x) = \sum_l |\omega_l^*| e^{ilx}, \quad \theta(x) = \sum |s|^2 |\bar{y}_s^*| e^{isx}. \tag{3.4.62}$$

Hence

$$\begin{aligned}
4\pi^2 \rho \sum_{l=s} |\omega_l^*||\bar{y}_s^*||s|^2 &\le \rho \left| \int_\Omega \xi(x)\theta(x)\mathrm{d}x \right| \\
&\le \rho \|\xi\|\|\theta\| = \rho \|e^{\tau A^{\frac{1}{2}}} \omega\| \|e^{\tau A^{\frac{1}{2}}} Ay\|,
\end{aligned} \tag{3.4.63}$$

$$\begin{aligned}
&4\pi^2 \sqrt{1+\mu^2} \sum_{j_1+k_1+\cdots+j_\sigma+k_\sigma+l=s} |u_{j_1}^*||v_{j_1}^*| \cdots |u_{j_\sigma}^*||v_{k_\sigma}^*||\omega_l^*||\bar{y}_s^*||s|^2 \\
&\quad = \sqrt{1+\mu^2} \int_\Omega \varphi_{j_1}(x)\Psi_{k_1}(x) \cdot \varphi_{j_\sigma}(x)\Psi_{k_\sigma}(x)\xi(x)\theta(x)\mathrm{d}x,
\end{aligned} \tag{3.4.64}$$

where $\xi(x)$, $\theta(x)$ are shown in equation (3.4.62) and

$$\begin{aligned}
\varphi_{j_1}(x) &= |u_{j_1}^*| e^{ij_1 x}, \quad \Psi_{k_1}(x) = |v_{k_1}^*| e^{ik_1 x}, \\
&\vdots \\
\varphi_{j_\sigma}(x) &= |u_{j_\sigma}^*| e^{ij_\sigma x}, \quad \Psi_{k_\sigma}(x) = |v_{k_\sigma}^*| e^{ik_\sigma x}.
\end{aligned} \tag{3.4.65}$$

From equation (3.4.64) we get

$$
\begin{aligned}
&4\pi^2\sqrt{1+\mu^2}\sum_{j_1+k_1+\cdots+j_\sigma+k_\sigma+l=s}|u_{j_1}^*||v_{k_1}^*|\cdots|u_{j_\sigma}^*||v_{k_\sigma}^*||\omega_l^*||\overline{y}_s^*||s|^2\\
&\le\sqrt{1+\mu^2}\left|\int_\Omega\varphi_{j_1}(x)\Psi_{k_1}(x)\cdot\varphi_{j_\sigma}(x)\Psi_{k_\sigma}(x)\xi(x)\theta(x)dx\right|\\
&\le\sqrt{1+\mu^2}\|\varphi_{j_1}\|_{4\sigma+2}\|\Psi_{k_1}\|_{4\sigma+2}\cdots\|\varphi_{j_\sigma}\|_{4\sigma+2}\|\Psi_{k_\sigma}\|_{4\sigma+2}\|\xi\|_{4\sigma+2}\|\theta\|\\
&\le\sqrt{1+\mu^2}\|\varphi_{j_1}\|_{H^1}\|\Psi_{k_1}\|_{H^1}\cdots\|\varphi_{j_\sigma}\|_{H^1}\|\Psi_{k_\sigma}\|_{H^1}\|\xi\|_{H^1}\|\theta\|\\
&\le C_1\|A^{\frac12}\varphi_{j_1}\|\|A^{\frac12}\Psi_{k_1}\|\cdots\|A^{\frac12}\varphi_{j_\sigma}\|\|A^{\frac12}\Psi_{k_\sigma}\|\|A^{\frac12}\xi\|\|\theta\|\\
&\le C_2\|e^{\tau A^{\frac12}}A^{\frac12}y\|^\sigma\|e^{\tau A^{\frac12}}A^{\frac12}v\|^\sigma\|e^{\tau A^{\frac12}}A^{\frac12}\omega\|\|e^{\tau A^{\frac12}}Ay\|.
\end{aligned}
\tag{3.4.66}
$$

By virtue of

$$
\begin{aligned}
&4\pi^2\alpha|\lambda_1|\sum_{j+k+l=s}|u_j^*||v_k^*||l||\omega_k^*||\overline{y}_s^*||s|^2\\
&=\alpha|\lambda_1|\int_\Omega\varphi(x)\Psi(x)\eta(x)\theta(x)dx,
\end{aligned}
\tag{3.4.67}
$$

where $\theta(x)$ is defined by equation (3.4.62), and since

$$
\varphi(x)=|u_j^*|e^{ijx},\quad\Psi(x)=|v_k^*|e^{ikx},\quad\eta(x)=|l||\omega_l^*|e^{ilx},\tag{3.4.68}
$$

by equation (3.4.67) we obtain

$$
\begin{aligned}
&4\pi^2\alpha|\lambda_1|\sum_{j+k+l=s}|u_j^*||v_k^*||l||\omega_k^*||\overline{y}_s^*||s|^2\\
&\le|\alpha||\lambda_1|\left|\int_\Omega\varphi(x)\Psi(x)\eta(x)\theta(x)dx\right|\\
&\le|\alpha||\lambda_1|\|\varphi\|_8\|\Psi\|_8\|\eta\|_4\|\theta\|\\
&\le C_3\|\varphi\|_{H^1}\|\Psi\|_{H^1}\|\eta\|^{\frac12}\|\eta\|_{H^1}^{\frac12}\|\theta\|\\
&\le C_4\|A^{\frac12}\varphi\|\|A^{\frac12}\Psi\|\|\eta\|\|A^{\frac12}\eta\|^{\frac12}\|\theta\|\\
&\le C_5\|e^{\tau A^{\frac12}}A^{\frac12}u\|\|e^{\tau A^{\frac12}}A^{\frac12}v\|\|e^{\tau A^{\frac12}}A^{\frac12}\omega\|^{\frac12}\\
&\quad\times\|e^{\tau A^{\frac12}}A\omega\|^{\frac12}\|e^{\tau A^{\frac12}}Ay\|.
\end{aligned}
\tag{3.4.69}
$$

Similar to equation (3.4.69), we have

$$
\begin{aligned}
&4\pi^2\alpha|\lambda_1|\sum_{j+k+l=s}|j||u_j^*||v_k^*||\omega_l^*||\overline{y}_s^*||s|^2\\
&\le C_6\|e^{\tau A^{\frac12}}A^{\frac12}u\|^{\frac12}\|e^{\tau A^{\frac12}}Au\|^{\frac12}\|e^{\tau A^{\frac12}}A^{\frac12}v\|\|e^{\tau A^{\frac12}}A^{\frac12}\omega\|\|e^{\tau A^{\frac12}}A^{\frac12}y\|,
\end{aligned}
\tag{3.4.70}
$$

$$\begin{aligned}
&4\pi^2\alpha|\lambda_1| \sum_{j+k+l=s} |u_j^*||k||v_k^*||\omega_l^*||\overline{y}_s^*||s|^2 \\
&\quad \le C_7\|e^{\tau A^{\frac{1}{2}}}A^{\frac{1}{2}}u\|\|e^{\tau A^{\frac{1}{2}}}A^{\frac{1}{2}}v\|^{\frac{1}{2}}\|e^{\tau A^{\frac{1}{2}}}A^{\frac{1}{2}}v\|^{\frac{1}{2}}\|e^{\tau A^{\frac{1}{2}}}A^{\frac{1}{2}}\omega\|\|e^{\tau A^{\frac{1}{2}}}Ay\|, \qquad (3.4.71)\\
&4\pi^2\beta \sum_{j+k+l=s} |u_j^*||v_k^*||l||\omega_l^*||\overline{y}_s^*||s|^2 \\
&\quad \le C_8\|e^{\tau A^{\frac{1}{2}}}A^{\frac{1}{2}}u\|\|e^{\tau A^{\frac{1}{2}}}A^{\frac{1}{2}}v\|\|e^{\tau A^{\frac{1}{2}}}A^{\frac{1}{2}}\omega\|^{\frac{1}{2}}\|e^{\tau A^{\frac{1}{2}}}A^{\frac{1}{2}}\omega\|^{\frac{1}{2}}\|e^{\tau A^{\frac{1}{2}}}Ay\|. \qquad (3.4.72)
\end{aligned}$$

From equations (3.4.60), (3.4.63), (3.4.66)–(3.4.69), we deduce the conclusion of the lemma. □

From Lemma 3.4.2, we get

$$\begin{aligned}
&|(e^{\tau A^{\frac{1}{2}}}R(u,u,u), e^{\tau A^{\frac{1}{2}}}Au)| \\
&\quad \le \rho\|e^{\tau A^{\frac{1}{2}}}u\|\|e^{\tau A^{\frac{1}{2}}}Au\| + C\|e^{\tau A^{\frac{1}{2}}}A^{\frac{1}{2}}u\|^{2\sigma+1}\|e^{\tau A^{\frac{1}{2}}}Au\| + C\|e^{\tau A^{\frac{1}{2}}}A^{\frac{1}{2}}u\|^{\frac{5}{2}}\|e^{\tau A^{\frac{1}{2}}}Au\|^{\frac{3}{2}} \\
&\quad \le \varepsilon\|e^{\tau A^{\frac{1}{2}}}Au\|^2 + C_1(\varepsilon)\|e^{\tau A^{\frac{1}{2}}}u\|^2 + C_2(\varepsilon)\|e^{\tau A^{\frac{1}{2}}}A^{\frac{1}{2}}u\|^{4\sigma+2} + C_3(\varepsilon)\|e^{\tau A^{\frac{1}{2}}}A^{\frac{1}{2}}u\|^{10} \\
&\quad \le \varepsilon\|e^{\tau A^{\frac{1}{2}}}Au\|^2 + C_4(\varepsilon)\|e^{\tau A^{\frac{1}{2}}}u\|^{4\sigma+2} + C_5(\varepsilon)\|e^{\tau A^{\frac{1}{2}}}u\|^{4\sigma+2} + C_6 \\
&\quad \le \varepsilon\|e^{\tau A^{\frac{1}{2}}}Au\|^2 + C_7(\varepsilon)\|e^{\tau A^{\frac{1}{2}}}A^{\frac{1}{2}}u\|^{4\sigma+2} + C_6. \qquad (3.4.73)
\end{aligned}$$

Theorem 3.4.3. *Suppose that conditions (3.4.4) is satisfied, $u_0 \in H^2(\Omega)$. Then there exists a constant k that depends on the initial value, such that each component of the solution for the problem (3.4.1)–(3.4.3) belongs to $D(A^{\frac{1}{2}}\exp(kA^{\frac{1}{2}}))$, the analytic continuation in the complex domain is given as follows:*

$$\Delta = \{t + se^{i\theta} : t \ge t_*,\ |\theta| \le \theta_0,\ 0 \le s \le T_0\} \qquad (3.4.74)$$

and

$$\|e^{kA^{\frac{1}{2}}}A^{\frac{1}{2}}u(Z)\| \le K, \quad Z \in \Delta, \qquad (3.4.75)$$

where θ_0, T_0 and K depend on the initial value, $|\theta_0| \le \pi/4$, t_* depends on the initial value and R, whenever $\|u_0\|_{H^1} \le R$.

Proof. From equations (3.4.6) and (3.4.73), we know Theorem 3.1 of [186] is valid. Hence the theorem has been proved. □

From Theorem 3.4.3 and Cauchy formula, we get

$$\left\|e^{kA^{\frac{1}{2}}}A^{\frac{1}{2}}\frac{\mathrm{d}}{\mathrm{d}t}u(t)\right\| \le K_1, \quad \forall t \ge t_1. \qquad (3.4.76)$$

Through equations (3.4.75) and (3.4.76) we get, when t is large enough, that

$$\|e^{kA^{\frac{1}{2}}}A^{\frac{1}{2}}q(t)\| \le K, \quad \left\|e^{kA^{\frac{1}{2}}}A^{\frac{1}{2}}\frac{\mathrm{d}}{\mathrm{d}t}q(t)\right\| \le K_1, \qquad (3.4.77)$$

where $q(t) = Q_m u(t)$. Hence when $t \geq t_*$,

$$\|A^{\frac{1}{2}} q(t)\| \leq K e^{-k\lambda_{m+1}^{\frac{1}{2}}}, \quad \|q(t)\| \leq K \lambda_{m+1}^{-\frac{1}{2}} e^{-k\lambda_{m+1}^{\frac{1}{2}}},$$
$$\left\|\frac{dq(t)}{dt}\right\| \leq K_1 \lambda_{m+1}^{-\frac{1}{2}} e^{-k\lambda_{m+1}^{\frac{1}{2}}}. \tag{3.4.78}$$

Using equation (3.4.78) instead of equation (3.4.25), similar to Theorem 3.4.2, we have

Theorem 3.4.4. *Suppose that condition (3.4.4) holds, $u_0 \in H^2(\Omega)$. Then there exists a constant E, depending on the initial value, such that*

$$\mathrm{dist}_H(\Sigma, u(t)) \leq E \lambda_{m+1}^{-1} e^{-k\lambda_{m+1}^{\frac{1}{2}}}, \quad t \geq t_*, \tag{3.4.79}$$

where $u(t)$ is the solution of problem (3.4.1)–(3.4.3), Σ is its approximate inertial manifold, t_* depends on the initial value and R, whenever $\|u_0\|_{H^1} \leq R$.

3.5 Bernard convection equation

Two-dimensional Newton–Boussinesq equation can be used to describe the famous Bernard convection:

$$\begin{cases} \partial_t \xi + u\partial_x \xi + v\partial_y \xi = \Delta\xi - \dfrac{R_\alpha}{P_y}\partial_x \theta, \\ \Delta\Psi = \xi, \quad u = \partial_y \Psi, \quad v = -\partial_x \Psi, \\ \partial_t \theta + u\partial_x \theta + v\partial_y \theta = \dfrac{1}{P_y}\Delta\theta, \end{cases}$$

where $(u, v) \triangleq u$ is the velocity vector; θ is the temperature; Ψ is the stream function; ξ is vorticity; $P_y > 0$ is the Prandtl constant; and $R_\alpha > 0$ is Rayleigh number. The above equation can be rewritten as

$$\frac{\partial}{\partial t}\Delta\Psi + J(\Psi, \Delta\Psi) = \Delta^2\Psi - \frac{R_\alpha}{P_y}\frac{\partial_\theta}{\partial_x}, \tag{3.5.1}$$

$$\frac{\partial_\theta}{\partial t} + J(\Psi, \theta) = \frac{1}{P_y}\Delta\theta, \tag{3.5.2}$$

where

$$J(u, v) = u_y v_x - u_x v_y.$$

The above equation can be endowed with the initial value

$$\Psi(x, y, 0) = \Psi_0(x, y), \quad \theta(x, y, 0) = \theta_0(x, y) \tag{3.5.3}$$

and periodic boundary conditions

$$\begin{aligned}\Psi(x+2D,y,t)&=\Psi(x,y,t), \quad \Psi(x,y+2D,t)=\Psi(x,y,t),\\ \theta(x+2D,y,t)&=\theta(x,y,t), \quad \theta(x,y+2D,t)=\theta(x,y,t).\end{aligned} \tag{3.5.4}$$

In 1987, Foias et al. [49] proved the existence and the finiteness of dimension of the global attractor. In [86] Guo proved the convergence of the spectral method and the existence and uniqueness of a global smooth solution. The nonlinear Galerkin method was proposed and its convergence was proved by Guo [90] in 1995. Guo and Wang [120] in 1996 proved the existence of approximate inertial manifolds.

In order to simplify the problem using functional formulation, we let $Au = -\Delta u$. The inner product on $H = L^2(\Omega)$ is $(\cdot)$, its norm is $\|\cdot\|$, $D(A) = \{u \in H^2(\Omega);\ u$ satisfies the equation (3.5.4)$\}$. Let $\Omega = (0, 2D)\times(0, 2D)$. Then equations (3.5.1)–(3.5.2) can be written as

$$\frac{\mathrm{d}}{\mathrm{d}t}A\Psi + J(\Psi, A\Psi) + A^2\Psi - \frac{R_\alpha}{P_y}B(\theta) = 0, \tag{3.5.5}$$

$$\frac{\mathrm{d}}{\mathrm{d}t}\theta + J(\Psi,\theta) + \frac{1}{P_y}A(\theta) = 0, \tag{3.5.6}$$

where $B(\theta) = \frac{\partial\theta}{\partial x}$ is a linear operator. From [90] we know that when $(\Psi_0, \theta_0) \in H^2 \times H^1$, the problem (3.5.1)–(3.5.4) has a unique solution (Ψ, θ),

$$\begin{aligned}&\Psi \in L^\infty(R^+; H^2(\Omega)), \quad \Delta\Psi \in L^2(0,T;H^1(\Omega)), \quad \forall T > 0,\\ &\theta \in L^\infty(R^+; H^1(\Omega)), \quad \Delta\theta \in L^2(0,T;H).\end{aligned}$$

One of the properties of equation (3.5.2) is that the average of the solution of the equations is conserved ($t > 0$):

$$\begin{aligned}m(\theta(t)) &= \frac{1}{|\Omega|}\iint_\Omega \theta(x,y,t)\mathrm{d}x\mathrm{d}y\\ &= \frac{1}{|\Omega|}\iint_\Omega \theta_0(x,y,t)\mathrm{d}x\mathrm{d}y = m(\theta_0).\end{aligned}$$

Hence, there exists an unbounded absorbing set in the whole space H. We introduce the subset of H:

$$H_\alpha = \{\theta \in H : |m(\theta)| \le \alpha\}, \quad \alpha \text{ is a fixed number},$$

Lemma 3.5.1 (Uniform Gronwall lemma, [197]). *Suppose g, h, y are three positive locally integrable functions on $[t_0, \infty)$, and y' is also locally integrable on the interval $[t_0, \infty)$. Let g, h, y satisfy the inequalities*

$$\frac{\mathrm{d}y}{\mathrm{d}t} \le gy + h, \quad t \ge t_0,$$

$$\int_t^{t+r} g(s)\mathrm{d}s \le a_1, \quad \int_t^{t+r} h(s)\mathrm{d}s \le a_2,$$
$$\int_t^{t+r} y(s)\mathrm{d}s \le a_3, \quad t \ge t_0.$$

where r, a_1, a_2, a_3 are positive constants. Then we have

$$y(t+r) \le \left(\frac{a_3}{r} + a_2\right)\exp(a_1), \quad t \ge t_0.$$

Lemma 3.5.2. *Let $\Psi_0 \in H^4$, $\theta_0 \in H_\alpha \cap H^2$. Then for the solution (Ψ, θ) of problem (3.5.2)–(3.5.4), we have the following estimates:*

$$\|A\Psi\|, \|A^{\frac{3}{2}}\Psi\|, \|\theta\|, \|A^{\frac{1}{2}}\theta\|, \left\|\frac{\mathrm{d}}{\mathrm{d}t}\theta\right\| \le M_0, \quad \forall t \ge t_*,$$

where M_0 depends on the parameters $(\alpha, \Omega, P_y, R_\alpha)$; t_* depends on $(\alpha, \Omega, P_y, R_\alpha)$ and R, whenever $\|\Psi_0\|_{H^4} \le R$ and $\|\theta_0\|_{H^2} \le R$.

Proof. Similarly as in Lemma 2.3 in [90], it is easy to see that there exists a constant C, which only depends on the initial value, such that

$$\|A\Psi\|, \|\theta\|, \|A^{\frac{1}{2}}\theta\|, \int_t^{t+1} \|A^{\frac{3}{2}}\Psi\|^2 \mathrm{d}t \le C, \quad \forall t \ge t_0, \tag{3.5.7}$$

where t_0 depends on the initial value and R, whenever $\|\Psi_0\|_{H^1} \le R$, $\|\theta_0\| \le R$; C here means the parameter depends only on the constants.

Taking the inner product of equation (3.5.5) with $A^2\Psi$, we obtain

$$\frac{1}{2}\frac{\mathrm{d}}{\mathrm{d}t}\|A^{\frac{3}{2}}\Psi\|^2 + (J(\Psi, A\Psi), A^2\Psi) + \|A^2\Psi\|^2 - \frac{R_\alpha}{P_y}(B(\theta), A^2\Psi) = 0, \tag{3.5.8}$$

$$\begin{aligned}|(J(\Psi, A\Psi), A^2\Psi)| &\le |J(\Psi, A\Psi)\|A^2\Psi\|| \\ &\le C\|A^{\frac{3}{2}}\Psi\|^2\|A^2\Psi\| \le \frac{1}{4}\|A^2\Psi\|^2 + C\|A^{\frac{3}{2}}\Psi\|^4,\end{aligned} \tag{3.5.9}$$

$$\begin{aligned}|(B(\theta), A^2\Psi)| &\le \|B(\theta)\|\|A^2\Psi\| \le \|A^{\frac{1}{2}}\theta\|\|A^2\Psi\| \\ &\le C\|A^2\Psi\| \le \frac{1}{4}\|A^2\Psi\|^2 + C.\end{aligned} \tag{3.5.10}$$

From equations (3.5.8)–(3.5.10) we get

$$\frac{\mathrm{d}}{\mathrm{d}t}\|A^{\frac{3}{2}}\Psi\|^2 + \|A^2\Psi\|^2 \le C\|A^{\frac{3}{2}}\Psi\|^4 + C,$$
$$\frac{\mathrm{d}}{\mathrm{d}t}\|A^{\frac{3}{2}}\Psi\|^2 \le C\|A^{\frac{3}{2}}\Psi\|^4 + C.$$

In order to use the claim of the uniform Gronwall lemma, we set

$$y = \|A^{\frac{3}{2}}\Psi\|^2, \quad g = C\|A^{\frac{3}{2}}\Psi\|^2, \quad h = C.$$

Then from equation (3.5.7), we know that conditions of Lemma 3.5.1 are satisfied, hence we have

$$\|A^{\frac{3}{2}}\Psi\|^2 \le C, \quad \forall t \ge t_0 + 1, \tag{3.5.11}$$

where t_0 is like in equation (3.5.7).

Differentiating equations (3.5.5)–(3.5.6) with t, we obtain

$$\frac{\mathrm{d}}{\mathrm{d}t}A\Psi_t + J(\Psi_t, A\Psi) + J(\Psi, A\Psi_t) + A^2\Psi_t - \frac{R_\alpha}{P_\gamma}B(\theta_t) = 0,$$

$$\frac{\mathrm{d}}{\mathrm{d}t}\theta_t + J(\Psi_t, \theta) + J(\Psi, \theta_t) + \frac{1}{P_\gamma}A(\theta_t) = 0.$$

Since

$$\Psi_0 \in H^4, \quad \theta_0 \in H^2,$$

using equations (3.5.5)–(3.5.6), we get

$$A\Psi_t(0) \in H, \quad \theta_t(0) \in H.$$

Repeating the proof in [90], we get

$$\left\|A\frac{\mathrm{d}}{\mathrm{d}t}\Psi\right\| \le C, \quad \left\|\frac{\mathrm{d}}{\mathrm{d}t}\theta\right\| \le C, \quad \forall t \ge t_0',$$

where t_0' depends on the initial value and R, whenever $\|\Psi_0\|_{H^4} \le R$, $\|\theta_0\|_{H^2} \le R$. Let $t_* = \max\{t_0', t_0 + 1\}$. When $t \ge t_*$, from the above inequalities (3.5.7) and (3.5.11), we get the claim of the lemma. □

Let $\{\omega_j(x,y)\}$, $j = 1, 2, \ldots$, be the periodic eigenvectors of A, which satisfy

$$A\omega_j = \lambda_j\omega_j, \quad j = 1, 2, \ldots,$$
$$\lambda_1 \le \lambda_2 \le \cdots, \quad \lambda_j \to +\infty, \quad j \to \infty.$$

Fixing a positive integer m, set $P = P_m$ to be the projection from H to the subspace $\operatorname{span}\{\omega_1, \omega_2, \ldots, \omega_m\}$. Let $Q = Q_m = I - P_m$. Then acting with P, Q on the formulas (3.5.5)–(3.5.6), we get the coupling equations for $\Psi_1 = P\Psi$, $\Psi_2 = Q\Psi$, $\theta_1 = P\theta$ and $\theta_2 = Q\theta$ as follows:

$$\begin{cases} \frac{\mathrm{d}}{\mathrm{d}t}A\Psi_1 + P_mJ(\Psi, A\Psi) + A^2\Psi_1 - \frac{R_\alpha}{P_\gamma}P_mB(\theta_t) = 0, \\ \frac{\mathrm{d}}{\mathrm{d}t}A\Psi_2 + Q_mJ(\Psi, A\Psi) + A^2\Psi_2 - \frac{R_\alpha}{P_\gamma}Q_mB(\theta_t) = 0, \end{cases} \tag{3.5.12}$$

$$\begin{cases} \dfrac{\mathrm{d}}{\mathrm{d}t}\theta_1 + P_m J(\Psi,\theta) + \dfrac{1}{P_y}A\theta_1 = 0, \\ \dfrac{\mathrm{d}}{\mathrm{d}t}\theta_2 + Q_m J(\Psi,\theta) + \dfrac{1}{P_y}A\theta_2 = 0. \end{cases} \tag{3.5.13}$$

Hereafter, we usually use the following inequality:

$$|J(u,v)| \le C\|A^{\frac{3}{2}}u\|\|A^{\frac{1}{2}}v\|, \quad u \in H^3,\ v \in H^1. \tag{3.5.14}$$

The proof of equation (3.5.14) is obvious, since

$$\begin{aligned} |J(u,v)| &= \Big(\iint_\Omega (u_y v_x - u_x v_y)^2 \mathrm{d}x\mathrm{d}y\Big)^{\frac{1}{2}} \\ &\le (\|u_y\|_\infty + \|u_x\|_\infty)\|\nabla v\| \\ &\le C(\|u_y\|^{\frac{1}{2}}\|\Delta u_y\|^{\frac{1}{2}} + \|u_x\|^{\frac{1}{2}}\|\Delta u_x\|^{\frac{1}{2}})\|\nabla v\| \\ &\le C\|\nabla u\|^{\frac{1}{2}}\|\nabla\Delta u\|^{\frac{1}{2}}\|\nabla v\| \\ &\le C\|A^{\frac{1}{2}}u\|^{\frac{1}{2}}\|A^{\frac{3}{2}}u\|^{\frac{1}{2}}\|A^{\frac{1}{2}}v\| \le C\|A^{\frac{3}{2}}u\|\|A^{\frac{1}{2}}v\|, \end{aligned}$$

thus we get equation (3.5.14).

Now we give the long-time estimate for Ψ_2, θ_2 and their derivatives.

Lemma 3.5.3. *Let $\Psi_0 \in H^4$, $\theta_0 \in H_0 \cap H^2$. Then there exists a constant M, which depends on parameter t_* and R, such that whenever $\|\Psi_0\|_{H^4} \le R$, $\|\theta_0\|_{H^2} \le R$,*

$$\|A^{\frac{3}{2}}\Psi_2(t)\| \le M\lambda_{m+1}^{-\frac{1}{2}}, \quad \Big\|\frac{\mathrm{d}}{\mathrm{d}t}A\Psi_2\Big\| \le M\lambda_{m+1}^{-1},$$
$$\|A^{\frac{3}{2}}\theta_2(t)\| \le M\lambda_{m+1}^{-\frac{1}{2}}, \quad \Big\|\frac{\mathrm{d}}{\mathrm{d}t}\theta_2\Big\| \le M\lambda_{m+1}^{-1}.$$

Proof. Taking the inner product of equation (3.5.12) with $A^2\Psi_2$ over H, we obtain

$$\frac{1}{2}\frac{\mathrm{d}}{\mathrm{d}t}|A^{\frac{3}{2}}\Psi_2|^2 + (J(\Psi,A\Psi),A^2\Psi_2) + |A^2\Psi_2|^2 - \frac{R_\alpha}{P_y}(B(\theta) - A^2\Psi_2) = 0. \tag{3.5.15}$$

From equation (3.5.14), we have

$$\begin{aligned} |(J(\Psi,A\Psi),A^2\Psi_2)| &\le \|J(\Psi,A\Psi)\|\|A^2\Psi_2\| \\ &\le C\|A^{\frac{3}{2}}\Psi\|^2\|A^2\Psi_2\| \le C\|A^2\Psi_2\|, \\ |(B(\theta),A^2\Psi_2)| &\le \|A^{\frac{1}{2}}\theta\|\|A^2\Psi_2\| \le C\|A^2\Psi_2\|. \end{aligned}$$

Hence by equation (3.5.15) we have

$$\frac{1}{2}\frac{\mathrm{d}}{\mathrm{d}t}|A^{\frac{3}{2}}\Psi_2|^2 + \|A^2\Psi_2\|^2 \le C\|A^2\Psi_2\| \le \frac{1}{2}\|A^2\Psi_2\|^2 + C.$$

Since $\|A^2\Psi_2\|^2 \geq \lambda_{m+1}\|A^{\frac{3}{2}}\Psi_2\|^2$, we get

$$\frac{1}{2}\frac{\mathrm{d}}{\mathrm{d}t}\left\|A^{\frac{3}{2}}\Psi_2\right\|^2 + \lambda_{m+1}\left\|A^{\frac{3}{2}}\Psi_2\right\|^2 \leq C.$$

Through Gronwall Lemma we know that

$$\left\|A^{\frac{3}{2}}\Psi_2(t)\right\|^2 \leq \left\|A^{\frac{3}{2}}\Psi_2(t_*)\right\|^2 e^{-\lambda_{m+1}(t-t_*)} + C\lambda_{m+1}^{-1}, \quad t \geq t_*,$$

where t_* is similar to that in Lemma 3.5.2.

Since

$$\left\|A^{\frac{3}{2}}\Psi_2(t_*)\right\| \leq \left\|A^{\frac{3}{2}}\Psi(t_*)\right\| \leq M,$$

we have

$$\left\|A^{\frac{3}{2}}\Psi_2(t)\right\|^2 \leq M^2 e^{-\lambda_{m+1}(t-t_*)} + C\lambda_{m+1}^{-1} \leq C\lambda_{m+1}^{-1}, \quad \forall t \geq t_*, \tag{3.5.16}$$

where $t_*' = \sup_m \max\{t_*, t_* + \frac{1}{\lambda_{m+1}} \lg \frac{M^2\lambda_{m+1}}{C}\}$.

Taking the inner product of equation (3.5.13) with $A\theta_2$ over H, we get

$$\frac{1}{2}\frac{\mathrm{d}}{\mathrm{d}t}\left\|A^{\frac{1}{2}}\theta_2\right\|^2 + (Q_m J(\Psi,\theta), A\theta_2) + \frac{1}{P_\gamma}\|A\theta_2\|^2 = 0.$$

Thus we have

$$\begin{aligned}\frac{1}{2}\frac{\mathrm{d}}{\mathrm{d}t}\left\|A^{\frac{1}{2}}\theta_2\right\|^2 + \frac{1}{P_\gamma}\|A\theta_2\|^2 &\leq |J(\Psi,\theta)||A\theta_2| \\ &\leq C\left\|A^{\frac{3}{2}}\Psi\right\|\left\|A^{\frac{1}{2}}\theta\right\|\|A\theta_2\| \leq C\|A\theta_2\| \leq \frac{1}{2P_\gamma}\|A\theta_2\|^2 + C,\end{aligned}$$

which yields

$$\frac{\mathrm{d}}{\mathrm{d}t}\left\|A^{\frac{1}{2}}\theta_2\right\|^2 + \frac{1}{P_\gamma}\|A\theta_2\|^2 \leq C,$$

$$\frac{\mathrm{d}}{\mathrm{d}t}\left\|A^{\frac{1}{2}}\theta_2\right\|^2 + \frac{\lambda_{m+1}}{P_\gamma}\left\|A^{\frac{1}{2}}\theta_2\right\|^2 \leq C.$$

By Gronwall lemma, we get

$$\begin{aligned}\left\|A^{\frac{1}{2}}\theta_2(t)\right\|^2 &\leq \left\|A^{\frac{1}{2}}\theta_2(t_*)\right\|^2 e^{-\frac{1}{P_\gamma}\lambda_{m+1}(t-t_*)} + C\lambda_{m+1}^{-1} \\ &\leq M^2 e^{-\frac{1}{P_\gamma}\lambda_{m+1}(t-t_*)} + C\lambda_{m+1}^{-1}, \quad \forall t \geq t_*,\end{aligned}$$

where t_* is similar to that in Lemma 3.5.2. This proves

$$\left\|A^{\frac{1}{2}}\theta_2(t)\right\|^2 \leq C\lambda_{m+1}^{-1}, \quad \forall t \geq t_*', \tag{3.5.17}$$

where $t_*' = \sup_m \max\{t_*, t_* + \frac{P_\gamma}{\lambda_{m+1}} \lg \frac{M^2\lambda_{m+1}}{C}\}$.

Differentiating equations (3.5.12) and (3.5.13) with respect to t and applying the above method, when t is large enough, we have

$$\left\|\frac{\mathrm{d}}{\mathrm{d}t}A\Psi_2(t)\right\|^2 \le M\lambda_{m+1}^{-1}, \quad \left\|\frac{\mathrm{d}}{\mathrm{d}t}\theta_2(t)\right\| \le M\lambda_{m+1}^{-1}. \tag{3.5.18}$$

Of course, at this time J not only involves Ψ, θ, but also includes $\frac{\mathrm{d}}{\mathrm{d}t}\Psi$, $\frac{\mathrm{d}}{\mathrm{d}t}\theta$. So we also need to use Lemma 3.5.2. From equations (3.5.16)–(3.5.18) we get the claim of the lemma. □

Since $\Psi_2, \theta_2 \in Q_m H$, by Lemma 3.5.3 we get

$$\|A\Psi_2\| \le M\lambda_{m+1}^{-1}, \quad \|\theta_2\| \le M\lambda_{m+1}^{-1}, \quad \forall t \ge t_*.$$

We now construct two-dimensional Newton–Boussinesq equations explicitly approximating inertial manifold.

In [90], the authors introduced the following nonlinear Galerkin method, and proved its convergence. The idea was to look for an approximate solution $(\Psi_1, \theta_1) \in P_m H \times P_m H$, satisfying

$$\begin{cases} \frac{\mathrm{d}}{\mathrm{d}t}A\Psi_1 + PJ(\Psi_1, A\Psi_1) + PJ(\varphi_1, A\Psi_1) + PJ(\Psi_1, A\Psi_1) + A^2\Psi_1 - \frac{R_\alpha}{P_\gamma}PB(\theta_1) = 0, \\ A^2\varphi_1 + QJ(\Psi_1, A\Psi_1) - \frac{R_\alpha}{P_\gamma}QB(\theta_1) = 0, \end{cases} \tag{3.5.19}$$

$$\begin{cases} \frac{\mathrm{d}}{\mathrm{d}t}\theta_1 + PJ(\Psi_1, \theta_1) + PJ(\varphi_1, \theta_1) + PJ(\Psi_1, \varphi_2) + \frac{1}{P_\gamma}A\theta_1 = 0, \\ \frac{1}{P_\gamma}A\varphi_2 + \alpha J(\Psi_1, \theta_1) = 0, \\ \Psi_1(0, x, y) = P\Psi_0(x, y), \theta_1(0, x, y) = P\theta_0(x, y), \end{cases} \tag{3.5.20}$$

where $\varphi_1, \varphi_2 \in Q_m H$. We notice that a nonlinear mapping F is as in the nonlinear Galerkin method. In the following, we prove that $\Sigma_1 = \operatorname{graph}(F)$ is an approximate inertial manifold. The mapping $F : P_m H \times P_m H \to Q_m H \times Q_m H$ is such that $F(\Psi_1, \theta_1) = (\varphi_1, \varphi_2)$ is valid for any $(\Psi_1, \theta_1) \in P_m H \times P_m H$. Here φ_1, φ_2 are determined by equations (3.5.19)–(3.5.20).

Theorem 3.5.1. *There exists a constant M, which only depends on parameters, such that*

$$\operatorname{dist}_{H^2 \times H}((\Psi, \theta), \Sigma_1) \le M\lambda_{m+1}^{-\frac{3}{2}}, \quad \forall t \ge t_*$$

where t_* depends on the parameters and R, whenever $\|\Psi_0\|_{H^4} \le R$, $\|\theta_0\|_{H^2} \le R$, and (Ψ, θ) is the solution of equations (3.5.1)–(3.5.4).

Proof. Taking the difference of equations (3.5.19) and (3.5.12), we get

$$\begin{aligned} &A^2(\varphi_1 - \Psi_2) + QJ(\Psi_1, A\Psi_1) - QJ(\Psi, A\Psi) \\ &\quad - \frac{R_\alpha}{P_\gamma}QB(\theta_1) + \frac{R_\alpha}{P_\gamma}QB(\theta) - \frac{\mathrm{d}}{\mathrm{d}t}A\Psi_2 = 0. \end{aligned}$$

By bilinear property of $J(u,v)$, we have

$$A^2(\varphi_1-\Psi_2)+QJ(-\Psi_2,A\Psi_1)+QJ(\Psi_1,-A\Psi_2)+\frac{R_\alpha}{P_\gamma}QB(\theta_2)-\frac{\mathrm{d}}{\mathrm{d}t}A\Psi_2=0, \quad (3.5.21)$$

$$|J(-\Psi_2,A\Psi_1)|\le C\|A^{\frac{3}{2}}\Psi_2\|\|A^{\frac{3}{2}}\Psi_1\|\le C\|A^{\frac{3}{2}}\Psi_2\|\|A^{\frac{3}{2}}\Psi\|\le C\lambda_{m+1}^{-\frac{1}{2}}, \quad (3.5.22)$$

$$|J(\Psi_1,-A\Psi_2)|\le C\|A^{\frac{3}{2}}\Psi\|\|A^{\frac{3}{2}}\Psi_2\|\le C\lambda_{m+1}^{-\frac{1}{2}}, \quad (3.5.23)$$

$$|B(\theta_2)|\le\|A^{\frac{1}{2}}\theta_2\|\le C\lambda_{m+1}^{-\frac{1}{2}}. \quad (3.5.24)$$

By equations (3.5.21)–(3.5.24) and Lemma 3.5.1, we get

$$\|A^2\varphi_1-A^2\Psi_2\|\le C\lambda_{m+1}^{-\frac{1}{2}}.$$

Hence we get

$$\|A\varphi_1-A\Psi_2\|\le C\lambda_{m+1}^{-\frac{3}{2}}. \quad (3.5.25)$$

Taking the difference of equations (3.5.20) and (3.5.13), it follows that

$$\frac{1}{P_\gamma}(A\varphi_2-A\theta_2)+QJ(\Psi_1,\theta_1)-QJ(\Psi_1,\theta)-\frac{\mathrm{d}}{\mathrm{d}t}\theta_2=0,$$

that is,

$$\frac{1}{P_\gamma}(A\varphi_2-A\theta_2)+QJ(-\Psi_2,\theta_1)+QJ(\Psi,-\theta_2)-\frac{\mathrm{d}}{\mathrm{d}t}\theta_2=0.$$

Then we get

$$\begin{aligned}\|A\varphi_2-A\theta_2\|&\le C\|A^{\frac{3}{2}}\Psi_2\|\|A^{\frac{1}{2}}\theta_1\|+C\|A^{\frac{3}{2}}\Psi\|\|A^{\frac{1}{2}}\theta_2\|+\left\|\frac{\mathrm{d}}{\mathrm{d}t}\theta_2\right\|\\&\le C\lambda_{m+1}^{-\frac{1}{2}}+C\lambda_{m+1}^{-1}\le C\lambda_{m+1}^{-\frac{1}{2}},\end{aligned}$$

which implies

$$\|\varphi_2-\theta_2\|\le C\lambda_{m+1}^{-\frac{3}{2}}. \quad (3.5.26)$$

For any solution $(\Psi(t),\theta(t))=(\Psi_1+\Psi_2,\theta_1+\theta_2)$ of problem (3.5.1)–(3.5.4), we have

$$\begin{aligned}\mathrm{dist}_{H^2\times H}((\Psi,\theta),\Sigma_1)&\le\|\Psi-(\Psi_1+\varphi_1)\|_{H^2}+\|\theta-(\theta_1+\varphi_2)\|\\&\le\|\Psi_2-\varphi_1\|_{H^2}+\|\theta_2-\varphi_2\|\\&\le C\|A\Psi_2-A\varphi_1\|+\|\theta_2-\varphi_2\|\le C\lambda_{m+1}^{-\frac{3}{2}},\quad\forall t\ge t_*,\end{aligned}$$

where t_* is chosen as in Lemmas 3.5.2 and 3.5.3. Theorem 3.5.1 has been proved. □

Now consider another implicit approximate inertial manifold. The manifold is determined by the contraction mapping, and provides a high-order approximation for the global attractor on Σ_1.

Introduce the balls:

$$\begin{aligned} B_m &= \{\Psi_1 \in P_m H : \|A^{\frac{3}{2}}\Psi_1\| \le 2M_0\}, \\ O_m &= \{\theta_1 \in P_m H : \|A^{\frac{1}{2}}\theta_1\| \le 2M_0\}, \\ B_m^{\perp} &= \{g \in Q_m H : \|A^{\frac{3}{2}}g\| \le 2M_0\}, \\ O_m^{\perp} &= \{h \in Q_m H : \|A^{\frac{1}{2}}h\| \le 2M_0\}, \end{aligned}$$

where M_0 is a constant as in Lemma 3.5.2.

Define the mapping $G : B_m \times O_m \to B_m^{\perp} \times O_m^{\perp}$ such that $G(\Psi_1, \theta_1) = (g, h)$ for any $(\Psi_1, \theta_1) \in B_m \times O_m$ and where (g, h) is determined by the following equations:

$$A^2 g + QJ(\Psi_1 + g, A\Psi_1 + Ag) - \frac{R_\alpha}{P_\gamma} QB(\theta_1 - h) = 0, \tag{3.5.27}$$

$$\frac{1}{P_\gamma} Ah + QJ(\Psi_1 + g, \theta_1 + h) = 0. \tag{3.5.28}$$

Firstly we prove that (g, h) is uniquely determined by equations (3.5.27)–(3.5.28).

Lemma 3.5.4. *There exists an integer m_0, which only depends on parameters, such that for $m \ge m_0$, equations (3.5.27)–(3.5.28) possess a unique solution $(g, h) \in B_m^{\perp} \times O_m^{\perp}$, $\forall (\Psi_1, \theta_1) \in B_m \times O_m$.*

Proof. Use the fixed point principle, by setting $(\Psi_1, \theta_1) \in B_m \times O_m$ and defining $\widetilde{G} : B_m^{\perp} \times O_m^{\perp} \to Q_m H \times Q_m H$ as follows: for $(g_1, h_1) \in B_m^{\perp} \times B_m^{\perp}$, $(g, h) = \widetilde{G}(g_1, h_1)$, which are determined by the following equations:

$$A^2 g + QJ(\Psi_1 + g_1, A\Psi_1 + Ag_1) - \frac{R_\alpha}{P_\gamma} QB(\theta_1 + h_1) = 0, \tag{3.5.29}$$

$$\frac{1}{P_\gamma} Ah + QJ(\Psi_1 + g_1, \theta_1 + h_1) = 0. \tag{3.5.30}$$

Obviously, $\widetilde{G}$ has a fixed point, namely the solution of equations (3.5.27)–(3.5.28). Indeed,

(1) When m is large enough, $\widetilde{G}$ maps $B_m^{\perp} \times O_m^{\perp}$ into itself. By equation (3.5.29), we get

$$\begin{aligned} \|A^2 g\| &\le \|J(\Psi_1 + g_1, A\Psi_1 + Ag_1)\| + \frac{R_\alpha}{P_\gamma}\|B(\theta_1 + h_1)\| \\ &\le C\|A^{\frac{3}{2}}\Psi_1 + A^{\frac{3}{2}}g_1\|^2 + \frac{R_\alpha}{P_\gamma}\|A^{\frac{1}{2}}\theta_1 + A^{\frac{1}{2}}h_1\| \le C. \end{aligned}$$

From the above, we get

$$\|A^{\frac{3}{2}}g\| \le C\lambda_{m+1}^{-\frac{1}{2}}.$$

Since $\lambda_{m+1} \to \infty$, there exists k_1, which depends on parameters, such that, when $m \ge k_1$, $g \in B_m^\perp$.

Through equation (3.5.28) we get

$$\begin{aligned}\frac{1}{P_y}\|Ah\| &\le \|J(\Psi_1 + g_1, \theta_1 + h_1)\| \\ &\le C\|A^{\frac{3}{2}}\Psi_1 + A^{\frac{3}{2}}g_1\|\|A^{\frac{1}{2}}\theta_1 + A^{\frac{1}{2}}h_1\| \\ &\le C(\Psi_1, \theta_1) \in B_m \times Q_m, \quad (g_1, h_1) \in B_m^\perp \times O_m^\perp.\end{aligned}$$

Hence if

$$\|A^{\frac{1}{2}}h\| \le C\lambda_{m+1}^{-\frac{1}{2}},$$

then there exists k_2, which depends on parameters, such that, when $m \ge k_2$, $h \in O_m^\perp$.

(2) $\widetilde{G}$ is contracting.

Let $(g_1, h_1), (g_2, h_2) \in B_m^\perp \times O_m^\perp$. By equation (3.5.29) we get

$$\begin{aligned}&A^2 g(g_1, h_1) - A^2 g(g_2, h_2) + QJ(\Psi_1 + g_1, A\Psi_1 + Ag_1) \\ &\quad - QJ(\Psi_1 + g_2, A\Psi_1 + Ag_2) - \frac{R_\alpha}{P_y}QB(\theta_1 + h_1) \\ &\quad + \frac{R_\alpha}{P_y}QB(\theta_1 + h_2) = 0,\end{aligned}$$

thus we have

$$\begin{aligned}&A^2 g(g_1, h_1) - A^2 g(g_2, h_2) + QJ(g_1 - g_2, A\Psi_1 + Ag_1) \\ &\quad + QJ(\Psi_1 + g_2, Ag_1 - Ag_2) - \frac{R_\alpha}{P_y}QB(h_1 - h_2) = 0.\end{aligned}$$

Then also

$$\begin{aligned}&\|A^2 g(g_1, h_1) - A^2 g(g_2, h_2)\| \\ &\quad \le \|QJ(g_1 - g_2, A\Psi_1 + Ag_1)\| + \|QJ(\Psi_1 + g_2, Ag_1 - Ag_2)\| + \frac{R_\alpha}{P_y}\|QB(h_1 - h_2)\| \\ &\quad \le C\|A^{\frac{3}{2}}g_1 - A^{\frac{3}{2}}g_2\|\|A^{\frac{3}{2}}\Psi_1 + A^{\frac{3}{2}}g_1\| + C\|A^{\frac{3}{2}}g_1 - A^{\frac{3}{2}}g_2\|\|A^{\frac{3}{2}}\Psi_1 + A^{\frac{3}{2}}g_2\| + C\|A^{\frac{1}{2}}(h_1 - h_2)\| \\ &\quad \le C\|A^{\frac{3}{2}}g_1 - A^{\frac{3}{2}}g_2\| + C\|A^{\frac{1}{2}}h_1 - A^{\frac{1}{2}}h_2\|,\end{aligned}$$

which yields

$$\begin{aligned}&\|A^{\frac{3}{2}}g(g_1, h_1) - A^{\frac{3}{2}}g(g_2, h_2)\| \\ &\quad \le C\lambda_{m+1}^{-\frac{1}{2}}\|A^{\frac{3}{2}}g_1 - A^{\frac{3}{2}}g_2\| + C\lambda_{m+1}^{-\frac{1}{2}}\|A^{\frac{1}{2}}h_1 - A^{\frac{1}{2}}h_2\|.\end{aligned} \tag{3.5.31}$$

Through equation (3.5.30), we get

$$\frac{1}{P_y}Ah(g_1,h_1)-\frac{1}{P_y}Ah(g_2,h_2)+QJ(\Psi_1+g_1,\theta_1+h_1)$$
$$-QJ(\Psi_2+g_2,\theta_2+h_2)=0,$$

that is,

$$\frac{1}{P_y}Ah(g_1,h_1)-\frac{1}{P_y}Ah(g_2,h_2)+QJ(g_1-g_2,\theta_1+h_1)$$
$$+QJ(\Psi_1+g_2,h_1-h_2)=0.$$

Therefore

$$\begin{aligned}
&\frac{1}{P_y}\|Ah(g_1,h_1)-Ah(g_2,h_2)\|\\
&\quad\le\|QJ(g_1-g_2,\theta_1+h_1)\|+\|QJ(\Psi_1+g_2,h_1-h_2)\|\\
&\quad\le C\|A^{\frac{3}{2}}g_1-A^{\frac{3}{2}}g_2\|\|A^{\frac{1}{2}}\theta_1+A^{\frac{1}{2}}h_1\|+C\|A^{\frac{3}{2}}\Psi_1+A^{\frac{3}{2}}g_2\|\|A^{\frac{1}{2}}h_1-A^{\frac{1}{2}}h_2\|\\
&\quad\le C\|A^{\frac{3}{2}}g_1-A^{\frac{3}{2}}g_2\|+C\|A^{\frac{1}{2}}h_1-A^{\frac{1}{2}}h_2\|,
\end{aligned}$$

which suggests

$$\begin{aligned}
&\|A^{\frac{1}{2}}h(g_1,h_1)-A^{\frac{1}{2}}h(g_2,h_2)\|\\
&\quad\le C\lambda_{m+1}^{-\frac{1}{2}}\|A^{\frac{3}{2}}g_1-A^{\frac{3}{2}}g_2\|+C\lambda_{m+1}^{-\frac{1}{2}}\|A^{\frac{1}{2}}h_1-A^{\frac{1}{2}}h_2\|.
\end{aligned}\tag{3.5.32}$$

Since $\lambda_{m+1}\to\infty$, by equations (3.5.31)–(3.5.32) we know that there exists $m_0\ge k_0$ such that whenever $m\ge m_0$, $\widetilde{G}$ is contracting.

By the preceding we know that $\widetilde{G}$ has a unique fixed point in $B_m^\perp\times Q_m^\perp$. Lemma 3.5.4 has been proved. □

Let $\Sigma_2=\mathrm{graph}(G)$. Then Σ_2 is an approximate inertial manifold. Indeed, we have

Theorem 3.5.2. *There exists $m_0\ge k_0$ such that for any $m\ge m_0$, there exists a constant M, for which we have*

$$\mathrm{dist}_{H^2\times H}((\Psi(t),\theta(t)),\Sigma_2)\le M\lambda_{m+1}^{-2},\quad\forall t\ge t_*,$$

where $(\Psi(t),\theta(t))$ is any solution of equations (3.5.1)–(3.5.4), M only depends on parameters, t_* depends on parameters and R, whenever $\|\Psi_0\|_{H^4}\le R$, $\|\theta_0\|_{H^2}\le R$.

Proof. From equations (3.5.27)–(3.5.32) we get

$$A^2g-A^2\Psi_2+QJ(\Psi_1+g,A\Psi_1+Ag)-QJ(\Psi_1,A\Psi)$$
$$-\frac{R_\alpha}{P_y}QB(\theta_1+h)+\frac{R_\alpha}{P_y}QB(\theta)-\frac{\mathrm{d}}{\mathrm{d}t}A\Psi_2=0,$$

that is,

$$A^2 g - A^2\Psi_2 + QJ(g - \Psi_2, A\Psi_1 + Ag) + QJ(\Psi, Ag - A\Psi_2) - \frac{R_\alpha}{P_\gamma} QB(h - \theta_2) - \frac{d}{dt} A\Psi_2 = 0.$$

Then we arrive at

$$\begin{aligned}
\|A^2 g - A^2\Psi_2\| &\le \|J(g - \Psi_2, A\Psi_1 + Ag)\| \\
&\quad + \|J(\Psi, Ag - A\Psi_2)\| + \frac{R_\alpha}{P_\gamma}\|B(h - \theta_2)\| + \left\|\frac{d}{dt}A\Psi_2\right\| \\
&\le C\|A^{\frac{3}{2}} g - A^{\frac{3}{2}}\Psi_2\|\|A^{\frac{3}{2}}\Psi_1 + A^{\frac{3}{2}} g\| \\
&\quad + C\|A^{\frac{3}{2}}\Psi\|\|A^{\frac{3}{2}} g - A^{\frac{3}{2}}\Psi_2\| + C\|A^{\frac{1}{2}} h - A^{\frac{1}{2}}\theta_2\| + \left\|\frac{d}{dt}A\Psi_2\right\| \\
&\le C\|A^{\frac{3}{2}} g - A^{\frac{3}{2}}\Psi_2\| + C\|A^{\frac{1}{2}} h - A^{\frac{1}{2}}\theta_2\| + C\lambda_{m+1}^{-1} \\
&\le C\lambda_{m+1}^{-\frac{1}{2}}\|A^2 g - A^2\Psi_2\| + C\|A^{\frac{1}{2}} h - A^{\frac{1}{2}}\theta_2\| + C\lambda_{m+1}^{-1}.
\end{aligned}$$

Thus there exists $m_0 \ge k_0$ such that for all $m \ge m_0$, we have

$$\begin{aligned}
\|A^2 g - A^2\Psi_2\| &\le C\|A^{\frac{1}{2}} h - A^{\frac{1}{2}}\theta_2\| + C\lambda_{m+1}^{-1} \\
&\le C\lambda_{m+1}^{-\frac{1}{2}}\|Ah - A\theta_2\| + C\lambda_{m+1}^{-1}.
\end{aligned} \tag{3.5.33}$$

Taking the difference of equations (3.5.28) and (3.5.13), it follows that

$$\frac{1}{P_\gamma}(Ah - A\theta_2) + QJ(\Psi_1 + g, \theta_1 + h) - QJ(\Psi_1, \theta) - \frac{d}{dt}\theta_2 = 0,$$

that is,

$$\frac{1}{P_\gamma}(Ah - A\theta_2) + QJ(g - \Psi_2, \theta_1 + h) + QJ(\Psi, h - \theta_2) - \frac{d}{dt}\theta_2 = 0,$$

which implies

$$\begin{aligned}
\|Ah - A\theta_2\| &\le C\|A^{\frac{3}{2}} g - A^{\frac{3}{2}}\Psi_2\|\|A^{\frac{1}{2}}\theta_1 + A^{\frac{1}{2}} h\| \\
&\quad + C\|A^{\frac{3}{2}}\Psi\|\|A^{\frac{1}{2}} h - A^{\frac{1}{2}}\theta_2\| + C\lambda_{m+1}^{-1} \\
&\le C\|A^{\frac{3}{2}} g - A^{\frac{3}{2}}\Psi_2\| + C\|A^{\frac{1}{2}} h - A^{\frac{1}{2}}\theta_2\| + C\lambda_{m+1}^{-1} \\
&\le C\lambda_{m+1}^{-\frac{1}{2}}\|A^2 g - A^2\Psi_2\| + C\lambda_{m+1}^{-\frac{1}{2}}\|Ah - A\theta_2\| + C\lambda_{m+1}^{-1}.
\end{aligned} \tag{3.5.34}$$

By virtue of (3.5.33)–(3.5.34), we deduce that, when $m \ge m_0$,

$$\|A^2 g - A^2\Psi_2\| + \|Ah - A\theta_2\| \le C\lambda_{m+1}^{-\frac{1}{2}}\|A^2 g - A^2\Psi_2\| + C\lambda_{m+1}^{-\frac{1}{2}}\|Ah - A\theta_2\| + C\lambda_{m+1}^{-1}.$$

Since $\lambda_{m+1} \to \infty$, we know that there exists $k_0 \geq m_0$, such that for all $m \geq k_0$ we have

$$\|A^2 g - A^2 \Psi_2\| + \|Ah - A\theta_2\| \leq C\lambda_{m+1}^{-1}.$$

Hence, we have

$$\begin{aligned}\|Ag - A\Psi_2\| &\leq C\lambda_{m+1}^{-2},\\ \|h - \theta_2\| &\leq C\lambda_{m+1}^{-2}.\end{aligned}$$

Then for any solution $(\Psi(t), \theta(t))$ of problem (3.5.1)–(3.5.4), we have

$$\begin{aligned}\operatorname{dist}_{H^2 \times H}((\Psi, \theta), \Sigma_2) &\leq \|\Psi(t) - (\Psi_1 + g)\|_{H^2} + \|\theta(t) - (\theta_1 + h)\|\\ &\leq \|\Psi_2 - g\|_{H^2} + \|\theta_2 - h\|\\ &\leq C\|A\Psi_2 - Ag\| + \|\theta_2 - h\| \leq C\lambda_{m+1}^{-2}.\end{aligned}$$

Theorem 3.5.2 has been proved. □

Finally, we introduce the nonlinear Galerkin method to approximate the inertial manifold Σ_2. To seek for the approximate solution $\Psi_m, \theta_m \in P_m H$, we need the following equations:

$$\begin{cases}\frac{\mathrm{d}}{\mathrm{d}t} A\Psi_m + PJ(\Psi_m, A\Psi_m) + PJ(g, A\Psi_m) + PJ(\Psi_m, Ag) + A^2\Psi_m - \frac{R_a}{P_y} PB(\theta_m) = 0,\\ \frac{\mathrm{d}\theta_m}{\mathrm{d}t} + PJ(\Psi_m, \theta_m) + PJ(g, \theta_m) + PJ(\Psi_m, h) + \frac{1}{P_y} A\theta_m = 0,\\ \Psi_m(0, x, y) = P_m \Psi_0(x, y), \quad \theta_m(0, x, y) = P_m \theta_0(x, y),\end{cases} \tag{3.5.35}$$

where $(g, h) \in Q_m H$ are determined by equations (3.5.27)–(3.5.28). To achieve this, (Ψ_1, θ_1) is replaced with (Ψ_m, θ_m). By the method in [90], we can prove the convergence of the approximate solution of the above form.

Theorem 3.5.3. *Let $\Psi_0(x, y) \in H^2$, and assume $\theta_0(x, y) \in H^1$ is a periodic function of x, y. Then the approximate solutions (Ψ_m, θ_m), determined by equation* (3.5.35), *when $m \to \infty$ converge to the generalized solution (Ψ, θ) of problem* (3.5.1)–(3.5.4).

3.6 Long wave–short wave (LS) equation

In 1977, Djordjevic et al. [48] studied two-dimensional gravity wave packet movement and first put forward the interaction of long and short waves. The general theory of interaction between long and short waves was studied by Grimshaw [82] and Denney [13]. In 1987, Guo [85] studied the existence and uniqueness of a smooth solution for the generalized LS equation. In 1991, Guo [87] studied the initial and periodic initial value problem for the generalized LS equations. 1994, Tsutsumi and Hatano [203] proved the existence and uniqueness of a solution in H^1. In 1996, Guo and Miao [107] considered an even more general LS equation and improved the results of [203] for the LS equation; [203] also contained an open problem. Guo and Chen [94] proved the orbital stability of a solitary wave for the LS equation. In 1996, Guo and Wang [118] studied the long time behavior for a nonlinear dissipative LS equation. In 1998, Guo and Wang [123] proved the existenc of attrators.

Consider the following generalized LS equations with the dissipative term [12]:

$$iu_t + u_{xx} - nu + i\alpha u + \beta g(|u|^2)u + h_1(x) = 0, \tag{3.6.1}$$

$$n_t + |u|_x^2 + \delta n + \gamma f(|u|^2)u + h_2(x) = 0, \tag{3.6.2}$$

with the initial value

$$u|_{t=0} = u_0(x), \quad n|_{t=0} = n_0(x), \quad x \in \Omega(-D, D),\ D > 0, \tag{3.6.3}$$

and periodic boundary condition

$$u(x-D,t) = u(x+D,t), \quad n(x-D,t) = n(x+D,t), \quad \forall x \in \mathbf{R},\ t \geq 0, \tag{3.6.4}$$

where $u = (u_1(x,t), \ldots, u_N(x,t))$ is an unknown complex vector; $n(x,t)$ is an unknown real function; $g(s)$ and $f(s)$, $0 \leq s < \infty$, are known real functions; $h_1(x)$ and $h_2(x)$ denote vector replication function and numerical function, respectively; α, β, γ and δ are real constants, $\alpha > 0$. In order to construct inertial manifolds of problem (3.6.1)–(3.6.4), we prove the existence and uniqueness of a global smooth solution for problem (3.6.1)–(3.6.4).

Lemma 3.6.1. *Let $u_0(x)$, $h_1(x)$ and $h_2(x) \in L^2(\Omega)$. Then for any solution $(u(t), n(t))$ of problem* (3.6.1)–(3.6.4), *we have*

$$\sup_{0 \leq t \leq T} \|u(t)\| \leq M_1, \quad \forall T > 0,$$

where $M_1 = M_1(\alpha, \beta, \gamma, \delta, f, g, h_1, h_2, T)$ is a positive constant.

Proof. Taking the inner product of equation (3.6.1) and u over H, we get

$$(iu_t + u_{xx} - nu + i\alpha u + \beta g(|u|^2)u + h_1(x), u) = 0. \tag{3.6.5}$$

Taking the imaginary part of equation (3.6.5), we obtain

$$\frac{1}{2}\frac{\mathrm{d}}{\mathrm{d}t}\|u(t)\|^2 + \alpha\|u(t)\|^2 + \mathrm{Im}(h, u) = 0. \tag{3.6.6}$$

By Gronwall lemma, we get the claim of Lemma 3.6.1. □

Lemma 3.6.2. *Suppose*

(1) $\rho g(s) \leq B_1 s^{2-\sigma} + C_1$, $s > 0$, $B_1 > 0$, $C_1 > 0$, $\sigma > 0$;

(2) $|f(s)| \leq B_2 s^{\frac{3}{2}} + C_2$, $s > 0$, $B_2 > 0$, $C_2 > 0$;

(3) $h_1(x), h_2(x) \in L^2(\Omega)$;
(4) $u_0(x) \in H^1(\Omega), n_0(x) \in L^2(\Omega)$.

Then we have

$$\sup_{0\le t\le T} \|u_x(t)\| + \sup_{0\le t\le T} \|n(t)\| \le M_2, \quad \forall T > 0,$$

where $M_2 = M_2(T, \|u_0\|_{H^1}, \|n_0\|)$ *is a positive constant.*

Proof. Taking the inner product of equation (3.6.1) and u_t over H, we get

$$(iu_t + u_{xx} - nu + i\alpha u + \beta g(|u|^2)u + h_1(x), u_t) = 0. \tag{3.6.7}$$

Taking the imaginary part of equation (3.6.7), we obtain

$$-\frac{1}{2}\frac{\mathrm{d}}{\mathrm{d}t}\|u_x\|^2 - \frac{1}{2}\int n\frac{\mathrm{d}}{\mathrm{d}t}|u|^2\mathrm{d}x + \mathrm{Re}(i\alpha u, u_t) + \frac{1}{2}\beta\int g(|u|^2)\frac{\mathrm{d}}{\mathrm{d}t}|u|^2\mathrm{d}x + \mathrm{Re}(h_1, u_t) = 0.$$

By virtue of

$$\begin{aligned}\int n\frac{\mathrm{d}}{\mathrm{d}t}|u|^2\mathrm{d}x &= \frac{\mathrm{d}}{\mathrm{d}t}\int n|u|^2\mathrm{d}x - \int |u|^2 n_t\mathrm{d}x \\ &= \frac{\mathrm{d}}{\mathrm{d}t}\int n|u|^2\mathrm{d}x + \delta\int n|u|^2\mathrm{d}x + \gamma\int f(|u|^2)|u|^2\mathrm{d}x + \int h_2|u|^2\mathrm{d}x,\end{aligned}$$

and letting

$$G(s) = \int_0^s g(\tau)\mathrm{d}\tau,$$

we get the following result:

$$\begin{aligned}&-\frac{1}{2}\frac{\mathrm{d}}{\mathrm{d}t}\|u_x\|^2 - \frac{1}{2}\frac{\mathrm{d}}{\mathrm{d}t}\int n|u|^2\mathrm{d}x - \frac{1}{2}\delta\int \frac{\mathrm{d}}{\mathrm{d}t}n|u|^2\mathrm{d}x \\ &- \frac{1}{2}\gamma\int f(|u|^2)|u|^2\mathrm{d}x - \frac{1}{2}\int h_2|u|^2\mathrm{d}x \\ &+ \frac{1}{2}\beta\frac{\mathrm{d}}{\mathrm{d}t}\int G(|u|^2)\mathrm{d}x + \mathrm{Re}(i\alpha u, u_t) + \frac{\mathrm{d}}{\mathrm{d}t}\mathrm{Re}(h_1, u) = 0.\end{aligned} \tag{3.6.8}$$

Integrating equation (3.6.8) over $t \in (0, T)$, we get

$$\begin{aligned}&\|u_x\|^2 + \int n|u|^2\mathrm{d}x + \delta\int_0^t\int n|u|^2\mathrm{d}x \\ &\quad + \gamma\int_0^t\int f(|u|^2)|u|^2\mathrm{d}x + \int_0^t\int h_2|u|^2\mathrm{d}x - \beta\int G(|u|^2)\mathrm{d}x\end{aligned}$$

$$
\begin{aligned}
&-2\int_0^t \mathrm{Re}(i\alpha u, u_t)\mathrm{d}t - 2\,\mathrm{Re}(h_1, u)\\
&= \|u_x(0)\|^2 + \int n(0)|u(0)|^2\mathrm{d}x\\
&\quad -\beta\int G(|u(0)|^2)\mathrm{d}x - 2\,\mathrm{Re}(h_1, u(0)).
\end{aligned} \tag{3.6.9}
$$

Now we estimate each term of equation (3.6.9). For all $\rho > 0$ we have

$$
\left|\int n|u|^2\mathrm{d}x\right| \le \rho\|n\|^2 + C\|u\|_4^4 \le \rho\|n\|^2 + \rho\|u_x\|^2 + M. \tag{3.6.10}
$$

Similarly we have

$$
\left|\delta\int n|u|^2\mathrm{d}x\right| \le \rho\|n\|^2 + \rho\|u_x\|^2 + M, \tag{3.6.11}
$$

$$
\begin{aligned}
\left|\gamma\int f(|u|^2)|u|^2\mathrm{d}x\right| &\le |\gamma|\int B_2|u|^5\mathrm{d}x + C_2|\gamma|\int |u|^2\mathrm{d}x\\
&\le M\|u\|_{H_1}^{\frac{3}{2}} + M \le M\|u_x\|^2 + M,
\end{aligned} \tag{3.6.12}
$$

$$
\left|\int h_2|u|^2\mathrm{d}x\right| \le \rho\|u_x\|^2 + M. \tag{3.6.13}
$$

Using assumption (1), we have

$$
\beta G(s) \le \frac{1}{3}B_1 s^{3-\sigma} + C_1 s, \quad \forall s > 0,
$$

and then

$$
\begin{aligned}
\beta\int G(|u|^2)\mathrm{d}x &\le \frac{1}{3}B_1\int |u|^{6-2\sigma}\mathrm{d}x + C_1\int |u|^2\mathrm{d}x\\
&\le M\|u_x\|^{2-\sigma} + M \le M\|u_x\|^2 + M,
\end{aligned} \tag{3.6.14}
$$

$$
|2\,\mathrm{Re}(h_1, u)| \le 2\|h_1\|\|u\| \le M. \tag{3.6.15}
$$

Taking the real part of (3.6.5), we obtain

$$
-\mathrm{Re}(iu, u_t) = \|u_x\|^2 + \int n|u|^2\mathrm{d}x - \beta\int g(|u|^2)|u|^2\mathrm{d}x - \mathrm{Re}(u, h_1).
$$

By virtue of assumption (1), we have

$$
\beta\int g(|u|^2)|u|^2\mathrm{d}x \le B_1\int |u|^{6-2\sigma}\mathrm{d}x + C_1\int |u|^2\mathrm{d}x \le \|u_x\|^2 + M,
$$

which yields

$$
-\mathrm{Re}(iu, u_t) \ge \int n|u|^2\mathrm{d}x - M,
$$

that is,

$$-2\alpha\,\mathrm{Re}(iu,u_t) \geq 2\alpha\int n|u|^2\mathrm{d}x - M. \tag{3.6.16}$$

From equations (3.6.9)–(3.6.16), we get

$$\begin{aligned}\|u_x\|^2 &\leq \rho\|n\|^2 + 2\rho\|u_x\|^2 + (2\alpha+1)\rho\int_0^t \|n\|^2\mathrm{d}t\\ &\quad + (2\alpha\rho + 2\rho + M)\int_0^t \|u_x\|^2\mathrm{d}t + Mt + M.\end{aligned} \tag{3.6.17}$$

Taking the inner product of equation (3.6.2) and n over H, we obtain

$$\frac{1}{2}\frac{\mathrm{d}}{\mathrm{d}t}\|n\|^2 + \int n|u|_x^2\mathrm{d}x + \delta\|n\|^2 + \gamma\int f(|u|^2)n\mathrm{d}x + \int h_2 n\mathrm{d}x = 0. \tag{3.6.18}$$

Since

$$\begin{aligned}\int n|u|_x^2\mathrm{d}x &= \int nu_x\bar{u}\mathrm{d}x + \int nu\bar{u}_x\mathrm{d}x\\ &= i\int(u_t\bar{u}_x - \bar{u}_t u_x)\mathrm{d}x + 2\,\mathrm{Re}\int i\alpha u\bar{u}_x\mathrm{d}x + 2\,\mathrm{Re}\int h_1\bar{u}_x\mathrm{d}x,\end{aligned} \tag{3.6.19}$$

$$\begin{aligned}\frac{\mathrm{d}}{\mathrm{d}t}\int(iu\bar{u}_x - iu_x\bar{u})\mathrm{d}x &= i\int(u_t\bar{u}_x + u\bar{u}_{xt} - u_{xt}\bar{u} - u_x\bar{u}_t)\mathrm{d}x\\ &= i\int(u_t\bar{u}_x - \bar{u}_t u_x + u_t\bar{u}_x - \bar{u}_t u_x)\mathrm{d}x\\ &= 2i\int(u_t\bar{u}_x - \bar{u}u_x)\mathrm{d}x,\end{aligned} \tag{3.6.20}$$

from equations (3.6.18)–(3.6.20), we get

$$\begin{aligned}&\frac{1}{2}\frac{\mathrm{d}}{\mathrm{d}t}\|n\|^2 + \frac{1}{2}\frac{\mathrm{d}}{\mathrm{d}t}\int(iu\bar{u}_x - iu_x\bar{u})\mathrm{d}x + 2\,\mathrm{Re}\int i\alpha u\bar{u}_x\mathrm{d}x\\ &\quad + 2\,\mathrm{Re}\int h_1\bar{u}_x\mathrm{d}x + \delta\|n\|^2 + \gamma\int f(|u|^2)n\mathrm{d}x + \int h_2 n\mathrm{d}x = 0.\end{aligned} \tag{3.6.21}$$

Integrating equation (3.6.21) over $t \in (0,t)$, we get

$$\begin{aligned}&\|n\|^2 + \int iu\bar{u}_x\mathrm{d}x - \int iu_x\bar{u}\mathrm{d}x + 4\,\mathrm{Re}\int_0^t\int i\alpha u\bar{u}_x\mathrm{d}x\\ &\quad + 4\,\mathrm{Re}\int_0^t\int h_1\bar{u}_x\mathrm{d}x + 2\delta\int_0^t\|n\|^2\mathrm{d}x + 2\gamma\int_0^t\int f(|u|^2)n\mathrm{d}x\mathrm{d}t\\ &\quad + 2\int_0^t\int h_2 n\mathrm{d}x\mathrm{d}t\\ &= \|n(0)\|^2 + i\int u(0)\bar{u}_x(0)\mathrm{d}x - i\int u_x(0)\bar{u}(0)\mathrm{d}x.\end{aligned} \tag{3.6.22}$$

Since

$$\left|i\int(u\bar{u}_x - u_x\bar{u})\mathrm{d}x\right| \le \rho\|u_x\|^2 + M, \tag{3.6.23}$$

$$\left|4\,\mathrm{Re}\int i\alpha u\bar{u}_x\mathrm{d}x\right| \le 4\alpha\int|u||u_x|^2\mathrm{d}x \le \rho\|u_x\|^2 + M, \tag{3.6.24}$$

$$\left|4\,\mathrm{Re}\int h_1\bar{u}_x\mathrm{d}x\right| \le 4\int|h_1||u_x|^2\mathrm{d}x \le \rho\|u_x\|^2 + M, \tag{3.6.25}$$

$$\begin{aligned}\left|2\gamma\int f(|u|^2)n\mathrm{d}x\right| &\le C\int|u|^3|n|\mathrm{d}x + C\int|n|\mathrm{d}x\\ &\le \|n\|^2 + M\|u_x\|^2 + M,\end{aligned} \tag{3.6.26}$$

together with equations (3.6.22)–(3.6.26), we get

$$\begin{aligned}\|n\|^2 \le \rho\|u_x\|^2 + (2\rho + M)\int_0^t\|u_x\|^2\mathrm{d}t\\ + 2(1+|\delta|)\int_0^t\|n\|^2\mathrm{d}t + Mt + M.\end{aligned} \tag{3.6.27}$$

From equations (3.6.17)–(3.6.27), we get

$$\begin{aligned}\|u_x\|^2 + \|n\|^2 \le \rho\|n\|^2 + 3\rho\|u_x\|^2 + (2\alpha\rho + 4\rho + M)\int\|u_x\|^2\mathrm{d}t\\ + (2\alpha\rho + \rho + 2 + 2|\delta|)\int_0^t\|n\|^2\mathrm{d}t + Mt + M.\end{aligned}$$

Choosing $\rho = \frac{1}{4}$, from the above equation we get

$$\|u_x\|^2 + \|n\|^2 \le M\int_0^t(\|u_x\|^2 + \|n\|^2)\mathrm{d}t + M, \quad \forall 0 \le t \le T.$$

By Gronwall lemma, we get

$$\|u_x\|^2 + \|n\|^2 \le M, \quad \forall 0 \le t \le T.$$

Lemma 3.6.2 has been proved. □

By virtue of Lemmas 3.6.1 and 3.6.2, we get

$$\sup_{0\le t\le T}\|u\|_{H^1} \le M, \quad \sup_{0\le t\le T}\|u\|_\infty \le M. \tag{3.6.28}$$

Lemma 3.6.3. *Let conditions of Lemma 3.6.2 be valid, and $g \in C^1[0,+\infty)$, $u_0 \in H^2$. Then we have*

$$\sup_{0\le t\le T} \|n_t\| + \sup_{0\le t\le T} \|u_t\| \le M_3, \quad \forall T>0,$$

where $M_3 = M_3(T, \|u_0\|_{H^2}, \|n_0\|)$ is a positive constant.

Proof. From equation (3.6.2) we get

$$\begin{aligned} \|n_t\| &\le \left\| \frac{\mathrm{d}}{\mathrm{d}t}(u\bar{u}) \right\| + |\delta|\|n\| + \|\gamma f(|u|^2)\| + \|h_2\| \\ &\le \|u_x\bar{u} + u\bar{u}_x\| + |\delta|\|n\| + \|\gamma f(|u|^2)\| + \|h_2\|. \end{aligned} \tag{3.6.29}$$

In virtue of inequality (3.6.28) we get

$$\|n_t\| \le M.$$

Differentiating (3.6.1) with respect to t, we get

$$\begin{aligned} &i\frac{\mathrm{d}}{\mathrm{d}t}u_t + u_{xxt} - n_t u - n u_t + i\alpha u_t + \beta g'(|u|^2)|u|^2 u_t \\ &\quad + \beta g'(|u|^2)|u|^2\bar{u}_t + \beta g(|u|^2)u_t = 0. \end{aligned} \tag{3.6.30}$$

Taking the inner product of equation (3.6.30) and u_t over H, and taking its imaginary part, we obtain

$$\frac{1}{2}\frac{\mathrm{d}}{\mathrm{d}t}\|u_t\|^2 + \alpha\|u_t\|^2 - \operatorname{Im}\int n_t u\bar{u}_t \mathrm{d}x + \operatorname{Im}\int \beta g'(|u|^2)u^2\bar{u}_t^2 \mathrm{d}x = 0.$$

From the above equation, we get

$$\sup_{0\le t\le T} \|u_t\| \le M. \qquad \square$$

Lemma 3.6.4. *Let the conditions of Lemma 3.6.3 be satisfied and consider $f \in C^1[0,+\infty)$, $n_0, h_2 \in H^1$. Then we have*

$$\sup_{0\le t\le T} \|n_x\| \le M_4,$$

where $M_4 = M_4(T, \|u_0\|_{H^2}, \|n_0\|_{H^2})$ is a positive constant.

Proof. Differentiating equation (3.6.2) with respect to x, we get

$$n_{xt} + u_{xx}\bar{u} + 2u_x\bar{u}_x + u\bar{u}_{xx} + \delta n_x + \gamma f'(|u|^2)|u|_x^2 + h_2' = 0. \tag{3.6.31}$$

Taking the inner product of equation (3.6.31) and n_x over H, we obtain

$$\begin{aligned} &\frac{1}{2}\frac{\mathrm{d}}{\mathrm{d}t}\|n_x\|^2 + \int u_{xx}\bar{u}n_x \mathrm{d}x + 2\int u_x\bar{u}_x n_x \mathrm{d}x \\ &\quad + \int u\bar{u}_{xx}n_x \mathrm{d}x + \delta\|n_x\|^2 + \gamma\int f'(|u|^2)u_x\bar{u}n_x \mathrm{d}x \\ &\quad + \gamma\int f'(|u|^2)u_x\bar{u}n_x \mathrm{d}x + \int h_2' n_x \mathrm{d}x = 0. \end{aligned}$$

After a detailed calculation, we get

$$\frac{\mathrm{d}}{\mathrm{d}t}\|n_x\|^2 \le M\|n_x\|^2 + M.$$

By Gronwall lemma, we get the claim of the lemma. □

Lemma 3.6.5. *Let the conditions of Lemma 3.6.4 be satisfied and consider $g \in C^2[0,+\infty)$, $h_1, n_0 \in H^1$, $u_0 \in H^3$. Then we have*

$$\sup_{0\le t\le T}\|n_{tx}\| + \sup_{0\le t\le T}\|u_{tx}\| \le M_5,$$

where $M_5 = M_5(T, \|u_0\|_{H^3}, \|n_0\|_{H^1})$ is a positive constant.

Proof. From equation (3.6.31), we get

$$\|n_{xt}\| \le \|u_{xx}\bar{u}\| + 2\|u_x\bar{u}_x\| + \|u\bar{u}_{xx}\| + \|\delta n_x\| + \|\gamma f'(|u|^2)|u|_x^2\| + \|h_2'\| \le M. \tag{3.6.32}$$

Differentiating equation (3.6.1) with respect to x and t, we get

$$\begin{aligned} &iu_{ttx} + u_{xxxt} - n_{xt}u - n_x u_t - n_t u_x - nu_{xt} + i\alpha u_{xt} \\ &\quad + \beta g''(|u|^2)|u|_t^2|u|_x^2 u + \beta g'(|u|^2)|u|_{tx}^2 u + \beta g'(|u|^2)|u|_x^2 u_t \\ &\quad + \beta g'(|u|^2)|u|_t^2 u_x + \beta g(|u|^2)u_{xt} = 0. \end{aligned}$$

Taking the inner product of above equation and u_{xt} over H, and then taking its imaginary part, we obtain

$$\begin{aligned} \frac{1}{2}\frac{\mathrm{d}}{\mathrm{d}t}\|u_{tx}\|^2 + \alpha\|u_{tx}\|^2 \le{}& \int |n_{xt}u\bar{n}_{xt}|\mathrm{d}x + \int |n_x u_t \bar{u}_{xt}|\mathrm{d}x + \int |n_t u_x \bar{u}_{xt}|\mathrm{d}x \\ &+ \int \big|\beta g''(|u|^2)u\bar{u}_x(u_x u_t \bar{u}^2 + \bar{u}_x u_t|u|^2 + u_t\bar{u}_t|u|^2 + \bar{u}_x\bar{u}_t u^2)\big|\mathrm{d}x \\ &+ \int \big|\beta g'(|u|^2)u\bar{u}_{xt}(u_{tx}\bar{u} + \bar{u}_x u_t + u_x\bar{u}_t + u\bar{u}_{tx})\big|\mathrm{d}x \\ &+ \int \big|\beta g'(|u|^2)(u_x\bar{u} + u\bar{u}_x)u_t\bar{u}_{tx}\big|\mathrm{d}x \\ &+ \int \big|\beta g'(|u|^2)(u_t\bar{u} + u\bar{u}_t)u_x\bar{u}_{xt}\big|\mathrm{d}x \le M\|u_{xt}\|^2 + M, \end{aligned}$$

and then we have

$$\frac{\mathrm{d}}{\mathrm{d}t}\|u_{xt}\|^2 \le M\|u_{xt}\|^2 + M. \tag{3.6.33}$$

Lemma 3.6.5 is proved. □

Lemma 3.6.6. *Let the conditions of Lemma 3.6.4 be satisfied and consider $u_0 \in H^4$, $n_0 \in H^2$. Then we have*

$$\sup_{0\le t\le T}\|n_{tt}\| + \sup_{0\le t\le T}\|u_{tt}\| \le M_6,$$

where $M_6 = M_6(T, \|u_0\|_{H^4}, \|n_0\|_{H^2})$ is a positive constant.

Proof. Differentiating (3.6.2) with respect to t, we get

$$\begin{aligned}\|n_{tt}\| &\le \|u_{xt}\bar{u}\| + \|u_x\bar{u}_t\| + \|u_t\bar{u}_x\| + \|u\bar{u}_{xt}\| + \|\delta n_t\| + \|\gamma f'(|u|^2)(u_t\bar{u} + u\bar{u}_t)\| \\ &\le 2\|u\|_\infty\|u_{xt}\| + 2\|u_x\|_\infty\|u_t\| + \|\delta n_t\| + 2\|\gamma f'(|u|^2)\|_\infty\|u\|_\infty\|u_t\| \le M. \end{aligned} \tag{3.6.34}$$

Differentiating equation (3.6.2) with respect to x twice, we get

$$\begin{aligned} &iu_{ttt} + u_{xxtt} - n_{tt}u - n_tu_t - nu_{tt} + i\alpha u_{tt} \\ &\quad + \beta g''(|u|^2)|u|^2|u|_t^2u + \beta g'(|u|^2)|u|_{tt}^2u + \beta g'(|u|^2)|u|_t^2u_t \\ &\quad + \beta g(|u|^2)u_{tt} = 0. \end{aligned}$$

Taking the inner product of the above equation and u_{tt} over H, and taking its imaginary part, we obtain

$$\begin{aligned} &\frac{1}{2}\frac{\mathrm{d}}{\mathrm{d}t}\|u_{tt}\|^2 + \alpha\|u_{tt}\|^2 - \operatorname{Im}\int n_{tt}u\bar{u}_{tt}\mathrm{d}x - 2\operatorname{Im}\int n_tu_t\bar{u}_{tt}\mathrm{d}x \\ &\quad + \operatorname{Im}\int \beta g''(|u|^2)u\bar{u}_{tt}(u_t\bar{u} + u\bar{u}_t)^2\mathrm{d}x \\ &\quad + \operatorname{Im}\int \beta g'(|u|^2)u\bar{u}_{tt}|u|_{tt}^2\mathrm{d}x \\ &\quad + 2\operatorname{Im}\int \beta g'(|u|^2)u_t\bar{u}_{tt}(u_t\bar{u} + u\bar{u}_t)\mathrm{d}x = 0, \end{aligned}$$

which yields

$$\begin{aligned} \frac{1}{2}\frac{\mathrm{d}}{\mathrm{d}t}\|u_{tt}\|^2 + \alpha\|u_{tt}\|^2 &\le \frac{1}{2}\|u\|_\infty(\|n_{tt}\|^2 + \|u_{tt}\|^2) + \|u_t\|_\infty(\|n_t\|^2 + \|u_{tt}\|^2) \\ &\quad + 2\|\beta g''(|u|^2)\|_\infty\|u\|_\infty^2\|u_t\|_\infty^2(\|u\|^2 + \|u_{tt}\|^2) \\ &\quad + \|\beta g'(|u|^2)\|_\infty\|u\|_\infty^2\|u_{tt}\|^2 \\ &\quad + 3\|\beta g'(|u|^2)\|_\infty\|u_t\|_\infty^2(\|u\|^2 + \|u_{tt}\|^2) \\ &\le M\|u_{tt}\|^2 + M. \end{aligned}$$

By Gronwall inequality, we have

$$\sup_{0\le t\le T}\|u_{tt}\| \le M, \quad \forall T > 0. \tag{3.6.35}$$

□

Lemma 3.6.7. *Let the conditions of Lemma 3.6.6 be satisfied and assume*

(1) $g(s) \in C^{2k-2}, f(s) \in C^{2k-3}, 0 \le s < \infty, k \ge 2$;

(2) $h_1 \in H^{2k-2}, h_2 \in H^{2k-3}$;

(3) $u_0 \in H^{2k}, n_0 \in H^{2k-2}$.

Then we have

$$\sup_{0\le t\le T}(\|D_t^{k-1}D_xn\| + \|D_t^{k-1}D_xu\| + \|D_t^ku\| + \|D_t^kn\|) \le M_7,$$

where $M_7 = M_7(T, \|u_0\|_{H^{2k}}, \|n_0\|_{H^{2k-2}})$ *is a positive constant.*

Proof. We can prove the claim by induction. The proof is similar to that of Lemma 9 in [87]. □

From Galerkin method and the technique used in [87], together with the a priori estimate of the above lemma, we obtain

Theorem 3.6.1. *Assume*
(1) $g(s) \in C^{2k-2}[0,\infty)$, $\beta g(s) \le B_1 s^{2-\sigma} + C_1$, $s > 0$, $B_1 > 0$, $C_1 > 0$, $\sigma > 0$;
(2) $f(s) \in C^{2k-3}[0,\infty)$, $|f(s)| \le B_2 s^{\frac{3}{2}} + C_2$, $s > 0$, $B_2 > 0$, $C_2 > 0$;
(3) $h_1 \in H^{2k-2}$, $h_2 \in H^{2k-3}$;
(4) $u_0 \in H^{2k}$, $n_0 \in H^{2k-2}$, $k \ge 2$.

Then there exists a unique global smooth solution $(u(x,t), n(x,t))$ *of problem* (3.6.1)–(3.6.4),

$$u(x,t) \in L^\infty((0,T); H^{2k}(\Omega)), \quad D_t^j u \in L^\infty((0,T); H^{2k-2j}(\Omega)),$$
$$n(x,t) \in L^\infty((0,T); H^{2k-2}(\Omega)), \quad D_t^3 n \in L^\infty((0,T); H^{2k-2j}(\Omega)).$$

Similarly, by Lemmas 3.6.1–3.6.4, we get an important case of approximate inertial manifolds.

Theorem 3.6.2. *Suppose that the conditions of Lemma* 3.6.2 *are satisfied and* $u_0 \in H^2(\Omega)$, $n_0 \in H^1(\Omega)$. *Then there exists a global unique solution* $(u(x,t), n(x,t))$ *of problem* (3.6.1)–(3.6.4),

$$u(x,t) \in L^\infty((0,T); H^2(\Omega)), \quad n(x,t) \in L^\infty((0,T); H^1(\Omega)).$$

We note that, if f satisfies

$$|f(s)| \le B_2 s^{\frac{3}{2}}, \quad s > 0,\ B_2 > 0,$$

then all the a priori estimates are consistent, all are independent of $\Omega = (-D, D)$. Applying again the tools from [87], and in the problem (3.6.1)–(3.6.4) letting $D \to \infty$, we get

Theorem 3.6.3. *Let the conditions of Lemma* 3.6.1 *be satisfied. Then there exists a unique global smooth solution* $(u(x,t), n(x,t))$ *such that*

$$u(x,t) \in L^\infty((0,T); H^{2k}(\mathbf{R})), \quad n(x,t) \in L^\infty((0,T); H^{2k-2}(\mathbf{R})).$$

For simplicity, we consider the simplified LS equations as follows:

$$iu_t - u_{xx} - nu + i\alpha u + h_1(x) = 0, \tag{3.6.36}$$
$$n_t + |u|_x^2 + \delta n + \gamma f(|u|^2) + h_2(x) = 0, \tag{3.6.37}$$

with initial value

$$u|_{t=0} = u_0(x), \quad n|_{t=0} = n_0(x), \quad x \in \Omega = (-D, D),\ D > 0, \tag{3.6.38}$$

and periodic boundary condition

$$u(x-D,t) = u(x+D,t), \quad n(x-D,t) = n(x+D,t), \quad \forall x \in \mathbf{R},\ t \ge 0. \tag{3.6.39}$$

In order to construct an approximate inertial manifold of problem (3.6.36)–(3.6.39), we set

(H_1) $\alpha > 0, \sigma > 0$;
(H_2) $f(s) \in C^2[0,\infty)$, $|f(s)| \le B_2 s^{\frac{3}{2}-\nu} + C_1$, $B_1 > 0$, $C_1 > 0$, $\nu > 0$;
(H_3) $h_1, h_2 \in H^1(\Omega)$;
(H_4) $u_0 \in H^3(\Omega)$, $n_0 \in H^1(\Omega)$.

Note that if conditions (H_1)–(H_4) are satisfied, then by Theorem 3.6.2 we know that there exists a unique global solution $(u(x,t), n(x,t))$ of problem (3.6.36)–(3.6.39), and the following lemma can be proved:

Lemma 3.6.8. *Suppose that (H_1)–(H_2) are satisfied. Then we have*

$$\|n_x\| + \|n_{tx}\| \le C, \quad \forall t \ge t_1,$$

where C_1 depends only on the initial value; t_1 depends on the initial value and R, whenever $\|(u_0, n_0)\|_{H^2\times H^1} \le R$.

Lemma 3.6.9. *Under assumptions (H_1)–(H_2), we have*

$$\|u_{xx}\| + \|u_{xxx}\| \le C_2, \quad \forall t \ge t_2,$$

where C_2 depends only on the initial value; t_2 depends on the initial value and R, whenever $\|(u_0, n_0)\|_{H^3\times H^1} \le R$.

In order to construct an approximate inertial manifold of problem (3.6.36)–(3.6.39), we write problem (3.6.36)–(3.6.37) in the form of abstract differential equations:

$$i\frac{du}{dt} - Au - B(u,n) + i\alpha u + h_1 = 0, \tag{3.6.40}$$

$$\frac{dn}{dt} + H(u) + \delta n + \gamma f(|u|^2) + h_2 = 0, \tag{3.6.41}$$

where $B(u,n) = nu$ is a bilinear operator, $H^1 \times H \to H$, $H(u) = |u|_x^2$ is a nonlinear operator, $H^1 \to H$, $A = -\partial_{xx}$ is an unbounded self-adjoint operator, and

$$D(A) = \{u \in H^2 : u(x+2D) = u(x),\ n(x+2D) = n(x)\}.$$

Then there exists an orthogonal basis $\{\omega_j\}_{j=1}^{\infty}$ composed of eigenvectors of A such that

$$\begin{aligned} &A\omega_j = \lambda_j\omega_j \\ &0 \le \lambda_1 \le \lambda_2 \le \cdots \le \lambda_j \to +\infty, \quad j \to \infty. \end{aligned}$$

Also $\forall m$, let $P = P_m$ be the projection of H onto the subspace spanned by $\{\omega_1, \dots, \omega_m\}$. Set $Q = Q_m = I - P_m$.

Acting with P_m and Q_m on equations (3.6.40) and (3.6.41), respectively, we have

$$\begin{cases} i\dfrac{dy}{dt} - Ay - P_mB(u,n) + i\alpha y + P_mh_1 = 0, \\ i\dfrac{dz}{dt} - Az - Q_mB(u,n) + i\alpha z + Q_mh_1 = 0, \end{cases} \tag{3.6.42}$$

$$\begin{cases} \dfrac{dp}{dt} + P_mH(u) + \delta p + \gamma P_mf(|u|^2) + P_mh_2 = 0, \\ \dfrac{dq}{dt} + Q_mH(u) + \delta q + \gamma Q_mf(|u|^2) + Q_mh_2 = 0, \end{cases} \tag{3.6.43}$$

where

$$y = P_mu, \quad z = Q_mu, \quad p = P_mn, \quad q = Q_mn.$$

By Lemmas 3.6.8 and 3.6.9, we get

$$\begin{aligned} \|A^{\frac{1}{2}}z\|, \|A^{\frac{1}{2}}z_t\|, \|A^{\frac{1}{2}}q\|, \|A^{\frac{1}{2}}q_t\| &\le C, \quad \forall t \ge t_*, \\ \|z\|, \|z_t\|, \|q\|, \|q_t\| &\le C\lambda_{m+1}^{-\frac{1}{2}}, \quad \forall t \ge t_* \end{aligned}$$

Define the mapping $\Phi : P_mH \times P_mH \to Q_mH \times Q_mH$ such that for any $(y,p) \in P_mH \times P_mH$, $\Phi(y,p) = (\Psi_1, \Psi_2)$, which satisfies

$$-A\Psi_1 - Q_mB(y,p) + Q_mh_1 = 0, \tag{3.6.44}$$

$$\delta\Psi_2 + Q_mH(y + \Psi_1) + \gamma Q_mf(|y|^2) + Q_mh_2 = 0. \tag{3.6.45}$$

Let $\Sigma_1 = \operatorname{graph}(\Phi)$. We will prove that it is an approximate inertial manifold of problem (3.6.36)–(3.6.39).

Theorem 3.6.4. *Suppose that (H_1)–(H_4) are satisfied. Then there exists a constant K, which depends on the initial value, such that*

$$\operatorname{dist}_{H^2 \times H}((u(t), n(t)), \Sigma_1) \le K\lambda_{m+1}^{-\frac{1}{2}}, \quad \forall t \ge t_*,$$

where $(u(t), n(t))$ is a global solution of problem (3.6.36)–(3.6.39); t_* only depends on the initial value and R, whenever $\|(u_0, n_0)\|_{H^3 \times H^1} \le R$.

Proof. Taking the difference of equation (3.6.44) and (3.6.42), it follows that

$$
\begin{aligned}
A\Psi_1 - Az &= Q_m B(u,n) - Q_m B(y,p) - i\alpha z - i\frac{dz}{dt} \\
&= Q_m B(u-y,n) + Q_m B(y,n-p) - i\alpha z - i\frac{dz}{dt} \\
&= Q_m B(z,n) + Q_m B(y,q) - i\alpha z - i\frac{dz}{dt}, \\
\|A\Psi_1 - Az\| &\le \|B(z,n)\| + \|B(z,q)\| + \alpha\|z\| + \left\|\frac{dz}{dt}\right\| \\
&\le \|n\|_\infty \|z\| + \|y\|_\infty \|q\| + \alpha\|z\| + \left\|\frac{dz}{dt}\right\| \\
&\le C\lambda_{m+1}^{-\frac{1}{2}}.
\end{aligned}
\tag{3.6.46}
$$

Taking away (3.6.45) from (3.6.43), we arrive at

$$
\begin{aligned}
\delta\Psi_2 - \delta q = {} & Q_m H(n) - Q_m H(y+\Psi_1) \\
& + \gamma Q_m f(|u|^2) - \gamma Q_m f(|y|^2) + \frac{dq}{dt}.
\end{aligned}
\tag{3.6.47}
$$

Since

$$
\begin{aligned}
H(u) - H(u-\Psi_1) &= |u|_x^2 - |y+\Psi|_x^2 \\
&= \left(u_x - y_x - \frac{d}{dt}\Psi_1\right)\overline{u} + \left(y_x + \frac{d}{dt}\Psi_1\right)(\overline{u} - \overline{y} - \overline{\Psi}_1) \\
&\quad + u\left(\overline{u}_x - \overline{y}_x - \frac{d}{dt}\overline{\Psi}_1\right) + (u - y - \Psi_1)\left(\overline{y}_x + \frac{d}{dt}\overline{\Psi}_1\right),
\end{aligned}
$$

we have

$$
\begin{aligned}
\|H(u) - H(u+\Psi_1)\| \le {} & 2\|u\|_\infty \|A^{\frac{1}{2}}\Psi_1 - A^{\frac{1}{2}}z\| \\
& + 2\|\Psi_1 - z\|_\infty \|A^{\frac{1}{2}}y + A^{\frac{1}{2}}\Psi_1\|.
\end{aligned}
\tag{3.6.48}
$$

By equation (3.6.44) we have

$$
\begin{aligned}
\|A\Psi_1\| &\le \|B(y,p)\| + \|h_1\| \le \|y\|_\infty \|p\| + \|h_1\| \\
&\le C\|y\|_{H^1}\|n\| + \|h_1\| \le C,
\end{aligned}
$$

and then

$$
\|A^{\frac{1}{2}}\Psi_1\| \le C. \tag{3.6.49}
$$

From equations (3.6.48)–(3.6.49), we obtain

$$
\begin{aligned}
\|H(u) - H(y+\Psi_1)\| &\le C\|A^{\frac{1}{2}}\Psi_1 - A^{\frac{1}{2}}z\| + C(\|A^{\frac{1}{2}}y\| + \|A^{\frac{1}{2}}\Psi_1\|)\|\Psi_1 - z\|_{H^1} \\
&\le C\|A^{\frac{1}{2}}\Psi - A^{\frac{1}{2}}z\| \le C\lambda_{m+1}^{\frac{1}{2}},
\end{aligned}
\tag{3.6.50}
$$

$$\begin{aligned}\|f(|u|^2)-f(|y|^2)\| &= C\|f'(\xi)(|u|^2-|y|^2)\|\\ &\le C\|2(u+y)\|\|z\| \le C(\|u\|_\infty+\|y\|_\infty)\|z\|\\ &\le C\|z\| \le C\lambda_{m+1}^{-1}\|Az\| \le C\lambda_{m+1}^{-1}.\end{aligned} \tag{3.6.51}$$

From equation (3.6.47), inequality (3.6.50) and equation (3.6.50) we have

$$\|\Psi_2 - q\| \le C\lambda_{m+1}^{-1} + \left\|\frac{\mathrm{d}q}{\mathrm{d}t}\right\| \le C\lambda_{m+1}^{-\frac{1}{2}}. \tag{3.6.52}$$

Therefore

$$\begin{aligned}&\operatorname{dist}_{H^2\times H}((u(t),n(t)),\Sigma_1)\\ &\le \|u(t)-(y(t)+\Psi_1)\|_{H^2} + \|n(t)-(p(t)+\Psi_2)\|\\ &\le \|A\Psi_1 - Az\| + \|\Psi_2 - q\| \le C\lambda_{m+1}^{\frac{1}{2}}.\end{aligned}$$

Thus Theorem 3.6.4 has been proved. □

3.7 One-dimensional ferromagnetic chain equation

Consider the following one-dimensional ferromagnetic chain equation:

$$\partial_t u = -\alpha u \times (u \times u_{xx}) + \beta(u \times u_{xx}), \quad (x,t) \in \Omega \times \mathbf{R}_+ \tag{3.7.1}$$

with initial value

$$u(x,0) = u_0(x), \quad |u_0(x)| = 1, \quad x \in \Omega = (-D,D),\ D > 0, \tag{3.7.2}$$

and periodic boundary condition

$$u(x-D,t) = u(x+D,t), \quad \forall x \in \mathbf{R},\ t \ge 0, \tag{3.7.3}$$

where $\times$ denotes the vector cross-product in $\mathbf{R}^3$; $\mathbf{u} = (u^1,u^2,u^3) : \mathbf{R}\times\mathbf{R}_+ \to \mathbf{R}^3$ is a rotating vector field; α, β are constants; $\alpha > 0$ is Gilbert damping constant. In 1995, Guo and Wang [115] studied the existence of an approximate inertial manifold of problem (3.7.1)–(3.7.3). From [217], we know that if $u_0 \in H^2(\Omega)$, $|u_0(x)| = 1$, then problem (3.7.1)–(3.7.3) possesses a unique global smooth solution $u(x,t)$ such that

$$u(x,t) \in L^\infty(\mathbf{R}^+, H^2(\Omega)).$$

In order to construct an approximate inertial manifold, we need to establish a uniform in t estimate.

Lemma 3.7.1. *Let* $|u_0(x)| = 1,\ x \in \mathbf{R}$. *Then for any smooth solution of problem* (3.7.1)–(3.7.3), *we have*

$$|u(x,t)|^2 = 1, \quad (x,t) \in \mathbf{R}\times\mathbf{R}^+, \tag{3.7.4}$$

$$\|u_x(t)\|^2 \le \|u_x(0)\|^2, \quad t \in \mathbf{R}^+. \tag{3.7.5}$$

Proof. Taking the dot product of equation (3.7.1) with u, we get

$$\frac{\partial}{\partial t}|u(x,t)|^2 = 0, \quad (x,t) \in \mathbf{R}\times\mathbf{R}^+,$$

and thus equation (3.7.4) is established. Taking the inner product of equation (3.7.1) and u_{xx}, we get

$$\begin{aligned} -\frac{1}{2}\frac{\mathrm{d}}{\mathrm{d}t}\|u_x\|^2 &= -\alpha\int_\Omega (u\times(u\times u_{xx}))\cdot u_{xx}\mathrm{d}x \\ &= -\alpha\int_\Omega (u_{xx}\times u)\cdot(u\times u_{xx})\mathrm{d}x = \alpha\int_\Omega |u\times u_{xx}|^2\mathrm{d}x. \end{aligned}$$

Hence

$$\frac{\mathrm{d}}{\mathrm{d}t}\|u_x\|^2 \le 0,$$

which yields equation (3.7.5). □

Introduce the subset H_ρ of H as

$$H_\rho = \{u \in H : |u(x)| = 1,\ \|u_x\| \le \rho\}.$$

Lemma 3.7.1 claims that H_ρ is a subspace of H. Hereafter we assume $u_0 \in H_\rho$.

Lemma 3.7.2. *Suppose that the conditions of Lemma* 3.7.1 *are satisfied. Then* u *is a smooth solution of problem* (3.7.1)–(3.7.3) *if and only if* u *is a smooth solution of the following problem:*

$$\frac{\partial u}{\partial t} = \alpha u_{xx} + \beta u\times u_{xx} + \alpha|u_x|^2 u, \tag{3.7.6}$$

$$u(x,0) = u_0(x), \quad |u_0(x)| = 1, \quad x \in \Omega, \tag{3.7.7}$$

$$u(x-D,0) = u_0(x-D,t), \quad x \in \mathbf{R},\ t \in \mathbf{R}^+. \tag{3.7.8}$$

Proof. See [217, Lemma 2]. □

Based on Lemma 3.7.2, hereafter we merely need to study the problem (3.7.6)–(3.7.8).

Lemma 3.7.3. *Let* $u_0 \in H^2 \cap H_\rho$. *Then for any solution of problem* (3.7.6)–(3.7.8) *we have*

$$\|u_{xx}(t)\| \le C_1, \quad \int_t^{t+1} \|u_{xxx}\|^2\mathrm{d}t \le C_1, \quad \forall t \ge t_1,$$

where C_1 *is a constant which depends on parameters* $(\alpha,\beta,\rho,\Omega)$; t_1 *depends on parameters* $(\alpha,\beta,\rho,\Omega)$ *and* R, *whenever* $\|u_0\|_{H^2} \le R$.

Proof. For simplicity, hereafter C denotes as arbitrary constant that merely depends on parameters $(\alpha, \beta, \rho, \Omega)$.

From $|u(x,t)| = 1$, we know that if $|u_x| \neq 0$, then u, u_x and $u \times u_x$ form an orthogonal basis of $\mathbf{R}^3$. Let $u_{xx} = \alpha_1 u + \alpha_2 u_x + \alpha_3 u \times u_x$. By a simple calculation, we get

$$\alpha_1 = -|u_x|^2, \quad \alpha_2 = \frac{u_x \cdot u_{xx}}{|u_x|^2}, \quad \alpha_3 = \frac{(u \times u_x) \cdot u_{xx}}{|u_x|^2}.$$

Differentiating equation (3.7.6) with respect to x twice, and taking the inner product with u_{xx}, we get

$$\begin{aligned}\frac{1}{2}\frac{\mathrm{d}}{\mathrm{d}t}\|u_{xx}\|^2 &= \int_\Omega u_{xx} \cdot (\alpha u_{xx} + \beta u \times u_{xx} + \alpha |u_x|^2 u)_{xx} \mathrm{d}x \\ &= -\alpha \|u_{xxx}\|^2 - \beta \int_\Omega (u_x \times u_{xx}) \cdot u_{xxx} \mathrm{d}x \\ &\quad - \alpha \int_\Omega u_{xxx} \cdot (|u_x|^2 u_x + 2u(u_x \cdot u_{xx}))\mathrm{d}x. \end{aligned} \tag{3.7.9}$$

Differentiating equation (3.7.4) with respect to x thrice, we obtain

$$\mathbf{u} \cdot \mathbf{u}_{xxx} = -\frac{3}{2}(|u_x|^2)_x. \tag{3.7.10}$$

Using equation (3.7.10), all terms on the right-hand side of estimate (3.7.9) can be written as follows:

$$\int_\Omega u_{xxx} \cdot (|u_x|^2 u_x)\mathrm{d}x = -\int_\Omega |u_x|^2 |u_{xx}|^2 \mathrm{d}x - 2\int_\Omega |u_x \cdot u_{xx}|^2 \mathrm{d}x, \tag{3.7.11}$$

$$\begin{aligned}\int_\Omega u_{xxx} \cdot (u(u_x \cdot u_{xx}))\mathrm{d}x &= \int_\Omega (u_{xxx} \cdot u)(u_x \cdot u_{xx})\mathrm{d}x \\ &= -3\int_\Omega |u_x \cdot u_{xx}|^2 \mathrm{d}x, \end{aligned} \tag{3.7.12}$$

$$\begin{aligned}\int_\Omega (u_x \times u_{xx}) \cdot u_{xxx} \mathrm{d}x &= \int_\Omega \left[u_x \times \left(-|u_x|^2 u + \frac{(u \times u_x) \cdot u_{xx}}{|u_x|^2}(u \times u_x)\right)\right] \cdot u_{xxx} \mathrm{d}x \\ &= \int_\Omega |u_x|^2 (u \times u_x) \cdot u_{xxx} \mathrm{d}x \\ &\quad + \int_\Omega \frac{(u \times u_x) \cdot u_{xx}}{|u_x|^2}(u_x \times (u \times u_x)) \cdot u_{xxx} \mathrm{d}x \\ &= \int_\Omega |u_x|^2 (u \times u_x) \cdot u_{xxx} \mathrm{d}x - \frac{3}{2}\int_\Omega (|u_x|^2)_x (u \times u_x) \cdot u_{xx} \mathrm{d}x \\ &= \int_\Omega |u_x|^2 (u \times u_x) \cdot u_{xxx} \mathrm{d}x + \frac{3}{2}\int_\Omega |u_x|^2 (u \times u_x) \cdot u_{xx} \mathrm{d}x \end{aligned}$$

$$+\frac{3}{2}\int_\Omega |u_x|^2(u\times u_x)\cdot u_{xxx}\mathrm{d}x$$
$$=\frac{5}{2}\int_\Omega |u_x|^2(u\times u_x)\cdot u_{xxx}\mathrm{d}x. \tag{3.7.13}$$

By equations (3.7.9), (3.7.11)–(3.7.13), we obtain

$$\begin{aligned}\frac{1}{2}\frac{\mathrm{d}}{\mathrm{d}t}\|u_{xx}\|^2+\alpha\|u_{xxx}\|^2&=\alpha\int_\Omega |u_x|^2|u_{xx}|^2\mathrm{d}x\\&\quad+8\alpha\int_\Omega |u_x\cdot u_{xx}|^2\mathrm{d}x-\frac{5}{2}\beta\int_\Omega |u_x|^2(u\times u_x)\times u_{xxx}\mathrm{d}x.\end{aligned} \tag{3.7.14}$$

In addition,

$$\begin{aligned}\frac{1}{4}\frac{\mathrm{d}}{\mathrm{d}t}\int_\Omega |u_x|^4\mathrm{d}x&=\int_\Omega |u_x|^2u_x\cdot(\alpha u_{xx}+\beta u\times u_{xx}+\alpha|u_x|^2u)_x\mathrm{d}x\\&=\int_\Omega |u_x|^2u_x\cdot(\alpha u_{xxx}+\beta u_x\times u_{xx}+\beta u\times u_{xxx}\\&\quad+\alpha|u_x|^2u_x+2\alpha u(u_x\cdot u_{xx}))\mathrm{d}x\\&=-\alpha\int_\Omega |u_x|^2|u_{xx}|^2\mathrm{d}x-2\alpha\int_\Omega |u_x\cdot u_{xx}|^2\mathrm{d}x\\&\quad+\alpha\int_\Omega |u_x|^6\mathrm{d}x-\beta\int_\Omega |u_x|^2(u_x\times u_x)\cdot u_{xxx}\mathrm{d}x,\end{aligned}$$

that is, we have

$$\begin{aligned}&-\beta\int_\Omega |u_x|^2(u\times u_x)\cdot u_{xxx}\mathrm{d}x\\&=\frac{1}{4}\frac{\mathrm{d}}{\mathrm{d}t}\int_\Omega |u_x|^4\mathrm{d}x+\alpha\int_\Omega |u_x|^2|u_{xx}|^2\mathrm{d}x+2\alpha\int_\Omega |u_x\cdot u_{xx}|^2\mathrm{d}x-\alpha\int_\Omega |u_x|^6\mathrm{d}x.\end{aligned} \tag{3.7.15}$$

Substituting equation (3.7.15) into equation (3.7.14), we get

$$\begin{aligned}&\frac{\mathrm{d}}{\mathrm{d}t}\|u_{xx}\|^2-\frac{5}{4}\frac{\mathrm{d}}{\mathrm{d}t}\int_\Omega |u_x|^4\mathrm{d}x+2\alpha\|u_{xx}\|^2+5\alpha\int_\Omega |u_x|^6\mathrm{d}x\\&=7\alpha\int_\Omega |u_x|^2|u_{xx}|^2\mathrm{d}x+26\alpha\int_\Omega |u_x\cdot u_{xx}|^2\mathrm{d}x.\end{aligned} \tag{3.7.16}$$

By virtue of

$$\begin{aligned}&7\alpha\int_\Omega |u_x|^2|u_{xx}|^2\mathrm{d}x+26\alpha\int_\Omega |u_x\cdot u_{xx}|^2\mathrm{d}x\\&\le 33\alpha\int_\Omega |u_x|^2|u_{xx}|^2\mathrm{d}x\end{aligned}$$

$$\le 5\alpha \int_\Omega |u_x|^6 \mathrm{d}x + C \int_\Omega |u_{xx}|^3 \mathrm{d}x \le 5\alpha \int_\Omega |u_x|^6 \mathrm{d}x + C\|u_x\|^{\frac{5}{4}} \|u_{xxx}\|^{\frac{7}{4}}$$

$$\le 5\alpha \int_\Omega |u_x|^6 \mathrm{d}x + \frac{\alpha}{2}\|u_{xxx}\|^2 + C,$$

and equation (3.7.16), we deduce

$$\frac{\mathrm{d}}{\mathrm{d}t}\|u_{xx}\|^2 - \frac{5}{4}\frac{\mathrm{d}}{\mathrm{d}t}\int_\Omega |u_x|^4 \mathrm{d}x + \frac{3}{2}\alpha\|u_{xxx}\|^2 \le C. \tag{3.7.17}$$

Since u_x is periodic, $\int_\Omega u_{xx}(x)\mathrm{d}x = 0$. Using Poincaré inequality, we get

$$\frac{\alpha}{2}\|u_{xxx}\|^2 \ge K\|u_{xx}\|^2. \tag{3.7.18}$$

Hence from equation (3.7.17) we get

$$\frac{\mathrm{d}}{\mathrm{d}t}\left(\|u_{xx}\|^2 - \frac{5}{4}\int_\Omega |u_x|^4 \mathrm{d}x\right) + K\left(\|u_{xx}\|^2 - \frac{5}{4}\int_\Omega |u_x|^4 \mathrm{d}x\right) + \alpha\|u_{xxx}\|^2 \le C, \tag{3.7.19}$$

that is,

$$\frac{\mathrm{d}}{\mathrm{d}t}\left(\|u_{xx}\|^2 - \frac{5}{4}\int_\Omega |u_x|^4 \mathrm{d}x\right) + K\left(\|u_{xx}\|^2 - \frac{5}{4}\int_\Omega |u_x|^4 \mathrm{d}x\right) \le C.$$

By Gronwall lemma, we get

$$\begin{aligned} &\left\|u_{xx}(t)\right\|^2 - \frac{5}{4}\int_\Omega |u_x|^4 \mathrm{d}x \\ &\quad\le \left(\left\|u_{xx}(0)\right\|^2 - \frac{5}{4}\int_\Omega |u_x(0)|^4 \mathrm{d}x\right)e^{-Kt} + \frac{C}{K} \\ &\quad\le \left\|u_{xx}(0)\right\|^2 e^{-Kt} + C \le R^2 e^{-Kt} + C, \quad t \ge 0 \le 2C,\ t \ge t_*, \end{aligned} \tag{3.7.20}$$

where $t_* = \frac{1}{K}\ln\frac{R^2}{C}$. Based on the interpolation inequality

$$\begin{aligned} \|u_x\|_4 &\le C|u_x|^{\frac{3}{4}}\|u_{xx}\|^{\frac{1}{4}}\frac{5}{4}\|u_x\|_4^2 \\ &\le C\|u_x\|^3\|u_{xx}\| \le C\|u_{xx}\| \le \frac{1}{2}\|u_{xx}\|^2 + C, \end{aligned} \tag{3.7.21}$$

from equations (3.7.20)–(3.7.21) we get

$$\left\|u_{xx}(t)\right\| \le C, \quad \forall t \ge t_*. \tag{3.7.22}$$

Integrating equations (3.7.19)–(3.7.22) with respect to t over $(t, t+1)$, we get

$$\int_t^{t+1} \|u_{xxx}\|^2 \mathrm{d}t \le C, \quad \forall t \ge t_*.$$

Thus Lemma 3.7.3 is proved. □

Lemma 3.7.4. *Let $u_0 \in H^{k+1} \cap H_p$, $k \ge 1$. Then we have*

$$\|D_x^{k+1} u(t)\|^2 \le C_k, \quad t \ge t_k,$$

$$\int_t^{t+1} \|D_x^{k+2} u(t)\|^2 \mathrm{d}t \le C_k, \quad t \ge t_k,$$

where constant C_k only depends on parameter k; t_k depend on parameter k and R; whenever $\|u_0\|_{H^2} \le R$.

Proof. We use induction to prove the lemma:

(1) When $k = 1$, the claim of Lemma 3.7.4 boils down to that of Lemma 3.7.3, which has been proved.

(2) Suppose the lemma is true up to $k-1$. Now we prove it is true also for k, $k \ge 2$. Differentiating $(k+1)$ times formula (3.7.6) with respect to x, and taking the inner product with D_x^{k+1}, we get

$$\begin{aligned} \frac{1}{2}\frac{\mathrm{d}}{\mathrm{d}t}\|D_x^{k+1}u\|^2 + \alpha\|D_x^{k+2}u\|^2 &= \beta \sum_{i=0}^{k+1} C_{k+1}^i \int_\Omega (D_x^{k+1-i}u \times D_x^i u_{xx}) \cdot D_x^{k+1} u \mathrm{d}x \\ &\quad + \alpha \sum_{i=0}^{k+1} C_{k+1}^i \int_\Omega D_x^i |u_0|^2 (D_x^{k+1-i} u \cdot D_x^{k+1} u) \mathrm{d}x. \end{aligned} \tag{3.7.23}$$

Through simple calculation we get

$$\begin{aligned} &\beta \sum_{i=0}^{k+1} C_{k+1}^i \int_\Omega (D_x^{k+1-i}u \times D_x^i u_{xx}) \cdot D_x^{k+1} u \mathrm{d}x \\ &\le C\|D_x^{k+1}u\|\|D_x^{k+2}u\| + C\|D_x^{k+1}u\|^2 \\ &\le \frac{\alpha}{4}\|D_x^{k+2}u\|^2 + C\|D_x^{k+1}u\|^2, \end{aligned} \tag{3.7.24}$$

$$\begin{aligned} &\alpha \sum_{i=0}^{k+1} C_{k+1}^i \int_\Omega D_x^i |u_0|^2 (D_x^{k+1-i} u \cdot D_x^{k+1} u) \mathrm{d}x \\ &\le \|D_x^{k+1}u\|^2 + C\|D_x^{k+1}u\|^2\|D_x^k u\|^2 + C\|D_x^{k+1}u\|\|D_x^{k+2}u\| \\ &\le \frac{\alpha}{4}\|D_x^{k+2}u\|^2 + C\|D_x^{k+1}u\|^2. \end{aligned} \tag{3.7.25}$$

From equations (3.7.23)–(3.7.25), we get

$$\frac{\mathrm{d}}{dt}\|D_x^{k+1}u\|^2+\alpha\|D_x^{k+2}u\|^2\leq C\|D_x^{k+1}u\|^2\leq\frac{\alpha}{2}\|D_x^{k+2}u\|^2+C,\tag{3.7.26}$$

which yields

$$\frac{\mathrm{d}}{\mathrm{d}t}\|D_x^{k+1}u\|^2+\frac{\alpha}{2K}\|D_x^{k+1}u\|^2\leq C,\tag{3.7.27}$$

where

$$\|D_x^{k+1}u\|\leq K\|D_x^{k+2}u\|\quad\left(\int D_x^{k+1}u\mathrm{d}x=0\right).$$

Then we get

$$\|D_x^{k+1}u\|^2\leq C,\quad t\geq t_*,\tag{3.7.28}$$

$$\int_t^{t+1}\|D_x^{k+1}u\|\mathrm{d}t\leq C,\quad t\geq t_*.\tag{3.7.29}$$

Thus Lemma 3.7.4 has been proved. □

Lemma 3.7.5. *Let $u_0\in H^{k+2}\cap H_\rho, k\geq 0$. Then there exists a constant C_k, depending only on parameter k, such that*

$$\|D_x^k u_t\|\leq C_k,\quad t\geq t_k,$$

where t_k depends on parameter k and R, whenever $\|u_0\|_{H^2}\leq R$.

Proof. Differentiating k times the equation with respect to x and applying Lemma 3.7.4, we get the claim. □

Now to construct an approximate inertial manifold of the ferromagnetic chain equation for the initial value problem (3.7.6)–(3.7.8), write (3.7.6)–(3.7.8) in an abstract form:

$$\frac{du}{\mathrm{d}t}+\alpha Au+B(u,u)+R(u)=0,\tag{3.7.30}$$

where $A=-\partial_{xx}$ is an unbounded self-adjoint operator, $D(A)=\{u\in H^2:\mathbf{u}\text{ satisfies equation (3.7.8)}\}$, $B(u,v)=-\beta u\times v_{xx}$ is a bilinear operator, $R(u)=-\alpha|u_x|^2$ is a nonlinear operator, $D(A)\to H$. Let $\{\omega_j\}_{j=1}^{\infty}$ be the orthogonal basis of H, which is composed of eigenvectors of A:

$$A\omega_j=\lambda_j\omega_j,\quad 0=\lambda_1<\lambda_2\leq\cdots\leq\lambda_j\to+\infty,\quad j\to\infty.$$

For a given m, let $P=P_m:H\to\operatorname{span}\{\omega_1,\omega_2,\dots,\omega_m\}$ be the projection, $Q=Q_m=I-P_m$. Then acting with P_m and Q_m on equation (3.7.30), we get

$$\begin{cases} \dfrac{dp}{dt} + \alpha Ap + P_m B(p+q, p+q) + P_m R(p+q) = 0, \\ \dfrac{dq}{dt} + \alpha Aq + Q_m B(p+q, p+q) + Q_m R(p+q) = 0, \end{cases} \tag{3.7.31}$$

where $p = P_m u$, $q = Q_m u$.

Using Lemmas 3.7.1, 3.7.4 and 3.7.5, we have

$$\|u\|, \|A^{\frac{3}{2}} u\| \le C, \quad t \ge t_*, \tag{3.7.32}$$

$$\|A^k u\|, \|A^{k+1} u\|, \|A^{k+\frac{3}{2}} u\|, \left\|A^k \frac{\partial u}{\partial t} u\right\| \le C, \quad t \ge t_*, \tag{3.7.33}$$

where constants C_k, t_k are defined as before. Note that

$$\begin{aligned} &\|A^{\frac{1}{2}} u\| = \left\|\frac{\partial u}{\partial x}\right\|, \quad u \in H^1, \\ &\|P_m u\| \le \|u\|, \quad \|Q_m u\| \le \|u\|, \quad u \in H, \\ &\|A^\alpha u\| \ge \lambda_{m+1}^\alpha \|u\|, \quad \alpha > 0,\ u \in Q_m D(A^\alpha). \end{aligned} \tag{3.7.34}$$

By equations (3.7.33)–(3.7.34) we deduce that

$$\|q\|, \|Aq\|, \|A^{\frac{3}{2}} q\|, \left\|\frac{d}{dt} q\right\| \le C_k \lambda_{m+1}^{-k}, \quad t \ge t_k. \tag{3.7.35}$$

Define the mapping $\Phi : P_m H \to Q_m H$ such that $p \in P_m H$, $\Phi(p) = \Psi$ is determined by

$$\alpha A\Psi + Q_m B(p,p) + Q_m R(p) = 0. \tag{3.7.36}$$

Letting $\Sigma = \text{graph}(\Phi)$, we can prove that Σ is an approximate inertial manifold of problem (3.7.6)–(3.7.8).

Theorem 3.7.1. *Let $u_0 \in H^{2k+2} \cap H_p$. Then for any positive integer k, there exists a constant C_k, depending on parameter k, such that*

$$\text{dist}_{H^2}(u(t), \Sigma) \le C_k \lambda_{m+1}^{-k}, \quad t \ge t_k,$$

where $u(t)$ is a solution of problem (3.7.6)–(3.7.8); t_k depends on parameter k and R; whenever $\|u_0\|_{H^2} \le R$.

Proof. Taking the difference of equation (3.7.36) and (3.7.31), it follows that

$$\begin{aligned} \alpha A\Psi - \alpha Aq &= \frac{dq}{dt} + Q_m B(p+q, p+q) - Q_m B(p,q) + Q_m R(p+q) - Q_m R(p) \\ &= \frac{dq}{dt} + Q_m B(p+q, p+q) \\ &\quad + Q_m B(p,q) + Q_m R(p+q) - Q_m R(p). \end{aligned} \tag{3.7.37}$$

Since

$$\begin{aligned}\|B(q,p+q)-B(p,q)\| &\le \beta\|q\times Au\| + \beta\|p\times Aq\| \\ &\le \beta\|Au\|_\infty\|q\| + \beta\|p\|\|Aq\|_\infty \\ &\le \beta\|A^{\frac{3}{2}}u\|\|q\| + \beta\|p\|\|A^{\frac{3}{2}}q\| \le C_k\lambda_{m+1}^{-k}, \end{aligned} \tag{3.7.38}$$

$$\begin{aligned}\|R(p+q)-R(p)\| &= \alpha\||u_x|^2u - |p_x|^2p\| \\ &\le \alpha\|(|u_x|^2-|p_x|^2)u\| + \alpha\||p_x|^2(u-p)\| \\ &\le \alpha\|(|u_x|+|p_x|)(|u_x|-|p_x|)u\| + \alpha\||p_x|^2q\| \\ &\le \alpha(\|u_x\|_\infty + \|p_x\|_\infty)\|q_x\|_\infty\|u\| + \alpha\|p_x\|_\infty^2\|q\| \\ &\le 2\alpha\|Au\|\|Aq\|\|u\| + \alpha\|Au\|\|q\| \le C_k\lambda_{m+1}^{-k}, \end{aligned} \tag{3.7.39}$$

by virtue of equations (3.7.37)–(3.7.39), we deduce

$$\|A\Psi - Aq\| \le C_k\lambda_{m+1}^{-k}, \quad t \ge t_k.$$

Therefore

$$\begin{aligned}\operatorname{dist}_{H^2}(u(t),\Sigma) &\le \|u(t)-(p(t)+\Phi(p(t)))\|_{H^2} \\ &\le \|\Phi(p(t))-q(t)\|_{H^2} \\ &\le \|A\Psi - Aq\| \le C_k\lambda_{m+1}^{-k}, \quad t\ge t_k.\end{aligned}$$

Finally, Theorem 3.7.1 has been proved. □

3.8 Nonlinear Schrödinger equation

In 1988, Ghidaglia [74] studied the existence of a global attractor and its dimension estimate for a class of nonlinear Schrödinger equations with damping. In 1996, Guo and Wang constructed two kinds of approximate inertial manifold [119] for such equations.

Consider the following nonlinear Schrödinger equation [20, 21, 126, 195]:

$$i\frac{\partial u}{\partial t} + \frac{\partial^2 u}{\partial x^2} + g(|u|^2)u + i\alpha u + h = 0, \tag{3.8.1}$$

with initial value condition

$$u(x,0) = u_0(x), \quad x\in\Omega = (-D,D) \tag{3.8.2}$$

and one of the following boundary conditions:

$$\begin{aligned} &\text{(Dirichlet)} && u(-D,t) = u(D,t) = 0, \quad t\in\mathbf{R}; \\ &\text{(Neumann)} && \frac{\partial u}{\partial x}(-D,t) = \frac{\partial u}{\partial x}(D,t) = 0, \quad t\in\mathbf{R}; \\ &\text{(periodic boundary)} && u(x+2D,t) = u(x,t), \quad x\in\mathbf{R},\ t\in\mathbf{R}_+, \end{aligned} \tag{3.8.3}$$

where u is an unknown complex-valued function; $\alpha > 0$ is constant. Also $h(x) \in L^2(\Omega)$, while $g(s)$, $0 \le s < \infty$, is a real-valued smooth function which satisfies

$$\lim_{s\to+\infty} \frac{G_+(s)}{s^3} = 0, \tag{3.8.4}$$

and $\exists \omega > 0$ such that

$$\lim_{s\to+\infty} \frac{f(s) - \omega G_(s)}{s^3} \le 0, \tag{3.8.5}$$

where $f(s) = sg(s)$, $G(s) = \int_0^s g(\tau)d\tau$, $G_+(s) = \max\{G(s), 0\}$. In [74] we have proved that under condition (3.8.4)–(3.8.5), for any $u_0 \in H^1(\Omega)$, problem (3.8.1)–(3.8.3) has a unique global solution $u(x, t)$ such that

$$u(x,t) \in L^\infty(\mathbf{R}^+; H^1(\Omega)),$$

and if $u_0(x) \in H^2(\Omega)$ then $u(x, t)$ satisfies

$$\|u\|, \|u_x\|, \|u_{xx}\|, \|u_t\|, \|u\|_\infty, \|u_x\|_\infty \le K, \quad 0 \le t \le T, \tag{3.8.6}$$

and

$$\|u\|, \|u_x\|, \|u_{xx}\|, \|u_t\|, \|u\|_\infty, \|u_x\|_\infty \le C, \quad t \ge t_*, \tag{3.8.7}$$

where K is a constant, which depends on parameters (α, Ω, g) and T, R; $\|u_0\|_{H^2} \le R$; t_* depends on parameters (α, Ω, g) and R; here and hereafter C denotes arbitrary constant only depending on parameters (α, Ω, g).

In order to construct an approximate inertial manifold, we need to provide a higher order a priori estimate.

Lemma 3.8.1. *Suppose the conditions* (3.8.4)–(3.8.5) *are satisfied and* $u_0 \in H^3(\Omega)$, $h(x) \in H^1(\Omega)$. *Then for any solution* $u(x, t)$ *of problem* (3.8.1)–(3.8.3) *we have*

$$\|u_{xxx}(\cdot, t)\| \le C, \quad t \ge t_*,$$

where the constant C only depends on parameters (α, Ω, g), t_* depends on parameters (α, Ω, g) and R, whenever $\|u_0\|_{H^2} \le R$.

Proof. Differentiating equation (3.8.1) with respect to x, we get

$$iu_{tx} + u_{xxx} + i\alpha u_x + g'(|u|^2)|u|_x^2 u + g(|u|^2)u_x + h'(x) = 0. \tag{3.8.8}$$

Taking the inner product of equation (3.8.8) and $u_{xxxt} + \alpha u_{xxx}$, we get

$$(iu_{tx} + u_{xxx} + i\alpha u_x + g'(|u|^2)|u|_x^2 u + g(|u|^2)u_x + h'(x), u_{xxxt} + \alpha u_{xxx}) = 0.$$

Taking the real part of the above equation, we arrive at

$$\begin{aligned}&\frac{1}{2}\frac{\mathrm{d}}{\mathrm{d}t}\|u_{xxx}\|^2+\alpha\|u_{xxx}\|^2+\mathrm{Re}(h',u_{xxxt}+\alpha u_{xxx})\\&\quad+\mathrm{Re}(g'(|u|^2)|u|_x^2u+g(|u|^2)u_x,u_{xxxt}+\alpha u_{xxx})=0.\end{aligned}\tag{3.8.9}$$

Since

$$\begin{aligned}&|(g'(|u|^2)|u|_x^2u+g(|u|^2)u_x,\alpha u_{xxx})|\\&\quad\le C\|g'(|u|^2)\|_\infty\|(u_x\overline{u}+u\overline{u}_x)\|_\infty\|u_{xxx}\|+C\|g(|u|^2)\|_\infty\|u_x\|_\infty\|u_{xxx}\|\\&\quad\le C\|u_{xxx}\|,\end{aligned}\tag{3.8.10}$$

$$\mathrm{Re}(h',u_{xxxt}+\alpha u_{xxx})=\frac{\mathrm{d}}{\mathrm{d}t}\mathrm{Re}\int h'\overline{u}_{xxx}\mathrm{d}x+\mathrm{Re}(h',\alpha u_{xxx})\tag{3.8.11}$$

$$|\mathrm{Re}(h',\alpha u_{xxx})|\le C\|h'\|\|u_{xxx}\|\le C\|u_{xxx}\|,\tag{3.8.12}$$

and due to equations (3.8.11)–(3.8.12), we have

$$\mathrm{Re}(h',u_{xxxt}+\alpha u_{xxx})\ge\frac{\mathrm{d}}{\mathrm{d}t}\mathrm{Re}\int h'\overline{u}_{xxx}\mathrm{d}x-C\|u_{xxx}\|.\tag{3.8.13}$$

Then by equations (3.8.9), (3.8.10) and (3.8.13) we have

$$\begin{aligned}&\frac{1}{2}\frac{\mathrm{d}}{\mathrm{d}t}\|u_{xxx}\|^2+\alpha\|u_{xxx}\|^2+\frac{\mathrm{d}}{\mathrm{d}t}\mathrm{Re}\int h'\overline{u}_{xxx}\mathrm{d}x\\&\quad+\mathrm{Re}(g'(|u|^2)|u|_x^2u+g(|u|^2)u_x,u_{xxxt})\le C\|u_{xxx}\|.\end{aligned}\tag{3.8.14}$$

Obviously,

$$\begin{aligned}&\mathrm{Re}(g'(|u|^2)|u|_x^2u+g(|u|^2)u_x,u_{xxxt})\\&\quad=\frac{\mathrm{d}}{\mathrm{d}t}\mathrm{Re}\int(g'(|u|^2)|u|_x^2u+g(|u|^2)u_x)\overline{u}_{xxx}\mathrm{d}x\\&\qquad-\mathrm{Re}\int\frac{\mathrm{d}}{\mathrm{d}t}(g'(|u|^2)|u|_x^2u+g(|u|^2)u_x)\overline{u}_{xxx}\mathrm{d}x\\&\quad=\frac{\mathrm{d}}{\mathrm{d}t}\mathrm{Re}\int(g'(|u|^2)|u|_x^2u+g(|u|^2)u_x)\overline{u}_{xxx}\mathrm{d}x\\&\qquad-\mathrm{Re}\int g''(|u|^2)|u|_t^2|u|_x^2u\cdot\overline{u}_{xxx}\mathrm{d}x\\&\qquad-\mathrm{Re}\int g'(|u|^2)|u|_{xt}^2u\cdot\overline{u}_{xxx}\mathrm{d}x\\&\qquad-\mathrm{Re}\int g'(|u|^2)|u|_x^2u_t\cdot\overline{u}_{xxx}\mathrm{d}x\\&\qquad-\mathrm{Re}\int g'(|u|^2)|u|_t^2u_x\cdot\overline{u}_{xxx}\mathrm{d}x\\&\qquad-\mathrm{Re}\int g(|u|^2)u_{xt}\cdot\overline{u}_{xxx}\mathrm{d}x.\end{aligned}\tag{3.8.15}$$

From equations (3.8.14)–(3.8.15), we have

$$\begin{aligned}\frac{1}{2}\frac{\mathrm{d}}{\mathrm{d}t}\|u_{xxx}\|^2 + \alpha\|u_{xxx}\|^2 + \frac{\mathrm{d}}{\mathrm{d}t}\mathrm{Re}\int h'\cdot\overline{u}_{xxx}\mathrm{d}x \\ + \frac{\mathrm{d}}{\mathrm{d}t}\mathrm{Re}\int (g'(|u|^2)|u|_x^2u + g(|u|^2)u_x)\cdot\overline{u}_{xxx}\mathrm{d}x \\ \le C\|u_{xxx}\| + \mathrm{Re}\int g''(|u|^2)|u|_t^2|u|_x^2u\cdot\overline{u}_{xxx}\mathrm{d}x \\ + \mathrm{Re}\int g'(|u|^2)|u|_{xt}^2u\cdot\overline{u}_{xxx}\mathrm{d}x \\ + \mathrm{Re}\int g'(|u|^2)|u|_x^2u_t\cdot\overline{u}_{xxx}\mathrm{d}x \\ + \mathrm{Re}\int g'(|u|^2)|u|_t^2u_x\cdot\overline{u}_{xxx}\mathrm{d}x \\ + \mathrm{Re}\int g(|u|^2)u_{xt}\cdot\overline{u}_{xxx}\mathrm{d}x.\end{aligned} \tag{3.8.16}$$

Now we estimate each term on the right-hand side of equation (3.8.16):

$$\begin{aligned}&\left|\int g''(|u|^2)|u|_t^2|u|_x^2u\cdot\overline{u}_{xxx}\mathrm{d}x\right| \\ &\quad\le\left|\int g''(|u|^2)(u_t\overline{u}+u\overline{u}_t)(u_x\overline{u}+u\overline{u}_x)u\cdot\overline{u}_{xxx}\mathrm{d}x\right| \\ &\quad\le 4\|g''(|u|^2)\|_\infty\|u\|_\infty^3\|u_x\|_\infty\int|u_t\overline{u}_{xxx}|\mathrm{d}x \\ &\quad\le C\|u_t\|\|u_{xxx}\|\le C\|u_{xxx}\|,\end{aligned} \tag{3.8.17}$$

$$\begin{aligned}&\left|\int g'(|u|^2)|u|_x^2u_t\cdot\overline{u}_{xxx}\mathrm{d}x\right| \\ &\quad\le\left|\int g'(|u|^2)(u_x\overline{u}+u\overline{u}_x)u_t\overline{u}_{xxx}\mathrm{d}x\right| \\ &\quad\le 2\|g'(|u|^2)\|_\infty\|u\|_\infty\|u_x\|_\infty\int|u_t\overline{u}_{xxx}|\mathrm{d}x \\ &\quad\le C\|u_t\|\|u_{xxx}\|\le C\|u_{xxx}\|,\end{aligned} \tag{3.8.18}$$

$$\begin{aligned}&\left|\int g'(|u|^2)|u|_t^2u_x\cdot\overline{u}_{xxx}\mathrm{d}x\right| \\ &\quad\le 2\|g'(|u|^2)\|_\infty\|u\|_\infty\|u_x\|_\infty\int|u_t\overline{u}_{xxx}|\mathrm{d}x \\ &\quad\le C\|u_{xxx}\|,\end{aligned} \tag{3.8.19}$$

$$\begin{aligned}&\mathrm{Re}\int g(|u|^2)u_{xt}\overline{u}_{xxx}\mathrm{d}x \\ &\quad=-\mathrm{Re}\int g'(|u|^2)|u|_x^2u_{xt}\overline{u}_{xx}\mathrm{d}x-\frac{1}{2}\frac{\mathrm{d}}{\mathrm{d}t}\int g(|u|^2)|u_{xx}|^2\mathrm{d}x \\ &\qquad+\frac{1}{2}\int g'(|u|^2)|u|_t^2|u_{xx}|\mathrm{d}x \\ &\quad\le-\frac{1}{2}\frac{\mathrm{d}}{\mathrm{d}t}\int g(|u|^2)|u_{xx}|^2\mathrm{d}x\end{aligned}$$

$$
\begin{aligned}
&+ 2\|g'(|u|^2)\|_\infty \|u\|_\infty \|u_x\|_\infty \int |u_{xt}\overline{u}_{xx}|\mathrm{d}x \\
&+ \|g'(|u|^2)\|_\infty \|u\|_\infty \|u_{xx}\|_\infty \int |u_t u_{xxx}|\mathrm{d}x \\
\le{}& -\frac{1}{2}\frac{\mathrm{d}}{\mathrm{d}t}\int g(|u|^2)|u_{xx}|^2\mathrm{d}x + C\|u_{xxx}\| + C,
\end{aligned} \tag{3.8.20}
$$

$$
\begin{aligned}
\operatorname{Re}&\int g'(|u|^2)u|u|^2_{xt}u\overline{u}_{xxx}\mathrm{d}x \\
&= -\operatorname{Re}\int g''(|u|^2)|u|^2_x|u|^2_{xt}u\overline{u}_{xx}\mathrm{d}x - \operatorname{Re}\int g'(|u|^2)|u|^2_{xxt}u\overline{u}_{xx}\mathrm{d}x \\
&\quad - \operatorname{Re}\int g'(|u|^2)|u|^2_{xt}u_x\overline{u}_{xx}\mathrm{d}x,
\end{aligned} \tag{3.8.21}
$$

$$
\begin{aligned}
&\left|\int g''(|u|^2)|u|^2_x|u|^2_{xt}u\overline{u}_{xx}\mathrm{d}x\right| \\
&\le \left|\int g''(|u|^2)(u_x\overline{u} + u\overline{u}_x)(u_{xt}\overline{u} + u_x\overline{u}_t + u_t\overline{u}_x + u\overline{u}_{xt})u\overline{u}_{xx}\mathrm{d}x\right| \\
&\le C\|u_{xt}\|\|u_{xx}\| + C\|u_t\|\|u_{xx}\| \\
&\le C\|u_{xxx}\| + C.
\end{aligned} \tag{3.8.22}
$$

Similarly, we have

$$
\left|\int g'(|u|^2)|u|^2_{xt}u_x\overline{u}_{xx}\mathrm{d}x\right| \le C\|u_{xxx}\| + C. \tag{3.8.23}
$$

Since

$$
\begin{aligned}
-\operatorname{Re}\int g'(|u|^2)|u|^2_{xxt}u\overline{u}_{xxx}\mathrm{d}x = -\operatorname{Re}\int g'(|u|^2)u\overline{u}_{xx}\{&u_{xxt}\overline{u} + u_{xx}\overline{u}_t + 2u_{xt}\overline{u}_x \\
&+ 2u_x\overline{u}_{xt} + u_t\overline{u}_{xx} + u\overline{u}_{xxt}\}\mathrm{d}x,
\end{aligned} \tag{3.8.24}
$$

we can bound each term on the right-hand side of equation (3.8.24):

$$
\begin{aligned}
&\left|\int g'(|u|^2)u\overline{u}_{xx}(u_{xx}\overline{u}_t + 2u_{xt}\overline{u}_x + 2u_x\overline{u}_{xt} + u_t\overline{u}_{xx})\mathrm{d}x\right| \\
&\le 2\|g'(|u|^2)\|_\infty \|u\|_\infty \|u_{xx}\|_\infty \int |u_t\overline{u}_{xx}|\mathrm{d}x \\
&\quad + 4\|g'(|u|^2)\|_\infty \|u\|_\infty \|u_x\|_\infty \int |\overline{u}_{xx}u_{xt}|\mathrm{d}x \\
&\le C\|u_x\|_{H^1}\|u_t\|\|u_{xx}\| + C\|u_{xt}\|\|u_{xx}\| \le C\|u_{xxx}\| + C,
\end{aligned} \tag{3.8.25}
$$

$$
\begin{aligned}
-\operatorname{Re}&\int g'(|u|^2)|u|^2\overline{u}_{xx}u_{xxt}\mathrm{d}x \\
&= -\frac{1}{2}\frac{\mathrm{d}}{\mathrm{d}t}\int g'(|u|^2)|u|^2|u_{xx}|^2\mathrm{d}x + \frac{1}{2}\int g''(|u|^2)|u|^2_t|u|^2|u_{xx}|^2\mathrm{d}x \\
&\quad + \frac{1}{2}\int g'(|u|^2)|u|^2_t|u_{xx}|^2\mathrm{d}x \\
&\le -\frac{1}{2}\frac{\mathrm{d}}{\mathrm{d}t}\int g'(|u|^2)|u|^2|u_{xx}|^2\mathrm{d}x + \|g''(|u|^2)\|_\infty \|u\|^3_\infty \|u_{xx}\|_\infty \int |u_t u_{xx}|\mathrm{d}x
\end{aligned}
$$

$$
\begin{aligned}
&+\|g'(|u|^2)\|_\infty\|u\|_\infty\|u_{xx}\|_\infty\int|u_tu_{xx}|\mathrm{d}x\\
&\le-\frac{1}{2}\frac{\mathrm{d}}{\mathrm{d}t}\int g'(|u|^2)|u|^2|u_{xx}|^2\mathrm{d}x+C\|u_{xx}\|_{H^1}\|u_t\|\|u_{xx}\|\\
&\le-\frac{1}{2}\frac{\mathrm{d}}{\mathrm{d}t}\int g'(|u|^2)|u|^2|u_{xx}|^2\mathrm{d}x+C\|u_{xxx}\|+C, \qquad (3.8.26)\\
-\operatorname{Re}&\int g'(|u|^2)u^2\overline{u}_{xx}\overline{u}_{xxt}\mathrm{d}x\\
&=-\frac{\mathrm{d}}{\mathrm{d}t}\operatorname{Re}\int g'(|u|^2)u^2\overline{u}_{xx}^2\mathrm{d}x+\operatorname{Re}\int g''(|u|^2)|u|_t^2u^2\overline{u}_{xx}^2\mathrm{d}x\\
&\quad+\operatorname{Re}\int g'(|u|^2)2uu_t\overline{u}_{xx}^2\mathrm{d}x+\operatorname{Re}\int g'(|u|^2)u^2\overline{u}_{xxt}\overline{u}_{xx}\mathrm{d}x.
\end{aligned}
$$

Then

$$
\begin{aligned}
-\operatorname{Re}&\int g'(|u|^2)u^2\overline{u}_{xx}\overline{u}_{xxt}\mathrm{d}x\\
&=-\frac{1}{2}\frac{\mathrm{d}}{\mathrm{d}t}\operatorname{Re}\int g'(|u|^2)u^2\overline{u}_{xx}^2\mathrm{d}x\\
&\quad+\frac{1}{2}\operatorname{Re}\int g''(|u|^2)|u|_t^2u^2\overline{u}_{xx}^2\mathrm{d}x+\operatorname{Re}\int g'(|u|^2)uu_t\overline{u}_{xx}\mathrm{d}x\\
&\le-\frac{1}{2}\frac{\mathrm{d}}{\mathrm{d}t}\operatorname{Re}\int g'(|u|^2)u^2\overline{u}_{xx}^2\mathrm{d}x\\
&\quad+\|g''(|u|^2)\|_\infty\|u\|_\infty^3\|u_{xx}\|_\infty\int|u_tu_{xx}|\mathrm{d}x\\
&\quad+\|g'(|u|^2)\|_\infty\|u\|_\infty\int|u_tu_{xx}|\mathrm{d}x\\
&\le-\frac{1}{2}\frac{\mathrm{d}}{\mathrm{d}t}\operatorname{Re}\int g'(|u|^2)u^2|u_{xx}|^2\mathrm{d}x+C\|u_{xx}\|_{H^1}\|u_t\|\|u_{xx}\|\\
&\le-\frac{1}{2}\frac{\mathrm{d}}{\mathrm{d}t}\operatorname{Re}\int g'(|u|^2)u^2u_{xx}^2\mathrm{d}x+C\|u_{xxx}\|+C. \qquad (3.8.27)
\end{aligned}
$$

From equations (3.8.21)–(3.8.27) we get

$$
\begin{aligned}
\operatorname{Re}&\int g'(|u|^2)|u|_{xt}^2u\overline{u}_{xxx}^2\mathrm{d}x\\
&\le-\frac{1}{2}\frac{\mathrm{d}}{\mathrm{d}t}\operatorname{Re}\int g'(|u|^2)|u^2||u_{xx}|^2\mathrm{d}x-\frac{1}{2}\frac{\mathrm{d}}{\mathrm{d}t}\operatorname{Re}\int g'(|u|^2)u^2\overline{u}_{xx}^2\mathrm{d}x+C\|u_{xxx}\|+C.
\end{aligned}
\tag{3.8.28}
$$

Moreover, from equations (3.8.16)–(3.8.20) and (3.8.28) we get

$$
\begin{aligned}
&\frac{\mathrm{d}}{\mathrm{d}t}\|u_{xxx}\|^2+\frac{\mathrm{d}}{\mathrm{d}t}2\operatorname{Re}\int h'\overline{u}_{xxx}\mathrm{d}x+\frac{\mathrm{d}}{\mathrm{d}t}\int g'(|u|^2)|u|^2|u_{xx}|^2\mathrm{d}x\\
&\quad+\frac{\mathrm{d}}{\mathrm{d}t}2\operatorname{Re}\int(g'(|u|^2)|u|_x^2u+g(|u|^2)u_x)\overline{u}_{xxx}\mathrm{d}x\\
&\quad+\frac{\mathrm{d}}{\mathrm{d}t}\operatorname{Re}\int g'(|u|^2)u^2\overline{u}_{xx}^2\mathrm{d}x
\end{aligned}
$$

$$+\frac{\mathrm{d}}{\mathrm{d}t}\int g(|u|^2)|u_{xx}|^2\mathrm{d}x + 2\alpha\|u_{xxx}\|^2$$
$$\le C\|u_{xxx}\| + C \le \frac{\alpha}{2}\|u_{xxx}\|^2 + C. \tag{3.8.29}$$

Let

$$\begin{aligned}E(t) = {} & \|u_{xxx}\|^2 + 2\,\mathrm{Re}\int h'\bar{u}_{xxx}^2\mathrm{d}x \\ & + 2\,\mathrm{Re}\int (g'(|u|^2)|u|_x^2 u + g(|u|^2)u_x)\bar{u}_{xxx}\mathrm{d}x \\ & + \int g'(|u|^2)|u|^2|u_{xx}|^2\mathrm{d}x \\ & + \mathrm{Re}\int g'(|u|^2)u^2\bar{u}_{xx}^2\mathrm{d}x + \int g(|u|^2)|u_{xx}|^2\mathrm{d}x.\end{aligned} \tag{3.8.30}$$

Then from equation (3.8.29) we get

$$\begin{aligned}\frac{\mathrm{d}E}{\mathrm{d}t} + \alpha E + \frac{\alpha}{2}\|u_{xxx}\|^2 \le {} & 2\,\mathrm{Re}\int h'\bar{u}_{xxx}\mathrm{d}x \\ & + 2\alpha\,\mathrm{Re}\int (g'(|u|^2)|u|_x^2 u + g(|u|^2)u_x)\bar{u}_{xxx}\mathrm{d}x \\ & + \alpha\int g'(|u|^2)|u|^2|u_{xxx}|^2\mathrm{d}x + \alpha\,\mathrm{Re}\int g'(|u|^2)u^2\bar{u}_{xx}^2\mathrm{d}x \\ & + \alpha\int g(|u|^2)|u_{xx}|^2\mathrm{d}x + C.\end{aligned} \tag{3.8.31}$$

We also estimate each term on the right-hand side of formula (3.8.31):

$$2\left|\mathrm{Re}\int h'\bar{u}_{xxx}\mathrm{d}x\right| \le 2\|h'\|\|u_{xxx}\| \le C\|u_{xxx}\| \tag{3.8.32}$$

$$\begin{aligned}&\left|2\,\mathrm{Re}\int (g'(|u|^2)|u|_x^2 u + g(|u|^2)u_x)\bar{u}_{xxx}\mathrm{d}x\right| \\ &\quad \le C\|u_x\|\|u_{xxx}\| \le C\|u_{xxx}\|,\end{aligned} \tag{3.8.33}$$

$$\left|\int g'(|u|^2)|u|^2|u_{xx}|^2\mathrm{d}x\right| \le \|g'(|u|^2)\|_\infty\|u\|_\infty^2\|u_{xx}\|^2 \le C. \tag{3.8.34}$$

Similarly, we have

$$\left|\mathrm{Re}\int g'(|u|^2)u^2\bar{u}_{xx}\mathrm{d}x\right| \le C, \tag{3.8.35}$$

$$\left|\int g(|u|^2)|u_{xx}|^2\mathrm{d}x\right| \le C. \tag{3.8.36}$$

By equations (3.8.31)–(3.8.36) we get

$$\frac{\mathrm{d}E}{\mathrm{d}t} + \alpha E + \frac{\alpha}{2}\|u_{xxx}\|^2 \le C\|u_{xxx}\| + C \le \frac{\alpha}{2}\|u_{xxx}\|^2 + C,$$

thus

$$\frac{dE}{dt} + \alpha E \le C. \tag{3.8.37}$$

From Gronwall lemma we have

$$E(t) \le E(t_*)e^{-\alpha(t-t_*)} + \frac{C}{\alpha}, \quad t \ge t_*. \tag{3.8.38}$$

Similarly, replacing equation (3.8.6) with equation (3.8.7), we have

$$\frac{dE}{dt} + \alpha E \le K, \quad 0 \le t \le T, \tag{3.8.39}$$

where the constant K depends on parameters (α, Ω, g), T and R, whenever $\|u_0\|_{H^3} \le R$. Thus we get

$$E(t_*) \le E(0)e^{-\alpha t_*} + \frac{C}{\alpha} \le C(R), \tag{3.8.40}$$

where $C(R)$ depends on parameters (α, Ω, g), R and $\|u_0\|_{H^3}$.

By equations (3.8.38) and (3.8.40) we get

$$E(t) \le \begin{cases} C(R))e^{-\alpha(t-t_*)} + \frac{C}{\alpha}, & t \ge t_*, \\ \frac{2C}{\alpha} & t \ge t_*, \end{cases} \tag{3.8.41}$$

where $t_* = \max\{t_*, t_* + \frac{1}{\alpha}\ln\frac{\alpha C(R)}{C}\}$.

Based on equation (3.8.30) and equations (3.8.32)–(3.8.36), we get

$$\|u_{xxx}\|^2 \le C\|u_{xxx}\| + C \le \frac{1}{2}\|u_{xxx}\|^2 + C, \quad t \ge t_*'.$$

From this we get the claim. □

Lemma 3.8.2. *Under the conditions of Lemma* 3.8.1, *we have the estimate*

$$\|u_{tx}\| \le C, \quad t \ge t_*.$$

Proof. By equation (3.8.8) we get

$$\begin{aligned} \|u_{tx}\| \le \|u_{xxx}\| + \alpha\|u_x\| + 2\|g'(|u|^2)\|_\infty \|u\|_\infty^2 \|u_x\| \\ + \|g(|u|^2)\|_\infty \|u_x\| + \|h\|_{H^1} \le C. \end{aligned}$$

Thus Lemma 3.8.2 has been proved. □

Now we construct an approximate inertial manifold of problem (3.8.1)–(3.8.3). Write equation (3.8.1) in an abstract differential equation form:

$$i\frac{du}{dt} - Au + g(|u|^2)u + i\alpha u + h = 0, \tag{3.8.42}$$

where $A = -\partial_{xx}$ is an unbounded self-adjoint operator

$$D(A) = \{u \in H^2 : u \text{ satisfies boundary condition (3.8.3)}\}.$$

Let $\{\omega_j\}_{j=1}^{\infty}$ be the orthogonal basis of H, which is composed of eigenvectors of A, that is,

$$\begin{aligned} &A\omega_j = \lambda_j \omega_j, \\ &0 \le \lambda_1 < \lambda_2 \le \cdots \le \lambda_j \to +\infty,\ j \to \infty. \end{aligned}$$

For a given m, set $P = P_m : H \to \operatorname{span}\{\omega_1, \dots, v_m\}$ to be the projection, $Q = Q_m = I - P_m$. Acting with P_m and Q_m on equation (3.6.42), we have

$$\begin{cases} i\dfrac{dp}{dt} - Ap + P_m(g(|u|^2)u) + i\alpha p + P_m h = 0, \\ i\dfrac{dq}{dt} - Aq + Q_m(g(|u|^2)u) + i\alpha q + Q_m h = 0, \end{cases} \tag{3.8.43}$$

where $p = P_m u$, $q = Q_m u$.

Note that

$$\begin{cases} \|A^{\frac{1}{2}}u\| = \left\|\dfrac{\partial u}{\partial x}\right\|, & u \in H^1, \\ \|P_m u\| \le \|u\|, \quad \|Q_m u\| \le \|u\|, & u \in H, \\ \|A^\alpha u\| \ge \lambda_{m+1}^\alpha \|u\|, & \alpha > 0,\ u \in Q_m D(A^\alpha). \end{cases} \tag{3.8.44}$$

Then from equation (3.8.7) and Lemma 3.8.2 we deduce that

$$\begin{cases} \|A^{\frac{1}{2}}q\|, \|A^{\frac{1}{2}}q_t\| \le C, & t \ge t_*, \\ \|q\|, \|q_t\| \le \lambda_{m+1}^{-\frac{1}{2}}, & t \ge t_*. \end{cases} \tag{3.8.45}$$

Now we define the mapping $\Phi : P_m H \to Q_m H$ such that for any $p \in P_m H$, $\Phi(p) = \Psi$ is given by the following equation:

$$-A\Psi + Q_m g(|p|^2)p + i\alpha\Psi + Q_m h = 0. \tag{3.8.46}$$

We first prove the existence of a unique solution $\Psi \in Q_m H$ of equation (3.8.46).

Lemma 3.8.3. *There is an integer m_0 depending on α and Ω such that when $m \ge m_0$, there exists a unique solution $\Psi \in Q_m H$ of equation (3.8.46) for any $p \in P_m H$.*

Proof. We will use the fixed point principle to prove the theorem. Let $p \in P_m H$ be fixed and introduce the mapping $G : Q_m H \to Q_m H$ such that for any $\varphi \in Q_m H$, $G(\varphi) = \Psi$ is given by the following equation:

$$-A\Psi + Q_m g(|p|^2)p + i\alpha\varphi + Q_m h = 0. \tag{3.8.47}$$

Obviously, any fixed point of G is a solution of equation (3.8.46).

In the following we prove that there exists an integer m_0, which depends on α and Ω, such that when $m \ge m_0$, $G : Q_mH \to Q_mH$ is a compact mapping. Let $\varphi_1, \varphi_2 \in Q_mH$, then by equation (3.8.47) we have

$$A\Psi_1 - A\Psi_2 = i\alpha(\varphi_1 - \varphi_2).$$

Hence

$$\|A\Psi_1 - A\Psi_2\| = \alpha\|\varphi_1 - \varphi_2\|. \tag{3.8.48}$$

From equations (3.8.44) and (3.8.48), we have

$$\|\Psi_1 - \Psi_2\| = \alpha\lambda_{m+1}^{-1}\|\varphi_1 - \varphi_2\|.$$

Since $\lambda_{m+1} \to \infty$, we can find an m_0, merely depending on α and Ω, such that when $m \ge m_0$, Ψ is compact. Hence G is the unique fixed point in Q_mH. This finishes the proof of the lemma. □

Letting $\Sigma_1 = \text{graph}(\Phi)$, we can prove that Σ_1 is an approximate inertial manifold of problem (3.8.1)–(3.8.3). In fact, we have

Theorem 3.8.1. *Suppose conditions (3.8.4)–(3.8.5) hold and $u_0 \in H^3, h \in H^1$. Then there exists a constant m_0, which depends on parameters (α, Ω, g), such that when $m \ge m_0$, we have*

$$\text{dist}_{H^2}(u(t), \Sigma_1) \le K\lambda_{m+1}^{-\frac{1}{2}}, \quad t \ge t_*,$$

where the constant K depends on parameters (α, Ω, g); $u(t)$ is the solution of problem (3.8.1)–(3.8.3); t_* only depends on parameters (α, Ω, g) and R, whenever $\|u_0\|_{H^2} \le R$.

Proof. Subtracting equation (3.8.46) from (3.8.43), we obtain

$$A\Psi - Aq = Q_m g(|p|^2)p - Q_m g(|u|^2)u + i\alpha(\Psi - q) - iq_t. \tag{3.8.49}$$

Furthermore, we have

$$\begin{aligned}
&\|g(|p|^2)p - g(|u|^2)u\| \\
&\quad \le \|(g(|p|^2) - g(|u|^2))p\| + \|g(|u|^2)q\| \\
&\quad \le \|g'(\xi)(|p|^2 - |u|^2)p\| + \|g(|u|^2)q\| \\
&\quad \le \|g'(\xi)(p\bar{p}q - p\bar{q}u)\| + \|g(|u|^2)q\| \\
&\le \|g'(\xi)\|_\infty\|p\|_\infty^2\|q\| + \|g'(\xi)\|_\infty\|p\|_\infty\|u\|_\infty\|q\| + \|g(|u|^2)\|_\infty\|q\| \\
&\quad \le C\|q\| \le C\lambda_{m+1}^{-\frac{1}{2}}.
\end{aligned} \tag{3.8.50}$$

From equations (3.8.49)–(3.8.50), we have

$$\begin{aligned}\|A\Psi - Aq\| &\le C\lambda_{m+1}^{-\frac{1}{2}} + \alpha\|\Psi - q\| + \|q_t\| \\ &\le C\lambda_{m+1}^{-\frac{1}{2}} + \alpha\lambda_{m+1}^{-1}\|A\Psi - Aq\|.\end{aligned}$$

Since $\lambda_{m+1} \to \infty$, we can find a positive integer m_0, merely depending on parameters (α, Ω, g), such that when $m \ge m_0$, we have

$$\|A\Psi - Aq\| \le C\lambda_{m+1}^{-\frac{1}{2}}, \quad t \ge t_*.$$

Hence

$$\begin{aligned}\operatorname{dist}_{H^2}(u(t), \Sigma_1) &\le \big\|u(t) - (p(t) + \Psi(t))\big\|_{H^2} \\ &\le \big\|\Psi(t) - q(t)\big\|_{H^2} = \big\|A\Psi(t) - Aq\big\| \le C\lambda_{m+1}^{-\frac{1}{2}}, \quad t \ge t_*.\end{aligned}$$

Theorem 3.8.1 has been proved. □

Note that, if $u_0 \in H^2$, by equation (3.8.7) we have

$$\|q\|, \|q_t\| \le C, \quad t \ge t_*. \tag{3.8.51}$$

From this, we get

Theorem 3.8.2. *Let the conditions* (3.8.4)–(3.8.5) *be satisfied and $u_0 \in H^2$, $h \in L^2$. Then there exists a constant m_0, which depends on parameters (α, Ω, g), such that when $m \ge m_0$, we have*

$$\operatorname{dist}_{H^1}(u(t), \Sigma_1) \le K\lambda_{m+1}^{-\frac{1}{2}}, \quad t \ge t_*,$$

where the constant K depends on parameters (α, Ω, g); t_* only depends on parameters (α, Ω, g) and R, whenever $\|u_0\|_{H^2} \le R$; and $u(t)$ is the solution of problem (3.8.1)–(3.8.3).

In the following, we introduce a simpler explicit inertial manifold $\Sigma_1 = \text{graph}(\Phi^*)$, in which Φ^* is the mapping $P_mH \to Q_mH$ such that for any $p \in P_mH$, $\Phi^*(p) = \varphi$ is given by the following equation:

$$-A\varphi + Q_m g(|p|^2)p + Q_m h = 0.$$

We can prove

Theorem 3.8.3. *Suppose conditions* (3.8.4)–(3.8.5) *are satisfied and $u_0 \in H^3$, $h \in H^1$. Then there exists a constant K which depends only on parameters (α, Ω, g) such that*

$$\operatorname{dist}_{H^2}(u(t), \Sigma_2) \le K\lambda_{m+1}^{-\frac{1}{2}}, \quad t \ge t_*,$$

where $u(t)$ is the solution of problem (3.8.1)–(3.8.3)*; t_* only depends on parameters (α, Ω, g) and R, whenever $\|u_0\|_{H^3} \le R$.*

Theorem 3.8.4. *Suppose that conditions (3.8.4)–(3.8.5) are satisfied and* $u_0 \in H^2$, $h \in L^2$. *Then there exists a constant K which depends only on parameters* (α, Ω, g) *such that*

$$\mathrm{dist}_{H^1}(u(t), \Sigma_2) \le K\lambda_{m+1}^{-\frac{1}{2}}, \quad t \ge t_*,$$

where $u(t)$ *is the solution of problem (3.8.1)–(3.8.3);* t_* *depends only on parameters* (α, Ω, g) *and R, whenever* $\|u_0\|_{H^2} \le R$.

3.9 The convergence of approximate inertial manifolds

Approximate inertial manifold plays a vital role in approximation calculation, but whether or not we can prove the existence of an inertial manifold by constructing approximate inertial manifolds is not clear. The problem had not been solved for a long time. In 1994, Debussche and Temam [44] constructed a sequence of approximate inertial manifolds and proved that it converges to the inertial manifold. Of course, the spectral gap conditions were still required.

Consider a class of nonlinear evolution equations in a Banach space $\mathscr{E}$ of the form

$$\begin{cases} \dfrac{du}{dt} + Au = f(u), \\ u(0) = u_0, \end{cases} \tag{3.9.1}$$

where A is a linear density operator in $\mathscr{E}$; $f \in C^1$ is a nonlinear operator, $E \to F$, where E and F are two Banach spaces,

$$E \subset F \subset \mathscr{E},$$

and the mapping is continuous. In $\mathscr{E}$, E, F, their norms are denoted by $|\cdot|_{\varepsilon}$, $|\cdot|_E$, and $|\cdot|_F$, respectively. Assume that the function f is globally Lipschitz, $E \to F$.

Let M_1 be a positive constant such that for any $x, y \in E$,

$$\begin{cases} |f(x) - f(y)|_F \le M_1 |x - y|_E, \\ |f(x)|_F \le M_1(1 + |x|_E). \end{cases} \tag{3.9.2}$$

For the operator A, suppose the linear equation is

$$\begin{cases} \dfrac{du}{dt} + Au = 0, \\ u(0) = u_0. \end{cases}$$

In ε, we define a strong continuous semigroup $(e^{-At})_{t \ge 0}$ such that

$$e^{-At} F \subset E, \quad \forall t > 0.$$

Suppose there exists an eigenvector sequence $(P_n)_{n\in N}$ of A and two series $(\lambda_n)_{n\in N}$, $(\Lambda_n)_{n\in N}$ such that

$$\Lambda_n \geq \lambda_n \geq 0, \quad \forall n \geq 0, \tag{3.9.3a}$$

$$\lambda_n \to \infty, \quad n \to \infty, \tag{3.9.3b}$$

$$\frac{\Lambda_n}{\lambda_n} \text{ is bounded as } n \to \infty. \tag{3.9.3c}$$

Also assume that for $Q_n = I - P_n$ we have

(∗) $P_n\mathscr{E}$ and $Q_n\mathscr{E}$ is invariant under the action of e^{-At}, $\forall t \geq 0$;

(∗∗) $(e^{-At})_{t\geq 0}$ can be extended to a semigroup $(e^{-At})_{t\in \mathbf{R}}$ in $P_n\mathscr{E}$, where these projections can define the exponential dichotomy of $(e^{-At})_{t\geq 0}$: there exist $K_1, K_2 > 0$, $\alpha \in (0,1)$, which are independent of n, such that

(H_1) for $t \leq 0$,

$$|e^{-At}P_n|_{\mathscr{L}(E)} \leq K_1 e^{-\lambda_n t},$$

$$|e^{-At}P_n|_{\mathscr{L}(F,E)} \leq K_1 \lambda_n^{\alpha} e^{-\lambda_n t};$$

(H_2) for $t > 0$,

$$|e^{-At}Q_n|_{\mathscr{L}(F,E)} \leq K_2\left(\frac{1}{t^{\alpha}} + \Lambda_n^{\alpha}\right)e^{-\Lambda_n t},$$

$$|A^{-1}e^{-At}Q_n|_{\mathscr{L}(F,E)} \leq K_2 \Lambda_n^{\alpha-1} e^{-\Lambda_n t}.$$

Finally, suppose that equation (3.9.1) defines a continuous semigroup $(S(t))_{t\geq 0}$ in E.

An approximate inertial manifold is constructed in the following. A sequence of approximate inertial manifolds is obtained by constructing the inertial manifold through the approximate Lyapunov–Perron method as the fixed point of a mapping defined by an integral equation. As is well known, the Lyapunov–Perron method consists in looking for Ψ as the fixed point of the mapping $\mathscr{J}$ given by

$$\mathscr{J}\Psi(y_0) = \int_{-\infty}^{0} e^{As} Q_n f(y(s)) + \Psi(y(s)) ds, \tag{3.9.4}$$

where y satisfies

$$\begin{cases} \dfrac{dy}{dt} + Ay = P_n(y + \Psi(y)), \\ y(0) = y_0. \end{cases} \tag{3.9.5}$$

Now we select the time step τ and positive integer N. Approximate formulas (3.9.4)–(3.9.5) as follows:

$$y_{k+1} = R_\tau y_k + S_\tau P_n f(y_k + \Psi(y_k)) \tag{3.9.6}$$

where $k = 0, \dots, N-1$, R_τ and S_τ are linear operators, which satisfy

$$(H_3) \quad \begin{cases} |R_\tau P_n|_{\mathscr{L}(E)} \le e^{\tau\lambda_n}, \\ |S_\tau P_n|_{\mathscr{L}(F,E)} \le K_3 \lambda_n^{\alpha-1}(e^{\tau\lambda_n} - 1), \end{cases}$$

and where positive constant K_3 does not depend on n. Define the mapping T_N^τ by

$$\begin{aligned} T_N^\tau \Psi(y) = {} & A^{-1}(I - e^{-A\tau}) \sum_{k=0}^{N-1} e^{-kA\tau} Q_n f(y_k + \Phi(y_k)) \\ & + A^{-1} e^{-NA\tau} Q_n f(y_N + \Phi(y_N)), \end{aligned}$$

where $(y_k)_{k=0,1,\dots,N}$ are calculated by equation (3.9.6).

For any $\Psi \in \mathscr{F}_{l,b}$,

$$\mathscr{F}_{l,b} = \left\{ \Psi : P_n E \to Q_n E \mid \operatorname{Lip} \Psi < l, \ \sup_{y \in P_n E} \frac{|\Psi(y)|_E}{1 + |y|_E} \le b \right\}, \tag{3.9.7}$$

$y_n \in P_n E$, y is the solution of equation (3.9.5), $\tilde{y}_k = y(-k\tau)$. Let

$$(H_4) \quad \begin{cases} \tilde{y}_{k+1} = R_\tau \tilde{y}_k + S_\tau P_n f(\tilde{y}_k + \Psi(\tilde{y}_k)) + \varepsilon_k, \\ |\varepsilon_k|_E \le \alpha_1(\lambda_n)\tau^2 (1 + |y_0|_E) e^{\tau(k+1)(\lambda_n + K_4 \lambda_n^\alpha)} \end{cases}$$

and suppose that the derivative with respect to y does not increase too fast, that is,

$$(H_5) \quad \left| \frac{dy}{dt} \right|_E \le \alpha_2(\lambda_n)(1 + |y_0|_E) e^{-(\lambda_n + K_5 \lambda_n^\alpha)t}, \quad t \in (-\infty, 0],$$

where K_4, K_5 do not depend on N, τ or n; $\alpha_1(\lambda_n), \alpha_2(\lambda_n)$ depends on λ_n, but not on N, τ.

For many partial differential equations with dissipation, the semigroup defined by equation (3.9.1) has a global attractor $\mathscr{A}$. In general it is embedded in a Banach subspace of E. From the definition of the attractor, $u_0 \in \mathscr{A}$. The solution of equation (3.9.1) is defined on $\mathbf{R}$ and is still in $\mathscr{A}$. As a result, u is uniformly bounded and its time derivative is uniformly bounded. For the real number field, the conditions (H_4), (H_5) can be replaced with

$$(H_4)' \quad \begin{cases} \tilde{y}_{k+1} = R_\tau \tilde{y}_k + S_\tau P_n f(u(-k\tau)) + \varepsilon_k, \\ |\varepsilon_k|_E \le \tau^2 \beta_1, \quad \forall k \le N, \end{cases}$$

$$(H_5)' \quad \left| \frac{dy}{dt} \right| \le \beta_2, \quad \forall t \le 0,$$

respectively, where $y = P_n u$ is the projection of solution of the equation (3.9.1) on $P_n E$ in $\mathscr{A}$; $\tilde{y}_k = y(-k\tau) = Pu(-k\tau)$, β_1, β_2 do not depend on N, τ or n.

Now we construct approximate inertial manifolds. Take a strictly positive sequence $(\tau_n)_{n \in \mathbf{N}}$, define a sequence $(\Phi_n)_{n \in \mathbf{N}}$ such that

$$\begin{cases} \Phi_0 = 0, \\ \Phi_{N+1} = T_N^{\tau N}(\Phi_N), \quad N \ge 0, \end{cases} \tag{3.9.8}$$

where T_0^τ is defined as

$$T_0^\tau \Phi(y_0) = A^{-1} Q_n f(y_0 + \Phi(y_0)).$$

Then graph(Φ_1) is an approximate inertial manifold. Letting M_1 = graph(Φ_N), we now estimate y_k and T_N^τ.

Lemma 3.9.1. *Let* (H_3) *be satisfied. Then the following is true:*

(i) *Assume* $\Psi \in \mathscr{F}_{l,b}$, $y_0 \in P_n E$, *and using equation* (3.9.6) *define* $(y_k)_{k=0,1,\dots,N}$. *Then we have*

$$|y_k|_E \le e^{k\tau(\lambda_n + K_3 M_1 (1+b)\lambda_n^\alpha)}(|y_0|_E + 1), \quad \forall k \in \mathbf{N},$$

and $\lambda_n^{1-\alpha} \ge K_3 M_1 (1+b)$.

(ii) *Assume* $\Psi^i \in \mathscr{F}_{l,b}$, $y_0^i \in P_n E$, *and by equation* (3.9.6) *define* $(y_k^i)_{k=0,\dots,N}$, $y_0 = y_0^i$, $i = 1, 2$. *Then we have*

$$\begin{aligned}|y_k^1 - y_k^2| &\le e^{k\tau(\lambda_n - K_3 M_1 (1+l)\lambda_n^\alpha)}|y_0^1 - y_0^2|_E \\ &\quad + \lambda_n^\alpha k\tau e^{k\tau(\lambda_n + K_6\lambda_n^\alpha)} K_3 M_1 |\Psi^1 - \Psi^2|_\infty (1 + |y_0|_E),\end{aligned}$$

where $K_5 = K_3 M_1 \max(1 + l, 1 + b)$.

Proof. From (H_3) and equations (3.9.2)–(3.9.7), we get

$$\begin{aligned}|y_{k+1}|_E &\le e^{\tau\lambda_n}|y_k|_E + K_3\lambda_n^{\alpha-1}(e^{\lambda_n} - 1)(M_1(1+b)(1+|y_k|_E)) \\ &\le e^{\tau(\lambda_n + K_3 M_1(1+b)\lambda_n^\alpha)}|y_k|_E + K_3\lambda_n^{\alpha-1}M_1(1+b)(e^{\lambda_n} - 1),\end{aligned}$$

where the following basic inequalities were used:

$$\begin{aligned}e^{\tau\lambda_n} - 1 &\le \tau\lambda_n e^{\tau\lambda_n}, \\ 1 + \tau K_1 M_1 (1+b)\lambda_n^\alpha &\le e^{\tau K_3 M_1 (1+b)\lambda_n^\alpha}.\end{aligned}$$

By successive iteration, we get

$$|y_k|_E \le e^{k\tau(\lambda_n + K_3 M_1(1+b)\lambda_n^\alpha)}(|y_0|_E + K_3 M_1 (1+b)\lambda_n^{\alpha-1}).$$

Hence (i) is proved. Take $(y_k^i)_{k=0,\dots,N}$, $i = 1, 2$, as in (ii) and denote $y_k = y_k^1 - y_k^2$. Then we have

$$\begin{aligned}|y_{k+1}|_E &\le e^{\tau\lambda_n}|y_k|_E + K_3\lambda_n^{\alpha-1}(e^{\tau\lambda_n} - 1)(|y_k|_E + |\Psi^1(y_k^1) - \Psi^2(y_k^2)|_E) \\ &\le e^{\tau\lambda_n}|y_k|_E + K_3 M_1 \lambda_n^{\alpha-1}(e^{\tau\lambda_n} - 1)(1+l)|y_k|_E + |\Psi^1 - \Psi^2|_\infty (1 + |y_k^1|_E).\end{aligned}$$

Applying (i) to $(y_k^i)_{k=0,\dots,N}$, we have:

$$|y_{k+1}|_E \le e^{\tau(\lambda_n + K_3 M_1 (1+l)\lambda_n^\alpha)}|y_k|_E + \tau K_3 M_1 \lambda_n^\alpha |\Psi^1 - \Psi^2|_\infty (1 + |y_0^1|_E) e^{(k+1)\tau(\lambda_n + K_3 M_1 (1+b)\lambda_n^\alpha)}.$$

Again by the circle principle, we deduce (ii). This completes the proof. □

Lemma 3.9.2. *Let (H_3) be satisfied. Assume $\Psi \in \mathscr{F}_{l,b}$, $y_0 \in P_nE$, let $(e_k)_{k=0,\dots,N}$ be a sequence in P_nE and $(y_k)_{k=0,\dots,N}$ be defined by equation (3.9.6). Consider the sequence $(\tilde{y}_k)_{k=0,\dots,N}$ defined as follows:*

$$\tilde{y}_{k+1} = R_\tau \tilde{y}_k + S_\tau P_n f(\tilde{y}_k + \Psi(\tilde{y}_k)) + e_k,$$
$$\tilde{y}_0 = y_0.$$

Then we have

$$|\tilde{y}_k - y_k|_E \le \sum_{j=0}^{k-1} e^{(k-1-j)\tau(\lambda_n + K_3 M_1 (1+l)\lambda_n^\alpha)} |e_j|_E, \quad k = 0, \dots, N.$$

Suppose ρ is bounded, $M_0 = \sup_{u\in E} |f(u)|_F$, and use $\widetilde{\mathscr{F}}_{l,b}$ to replace $\mathscr{F}_{l,b}$:

$$\widetilde{\mathscr{F}}_{l,b} = \{\Psi : P_nE \to Q_nE \mid |\Psi(y)|_E \le b,\ \forall y \in P_nE,$$
$$|\Psi(x) - \Psi(y)|_E \le l|x-y|_E,\ \forall x, y \in P_nE\}, \tag{3.9.9}$$

which leads to a simpler calculation.

Proposition 3.9.1. *Assume that (H_2) and (H_3) hold. Then there exist two constants C_1 and C_2 such that if*

$$(N+1)\tau \le \frac{C_1}{\lambda_n^\alpha}, \quad \lambda_n \ge C_2,$$

then there exist l, b_0 such that $\mathscr{F}_N^\tau : \widetilde{\mathscr{F}}_{l,b} \to \widetilde{\mathscr{F}}_{l,b}$, $\forall b \ge b_0$.

Proof. Consider $\Psi \in \widetilde{\mathscr{F}}_{l,b}$, $y_0 \in P_nE$, and define function $\bar{y}(s)$, $s \in (-\infty, 0]$ by

$$\begin{cases} \bar{y}(s) = y_k, & s \in ((-k+1)\tau, -k\tau], \quad k = 0, \dots, N-1, \\ \bar{y}(s) = y_k, & s \in (-\infty, -N\tau), \end{cases} \tag{3.9.10}$$

where $(y_k)_{k=0,\dots,N}$ is calculated by formula (3.9.6). Then $\mathscr{F}_N^\tau \Psi(y_0)$ can be written as

$$\mathscr{F}_N^\tau \Psi(y_0) = \int_{-\infty}^{0} e^{As} Q_n f(\bar{y}(s) + \Psi(\bar{y}(s)))ds$$
$$= \int_{-(N+1)\tau}^{0} e^{As} Q_n f(\bar{y}(s) + \Psi(\bar{y}(s)))ds + A^{-1} e^{-(N+1)A\tau} Q_n f(y_N + \Psi(y_N)).$$

From the bound of (H_2) and f, we get

$$|\mathscr{F}_N^\tau \Psi(y_0)| \le \int_{-\infty}^{0} K_2 M_0 \left(\frac{1}{|s|^\alpha} + \Lambda_n^\alpha \right) e^{\Lambda_n s} ds$$
$$\le K_2 M_0 \left(\int_{-\infty}^{0} \frac{e^y}{|y|^\alpha} dy + 1 \right) \Lambda_n^{\alpha-1}$$
$$\le K_2 M_0 (\Gamma(-1-\alpha) + 1) \Lambda_n^{\alpha-1} \le b.$$

If

$$b \geq b_0 = K_2 M_0 (\Gamma(-1-\alpha)+1) C_2^{\alpha-1}, \tag{3.9.11}$$

choosing $y_0^1, y_0^2 \in P_n E$, by equation (3.9.6) defining $(y_k^i)_{k=0,\dots,N}$, $y_0 = y_0^i$, and constructing $\overline{y}^i$ as before $i = 1, 2$, then

$$\begin{aligned}
&\mathscr{F}_N^\tau \Psi(y_0^1) - \mathscr{F}_N^\tau \Psi(y_0^2) \\
&\quad = \int_{-(N+1)\tau}^{0} e^{As} Q_n f(\overline{y}^1(s) + \Psi(\overline{y}^1(s))) - f(\overline{y}^2(s) + \Psi(\overline{y}^2(s))) \mathrm{d}s \\
&\qquad + A^{-1} e^{-A(N+1)\tau} Q_n (f(y_N^1 + \Psi(y_N^1)) - f(y_N^2 + \Psi(y_N^2))).
\end{aligned}$$

By (H_2), through equations (3.9.9) and (3.9.2) we get

$$\begin{aligned}
&|\mathscr{F}_N^\tau \Psi(y_0^1) - \mathscr{F}_N^\tau \Psi(y_0^2)|_E \\
&\quad \leq K_2 M_1 (1+l) \int_{-(N+1)\tau}^{0} \left(\frac{1}{|s|^\alpha} + \Lambda_n^\alpha\right) |\overline{y}^1(s) - \overline{y}^2(s)|_E \mathrm{d}s \\
&\qquad + K_2 M_1 (1+l) \Lambda_n^{\alpha-1} e^{-\Lambda_n^\alpha (N+1)\tau} |y_N^1 - y_N^2|_E .
\end{aligned}$$

Using Lemma 3.9.1(ii), we deduce that for $\Psi_1, \Psi_2 \in \widetilde{\mathscr{F}}_{l,b} \subset \mathscr{F}_{l,b}$,

$$\begin{aligned}
|y_k^1 - y_k^2|_E &\leq e^{k\tau(\lambda_n - K_3 M_1 (1+l)\lambda_n^\alpha)} |y_0^1 - y_0^2|_E \\
|\overline{y}^1(s) - \overline{y}^2(s)|_E &\leq e^{-s(\lambda_n + K_3 M_1 (1+l)\lambda_n^\alpha)} |y_0^1 - y_0^2|_E .
\end{aligned}$$

Hence, for s we have

$$\begin{aligned}
&|\mathscr{F}_N^\tau \Psi(y_0^1) - \mathscr{F}_N^\tau \Psi(y_0^2)|_E \\
&\quad \leq K_2 M_1 (1-l) |y_0^1 - y_0^2|_E \int_{-(N+1)\tau}^{0} \left(\frac{1}{|s|^\alpha} + \Lambda_n^\alpha\right) e^{(\Lambda_n - \lambda_n - K_3 M_1 (1+l)\lambda_n^\alpha)s} \mathrm{d}s \\
&\qquad + K_2 M_1 (1+l) |y_0^1 - y_0^2|_E \Lambda_n^{\alpha-1} e^{-(\Lambda_n - \lambda_n - K_3 M_1 (1+l)\lambda_n^\alpha)(N+1)\tau} .
\end{aligned}$$

From equation (3.9.3) and inequality $(N+1)\tau \leq \frac{C_1}{\lambda_n^\alpha}$, we have

$$\begin{aligned}
&\int_{-(N+1)\tau}^{0} \frac{1}{|s|^\alpha} e^{(\Lambda_n - \lambda_n - K_3 M_1 (1+l)\lambda_n^\alpha)s} \mathrm{d}s \\
&\quad \leq e^{K_3 M_1 C_1 (1+l)} \int_{-(N+1)\tau}^{0} \frac{1}{|s|^\alpha} \mathrm{d}s \leq e^{K_3 M_1 C_1 (1+l)} \frac{1}{1-\alpha} \lambda_n^{\alpha(\alpha-1)} C_1^{1-\alpha}
\end{aligned}$$

$$
\begin{aligned}
&|\mathcal{F}_N^\tau \Psi(y_0^1) - \mathcal{F}_N^\tau \Psi(y_0^2)|_E \\
&\quad \le K_2 M_1 (1+l)|y_0^1 - y_0^2|_E e^{K_3 M_1 (1+l)} \\
&\quad\quad \times \left(\frac{1}{1-\alpha} \lambda_n^{\alpha(\alpha-1)} C_1^{1-\alpha} + \frac{\Lambda_n^\alpha}{K_3 M_1 (1+l) \lambda_n^\alpha} + \Lambda_n^{\alpha-1} \right) \le l |y_0^1 - y_0^2|_E.
\end{aligned}
$$

If

$$
\left(K_2 M_1 \left(\frac{1}{1-\alpha} \lambda_n^{\alpha(\alpha-1)} C_1^{1-\alpha} + \Lambda_n^{\alpha-1} \right)(1+l) + \frac{K_2}{K_3} \left(\frac{\Lambda_n}{\lambda_n} \right)^\alpha \right) e^{K_3 M_1 C_1 (1+l)} \le l, \tag{3.9.12}
$$

we can choose C_2 and δ_0 such that, when $\lambda_n \ge C_2, C_1 \le \delta_0$,

$$
K_2 M_1 \left(\frac{1}{1-\alpha} \lambda_n^{\alpha(\alpha-1)} C_1^{1-\alpha} + \Lambda_n^{\alpha-1} \right) e^{K_3 M_1 C_1 (1+l)} \le \frac{1}{2}.
$$

Therefore, when

$$
\begin{aligned}
&\left(\frac{1}{2} + C_3' \right) e^{K_3 M_1 C_1 l} \le l, \\
&C_3' = \frac{K_2}{K_3} \sup_k \left(\frac{\Lambda_k}{\lambda_n} \right)^\alpha e^{K_3 M_1 \delta_0} + \frac{1}{2}, \\
&C_1 = \min\left(\delta_0, \frac{\ln \frac{3}{2}}{6 K_3 M_1 C_3'} \right), \quad l = 6 C_3',
\end{aligned}
$$

equation (3.9.12) is true. This proves the proposition. □

The result implies that if the sequence $(\tau_N)_{N \in \mathbf{N}}$ is chosen so that

$$
\tau_N \le \frac{C_1}{(N+1)\lambda_n^\alpha}, \quad \forall N, \tag{3.9.13}
$$

then the sequence $(\Phi_N)_{N\in\mathbf{N}} \subset \mathcal{F}_{l,b}$, where b, l are given in Proposition 3.9.1.

Now we give an estimate of thickness of an approximate manifold, which involves the attractor in a neighborhood of M_N.

Proposition 3.9.2. *Let the assumptions of Proposition 3.9.1, and conditions $(H_4)'$, $(H_5)'$ be satisfied. Then there exist three constants C_3, C_4, C_5 such that for any $\Psi \in \mathcal{F}_{l,b}$, we have*

$$
\begin{aligned}
\max_{u=y+z\in A} |\mathcal{F}_N^\tau \Psi(y) - z|_E &\le C_3 \Lambda_n^{\alpha-1} \max_{y+z\in A} |\Psi(y) - z|_E \\
&\quad + C_4 (\Lambda_n^{\alpha-1} \beta_2 + \Lambda_n^{-1} \beta_1)\tau + C_5 \left(\frac{((N+1)\tau)^{-\alpha}}{\Lambda_n} + \Lambda_n^{\alpha-1} \right) e^{-\Lambda_n (N+1)\tau}.
\end{aligned}
$$

Proof. Suppose $\Psi \in \mathcal{F}_{l,b}$, $u_0 = y_0 + z_0 \in \mathcal{A}$. Let $(y_k^i)_{i=0,\ldots,N}$ be the sequence constructed by equation (3.9.6) and let $\bar{y}(s)$ be defined by (3.9.10), $s \in (-\infty, 0]$, $y = P_n u$, $z = Q_n u$,

$\tilde{y}_k = y(-k\tau)$. Then

$$\mathscr{F}_N^\tau P_n(y_0) = \int_{-\infty}^{0} e^{As} Q_n f(\tilde{y}(s) + \Psi(\tilde{y}(s)))ds$$

$$z_0 = \int_{-\infty}^{0} e^{As} Q_n f(y(s) + z(s))ds.$$

Hence

$$|\mathscr{F}_N^\tau P_n(y_0) - z_0| \le \int_{-\infty}^{0} |e^{As} Q_n f(\tilde{y}(s) + \Psi(\tilde{y}(s))) - f(y(s) + z(s))|_E ds.$$

Using condition (H_2) and equations (3.9.2), (3.9.9) we get

$$\begin{aligned}
&|\mathscr{F}_N^\tau P_n(y_0) - z_0| \\
&\le K_2 M_1 \int_{-(N+1)\tau}^{0} \left(\frac{1}{|s|^\alpha} + \Lambda_n^\alpha\right) e^{\Lambda_n^s} ((1+l)|y(s) - \tilde{y}(s)|_E + |\Psi(y(s)) - z(s)|_E) ds \\
&\quad + 2M_0 K_2 \int_{-\infty}^{-(N+1)\tau} \left(\frac{1}{|s|^\alpha} + \Lambda_n^\alpha\right) e^{\Lambda_n^s} ds. \qquad (3.9.14)
\end{aligned}$$

Since $(H_4)'$ is assumed, we have

$$\tilde{y}_{k+1} = R_\tau \tilde{y}_k + S_\tau P_n f(\tilde{y}_k + z(-k\tau)) + \varepsilon_k.$$

By Lemma 3.9.2, we have

$$\begin{aligned}
|y_k - \tilde{y}_k| &\le \sum_{j=0}^{k-1} e^{(k-l-j)\tau(\lambda_n + K_3 M_1 (1+l)\lambda_n^\alpha)} \\
&\quad \times (|\varepsilon_j|_E + |S_\tau P_n (f(\tilde{y}_j + z(-j\tau)) - f(\tilde{y}_j + \Psi(\tilde{y}_j)))|_E) \\
&\le \left[\tau^2 \beta_1 - K_3 M_1 \lambda_n^{\alpha-1} (e^{\tau\lambda_n} - 1) \max_{y+z\in\mathscr{A}} |\Psi(y) - z|_E\right] \\
&\quad \times e^{k\tau(\lambda_n + K_3 M_1 (1+l)\lambda_n^\alpha)} / [e^{\tau(\lambda_n + K_3 M_1 (1+l)\lambda_n^\alpha)} - 1] \\
&\le \left[\frac{\tau\beta_1}{\lambda_n + K_3 M_1 (1+l)\lambda_n^\alpha} + K_3 M_1 \lambda_n^{\alpha-1} \max_{y+z\in\mathscr{A}} |\Psi(y) - z|_E\right] \\
&\quad \times e^{k\tau(\lambda_n + K_3 M_1 (1+l)\lambda_n^\alpha)}.
\end{aligned}$$

Since $s \in (-(k+1)\tau, -k\tau)$, it follows from $(H_5)'$ that

$$\begin{aligned}
|y(s) - \overline{y}(s)|_E &\le \tau\beta_2 + |y_k - \overline{y}_k|_E \\
&\le \tau\beta_2 + \left[\frac{\tau\beta_1}{\lambda_n + K_3 M_1 (1+l)\lambda_n^\alpha} + K_3 M_1 \lambda_n^{\alpha-1} \max_{y+z\in\mathscr{A}} |\Psi(y) - z|_E\right] \\
&\quad \times e^{-sk\tau(\lambda_n + K_3 M_1 (1+l)\lambda_n^\alpha)}.
\end{aligned}$$

Plugging the above equation into equation (3.9.14), we get

$$\begin{aligned}|\mathscr{F}_N^{\tau}\Phi(y_0) - z_0|_E &\le K_2M_1\beta_2(1+l)\left(\int_{-\infty}^{0}\frac{e^u}{|u|^\alpha}du + 1\right)\Lambda_n^{\alpha-1}\tau \\ &\quad + K_2M_1(1+l)\left[\frac{\tau\beta_1}{\lambda_n + K_3M_1(1+l)\lambda_n^\alpha} + K_3M_1\lambda_n^{\alpha-1}\max_{y+z\in\mathscr{A}}|\Psi(y) - z|_E\right] \\ &\quad \times \int_{-(N+1)\tau}^{0}\left(\frac{1}{|s|^\alpha} + \Lambda_n^\alpha\right)e^{-K_3M_1(1+l)\lambda_n^\alpha s}ds \\ &\quad + K_2M_1(\Gamma(-1-\alpha)+1)\Lambda_n^{\alpha-1}\max_{x+z\in\mathscr{A}}|\Psi(y) - z|_E \\ &\quad + 2K_2M_0\frac{((N+1)\tau)^{-\alpha} + \Lambda_n^\alpha}{\Lambda_n}e^{-\Lambda_n(N+1)\tau}.\end{aligned}$$

Through the proof of Proposition 3.9.1, we get

$$K_2M_1(1+l)\int_{-(N+1)\tau}^{0}\left(\frac{1}{|s|^\alpha} + \Lambda_n^\alpha\right)e^{-K_3M_1(1+l)s}ds \le l.$$

Hence

$$\begin{aligned}|\mathscr{F}_N^{\tau}\Phi(y_0) - z_0| &\le C_4(\Lambda_n^{\alpha-1}\beta_2 + \Lambda_n^{-1}\beta_1)\tau \\ &\quad + C_3\Lambda_n^{\alpha-1}\max_{y+z\in\mathscr{A}}|\Psi(y) - z|_E \\ &\quad + 2K_2M_0\frac{((N+1)\tau)^{-\alpha} + \Lambda_n^\alpha}{\Lambda_n}e^{-\Lambda_n(N+1)\tau},\end{aligned}$$

where C_3, C_4 are independent of N, τ and n. Thus, if τ_N satisfies equation (3.9.13), $\lambda_n \ge C_2$, then we have

$$\begin{aligned}\max_{u=y+z\in\mathscr{A}}|\Psi_{N+1}(y) - z|_E &\le C_3\Lambda_n^{\alpha-1}\max_{y+z\in\mathscr{A}}|\Psi_N(y) - z|_E \\ &\quad + C_4(\Lambda_n^{\alpha-1}\beta_2 + \Lambda_n^{-1}\beta_1)\tau_N + C_5\frac{((N+1)\tau_N)^{-\alpha} + \Lambda_n^\alpha}{\Lambda_n}e^{-\Lambda_n(N+1)\tau_N}.\end{aligned}$$

Proposition 3.9.2 has been proved. □

Theorem 3.9.1. *Let* (H_2), (H_3), $(H_4)'$, $(H_5)'$ *be satisfied. Assume the sequence* $(\tau_N)_{N\in\mathbf{N}}$ *satisfies*

$$C_6 \le \tau_N(N+1)\lambda_n^\alpha \le C_1, \quad \forall N,$$

where C_6 *is an appropriate constant, and constant* C_1 *is given in Proposition 3.9.2. Then the sequence* $(\Phi_N)_{N\in\mathbf{N}}$ *defined by equation* (3.9.8) *satisfies:*

$$\max_{u=y+z\in\mathscr{A}} |\Psi_N(y)-z|_E$$

$$\leq (C_3\Lambda_n^{\alpha-1})^N \max_{u\in\mathscr{A}} |Q_n u|_E + C_4(\Lambda_n^{\alpha-1}\beta_2 + \Lambda_n^{-1}\beta_1)\sum_{j=0}^{N-1}(C_3\Lambda_n^{\alpha-1})^j\tau_{N-1-j} + 4C_5\Lambda_n^{\alpha-1}e^{-C_6\Lambda_n^{1-\alpha}},$$

provided $\lambda_n \geq C_7$, *where* C_7 *is another constant.*

This yields an estimate of distance from M_N to $\mathscr{A}$:

$$\begin{aligned} d_E(\mathscr{A}, M_N) &= \sup_{v\in\mathscr{A}} \inf_{\omega\in M_N} |v-\omega|_E \\ &\leq (C_3\Lambda_n^{\alpha-1})^N \max_{u=y+z\in\mathscr{A}} |Q_n u|_E + C_4(\Lambda_n^{\alpha-1}\beta_2 + \Lambda_n^{-1}\beta_1) \\ &\quad \times \sum_{j=0}^{N-1}(C_3\Lambda_n^{\alpha-1})^j\tau_{N-1-j} + 4C_5\Lambda_n^{\alpha-1}e^{-C_6\Lambda_n^{1-\alpha}}. \end{aligned} \tag{3.9.15}$$

The first and second terms on right-hand side of inequality (3.9.15) tend to zero when $N \to \infty$. So the above distance decreases and comes close to a very small number

$$4C_5\Lambda_n^{\alpha-1}e^{-C_6\Lambda_n^{1-\alpha}}.$$

Hence, if N is large enough, M_N is an explicit inertial manifold of order

$$8C_5\Lambda_n^{\alpha-1}e^{-C_6\Lambda_n^{1-\alpha}}.$$

Now we prove the convergence of $\mathscr{F}_N^\tau$ when $N \to \infty$.

Proposition 3.9.3. *Let* (H_2) *and* (H_3) *be satisfied. If there exists a constant* C_8 *such that* $\Lambda_n - \lambda_n \geq C_8(\Lambda_n^\alpha + \lambda_n^\alpha)$, *then we have: for any* N *and* $\tau > 0$, $\mathscr{F}_N^\tau$ *maps* $\mathscr{F}_{l,b}$ *into itself, and* $\mathscr{F}_N^\tau$ *is a strict compact mapping, its compression constant is less than* $\frac{1}{2}$.

Proof. Let $\Psi \in \mathscr{F}_{l,b}$, $y_0 \in P_n E$, and by equation (3.9.6) construct $(y_k^i)_{k=0,\dots,N}$. By equation (3.9.10) define $\overline{y}$. From Lemma 3.9.1 we get that if C_8 is large enough, $C_8 \geq K_3M_1(1+l)(\sup_k \frac{\Lambda_k}{\lambda_k})^{1-\alpha}$, then

$$|\overline{y}(s)|_E \leq e^{-s(\lambda_n + K_3M_1(1+b)\lambda_n^\alpha)}(|y_0|_E + 1), \quad s \leq 0.$$

Denote

$$\mathscr{F}_N^\tau\Psi(y_0) = \int_{-\infty}^{0} e^{As}Q_n f(\overline{y}(s) + \Psi(\overline{y}(s)))dx.$$

Using (H_2) and equations (3.9.2), (3.9.7), we then get

$$|\mathscr{F}_N^\tau\Psi(y_0)|_E \leq \int_{-\infty}^{0} K_2M_1(1+b)(1+|y_0|_E)$$

$$\times \left(\frac{1}{|s|^{\alpha}} + \Lambda_n^{\alpha}\right) e^{(\Lambda_n - \lambda_n - K_3 M_1 (1+b)\lambda_n^{\alpha})s} ds$$

$$+ \int_{-\infty}^{0} K_2 M_1 (1+b) \left(\frac{1}{|s|^{\alpha}} + \Lambda_n^{\alpha}\right) e^{\Lambda_n s} ds$$

$$\leq K_2 M_1 (1+b)(\Gamma(-1-\alpha)+1)$$

$$\times \left(\frac{\Lambda_n^{\alpha}}{\Lambda_n - \lambda_n - K_3 M_1 (1+b)\lambda_n^{\alpha}} + \Lambda_n^{\alpha-1}\right)(1 + |y_0|_E)$$

$$\leq b(1 + |y_0|_E),$$

where

$$C_8 \geq \max\left(K_3 M_1 (1+b), \frac{2K_2 M_1 (1+b}{b}\right)(\Gamma(-1-\alpha)+1).$$

Now we take $y_0^1, y_0^2 \in P_n E$, and using equation (3.9.6) define $(y_k^i)_{k=0,\dots,N}$, $y_0 = y_0^i$, and $\overline{y}^i$ as before for $i = 1, 2$. From Lemma 3.9.1 we get

$$|\overline{y}^1(s) - \overline{y}^2(s)|_E \leq e^{-s(\lambda_n + K_6 \lambda_n^{\alpha})} |y_0^1 - y_0^2|_E, \quad s \leq 0.$$

Hence from (H_2) and equations (3.9.2) (3.9.3), we get

$$|\mathcal{F}_N^{\tau} \Psi(y_0^1) - \mathcal{F}_N^{\tau} \Psi(y_0^2)|_E$$

$$\leq \int_{-\infty}^{0} |e^{As} Q_n f(\overline{y}^1(s) + \Psi(\overline{y}^1(s))) - f(\overline{y}^2(s) + \Psi(\overline{y}^2(s)))|_E ds$$

$$\leq K_2 M_1 (1+l) |y_0^1 - y_0^2|_E (\Gamma(-1-\alpha)+1) \frac{\Lambda_n^{\alpha}}{\Lambda_n - \lambda_n - K_6 \lambda_n^{\alpha}}$$

$$\leq l |y_0^1 - y_0^2|_E,$$

where $C_8 \geq \max(K_6, \frac{K_2 M_1 (1+l)}{l})$.

Then we prove that $\mathcal{F}_N^{\tau}$ is strictly compact. Taking $\Psi^1, \Psi^2 \in \mathcal{F}_{l,b}$, $y_0 \in P_n E$, we use $P_n = \Psi^1$, $P_n = \Psi^2$ and equation (3.9.6) to construct $(y_k^1)_{k=0,\dots,N}$, $(y_k^2)_{k=0,\dots,N}$, and $\overline{y}^1, \overline{y}^2$, respectively, as before. By Lemma 3.9.1, since $y_0 = y_0^1 = y_0^2$, we get

$$|\overline{y}^1(s) - \overline{y}^2(s)|_E \leq K_3 M_1 \lambda_n^{\alpha} |\Psi_1 - \Psi_2|_{\infty} (1 + |y_0|_E) |s| e^{-s(\lambda_n + K_6 \lambda_n^{\alpha})}.$$

Writing

$$\mathcal{F}_N^{\tau} \Psi^1(y_0) - \mathcal{F}_N^{\tau} \Psi^2(y_0) \leq \int_{-\infty}^{0} e^{As} Q_n f(\overline{y}^1(s) + \Psi(\overline{y}^1(s)))$$

$$- f(\overline{y}^2(s) + \Psi(\overline{y}^2(s))) ds,$$

from (H_2) and equations (3.9.2), (3.9.7) we get

$$\begin{aligned}
&|\mathscr{F}_N^\tau \Psi^1(y_0) - \mathscr{F}_N^\tau \Psi^2(y_0)|_E \\
&\le K_2 M_1 \int_{-\infty}^{0} \Big(\frac{1}{|s|^\alpha} + \Lambda_n^\alpha\Big) e^{\Lambda_n s}(1+l)|\bar{y}_0^1(s) - \bar{y}_0^2(s)|_E \\
&\quad + |\Psi_1 - \Psi_2|_\infty (1 + |\bar{y}^1(s)|_E) ds \\
&\le K_2 M_1 |\Psi_1 - \Psi_2|_\infty \Bigg[K_3 M_1 (1+l) \lambda_n^\alpha (1 + |y_0|_E) \\
&\quad \times \int_{-\infty}^{0} (|s|^{1-\alpha} + \Lambda_n^\alpha |s|) e^{(\Lambda_n - \lambda_n - K_6 \lambda_n^\alpha) s} ds + \int_{-\infty}^{0} \Big(\frac{1}{|s|^\alpha} + \Lambda_n^\alpha\Big) e^{\Lambda_n s} ds \\
&\quad + (1 + |y_0|_E) \int_{-\infty}^{0} (|s|^{1-\alpha} + \Lambda_n^\alpha |s|) \times e^{(\Lambda_n - \lambda_n - K_3 M_4 (1+b) \lambda_n^\alpha) s} ds \Bigg].
\end{aligned}$$

Then for $\bar{y}^1$ using Lemma 3.9.1, by direct calculation we get

$$\begin{aligned}
&|\mathscr{F}_N^\tau \Psi^1(y_0) - \mathscr{F}_N^\tau \Psi^2(y_0)|_E \\
&\quad \le K_2 M_1 (1 + |y_0|_E) |\Psi_1 - \Psi_2|_\infty \\
&\quad \times \Bigg[\Lambda_n^{\alpha-1} (\Gamma(-1-\alpha) + 1) + K_3 M_1 (1+l) \frac{\Lambda_n^\alpha \lambda_n^\alpha}{(\Lambda_n - \lambda_n - K_6 \lambda_n^\alpha)^2} \\
&\quad \times ((1-\alpha)\Gamma(-1-\alpha) + 1) + \frac{\Lambda_n^\alpha}{(\Lambda_n - \lambda_n - K_3 M_1 (1+b) \lambda_n^\alpha)} (\Gamma(-1-\alpha) + 1) \Bigg] \\
&\quad \le \frac{1}{2} (1 + |y_0|_E) |\Psi_1 - \Psi_2|_\infty,
\end{aligned}$$

where we assume that C_8 is greater than an appropriate constant which is independent of n. Thus Proposition 3.9.3 has been proved. □

Below, in order to compare the distance between $\mathscr{F}_N^\tau \Psi_N$ and Φ or Ψ, we first state a theorem on the existence of inertial manifolds for equation (3.9.1) in [45]. Let

$$J\Psi(y_0) = \int_{-\infty}^{0} e^{As} Q_n f(y(s) + P_n(y(s))) ds.$$

Theorem 3.9.2. *If there exists a constant C_0, depending on f, K_1, K_2, l and b, such that $\Lambda_n - \lambda_n \ge C_0(\Lambda_n^\alpha + \lambda_n^\alpha)$, then the functional $\mathscr{J}$ maps $\mathscr{F}_{l,b}$ into itself, and $\mathscr{J}$ is a strictly compact mapping. Hence it has a fixed point, that is, $\mathscr{J}\Phi = \Phi$, graph Φ is a C_1-inertial manifold of equation (3.9.1).*

Proposition 3.9.4. *Under conditions of Proposition 3.9.3, Theorem 3.9.2, and assuming (H_4) and (H_5), there exist two constants C_9, C_{10} such that if $\frac{\Lambda_n - \lambda_n}{\Lambda_n^\alpha + \lambda_n^\alpha} \ge C_9 \ge C_8$, then for all*

$\Psi \in \mathscr{F}_{l,b}$ *and all* N, τ, *we have*

$$|\mathscr{F}_N^\tau P_n - \Phi|_\infty \le \frac{1}{2}|\Psi - \Phi|_\infty + \varepsilon(N,\tau),$$

where

$$\varepsilon(N,\tau) = C_{10}\left(\left(\alpha_2(\lambda_n) + \frac{\alpha_1(\lambda_n)}{\lambda_n^\alpha}\right)\tau + \frac{\alpha_2(\lambda_n)}{\Lambda_n^\alpha} e^{-\Lambda_n^\alpha N\tau}\right).$$

Proof. From Proposition 3.9.3, we have

$$|\mathscr{F}_N^\tau \Psi - \Phi|_\infty \le \frac{1}{2}|\Psi - \Phi|_\infty + |\mathscr{F}_N^\tau \Phi - \Phi|_\infty.$$

We estimate $|\mathscr{F}_N^\tau \Phi - \Phi|_\infty$ successively. Suppose that y is the solution of the following equation:

$$\begin{cases} \dfrac{dy}{dt} + Ay = P_n f(y + \Phi(y)), \\ y(0) = y_0. \end{cases}$$

Taking $\overline{y}_k = y(-k\tau)$, by equation (3.9.6) we construct $(y_k^i)_{k=0,\dots,N}$, $\Psi = \Phi$, and then by (H_4) and Lemma 3.9.2, with $e_k = \varepsilon_k$, we have

$$\begin{aligned} |\overline{y}_k - y_k|_E &\le \sum_{j=0}^{k-1} e^{(k-1-j)\tau(\lambda_n + K_3 M_1(1+l)\lambda_n^\alpha)} |\varepsilon_j|_E \\ &\le \alpha_1(\lambda_n)\tau^2(1 + |y_0|_E) k e^{k\tau(\lambda_n + C_4'\lambda_n^\alpha)}, \end{aligned} \tag{3.9.16}$$

where $C_4' = \max(K_4, K_3 M_1(1+l))$. Using equation (3.9.10) we define $\overline{y}$. Then by employing (H_5) and equation (3.9.16), we arrive at

$$\begin{aligned} |y(s) - \overline{y}(s)|_E &\le \tau\alpha_2(\lambda_n)(1 + |y_0|_E)e^{-s(\lambda_n + K_5\lambda_n^\alpha)} \\ &\quad + \tau\alpha_1(\lambda_n)(1 + |y_0|_E)|s|e^{-s(\lambda_n + C_4'\lambda_n^\alpha)}, \\ &\quad s \in (-(k+1)\tau, -k\tau]. \end{aligned} \tag{3.9.17}$$

Similarly, we can deduce that

$$\begin{aligned} |y(s) - \overline{y}(s)|_E &\le |s + N\tau|\alpha_2(\lambda_n)(1 + |y_0|_E)e^{-s(\lambda_n + K_5\lambda_n^\alpha)} \\ &\quad + \tau\alpha_1(\lambda_n)(1 + |y_0|_E)|s|e^{-s(\lambda_n + C_4'\lambda_n^\alpha)}, \\ &\quad s \in (-\infty, -N\tau]. \end{aligned} \tag{3.9.18}$$

Since

$$\mathscr{F}_N^\tau \Phi(y_0) - \Phi(y_0) = \int_{-\infty}^{0} e^{As} Q_n f(\overline{y}(s) + \Phi(\overline{y}(s))) - f(y(s) + \Phi(y(s)))ds,$$

together with (H_2) and equations (3.9.2), (3.9.7), we obtain

$$
\begin{aligned}
&|\mathscr{F}_N^\tau \Phi(y_0) - \Phi(y_0)|_E \\
&\le K_2 M_1 (1+l) \int_{-\infty}^{0} \left(\frac{1}{|s|^\alpha} + \Lambda_n^\alpha\right) e^{\Lambda_n s} |\bar{y}(s) - y(s)|_E \, sds \\
&\le K_2 M_1 (1+l)(1 + |y_0|_E) \\
&\quad \times \left[\alpha_2(\lambda_n)\tau \int_{-\infty}^{0} \left(\frac{1}{|s|^\alpha} + \Lambda_n^\alpha\right) e^{\Lambda_n - \lambda_n - K_5 \lambda_n^\alpha} sds \right. \\
&\quad + \alpha_1(\lambda_n) \int_{-\infty}^{0} (|s|^{\alpha-1} + \Lambda_n^\alpha) e^{(\Lambda_n - \lambda_n - C_4' \lambda_n^\alpha)} sds \\
&\quad \left. + \alpha_2(\lambda_n) \int_{-\infty}^{-N\tau} |s + N\tau| \left(\frac{1}{|s|^\alpha} + \Lambda_n^\alpha\right) e^{\Lambda_n - \lambda_n - K_5 \lambda_n^\alpha} sds \right] \\
&\le (1 + |y_0|_E) C_{10} \left(\alpha_2(\lambda_n) \frac{\Lambda_n^\alpha}{\Lambda_n - \lambda_n - K_5 \lambda_n^\alpha} \tau \right. \\
&\quad + \alpha_1(\lambda_n) \frac{\Lambda_n^\alpha}{(\Lambda_n - \lambda_n - C_4' \lambda_n^\alpha)^2} \tau \\
&\quad \left. + \alpha_2(\lambda_n) \frac{\Lambda_n^\alpha}{(\Lambda_n - \lambda_n - K_5 \lambda_n^\alpha)^2} e^{-(\Lambda_n - \lambda_n - K_5 \lambda_n^\alpha) N\tau} \right) \\
&\le C_{10} (1 + |y_0|_E) \left(\alpha_2(\lambda_n)\tau + \frac{\alpha_1(\lambda_n)}{\Lambda_n^\alpha} + \frac{\alpha_2(\lambda_n)}{\Lambda_n^\alpha} e^{-\Lambda_n^\alpha N\tau} \right),
\end{aligned}
$$

where

$$C_9 \ge \max(K_5, 1, C_8, C_4'). \qquad \square$$

The constant C_{10} can be calculated in detail, and does not depend on N, τ or n. Since, when $\tau \to 0$, $N\tau \to \infty$, this leads to $\varepsilon(N,\tau) \to 0$. It is then easy to get

Theorem 3.9.3. *Under the assumptions of Proposition 3.9.4, defining the sequence $(\Phi_N)_{N\in\mathbf{N}}$ by equation (3.9.8) which, it converges to Φ with respect to $|\cdot|_\infty$, where $(\tau_N)_{N\in\mathbf{N}}$ is satisfied that $N\tau_N \to \infty$ as $N \to \infty$.*

Below we prove that the sequence $(\Phi_N)_{N\in\mathbf{N}}$ converge to Φ under very few conditions in the C-topology. Introduce the set

$$
\begin{aligned}
\mathscr{G}_l &= \left\{\Delta : P_n E \to \mathscr{L}\left(P_n E, Q_n E \mid \sup_{y\in P_n E} |\Delta(y)|_{\mathscr{L}(P_n E, Q_n E)} \le l\right)\right\}, \\
|\Delta|_\infty &= \sup_{y\in P_N E} |\Delta(y)|_{\mathscr{L}(P_n E, Q_n E)}.
\end{aligned}
$$

For any $\Psi \in \mathscr{F}_{l,b}$, define the mapping $T_\Psi : \Delta \in \mathscr{G} \to T_\Psi(\Delta)$ by

$$T_\Psi(\Delta)(y_0)\eta_0 = \int_{-\infty}^{0} e^{As} Q_n Df(y(s) + \Psi(y(s)))(\eta(s) + \Delta(y(s))\eta(s))ds, \tag{3.9.19}$$

where y is the solution of the following problem:

$$\begin{cases} \dfrac{dy}{dt} + Ay = P_n f(y + \Psi(y)), \\ y(0) = y_0, \end{cases} \tag{3.9.20}$$

and assume η satisfies

$$\begin{cases} \dfrac{d\eta}{dt} + A\eta = P_n f(y + \Psi(y))(\eta + \Delta(y)\eta), \\ \eta(0) = \eta_0. \end{cases} \tag{3.9.21}$$

Under the assumption of Theorem 3.9.2, we can prove for $\Psi \in \mathscr{F}_{l,b}$, that T_Ψ maps $\mathscr{F}_l$ into itself. Moreover, it is a strictly compact mapping, and this holds uniformly with respect to Ψ. Hence, we can prove $\Phi \in C^1$.

Now by the construction method of $\mathscr{F}_N^\tau$ we come up with an approximation of T_Ψ. For all $N \in \mathbf{N}$, $\tau > 0$, $\Psi \in \mathscr{F}_{l,b}$, we define $T_{N,\Psi}^\tau$ in $\mathscr{F}_l$ as follows:

$$\begin{aligned} T_{N,\Psi}^\tau(\Delta)(y_0)\eta_0 = {} & A^{-1}(I - e^{-\tau A}) \sum_{k=0}^{N-1} e^{-k\tau A} Q_n Df(y_k + \Psi(y_k))(\eta_k + \Delta(y_k)\eta_k) \\ & + A^{-1} e^{-N\tau A} Q_n Df(y_N + \Psi(y_N))(\eta_k + \Delta(y_N)\eta_N), \end{aligned}$$

where $(y_k^i)_{k=0,\dots,N}$ and $(\eta_k)_{k=0,\dots,N}$ can be calculated by the following equations:

$$y_{k+1} = R_\tau y_k + S_\tau P_n f(y_k + \Psi(y_k)), \tag{3.9.22}$$

$$\eta_{k+1} = R_\tau \eta_k + S_\tau P_n Df(y_k + \Psi(y_k))(\eta_k + \Delta(y_k)\eta_k). \tag{3.9.23}$$

Suppose that $(\Phi_N)_{N\in\mathbf{N}}$ is defined by equation (3.9.8). Then it is easy to get

$$D\Phi_{N+1} = T_{N,\Phi_N}^{\tau_N}(D\Phi_N).$$

Under some additional assumptions, we prove that $T_{N,\Phi}^\tau$ is approaching T_Ψ in a certain sense, and thus $D\Phi_N$ is close to $D\Phi$.

Firstly, because of the technical reasons, we suppose f has finite support. This is valid in some application expamples. Because the function F can be truncated, suppose

$$(H_6) \quad \begin{cases} |R_\tau y|_E \geq |y|_E, & \tau > 0,\ t > 0, \\ |e^{-At} y|_E \geq |y|_E, & y \in P_n E, \\ |y + z|_E \geq |y|_E, & z \in Q_n E. \end{cases} \tag{3.9.24}$$

We say that equations (3.9.22)–(3.9.23) approximate (3.9.20)–(3.9.21) if for $\overline{y}(\overline{\eta})$ defined on the interval $(-\infty, 0]$, equal to $y_k(\eta_k)$ on $(-(k+1)\tau, -k\tau]$, and equal to $y_N(\eta_N)$ on $(-\infty, -N\tau]$, respectively, we have

(H_7) for all $T > 0$, $\overline{y}$, $\overline{\eta}$ converge to the solution y, η of equation (3.9.20)–(3.9.21) in $[-T, 0]$ consistently, where $y_0, \eta_0 \in P_nE$, for any bounded set, $\tau \to 0$, $N\tau \to \infty$.

Theorem 3.9.4. *Suppose the assumptions of Theorem 3.9.3 hold, (H_6) and (H_7) are satisfied, f has compact support, and Df is consistently continuous. Then the sequence $(\Phi_N)_{N\in\mathbf{N}}$ converges to Φ in the C^1-topology, when $\tau_N \to 0$, $N\tau_N \to \infty$.*

Proof. From Theorem 3.9.3, Φ_N consistently converges to Φ in any bounded set of P_nE. Suppose that B is a ball in the support of f, and R_1 is its radius. Choosing any $y_0 \in P_nE$ such that

$$|y_0|_E \geq R_1,$$

by (H_6) we deduce that when y is the solution of equation (3.9.5) and $(y_k^i)_{k=0,\dots,N}$ is defined by equation (3.9.6), we have

$$\begin{aligned} |y(s)|_E &\geq R_1, \quad s \leq 0, \\ |y_k|_E &\geq R_1, \quad k = 0, \dots, N. \end{aligned}$$

From equations (3.9.4) and (3.9.7), we deduce that

$$\Phi(y_0) = \Phi_N(y_0) = 0.$$

Hence, for all N, Φ_N has its support included in B and Φ_N is uniformly convergent to Φ in P_nE.

Denote

$$\begin{aligned} D\Phi_{N+1} - D\Phi = {} & T^{\tau_N}_{N,\Phi_N}(D\Phi_N) - T^{\tau_N}_{N,\Phi_N}(D\Phi_N) \\ & + T^{\tau_N}_{N,\Phi_N}(D\Phi_N) - T_\Phi(D\Phi). \end{aligned} \tag{3.9.25}$$

Using the method of Proposition 3.9.3, we get

$$\left|T^{\tau_N}_{N,\Phi_N}(D\Phi_N) - T^{\tau_N}_{N,\Phi_N}(D\Phi)\right|_\infty \leq \frac{1}{2}|D\Phi_N - D\Phi|_\infty. \tag{3.9.26}$$

Consider $y_0, \eta_0 \in P_nE$, $|\eta_0| = 1$, and let $(y_k^i)_{k=0,\dots,N}$ and $(\eta_k)_{k=0,\dots,N}$, $((y_k^N)_{k=0,\dots,N}$ and $(\eta_k^N)_{k=0,\dots,N})$ be defined by equations (3.9.22)–(3.9.23). Letting $\Psi = \Phi$, $\Delta = D\Phi$ ($\Psi = \Phi_N$, $\Delta = D\Phi_N$), define $\overline{y}$, $\overline{\eta}$ ($\overline{y}^N$, $\overline{\eta}^N$) as in (H_7). Then if y, η are the solutions of equations (3.9.20)–(3.9.21) with $\Psi = \Phi$, $\Delta = D\Phi$, we have

$$\begin{aligned} & T^{\tau_N}_{N,\Phi_N}(D\Phi)(y_0)(\eta_0) - T_\Phi(D\Phi)(y_0)(\eta_0) \\ & = \int_{-\infty}^{0} e^{As} Q_n(Df(\overline{y}^N(s) + \Phi_N(\overline{y}^N(s))) \end{aligned}$$

$$\times (\overline{\eta}^N(s) + D\Phi(\overline{y}^N(s))\overline{\eta}^N(s))$$
$$- Df(y(s) + \Phi(y(s)))(\eta(s) + D\Phi(y(s))\eta(s))ds. \quad (3.9.27)$$

If $|y_0|_E \geq R_1$, then we can prove

$$T_\Phi(D\Phi)(y_0)(\eta_0) = T^{\tau_N}_{N,\Phi_N}(D\Phi)(y_0)(\eta_0) = 0.$$

There exists a constant C_5' such that the integral of equation (3.9.27) is less than

$$C_5'\left(\frac{1}{|s|^\alpha} + \Lambda_n^\alpha\right)|\eta_0|_E e^{(\Lambda_n - \lambda_n - K_2 M_1 (1+l)\lambda_n^\alpha)s}.$$

Hence, the integral for y_0, η_0 is uniformly convergent. Since $|\eta_0|_E = 1$, for $\varepsilon > 0$, there exists T, which is independent of y_0, η_0, such that

$$\int_{-\infty}^{-T} |e^{As}Q_n(Df(\overline{y}^N(s) + \Phi_N(\overline{y}^N(s)))$$
$$\times (\overline{\eta}^N(s) + D\Phi(\overline{y}^N(s))\overline{\eta}^N(s))$$
$$- Df(y(s) + \Phi(y(s)))(\eta(s) + D\Phi(y(s))\eta(s))|_E ds \leq \frac{\varepsilon}{2}.$$

The remainder of the integral (3.9.27) is divided into the following sum of integrals:

$$I_1 = \int_{-T}^{0} e^{As}Q_n(Df(\overline{y}^N(s) + \Phi_N(\overline{y}^N(s)))$$
$$\times (\overline{\eta}^N(s) + D\Phi(\overline{y}^N(s))\overline{\eta}^N(s))$$
$$- Df(\overline{y}(s) + \Phi(\overline{y}(s)))(\overline{\eta}(s) + D\Phi(\overline{y}(s))\overline{\eta}(s))ds,$$

$$I_2 = \int_{-T}^{0} e^{As}Q_n(Df(\overline{y}(s) + \Phi_N(\overline{y}(s)))$$
$$\times (\overline{\eta}(s) + D\Phi(\overline{y}(s))\overline{\eta}(s))$$
$$- Df(y(s) + \Phi(y(s)))(\eta(s) + D\Phi(y(s))\eta(s))ds.$$

In the proof of Proposition 3.9.3, we have shown that there exists a constant $C_6'(T, n)$ such that

$$|\overline{y}(s) - \overline{y}^N(s)|_E \leq C_6'(T, n)|\Phi_N - \Phi|_\infty(1 + |y_0|_E)$$
$$\leq C_6'(T, n)(1 + R_1)|\Phi_N - \Phi|_\infty. \quad (3.9.28)$$

We can find a bounded set B_T, which depends on n, R_1 and T, such that $\overline{y}(s)$ and $\overline{y}^N(s)$ are in B_T, $\forall s \in [-T, 0]$. Therefore for any $\alpha > 0$, if we choose α_n such that

$$|\overline{y}(s) - \overline{y}^N(s)|_E + |\Phi(\overline{y}(s)) - \Phi^N(\overline{y}^N(s))|_E \leq \alpha_N,$$
$$\forall s \in [-T, 0], \ |y_0| \leq R_1,$$

we can prove that there exists $C_7'(T,n)$ such that

$$\begin{aligned}
&\left|\bar{\eta}^N(s)-\bar{\eta}(s)\right|_E \le C_7'(T,n)(M_1(\alpha_N)+M_2(\alpha_N)) \\
&s\in[-T,0],\ |y_0|_E\le R_1,\ |\eta_0|_E=1,
\end{aligned} \tag{3.9.29}$$

where

$$M_1(\alpha)=\sup_{|x-y|\le\alpha}\left|Df(x)-Df(y)\right|_{\mathscr{L}(E,F)},$$

$$M_2(\alpha)=\sup_{x,y\in B_T,\ |x-y|_E\le\alpha}\left|D\Phi(x)-D\Phi(y)\right|_{\mathscr{L}(E)}.$$

From equations (3.9.28)–(3.9.29), Df in P_nE and $D\Phi$ in B_T are uniformly continuous, which implies that I_1 uniformly converges to zero for given y_0, η_0, which satisfy $|y_0|_E\le R_1$, $|\eta_0|_E=1$. Then by (H_7), we know that I_2 uniformly converges to zero for y_0, η_0, since $|y_0|_E\le R_1$, $|\eta_0|_E=1$.

Hence, $\mathscr{I}^{\tau N}_{N,\Phi_N}(D\Phi)(y_0)\eta_0 - T_\Phi(D\Phi)(y_0)\eta_0$ uniformly converges to zero for y_0, η_0 satisfying $|y_0|_E\le R_1$, $|\eta_0|_E=1$. But since $|y_0|\le R_1$, it is zero. Hence

$$\left|T^{\tau_N}_{N,\Phi}(D\Phi_N)-T_\Phi(D\Phi)\right|_\infty\to 0,\quad N\to\infty.$$

From this and equations (3.9.25)–(3.9.26) we get

$$|D\Phi-D\Phi_N|_\infty\to 0,\quad N\to\infty.$$

Thus the theorem has been proved. □

Now we give three examples to illustrate the established results.

Example 1

Consider the classic parabolic semilinear equation with symmetric linear part, namely

$$\begin{cases}\dfrac{du}{dt}+Au=g(u),\\ u(0)=u_0,\end{cases} \tag{3.9.30}$$

where the operator A is a dense linear positive unbounded self-adjoint operator in a Hilbert space, which has norm $|\cdot|$. This equation has a compact pre-solution set, which is composed of an orthogonal basis $\{\omega_j\}_{j\in\mathbf{N}}$ of eigenvectors with the corresponding eigenvalues

$$0<\mu_0\le\mu_1\le\cdots\le\mu_j\to+\infty.$$

Space $D(A^s)$, $s>0$ is defined as usual, with the norm $|\cdot|_s=|A^s\cdot|$. Let $g\in C^1: D(A^{\alpha+\gamma})\to D(A^\gamma)$, $\gamma\ge0$, $\alpha\in[0,1)$. Take

$$\mathscr{E}=H,\quad F=D(A^\gamma),\quad E=D(A^{\alpha+\gamma}).$$

Define P_n as the characteristic projection of A. Then P_nH and Q_nH are invariant under the action of e^{-At} $(t \geq 0)$. The semigroup $(e^{-At})_{t\geq 0}$ can be extended to a group in P_nH, $e^{-At}D(A^{\gamma}) \subset D(A^{\alpha+\gamma})$, $\forall t > 0$.

Assume that conditions (H_1) and (H_2) are easy to verify that they are satisfied, and just take

$$K_1 = K_2 = 1, \quad \lambda_n = \mu_n, \quad \Lambda_n = \mu_{n+1}.$$

Under suitable assumptions, equation (3.9.30) in $D(A^{\alpha+\gamma})$ defines a continuous semigroup $(S(t))_{t\geq 0}$.

Generally speaking, the function g is not globally Lipschitz. In order to overcome this difficulty, we can set $(S(t))_{t\geq 0}$ to be a bounded absorbing set in $D(A^{\alpha+\gamma})$. Take $R \geq 0$ appropriately large, so that the ball with a radius of R in $D(A^{\alpha+\gamma})$ contains this absorbing set. The function f is defined as:

$$f(u) = \theta\left(\frac{|u|^2_{\alpha-\gamma}}{R^2}\right)g(u),$$

where $\theta(x) \in C^1$ is such that

$$\begin{cases} \theta(x) = 1, & x \leq 1, \\ \theta(x) \leq 1, & \forall x, \\ \theta(x) = 0, & x \geq 2. \end{cases}$$

In general, we consider the truncated equation

$$\begin{cases} \dfrac{du}{dt} + Au = f(u), \\ u(0) = u_0, \end{cases} \tag{3.9.31}$$

Consider the setting of Proposition 3.9.3, and the gap condition is given by

$$\limsup_{n\to\infty} \frac{\mu_{n+1} - \mu_n}{\mu^{\alpha}_{n+1} + \mu^{\alpha}_n} = \infty. \tag{3.9.32}$$

When A is an elliptic operator on a bounded domain $\Omega \subset \mathbf{R}^n$, its eigenvalues possess the asymptotical property

$$\mu_n \sim Cn^p, \tag{3.9.33}$$

which can justify the assumption of equation (3.9.32). Condition

$$p(\alpha - 1) > 1 \tag{3.9.34}$$

is not implied by equation (3.9.32). For example, let $A = -\Delta$, with periodic boundary conditions on $[0, L_1] \times [0, L_2] \subset \mathbf{R}^2$. When $\frac{L_1}{L_2}$ is a rational number and $\alpha = 0$, equation (3.9.32) is satisfied, but $p(1 - \alpha) = 1$, and so inequality (3.9.34) does not hold.

The approximation formula (3.9.6) for equation (3.9.5) can be discretized using different methods. First, consider the simple Euler scheme:

$$\frac{y_k - y_{k-1}}{\tau} + Ay_k = P_n f(y_k + \Psi(y_k)),$$

which gives

$$\begin{cases} R_\tau = I + \tau A, \\ S_\tau = -\tau I. \end{cases} \tag{3.9.35}$$

We verify that (H_3), (H_4), (H_5), $(H_4)'$, $(H_5)'$ hold. First, (H_3) is obvious if we take

$$K_3 = 1.$$

And it is easy to verify (H_5). By equation (3.9.5) we get

$$\left|\frac{\mathrm{d}y}{\mathrm{d}t}\right|_{\alpha+\gamma} \le (\lambda_n + M_1(1+b)\lambda_n^\alpha)|y|_{\alpha+1} + \lambda_n^\alpha M_1(1+b).$$

Denote

$$y(t) = e^{-At}y(0) + \int_0^t e^{-A(t-s)}P_n f(y(s) + \Psi(y(s)))\mathrm{d}s.$$

Then we get

$$\begin{aligned} |y(t)|_{\alpha+\gamma} &\le e^{-\lambda_n t}|y(0)|_{\alpha+\gamma} + \lambda_n^\alpha \int_0^t e^{-\lambda_n(t-s)}|f(y(s)+\Psi(s))|_\gamma \mathrm{d}s \\ &\le e^{-\lambda_n t}|y(0)|_{\alpha+\gamma} + \lambda_n^{\alpha-1}M_1(1+b) \\ &\quad + M_1(1+b)\lambda_n^\alpha \int_0^t e^{-\lambda_n(t-s)}|y(s)|_{\alpha+\gamma}\mathrm{d}s. \end{aligned}$$

Using Gronwall lemma, we arrive at

$$|y(t)|_{\alpha+\gamma} \le 2(|y(0)|_{\alpha+\gamma} + \lambda_n^{\alpha-1}M_1(1+b))e^{-(\lambda_n+M_1(1+b)\lambda_n^\alpha)t}.$$

Therefore,

$$\begin{aligned} \left|\frac{\mathrm{d}y}{\mathrm{d}t}\right|_{\alpha+\gamma} &\le 3(\lambda_n + M_1(1+b)\lambda_n^\alpha) \\ &\quad \times (|y(0)|_{\alpha+\gamma} + \lambda_n^{\alpha-1}M_1(1+b))e^{-(\lambda_n+M_1(1+b)\lambda_n^\alpha)t}, \end{aligned}$$

and so (H_5) is established, where

$$\begin{aligned} \alpha_2(\lambda) &= 3(\lambda_n + M_1(1+b)\lambda_n^\alpha)\max\Big(1, M_1(1+b)\sup_n \lambda_n^{\alpha-1}\Big), \\ K_5 &= M_1(1+b). \end{aligned}$$

Now we prove (H_4). Let y be the solution of equation (3.9.5). If we set $\tilde{y}_k = y(-k\tau)$, then

$$\frac{\tilde{y}_k - \tilde{y}_{k+1}}{\tau} + A\tilde{y}_k = P_n f(\tilde{y}_k + \Psi(\tilde{y}_k)) + \varepsilon_k',$$

where

$$\varepsilon_k' = \frac{1}{\tau} \int_{-(k+1)\tau}^{-k\tau} \left(\frac{dy}{dt}(s) - \frac{dy}{dt}(-k\tau) \right) ds.$$

Hence, $\varepsilon_k = \tau\varepsilon_k'$. Using equation (3.9.5),

$$\begin{aligned} \left| \frac{dy}{dt}(s) - \frac{dy}{dt}(-k\tau) \right|_{\alpha+\gamma} &\leq |Ay(s) - Ay(-ks)|_{\alpha+\gamma} \\ &\quad + |P_n(f(y(s) + \Psi(y(s))) - f(\tilde{y}(s)_k + \Psi(\tilde{y}_k)))|_{\alpha+\gamma}. \end{aligned}$$

Since $\Psi \in \mathscr{F}_{l,b}$ and due to equation (3.9.2), we have

$$\begin{aligned} \left| \frac{dy}{dt}(s) - \frac{dy}{dt}(-k\tau) \right|_{\alpha+\gamma} &\leq (\lambda_n + M_1(1+l)\lambda_n^\alpha)|y(s) - y(-k\tau)|_{\alpha+\gamma} \\ &\leq \tau(\lambda_n + M_1(1+l)\lambda_n^\alpha) \int_{-k\tau}^{s} \left| \frac{dy}{dt}(\theta) \right|_{\alpha+\gamma} d\theta \\ &\leq \tau(\lambda_n + M_1(1+l)\lambda_n^\alpha)(k\tau + s)\alpha_2(\lambda)e^{-(\lambda_n + K_5\lambda_n^\alpha)t}. \end{aligned}$$

From this we can deduce (H_4), where

$$\begin{aligned} \alpha_1(\lambda) &= (\lambda_n + M_1(1+l)\lambda_n^\alpha)(k\tau + s)\alpha_2(\lambda), \\ K_4 &= K_5. \end{aligned}$$

In order to verify the $(H_4)'$, $(H_5)'$, each specific equation requires a special proof. First of all, it is necessary to prove the existence of an attractor and show that it is bounded in $D(A^{\alpha+\gamma})$. Secondly, the bounds of time derivatives can be obtained from the time analyticity of solution and the Cauchy inequality. In the end, the proof of (H_6) is direct. Condition (H_7) can be obtained from the classic, but cumbersome calculation. In summary, the inertial manifolds exist, and $\{\Phi_N\}_{N\in\mathbf{N}}$ converges in the C^1-topology, as long as equation (3.9.32) holds when n is sufficiently large.

Another possible approximation of equation (3.9.5) is

$$y_{k+1} = e^{A\tau}y_k + A^{-1}(e^{A\tau} - 1)P_n f(y_k + \Psi(y_k)),$$

where

$$R_\tau = e^{A\tau}, \quad S_\tau = A^{-1}(e^{A\tau} - I). \tag{3.9.36}$$

All assumptions can be verified similarly.

Example 2
Consider the steady solution $\overline{u}$ of equation (3.9.31), namely

$$A\overline{u} = f(\overline{u}).$$

Set

$$\begin{aligned} v &= u - \overline{u}, \\ \mathscr{A}v &= Au - Df(\overline{u})v + \gamma v, \\ h(v) &= f(\overline{u} + v) - f(\overline{u}) - Df(\overline{u})v + \gamma v, \end{aligned}$$

where the parameter γ is chosen such that $\mathscr{A} \geq 0$ (that is, $(\mathscr{A}v, v) > 0$, $\forall v \in D(\mathscr{A}) = D(A)$). Then equation (3.9.31) can be changed into

$$\begin{cases} \dfrac{dv}{dt} + \mathscr{A}v = h(v), \\ v(0) = u_0 - \overline{u}. \end{cases} \tag{3.9.37}$$

If we set

$$\begin{aligned} &\mu_j \sim Cj^p, \quad p > 0, \\ &\mathscr{E} = H, \quad F = D(A^\gamma) = D(\mathscr{A}^\gamma), \quad E = D(A^{\alpha+\gamma}) = E(\mathscr{A}^{\alpha+\gamma}), \end{aligned}$$

and if $p(1-\alpha) > 1$, then there exists the eigenvector sequence $\{\mathscr{P}_{n_k}\}_{k\in\mathbf{N}}$ of $\mathscr{A}$ such that (H_1), (H_2) hold, where

$$\begin{aligned} &P_k = \mathscr{P}_{nk}, \quad \lambda_k = \mu_{nk} + C'\mu_{nk}^\alpha, \\ &\Lambda_k = \mu_{n_{k+1}} - C'\mu_{n_{k+1}}^\alpha, \end{aligned}$$

where C' is a constant, which depends on α and f. Function h is globally Lipschitz and satisfies equation (3.9.2). Since equation (3.9.31) implies the existence of an absorbing ball in $D(A^{\alpha+\gamma})$, this leads to a similar result for equation (3.9.37).

Now we construct inertial manifolds of equation (3.9.37). Since

$$\begin{aligned} &\Lambda_k - \lambda_k = \mu_{n_{k+1}} - \mu_{nk} - C'(\mu_{n_k}^\alpha + \mu_{n_{k+1}}^\alpha), \\ &\mu_{n_k} \sim n_k^p, \end{aligned}$$

and condition $p(1-\alpha) > 1$ holds, this implies that Theorem 9.6 and spectral gap condition are satisfied. As an approximation of equation (3.9.5) one can consider

$$\frac{y_k - y_{k+1}}{\tau} + \mathscr{A}y_k = P_nf(y_k + \Psi(y_k)).$$

For

$$\begin{aligned} R_\tau &= I + \tau\mathscr{A}, \\ S_\tau &= -\tau I, \end{aligned}$$

we need to change the norm on $P_kD(A^{\alpha+\gamma})$ to ensure (H_3) is satisfied. For any $y \in D(A^{\alpha+\gamma})$, we define

$$\|y\|_H = \sup_{l\geq 0} \frac{|\mathscr{A}^l y|}{\lambda_k^l},$$
$$\|y\|_{\alpha+\gamma} = \|\mathscr{A}^{\alpha+\gamma} y\|_H,$$
$$\|y\|_\gamma = \|\mathscr{A}^\gamma y\|_H,$$

which can prove that there exists a C'' such that

$$|\mathscr{A}^l y| \leq C'' \lambda_k^l |y|, \quad \forall l \geq 0,\ k \geq 0,\ y \in P_kD(A^{\alpha+\gamma}).$$

Introduce the norms

$$\|y+z\|_H = \|y\|_h + |z|,$$
$$\|y+z\|_{\alpha+\gamma} = \|y\|_{\alpha+\gamma} + |z|_{\alpha+\gamma},$$
$$\|y+z\|_\gamma = \|y\|_\gamma + |z|_\gamma,$$

which are equivalent to the previous respective norms. And their constants do not depend on k. It is easy to prove that (H_3) is established for these norms. Conditions (H_4), (H_5), (H_4'), (H_5') are checked as in the previous estimation, while (H_6), (H_7) are also easy to verify.

An approximation of equation (3.9.5) can be considered as follows:

$$y_{k+1} = e^{\mathscr{A}\tau} y_k + \mathscr{A}^{-1}(e^{\mathscr{A}\tau} - I)P_n f(y_k + \Psi(y_k)).$$

At this time we can use different forms of the norm to meet the first part of (H_3), the second part can be obtained from the following:

$$\mathscr{A}^{-1}(e^{\mathscr{A}\tau} - I)P_n = \int_0^\tau e^{\mathscr{A}\tau} P_n \mathrm{d}s.$$

Example 3

A class of equations with linear antisymmetric operators is the following:

$$\begin{cases} \dfrac{\mathrm{d}u}{\mathrm{d}t} + A_0 u + Cu + f(u) = 0, \\ u(0) = u_0, \end{cases}$$

where A_0 satisfies the assumption of A discussed in Example 1, C is a linear bounded antisymmetric operator, $D(A^{s_0}) \to H$, $s_0 > 0$; F is a C^1 function, $D(A_0^{\alpha+\gamma}) \to D(A_0^\gamma)$, $\gamma \geq 0$, $\alpha \in (0,1]$. Suppose that C and A_0 commute. Take

$$A = A_0 + C, \quad E = D(A_0^{\alpha+\gamma}), \quad F = D(A_0), \quad \mathscr{E} = H,$$

$(P_n)_{n\in\mathbf{N}}$, $(\lambda_n)_{n\in\mathbf{N}}$, $(\Lambda_n)_{n\in\mathbf{N}}$ as in Example 1. Conditions (H_1), (H_2) are then easy to verify.

The first example of such equation is the laser equation. In this case,

$$\begin{aligned} H &= L^2(\Omega), \quad \Omega \subset \mathbf{R}^m \text{ is a bounded open set,} \\ A_0 u &= -\Delta u + u, \end{aligned}$$

with Dirichlet or Neumann boundary condition,

$$\begin{aligned} Cu &= i\alpha(A_0 u - u), \\ F(u) &= (1+\gamma)u + (1+i\beta)f(|u|^2)u, \end{aligned}$$

$f(s) = \frac{s}{1+\delta s}$, $\delta > 0$. Ginzburg–Landau equation is the second example. The nonlinear function of the GL equation is not globally Lipschitz. But it has an absorbing sphere. Proposition 3.9.3 is applicable for both equations, as long as the spectral gap condition is satisfied. When the space dimension is 1 or $\Omega = (0, L_1) \times (0, L_2) \subset \mathbf{R}^2$, and $\frac{L_1}{L_2}$ is a finite number, then this gap condition holds.

An approximation of equation (3.9.5) is

$$y_{k+1} = e^{(A_0+C)\tau} y_k + (A_0 + C)^{-1}(e^{(A_0+C)\tau} - 1)P_n f(y_k + \Psi(y_k)).$$

Using the equality

$$(A_0 + C)^{-1}(e^{(A_0+C)\tau} - I) = \int_0^\tau e^{(A_0+C)s} ds,$$

condition (H_3) is verified.

The laser equation is not dissipative, and does not have attractors. Proposition 3.9.3 and Theorem 3.9.3 are applicable. For the GL equation, the existence of the attractor was proved. And the solution is time analytic. Therefore, $(H_4)'$ and $(H_5)'$ are established, while (H_6) and (H_7) are also easy to verify.

Bibliography

[1] Ablowitz, M. J. and Fokas, A. S., On the inverse scattering transform of multidimension nonlinear equations related to first order systems in the plane, J. Math. Phys., 1984, 25:2494–2505

[2] Ablowitz, M. J. and Haberman, R., Nonlinear evolution equations in two and three dimensions, Phys. Rev. Lett., 1975, 35:1185–1188

[3] Aimer, M. A. and Denar, P., Universite de TULON et da VAR, 1982

[4] Anker, D. and Freeman, N. C., On the soliton solutions of the Davey–Stewartson equation for long waves, Proc. R. Soc. Lond. A, 1978, 360:529–540

[5] Appert, K. and Vaclavik, J., Dynamics of coupled solitons, J. Phys. Fluids, 1977, 20(11):1845–1849

[6] Aubin, T., Nonlinear Analysis on Manifolds. Monge–Ampere Equations, Spring-Verlag, New York, Heidelberg, Berlin, 1982

[7] Babin, A. and Vishik, M. I., Attractors of partial differential equations and estimates of their dimension, Usp. Mat. Nauk, 1983, 38:133–187

[8] Babin, A. V. and Vishik, M. I., Attractor of Evolution Equations, Stud. Math. Appl., vol. 25, North-Holland, Amsterdam, London, New York, Tokyo, 1992

[9] Baillon, J. B. and Chadam, J. M., The Cauchy problem for the coupled Schrödinger–Klein–Gordon equations, in: Contemporary Developments in Continuum Mechanics and Partial Differential Equation, North-Holland, Amsterdam, New York, 1978, pp. 37–44

[10] Bartuccelli, M., Constantin, P., Doering, C. R., et al., On the possibility of soft and hard turbulence in the complex Ginzburg–Landau equation, Phys. D, Nonlinear Phenom., 1990, 44(3):421–444

[11] Bates, P. W. and Zheng, S., Inertial manifolds and inertial sets for the phase-field equations, J. Dyn. Differ. Equ., 1992, 4(2):375–398

[12] Benjamin, T. B., Bona, J. L., and Mahony, J. J., Model equations for long waves in nonlinear dispersive systems, Philos. Trans. R. Soc. Lond. A, Math. Phys. Eng. Sci., 1972, 272(1220):47–78

[13] Benny, D. J., A General theory for interactions between short and long waves, Stud. Appl. Math., 1977, 56:81–94

[14] Berger, M. S. and Chen, Y. Y., Symmetric vortices for the Ginzburg–Landau equations of superconductivity and the nonlinear desingularization phenomenon, J. Funct. Anal., 1989, 82(2):259–295

[15] Berkaliev, Z. B., Attractor of some quasilinear system of differential equations with viscoelastic terms, Russ. Math. Surv., 1985, 40(1):209–210

[16] Berkaliev, Z. B., Attractor of nonlinear evolutionary equation of viscoelasticity, Mosc. Univ. Math. Bull., 1985, 40(5):61–63

[17] Bernal, A. R., Inertial manifolds for dissipative semiflows in Banach spaces, Appl. Anal., 1990, 37:95–141

[18] Biler, P., Attractors for the system of Schrödinger and Klein–Gordon equations with Yukawa coupling, SIAM J. Math. Anal., 1990, 21(5): 1190–1212

[19] Brand, H. R. and Deissler, R. J., Interaction of localized solutions for subcritical bifurcations, Phys. Rev. Lett., 1989, 63(26):2801

[20] Cazenave, T., An Introduction to Nonlinear Schrödinger Equations, Textos Métodos Mat., vol. 22, 1889

https://doi.org/10.1515/9783110549652-004

[21] Cazenave, T, Haraux, A, and Martel, Y., An Introduction to Semilinear Evolution Equations, Blackwell Publishing Ltd, Oxford, 1998

[22] Chang, Q. and Guo, B., Finite difference method for a nonlinear wave equation, J. Comput. Math., 1984, 2(4):297–304

[23] Chapman, S. J., Howison, S. D., and Ockendon, J. R., Macroscopic models for superconductivity, SIAM Rev., 1992, 34(4):529

[24] Chen, Y. and Guo, B., Two dimensional Landau–Lifshitz equations, J. Partial Differ. Equ., 1996, 9:313–322

[25] Chen, Y., Ding, S., and Guo, B., Partial regularity for two dimensional Landau–Lifshitz equations, Acta Math. Sin., 1998, 14(3):423–432

[26] Chen, F., Guo, B., and Wang, P., Long time behavior of strongly damped nonlinear wave equations, J. Differ. Equ., 1998, 147(2):231–241

[27] Chidaglia, J. M., A note on the strong convergence towards attractors for damped forced KdV equation, J. Differ. Equ., 1994, 110:356–359

[28] Chow, S. N. and Lu, K., Invariant for flows in Banach spaces, J. Differ. Equ., 1988, 74:285–317

[29] Cipolatti, R., On the existence of standing waves for a Davey–Stewartson system, Commun. Partial Differ. Equ., 1992, 17(5–6):967–988

[30] Collet, P., Eaknam, J. P., Epstein, H., et al., Analyticity for the Kuramoto–Sivashinsky equation, Physica D, 1993, 67:321–326

[31] Constantin, P. and Foias, C., Global Lyapunov exponents, Kaplan–Yorke formula and the dimensions of attractors for two Navier–Stokes equations, Commun. Pure Appl. Math., 1985, 38:1–27

[32] Constantin, P., Foias, C., Manley, O., et al., Determining modes and fractal dimension of turbulent flows, J. Fluid Mech., 1985, 150:427–440

[33] Constantin, P., Foias, C., and Temam, R., Attractors Representing Turbulent Flow, Mem. Am. Math. Soc., vol. 53(314), 1985

[34] Constantin, P., Foias, C., Nicolaenko, B., et al., Integral and Inertial Manifolds for Dissipative Partial Differential Equations, Appl. Math. Sci., vol. 70, Springer-Verlag, New York, 1988

[35] Constantin, P., Foias, C., Nicolaenko, B., et al., Spectral barriers and inertial manifolds for dissipative partial differential equations, J. Dyn. Differ. Equ., 1988, 1:41–53

[36] Constantin, P., Foias, C., and Temam, R., On the dimension, of the attractors in two-dimensional turbulence, Physica D, 1988, 30:284–286

[37] Conway, E., Hoff, D., and Smoller, J., Large time behavior of solution of nonlinear reaction diffusion equations, SIAM J. Appl. Math., 1978, 35:1–16

[38] Davey, A. and Stewartson, K., On three-dimensional packets of surface of waves, J. Fluid Mech., 1977, 79:703–714

[39] Davey, A., Hocking, L. M., and Stewartson, K., On the nonlinear evolution of three-dimensional disturbances in plane Poiseuille flow, J. Fluid Mech., 1974, 63(3):529–536

[40] Debussche, A., Inertial manifolds and Sacker's equation, Differ. Integral Equ., 1990, 3(3):457–486

[41] Debussche, A. and Marion, M., On the construction of families of approximate inertial manifolds, J. Differ. Equ., 1992, 100(1):173–201

[42] Debussche, A, Temam, R., Inertial manifolds and the slow manifolds in meteorology, Differ. Integral Equ., 1991, 4(5):897–931

[43] Debussche, A. and Temam, R., Inertial manifolds and their dimension, in: Anderson, S. I., Anderson, A. E., and Ottason, O., eds., Dynamical Systems, Theory and Applications, World Scientific Publishing Co., Singapore, 1993

[44] Debussche, A and Temam, R., Convergent families of approximate inertial manifolds, J. Math. Pures Appl., 1994, 73:489–522

[45] Debussche, A. and Temam, R., Inertial manifolds and their dimension, in: Proceedings of the Soderstorns Sommar Universitet, Huddinge, Stockholm, 1992, pp. 15–20
[46] Deissler, R. J. and Brand, H. R., Generation of counter propagating nonlinear interacting traveling waves by localized noise, Phys. Lett. A, 1988, 130(4–5):293–298
[47] Demengel, F. and Ghidaglia, J. M., Inertial manifolds for partial differential evolution equations under time discretization: existence, convergence and applications, J. Math. Anal. Appl., 1991, 155:177–225
[48] Djordjevic, V. D. and Redekopp, L. G., On two-dimensional packets of capillary-gravity waves, J. Fluid Mech., 1997, 79:703–714
[49] Doering, C. R., Gibbon, J. D., and Levermore,C. D., Weak and strong solutions of the complex Ginzburg–Landau equation, Physica D, 1994, 71(3):285–318
[50] Duan, J. and Holmes, P., On the Cauchy problem of generalized Ginzburg Laudau equation, Nonlinear Anal. TMA, 1994, 22:1033–1040
[51] Duan, J., Holmes, P., and Titi, E. S., Global existence theory for a generalized Ginzburg–Landau equation, Nonlinearity, 1992, 5(6):1303–1314
[52] Dubois, T., Jauberteau, F., and Temam, R., Solution of the incompressible Navier–Stokes equations by the nonlinear Galerkin method, J. Sci. Comput., 1993, 8(2):167–194
[53] Eckmann, J. P. and Gallay, Th. Front solutions for the Ginzburg–Landau equation, Commun. Math. Phys., 1993, 152:221–248
[54] Fabes, E., Luskin, M., and Sell, G. R., Construction of inertial manifolds by elliptic regularization, J. Differ. Equ., 1991, 88(2):335–381
[55] Federer, H., Geometric Measure Theory, Springer-Verlag, Berlin, Heidelberg, New York, 1969, 676 pp
[56] Fenichel, N., Geometric singular perturbation theory for ordinary differential equations, J. Differ. Equ., 1979, 31:53–98
[57] Fitzgibbon, W. E., Strongly damped quasilinear evolution equations, J. Math. Anal. Appl., 1981, 79(2):536–550
[58] Flahaut, I., Attractor for the dissipative Zakharov system, Nonlinear Anal. TMA, 1991, 16(7/8):599–633
[59] Foias, C. and Kukavica, I., Determining nodes for the Kuramoto–Sivashinsky equation, J. Dyn. Differ. Equ., 1995, 7(2):365–373
[60] Foias, C. and Temam, R., Remarques sur les equations de Navier–Stokes stationnaires et les phénomênēs Successifs de bifurcation, Ann. Sc. Norm. Super. Pisa, 1978, 4(5):29–63
[61] Foias, C and Temam, R., Determination of the solution for Navier–Stokes equations of nodal value, Math. Comput., 1984, 43:117–133
[62] Foias, C. and Temam, R., Gevrey class regularity for the solutions of the Navier–Stokes equations, J. Funct. Anal., 1989, 87:359–369
[63] Foias, C. and Temam, R., Approximation of attractors by algebraic or analytic set, SIAM J. Math. Anal., 1994, 25(5):1269–1302
[64] Foias, C. and Tewanm, R., The connection between the Navier–Stokes equations dynamical systems and turbulence, in: Grandall, M. G., Rabindowitz, P. H., and Turner, E. E. L., eds., Directions in Partial Differential Equation, 1987, pp. 55–73
[65] Foias, C., Sell, G. R., and Temam, R., Varites inertilles des equations differentielles dissipatives, C. R. Acad. Sci. Paris, Ser. I Math., 1985, 301:139–142
[66] Foias, C., Manley, O. P., and Temam, R., Sur l'interaction des petits et grands tourbillons dans les écoulements turbulents, C. R. Acad. Sci., Ser. I Math., 1987:497–500
[67] Foias, C., Manley, G., and Temam, R., Attractors for the Benard problem existence and physical bound on their fractal dimension, Nonlinear Anal. TMA 1987, 11:939–967

[68] Foias, C., Manley, O., and Temam, R., On the interaction of small and large eddies in two dimensional turbulent flows, Math. Model. Numer. Anal., 1988, 22:93–114

[69] Foias, C., Nicolaenko, B., Sell, G. R., et al., Inertial manifolds for the Kuramoto–Sivashinsky equation and an estimate of their lowest dimension, J. Math. Pures Appl., 1988, 67:197–225

[70] Fukuda, I. and Tsutsumi, M., On coupled Klein–Gordon–Schrödinger equations II, Math. Appl., 1978, 66:358–378

[71] Gao, H. and Duan, J., On the initial-value problem for the generalized two-dimensional Ginzburg–Landau equation, J. Math. Anal. Appl., 1997, 216(2):536–548

[72] Gao, H. and Guo, B., On the number of determining nodes for the generalized Ginzburg–Landau equations, J. Partial Differ. Equ., 1997, 10:97–106

[73] Gao, H. and Guo, B., Global dynamics and control of a nonlinear body equation with strong structural damping, Acta Math. Sin., 1998, 14:(2):183–190

[74] Ghidaglia, J. M., Finite dimensional behavior for weakly and damped driven Schrödinger equation, Ann. Inst. Henri Poincaré, Anal. Non Linéaire, 1988, 4:365–405

[75] Ghidaglia, J. M. and Satit, J. C., On the initial value problem for the Davey–Stewartson systems, Nonlinearity, 1990, 3:475–506

[76] Ghidaglia, J. M. and Temam, R., Lower bound on the dimension of the attractor for the Navier–Stokes equations in space dimension 3, in: Francaviglia, M., ed., Mechanics. Analysis and Geometry: 200 Years after Lagrange, Elsevier, Amsterdam, 1990

[77] Ghidaglia, J. M., Marison, M., and Temam, R., Generalization of the Sobolev–Lieb–Thirring inequalities and applications to the dimensions of attractor, Differ. Integral Equ., 1988, 1(1):1–21

[78] Gibbons, J., Thornhill, S. G., Wardrop, M. J., et al., On the theory of Langmuir solitons, J. Plasma Phys., 1977, 17(2):153–170

[79] Glazier, J. A. and Kolodner, P., Interaction of nonlinear pules in convection in binary fluids, Phys. Rev. A, 1990, 43(4):4269–4280

[80] Greenberg, J. M., On the existence, uniqueness and stability of the equation $\rho_0 X_{tt} = E(X_x)X_{xx} + \lambda X_{xxt}$, J. Math. Anal. Appl., 1969, 25:575–591

[81] Greenberg, J. M., Camy, R. C. M., and Mizel, V. J., On the existence uniqueness, and stability of solutions of the equation $\sigma'(u_x)u_{xx} + \lambda u_{xtx} = \rho_0 u_{tt}$, J. Math. Mech., 1967/1968, 17:707–728

[82] Grimshaw, R. H. J., The modulation of an internal gravity wave packet, and the resonance with the mean motion, Stud. Appl. Math., 1977, 56(3):241–266

[83] Guo, B., The global solution of the system of equations for complex Schrödinger field coupled with Boussinesq type self-consistent field, Acta Math. Sin., 1983, 26:297–306

[84] Guo, B., Existence and uniqueness of the global solution of the Cauchy problem and the periodic initial value problem for a class of the coupled system of KdV-nonlinear Schrödinger equations, Acta Math. Sinica 1983, 26(5):513–532

[85] Guo, B., The global solution for one class of the system of LS nonlinear wave interaction, J. Math. Res. Expo., 1987, 1:59–76

[86] Guo, B., Spectral method for solving two dimensional Newton–Boussinesq equations, Acta Math. Appl. Sin., 1989, 5(2):208–218

[87] Guo, B., The periodic initial value problems and initial value problem for one clas of generalized I. S type equations, J. Eng. Math., 1991, 8(1):47–53

[88] Guo, B., The global attractors for the periodic initial value problem of generalized Kuramoto–SivaShinsky type equations, Prog. Nat. Sci., 1993, 3(4):327

[89] Guo, B., Initial manifolds for the generalized Kuramoto–Sivashinsky type equations, J. Math. Study, 1995, 28(3):50–62

[90] Guo, H., Nonlinear Galerkin methods for solving two dimensional Newton–Boussinesq equations, Chin. Ann. Math., Ser. B, 1995, 16(3):379–390
[91] Guo, B., Existence of the initial manifolds for generalized Kuramoto–Sivashinsky equation, J. Math. Study, 1996, 29(1):38–51
[92] Guo, B. and Chang, Q., Attractors and dimensions for discretizations of a generalized Ginzburg–Landau equations, J. Partial Differ. Equ., 1996, 9(4):365–383
[93] Guo, B. and Chen, F., Finite dimensional behavior of global attractors for weakly damped and forced KdV equations coupling with nonlinear Schrödinger equations, Nonlinear Anal. TMA, 1997, 29(5):569–584
[94] Guo, B. and Chen, L., Orbital stability of solitary waves of the long-short resonance equations, Math. Methods Appl. Sci., 1998, 21(10):883–894
[95] Guo, B. and Ding, S., Initial-boundary value problem of Landau–Lifshitz system (I), Prog. Nat. Sci., 1998, 8(1):11–23
[96] Guo, B. and Ding, S., Initial-boundary value problem of Landau–Lifshitz system (II), Prog. Nat. Sci., 1998, 8(2):147–151
[97] Guo, B. and Gao, H., Finite dimensional behavior of generalized Ginzburg–Landau equation, Prog. Nat. Sci., 1995, 5(6):599–610
[98] Guo, H. and Han, Y., Remarks on the generalized Kadomtsev–Petviashvili equations and two dimensional Benjamin–Ono equations, Proc. R. Soc. Lond. A, 1996, 452:1585–1595
[99] Guo, B. and Han, Y., Cauchy problem of nonlinear Schrödinger–Boussinesq equation in L^2 and H^1, in: Proceedings of the Conference on Nonlinear Evolution Equations and Infinite-dimensional Dynamical Systems (Shanghai, 1995), World Sci. Publ., River Edge, NJ, 1997, pp. 66–72
[100] Guo, B. and Hong, M., The Landau–Lifshitz equations of the ferromagnetic spin chain and harmonic maps, Calc. Var. Partial Differ. Equ., 1993, 1(4):311–334
[101] Guo, B., and Jing, Z., On the generalized Kuramoto–Sivashinsky type equations with the dispersive effects, Ann. Math. Res., 1892, 25(2):1–24
[102] Guo, B. and Jing, Z., Slow time-periodic solutions of cubic-quintic Ginzburg–Landau equations (I), Prog. Nat. Sci., 1998, 8(4):403–415
[103] Guo, B. and Jing, Z., Slow time-periodic solutions of cubic-quintic Ginzburg–Landau equations (II), Prog. Nat. Sci., 1998, 8(5):539–547
[104] Guo, B. and Li, Y., Attractor for dissipative Klein–Gordon–Schrödinger equations in R^3, J. Differ. Equ., 1997, 136(2):356–377
[105] Guo, B. and Li, Y., Long time behavior of solutions of Davey–Stewartson equations, Acta Math. Appl. Sin., 2001, 17(1):86–97
[106] Guo, B. and Miao, C., Asymptotic behavior of coupled Klein–Gordon–Schrödinger equations, Sci. China Ser. A, 1995, 25(7):705–714
[107] Guo, B. and Miao C., Well posedness of Cauchy problem for coupled system of the long-short wave equations, J. Partial Differ. Equ., 1998, 11(1):83–96
[108] Guo, B. and Shen, L., The global solution of initial value equation for nonlinear Schrödinger–Boussinesq equation in 3-dimensions, Acta Math. Appl. Sin., 1993, 6:223–233
[109] Guo, B. and Su, F., The global attractors for the periodic initial value problem of generalized Kuramoto–Sivashinsky type equations in multi dimensions, J. Partial Differ. Equ., 1993, 6(3):217–236
[110] Guo, B. and Su, F., The attractors for Landau–Lifshitz–Maxwell equations, J. Partial Differ. Equ., 2000, 4:320–340
[111] Guo, B. and Tan, S., Global smooth solution for nonlinear equation of Hirota type, Sci. China Ser. A 1992, 35(2):1425–1433

[112] Guo, B. and Tan, S., Long-time behavior for the equation of finite-depth fluids, Commun. Math. Phys., 1994, 163(1):1–15
[113] Guo, B. and Tan, S., Global smooth solution for a coupled non-linear wave equations, Math. Methods Appl. Sci., 1991, 14(6):419–425
[114] Guo, B. and Wang, B., Finite dimensional behavior for the derivative Ginzburg–Landau equation in two spatial dimensions, Physica D, 1995, 89:83–90
[115] Guo, B. and Wang, B., Approximation to the global attractor for Landau–Lifshitz equation of the ferromagnetic spin chain, Beijing Math., 1995, 1:164–176
[116] Guo, B. and Wang, Y., Generalized Landau–Lifshitz systems of ferromagnetic spin chain type and harmonic maps, Sci. China Ser. A, 1996, 26:800–810
[117] Guo, B. and Wang, B., Gevrey regularity and approximate inertial manifolds for the derivative Ginzburg–Landau equation in two space dimensions, Discrete Contin. Dyn. Syst., 1996, 3(4):455–466
[118] Guo, B. and Wang, B., The global solution and its long time behavior for a class of generalized LS type equations, Prog. Nat. Sci., 1996, 6(5):533–546
[119] Guo, B. and Wang, B., Approximation to the global attractor for a nonlinear Schrödinger equation, Appl. Math. B, 1996, 11:125–136
[120] Guo, B. and Wang, B., Approximate inertial manifolds to the Newton–Boussinesq equations, J. Partial Differ. Equ., 1996, 9:237–250
[121] Guo, B. and Wang, B., Upper semicontinuity of attractors for the reaction–diffusion equation, Acta Math. Sci., 1998, 18(2):146–157
[122] Guo, B. and Wang, B., Gevrey class regularity and approximate inertial manifolds for the Newton–Boussinesq equations, Chin. Ann. Math., Ser. B, 1998, 2:179–188
[123] Guo, B. and Wang, B., Attractor for the long-short wave equations, J. Partial Differ. Equ., 1998, 11:361–383
[124] Guo, B. and Wu, Y., Remarks on the global attractor of the weakly dissipative Benjamin–Ono equation, Northeast. Math. J., 1995, 11(4):489–496
[125] Guo, B. and Wu, Y., Finite-dimensional behavior of the Ginzburg–Landau for superconductivity, Prog. Nat. Sci., 1995, 5(6):658–667
[126] Guo, B. and Wu, Y., Global existence and nonexistence of solution of a forced nonlinear Schrödinger equation, J. Math. Phys., 1995, 36(7):3479–3484
[127] Guo, B. and Wu, Y., Global attractor and its dimension estimates for the generalized dissipative KdV equation, Acta Math. Appl. Sin., 1998, 14(3):252–259
[128] Guo, B. and Yang, L., The global attractors for the periodic initial value problem for a coupled nonlinear wave equation, Math. Methods Appl. Sci., 1994, 19:131–144
[129] Guo, B. and Yuan, G., Global smooth solution for the Klein–Gordon–Zakharov equations, J. Math. Phys., 1995, 36(8):4110–4124
[130] Guo, B. and Yuan, G., On the suitable weak solutions for the Cauchy problems of the Boussinesq equations, Nonlinear Anal. TMA, 1996, 26(8):1367–1385
[131] Guo, B. and Yuan, G., The Cauchy problem for the system of Zakharov equations arising from ion-acoustic models, Proc. R. Soc. Edinb. A, 1996, 126:811–820
[132] Guo, B.and Zhang, L., Decay of solution to magneto-hydrodynamics equations in two space dimensions, Proc. R. Soc. Lond. A, 1995, 449:79–91
[133] Guo, B., Jing, Z., and Lu, B., Spatiotemporal complexity of the cubic Ginzburg–Landau equation, Commun. Nonlinear Sci. Numer. Simul., 1996, 1(4):12–17
[134] Guo, B., Miao, C., and Huang, H., Global flow generated by coupled system of Schrödinger–BM system equations, Sci. China Ser. A, 1998, 41(2):131–138
[135] Haken, H., Synergetics – An Introduction, Springer, New York, 1987

[136] Hale, J., Ordinary Differential Equations, Robert E. Krieger Publishing Co., Inc., Huntington, N. Y., 1980
[137] Hale, J. K., Asymptotic Behavior of Dissipative Systems, Math. Surv. Monogr., vol. 25, Am. Math. Soc., Providence, RI, 1988
[138] Hayashi, N. and Saut, J. C., Global existence of small solutions to Davey–Stewartson and the Ishimori systems, Differ. Integral Equ., 1995, 8:1657–1675
[139] Hayashi, N. and Von Wahl, W., On the global strong solution of coupled Klein–Schrödinger equations, J. Math. Soc. Jpn., 1987, 39:489–497
[140] He, Y. and Li, K., Nonlinear Galerkin approximation of the two dimensional exterior Navier–Stokes problem, Discrete Contin. Dyn. Syst., 1998, 2(4):467–482
[141] Henry, D., Geometric Theory of Semilinear Parabolic Equations, Lect. Notes Math., vol. 840, Springer-Verlag, Berlin, 1981
[142] Holmes, C. A., Bounds solutions of the nonlinear parabolic amplitude equation for plane Poiseuille flow, Proc. R. Soc. Lond. A, 1985, 402:299–322
[143] Iorio, R. J., Jr, On the Cauchy problem for the Benjamin–Ono equation, Commun. Partial Differ. Equ., 1991, 11(2):1031–1081
[144] Ito, M., Symmetries and conservation laws of a coupled nonlinear wave equation, Phys. Lett. A, 1982, 91(7):335–338
[145] Jolly, M. S., Explicit Construction of an inertial manifold for a reaction diffusion equation, J. Differ. Equ., 1989, 38(2):220–261
[146] Jolly, M. S., Kevrekidis, I. G., and Tid, E. S., Approximate inertial for the Kuramoto Sivashinsky equation: analysis and computations, Physica D 1990, 44:38–60
[147] Joseph, R. I., Solitary waves in a finite depth fluid, J. Phys. A, Gen. Phys., 1977, 10(12):225–227
[148] Kato, T., in: Studies in Applied Mathematics, Adv. Math. Suppl. Stud., vol. 8, 1983, pp. 93–128
[149] Keefe, L. R., Dynamics of perturbed wavetrain solutions the Ginzburg–Landau equation, Stud. Appl. Math., 1985, 73:91–153
[150] Kovacic, G., Singular perturbation theory for homoclinic orbits in a class of near integrable dissipative systems, SIAM J. Math. Anal. 1995:1511–1642
[151] Kovacic, G. and Wiggins, S., Orbits homoclinic to resonance, with in applications to chaos in a model of the forced and damped Sine–Gordon equation, Physica D, 1992, 57:185–225
[152] Kraichnan, R. H., Inertial ranges in two-dimensional turbulence, Phys. Fluids, 1967, 10:1417–1423
[153] Kukavica, I., Hausdorff length of level sets for solutions for the Ginzburg–Landau equations, Nonlinearity, 1991, 8:113–129
[154] Kukavica, I., On the number of determining Dodes for the Ginzburg–Landau equations, Nonlinearity, 1992:997–1006
[155] Kukavica, I., An upper bound for the winding number for the solutions of the Ginzburg–Landau equations, Indiana Univ. Math. J., 1992, 11(3):825–836
[156] Kupershmidt, B. A., A coupled KdV equation with dispersion, Phys. Lett. A 1985, 18(10):571–572
[157] Kuramoto, Y., Diffusion-induced chaos in reaction systems, Prog. Theor. Phys. Suppl., 1978, 64:346–367
[158] Kwak, M., Finite dimensional inertial forms for the 2D Navier–Stokes equations, Indiana Univ. Math. J., 1992, 41:927–981
[159] Lakshmanan, M. and Nakauma, K., Landau–Lifshitz equations of ferromagnetism: Exact treatment of the gilbert damping, Phys. Rev. Lett. 1984, 53(6):2497–2499
[160] Landau, L. and Lifshitz, E., Electrodynamique des Milieux Continus, Course de Physique Theorique, Tome VI, Mir, Moscow, 1969

[161] Li, Y., Finite dimension of global attractor for weakly dissipative Klein–Gordon–Schrödinger Equation, Nonlinear World, 1997, 4:573–595
[162] Li, Y. and Guo, B., Attractor for dissipative Zakharov equation an un-bounded domain, Rev. Math. Phys., 1997, 9(6):675–687
[163] Li, Y. and Guo, B., Global attractor for generalized 2D Ginzburg–Landau equation, in: Partial Differential Equations and Spectral Theory, vol. 126, Birkhäuser, Basel, 2001, pp. 197–204
[164] Li, Y., McLaughlin, D. W., Shatah, J., et al., Persistent homoclinic orbits for a perturbed nonlinear Schrödinger equation, Commun. Pure Appl. Math., 1996, 49(11):1175–1255
[165] Linares, F. and Ponce, G., On the Davey–Stewartson systems, Ann. Inst. Henri Poincaré, Anal. Non Linéaire, 1993, 10(5):523–548
[166] Lions, P. L., The concentration-compactness principle in the calculus of variations. The limit case, Part I, Rev. Mat. Iberoam., 1985, 1(1): 145–201
[167] Lu, B., The mathematical analysis and numerical analysis for Landau Lifshitz and Ginzburg Landau equation, PhD thesis, China Academy of Engineering Physics, 1998
[168] Lu, B. and Fang, S., Spectral and pseudospectral methods for the ferro-magnetic chain equations, Math. Numer. Sin., 1998, 20(1):1–15
[169] Makhankov, V. G., On stationary solutions of Schrödinger equations with a self-consistent potential satisfying Boussinesq's equations, Phys. Lett. A, 1974, 500:42–44
[170] Makhankov, V. G., Dynamics of classical solitons, Phys. Rep., 1978, 35(1):1–128
[171] Mallet-Paret, J. and Sell, G. R., Inertial manifolds for reaction–diffusion equations in higher space dimensions, J. Am. Math. Soc., 1988, 1:805–866
[172] Mane, R., Reduction of semilinear parabolic equations to finite dimensional C^1 flows, in: Geometry and Topology, Lect. Notes Math., vol. 597, Springer-Verlag, New York, 1977, pp. 361–378
[173] Marion, M., Approximate inertial manifolds for reaction diffusion equation in higher space dimension, J. Am. Math. Soc., 1988, 1:805–866
[174] Marion, M., Approximate inertial manifolds for the patterns formation Cahn–Hilliard equation, Math. Model. Numer. Anal. 1989, 23:463–480
[175] Massatt, P., Limiting behavior for strongly damped nonlinear waves equations, J. Differ. Equ., 1983, 48:334–349
[176] McLaughlin, D. and Overman II, E. A., Whiskered tori for integrable PDE's: chaotic behavior in near integrable PDE's, in: Surveys in Applied Mathematics, vol. I, Plenum, New York, 1995, pp. 83–203
[177] Mora, X. Finite dimensional attracting manifold for damped semilinear wave equations, in: Diaz, I. and Lions, L., eds., Contr. to Nonlinear Partial Diff. Eqs., Longmans Green, New York, 1983, pp. 172–183
[178] Nakamura, K. and Sasada, T., Solitons and wave trains in ferromagnets, Phys. Lett. A, 1974, 48(5):321–322
[179] Nicolaenko, B., Large scale spatial structures in two-dimensional turbulent flows, Nucl. Phys. B, 1987, 2(87):453–484
[180] Nicolaenko, B., Inertial manifolds for models of compressible gas dynamics, in: The Connection between Infinite Dim. and Finite Dim., Dyn. Syst, Contemp. Math., vol. 99, 1989, pp. 165–180
[181] Nicolaenko, B., Scheurer, B., and Temam, R., Some global dynamical properties of the Kuramoto–Sivashinsky equation: Nonlinear stability and attractors, Physica D, 1985, 16:155–183
[182] Nicolaenko, B., Shearor, B., and Temam, R., Attractors for the Kuramoto–Sivashinsky equations, Lect. Appl. Math., 1986, 23:149–170

[183] Nicolaenko, B., Scheurer, B., and Temam, R., Some global dynamical properties of a class of pattern formation equations, Commun. Partial Differ. Equ., 1987, 14(14):245–297
[184] Nishikawa, K., Hejo, H., Mima, K., et al., Coupled nonlinear electron-plasma and ion-acoustic waves, Phys. Rev. Lett. A, 1974, 33:148–150
[185] Pazy, A., Semigroups of Linear Operators and Applications to Partial Differential Equation, Springer-Verlag, Berlin, 1983
[186] Promislow, K., Time analyticity and Gevrey regularity for solutions of a class of dissipative partial differential equations, Nonlinear Anal. TMA, 1991, 16:959–980
[187] Qian, M., Qin, W. X., and Zhu, S., One dimensional global attractor for discretization of the damped driven Sine–Gordon equation, Nonlinear Anal. TMA, 1998, 34(7):941–951
[188] Robinson, J. C., Inertial manifolds and the cone condition, Dyn. Syst. Appl. 1993, 2:311–330
[189] Rosa, R. and Temam, R., Initial manifolds and normal hyperbolicity, Acta Appl. Math., 1996, 45:1–50
[190] Schoen, R. and Yan, S. T., Differential Geom., Academic Press, Sinica, 1998
[191] Sell, G. R. and You, Y., Inertial manifolds: the nonself adjoint case, J. Differ. Equ., 1992, 96:203–255
[192] Shang, T., Vivashinsky, G., Irregular flow of a liquid film down a vertical column, J. Phys. France, 1982, 43(3):459–466
[193] Sivashinsky, G. I., Nonlinear analysis of hydrodynamic instability in laminar flames. I. Derivation of basic equations, Acta Astronaut., 1977, 4:1177–1206
[194] Su, F. and Guo, B., The global smooth solution for Landau–Lifshitz–Maxwell equation without dissipation, J. Partial Differ. Equ., 1998, 3:193–208
[195] Sulem, C. and Sulem, P. L., Quelques results de regularite pour les equations de la turbulence de Langmuir, C. R. Acad. Sci. Paris, 1979, 289:173–176
[196] Taboada, M., Finite dimensional asymptotic behavior for the Swift–Hohenberg model of convection, Nonlinear Anal. TMA, 1990, 1:43–54
[197] Temam, R., Differential Dynamical Systems in Mechanics and Physics, Springer-Verlag, Berlin, 1988, second edition in 1997
[198] Temam, R., Attractors for the Navier–Stokes equations, localization and approximation, J. Fac. Sci., Univ. Tokyo, Sect. 1A 1989, 36:629–647
[199] Temam, R. and Wang, X., Estimates on the lowest dimension of inertial manifolds for KS equation in the general case, Differ. Integral Equ., 1994, 7(4):1095–1108
[200] Todd, K., Existence and stability of singular heteroclinic orbits for the Ginzburg–Landau equation, Nonlinearity, 1996, 9:669–685
[201] Tsutsumi, M., Deacy of weak solutions to the Davey–Stewartson systems, J. Math. Anal. Appl., 1994, 182(3):680–704
[202] Tsutsumi, M. and Fukuda, I., On solutions of the derivative nonlinear Schrödinger equation: existence and uniqueness theorem, Funkc. Ekvacioj, 1980, 23(3):259–277
[203] Tsutsumi, M. and Hatano, S., Well posedness of the Cauchy problem for the long-short wave resonance equation, Nonlinear Anal. TMA, 1994, 22:151–171
[204] van Saarloos, W. and Hohenberg, P. C., Fronts, pulses, sources and sinks in generalized complex Ginzburg–Landau equations, Physica D, 1992, 56:303–367
[205] Vanderbauwhede, A. and Iooss, G., Center manifold theory in infinite dimensions, in: Dynamics Reported, vol. 1, Springer-Verlag, Berlin, Heidelberg, 1992, pp. 125–162
[206] Wang, B. and Guo, B., Attractors for the Davey–Stewartson systems in R^2, J. Math. Phys. 1997, 38(5):2524–2534
[207] Yan, Y., Attractor and dimensions for discretizations of a weakly damped Schrödinger equation and a Sine–Gordon equation, Nonlinear Anal. TMA, 1993, 20(12):1417–1452

[208] Yang, Y., Existence, regularity, and asymptotic behavior of the solutions to the Ginzburg–Landau equations on R^3, Commun. Math. Phys., 1989, 123(1):147–161
[209] Yang, Y., Global spatially-periodic solutions to the Ginzburg–Landau equation, Proc. R. Soc. Edinb. A, 1994, 110:263–273
[210] Yang, Linge and Guo, B., Initial-boundary value problem for the Davey–Stewartson system, Prog. Nat. Sci., 1997, 7(3):272–279
[211] Zaharov, V. E. and Takhtajan, L. A., Equivalence between nonlinear Schrödinger equation and Heisenberg ferromagnet equation, Theor. Math. Phys., 1979, 38(1):17–23
[212] Zakharov, V. E. and Shabat, A. B., Integrable system of nonlinear equations in mathematical physics, Funct. Anal. Appl., 1974, 83:43–53
[213] Zhao, Y., The approximate weak inertial manifolds of a class of nonlinear hyperbolic dynamical system, Sci. China, 1996, 39(7):694–708
[214] Zhou, Y., Applications of Discrete Functional Analysis to the Finite Difference Method, International Academic Publishers, Bern, 1990
[215] Zhou, Y and Guo, B., Initial value problems for a nonlinear singular integral–differential equation of deep water, in: Partial Differential Equations, Springer, Berlin, Heidelberg, 1988, pp. 1275–1288
[216] Zhou, Y. and Guo, B., Nonlinear partial differential equations in physics and mechanics, in: Gu, C., Ding, X., and Yang, C.-C., eds., Partial Differential Equation in China, Kluwer Academic Publishers, Dordrecht, 1994, pp. 127–159
[217] Zhou, Y., Guo, B., and Tan, S., Existence and uniqueness of smooth solution of system of ferromagnetic chain, Sci. China Ser. A. 1991, 34:257–266

Index